T0181841

# Lecture Notes in Artificial Intelligence     12459

## Subseries of Lecture Notes in Computer Science

More information about this subseries at http://www.springer.com/series/1244

Frank Hutter · Kristian Kersting ·
Jefrey Lijffijt · Isabel Valera (Eds.)

# Machine Learning and Knowledge Discovery in Databases

European Conference, ECML PKDD 2020
Ghent, Belgium, September 14–18, 2020
Proceedings, Part III

 Springer

*Editors*
Frank Hutter 🆔
Albert-Ludwigs-Universität
Freiburg, Germany

Kristian Kersting 🆔
TU Darmstadt
Darmstadt, Germany

Jefrey Lijffijt 🆔
Ghent University
Ghent, Belgium

Isabel Valera 🆔
Saarland University
Saarbrücken, Germany

ISSN 0302-9743          ISSN 1611-3349   (electronic)
Lecture Notes in Artificial Intelligence
ISBN 978-3-030-67663-6          ISBN 978-3-030-67664-3   (eBook)
https://doi.org/10.1007/978-3-030-67664-3

LNCS Sublibrary: SL7 – Artificial Intelligence

This Springer imprint is published by the registered company Springer Nature Switzerland AG
The registered company address is: Gewerbestrasse 11, 6330 Cham, Switzerland

# Preface

This edition of the European Conference on Machine Learning and Principles and Practice of Knowledge Discovery in Databases (ECML PKDD 2020) is one that we will not easily forget. Due to the emergence of a global pandemic, our lives changed, including many aspects of the conference. Because of this, we are perhaps more proud and happy than ever to present these proceedings to you.

ECML PKDD is an annual conference that provides an international forum for the latest research in all areas related to machine learning and knowledge discovery in databases, including innovative applications. It is the leading European machine learning and data mining conference and builds upon a very successful series of ECML PKDD conferences.

Scheduled to take place in Ghent, Belgium, due to the SARS-CoV-2 pandemic, ECML PKDD 2020 was the first edition to be held fully virtually, from the 14th to the 18th of September 2020. The conference attracted over 1000 participants from all over the world. New this year was a joint event with local industry on Thursday afternoon, the AI4Growth industry track. More generally, the conference received substantial attention from industry through sponsorship, participation, and the revived industry track at the conference.

The main conference programme consisted of presentations of 220 accepted papers and five keynote talks (in order of appearance): Max Welling (University of Amsterdam), Been Kim (Google Brain), Gemma Galdon-Clavell (Eticas Research & Consulting), Stephan Günnemann (Technical University of Munich), and Doina Precup (McGill University & DeepMind Montreal).

In addition, there were 23 workshops, nine tutorials, two combined workshop-tutorials, the PhD Forum, and a discovery challenge.

Papers presented during the three main conference days were organized in four different tracks:

- Research Track: research or methodology papers from all areas in machine learning, knowledge discovery, and data mining;
- Applied Data Science Track: papers on novel applications of machine learning, data mining, and knowledge discovery to solve real-world use cases, thereby bridging the gap between practice and current theory;
- Journal Track: papers that were published in special issues of the journals *Machine Learning* and *Data Mining and Knowledge Discovery*;
- Demo Track: short papers that introduce a new system that goes beyond the state of the art, accompanied with a video of the demo.

We received a record number of 687 and 235 submissions for the Research and Applied Data Science Tracks respectively. We accepted 130 (19%) and 65 (28%) of these. In addition, there were 25 papers from the Journal Track, and 10 demo papers (out of 25 submissions). All in all, the high-quality submissions allowed us to put together an exceptionally rich and exciting program.

The Awards Committee selected research papers that were considered to be of exceptional quality and worthy of special recognition:

- Data Mining best paper award: "Revisiting Wedge Sampling for Budgeted Maximum Inner Product Search", by Stephan S. Lorenzen and Ninh Pham.
- Data Mining best student paper award: "SpecGreedy: Unified Dense Subgraph Detection", by Wenjie Feng, Shenghua Liu, Danai Koutra, Huawei Shen, and Xueqi Cheng.
- Machine Learning best (student) paper award: "Robust Domain Adaptation: Representations, Weights and Inductive Bias", by Victor Bouvier, Philippe Very, Clément Chastagnol, Myriam Tami, and Céline Hudelot.
- Machine Learning best (student) paper runner-up award: "A Principle of Least Action for the Training of Neural Networks", by Skander Karkar, Ibrahim Ayed, Emmanuel de Bézenac, and Patrick Gallinari.
- Best Applied Data Science Track paper: "Learning to Simulate on Sparse Trajectory Data", by Hua Wei, Chacha Chen, Chang Liu, Guanjie Zheng, and Zhenhui Li.
- Best Applied Data Science Track paper runner-up: "Learning a Contextual and Topological Representation of Areas-of-Interest for On-Demand Delivery Application", by Mingxuan Yue, Tianshu Sun, Fan Wu, Lixia Wu, Yinghui Xu, and Cyrus Shahabi.
- Test of Time Award for highest-impact paper from ECML PKDD 2010: "Three Naive Bayes Approaches for Discrimination-Free Classification", by Toon Calders and Sicco Verwer.

We would like to wholeheartedly thank all participants, authors, PC members, area chairs, session chairs, volunteers, co-organizers, and organizers of workshops and tutorials for their contributions that helped make ECML PKDD 2020 a great success. Special thanks go to Vicky, Inge, and Eneko, and the volunteer and virtual conference platform chairs from the UGent AIDA group, who did an amazing job to make the online event feasible. We would also like to thank the ECML PKDD Steering Committee and all sponsors.

October 2020

Tijl De Bie
Craig Saunders
Dunja Mladenić
Yuxiao Dong
Frank Hutter
Isabel Valera
Jefrey Lijffijt
Kristian Kersting
Georgiana Ifrim
Sofie Van Hoecke

# Organization

## General Chair

Tijl De Bie                    Ghent University, Belgium

## Research Track Program Chairs

Frank Hutter                   University of Freiburg & Bosch Center for AI,
                                   Germany
Isabel Valera                  Max Planck Institute for Intelligent Systems, Germany
Jefrey Lijffijt                Ghent University, Belgium
Kristian Kersting              TU Darmstadt, Germany

## Applied Data Science Track Program Chairs

Craig Saunders                 Amazon Alexa Knowledge, UK
Dunja Mladenić                 Jožef Stefan Institute, Slovenia
Yuxiao Dong                    Microsoft Research, USA

## Journal Track Chairs

Aristides Gionis               KTH, Sweden
Carlotta Domeniconi            George Mason University, USA
Eyke Hüllermeier               Paderborn University, Germany
Ira Assent                     Aarhus University, Denmark

## Discovery Challenge Chair

Andreas Hotho                  University of Würzburg, Germany

## Workshop and Tutorial Chairs

Myra Spiliopoulou              Otto von Guericke University Magdeburg, Germany
Willem Waegeman                Ghent University, Belgium

## Demonstration Chairs

Georgiana Ifrim                University College Dublin, Ireland
Sofie Van Hoecke               Ghent University, Belgium

## Nectar Track Chairs

| | |
|---|---|
| Jie Tang | Tsinghua University, China |
| Siegfried Nijssen | Université catholique de Louvain, Belgium |
| Yizhou Sun | University of California, Los Angeles, USA |

## Industry Track Chairs

| | |
|---|---|
| Alexander Ypma | ASML, the Netherlands |
| Arindam Mallik | imec, Belgium |
| Luis Moreira-Matias | Kreditech, Germany |

## PhD Forum Chairs

| | |
|---|---|
| Marinka Zitnik | Harvard University, USA |
| Robert West | EPFL, Switzerland |

## Publicity and Public Relations Chairs

| | |
|---|---|
| Albrecht Zimmermann | Université de Caen Normandie, France |
| Samantha Monty | Universität Würzburg, Germany |

## Awards Chairs

| | |
|---|---|
| Danai Koutra | University of Michigan, USA |
| José Hernández-Orallo | Universitat Politècnica de València, Spain |

## Inclusion and Diversity Chairs

| | |
|---|---|
| Peter Steinbach | Helmholtz-Zentrum Dresden-Rossendorf, Germany |
| Heidi Seibold | Ludwig-Maximilians-Universität München, Germany |
| Oliver Guhr | Hochschule für Technik und Wirtschaft Dresden, Germany |
| Michele Berlingerio | Novartis, Ireland |

## Local Chairs

| | |
|---|---|
| Eneko Illarramendi Lerchundi | Ghent University, Belgium |
| Inge Lason | Ghent University, Belgium |
| Vicky Wandels | Ghent University, Belgium |

## Proceedings Chair

| | |
|---|---|
| Wouter Duivesteijn | Technische Universiteit Eindhoven, the Netherlands |

## Sponsorship Chairs

Luis Moreira-Matias        Kreditech, Germany
Vicky Wandels             Ghent University, Belgium

## Volunteering Chairs

Junning Deng              Ghent University, Belgium
Len Vande Veire           Ghent University, Belgium
Maarten Buyl              Ghent University, Belgium
Raphaël Romero            Ghent University, Belgium
Robin Vandaele            Ghent University, Belgium
Xi Chen                   Ghent University, Belgium

## Virtual Conference Platform Chairs

Ahmad Mel                 Ghent University, Belgium
Alexandru Cristian Mara   Ghent University, Belgium
Bo Kang                   Ghent University, Belgium
Dieter De Witte           Ghent University, Belgium
Yoosof Mashayekhi         Ghent University, Belgium

## Web Chair

Bo Kang                   Ghent University, Belgium

## ECML PKDD Steering Committee

Andrea Passerini          University of Trento, Italy
Francesco Bonchi          ISI Foundation, Italy
Albert Bifet              Télécom Paris, France
Sašo Džeroski             Jožef Stefan Institute, Slovenia
Katharina Morik           TU Dortmund, Germany
Arno Siebes               Utrecht University, the Netherlands
Siegfried Nijssen         Université catholique de Louvain, Belgium
Michelangelo Ceci         University of Bari Aldo Moro, Italy
Myra Spiliopoulou         Otto von Guericke University Magdeburg, Germany
Jaakko Hollmen            Aalto University, Finland
Georgiana Ifrim           University College Dublin, Ireland
Thomas Gärtner            University of Nottinghem, UK
Neil Hurley               University College Dublin, Ireland
Michele Berlingerio       IBM Research, Ireland
Elisa Fromont             Université de Rennes 1, France
Arno Knobbe               Universiteit Leiden, the Netherlands
Ulf Brefeld               Leuphana Universität Luneburg, Germany
Andreas Hotho             Julius-Maximilians-Universität Würzburg, Germany

# Program Committees

## Guest Editorial Board, Journal Track

| | |
|---|---|
| Michael Kamp | Monash University |
| Mehdi Kaytoue | Infologic |
| Marius Kloft | TU Kaiserslautern |
| Dragi Kocev | Jožef Stefan Institute |
| Peer Kröger | Ludwig-Maximilians-Universität Munich |
| Meelis Kull | University of Tartu |
| Ondrej Kuzelka | KU Leuven |
| Mark Last | Ben-Gurion University of the Negev |
| Matthijs van Leeuwen | Leiden University |
| Marco Lippi | University of Modena and Reggio Emilia |
| Claudio Lucchese | Ca' Foscari University of Venice |
| Brian Mac Namee | University College Dublin |
| Gjorgji Madjarov | Ss. Cyril and Methodius University of Skopje |
| Fabrizio Maria Maggi | Free University of Bozen-Bolzano |
| Giuseppe Manco | ICAR-CNR |
| Ernestina Menasalvas | Universidad Politécnica de Madrid |
| Aditya Menon | Google Research |
| Katharina Morik | TU Dortmund |
| Davide Mottin | Aarhus University |
| Animesh Mukherjee | Indian Institute of Technology Kharagpur |
| Amedeo Napoli | LORIA |
| Siegfried Nijssen | Université catholique de Louvain |
| Eirini Ntoutsi | Leibniz University Hannover |
| Bruno Ordozgoiti | Aalto University |
| Pance Panov | Jožef Stefan Institute |
| Panagiotis Papapetrou | Stockholm University |
| Srinivasan Parthasarathy | Ohio State University |
| Andrea Passerini | University of Trento |
| Mykola Pechenizkiy | Technische Universiteit Eindhoven |
| Charlotte Pelletier | Univ. Bretagne Sud/IRISA |
| Ruggero Pensa | University of Turin |
| Francois Petitjean | Monash University |
| Nico Piatkowski | TU Dortmund |
| Evaggelia Pitoura | Univ. of Ioannina |
| Marc Plantevit | Claude Bernard University Lyon 1 |
| Kai Puolamäki | University of Helsinki |
| Chedy Raïssi | Inria |
| Matteo Riondato | Amherst College |
| Joerg Sander | University of Alberta |
| Pierre Schaus | UCLouvain |
| Lars Schmidt-Thieme | University of Hildesheim |
| Matthias Schubert | LMU Munich |
| Thomas Seidl | LMU Munich |
| Gerasimos Spanakis | Maastricht University |
| Myra Spiliopoulou | Otto von Guericke University Magdeburg |
| Jerzy Stefanowski | Poznań University of Technology |

| | |
|---|---|
| Giovanni Stilo | Università degli Studi dell'Aquila |
| Mahito Sugiyama | National Institute of Informatics |
| Andrea Tagarelli | University of Calabria |
| Chang Wei Tan | Monash University |
| Nikolaj Tatti | University of Helsinki |
| Alexandre Termier | Univ. Rennes 1 |
| Marc Tommasi | University of Lille |
| Ivor Tsang | University of Technology Sydney |
| Panayiotis Tsaparas | University of Ioannina |
| Steffen Udluft | Siemens |
| Celine Vens | KU Leuven |
| Antonio Vergari | University of California, Los Angeles |
| Michalis Vlachos | University of Lausanne |
| Christel Vrain | LIFO, Université d'Orléans |
| Jilles Vreeken | Helmholtz Center for Information Security |
| Willem Waegeman | Ghent University |
| Marcel Wever | Paderborn University |
| Stefan Wrobel | Univ. Bonn and Fraunhofer IAIS |
| Yinchong Yang | Siemens AG |
| Guoxian Yu | Southwest University |
| Bianca Zadrozny | IBM |
| Ye Zhu | Monash University |
| Arthur Zimek | University of Southern Denmark |
| Albrecht Zimmermann | Université de Caen Normandie |
| Marinka Zitnik | Harvard University |

**Area Chairs, Research Track**

| | |
|---|---|
| Cuneyt Gurcan Akcora | The University of Texas at Dallas |
| Carlos M. Alaíz | Universidad Autónoma de Madrid |
| Fabrizio Angiulli | University of Calabria |
| Georgios Arvanitidis | Max Planck Institute for Intelligent Systems |
| Roberto Bayardo | Google |
| Michele Berlingerio | IBM |
| Michael Berthold | University of Konstanz |
| Albert Bifet | Télécom Paris |
| Hendrik Blockeel | Katholieke Universiteit Leuven |
| Mario Boley | MPI Informatics |
| Francesco Bonchi | Fondazione ISI |
| Ulf Brefeld | Leuphana Universität Lüneburg |
| Michelangelo Ceci | Università degli Studi di Bari Aldo Moro |
| Duen Horng Chau | Georgia Institute of Technology |
| Nicolas Courty | Université de Bretagne Sud/IRISA |
| Bruno Cremilleux | Université de Caen Normandie |
| Andre de Carvalho | University of São Paulo |
| Patrick De Causmaecker | Katholieke Universiteit Leuven |

| | |
|---|---|
| Nicola Di Mauro | Università degli Studi di Bari Aldo Moro |
| Tapio Elomaa | Tampere University |
| Amir-Massoud Farahmand | Vector Institute & University of Toronto |
| Ángela Fernández | Universidad Autónoma de Madrid |
| Germain Forestier | Université de Haute-Alsace |
| Elisa Fromont | Université de Rennes 1 |
| Johannes Fürnkranz | Johannes Kepler University Linz |
| Patrick Gallinari | Sorbonne University |
| Joao Gama | University of Porto |
| Thomas Gärtner | TU Wien |
| Pierre Geurts | University of Liège |
| Manuel Gomez Rodriguez | MPI for Software Systems |
| Przemyslaw Grabowicz | University of Massachusetts Amherst |
| Stephan Günnemann | Technical University of Munich |
| Allan Hanbury | Vienna University of Technology |
| Daniel Hernández-Lobato | Universidad Autónoma de Madrid |
| Jose Hernandez-Orallo | Universitat Politècnica de València |
| Jaakko Hollmén | Aalto University |
| Andreas Hotho | University of Würzburg |
| Neil Hurley | University College Dublin |
| Georgiana Ifrim | University College Dublin |
| Alipio M. Jorge | University of Porto |
| Arno Knobbe | Universiteit Leiden |
| Dragi Kocev | Jožef Stefan Institute |
| Lars Kotthoff | University of Wyoming |
| Nick Koudas | University of Toronto |
| Stefan Kramer | Johannes Gutenberg University Mainz |
| Meelis Kull | University of Tartu |
| Niels Landwehr | University of Potsdam |
| Sébastien Lefèvre | Université de Bretagne Sud |
| Daniel Lemire | Université du Québec |
| Matthijs van Leeuwen | Leiden University |
| Marius Lindauer | Leibniz University Hannover |
| Jörg Lücke | University of Oldenburg |
| Donato Malerba | Università degli Studi di Bari "Aldo Moro" |
| Giuseppe Manco | ICAR-CNR |
| Pauli Miettinen | University of Eastern Finland |
| Anna Monreale | University of Pisa |
| Katharina Morik | TU Dortmund |
| Emmanuel Müller | University of Bonn |
| Sriraam Natarajan | Indiana University Bloomington |
| Alfredo Nazábal | The Alan Turing Institute |
| Siegfried Nijssen | Université catholique de Louvain |
| Barry O'Sullivan | University College Cork |
| Pablo Olmos | University Carlos III of Madrid |
| Panagiotis Papapetrou | Stockholm University |

| Andrea Passerini | University of Turin |
| Mykola Pechenizkiy | Technische Universiteit Eindhoven |
| Ruggero G. Pensa | University of Torino |
| Francois Petitjean | Monash University |
| Claudia Plant | University of Vienna |
| Marc Plantevit | Université Claude Bernard Lyon 1 |
| Philippe Preux | Université de Lille |
| Rita Ribeiro | University of Porto |
| Celine Robardet | INSA Lyon |
| Elmar Rueckert | University of Lübeck |
| Marian Scuturici | LIRIS-INSA de Lyon |
| Michèle Sebag | Univ. Paris-Sud |
| Thomas Seidl | Ludwig-Maximilians-Universität Muenchen |
| Arno Siebes | Utrecht University |
| Alessandro Sperduti | University of Padua |
| Myra Spiliopoulou | Otto von Guericke University Magdeburg |
| Jerzy Stefanowski | Poznań University of Technology |
| Yizhou Sun | University of California, Los Angeles |
| Einoshin Suzuki | Kyushu University |
| Acar Tamersoy | Symantec Research Labs |
| Jie Tang | Tsinghua University |
| Grigorios Tsoumakas | Aristotle University of Thessaloniki |
| Celine Vens | KU Leuven |
| Antonio Vergari | University of California, Los Angeles |
| Herna Viktor | University of Ottawa |
| Christel Vrain | University of Orléans |
| Jilles Vreeken | Helmholtz Center for Information Security |
| Willem Waegeman | Ghent University |
| Wendy Hui Wang | Stevens Institute of Technology |
| Stefan Wrobel | Fraunhofer IAIS & Univ. of Bonn |
| Han-Jia Ye | Nanjing University |
| Guoxian Yu | Southwest University |
| Min-Ling Zhang | Southeast University |
| Albrecht Zimmermann | Université de Caen Normandie |

## Area Chairs, Applied Data Science Track

| Michelangelo Ceci | Università degli Studi di Bari Aldo Moro |
| Tom Diethe | Amazon |
| Faisal Farooq | IBM |
| Johannes Fürnkranz | Johannes Kepler University Linz |
| Rayid Ghani | Carnegie Mellon University |
| Ahmed Hassan Awadallah | Microsoft |
| Xiangnan He | University of Science and Technology of China |
| Georgiana Ifrim | University College Dublin |
| Anne Kao | Boeing |

| Javier Latorre | Apple |
| Hao Ma | Facebook AI |
| Gabor Melli | Sony PlayStation |
| Luis Moreira-Matias | Kreditech |
| Alessandro Moschitti | Amazon |
| Kitsuchart Pasupa | King Mongkut's Institute of Technology Ladkrabang |
| Mykola Pechenizkiy | Technische Universiteit Eindhoven |
| Julien Perez | NAVER LABS Europe |
| Xing Xie | Microsoft |
| Chenyan Xiong | Microsoft Research |
| Yang Yang | Zhejiang University |

## Program Committee Members, Research Track

| Moloud Abdar | Deakin University |
| Linara Adilova | Fraunhofer IAIS |
| Florian Adriaens | Ghent University |
| Zahra Ahmadi | Johannes Gutenberg University Mainz |
| M. Eren Akbiyik | IBM Germany Research and Development GmbH |
| Youhei Akimoto | University of Tsukuba |
| Ömer Deniz Akyildiz | University of Warwick and The Alan Turing Institute |
| Francesco Alesiani | NEC Laboratories Europe |
| Alexandre Alves | Universidade Federal de Uberlândia |
| Maryam Amir Haeri | Technische Universität Kaiserslautern |
| Alessandro Antonucci | IDSIA |
| Muhammad Umer Anwaar | Mercateo AG |
| Xiang Ao | Institute of Computing Technology, Chinese Academy of Sciences |
| Sunil Aryal | Deakin University |
| Thushari Atapattu | The University of Adelaide |
| Arthur Aubret | LIRIS |
| Julien Audiffren | Fribourg University |
| Murat Seckin Ayhan | Eberhard Karls Universität Tübingen |
| Dario Azzimonti | Istituto Dalle Molle di Studi sull'Intelligenza Artificiale |
| Behrouz Babaki | Polytechnique Montréal |
| Rohit Babbar | Aalto University |
| Housam Babiker | University of Alberta |
| Davide Bacciu | University of Pisa |
| Thomas Baeck | Leiden University |
| Abdelkader Baggag | Qatar Computing Research Institute |
| Zilong Bai | University of California, Davis |
| Jiyang Bai | Florida State University |
| Sambaran Bandyopadhyay | IBM |
| Mitra Baratchi | University of Twente |
| Christian Beecks | University of Münster |
| Anna Beer | Ludwig Maximilian University of Munich |

| | |
|---|---|
| Adnene Belfodil | Munic Car Data |
| Aimene Belfodil | INSA Lyon |
| Ines Ben Kraiem | UT2J-IRIT |
| Anes Bendimerad | LIRIS |
| Christoph Bergmeir | Monash University |
| Max Berrendorf | Ludwig Maximilian University of Munich |
| Louis Béthune | ENS de Lyon |
| Anton Björklund | University of Helsinki |
| Alexandre Blansché | Université de Lorraine |
| Laurens Bliek | Delft University of Technology |
| Isabelle Bloch | ENST - CNRS UMR 5141 LTCI |
| Gianluca Bontempi | Université Libre de Bruxelles |
| Felix Borutta | Ludwig-Maximilians-Universität München |
| Ahcène Boubekki | Leuphana Universität Lüneburg |
| Tanya Braun | University of Lübeck |
| Wieland Brendel | University of Tübingen |
| Klaus Brinker | Hamm-Lippstadt University of Applied Sciences |
| David Browne | Insight Centre for Data Analytics |
| Sebastian Bruckert | Otto Friedrich University Bamberg |
| Mirko Bunse | TU Dortmund University |
| Sophie Burkhardt | University of Mainz |
| Haipeng Cai | Washington State University |
| Lele Cao | Tsinghua University |
| Manliang Cao | Fudan University |
| Defu Cao | Peking University |
| Antonio Carta | University of Pisa |
| Remy Cazabet | Université Lyon 1 |
| Abdulkadir Celikkanat | CentraleSupelec, Paris-Saclay University |
| Christophe Cerisara | LORIA |
| Carlos Cernuda | Mondragon University |
| Vitor Cerqueira | LIAAD-INESCTEC |
| Mattia Cerrato | Università di Torino |
| Ricardo Cerri | Federal University of São Carlos |
| Laetitia Chapel | IRISA |
| Vaggos Chatziafratis | Stanford University |
| El Vaigh Cheikh Brahim | Inria/IRISA Rennes |
| Yifei Chen | University of Groningen |
| Junyang Chen | University of Macau |
| Jiaoyan Chen | University of Oxford |
| Huiyuan Chen | Case Western Reserve University |
| Run-Qing Chen | Xiamen University |
| Tianyi Chen | Microsoft |
| Lingwei Chen | The Pennsylvania State University |
| Senpeng Chen | UESTC |
| Liheng Chen | Shanghai Jiao Tong University |
| Siming Chen | Frauenhofer IAIS |

| | |
|---|---|
| Liang Chen | Sun Yat-sen University |
| Dawei Cheng | Shanghai Jiao Tong University |
| Wei Cheng | NEC Labs America |
| Wen-Hao Chiang | Indiana University - Purdue University Indianapolis |
| Feng Chong | Beijing Institute of Technology |
| Pantelis Chronis | Athena Research Center |
| Victor W. Chu | The University of New South Wales |
| Xin Cong | Institute of Information Engineering, Chinese Academy of Sciences |
| Roberto Corizzo | UNIBA |
| Mustafa Coskun | Case Western Reserve University |
| Gustavo De Assis Costa | Instituto Federal de Educação, Ciência e Tecnologia de Goiás |
| Fabrizio Costa | University of Exeter |
| Miguel Couceiro | Inria |
| Shiyao Cui | Institute of Information Engineering, Chinese Academy of Sciences |
| Bertrand Cuissart | GREYC |
| Mohamad H. Danesh | Oregon State University |
| Thi-Bich-Hanh Dao | University of Orléans |
| Cedric De Boom | Ghent University |
| Marcos Luiz de Paula Bueno | Technische Universiteit Eindhoven |
| Matteo Dell'Amico | NortonLifeLock |
| Qi Deng | Shanghai University of Finance and Economics |
| Andreas Dengel | German Research Center for Artificial Intelligence |
| Sourya Dey | University of Southern California |
| Yao Di | Institute of Computing Technology, Chinese Academy of Sciences |
| Stefano Di Frischia | University of L'Aquila |
| Jilles Dibangoye | INSA Lyon |
| Felix Dietrich | Technical University of Munich |
| Jiahao Ding | University of Houston |
| Yao-Xiang Ding | Nanjing University |
| Tianyu Ding | Johns Hopkins University |
| Rui Ding | Microsoft |
| Thang Doan | McGill University |
| Carola Doerr | Sorbonne University, CNRS |
| Xiao Dong | The University of Queensland |
| Wei Du | University of Arkansas |
| Xin Du | Technische Universiteit Eindhoven |
| Yuntao Du | Nanjing University |
| Stefan Duffner | LIRIS |
| Sebastijan Dumancic | Katholieke Universiteit Leuven |
| Valentin Durand de Gevigney | IRISA |

| | |
|---|---|
| Saso Dzeroski | Jožef Stefan Institute |
| Mohamed Elati | Université d'Evry |
| Lukas Enderich | Robert Bosch GmbH |
| Dominik Endres | Philipps-Universität Marburg |
| Francisco Escolano | University of Alicante |
| Bjoern Eskofier | Friedrich-Alexander University Erlangen-Nürnberg |
| Roberto Esposito | Università di Torino |
| Georgios Exarchakis | Institut de la Vision |
| Melanie F. Pradier | Harvard University |
| Samuel G. Fadel | Universidade Estadual de Campinas |
| Evgeniy Faerman | Ludwig Maximilian University of Munich |
| Yujie Fan | Case Western Reserve University |
| Elaine Faria | Federal University of Uberlândia |
| Golnoosh Farnadi | Mila/University of Montreal |
| Fabio Fassetti | University of Calabria |
| Ad Feelders | Utrecht University |
| Yu Fei | Harbin Institute of Technology |
| Wenjie Feng | The Institute of Computing Technology, Chinese Academy of Sciences |
| Zunlei Feng | Zhejiang University |
| Cesar Ferri | Universitat Politècnica de València |
| Raul Fidalgo-Merino | European Commission Joint Research Centre |
| Murat Firat | Technische Universiteit Eindhoven |
| Francoise Fogelman-Soulié | Tianjin University |
| Vincent Fortuin | ETH Zurich |
| Iordanis Fostiropoulos | University of Southern California |
| Eibe Frank | University of Waikato |
| Benoît Frénay | Université de Namur |
| Nikolaos Freris | University of Science and Technology of China |
| Moshe Gabel | University of Toronto |
| Ricardo José Gabrielli Barreto Campello | University of Newcastle |
| Esther Galbrun | University of Eastern Finland |
| Claudio Gallicchio | University of Pisa |
| Yuanning Gao | Shanghai Jiao Tong University |
| Alberto Garcia-Duran | Ecole Polytechnique Fédérale de Lausanne |
| Eduardo Garrido | Universidad Autónoma de Madrid |
| Clément Gautrais | KU Leuven |
| Arne Gevaert | Ghent University |
| Giorgos Giannopoulos | IMSI, "Athena" Research Center |
| C. Lee Giles | The Pennsylvania State University |
| Ioana Giurgiu | IBM Research - Zurich |
| Thomas Goerttler | TU Berlin |
| Heitor Murilo Gomes | University of Waikato |
| Chen Gong | Shanghai Jiao Tong University |
| Zhiguo Gong | University of Macau |

| | |
|---|---|
| Hongyu Gong | University of Illinois at Urbana-Champaign |
| Pietro Gori | Télécom Paris |
| James Goulding | University of Nottingham |
| Kshitij Goyal | Katholieke Universiteit Leuven |
| Dmitry Grishchenko | Université Grenoble Alpes |
| Moritz Grosse-Wentrup | University of Vienna |
| Sebastian Gruber | Siemens AG |
| John Grundy | Monash University |
| Kang Gu | Dartmouth College |
| Jindong Gu | Siemens |
| Riccardo Guidotti | University of Pisa |
| Tias Guns | Vrije Universiteit Brussel |
| Ruocheng Guo | Arizona State University |
| Yiluan Guo | Singapore University of Technology and Design |
| Xiaobo Guo | University of Chinese Academy of Sciences |
| Thomas Guyet | IRISA |
| Jiawei Han | University of Illinois at Urbana-Champaign |
| Zhiwei Han | fortiss GmbH |
| Tom Hanika | University of Kassel |
| Shonosuke Harada | Kyoto University |
| Marwan Hassani | Technische Universiteit Eindhoven |
| Jianhao He | Sun Yat-sen University |
| Deniu He | Chongqing University of Posts and Telecommunications |
| Dongxiao He | Tianjin University |
| Stefan Heidekrueger | Technical University of Munich |
| Nandyala Hemachandra | Indian Institute of Technology Bombay |
| Till Hendrik Schulz | University of Bonn |
| Alexander Hepburn | University of Bristol |
| Sibylle Hess | Technische Universiteit Eindhoven |
| Javad Heydari | LG Electronics |
| Joyce Ho | Emory University |
| Shunsuke Horii | Waseda University |
| Tamas Horvath | University of Bonn and Fraunhofer IAIS |
| Mehran Hossein Zadeh Bazargani | University College Dublin |
| Robert Hu | University of Oxford |
| Weipeng Huang | Insight |
| Jun Huang | University of Tokyo |
| Haojie Huang | The University of New South Wales |
| Hong Huang | UGoe |
| Shenyang Huang | McGill University |
| Vân Anh Huynh-Thu | University of Liège |
| Dino Ienco | INRAE |
| Siohoi Ieng | Institut de la Vision |
| Angelo Impedovo | Università "Aldo Moro" degli studi di Bari |

| | |
|---|---|
| Muhammad Imran Razzak | Deakin University |
| Vasileios Iosifidis | Leibniz University Hannover |
| Joseph Isaac | Indian Institute of Technology Madras |
| Md Islam | Washington State University |
| Ziyu Jia | Beijing Jiaotong University |
| Lili Jiang | Umeå University |
| Yao Jiangchao | Alibaba |
| Tan Jianlong | Institute of Information Engineering, Chinese Academy of Sciences |
| Baihong Jin | University of California, Berkeley |
| Di Jin | Tianjin University |
| Wei Jing | Xi'an Jiaotong University |
| Jonathan Jouanne | ARIADNEXT |
| Ata Kaban | University of Birmingham |
| Tomasz Kajdanowicz | Wrocław University of Science and Technology |
| Sandesh Kamath | Chennai Mathematical Institute |
| Keegan Kang | Singapore University of Technology and Design |
| Bo Kang | Ghent University |
| Isak Karlsson | Stockholm University |
| Panagiotis Karras | Aarhus University |
| Nikos Katzouris | NCSR Demokritos |
| Uzay Kaymak | Technische Universiteit Eindhoven |
| Mehdi Kaytoue | Infologic |
| Pascal Kerschke | University of Münster |
| Jungtaek Kim | Pohang University of Science and Technology |
| Minyoung Kim | Samsung AI Center Cambridge |
| Masahiro Kimura | Ryukoku University |
| Uday Kiran | The University of Tokyo |
| Bogdan Kirillov | ITMO University |
| Péter Kiss | ELTE |
| Gerhard Klassen | Heinrich Heine University Düsseldorf |
| Dmitry Kobak | Eberhard Karls University of Tübingen |
| Masahiro Kohjima | NTT |
| Ziyi Kou | University of Rochester |
| Wouter Kouw | Technische Universiteit Eindhoven |
| Fumiya Kudo | Hitachi, Ltd. |
| Piotr Kulczycki | Systems Research Institute, Polish Academy of Sciences |
| Ilona Kulikovskikh | Samara State Aerospace University |
| Rajiv Kumar | IIT Bombay |
| Pawan Kumar | IIT Kanpur |
| Suhansanu Kumar | University of Illinois, Urbana-Champaign |
| Abhishek Kumar | University of Helsinki |
| Gautam Kunapuli | The University of Texas at Dallas |
| Takeshi Kurashima | NTT |
| Vladimir Kuzmanovski | Jožef Stefan Institute |

Shiwei Liu                          Technische Universiteit Eindhoven
Shenghua Liu                        Institute of Computing Technology, Chinese Academy
                                       of Sciences
Corrado Loglisci                    University of Bari Aldo Moro
Andrey Lokhov                       Los Alamos National Laboratory
Yijun Lu                            Alibaba Cloud
Xuequan Lu                          Deakin University
Szymon Lukasik                      AGH University of Science and Technology
Phuc Luong                          Deakin University
Jianming Lv                         South China University of Technology
Gengyu Lyu                          Beijing Jiaotong University
Vijaikumar M.                       Indian Institute of Science
Jing Ma                             Emory University
Nan Ma                              Shanghai Jiao Tong University
Sebastian Mair                      Leuphana University Lüneburg
Marjan Mansourvar                   University of Southern Denmark
Vincent Margot                      Advestis
Fernando Martínez-Plumed            Joint Research Centre - European Commission
Florent Masseglia                   Inria
Romain Mathonat                     Université de Lyon
Deepak Maurya                       Indian Institute of Technology Madras
Christian Medeiros Adriano          Hasso-Plattner-Institut
Purvanshi Mehta                     University of Rochester
Tobias Meisen                       Bergische Universität Wuppertal
Luciano Melodia                     Friedrich-Alexander Universität Erlangen-Nürnberg
Ernestina Menasalvas                Universidad Politécnica de Madrid
Vlado Menkovski                     Technische Universiteit Eindhoven
Engelbert Mephu Nguifo              Université Clermont Auvergne
Alberto Maria Metelli               Politecnico di Milano
Donald Metzler                      Google
Anke Meyer-Baese                    Florida State University
Richard Meyes                       University of Wuppertal
Haithem Mezni                       University of Jendouba
Paolo Mignone                       Università degli Studi di Bari Aldo Moro
Matej Mihelčić                      University of Zagreb
Decebal Constantin Mocanu           University of Twente
Christoph Molnar                    Ludwig Maximilian University of Munich
Lia Morra                           Politecnico di Torino
Christopher Morris                  TU Dortmund University
Tadeusz Morzy                       Poznań University of Technology
Henry Moss                          Lancaster University
Tetsuya Motokawa                    University of Tsukuba
Mathilde Mougeot                    Université Paris-Saclay
Tingting Mu                         The University of Manchester
Andreas Mueller                     NYU
Tanmoy Mukherjee                    Queen Mary University of London

| | |
|---|---|
| Ksenia Mukhina | ITMO University |
| Peter Müllner | Know-Center |
| Guido Muscioni | University of Illinois at Chicago |
| Waleed Mustafa | TU Kaiserslautern |
| Mohamed Nadif | University of Paris |
| Ankur Nahar | Indian Institute of Technology Jodhpur |
| Kei Nakagawa | Nomura Asset Management Co., Ltd. |
| Haïfa Nakouri | University of Tunis |
| Mirco Nanni | KDD-Lab ISTI-CNR Pisa |
| Nicolo' Navarin | University of Padova |
| Richi Nayak | Queensland University of Technology |
| Mojtaba Nayyeri | University of Bonn |
| Daniel Neider | MPI SWS |
| Nan Neng | Institute of Information Engineering, Chinese Academy of Sciences |
| Stefan Neumann | University of Vienna |
| Dang Nguyen | Deakin University |
| Kien Duy Nguyen | University of Southern California |
| Jingchao Ni | NEC Laboratories America |
| Vlad Niculae | Instituto de Telecomunicações |
| Sofia Maria Nikolakaki | Boston University |
| Kun Niu | Beijing University of Posts and Telecommunications |
| Ryo Nomura | Waseda University |
| Eirini Ntoutsi | Leibniz University Hannover |
| Andreas Nuernberger | Otto von Guericke University of Magdeburg |
| Tsuyoshi Okita | Kyushu Institute of Technology |
| Maria Oliver Parera | GIPSA-lab |
| Bruno Ordozgoiti | Aalto University |
| Sindhu Padakandla | Indian Institute of Science |
| Tapio Pahikkala | University of Turku |
| Joao Palotti | Qatar Computing Research Institute |
| Guansong Pang | The University of Adelaide |
| Pance Panov | Jožef Stefan Institute |
| Konstantinos Papangelou | The University of Manchester |
| Yulong Pei | Technische Universiteit Eindhoven |
| Nikos Pelekis | University of Piraeus |
| Thomas Pellegrini | Université Toulouse III - Paul Sabatier |
| Charlotte Pelletier | Univ. Bretagne Sud |
| Jaakko Peltonen | Aalto University and Tampere University |
| Shaowen Peng | Kyushu University |
| Siqi Peng | Kyoto University |
| Bo Peng | The Ohio State University |
| Lukas Pensel | Johannes Gutenberg University Mainz |
| Aritz Pérez Martínez | Basque Center for Applied Mathematics |
| Lorenzo Perini | KU Leuven |
| Matej Petković | Jožef Stefan Institute |

| | |
|---|---|
| Bernhard Pfahringer | University of Waikato |
| Weiguo Pian | Chongqing University |
| Francesco Piccialli | University of Naples Federico II |
| Sebastian Pineda Arango | University of Hildesheim |
| Gianvito Pio | University of Bari "Aldo Moro" |
| Giuseppe Pirrò | Sapienza University of Rome |
| Anastasia Podosinnikova | Massachusetts Institute of Technology |
| Sebastian Pölsterl | Ludwig Maximilian University of Munich |
| Vamsi Potluru | JP Morgan AI Research |
| Rafael Poyiadzi | University of Bristol |
| Surya Prakash | University of Canberra |
| Paul Prasse | University of Potsdam |
| Rameshwar Pratap | Indian Institute of Technology Mandi |
| Jonas Prellberg | University of Oldenburg |
| Hugo Proenca | Leiden Institute of Advanced Computer Science |
| Ricardo Prudencio | Federal University of Pernambuco |
| Petr Pulc | Institute of Computer Science of the Czech Academy of Sciences |
| Lei Qi | Iowa State University |
| Zhenyue Qin | The Australian National University |
| Rahul Ragesh | PES University |
| Tahrima Rahman | The University of Texas at Dallas |
| Zana Rashidi | York University |
| S. S. Ravi | University of Virginia and University at Albany – SUNY |
| Ambrish Rawat | IBM |
| Henry Reeve | University of Birmingham |
| Reza Refaei Afshar | Technische Universiteit Eindhoven |
| Navid Rekabsaz | Johannes Kepler University Linz |
| Yongjian Ren | Shandong University |
| Zhiyun Ren | The Ohio State University |
| Guohua Ren | LG Electronics |
| Yuxiang Ren | Florida State University |
| Xavier Renard | AXA |
| Martí Renedo Mirambell | Universitat Politècnica de Catalunya |
| Gavin Rens | Katholiek Universiteit Leuven |
| Matthias Renz | Christian-Albrechts-Universität zu Kiel |
| Guillaume Richard | EDF R&D |
| Matteo Riondato | Amherst College |
| Niklas Risse | Bielefeld University |
| Lars Rosenbaum | Robert Bosch GmbH |
| Celine Rouveirol | Université Sorbonne Paris Nord |
| Shoumik Roychoudhury | Temple University |
| Polina Rozenshtein | Aalto University |
| Peter Rubbens | Flanders Marine Institute (VLIZ) |
| David Ruegamer | LMU Munich |

| | |
|---|---|
| Matteo Ruffini | ToolsGroup |
| Ellen Rushe | Insight Centre for Data Analytics |
| Amal Saadallah | TU Dortmund |
| Yogish Sabharwal | IBM Research - India |
| Mandana Saebi | University of Notre Dame |
| Aadirupa Saha | IISc |
| Seyed Erfan Sajjadi | Brunel University |
| Durgesh Samariya | Federation University |
| Md Samiullah | Monash University |
| Mark Sandler | Google |
| Raul Santos-Rodriguez | University of Bristol |
| Yucel Saygin | Sabancı University |
| Pierre Schaus | UCLouvain |
| Fabian Scheipl | Ludwig Maximilian University of Munich |
| Katerina Schindlerova | University of Vienna |
| Ute Schmid | University of Bamberg |
| Daniel Schmidt | Monash University |
| Sebastian Schmoll | Ludwig Maximilian University of Munich |
| Johannes Schneider | University of Liechtenstein |
| Marc Schoenauer | Inria Saclay Île-de-France |
| Jonas Schouterden | Katholieke Universiteit Leuven |
| Leo Schwinn | Friedrich-Alexander-Universität Erlangen-Nürnberg |
| Florian Seiffarth | University of Bonn |
| Nan Serra | NEC Laboratories Europe GmbH |
| Rowland Seymour | Univeristy of Nottingham |
| Ammar Shaker | NEC Laboratories Europe |
| Ali Shakiba | Vali-e-Asr University of Rafsanjan |
| Junming Shao | University of Science and Technology of China |
| Zhou Shao | Tsinghua University |
| Manali Sharma | Samsung Semiconductor Inc. |
| Jiaming Shen | University of Illinois at Urbana-Champaign |
| Ying Shen | Sun Yat-sen University |
| Hao Shen | fortiss GmbH |
| Tao Shen | University of Technology Sydney |
| Ge Shi | Beijing Institute of Technology |
| Ziqiang Shi | Fujitsu Research & Development Center |
| Masumi Shirakawa | hapicom Inc./Osaka University |
| Kai Shu | Arizona State University |
| Amila Silva | The University of Melbourne |
| Edwin Simpson | University of Bristol |
| Dinesh Singh | RIKEN Center for Advanced Intelligence Project |
| Jaspreet Singh | L3S Research Centre |
| Spiros Skiadopoulos | University of the Peloponnese |
| Gavin Smith | University of Nottingham |
| Miguel A. Solinas | CEA |
| Dongjin Song | NEC Labs America |

| Arnaud Soulet | Université de Tours |
|---|---|
| Marvin Ssemambo | Makerere University |
| Michiel Stock | Ghent University |
| Filipo Studzinski Perotto | Institut de Recherche en Informatique de Toulouse |
| Adisak Sukul | Iowa State University |
| Lijuan Sun | Beijing Jiaotong University |
| Tao Sun | National University of Defense Technology |
| Ke Sun | Peking University |
| Yue Sun | Beijing Jiaotong University |
| Hari Sundaram | University of Illinois at Urbana-Champaign |
| Gero Szepannek | Stralsund University of Applied Sciences |
| Jacek Tabor | Jagiellonian University |
| Jianwei Tai | IIE, CAS |
| Naoya Takeishi | RIKEN Center for Advanced Intelligence Project |
| Chang Wei Tan | Monash University |
| Jinghua Tan | Southwestern University of Finance and Economics |
| Zeeshan Tariq | Ulster University |
| Bouadi Tassadit | IRISA-Université de Rennes 1 |
| Maryam Tavakol | TU Dortmund |
| Romain Tavenard | Univ. Rennes 2/LETG-COSTEL/IRISA-OBELIX |
| Alexandre Termier | Université de Rennes 1 |
| Janek Thomas | Fraunhofer Institute for Integrated Circuits IIS |
| Manoj Thulasidas | Singapore Management University |
| Hao Tian | Syracuse University |
| Hiroyuki Toda | NTT |
| Jussi Tohka | University of Eastern Finland |
| Ricardo Torres | Norwegian University of Science and Technology |
| Isaac Triguero Velázquez | University of Nottingham |
| Sandhya Tripathi | Indian Institute of Technology Bombay |
| Holger Trittenbach | Karlsruhe Institute of Technology |
| Peter van der Putten | Leiden University & Pegasystems |
| Elia Van Wolputte | KU Leuven |
| Fabio Vandin | University of Padova |
| Titouan Vayer | IRISA |
| Ashish Verma | IBM Research - US |
| Bouvier Victor | Sidetrade MICS |
| Julia Vogt | University of Basel |
| Tim Vor der Brück | Lucerne University of Applied Sciences and Arts |
| Yb W. | Chongqing University |
| Krishna Wadhwani | Indian Institute of Technology Bombay |
| Huaiyu Wan | Beijing Jiaotong University |
| Qunbo Wang | Beihang University |
| Beilun Wang | Southeast University |
| Yiwei Wang | National University of Singapore |
| Bin Wang | Xiaomi AI Lab |

| Jiong Wang | Institute of Information Engineering, Chinese Academy of Sciences |
| Xiaobao Wang | Tianjin University |
| Shuheng Wang | Nanjing University of Science and Technology |
| Jihu Wang | Shandong University |
| Haobo Wang | Zhejiang University |
| Xianzhi Wang | University of Technology Sydney |
| Chao Wang | Shanghai Jiao Tong University |
| Jun Wang | Southwest University |
| Jing Wang | Beijing Jiaotong University |
| Di Wang | Nanyang Technological University |
| Yashen Wang | China Academy of Electronics and Information Technology of CETC |
| Qinglong Wang | McGill University |
| Sen Wang | University of Queensland |
| Di Wang | State University of New York at Buffalo |
| Qing Wang | Information Science Research Centre |
| Guoyin Wang | Chongqing University of Posts and Telecommunications |
| Thomas Weber | Ludwig-Maximilians-Universität München |
| Lingwei Wei | University of Chinese Academy of Sciences; Institute of Information Engineering, CAS |
| Tong Wei | Nanjing University |
| Pascal Welke | University of Bonn |
| Yang Wen | University of Science and Technology of China |
| Yanlong Wen | Nankai University |
| Paul Weng | UM-SJTU Joint Institute |
| Matthias Werner | ETAS GmbH, Bosch Group |
| Joerg Wicker | The University of Auckland |
| Uffe Wiil | University of Southern Denmark |
| Paul Wimmer | University of Lübeck; Robert Bosch GmbH |
| Martin Wistuba | University of Hildesheim |
| Feijie Wu | The Hong Kong Polytechnic University |
| Xian Wu | University of Notre Dame |
| Hang Wu | Georgia Institute of Technology |
| Yubao Wu | Georgia State University |
| Yichao Wu | SenseTime Group Limited |
| Xi-Zhu Wu | Nanjing University |
| Jia Wu | Macquarie University |
| Yang Xiaofei | Harbin Institute of Technology, Shenzhen |
| Yuan Xin | University of Science and Technology of China |
| Liu Xinshun | VIVO |
| Taufik Xu | Tsinghua University |
| Jinhui Xu | State University of New York at Buffalo |
| Depeng Xu | University of Arkansas |
| Peipei Xu | University of Liverpool |

| | |
|---|---|
| Yichen Xu | Beijing University of Posts and Telecommunications |
| Bo Xu | Donghua University |
| Hansheng Xue | Harbin Institute of Technology, Shenzhen |
| Naganand Yadati | Indian Institute of Science |
| Akihiro Yamaguchi | Toshiba Corporation |
| Haitian Yang | Institute of Information Engineering, Chinese Academy of Sciences |
| Hongxia Yang | Alibaba Group |
| Longqi Yang | HPCL |
| Xiaochen Yang | University College London |
| Yuhan Yang | Shanghai Jiao Tong University |
| Ya Zhou Yang | National University of Defense Technology |
| Feidiao Yang | Institute of Computing Technology, Chinese Academy of Sciences |
| Liu Yang | Tianjin University |
| Chaoqi Yang | University of Illinois at Urbana-Champaign |
| Carl Yang | University of Illinois at Urbana-Champaign |
| Guanyu Yang | Xi'an Jiaotong - Liverpool University |
| Yang Yang | Nanjing University |
| Weicheng Ye | Carnegie Mellon University |
| Wei Ye | Peking University |
| Yanfang Ye | Case Western Reserve University |
| Kejiang Ye | SIAT, Chinese Academy of Sciences |
| Florian Yger | Université Paris-Dauphine |
| Yunfei Yin | Chongqing University |
| Lu Yin | Technische Universiteit Eindhoven |
| Wang Yingkui | Tianjin University |
| Kristina Yordanova | University of Rostock |
| Tao You | Northwestern Polytechnical University |
| Hong Qing Yu | University of Bedfordshire |
| Bowen Yu | Institute of Information Engineering, Chinese Academy of Sciences |
| Donghan Yu | Carnegie Mellon University |
| Yipeng Yu | Tencent |
| Shujian Yu | NEC Laboratories Europe |
| Jiadi Yu | Shanghai Jiao Tong University |
| Wenchao Yu | University of California, Los Angeles |
| Feng Yuan | The University of New South Wales |
| Chunyuan Yuan | Institute of Information Engineering, Chinese Academy of Sciences |
| Sha Yuan | Tsinghua University |
| Farzad Zafarani | Purdue University |
| Marco Zaffalon | IDSIA |
| Nayyar Zaidi | Monash University |
| Tianzi Zang | Shanghai Jiao Tong University |
| Gerson Zaverucha | Federal University of Rio de Janeiro |

| | |
|---|---|
| Javier Zazo | Harvard University |
| Albin Zehe | University of Würzburg |
| Yuri Zelenkov | National Research University Higher School of Economics |
| Amber Zelvelder | Umeå University |
| Mingyu Zhai | NARI Group Corporation |
| Donglin Zhan | Sichuan University |
| Yu Zhang | Southeast University |
| Wenbin Zhang | University of Maryland |
| Qiuchen Zhang | Emory University |
| Tong Zhang | PKU |
| Jianfei Zhang | Case Western Reserve University |
| Nailong Zhang | MassMutual |
| Yi Zhang | Nanjing University |
| Xiangliang Zhang | King Abdullah University of Science and Technology |
| Ya Zhang | Shanghai Jiao Tong University |
| Zongzhang Zhang | Nanjing University |
| Lei Zhang | Institute of Information Engineering, Chinese Academy of Sciences |
| Jing Zhang | Renmin University of China |
| Xianchao Zhang | Dalian University of Technology |
| Jiangwei Zhang | National University of Singapore |
| Fengpan Zhao | Georgia State University |
| Lin Zhao | Institute of Information Engineering, Chinese Academy of Sciences |
| Long Zheng | Huazhong University of Science and Technology |
| Zuowu Zheng | Shanghai Jiao Tong University |
| Tongya Zheng | Zhejiang University |
| Runkai Zheng | Jinan University |
| Cheng Zheng | University of California, Los Angeles |
| Wenbo Zheng | Xi'an Jiaotong University |
| Zhiqiang Zhong | University of Luxembourg |
| Caiming Zhong | Ningbo University |
| Ding Zhou | Columbia University |
| Yilun Zhou | MIT |
| Ming Zhou | Shanghai Jiao Tong University |
| Yanqiao Zhu | Institute of Automation, Chinese Academy of Sciences |
| Wenfei Zhu | King |
| Wanzheng Zhu | University of Illinois at Urbana-Champaign |
| Fuqing Zhu | Institute of Information Engineering, Chinese Academy of Sciences |
| Markus Zopf | TU Darmstadt |
| Weidong Zou | Beijing Institute of Technology |
| Jingwei Zuo | UVSQ |

**Program Committee Members, Applied Data Science Track**

| | |
|---|---|
| Deepak Ajwani | Nokia Bell Labs |
| Nawaf Alharbi | Kansas State University |
| Rares Ambrus | Toyota Research Institute |
| Maryam Amir Haeri | Technische Universität Kaiserslautern |
| Jean-Marc Andreoli | Naverlabs Europe |
| Cecilio Angulo | Universitat Politècnica de Catalunya |
| Stefanos Antaris | KTH Royal Institute of Technology |
| Nino Antulov-Fantulin | ETH Zurich |
| Francisco Antunes | University of Coimbra |
| Muhammad Umer Anwaar | Technical University of Munich |
| Cristian Axenie | Audi Konfuzius-Institut Ingolstadt/Technical University of Ingolstadt |
| Mehmet Cem Aytekin | Sabancı University |
| Anthony Bagnall | University of East Anglia |
| Marco Baldan | Leibniz University Hannover |
| Maria Bampa | Stockholm University |
| Karin Becker | UFRGS |
| Swarup Ranjan Behera | Indian Institute of Technology Guwahati |
| Michael Berthold | University of Konstanz |
| Antonio Bevilacqua | Insight Centre for Data Analytics |
| Ananth Reddy Bhimireddy | Indiana University Purdue University - Indianapolis |
| Haixia Bi | University of Bristol |
| Wu Bin | Zhengzhou University |
| Thibault Blanc Beyne | INP Toulouse |
| Andrzej Bobyk | Maria Curie-Skłodowska University |
| Antonio Bonafonte | Amazon |
| Ludovico Boratto | Eurecat |
| Massimiliano Botticelli | Robert Bosch GmbH |
| Maria Brbic | Stanford University |
| Sebastian Buschjäger | TU Dortmund |
| Rui Camacho | University of Porto |
| Doina Caragea | Kansas State University |
| Nicolas Carrara | University of Toronto |
| Michele Catasta | Stanford University |
| Oded Cats | Delft University of Technology |
| Tania Cerquitelli | Politecnico di Torino |
| Fabricio Ceschin | Federal University of Paraná |
| Jeremy Charlier | University of Luxembourg |
| Anveshi Charuvaka | GE Global Research |
| Liang Chen | Sun Yat-sen University |
| Zhiyong Cheng | Shandong Artificial Intelligence Institute |

| | |
|---|---|
| Silvia Chiusano | Politecnico di Torino |
| Cristian Consonni | Eurecat - Centre Tecnòlogic de Catalunya |
| Laure Crochepierre | RTE |
| Henggang Cui | Uber ATG |
| Tiago Cunha | University of Porto |
| Elena Daraio | Politecnico di Torino |
| Hugo De Oliveira | HEVA/Mines Saint-Étienne |
| Tom Decroos | Katholieke Universiteit Leuven |
| Himel Dev | University of Illinois at Urbana-Champaign |
| Eustache Diemert | Criteo AI Lab |
| Nat Dilokthanakul | Vidyasirimedhi Institute of Science and Technology |
| Daizong Ding | Fudan University |
| Kaize Ding | ASU |
| Ming Ding | Tsinghua University |
| Xiaowen Dong | University of Oxford |
| Sourav Dutta | Huawei Research |
| Madeleine Ellis | University of Nottingham |
| Benjamin Evans | Brunel University London |
| Francesco Fabbri | Universitat Pompeu Fabra |
| Benjamin Fauber | Dell Technologies |
| Fuli Feng | National University of Singapore |
| Oluwaseyi Feyisetan | Amazon |
| Ferdinando Fioretto | Georgia Institute of Technology |
| Caio Flexa | Federal University of Pará |
| Germain Forestier | Université de Haute-Alsace |
| Blaz Fortuna | Qlector |
| Enrique Frias-Martinez | Telefónica Research and Development |
| Zuohui Fu | Rutgers University |
| Takahiro Fukushige | Nissan Motor Co., Ltd. |
| Chen Gao | Tsinghua University |
| Johan Garcia | Karlstad University |
| Marco Gärtler | ABB Corporate Research Center |
| Kanishka Ghosh Dastidar | Universität Passau |
| Biraja Ghoshal | Brunel University London |
| Lovedeep Gondara | Simon Fraser University |
| Severin Gsponer | Science Foundation Ireland |
| Xinyu Guan | Xi'an Jiaotong University |
| Karthik Gurumoorthy | Amazon |
| Marina Haliem | Purdue University |
| Massinissa Hamidi | Laboratoire LIPN-UMR CNRS 7030, Sorbonne Paris Cité |
| Junheng Hao | University of California, Los Angeles |
| Khadidja Henni | Université TÉLUQ |

| | |
|---|---|
| Martin Holena | Institute of Computer Science Academy of Sciences of the Czech Republic |
| Ziniu Hu | University of California, Los Angeles |
| Weihua Hu | Stanford University |
| Chao Huang | University of Notre Dame |
| Hong Huang | UGoe |
| Inhwan Hwang | Seoul National University |
| Chidubem Iddianozie | University College Dublin |
| Omid Isfahani Alamdari | University of Pisa |
| Guillaume Jacquet | Joint Research Centre - European Commission |
| Nishtha Jain | ADAPT Centre |
| Samyak Jain | NIT Karnataka, Surathkal |
| Mohsan Jameel | University of Hildesheim |
| Di Jiang | WeBank |
| Song Jiang | University of California, Los Angeles |
| Khiary Jihed | Johannes Kepler Universität Linz |
| Md. Rezaul Karim | Fraunhofer FIT |
| Siddhant Katyan | IIIT Hyderabad |
| Jin Kyu Kim | Facebook |
| Sundong Kim | Institute for Basic Science |
| Tomas Kliegr | Prague University of Economics and Business |
| Yun Sing Koh | The University of Auckland |
| Aljaz Kosmerlj | Jožef Stefan Institute |
| Jitin Krishnan | George Mason University |
| Alejandro Kuratomi | Stockholm University |
| Charlotte Laclau | Laboratoire Hubert Curien |
| Filipe Lauar | Federal University of Minas Gerais |
| Thach Le Nguyen | The Insight Centre for Data Analytics |
| Wenqiang Lei | National University of Singapore |
| Camelia Lemnaru | Universitatea Tehnică din Cluj-Napoca |
| Carson Leung | University of Manitoba |
| Meng Li | Ant Financial Services Group |
| Zeyu Li | University of California, Los Angeles |
| Pieter Libin | Vrije Universiteit Brussel |
| Tomislav Lipic | Ruđer Bošković Institut |
| Bowen Liu | Stanford University |
| Yin Lou | Ant Financial |
| Martin Lukac | Nazarbayev University |
| Brian Mac Namee | University College Dublin |
| Fragkiskos Malliaros | Université Paris-Saclay |
| Mirko Marras | University of Cagliari |
| Smit Marvaniya | IBM Research - India |
| Kseniia Melnikova | Samsung R&D Institute Russia |

| | |
|---|---|
| João Mendes-Moreira | University of Porto |
| Ioannis Mitros | Insight Centre for Data Analytics |
| Elena Mocanu | University of Twente |
| Hebatallah Mohamed | Free University of Bozen-Bolzano |
| Roghayeh Mojarad | Université Paris-Est Créteil |
| Mirco Nanni | KDD-Lab ISTI-CNR Pisa |
| Juggapong Natwichai | Chiang Mai University |
| Sasho Nedelkoski | TU Berlin |
| Kei Nemoto | The Graduate Center, City University of New York |
| Ba-Hung Nguyen | Japan Advanced Institute of Science and Technology |
| Tobias Nickchen | Paderborn University |
| Aastha Nigam | LinkedIn Inc |
| Inna Novalija | Jožef Stefan Institute |
| Francisco Ocegueda-Hernandez | National Oilwell Varco |
| Tsuyoshi Okita | Kyushu Institute of Technology |
| Oghenejokpeme Orhobor | The University of Manchester |
| Aomar Osmani | Université Sorbonne Paris Nord |
| Latifa Oukhellou | IFSTTAR |
| Rodolfo Palma | Inria Chile |
| Pankaj Pandey | Indian Institute of Technology Gandhinagar |
| Luca Pappalardo | University of Pisa, ISTI-CNR |
| Paulo Paraíso | INESC TEC |
| Namyong Park | Carnegie Mellon University |
| Chanyoung Park | University of Illinois at Urbana-Champaign |
| Miquel Perelló-Nieto | University of Bristol |
| Nicola Pezzotti | Philips Research |
| Tiziano Piccardi | Ecole Polytechnique Fédérale de Lausanne |
| Thom Pijnenburg | Elsevier |
| Valentina Poggioni | Università degli Studi di Perugia |
| Chuan Qin | University of Science and Technology of China |
| Jiezhong Qiu | Tsinghua University |
| Maria Ramirez-Loaiza | Intel Corporation |
| Manjusha Ravindranath | ASU |
| Zhaochun Ren | Shandong University |
| Antoine Richard | Georgia Institute of Technology |
| Kit Rodolfa | Carnegie Mellon University |
| Mark Patrick Roeling | Technical University of Delft |
| Soumyadeep Roy | Indian Institute of Technology Kharagpur |
| Ellen Rushe | Insight Centre for Data Analytics |
| Amal Saadallah | TU Dortmund |
| Carlos Salort Sanchez | Huawei |
| Eduardo Hugo Sanchez | IRT Saint Exupéry |
| Markus Schmitz | University of Erlangen-Nuremberg/BMW Group |
| Ayan Sengupta | Optum Global Analytics (India) Pvt. Ltd. |
| Ammar Shaker | NEC Laboratories Europe |

| | |
|---|---|
| Manali Sharma | Samsung Semiconductor Inc. |
| Jiaming Shen | University of Illinois at Urbana-Champaign |
| Dash Shi | LinkedIn |
| Ashish Sinha | IIT Roorkee |
| Yorick Spenrath | Technische Universiteit Eindhoven |
| Simon Stieber | University of Augsburg |
| Hendra Suryanto | Rich Data Corporation |
| Raunak Swarnkar | IIT Gandhinagar |
| Imen Trabelsi | National Engineering School of Tunis |
| Alexander Treiss | Karlsruhe Institute of Technology |
| Rahul Tripathi | Amazon |
| Dries Van Daele | Katholieke Universiteit Leuven |
| Ranga Raju Vatsavai | North Carolina State University |
| Vishnu Venkataraman | Credit Karma |
| Sergio Viademonte | Vale Institute of Technology, Vale SA |
| Yue Wang | Microsoft Research |
| Changzhou Wang | The Boeing Company |
| Xiang Wang | National University of Singapore |
| Hongwei Wang | Shanghai Jiao Tong University |
| Wenjie Wang | Emory University |
| Zirui Wang | Carnegie Mellon University |
| Shen Wang | University of Illinois at Chicago |
| Dingxian Wang | East China Normal University |
| Yoshikazu Washizawa | The University of Electro-Communications |
| Chrys Watson Ross | University of New Mexico |
| Dilusha Weeraddana | CSIRO |
| Ying Wei | The Hong Kong University of Science and Technology |
| Laksri Wijerathna | Monash University |
| Le Wu | Hefei University of Technology |
| Yikun Xian | Rutgers University |
| Jian Xu | Citadel |
| Haiqin Yang | Ping An Life |
| Yang Yang | Northwestern University |
| Carl Yang | University of Illinois at Urbana-Champaign |
| Chin-Chia Michael Yeh | Visa Research |
| Shujian Yu | NEC Laboratories Europe |
| Chung-Hsien Yu | University of Massachusetts Boston |
| Jun Yuan | The Boeing Company |
| Stella Zevio | LIPN |
| Hanwen Zha | University of California, Santa Barbara |
| Chuxu Zhang | University of Notre Dame |
| Fanjin Zhang | Tsinghua University |
| Xiaohan Zhang | Sony Interactive Entertainment |
| Xinyang Zhang | University of Illinois at Urbana-Champaign |
| Mia Zhao | Airbnb |
| Qi Zhu | University of Illinois at Urbana-Champaign |

| | |
|---|---|
| Hengshu Zhu | Baidu Inc. |
| Tommaso Zoppi | University of Florence |
| Lan Zou | Carnegie Mellon University |

**Program Committee Members, Demo Track**

| | |
|---|---|
| Deepak Ajwani | Nokia Bell Labs |
| Rares Ambrus | Toyota Research Institute |
| Jean-Marc Andreoli | NAVER LABS Europe |
| Ludovico Boratto | Eurecat |
| Nicolas Carrara | University of Toronto |
| Michelangelo Ceci | Università degli Studi di Bari Aldo Moro |
| Tania Cerquitelli | Politecnico di Torino |
| Liang Chen | Sun Yat-sen University |
| Jiawei Chen | Zhejiang University |
| Zhiyong Cheng | Shandong Artificial Intelligence Institute |
| Silvia Chiusano | Politecnico di Torino |
| Henggang Cui | Uber ATG |
| Tiago Cunha | University of Porto |
| Chris Develder | Ghent University |
| Nat Dilokthanakul | Vidyasirimedhi Institute of Science and Technology |
| Daizong Ding | Fudan University |
| Kaize Ding | ASU |
| Xiaowen Dong | University of Oxford |
| Fuli Feng | National University of Singapore |
| Enrique Frias-Martinez | Telefónica Research and Development |
| Zuohui Fu | Rutgers University |
| Chen Gao | Tsinghua University |
| Thomas Gärtner | TU Wien |
| Derek Greene | University College Dublin |
| Severin Gsponer | University College Dublin |
| Xinyu Guan | Xi'an Jiaotong University |
| Junheng Hao | University of California, Los Angeles |
| Ziniu Hu | University of California, Los Angeles |
| Chao Huang | University of Notre Dame |
| Hong Huang | UGoe |
| Neil Hurley | University College Dublin |
| Guillaume Jacquet | Joint Research Centre - European Commission |
| Di Jiang | WeBank |
| Song Jiang | University of California, Los Angeles |
| Jihed Khiari | Johannes Kepler Universität Linz |
| Mark Last | Ben-Gurion University of the Negev |
| Thach Le Nguyen | The Insight Centre for Data Analytics |
| Vincent Lemaire | Orange Labs |
| Camelia Lemnaru | Universitatea Tehnică din Cluj-Napoca |
| Bowen Liu | Stanford University |

## Sponsors

# Contents – Part III

## Computer Vision and Image Processing

## Natural Language Processing

## Bioinformatics

# Combinatorial Optimization

# Algorithms for Optimizing the Ratio of Monotone $k$-Submodular Functions

Hau Chan[1(⊠)], Grigorios Loukides[2], and Zhenghui Su[1]

[1] University of Nebraska-Lincoln, Lincoln, NE, USA
hchan3@unl.edu, zsu@huskers.unl.edu
[2] King's College London, London, UK
grigorios.loukides@kcl.ac.uk

**Abstract.** We study a new optimization problem that minimizes the ratio of two monotone $k$-submodular functions. The problem has applications in sensor placement, influence maximization, and feature selection among many others where one wishes to make a tradeoff between two objectives, measured as a ratio of two functions (e.g., solution cost vs. quality). We develop three greedy based algorithms for the problem, with approximation ratios that depend on the curvatures and/or the values of the functions. We apply our algorithms to a sensor placement problem where one aims to install $k$ types of sensors, while minimizing the ratio between cost and uncertainty of sensor measurements, as well as to an influence maximization problem where one seeks to advertise $k$ products to minimize the ratio between advertisement cost and expected number of influenced users. Our experimental results demonstrate the effectiveness of minimizing the respective ratios and the runtime efficiency of our algorithms. Finally, we discuss various extensions of our problems.

**Keywords:** k-submodular function · Greedy algorithm · Approximation

## 1  Introduction

In many applications ranging from machine learning such as feature selection and clustering [1] to social network analysis [14], we want to select $k$ disjoint subsets of elements from a ground set that optimize a *k-submodular* function. A $k$-submodular function takes in $k$ disjoint subsets as argument and has a diminishing returns property with respect to each subset when fixing the other $k-1$ subsets [6]. For example, in sensor domain with $k$ types of sensors, a $k$-submodular function can model the diminishing cost of obtaining an extra sensor of a type when fixing the numbers of sensors of other types. Most recently, the problem of $k$-submodular function maximization has been studied in [6,7,14,18].

In this work, we study a new optimization problem which aims to find $k$ disjoint subsets of a ground set that minimize the ratio of two non-negative and monotone $k$-submodular functions. We call this the RS-$k$ minimization problem.

© Springer Nature Switzerland AG 2021
F. Hutter et al. (Eds.): ECML PKDD 2020, LNAI 12459, pp. 3–19, 2021.
https://doi.org/10.1007/978-3-030-67664-3_1

The problem can be used to model a situation where one needs to make a trade-off between two different objectives (e.g., solution cost and quality). For the exposition of our problem, we provide some preliminary definitions.

Let $V = \{u_1, ..., u_n\}$ be a non-empty set of $n$ elements, $k \geq 1$ be an integer, and $[k] = \{1, ..., k\}$. Let also $(k+1)^V = \{(X_1, ..., X_k) \mid X_i \subseteq V, \forall i \in [k], \text{ and } X_i \cap X_j = \emptyset, \forall i \neq j \in [k]\}$ be the set of $k$ (pairwise) disjoint subsets of $V$. A function $f : (k+1)^V \rightarrow \mathbb{R}$ is $k$-submodular [6] if and only if for any $\mathbf{x}, \mathbf{y} \in (k+1)^V$, it holds that:

$$f(\mathbf{x}) + f(\mathbf{y}) \geq f(\mathbf{x} \sqcap \mathbf{y}) + f(\mathbf{x} \sqcup \mathbf{y}), \qquad (1)$$

where

$$\mathbf{x} \sqcap \mathbf{y} = (X_1 \cap Y_1, ..., X_k \cap Y_k) \qquad \text{and}$$

$$\mathbf{x} \sqcup \mathbf{y} = ((X_1 \cup Y_1) \setminus (\bigcup_{i \in [k] \setminus \{1\}} X_i \cup Y_i), ..., (X_k \cup Y_k) \setminus (\bigcup_{i \in [k] \setminus \{k\}} X_i \cup Y_i)).$$

When $k = 1$, this definition coincides with the standard definition of a submodular function. Given any $\mathbf{x} = (X_1, ..., X_k) \in (k+1)^V$ and $\mathbf{y} = (Y_1, ..., Y_k) \in (k+1)^V$, we write $\mathbf{x} \preceq \mathbf{y}$ if and only if $X_i \subseteq Y_i, \forall i \in [k]$. The function $f : (k+1)^V \rightarrow \mathbb{R}$ is *monotone* if and only if $f(\mathbf{x}) \leq f(\mathbf{y})$ for any $\mathbf{x} \preceq \mathbf{y}$.

Our RS-$k$ minimization problem aims to find a subset of $(k+1)^V$ that minimizes the ratio of non-negative and monotone $k$-submodular functions $f$ and $g$: $\min_{0 \neq \mathbf{x} \in (k+1)^V} \frac{f(\mathbf{x})}{g(\mathbf{x})}$. The maximization version can be defined analogously.

**Applications.** We outline some specific applications of our problem below:

*1. Sensor placement*: Consider a set of locations $V$ and $k$ different types of sensors. Each sensor can be installed in a single location and has a different purpose (e.g., monitoring humidity or temperature). Installing a sensor of type $i \in [k]$ incurs a certain cost that diminishes when we install more sensors of that type. The cost diminishes, for example, because one can purchase sensors of the same type in bulks or reuse equipment for installing the sensors of that type for multiple sensor installations [8]. We want to select a vector $\mathbf{x} = (X_1, ..., X_k)$ containing $k$ subsets of locations, each corresponding to a different type of sensors, so that the sensors have measurements of low uncertainty and also have small cost. The uncertainty of the sensors in the selected locations is captured by the entropy function $H(\mathbf{x})$ (large $H(\mathbf{x})$ implies low uncertainty) and their cost is captured by the cost function $C(\mathbf{x})$; $H(\mathbf{x})$ and $C(\mathbf{x})$ are monotone $k$-submodular [8,14]. The problem is to select $\mathbf{x}$ that minimizes the ratio $\frac{C(\mathbf{x})}{H(\mathbf{x})}$.

*2. Influence maximization*: Consider a set of users (*seeds*) $V$ who receive incentives (e.g., free products) from a company, to influence other users to use $k$ products through word-of-mouth effects. The expected number of influenced users $I(\mathbf{x})$ and the cost of the products $C(\mathbf{x})$ for a vector $\mathbf{x}$ of seeds are monotone $k$-submodular functions [13,14]. We want to select a vector $\mathbf{x} = (X_1, ..., X_k)$ that contains $k$ subsets of seeds, so that each subset is given a different product to advertise and $\mathbf{x}$ maximizes the ratio $\frac{I(\mathbf{x})}{C(\mathbf{x})}$, or equivalently minimizes $\frac{C(\mathbf{x})}{I(\mathbf{x})}$.

*3. Coupled feature selection*: Consider a set of features $V$ which leads to an accurate pattern classifier. We want to predict $k$ variables $Z_1, \ldots, Z_k$ using features $Y_1, \ldots, Y_{|V|}$. Due to communication constraints [16], each feature can be used to predict only one $Z_i$. When $Y_1, \ldots Y_{|V|}$ are pairwise independent given $Z$, the monotone $k$-submodular function $F(\mathbf{x}) = H(X_1, \ldots, X_k) - \sum_{i \in [k]} \sum_{j \in X_i} H(Y_j | Z_i)$, where $H$ is the entropy function and $X_i$ is a group of features, captures the joint quality of feature groups [16], and the monotone $k$-submodular function $C(\mathbf{x})$ captures their cost. The problem is to select $\mathbf{x}$ that minimizes the ratio $\frac{C(\mathbf{x})}{F(\mathbf{x})}$.

The RS-1 minimization problem has been studied in [1,15,17] and its applications include maximizing the F-measure in information retrieval, as well as maximizing normalized cuts and ratio cuts [1]. However, the algorithms in these works cannot be directly applied to our problem.

**Contributions.** Our contributions can be summarized as follows:

*1.* We define RS-$k$ minimization problem, a new optimization problem that seeks to minimize the ratio of two monotone $k$-submodular functions and finds applications in influence maximization, sensor placement, and feature selection.

*2.* We introduce three greedy based approximation algorithms for the problem: $k$-GREEDRATIO, $k$-STOCHASTICGREEDRATIO, and Sandwich Approximation Ratio (SAR). The first two algorithms have an approximation ratio that depends on the curvature of $f$ and the size of the optimal solution. These algorithms generalize the result of [15] to $k$-submodular functions and improve the result of [1] for the RS-1 minimization problem. $k$-STOCHASTIC-GREEDRATIO differs from $k$-GREEDRATIO in that it is more efficient, as it uses sampling, and in that it achieves the guarantee of $k$-GREEDRATIO with a probability of at least $1 - \delta$, for any $\delta > 0$. SAR has an approximation ratio that depends on the values of $f$, and it is based on the sandwich approximation strategy [10] for non-submodular maximization, which we extend to a ratio of $k$-submodular functions.

*3.* We experimentally demonstrate the effectiveness and efficiency of our algorithms on the sensor selection problem and the influence maximization problem outlined above, by comparing them to three baselines. The solutions of our algorithms have several times higher quality compared to those of the baselines.

## 2 Related Work

The concept of $k$-submodular function was introduced in [6], and the problem of maximizing a $k$-submodular function has been considered in [6,7,14,18]. For example, [7] studied unconstrained $k$-submodular maximization and proposed a $\frac{k}{2k-1}$-approximation algorithm for monotone functions and a $\frac{1}{2}$-approximation algorithm for nonmonotone functions. The work of [14] studied constrained $k$-submodular maximization, with an upper bound on the solution size, and proposed $k$-GREEDY-$TS$, a $\frac{1}{2}$-approximation algorithm for monotone functions, and a randomized version of it. These algorithms cannot deal with our problem.

When $k = 1$, the RS-$k$ minimization problem coincides with the submodular ratio minimization problem studied in [1,15,17]. The work of [1] initiated the study of the latter problem and proposed GREEDRATIO, an $\frac{1}{1-e^{\kappa_f-1}}$-approximation algorithm, where $k_f$ is the curvature of a submodular function $f$ [9]. The work of [15] provided an improved approximation of $\frac{|X^*|}{1+(|X^*|-1)(1-\hat{\kappa}_f(X^*))}$ for GREEDRATIO, where $|X^*|$ is the size of an optimal solution $X^*$ and $\hat{\kappa}_f$ is an alternative curvature notion of $f$ [9]. The work of [17] considered the submodular ratio minimization problem where both of the functions may not be submodular and showed that GREEDRATIO provides approximation guarantees that depend on the submodularity ratio [5] and the curvatures of the functions. These previous results do not apply to RS-$k$ minimization with $k > 1$.

## 3  Preliminaries

We may use the word dimension when referring to a particular subset of $\mathbf{x} \in (k+1)^V$. We use $\mathbf{0} = (X_1 = \{\}, \ldots, X_k = \{\})$ to denote the vector of $k$ empty subsets and $\mathbf{x} = (X_i, \mathbf{x}_{-i})$ to highlight the set $X_i$ in dimension $i$ and $\mathbf{x}_{-i}$ the subsets of other dimensions except $i$.

We define the *marginal gain* of a $k$-submodular function $f$ when adding an element to a dimension $i \in [k]$ to be $\Delta_{u,i}f(\mathbf{x}) = f(X_1, \ldots, X_{i-1}, X_i \cup \{u\}, X_{i+1}, \ldots, X_k) - f(X_1, \ldots, X_k)$, where $\mathbf{x} \in (k+1)^V$ and $u \notin \bigcup_{l \in [k]} X_l$.

Thus, we can also define $k$-submodular functions as those that are monotone and have a diminishing returns property in each dimension [18].

**Definition 1 ($k$-submodular function).** *A function $f : (k+1)^V \to \mathbb{R}$ is $k$-submodular if and only if: (a) $\Delta_{u,i}f(\mathbf{x}) \geq \Delta_{u,i}f(\mathbf{y})$, for all $\mathbf{x}, \mathbf{y} \in (k+1)^V$ with $\mathbf{x} \preceq \mathbf{y}$, $u \notin \cup_{\ell \in [k]} Y_i$, and $i \in [k]$, and (b) $\Delta_{u,i}f(\mathbf{x}) + \Delta_{u,j}f(\mathbf{x}) \geq 0$, for any $\mathbf{x} \in (k+1)^V$, $u \notin \cup_{\ell \in [k]} X_i$, and $i, j \in [k]$ with $i \neq j$.*

Part (a) of Definition 1 is known as the *diminishing returns* property and part (b) as *pairwise monotonicity*. We define a $k$-modular function as follows.

**Definition 2 ($k$-modular function).** *A function $f$ is $k$-modular if and only if $\Delta_{u,i}f(\mathbf{x}) = \Delta_{u,i}f(\mathbf{y})$, for all $\mathbf{x}, \mathbf{y} \in (k+1)^V$ with $\mathbf{x} \preceq \mathbf{y}$, $u \notin \cup_{\ell \in [k]} Y_i$, and $i \in [k]$.*

Without loss of generality, we assume that the functions $f$ and $g$ in RS-$k$ minimization are normalized such that $g(\mathbf{0}) = f(\mathbf{0}) = 0$. Moreover, we assume that, for each $u \in V$, there are dimensions $i, i' \in [k]$ such that $f(\{u\}_i, \mathbf{0}_{-i}) > 0$ and $g(\{u\}_{i'}, \mathbf{0}_{-i'}) > 0$. Otherwise, we can remove each $u \in V$ such that $g(\{u\}_i, \mathbf{0}_{-i}) = 0$ for all $i \in [k]$ from the candidate solution of the problem, as adding it to any dimension will not decrease the value of the ratio. Also, if there is $u \in V$ with $f(\{u\}_i, \mathbf{0}_{-i}) = 0$ for all $i \in [k]$, we could add $u$ to the final solution as it will not increase the ratio. If there is more than one such element, we need to determine which dimensions to add the elements into, so that $g$ is maximized, which boils down to solving a $k$-submodular function maximization problem. Applying an existing $\alpha$-approximation algorithm [6,7,14,18] to that problem would yield an additional approximation multiplicative factor of $\alpha$ to the ratio of the final solution of our problem. Finally, we assume that the values of $f$ and $g$ are given by value oracles.

## 4 The $k$-GreedRatio Algorithm

We present our first algorithm for the RS-$k$ minimization problem, called $k$-GREEDRATIO. The algorithm iteratively adds an element that achieves the best marginal gain to the functions of the ratio $f/g$ and terminates by returning the subsets (created from each iteration) that have the smallest ratio. We show that $k$-GREEDRATIO has a bounded approximation ratio that depends on the curvature of the function $f$. We also show that the algorithm yields an optimal solution when $f$ and $g$ are $k$-modular.

---

**Algorithm:** $k$-GREEDRATIO
**Input:** $V$, $f : (k+1)^V \to \mathbb{R}_{\geq 0}$, $g : (k+1)^V \to \mathbb{R}_{\geq 0}$
**Output:** Solution $\mathbf{x} \in (k+1)^V$

1  $j \leftarrow 0$; $\mathbf{x}_j \leftarrow \mathbf{0}$; $R \leftarrow V$; $S \leftarrow \{\}$
2  **while** $R \neq \{\}$ **do**

3      $(u,i) \in \arg\min\limits_{u \in R, i \in [k]} \dfrac{\Delta_{u,i} f(\mathbf{x}_j)}{\Delta_{u,i} g(\mathbf{x}_j)}$

4      $\mathbf{x}_{j+1} \leftarrow (X_1, \ldots, X_i \cup \{u\}, \ldots X_k)$
5      $R \leftarrow \{u \in R \mid u \notin X_i, \forall i \in [k], \text{ and } \exists i \in [k] : \Delta_{u,i} g(\mathbf{x}_{j+1}) > 0\}$
6      $S \leftarrow S \cup \{\mathbf{x}_{j+1}\}$
7      $j \leftarrow j + 1$

8  $\mathbf{x} \leftarrow \arg\min\limits_{\mathbf{x}_j \in S} \dfrac{f(\mathbf{x}_j)}{g(\mathbf{x}_j)}$

9  **return** $\mathbf{x}$

---

### 4.1 The Ratio of $k$-Submodular Functions

Following [4,9], we define the curvature of a $k$-submodular function of dimension $i \in [k]$ for any $\mathbf{x} \in (k+1)^V$ as $\kappa_{f,i}(\mathbf{x}_{-i}) = 1 - \min_{u \in V \setminus \bigcup_{j \neq i} X_j} \frac{\Delta_{u,i} f(V \setminus \{u\}, \mathbf{x}_{-i})}{f(\{u\}, \mathbf{x}_{-i})}$, and $\kappa_{f,i}(X_i, \mathbf{x}_{-i}) = 1 - \min_{u \in X_i \setminus \bigcup_{j \neq i} X_j} \frac{\Delta_{u,i} f(X_i \setminus \{u\}, \mathbf{x}_{-i})}{f(\{u\}, \mathbf{x}_{-i})}$. We extend a relaxed version [9] of the above definition as $\hat{\kappa}_{f,i}(X_i, \mathbf{x}_{-i}) = 1 - \frac{\sum_{u \in X_i} \Delta_{u,i} f(X_i \setminus \{u\}, \mathbf{x}_{-i})}{\sum_{u \in X_i} f(\{u\}, \mathbf{x}_{-i})}$.

Note that, for a given $\mathbf{x} \in (k+1)^V$, $\hat{\kappa}_{f,i}(X_i, \mathbf{x}_{-i}) \leq \kappa_{f,i}(X_i, \mathbf{x}_{-i}) \leq \kappa_{f,i}(\mathbf{x}_{-i})$ when $f$ is monotone submodular in each dimension [9].

Let $\hat{\kappa}_{f,i}^{\max}(X_i) = \max_{(X_i, \bar{\mathbf{x}}_{-i}) \in (k+1)^V} \hat{\kappa}_{f,i}(X_i, \bar{\mathbf{x}}_{-i})$. The following lemma (whose proof easily follows from Lemma 3.1 of [9]) provides an upper bound on the sum of the individual elements of a given set of elements for a dimension.

**Lemma 1.** *Given any non-negative and monotone $k$-submodular function $f$, $\boldsymbol{x} \in (k+1)^V$ and $i \in [k]$, $\sum_{u \in X_i} f(\{u\}_i, \boldsymbol{x}_{-i}) \leq \frac{|X_i|}{1+(|X_i|-1)(1-\hat{\kappa}_{f,i}^{\max}(X_i))} f(X_i, \boldsymbol{x}_{-i})$.*

Notice that the inequality in Lemma 1 depends only on $X_i$. Thus, it holds for any $\bar{\mathbf{x}} \in (k+1)^V$ as long as $\bar{X}_i = X_i$. We now begin to prove the approximation guarantee of $k$-GREEDRATIO.

Let $\mathbf{x}^* = (X_1^*, ..., X_k^*) \in \arg\min_{\mathbf{0} \neq \mathbf{x} \in (k+1)^V} \frac{f(\mathbf{x})}{g(\mathbf{x})}$ be an optimal solution of the RS-$k$ minimization problem. Let $S(\mathbf{x}^*) = \{\mathbf{x} \in (k+1)^V \mid |X_i| = \mathbb{1}[|X_i^*| > 0] \ \forall i \in [k]\}$ be the subsets of $(k+1)^V$ in which each dimension contains at most one element (with respect to $\mathbf{x}^*$) where $\mathbb{1}[\cdot]$ is an indicator function. Given $S(\mathbf{x}^*)$, we let $\mathbf{x}' = (X_1', ..., X_k') \in \arg\min_{\mathbf{0} \neq \mathbf{x} \in S(\mathbf{x}^*)} \frac{f(\mathbf{x})}{g(\mathbf{x})}$. We first compare $\mathbf{x}'$ with $\mathbf{x}^*$ using a proof idea from [15].

**Theorem 1.** *Given two non-negative and monotone $k$-submodular functions $f$ and $g$, we have $\frac{f(\boldsymbol{x}')}{g(\boldsymbol{x}')} \leq \alpha \frac{f(\boldsymbol{x}^*)}{g(\boldsymbol{x}^*)}$, where $\alpha = \prod_{i \in [k] s.t. |X_i^*| > 0} \frac{|X_i^*|}{1+(|X_i^*|-1)(1-\hat{\kappa}_{f,i}^{\max}(X_i^*))}$.*

*Proof.* We have that

$$g(\mathbf{x}^*) = g(X_1^*, .., X_k^*) \leq \sum_{u_i \in X_i^*} g(\{u_i\}_i, \mathbf{x}_{-i}^*) \leq \sum_{u_1 \in X_1^*, ..., u_k \in X_k^*} g(\{u_1\}_1, ..., \{u_k\}_k)$$

$$\leq \sum_{u_1 \in X_1^*, ..., u_k \in X_k^*} f(\{u_1\}_1, ..., \{u_k\}_k) \frac{g(\mathbf{x}')}{f(\mathbf{x}')}$$

$$\leq \prod_{i \in [k] s.t. |X_i^*| > 0} \frac{|X_i^*|}{1 + (|X_i^*| - 1)(1 - \hat{\kappa}_{f,i}^{\max}(X_i^*))} \frac{g(\mathbf{x}')}{f(\mathbf{x}')} f(\mathbf{x}^*),$$

where the first inequality is from applying Definition 1(a) to dimension $i$, the second inequality is from applying Definition 1(a) to other dimensions successively, the third inequality is from noting that $\frac{f(\mathbf{x}')}{g(\mathbf{x}')} \leq \frac{f(\mathbf{x})}{g(\mathbf{x})} \iff g(\mathbf{x}) \leq f(\mathbf{x})\frac{g(\mathbf{x}')}{f(\mathbf{x}')}$ for any $\mathbf{x} \in S(\mathbf{x}^*)$ and summing up the corresponding terms, and the fourth inequality is from applying Lemma 1 repeatedly from $i = 1$ to $k$. □

Notice that $\alpha$ in Theorem 1 could be hard to compute[1]. However, the bound is tight. For instance, if $f$ is $k$-modular, then $\mathbf{x}'$ yields an optimal solution.

Theorem 1 shows that we can approximate the optimal solution, using the optimal solution $\mathbf{x}'$ where each dimension contains at most one element. However, computing any $\mathbf{x}'$ cannot be done efficiently for large $k$. Our next theorem shows that $k$-GREEDRATIO solution can be used to approximate any $\mathbf{x}'$, which, in turn can be used to approximate any $\mathbf{x}^*$. To begin, we let $\bar{x} \in \arg\min_{x \in V} \min_{i \in [k]} \frac{f(\{x\}_i, \mathbf{0}_{-i})}{g(\{x\}_i, \mathbf{0}_{-i})}$ and let $\bar{i}$ be the corresponding dimension.

---

[1] It is possible to obtain a computable bound by redefining the curvature related parameters with respect to $\mathbf{x}_{-i} = \mathbf{0}_{-i}$. The proof in Theorem 1 follows similarly up until the third inequality. However, the achieved approximation essentially depends on the product of the sizes of the sets (without all $k$ but one denominator term).

**Theorem 2.** *Given two non-negative and monotone $k$-submodular functions $f$ and $g$, we have $\frac{f(\{\overline{x}\}_{\overline{i}},\mathbf{0}_{-\overline{i}})}{g(\{\overline{x}\}_{\overline{i}},\mathbf{0}_{-\overline{i}})} \leq k\frac{f(\boldsymbol{x}'')}{g(\boldsymbol{x}'')}$, for any $\boldsymbol{x}'' \in S = \{\boldsymbol{x} \in (k+1)^V \mid |X_i| \leq 1 \ \forall i \in [k]\}$.*

*Proof.* We have that

$$g(\mathbf{x}'') = g(\{x_1''\}, ..., \{x_k''\}) \leq g(\{x_1''\}, \mathbf{0}_{-1}) + ... + g(\{x_k''\}, \mathbf{0}_{-k})$$

$$\leq [f(\{x_1''\}, \mathbf{0}_{-1}) + ... + f(\{x_k''\}, \mathbf{0}_{-k})]\frac{g(\{\overline{x}\}_{\overline{i}},\mathbf{0}_{-\overline{i}})}{f(\{\overline{x}\}_{\overline{i}},\mathbf{0}_{-\overline{i}})} \leq kf(\mathbf{x}'')\frac{g(\{\overline{x}\}_{\overline{i}},\mathbf{0}_{-\overline{i}})}{f(\{\overline{x}\}_{\overline{i}},\mathbf{0}_{-\overline{i}})},$$

where the first inequality is from applying the definition of $k$-submodularity according to Inequality 1 in Sect. 1 repeatedly, the second inequality is from noting that $\frac{f(\{\overline{x}\}_{\overline{i}},\mathbf{0}_{-\overline{i}})}{g(\{\overline{x}\}_{\overline{i}},\mathbf{0}_{-\overline{i}})} \leq \frac{f(\{u\}_i,\mathbf{0}_{-i})}{g(\{u\}_i,\mathbf{0}_{-i})} \longleftrightarrow g(\{u\}_j, \mathbf{0}_{-j}) \leq f(\{u\}_j, \mathbf{0}_{-j})\frac{g(\{\overline{x}\}_{\overline{i}},\mathbf{0}_{-\overline{i}})}{f(\{\overline{x}\}_{\overline{i}},\mathbf{0}_{-\overline{i}})}$ for any $u \in V$ and $j \in [k]$ and summing up the corresponding terms, and the third inequality is due to monotonicity. $\square$

Combining Theorems 1 and 2, we have the following result.

**Theorem 3.** *Given two non-negative and monotone $k$-submodular functions $f$ and $g$, $k$-GREEDRATIO finds a solution that is at most $k\alpha$ times of the optimal solution, where $\alpha = \prod_{i \in [k] s.t. |X_i^*| > 0} \frac{|X_i^*|}{1+(|X_i^*|-1)(1-\hat{\kappa}_{f,i}^{\max}(X_i^*))}$, in $O(|V|^2 k)$ time, assuming it is given (value) oracle access to $f$ and $g$.*

*Proof.* Let $\mathbf{x}$ be the output of $k$-GREEDRATIO. We have that $\frac{f(\mathbf{x})}{g(\mathbf{x})} \leq \frac{f(\{\overline{x}\}_{\overline{i}},\mathbf{0}_{-\overline{i}})}{g(\{\overline{x}\}_{\overline{i}},\mathbf{0}_{-\overline{i}})} \leq k\frac{f(\mathbf{x}')}{g(\mathbf{x}')} \leq k\alpha\frac{f(\mathbf{x}^*)}{g(\mathbf{x}^*)}$ where the first inequality holds because $\overline{x}_i$ is the first element selected by the algorithm, the second inequality is due to Theorem 2 (which holds for any $\mathbf{x}'' \in S$), and the third inequality is due to Theorem 1 and $\alpha = \prod_{i \in [k] s.t. |X_i^*| > 0} \frac{|X_i^*|}{1+(|X_i^*|-1)(1-\hat{\kappa}_{f,i}^{\max}(X_i^*))}$. $k$-GREEDRATIO needs $O(|V|^2 k)$ time, as step 3 needs $O(|V|k)$ time and the loop in step 2 is executed $O(|V|)$ times. $\square$

Our result generalizes the result of [15] to $k$-submodular functions and improves the result of [1] for the RS-1 minimization problem.

## 4.2 The Ratio of $k$-Modular Functions

**Theorem 4.** *Given two non-negative and monotone $k$-modular functions $f$ and $g$, $k$-GREEDRATIO finds an optimal solution $\mathbf{x} \in \arg\min_{\mathbf{x}' \in (k+1)^V} \frac{f(\mathbf{x}')}{g(\mathbf{x}')}$. There is an $O(|V|k+|V|\log|V|)$-time implementation of $k$-GREEDRATIO, assuming it is given (value) oracle access to $f$ and $g$.*

*Proof.* The proof follows a similar argument to [1]. From the definition of $k$-modular function, $f$ and $g$ satisfy $\Delta_{u,i}f(\mathbf{x}) = f(\{u\}_i, \mathbf{0}_{-i})$ and $\Delta_{u,i}g(\mathbf{x}) = g(\{u\}_i, \mathbf{0}_{-i})$ for all $\mathbf{x} \in (k+1)^V$, $u \notin \cup_{\ell \in [k]} X_\ell$, and $i \in [k]$.

As a result, we can provide an (efficient) alternative implementation of $k$-GREEDRATIO by computing $Q(u) = \min_{i \in [k]} \frac{f(\{u\}_i, \mathbf{0}_{-i})}{g(\{u\}_i, \mathbf{0}_{-i})}$ for each $u \in V$ and sorting $u$'s in increasing order of $Q(u)$ (breaking ties arbitrarily).

Without loss of generality, let $Q(u_1) \leq Q(u_2)... \leq Q(u_n)$ be such an ordering and let $i_1, i_2,..., i_n$ be the corresponding dimensions so that $\frac{f(\{u_1\}_{i_1}, \mathbf{0}_{-i_1})}{g(\{u_1\}_{i_1}, \mathbf{0}_{-i_1})} \leq$ $... \leq \frac{f(\{u_n\}_{i_n}, \mathbf{0}_{-i_n})}{g(\{u_n\}_{i_n}, \mathbf{0}_{-i_n})}$. It is not hard to see that $k$-GREEDRATIO picks the first $i$ elements according to the ordering (each in the $i$ iteration).

Let $\mathbf{x}^* \in \arg\min_{\mathbf{x} \in (k+1)^V} \frac{f(\mathbf{x})}{g(\mathbf{x})}$ and $r^* = \frac{f(\mathbf{x}^*)}{g(\mathbf{x}^*)}$. There must be some $u_j \in V$ such that $Q(u_j) \leq \tau^*$, otherwise $r^*$ cannot be obtained from the elements of $\mathbf{x}^*$.

Consider the set $\mathbf{0} \neq \mathbf{x}^{\tau^*} = (X_1, ..., X_k)$ where $X_l = \{u_j \in V \mid Q(u_j) \leq \tau^*$ and $i_j = l\}$ for each $l \in [k]$. First note that $\mathbf{x}^{\tau^*}$ is among the solutions $\{\mathbf{x}_i\}_{i=1}^n$ obtained by $k$-GREEDRATIO. Second, we have that $\tau^* \leq \frac{f(\mathbf{x}^{\tau^*})}{g(\mathbf{x}^{\tau^*})} \leq \tau^*$ (i.e., each of the ratio is bounded by $\tau^*$). Thus, $\frac{f(\mathbf{x}^{\tau^*})}{g(\mathbf{x}^{\tau^*})} = \tau^*$.

The above implementation takes $O(|V|k)$ and $O(|V|\log|V|)$ time to compute ratios for each element/dimension and to sort the $|V|$ elements, respectively. $\square$

# 5   $k$-StochasticGreedRatio

We introduce a more efficient randomized version of $k$-GREEDRATIO that uses a smaller number of function evaluations at each iteration of the algorithm. The algorithm is linear in $|V|$ and it uses sampling in a similar way as the algorithm of [12] for submodular maximization. That is, it selects elements to add into $\mathbf{x}$ based on a sufficiently large random sample of $V$ instead of $V$.

---

**Algorithm**: $k$-STOCHASTICGREEDRATIO

**Input**: $V$, $f : (k+1)^V \to \mathbb{R}_{\geq 0}$, $g : (k+1)^V \to \mathbb{R}_{\geq 0}$, $\delta > 0$

**Output**: Solution $\mathbf{x} \in (k+1)^V$

1  $j \leftarrow 0$; $\mathbf{x}_j \leftarrow \mathbf{0}$; $R \leftarrow V$; $S \leftarrow \{\}$

2  **while** $R \neq \{\}$ **do**

3        $Q \leftarrow$ a random subset of size $\min\{\lceil \log \frac{|V|}{\delta} \rceil, |V|\}$ uniformly sampled with replacement from $V \setminus S$

4        $(u, i) \in \arg\min_{u \in Q, i \in [k]} \dfrac{\Delta_{u,i} f(\mathbf{x}_j)}{\Delta_{u,i} g(\mathbf{x}_j)}$

5  the next steps are the same as steps 4 to 9 of $k$-GREEDRATIO

---

**Theorem 5.** *With probability at least $1 - \delta$, $k$-STOCHASTICGREEDRATIO outputs a solution that is: (a) at most $k\alpha$ times of the optimal solution when $f$ and $g$ are non-negative and monotone $k$-submodular, or (b) optimal when $f$ and $g$ are non-negative and monotone $k$-modular where $\alpha$ is the ratio in Theorem 3.*

*Proof.* (a) Let $Q^1$ be $Q$ of the first iteration and consider the first element selected by $k$-GREEDRATIO. If $|Q^1| = |V|$, then the probability that the first element is *not* contained in $Q^1$ is 0. Otherwise, this probability is

$\left(1 - \frac{1}{|V|}\right)^{|Q^1|} \le e^{-\log \frac{|V|}{\delta}} = \frac{\delta}{|V|}$. We have that with probability at least $1 - \delta$ $k$-STOCHASTICGREED-RATIO selects the first element. The claims follows from this and Theorem 3.

(b) It follows from the fact that $k$-STOCHASTICGREEDRATIO selects the first element with probability at least $1 - \delta$ and Theorem 4.     □

**Lemma 2.** *The time complexity of $k$-STOCHASTICGREEDRATIO is $O(k|V|\log\frac{|V|}{\delta})$ for $\delta \ge \frac{|V|}{e^{|V|}}$ and $O(k|V|^2)$ otherwise, where e is the base of the natural logarithm.*

*Proof.* Step 4 needs $O(k\min\{\lceil\log\frac{|V|}{\delta}\rceil,|V|\}) = O(k\min\{\log\frac{|V|}{\delta},|V|\})$ time and it is executed $O(|V|)$ times, once per iteration of the loop in step 2. If the sample size $\min\{\lceil\log\frac{|V|}{\delta}\rceil,|V|\} = \lceil\log\frac{|V|}{\delta}\rceil$, or equivalently if $\delta \ge \frac{|V|}{e^{|V|}}$, then the algorithm takes $O(k|V|\log\frac{|V|}{\delta})$ time. Otherwise, it takes $O(k|V|^2)$ time.     □

## 6    Sandwich Approximation Ratio (SAR)

We present SAR, a greedy based algorithm that provides an approximation guarantee based on the value of $f$, by extending the idea of [10] from non-submodular function maximization to RS-$k$ minimization problems. SAR uses an upper bound $k$-submodular function and a lower bound $k$-submodular function of the ratio function $f/g$, applies $k$-GREEDY-TS [14] with size constraint $|V|$ using each bound function as well as $f/g$, and returns the solution that maximizes the ratio among the solutions constructed by $k$-GREEDY-TS[2].

SAR uses the functions $\frac{g(x)}{2c}$ and $\frac{g(x)}{c'}$, where $2c \ge c^* = \max_{\mathbf{x}\in(k+1)^V} f(\mathbf{x})$ and $c' = \min_{x\in V}\min_{i\in[k]} f(\{x\}_i, \mathbf{0}_{-i})$. It is easy to see that these functions bound $h(\mathbf{x}) = \frac{g(\mathbf{x})}{f(\mathbf{x})}$ from below and above, respectively. While $c^*$ cannot be computed exactly, we have $2c \ge c^*$, where $c$ is the solution of the $k$-GREEDY-TS $\frac{1}{2}$-approximation algorithm for maximizing a monotone $k$-submodular function [14], when applied with function $f$ and solution size threshold $|V|$.

---

**Algorithm**: Sandwich Approximation Ratio (SAR)
**Input**: $V$, $f : (k+1)^V \to \mathbb{R}_{\ge 0}$, $g : (k+1)^V \to \mathbb{R}_{\ge 0}$, $c$, $c'$
**Output**: Solution $\mathbf{x}_{SAR} \in (k+1)^V$
1  $(\mathbf{x}_\ell^1, \ldots, \mathbf{x}_\ell^{|V|}) \leftarrow k$-GREEDY-TS with $g/2c$ and threshold $|V|$
2  $(\mathbf{x}_h^1, \ldots, \mathbf{x}_h^{|V|}) \leftarrow k$-GREEDY-TS with $h = g/f$ and threshold $|V|$
3  $(\mathbf{x}_u^1, \ldots, \mathbf{x}_u^{|V|}) \leftarrow k$-GREEDY-TS with $g/c'$ and threshold $|V|$
4  **return** $\mathbf{x}_{SAR} \leftarrow \arg\max_{\mathbf{x}\in\{\mathbf{x}_\ell^1,\ldots,\mathbf{x}_\ell^{|V|},\mathbf{x}_h^1,\ldots,\mathbf{x}_h^{|V|},\mathbf{x}_u^1,\ldots,\mathbf{x}_u^{|V|}\}} \frac{g(\mathbf{x})}{f(\mathbf{x})}$

---

[2] SAR can be easily modified to use other algorithms for monotone $k$-submodular maximization instead of $k$-GREEDY-TS, such as the algorithm of [7].

---

**Algorithm**: $k$-Greedy-TS
**Input**: $f : (k+1)^V \to \mathbb{R}_{\geq 0}$, solution size threshold $B$
**Output**: Vector of solutions $\mathbf{x}_f^1, \ldots, \mathbf{x}_f^B$ [3]

1  $\mathbf{x}_f^0 \leftarrow \mathbf{0}$
2  **for** $j = 1$ *to* $B$ **do**
3      $(u, i) \in \arg\max_{u \in V \setminus R, i \in [k]} \Delta_{u,i} f(\mathbf{x}_f)$
4      $\mathbf{x}_f^j \leftarrow \mathbf{x}_f^{j-1}$
5      $\mathbf{x}_f^j(u) \leftarrow i$
6      $R \leftarrow R \cup \{u\}$
7  **return** $(\mathbf{x}_f^1, \ldots, \mathbf{x}_f^B)$

---

To provide SAR's approximation guarantee, we define $\ell(\mathbf{x}) = \frac{g(\mathbf{x})}{2c}$, $h(\mathbf{x}) = \frac{g(\mathbf{x})}{f(\mathbf{x})}$, and $u(\mathbf{x}) = \frac{g(\mathbf{x})}{c'}$ for all $\mathbf{x} \in (k+1)^V$. Let $s(\mathbf{x}) = \sum_{i \in [k]} |X_i|$ be the size of any $\mathbf{x} \in (k+1)^V$. Let $\mathbf{x}_h^* \in \arg\max_{\mathbf{x} \in (k+1)^V} \frac{g(\mathbf{x})}{f(\mathbf{x})}$ be the optimal solution and $s = s(\mathbf{x}_h^*)$ be the size of the optimal solution. We let $\mathbf{x}_\ell^{j*} \in \arg\max_{\mathbf{x} \in (k+1)^V : s(\mathbf{x})=j} \ell(\mathbf{x})$, $\mathbf{x}_h^{j*} \in \arg\max_{\mathbf{x} \in (k+1)^V : s(\mathbf{x})=j} \frac{g(\mathbf{x})}{f(\mathbf{x})}$, and $\mathbf{x}_u^{j*} \in \arg\max_{\mathbf{x} \in (k+1)^V : s(\mathbf{x})=j} u(\mathbf{x})$.

**Theorem 6.** *Given two non-negative and monotone $k$-submodular functions $f$ and $g$, SAR finds a solution at most $2 / \max(\frac{c'}{f(\mathbf{x}_u^s)}, \frac{f(\mathbf{x}_h^{s*})}{2c})$ times the optimal solution in $O(|V|^2 k)$ time, assuming it is given (value) oracle access to $f$ and $g$.*

*Proof.* We first show that, for each $j \in [|V|]$, $\max_{\mathbf{x}^j \in \{\mathbf{x}_\ell^j, \mathbf{x}_h^j, \mathbf{x}_u^j\}} \frac{g(\mathbf{x}^j)}{f(\mathbf{x}^j)}$ from SAR approximates $\frac{g(\mathbf{x}_h^{j*})}{f(\mathbf{x}_h^{j*})}$. Since $s \in [|V|]$ and $\mathbf{x}_{SAR} \in \arg\max_{\mathbf{x}^j \in \{\mathbf{x}_\ell^j, \mathbf{x}_h^j, \mathbf{x}_u^j\}_{j \in [|V|]}} \frac{g(\mathbf{x}^j)}{f(\mathbf{x}^j)}$, our claimed approximation follows immediately. For a fixed $j \in [|V|]$, we have

$$h(\mathbf{x}_u^j) = \frac{h(\mathbf{x}_u^j)}{u(\mathbf{x}_u^j)} u(\mathbf{x}_u^j) \geq \frac{h(\mathbf{x}_u^j)}{u(\mathbf{x}_u^j)} \frac{1}{2} u(\mathbf{x}_u^{j*}) \geq \frac{h(\mathbf{x}_u^j)}{u(\mathbf{x}_u^j)} \frac{1}{2} u(\mathbf{x}_h^{j*}) \geq \frac{h(\mathbf{x}_u^j)}{u(\mathbf{x}_u^j)} \frac{1}{2} h(\mathbf{x}_h^{j*}),$$

where the first inequality is due to the use of $k$-Greedy-TS in [14] for a fixed size $j$, the second inequality follows from the definition of $\mathbf{x}_u^{j*}$, and the third inequality is from the fact that $u$ upper-bounds $h$.

We also have $h(\mathbf{x}_\ell^j) \geq \ell(\mathbf{x}_\ell^j) \geq \frac{1}{2}\ell(\mathbf{x}_\ell^{j*}) \geq \frac{1}{2}\ell(\mathbf{x}_h^{j*}) \geq \frac{\ell(\mathbf{x}_h^{j*})}{h(\mathbf{x}_h^{j*})} \frac{1}{2} h(\mathbf{x}_h^{j*})$, where the first inequality is due to the use of $k$-Greedy-TS in [14] for a fixed size $j$, the second inequality follows from the definition of $\mathbf{x}_\ell^{j*}$, and the third inequality is from the fact that $\ell$ lower-bounds $h$.

---

[3] We modify $k$-Greedy-*TS* to return every partial solution $\mathbf{x}_f^j$, instead of only $\mathbf{x}_f^B$.

From combining $h(\mathbf{x}_u^j) \geq \frac{h(\mathbf{x}_u^j)}{u(\mathbf{x}_u^j)} \frac{1}{2} h(\mathbf{x}_h^{j*})$ and $h(\mathbf{x}_\ell^j) \geq \frac{\ell(\mathbf{x}_h^{j*})}{h(\mathbf{x}_h^{j*})} \frac{1}{2} h(\mathbf{x}_h^{j*})$, we have

$$\max_{\mathbf{x}^j \in \{\mathbf{x}_\ell^j, \mathbf{x}_h^j, \mathbf{x}_u^j\}} h(\mathbf{x}^j) \geq \max\left(\frac{h(\mathbf{x}_u^j)}{u(\mathbf{x}_u^j)}, \frac{\ell(\mathbf{x}_h^{j*})}{h(\mathbf{x}_h^{j*})}\right) \frac{1}{2} h(\mathbf{x}_h^{j*}).$$

From the above for each $j$ and step 4 of SAR, we have

$$h(\mathbf{x}_{\mathrm{SAR}}) \geq \max_{j \in [|V|]}\left\{\max\left(\frac{h(\mathbf{x}_u^j)}{u(\mathbf{x}_u^j)}, \frac{\ell(\mathbf{x}_h^{j*})}{h(\mathbf{x}_h^{j*})}\right) \frac{1}{2} h(\mathbf{x}_h^{j*})\right\} \geq \left(\frac{h(\mathbf{x}_u^s)}{u(\mathbf{x}_u^s)}, \frac{\ell(\mathbf{x}_h^{s*})}{h(\mathbf{x}_h^{s*})}\right) \frac{1}{2} h(\mathbf{x}_h^{s*}).$$

It follows that: $\frac{f(\mathbf{x}_{\mathrm{SAR}})}{g(\mathbf{x}_{\mathrm{SAR}})} \leq 2 \Big/ \max\left(\frac{c'}{f(\mathbf{x}_u^s)}, \frac{f(\mathbf{x}_h^{s*})}{2c}\right) \arg\min_{\mathbf{x} \in (k+1)^V} \frac{f(\mathbf{x})}{g(\mathbf{x})}$. The time complexity of SAR follows from executing $k$-GREEDY-$TS$ three times.    □

# 7    Experimental Results

We experimentally evaluate the effectiveness and efficiency of our algorithms for cost-effective variants [13] of two problems on two publicly available datasets; a sensor placement problem where sensors have $k$ different types [14], and an influence maximization problem under the $k$-topic independent cascade model [14]. We compare our algorithms to three baseline algorithms, alike those in [14], as explained below. We implemented all algorithms in C++ and executed them on an Intel Xeon @ 2.60 GHz with 128 GB RAM. Our source code and the datasets we used are available at: https://bitbucket.org/grigorios_loukides/ksub.

## 7.1    Sensor Placement

**Entropy and Cost Functions.** We first define the entropy function for the problem, following [14]. Let the set of random variables $\Omega = \{X_1, \ldots, X_n\}$ and $H(\mathcal{S}) = -\sum_{\mathbf{s} \in \mathrm{dom}\ \mathcal{S}} Pr[\mathbf{s}] \cdot \log Pr[\mathbf{s}]$ be the entropy of a subset $\mathcal{S} \subseteq \Omega$ and dom $\mathcal{S}$ is the domain of $\mathcal{S}$. The conditional entropy of $\Omega$ after having observed $\mathcal{S}$ is $H(\Omega \mid \mathcal{S}) = H(\Omega) - H(\mathcal{S})$. Thus, an $\mathcal{S}$ with large entropy $H(\mathcal{S})$ has small uncertainty and is preferred. In our sensor placement problem, we want to select locations at which we will install sensors of $k$ types, one sensor per selected location. Let $\Omega = \{X_i^u\}_{i \in [k], u \in V}$ be the set of random variables for each sensor type $i \in [k]$ and each location $u \in V$. Each $X_i^u$ is the random variable representing the observation collected from a sensor of type $i$ that is installed at location $u$. Thus, $X_i = \{X_i^u\} \subseteq \Omega$ is the set representing the observations for all locations at which a sensor of type $i \in [k]$ is installed. Then, the entropy of a vector $\mathbf{x} = (X_1, \ldots, X_k) \in (k+1)^V$ is given by $H(\mathbf{x}) = H(\cup_{i \in [k]} X_i)$.

Let $c_i$ be the cost of installing any sensor of type $i \in [k]$. We selected each $c_i$ uniformly at random from $[1, 10]$, unless stated otherwise, and computed the cost of a vector $\mathbf{x} = (X_1, \ldots, X_k) \in (k+1)^V$ using the cost function $C(\mathbf{x}) = \sum_{i \in [k]} c_i \cdot |X_i|^\beta$, where $\beta \in (0, 1]$ is a user-specified parameter, similarly to [8].

This function models that the cost of installing an extra sensor of any type $i$ diminishes when more sensors of that type have been installed (i.e., when $|X_i|$ is larger) and that the total cost of installing sensors is the sum of the costs of installing all sensors of each type. The function $|X_i|^\beta$, for $\beta \in (0, 1]$, is monotone submodular, as a composition of a monotone concave function and a monotone modular function [3]. Thus, $C(\mathbf{x})$ is monotone $k$-submodular, as a composition of a monotone concave and a monotone $k$-modular function (the proof easily follows from Theorem 5.4 in [3]). The RS-$k$ minimization problem is to minimize $\frac{C(\mathbf{x})}{H(\mathbf{x})}$. We solve the equivalent problem of maximizing $\frac{H(\mathbf{x})}{C(\mathbf{x})}$.

**Setup.** We evaluate our algorithms on the Intel Lab dataset (http://db.csail.mit.edu/labdata/labdata.html) which is preprocessed as in [14]. The dataset is a log of approximately 2.3 million values that are collected from 54 sensors installed in 54 locations in the Intel Berkeley research lab. There are three types of sensors. Sensors of type 1, 2, and 3 collect temperature, humidity, and light values, respectively. Our $k$-submodular functions take as argument a vector: (1) $\mathbf{x} = (X_1)$ of sensors of type 1, when $k = 1$; (2) $\mathbf{x} = (X_1, X_2)$ of sensors of type 1 and of type 2, when $k = 2$, or (3) $\mathbf{x} = (X_1, X_2, X_3)$ of sensors of type 1 and of type 2 and of type 3, when $k = 3$.

We compared our algorithms to two baselines: 1. SINGLE($i$), which allocates only sensors of type $i$ to locations, and 2. RANDOM, which allocates sensors of any type randomly to locations. The baselines are similar to those in [14]; the only difference is that SINGLE($i$) is based on $\frac{H}{C}$. That is, SINGLE($i$) adds into the dimension $i$ of vector $\mathbf{x}$ the location that incurs the maximum gain with respect to $\frac{H}{C}$. We tested all different $i$'s and report results for SINGLE(1), which performed the best. Following [14], we used the lazy evaluation technique [11] in $k$-GREEDRATIO and SAR, for efficiency. For these algorithms, we maintain an upper bound on the gain of adding each $u$ in $X_i$, for $i \in [k]$, with respect to $\frac{H}{C}$ and apply the technique directly. For $k$-STOCHASTICGREEDRATIO, we used $\delta = 10^{-1}$ (unless stated otherwise) and report the average over 10 runs.

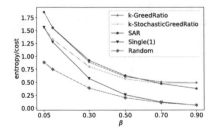

**Fig. 2.** Entropy to cost ratio for varying $\beta \in [0.05, 0.9]$ and $k = 2$.

**Results.** Figure 1 shows that our algorithms outperform the baselines with respect to entropy to cost ratio. In these experiments, we used $c_1 = c_2 = c_3 = 1$. Specifically, our algorithms outperform the best baseline, SINGLE(1), by 5.2, 5.1, and 3.6 times on average over the results of Fig. 1a and 1d. Our algorithms perform best when $C$ is close to being $k$-modular (i.e., in Figs. 1d, 1e and 1f where $\beta = 0.9$). In this case, they outperform SINGLE(1) by at least 6.2 times. The cost function $C(\mathbf{x})$ increases as $\beta$ increases and thus it affects the entropy to cost ratio $H(\mathbf{x})/C(\mathbf{x})$ more substantially when $\beta = 0.9$. Yet, our algorithms again selected sensors with smaller costs than the baselines (see Fig. 1f), achieving a solution with much higher ratio (see Fig. 1d). The good

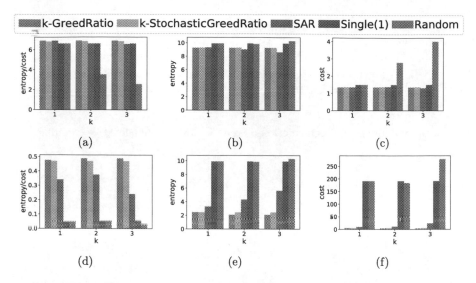

**Fig. 1.** (a) Entropy to cost ratio $\frac{H}{C}$ for varying $k \in [1,3]$ and $\beta = 0.1$. (b) Entropy $H$ for varying $k \in [1,3]$ and $\beta = 0.1$. (c) Cost $C$ for varying $k \in [1,3]$ and $\beta = 0.1$. (d) Entropy to cost ratio $\frac{H}{C}$ for varying $k \in [1,3]$ and $\beta = 0.9$. (b) Entropy $H$ for varying $k \in [1,3]$ and $\beta = 0.9$. (c) Cost $C$ for varying $k \in [1,3]$ and $\beta = 0.9$.

performance of $k$-GREEDRATIO and $k$-STOCHASTICGREEDRATIO when $\beta = 0.9$ is because $C(\mathbf{x})$ is "close" to $k$-modular (it is $k$-modular with $\beta = 1$) and these algorithms offer a better approximation guarantee for a $k$-modular function (see Theorem 3).

Figure 2 shows that our algorithms outperform the baselines with respect to entropy to cost ratio for different values of $\beta \in [0.05, 0.9]$. Specifically, $k$-GREEDRATIO, $k$-STOCHASTICGREEDRATIO, and SAR outperform the best baseline, SINGLE(1), by 3.4, 3.1, and 3 times on average, respectively. Also, note that $k$-GREEDRATIO and $k$-STOCHASTICGREEDRATIO perform very well for $\beta > 0.5$, as they again were able to select sensors with smaller costs.

Figure 3a shows the number of function evaluations ($\frac{H}{C}$, $H$ and $C$ for SAR and $\frac{H}{C}$ for all other algorithms) when $k$ varies in $[1,3]$. The number of function evaluations is a proxy for efficiency and shows the benefit of lazy evaluation [14]. As can be seen, the number of function evaluations is the largest for SAR, since it evaluates both $h = \frac{H}{C}$ and $g = C$, while it is zero for RANDOM, since it does not evaluate any function to select sensors. SINGLE(1) performs a small number of evaluations of $\frac{H}{C}$, since it adds all sensors into a fixed dimension. $k$-STOCHASTICGREEDRATIO performs fewer evaluations than $k$-GREEDRATIO due to the use of sampling. Figure 3b shows the runtime of all algorithms for the same experiment as that of Fig. 3a. Observe that all algorithms take more time as $k$ increases and that our algorithms are slower than the baselines, since they perform more function evaluations. However, the runtime of our algorithms increases sublinearly with $k$. $k$-STOCHASTICGREEDRATIO is the fastest, while SAR is the slowest among our algorithms.

Figures 3c and 3d show the impact of parameter $\delta$ on the entropy to cost ratio and on runtime of $k$-STOCHASTICGREEDRATIO, respectively. As can be seen, when $\delta$ increases, the algorithm finds a slightly worse solution but runs faster. This is because a smaller $\delta$ leads to a smaller sample size. In fact, the sample size was 30% for $\delta = 10^{-5}$ and 10.9% for $\delta = 0.2$.

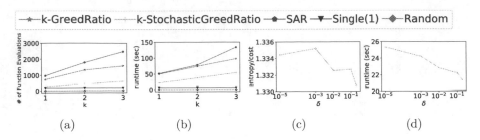

(a)            (b)            (c)            (d)

**Fig. 3.** (a) Number of evaluations of $\frac{H}{C}$, $H$, or $C$, for varying $k \in [1,3]$ and $\beta = 0.9$. (b) Runtime (sec) for varying $k \in [1,3]$ and $\beta = 0.9$. (c) Entropy to cost ratio $\frac{H}{C}$ for varying $\delta \in [10^{-5}, 0.2]$ used in $k$-STOCHASTICGREEDRATIO. (d) Runtime (sec) for varying $\delta \in [10^{-5}, 0.2]$ used in $k$-STOCHASTICGREEDRATIO.

## 7.2   Influence Maximization

**Influence and Cost Functions.** In the $k$-topic independent cascade model [14], $k$ different topics spread through a social network independently. At the initial time $t = 0$, there is a vector $\mathbf{x} = (X_1, \ldots, X_k)$ of influenced users called *seeds*. Each seed $u$ in $X_i$, $i \in [k]$, is influenced about topic $i$ and has a single chance to influence its out-neighbor $v$, if $v$ is not already influenced. The node $v$ is influenced at $t = 1$ by $u$ on topic $i$ with probability $p_{u,v}^i$. When $v$ becomes influenced, it stays influenced and has a single chance to influence each of its out-neighbors that is not already influenced. The process proceeds similarly until no new nodes are influenced. The expected number of influenced users is $I(\mathbf{x}) = \mathrm{E}[|\cup_{i \in [k]} A_i(X_i)|]$, where $A_i(X_i)$ is a random variable representing the set of users influenced about topic $i$ through $X_i$. The influence function $I$ is shown to be $k$-submodular [14]. The selection of a node $u$ as seed in $X_i$ incurs a cost $C(u, i)$, which we selected uniformly at random from $[2000, 20000]$. The cost of $\mathbf{x}$ is $C(\mathbf{x}) = \left( \sum_{u \in \cup_{i \in [k]} X_i} C(u, i) \right)^{\beta}$, where $\beta \in (0, 1]$. $C(\mathbf{x})$ is monotone $k$-submodular, as a composition of a monotone concave and a monotone $k$-modular function (the proof easily follows from Theorem 5.4 in [3]). The RS-$k$ minimization problem is to minimize $\frac{C(\mathbf{x})}{I(\mathbf{x})}$. We solve the equivalent problem of maximizing $\frac{I(\mathbf{x})}{C(\mathbf{x})}$.

**Setup.** We evaluate our algorithms on a dataset of a social news website (http://www.isi.edu/~lerman/downloads/digg2009.html) following the setup of [14]. The dataset consists of a graph and a log of user votes for stories. Each node represents a user and each edge $(u, v)$ represents that user $u$ can watch the

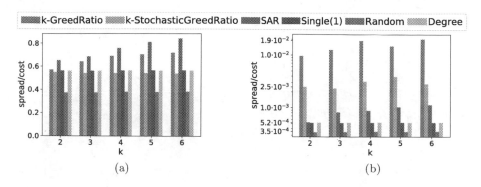

**Fig. 4.** Spread to cost ratio $\frac{I}{C}$ for varying: (a) $k \in [2,6]$ and $\beta = 0.5$, (b) $k \in [2,6]$ and $\beta = 0.9$

activity of $v$. The edge probabilities $p^i_{u,v}$ for each edge $(u,v)$ and topic $i$ were set using the method of [2]. We compared our algorithms to three baselines [14]: 1. SINGLE($i$); 2. RANDOM, and 3. DEGREE. SINGLE($i$) is similar to that used in Sect. 7.1 but it is based on $\frac{I}{C}$. Following [14], we used the lazy evaluation technique [11] on $k$-GREEDRATIO, SAR, and $k$-STOCHASTICGREEDRATIO. For the first two algorithms, we applied the technique similarly to Sect. 7.1. For $k$-STOCHASTICGREEDRATIO, we maintain an upper bound on the gain of adding each $u$ into $X_i$, for $i \in [1,k]$, w.r.t. $\frac{H}{C}$ and select the element in $Q$ with the largest gain in each iteration. We tested all $i$'s in SINGLE($i$) and report results for SINGLE(1) that performed best. DEGREE sorts all nodes in decreasing order of out-degree and assigns each of them to a random topic. We simulated the influence process based on Monte Carlo simulation. For $k$-STOCHASTICGREEDRATIO, we used $\delta = 10^{-1}$ and report the average over 10 runs.

**Results.** Figures 4a and 4b show that our algorithms outperform all three baselines, by at least 15.3, 3.3, and 1.5 times on average for $k$-GREEDRATIO, $k$-STOCHASTICGREEDRATIO, and SAR, respectively. The first two algorithms perform best when $C$ is close to being $k$-modular (i.e., in Fig. 4b where $\beta = 0.9$). This is because $C$ is $k$-modular when $\beta = 1$ and these algorithms offer a better approximation guarantee for a $k$-modular function (see Theorem 3).

Figure 5a shows that all algorithms perform similarly for $\beta < 0.5$. This is because in these cases $C$ has a much smaller value than $I$. Thus, the benefit of our algorithms in terms of selecting seeds with small costs is not significant. For $\beta \geq 0.5$, our algorithms substantially outperformed the baselines, with $k$-GREEDRATIO and $k$-STOCHASTICGREEDRATIO performing better as $\beta$ approaches 1 for the reason mentioned above.

Figure 5b shows the number of evaluations of the functions $\frac{I}{C}$ and $I$ for SAR, and of $\frac{I}{C}$ for all other algorithms, when $k$ varies in $[2,6]$. The number of evaluations is the largest for SAR, since SAR applies $k$-GREEDY-TS on both $h = \frac{I}{C}$ and $g = C$, and zero for RANDOM and DEGREE, since these algorithms select seeds without evaluating $\frac{I}{C}$. SINGLE(1) performs a small number of function

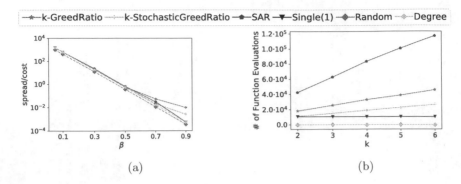

**Fig. 5.** (a) Spread to cost ratio for varying $\beta \in [0.05, 0.9]$ and $k = 6$. (b) Number of evaluations of $\frac{I}{C}$, $I$ or $C$, for varying $k \in [2,6]$ and $\beta = 0.7$.

evaluations of $\frac{I}{C}$, since it adds all nodes into a fixed dimension (i.e., dimension 1). $k$-STOCHASTICGREEDRATIO performs fewer function evaluations than $k$-GREEDRATIO, because it uses sampling. $k$-STOCHASTICGREEDRATIO was also 30% faster than SAR on average but 5 times slower than SINGLE(1).

## 8   Conclusion

In this paper, we studied RS-$k$ minimization, a new optimization problem that seeks to minimize the ratio of two monotone $k$-submodular functions. To deal with the problem, we developed $k$-GREEDRATIO, $k$-STOCHASTICGREEDRATIO, and Sandwich Approximation Ratio (SAR), whose approximation ratios depend on the curvatures of the $k$-submodular functions and the values of the functions. We also demonstrated the effectiveness and efficiency of our algorithms by applying them to sensor placement and influence maximization problems.

**Extensions.** One interesting question is to consider the RS-$k$ minimization problem with size constraints, alike those for $k$-submodular maximization in [14]. The constrained $k$-minimization problem seeks to select $k$ disjoint subsets that minimize the ratio and either all contain at most a specified number of elements, or each of them contains at most a specified number of elements. Our algorithms can be extended to tackle this constrained problem.

Another interesting question is to consider RS-$k$ minimization when $f$ and $g$ are not exactly $k$-submodular. A recent work [17] shows that the approximation ratios of algorithm for RS-1 minimization depend on the curvatures and submodularity ratios [5] of the functions $f$ and $g$, when $f$ and $g$ are not submodular. A similar idea can be considered for our algorithms, provided that we extend the notion of submodularity ratio to $k$-submodular functions.

**Acknowledgments.** We would like to thank the authors of [14] for providing the code of the baselines.

# References

1. Bai, W., Iyer, R., Wei, K., Bilmes, J.: Algorithms for optimizing the ratio of submodular functions. In: ICML, vol. 48, pp. 2751–2759 (2016)
2. Barbieri, N., Bonchi, F., Manco, G.: Topic-aware social influence propagation models. In: ICDM, pp. 81–90 (2012)
3. Bilmes, J.A., Bai, W.: Deep submodular functions. CoRR abs/1701.08939 (2017). http://arxiv.org/abs/1701.08939
4. Conforti, M., Cornuéjols, G.: Submodular set functions, matroids and the greedy algorithm: tight worst-case bounds and some generalizations of the Rado-Edmonds theorem. Discret. Appl. Math. **7**(3), 251–274 (1984)
5. Das, A., Kempe, D.: Submodular meets spectral: greedy algorithms for subset selection, sparse approximation and dictionary selection. In: ICML, pp. 1057–1064 (2011)
6. Huber, A., Kolmogorov, V.: Towards minimizing $k$-submodular functions. In: Mahjoub, A.R., Markakis, V., Milis, I., Paschos, V.T. (eds.) ISCO 2012. LNCS, vol. 7422, pp. 451–462. Springer, Heidelberg (2012). https://doi.org/10.1007/978-3-642-32147-4_40
7. Iwata, S., Tanigawa, S.I., Yoshida, Y.: Improved approximation algorithms for k-submodular function maximization. In: SODA, pp. 404–413 (2016)
8. Iyer, R., Bilmes, J.: Algorithms for approximate minimization of the difference between submodular functions, with applications. In: UAI, pp. 407–417 (2012)
9. Iyer, R.K., Jegelka, S., Bilmes, J.A.: Curvature and optimal algorithms for learning and minimizing submodular functions. In: NIPS, pp. 2742–2750 (2013)
10. Lu, W., Chen, W., Lakshmanan, L.V.S.: From competition to complementarity: comparative influence diffusion and maximization. PVLDB **9**(2), 60–71 (2015)
11. Minoux, M.: Accelerated greedy algorithms for maximizing submodular set functions. In: Stoer, J. (ed.) Optimization Techniques. Lecture Notes in Control and Information Sciences, vol. 7, pp. 234–243. Springer, Heidelberg (1978). https://doi.org/10.1007/BFb0006528
12. Mirzasoleiman, B., Badanidiyuru, A., Karbasi, A., Vondrák, J., Krause, A.: Lazier than lazy greedy. In: AAAI, pp. 1812–1818 (2015)
13. Nguyen, H., Zheng, R.: On budgeted influence maximization in social networks. IEEE J. Sel. Areas Commun. **31**(6), 1084–1094 (2013)
14. Ohsaka, N., Yoshida, Y.: Monotone k-submodular function maximization with size constraints. In: NIPS, pp. 694–702 (2015)
15. Qian, C., Shi, J.C., Yu, Y., Tang, K., Zhou, Z.H.: Optimizing ratio of monotone set functions. In: IJCAI, pp. 2606–2612 (2017)
16. Soma, T.: No-regret algorithms for online $k$-submodular maximization. In: PMLR, vol. 89, pp. 1205–1214 (2019)
17. Wang, Y.J., Xu, D.C., Jiang, Y.J., Zhang, D.M.: Minimizing ratio of monotone non-submodular functions. J. Oper. Res. Soc. China **7**(3), 449–459 (2019)
18. Ward, J., Živný, S.: Maximizing k-submodular functions and beyond. ACM Trans. Algorithm. **12**(4), 47:1–47:26 (2016)

# Mining Dense Subgraphs
# with Similar Edges

Polina Rozenshtein[1]([✉]), Giulia Preti[2], Aristides Gionis[3],
and Yannis Velegrakis[4]

[1] Institute of Data Science, National University of Singapore, Singapore, Singapore
idspoli@nus.edu.sg
[2] ISI Foundation, Turin, Italy
[3] KTH Royal Institute of Technology, Stockholm, Sweden
[4] Utrecht University, Utrecht, The Netherlands

**Abstract.** When searching for interesting structures in graphs, it is often important to take into account not only the graph connectivity, but also the metadata available, such as node and edge labels, or temporal information. In this paper we are interested in settings where such metadata is used to define a similarity between edges. We consider the problem of finding subgraphs that are *dense* and whose edges are *similar* to each other with respect to a given similarity function. Depending on the application, this function can be, for example, the Jaccard similarity between the edge label sets, or the temporal correlation of the edge occurrences in a temporal graph.

We formulate a Lagrangian relaxation-based optimization problem to search for dense subgraphs with high pairwise edge similarity. We design a novel algorithm to solve the problem through parametric MIN-CUT [15,17], and provide an efficient search scheme to iterate through the values of the Lagrangian multipliers. Our study is complemented by an evaluation on real-world datasets, which demonstrates the usefulness and efficiency of the proposed approach.

## 1 Introduction

Searching for densely-connected structures in graphs is a task with numerous applications [1,7,10,23] and extensive theoretical work [4,16,19]. A densely-connected subset of nodes may represent a community in a social network, a set of interacting proteins, or a group of related entities in a knowledge base. Given the relevance of the problem in different applications, a number of measures have been used to capture graph density, including average degree [19], quasi-cliques [23], and $k$-clique subgraphs [22].

Often, however, real-world graphs have attributes associated with their edges, which describe how nodes are related with each other. This is common in social

---

**Electronic supplementary material** The online version of this chapter (https://doi.org/10.1007/978-3-030-67664-3_2) contains supplementary material, which is available to authorized users.

F. Hutter et al. (Eds.): ECML PKDD 2020, LNAI 12459, pp. 20–36, 2021.
https://doi.org/10.1007/978-3-030-67664-3_2

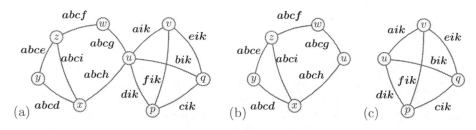

**Fig. 1.** (a) Input graph $G$ with edge labels; (b) subgraph $G_B$ of users $B = \{x, y, z, w, u\}$, where each edge pair shares 3 out of 4 labels; (c) clique $G_C$ of users $C = \{u, v, p, q\}$, where each edge pair shares 2 out of 3 labels.

networks, where we can distinguish multiple types of relationships between individuals (friends, family, class-mates, work, etc.), as well as several types of interactions (likes, messages, and comments). Similarly, a communication network records information that describes the communication patterns between its nodes, the volume of data exchanged between two nodes, or the level of congestion at a given link, as a function of time.

Incorporating this rich information into standard graph-mining tasks, such as dense-subgraph mining, can provide a better understanding of the graph, and enable the discovery of clear, cohesive, and homogeneous groups and patterns [14]. For instance, a group of hashtags that form a dense subgraph in the Twitter's hashtag co-occurence network becomes more meaningful for a social scientist if those hashtags are also correlated in time, as they likely indicate a recurrent topic of discussion, or an emerging trend.

In this paper, we study a general graph-mining problem where the input is a graph $G = (V, E)$ and a function $s : E \times E \rightarrow \mathbb{R}_{\geq 0}$ that measures similarity between pairs of edges. We do not restrict the choice of edge similarity $s$, meaning that it can be defined using any type of information that is available about the edges. For example, for a graph with edge labels, the similarity of two edges can be defined as the Jaccard similarity between their label sets, while for a temporal graph, where edges are active in some timestamps and inactive in others, the similarity can be defined as the temporal correlation between the edge time series. Given a similarity function, we are interested in finding *dense* subgraphs whose edges are *similar* to each other. Consider the following example.

**Example.** As a toy example, Fig. 1(a) illustrates a portion of a social network, where a set of labels is available for each connection, describing the topics on which the two users have interacted with each other. Figures 1(b) and 1(c) highlight two dense subgraphs $G_B$ and $G_C$, represented by the sets of users $B = \{x, y, z, w, u\}$ and $C = \{u, v, p, q\}$, respectively. The graph $G_C$ is denser than $G_B$ ($G_C$ is a clique), meaning that the users in $C$ have interacted more. However, the edges of $G_B$ have more labels in common than those of $G_C$ (3 out of 4 per edge pair, versus 2 out of 3), meaning that the users in $B$ share more topics of interest. This example shows that when multiple metrics of interest are taken into consideration, some solutions may optimize some of the metrics, while other solutions may optimize the other metrics. For example, an advertiser may be interested in finding both tighter groups of users and highly similar groups of users, because the first ones have more connections and thus they can influence more other users in the group, while the second ones have more interests in common and thus they are more likely to like similar products.    □

The previous example brings an interesting trade-off: some subgraphs have higher density, while other have higher edge similarity. This is a typical situation in *bi-criteria optimization* [12]. A common approach to study such problems is by using a Lagrangian relaxation, i.e., combining the two objectives into a weighted sum and solving the resulting optimization problem for different weights. We adopt this approach and combine the density and the edge-similarity of the subgraph induced by an edge set. Then, we reformulate the problem and design a novel efficient algorithm to solve the relaxation based on *parametric minimum cut* [15,17]. We explore possible density-similarity trade-offs and provide an efficient search procedure through the values of the Lagrangian multipliers.

We demonstrate experimentally that our method finds efficiently a set of solutions on real-world datasets. A wide range of the weighting parameter effectively controls the trade-off between similarity and density. Additionally, we present a case study where we explore the properties of the discovered subgraphs.

All omitted proofs can be found in the Supplementary Material.

## 2    Problem Formulation

We consider an undirected graph $G = (V, E)$ with node set $V$ and edge set $E$. All our algorithms extend to weighted graphs, but for simplicity of presentation we discuss the unweighted case. To avoid degenerate cases, we assume that $G$ has at least 2 edges. We consider subsets of edges and edge-induced subgraphs:

**Definition 1 (Edge-induced subgraph).** *Let $G = (V, E)$ be an undirected graph and $X$ a subset of edges. The subgraph $G(X) = (V(X), X)$ of $G$ is induced by $X$, where $V(X)$ contains all the nodes that are endpoints of edges in $X$.*

We define the density of an edge-induced subgraph as the standard half of average degree or the number of edges divided by the number of nodes [9,19]:

**Definition 2 (Density).** *Given an undirected graph $G = (V, E)$ and a set of edges $X \subseteq E$, the density of the edge-induced graph $G(X) = (V(X), X)$ is defined as*

$$D(G(X)) = \frac{1}{2} \frac{\sum_{u \in V(X)} deg(u)}{|V(X)|} = \frac{|X|}{|V(X)|},$$

*where $deg(u)$ denotes the degree of a node $u \in V$. We refer to $D(G(X))$ as the density of the set of edges $X \subseteq E$, and denote it by $D(X) = D(G(X))$.*

We assume that the graph $G$ is equipped with a non-negative *edge similarity function* $s : E \times E \to \mathbb{R}_{\geq 0}$. We define the *total edge similarity* of an edge $e$ as $s_{total}(e, X) = \sum_{e_i \in X \wedge e \neq e_i} s(e, e_i)$. We then define the *subgraph edge similarity* of an edge-induced subgraph as half of the average total edge similarity:

**Definition 3 (Subgraph edge similarity).** *The similarity of a set of edges $X$ with at least 2 edges is defined to be*

$$S(X) = \frac{1}{2} \frac{\sum_{e \in X} s_{total}(e, X)}{|X|} = \frac{1}{|X|} \sum_{\{e_i, e_j\} \in X^2} s(e_i, e_j),$$

*where $X^2$ is the set of all the unordered pairs of edges in $X$, i.e., $X^2 = \{\{e_i, e_j\} \mid e_i, e_j \in X$ with $e_i \neq e_j\}$. If $|X| \leq 1$, we set $S(X) = 0$.*

In this paper we look for edge-induced subgraphs that have *high density* and *high subgraph edge similarity*. Note that the more common definition of node-induced subgraphs is not suitable for our problem setting, because a solution to our problem is not defined by a node set. Indeed, excluding some edges from a node-induced subgraph may lead to a subgraph, which is less dense, but have edges more similar to each other.

As shown in Fig. 1, there may not exist solutions that optimize the two objectives at the same time. One possible approach is to search for subgraphs whose density and subgraph edge similarity exceed given thresholds. However, setting meaningful thresholds requires domain knowledge, which may be expensive to acquire. Here, we rely on a common approach to cope with bi-criteria optimization problems, namely to formulate and solve a Lagrangian relaxation:

*Problem 1 (DSS).* Given an undirected graph $G = (V, E)$ with an edge-similarity function $s : E \times E \to \mathbb{R}_{\geq 0}$ and a non-negative real number $\mu \geq 0$, find a subset of edges $X \subseteq E$, that maximizes the objective $O_\mu(X \mid \mu) = S(X) + \mu\, D(X)$.

## 3   Proposed Method

We start describing our solution by reformulating the DSS problem. The reformulation will allow us to use efficient algorithmic techniques. We alter the DSS objective by substituting the density term $D$ with the inverse negated term $-1/D$. Without loss of generality, we require that the solution edge set $X$ contains at least one edge. The resulting problem is the following.

*Problem 2 (DSS-INV).* Given an undirected graph $G = (V, E)$ and a non-negative real number $\lambda \geq 0$, find a subset of edges $X \subseteq E$, with $|X| \geq 1$, that maximizes the objective $O_\lambda(X \mid \lambda) = S(X) - \lambda/D(X)$.

For shorthand, we denote $-1/D$ as $\bar{D}$. We first show that DSS can be mapped to DSS-INV, so that optimal solutions for the one problem can be found by solving the other, with parameters $\mu$ and $\lambda$ appropriatelly chosen. Then, we focus on solving the DSS-INV problem.

**Proposition 1.** *An edge set $X^*$ is an optimal solution for DSS with parameter $\mu$ if and only if $X^*$ is an optimal solution for DSS-INV with $\lambda = D^2(X^*)\mu$.*

The mapping provided in Proposition 1 guarantees that a solution to DSS with a parameter $\mu$ can be found by solving DSS-INV with a corresponding parameter $\lambda$. A drawback is that to construct an DSS-INV instance for a given DSS instance with a fixed $\mu$ we need to know the density of DSS's solution $D(X^*)$. However, in general, the Lagrangian multiplier $\mu$ is often not known in advance, and the user needs to experiment with several values and select the setting leading to an interesting solution. In such cases, arguably, there is no difference between experimenting with $\mu$ for DSS or with $\lambda$ for DSS-INV. Furthermore, if the value of $\mu$ is given, we will show that our solution provides an efficient framework to explore the solution space of DSS-INV for all possible values of $\lambda$, and identify the solutions for the given value of $\mu$.

## 3.1 Fractional Programming

Following the connection established in the previous section, our goal is therefore to solve problem DSS-INV for a given value of $\lambda$. We use the technique of *fractional programming*, based on the work by Gallo et al. [15]. For completeness, we review the technique. We first define the *fractional programming* (FP) problem:

*Problem 3 (FP).* Given an undirected graph $G = (V, E)$, and edge set functions $F_1 : 2^E \rightarrow \mathbb{R}$ and $F_2 : 2^E \rightarrow \mathbb{R}_{\geq 0}$, find a subset of edges $X \subseteq E$ so that $c(X) = \frac{F_1(X)}{F_2(X)}$ is maximized.

The following problem, which we call Q, is closely related to the FP problem:

*Problem 4 (Q).* Given an undirected graph $G = (V, E)$, edge set functions $F_1 : 2^E \rightarrow \mathbb{R}$ and $F_2 : 2^E \rightarrow \mathbb{R}_{\geq 0}$, and a real number $c \in \mathbb{R}$, find a subset of edges $X \subseteq E$ so that $Q(X \mid c) = F_1(X) - cF_2(X)$ is maximized.

The key result of fractional programming [15] states that:

**Proposition 2 (Gallo et al. [15]).** *A set $X^*$ is an optimal solution to an instance of the FP problem with solution value $c(X^*)$, if and only if $X^*$ is an optimal solution to the corresponding Q problem with $c = c(X^*)$ and $Q(X^* \mid c(X^*)) = 0$.*

Proposition 2 provides the basis for the following iterative algorithm (FP-algo), which finds a solution to FP by solving a series of instances of Q problems [15].

Algorithm FP-algo:

1. Select some $X_0 \subseteq E$. Set $c_0 \leftarrow F_1(X_0)/F_2(X_0)$, and $k \leftarrow 0$.
2. Compute $X_{k+1}$ by solving the Q problem:

$$Q(X_{k+1} \mid c_k) \leftarrow \max_{X \subseteq E}\{F_1(X) - c_k F_2(X)\}.$$

3. If $Q(X_{k+1} \mid c_k) = 0$, then return $X^* \leftarrow X_k$.
   Otherwise, set $c_{k+1} \leftarrow F_1(X_{k+1})/F_2(X_{k+1})$, $k \leftarrow k + 1$, and go to Step (2).

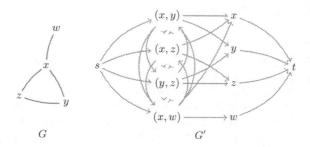

**Fig. 2.** A graph $G$ (left), and the corresponding flow graph $G'$ (right) used for solving Q with parametric MIN–CUT techniques, as described in Sect. 3.2. The edge weights in $G'$ are not shown to avoid clutter.

It can be shown [15] that the sequence $(c_k)$ generated by FP-algo is increasing, and that if $F_2$ is an integer-valued set function (and we will see that this is our case), then the number of iterations of FP-algo is bounded by the number of elements in the underlying set, which in our case is the edgeset $E$.

We formulate DSS-INV as an instance of FP. As DSS-INV is parameterized by $\lambda \geq 0$, we introduce such parameter in FP and set

$$F_1(X \mid \lambda) = \sum_{\{e_i, e_j\} \in X^2} s(e_i, e_j) - \lambda |V(X)|, \text{ and } F_2(X) = |X|.$$

Now, DSS-INV becomes an instance of FP and algorithm FP-algo can be applied. As $F_2(X)$ is the number of edges in the solution, the algorithm FP-algo is guaranteed to halt after solving $\mathcal{O}(|E|)$ instances of the Q problem.

Each instance of the Q problem at Step (2) of FP-algo can be solved efficiently by a parametric preflow/minimum cut algorithm [15]. The construction of the flow graph is presented in the next section.

Since we introduced the parameter $\lambda$, we need to write the objectives of FP and Q as $c(X \mid \lambda)$ and $Q(X \mid c, \lambda)$, respectively, but we will omit the dependency on $\lambda$ when it is clear. We denote the optimal values of FP and Q as $c^*(\lambda) = c(X^* \mid \lambda)$ and $Q^*(c, \lambda) = Q(X^* \mid c, \lambda)$, respectively.

## 3.2 Parametric MIN–CUT

In this section we show how to solve instances of the Q problem by using a mapping to the MIN–CUT problem. A similar approach has been used, among others, by Goldberg [16] to solve the densest-subgraph problem.

Let the input of Q be a graph $G = (V, E)$ with edge similarity $S$ and parameters $c \in \mathbb{R}$ and $\lambda \in \mathbb{R}_{\geq 0}$. We construct the following directed weighed network $G' = (U', E', w')$. The set $U'$ is defined as $U' = U_E \cup U_V \cup \{s, t\}$, where $U_E = \{u_e \mid e \in E\}$ contains a node $u_e$ for each edge $e \in E$, $U_V = \{u_v \mid v \in V\}$

contains a node $u_v$ for each node $v \in V$, and the nodes $s$ and $t$ are additional source and sink nodes. The nodes in $U_E$ are pairwise connected by bidirectional edges $(u_e, u_d)$ with weight $\frac{1}{2}S(e,d)$, whereas the nodes in $U_V$ are not connected to each other. Additionally, there is a directed edge $(u_e, u_v)$ for each $v \in V$ that is an endpoint of $e \in E$ with weight $w'(u_e, u_v) = +\infty$. Finally, the source $s$ is connected to all the nodes in $U_E$ by directed edges with weight $w'(s, u_e) = \frac{1}{2}\sum_{d \in E, d \neq e} s(e,d) - c$, and each node in $U_V$ is connected to $t$ by a directed edge with weight $\lambda$. The construction of $G'$ for a given $G$ is clearly polynomial. An example of the construction of $G'$ is shown in Fig. 2.

We now solve the $(s,t)$-MIN-CUT problem on the graph $G' = (U', E', w')$, parameterized with $c$ and $\lambda$. Let $(\{s\} \cup U^*, \{t\} \cup \overline{U}^*)$ be the minimum cut in $G'$, and let $\mathsf{C}^*(c, \lambda)$ be its value. The next proposition establishes the connection between the optimal values of MIN-CUT on $G'$ and the Q problem on $G$, and describes how the solution edge set for the Q problem can be derived from the solution cut set of MIN-CUT.

**Proposition 3.** *The value of the $(s,t)$-MIN-CUT in the graph $G' = (U', E', w')$ for given parameters $c$ and $\lambda$ corresponds to the optimum value for the Q problem with the same values of $c$ and $\lambda$. The solution edge set $X^*$ for Q problem on $G$ can be reconstructed from the minimum-cut set $\{s\} \cup U^* \subseteq U'$ in $G'$ as $X^* = U^* \cap U_E$.*

To summarize, in the previous sections we have established the following:

**Proposition 4.** *An instance of DSS-INV for a given parameter $\lambda$ can be solved by mapping it to Problem FP and applying the FP-algo. Problem Q in the iterative step of FP-algo can be solved by mapping it to the parametric MIN-CUT problem, as shown in Proposition 3.*

Let us evaluate the time and space complexity of the proposed solution. In FP-algo we iteratively search for optimal values in the Q problem by solving MIN-CUT problems. In each iteration, only the source link capacities are updated as $c_k$ changes, and, as mentioned before, sequence $(c_k)$ grows monotonically. This setting can be handled efficiently in the parametric MIN-CUT framework, which incrementally updates the solution from the previous iteration. The state-of-the-art algorithm for parametric MIN-CUT [17] requires $\mathcal{O}(mn \log n + kn)$ time and $\mathcal{O}(m)$ space for a graph with $n$ nodes, $m$ edges, and $k$ updates of edge capacities (iterations in FP-algo). Recall that the number of iterations is bounded by $\mathcal{O}(|E|)$, and thus, solving DSS-INV for a fixed $\lambda$ requires $\mathcal{O}(|E|^3 \log |E|)$ time and $\mathcal{O}(|E|^2)$ space.

### 3.3    $\lambda$-Exploration

Having discussed how to solve the DSS-INV problem for a fixed $\lambda$, we now introduce a framework to efficiently enumerate the solutions for all possible values of $\lambda$. The goal is to identify the ranges of values of $\lambda$ that yield identical solutions and exclude them from the search.

First, we show the monotonicity of the optimal solution value of DSS-INV, the optimal subgraph similarity and density values with respect to $\lambda$.

**Proposition 5.** *The optimal solution value of* DSS-INV *is a monotonically non-increasing function of* $\lambda$. *The density of the optimal edge set is a monotonically non-decreasing function of* $\lambda$. *The subgraph edge similarity of the optimal edge set is a monotonically non-increasing function of* $\lambda$.

From the definition of optimality and Proposition 5, it follows that:

**Corollary 1.** *Given two solutions* $X_1$ *and* $X_2$ *to* DSS-INV *for* $\lambda_1$ *and* $\lambda_2$ *with* $\lambda_1 < \lambda_2$, *either (i)* $S(X_1) = S(X_2)$ *and* $D(X_1) = D(X_2)$ *or (ii)* $S(X_1) > S(X_2)$ *and* $D(X_1) < D(X_2)$.

The monotonicity of the optimal values of the objective functions and Corollary 1 will guide our exploration of the $\lambda$ ranges.

Note that the shown monotonic properties are not strict and it is indeed easy to construct an example input graph, where different values of $\lambda$ lead to solutions to DSS-INV with the same values of subgraph edge similarity and density. To avoid a redundant search, we would like to solve DSS-INV for all the values of $\lambda$ that lead to distinct combinations of density and similarity values. Such redundant values of $\lambda$ can be pruned by observing that when two values $\lambda_1$ and $\lambda_2$ give solutions with the same values $S_1 = S_2$ and $D_1 = D_2$, then all $\lambda \in [\lambda_1, \lambda_2]$ must also lead to the same optimal density and subgraph edge similarity, and thus the interval $[\lambda_1, \lambda_2]$ can be discarded from further search. This result follows from the monotonicity of the optimal values of density and similarity (Proposition 5).

The proposed approach to search for different values of $\lambda$ is a breadth-first iterative algorithm. At the beginning, the set of distinct solutions $\mathcal{P}$ is empty, and $\lambda_\ell = \lambda_{\min}$ and $\lambda_u = \lambda_{\max}$. The algorithm maintains a queue of candidate search intervals $T$, which is initially empty. To avoid clutter, we denote the solution values of subgraph edge similarity and density $(S, D)$ for a given $\lambda$ as $t(\lambda)$.

Algorithm $\lambda$-exploration:

1. Compute set $X^*(\lambda_\ell)$ and add it to $\mathcal{P}$.
2. Compute set $X^*(\lambda_u)$.
3. If $t(\lambda_u) \neq t(\lambda_\ell)$, then push $(\lambda_\ell, \lambda_u, t(\lambda_\ell), t(\lambda_u))$ to the queue $T$ and add $X^*(\lambda_u)$ to $\mathcal{P}$.
4. While $Q$ is not empty:
    (a) Pop $(\lambda_\ell, \lambda_u, t(\lambda_\ell), t(\lambda_u))$ from $T$.
    (b) Set $\lambda_m = (\lambda_\ell + \lambda_u)/2$ and compute $X^*(\lambda_m)$.
    (c) If $t(\lambda_m) \neq t(\lambda_\ell)$, then push $(\lambda_\ell, \lambda_m, t(\lambda_\ell), t(\lambda_m))$ to $T$.
    (d) If $t(\lambda_m) \neq t(\lambda_u)$, then push $(\lambda_m, \lambda_u, t(\lambda_m), t(\lambda_u))$ to $T$.
    (e) If $t(\lambda_m) \neq t(\lambda_\ell)$ and $t(\lambda_m) \neq t(\lambda_u)$, add $X^*(\lambda_m)$ to $\mathcal{P}$.

To bound the number of calls of $\lambda$-search, we need to lower bound the difference between two consecutive values of $\lambda$ that lead to two different solutions. This lower bound is given in the next proposition, together with upper and lower bounds for $\lambda$ values.

**Proposition 6.** *To obtain all the distinct solutions in the $\lambda$-exploration algorithm, a lower bound for a value of $\lambda$ is $\lambda_{min} = s_{min}/2|E|$, an upper bound is $\lambda_{max} = s_{max}|E|^2/2$, and a lower bound for the difference between two values of $\lambda$ leading to solutions with distinct density and subgraph edge similarity values is $\delta_\lambda = s_{min}/2|E|$. Here $s_{min} = \min_{\{e_1,e_2\}\in X^2} s(e_1,e_2)$ and $s_{max} = \max_{\{e_1,e_2\}\in X^2} s(e_1,e_2)$.*

Given the bounds in Proposition 6, an upper bound on the number of different values of $\lambda$ that we need to try is $I_\lambda = (\lambda_{max} - \lambda_{min})/\delta_\lambda \leq |E|^3 \frac{s_{max}}{s_{min}}$, where $s_{max}$ and $s_{min}$ are the largest and the smallest non-zero values of edge similarity between two edges in the input graph. Thus, for a complete exploration of all the possible $\lambda$ values leading to different values of subgraph edge similarity and density of the solution graph, we need $\mathcal{O}(|E|^3)$ iterations. This estimate is pessimistic and assumes no subranges of $\lambda$ are pruned during the exploration. As we will see later, on practice the exploration typically requires around $|E|$ number of iterations.

# 4   Related Work

In this paper we consider the problem of finding subgraphs that maximize both a density measure and a similarity measure. The problem of finding dense structures in graphs has been extensively studied in the literature, as it finds applications in many domains such as community detection [10,13], event detection [1], and fraud detection [18]. Existing works have addressed the task of finding the best solution that satisfies the given constraints, such as, the densest subgraph [16,18], the densest subgraph of $k$ vertices [4], the densest subgraph in a dual network [25], or the best $\alpha$-quasi-clique [23]. Other works have aimed at retrieving a set of good solutions, such as top-$k$ densest subgraphs in a graph collection [24], $k$ diverse subgraphs with maximum total aggregate density [2], or $k$-cores with maximum number of common attributes [14]. However, these works optimize a single measure, i.e., the density, thus ignoring other properties of the graph, or find a solution that depends on an input query.

There are a few works focusing on edge similarity. The closest to our work, Boden et al. [5], considers edge-labeled multilayer graph and looks for vertex sets that are densely connected by edges with similar labels in a subset of the graph layers. They set a pairwise edge similarity threshold for a layer and look for 0.5 quasi-cliques, which persist for at least 2 layers. In contrast, our approach does not require any preset parameters and offers a comprehensive exploration of the space of dense and similar subgraphs.

Motivated by applications in fraud detection, Shin et al. [18] propose a greedy algorithm that detects the $k$ densest blocks in a tensor with $N$ attributes, with guarantees on the approximation. The framework outputs $k$ blocks by greedy iterative search. Yikun et al. [26] propose a novel model and a measure for dense fraudulent blocks detection. The measure is tailored for the fraud detection in multi-dimensional entity data, such as online product reviews, but could be possibly adapted to capture other types of node and edge similarities. They propose an efficient algorithm, which outputs several graph with some approximation guarantees. In contrast to the approaches above, our work offers exploration of *exact* solutions with different trade-offs.

Multi-objective optimization for interesting graph structures search was studied in the context of frequent pattern mining [6,21] and graph partitioning [3]. However, most of the frequent pattern works do not consider the density as an objective function, but depend on the notion of frequency in the graph and cannot be extended to our case. Carranza et al. [6], instead, define a conductance measure in terms of an input pattern and then recursively cut the graph into partitions with minimum conductance. The result, however, depends on the input. Graph partitioning approaches optimize quality functions based on modularity and node/edge attributes, but focus on a complete partition of the input graph [11,20] and do not guarantee to the quality of each individual partition.

## 5    Experimental Evaluation

We evaluate the proposed method using real-world multiplex networks from the CoMuNe lab database[1] and the BioGRID datasets[2]. An implementation of our method is publicly available[3]. In the following, we refer to our approach as DenSim. For the parametric MIN–CUT problem, we use Hochbaum's algorithm [17] and its open-source C implementation [8]. The experiments are conducted on a Xeon Gold 6148 2.40 GHz machine.

**Datasets.** We use the following real-world datasets: *CS-Aarhus* is a multiplex social network consisting of five kinds of online and offline relationships (Facebook, leisure, work, co-authorship, and lunch) between the employees of the Computer Science department at Aarhus University. *EU-Air* is a multilayer network composed by 37 layers, each one corresponding to a different airline operating in Europe. *Neuronal C.e.* is the C. elegans connectome multiplex that consists of layers corresponding to different synaptic junctions: electric, chemical monadic, and polyadic. *Genetic C.e.* and *Genetic A.th.* are multiplex networks of genetic interactions for C. elegans and Arabidopsis thaliana. Table 1 summarizes the main characteristics of the datasets.

All datasets are multilayer networks $G_M = (V, E, \ell)$ with a labeling function $\ell : E \rightarrow 2^L$, where $L$ is the set of all possible labels. We omit edge directionality if it is present in the dataset. The edge similarity function is defined as the Jaccard coefficient of the labellings: $s(e_1, e_2) = |\ell(e_1) \cap \ell(e_2)| / |\ell(e_1) \cup \ell(e_2)|$. If the pairwise edge similarity is 0 for a pair of edges, we do not materialize the corresponding edge in the MIN–CUT problem graph.

We note that, while the datasets are not large in terms of number of nodes and edges, the number of edges co-appearing in at least one layer is significant. Furthermore, the subgraphs edge similarity is high.

---

[1] https://comunelab.fbk.eu/data.php.
[2] https://thebiogrid.org.
[3] https://github.com/polinapolina/dense-subgraphs-with-similar-edges.

**Table 1.** Network characteristics. $|V|$: number of vertices; $|E|$: the number of edges; $L$: the number of layers; $|E|_{avg}$: number of edges per layer; $|E_{mult}|$: number of edges in the multiplex (same edges on different layers are counted as distinct); $|E_{meta}|$: number of unordered edge pairs co-appeared at least in one layer; $D$: density of the network; $D_{avg}$: average density across the layers; $S$: similarity of the network's edge set $S(E)$; $\ell_{avg}$: average participation of an edge to a layer.

| Dataset | $|V|$ | $|E|$ | $L$ | $|E|_{avg}$ | $|E_{mult}|$ | $|E_{meta}|$ | $D$ | $D_{avg}$ | $S$ | $\ell_{avg}$ |
|---|---|---|---|---|---|---|---|---|---|---|
| CS-Aarhus | 61 | 353 | 5 | 124.00 | 620 | 39565 | 5.78 | 2.60 | 57.44 | 1.75 |
| EU-Air | 417 | 2953 | 37 | 96.97 | 3588 | 360082 | 7.08 | 1.56 | 94.64 | 1.21 |
| Neuronal C.e. | 279 | 2290 | 3 | 1036.0 | 5863 | 1762756 | 8.20 | 3.86 | 534.70 | 1.35 |
| Genetic C.e. | 3879 | 7908 | 6 | 1338.66 | 8182 | 17249444 | 2.03 | 1.23 | 2141.88 | 1.01 |
| Genetic A.th. | 6980 | 16713 | 7 | 2499.57 | 18655 | 91782863 | 2.39 | 1.18 | 5156.65 | 1.04 |

**Baselines.** We compare DenSim with two baselines, BLDen and BLSim.

Algorithm BLDen optimizes the density directly, but takes into account edge-similarity indirectly. It outputs the set of edges of the densest weighted subgraph in the complete graph $G_D = (V, E_D, w^D)$, which has the same nodes as the input multiplex network $G_M$. The edge-weighting function $w^D$ has two components, i.e., $w^D = (w_1^D, w_2^D)$. The weight $w_1^D(u, v)$ captures the graph topology, i.e., $w_1^D(u, v) = 1$ if $(u, v) \in E$ for some layer and 0 otherwise. The weight $w_2^D(u, v)$ captures the similarity of node activity across layers: we first define the node labels as the set of all layers where a node appears in some edge $\ell(u) = \cup_{u \in e} \ell(e)$, and then define $w_2^D(u, v)$ to be the Jaccard index between the sets $\ell(u)$ and $\ell(v)$. The final weight of an edge is a weighted sum $w^D(u, v) = w_1^D(u, v) + \gamma w_2^D(u, v)$, where $\gamma$ regulates the importance of the components. By tuning $\gamma$ we can obtain a trade-off between topological density and subgraph edge similarity.

Algorithm BLSim is the counterpart of BLDen, which optimizes the edge-similarity directly, but accounts for density indirectly. BLSim finds the densest weighted subgraph of the complete graph $G_S = (E, E_S, w^S)$ with $w^S = (w_1^S, w_2^S)$. Here weight $w_1^S(e_1, e_2)$ is the edge similarity in the multiplex network $G_M$, i.e., $w_1(e_1, e_2) = s(e_1, e_2)$ and $w_2^S(e_1, e_2)$ represents the topological information, i.e., $w_2^S(e_1, e_2) = 1$ if $e_1$ and $e_2$ have a common node in the original graph, and 0 otherwise. Again, the final edge weight is a weighted sum $w^S(e_1, e_2) = w_1^S(e_1, e_2) + \gamma w_2^S(e_1, e_2)$. When $\gamma = 0$, finding the densest weighted subgraph is equivalent to finding the set of edges in the original graph with the largest similarity. We tune $\gamma$ to obtain a trade-off between subgraph edge similarity and topological density.

As with DenSim, we do not materialize 0-weight edges in the baselines. Both baselines search for the densest subgraph. Similarly to DenSim, we use the parametric MIN–CUT framework [15].

**Experimental Results.** Figure 3 shows the different characteristics of solutions discovered in the datasets during $\lambda$-exploration. We observe that the baselines are extremely sensitive to the values of $\gamma$: it is hard to find a set of $\gamma$ values that

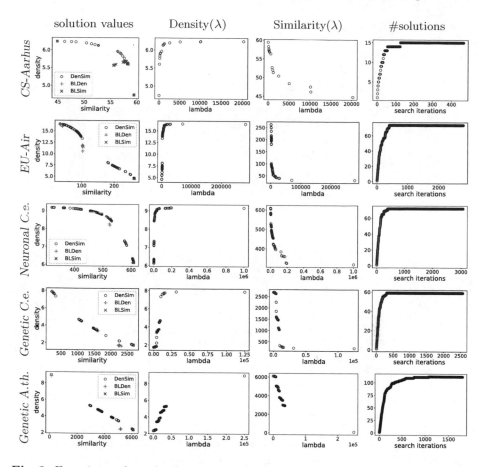

**Fig. 3.** Experimental results for our method DenSim, and the baselines BLDen and BLSim on real-world datasets. Each row represents one dataset. The first column shows the values of subgraph edge similarity and density of the discovered solutions. Column 1 also show the values of solutions, discovered by the baselines. Columns 2 and 3 show $D$ and $S$ as a function of $\lambda$. Column 4 shows how the number of discovered unique optimal solutions grows with the number of iterations in $\lambda$-exploration.

lead to distinct solutions. Moreover, the range and granularity of $\gamma$ depend on the datasets, and it is up to the end-user to decide their values. To provide a somewhat unified comparison, we allow $\gamma$ to range from 0 to 10 with step 0.1.

The first column in Fig. 3 shows the values of density and subgraph edge similarity of the solutions found. The solutions discovered by DenSim cover the space of possible values of similarity and density rather uniformly, providing a range of trade-offs. The solutions discovered by the baselines are mostly grouped around the same values and often dominated by solutions of DenSim. Note that the baselines successfully find the points with the largest density or similarity. By design, these solutions correspond to values of $\gamma = 0$, and they also correspond to the solutions of DenSim for $\lambda = \lambda_{min}$ and $\lambda = \lambda_{max}$.

**Table 2.** Number of subgraphs and running time characteristics. $|\mathcal{P}|$: number of discovered optimal solutions; $t$(s): total time (seconds) for the search; $I_\lambda$: number of tested values of $\lambda$; $t_\lambda$(s): average time to test one value of $\lambda$; $I_{MC}$: average number of MIN-CUT problems solved for one $\lambda$ (i.e., average number of iterations in FP-algo).

| Dataset | $|\mathcal{P}|$ | $t$(s) | $I_\lambda$ | $t_\lambda$(s) | $I_{MC}$ |
|---|---|---|---|---|---|
| CS-Aarhus | 15 | 2.26 | 465 | 0.003 | 2.89 |
| EU-Air | 74 | 314 | 2770 | 0.069 | 6.00 |
| Neuronal C.e. | 72 | 1015 | 3064 | 0.244 | 4.60 |
| Genetic C.e. | 59 | 10159 | 2561 | 3.075 | 4.72 |
| Genetic A.th. | 112 | 43200 | 1794 | 20.540 | 5.68 |

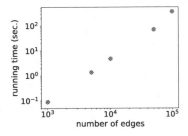

**Fig. 4.** Running time in seconds to calculate 10 first solutions. The input graph is a $G_{n,m}$ random graph with $n = 1000$ nodes, $m$ edges, the probability that a pair of edges has a non-zero similarity is set to 0.001.

Columns 2 and 3 show optimal density $D$ and subgraph edge similarity $S$ as functions of $\lambda$. As expected, larger values of $\lambda$ correspond to solutions with larger density and smaller similarity. The range of $\lambda$ that gives unique optimal solutions is dataset-dependent and not uniform. However, due to the monotonicity property we can search for these values efficiently, in contrast to the naïve search for the baselines. The last column of Fig. 3 shows the efficiency of $\lambda$-exploration. All unique solutions are found after 200 to 1000 iterations.

In Table 2 we report the number of the unique optimal solutions and running times. The total running time varies from seconds to hours. We should highlight, however, that finding a solution for a single value of $\lambda$ takes on average 20 s for the largest dataset. Thus, if the search progresses fast (as shown in the last row of Fig. 3) and a sufficient number of optimal solutions have been found, we can terminate the search. As we discussed before, we implement $\lambda$-exploration as a BFS, so that at any point we have a diverse set of $\lambda$ tested. It is worth noting that FP-algo converges in about 5 iterations on average, and thus the MIN-CUT algorithm is not run many times.

**Scalability.** In order to test the scalability of DenSim we generate a number of random graphs with $n = 1000$ nodes and varying $m$ number of edges. We draw the graphs from a random graph model $G_{n,m}$. The similarities between the

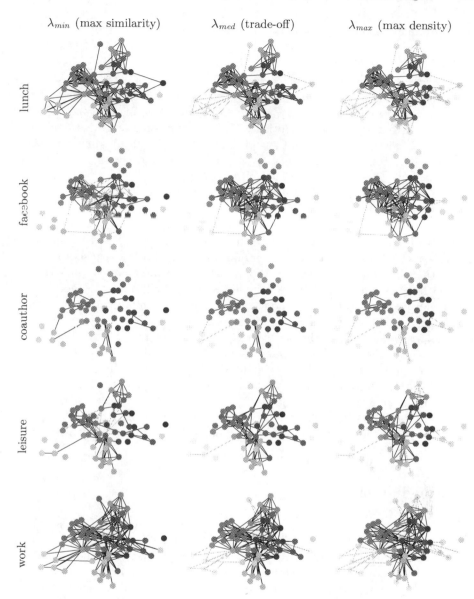

**Fig. 5.** Some solutions output by DenSim for *CS-Aarhus* dataset. Each column corresponds to a solution with the smallest (left), largest (right), and median (center) value of λ. Each row corresponds to a layer in the dataset. The nodes and edges, included into the solution are shown with the bright colors, the rest of the nodes and edges on each layer are drawn transparently. (Color figure online)

edges are random values from $(0, 1]$, and the probability that a pair of edges has a non-zero similarity is set to 0.001. We run DenSim until it found 10 solutions, and the running time is reported in Fig. 4. It took 6 minutes to find 10 solutions for the largest graph with 100000 edges.

**Case Study.** We run DenSim on the *CS-Aarhus* dataset. We pick three of the solutions discovered: one for $\lambda_{min}$, one for $\lambda_{max}$, and one for the median value $\lambda_{med}$. Recall that $\lambda_{min}$ gives a solution with maximum subgraph edge similarity, while density is ignored; while $\lambda_{max}$ gives a solution with maximum density, ignoring edge similarity. Any other $\lambda_{med}$ should provide some balance between these extremes. The solutions are visualized in Fig. 5.

The graph maximizing the subgraph similarity ($\lambda_{min}$) includes all the edges from the layers of "work" and "lunch." Since the dataset contains relationships between the employees of the same university department, it is intuitive that these two layers define the edge set with the largest subgraph edge similarity. All network nodes are included in this solution, as all these people share similar interactions at work and lunch. Facebook and leisure interactions, not overlapping with "work" and "lunch", are excluded, as they are localized in their layers. The resulting graph contains 61 nodes and 289 edges, while the subgraph edge similarity is 59.43 and the density is 4.73.

The graph maximizing the density ($\lambda_{max}$) includes the edges of the densest subgraph from the "work" layer, and reinforces it by adding edges from other layers. The graph contains 45 nodes and 281 edges, and it is the smallest of the three. The subgraph edge similarity is 44.83 and the density is 6.24.

The trade-off graph ($\lambda_{med}$) selects 325 edges, more than the other two, while it has 53 nodes. Its subgraph similarity is 52.64 and its density 6.13. The graph resembles the one for $\lambda_{max}$, but adds interactions that decrease the density while increasing the subgraph similarity.

# 6  Concluding Remarks

In this paper we study a novel graph-mining problem, where the goal is to find a set of edges that maximize the density of the edge-induced subgraph and the subgraph edge similarity. We reformulate the problem as a non-standard Lagrangian relaxation and develop a novel efficient algorithm to solve the relaxation based on parametric minimum-cut [15,17]. We provide an efficient search strategy through the values of Lagrangian multipliers. The approach is evaluated on real-world datasets and compared against intuitive baselines.

**Acknowledgments.** This research was partially supported by the National Research Foundation, Singapore under its AI Singapore Programme (AISG Award No: AISG-GC-2019-001). Aristides Gionis is supported by three Academy of Finland projects (286211, 313927, 317085), the ERC Advanced Grant REBOUND (834862), the EC H2020 RIA project "SoBigData++" (871042), and the Wallenberg AI, Autonomous Systems and Software Program (WASP). The funders had no role in study design, data collection and analysis, decision to publish, or preparation of the manuscript.

# References

1. Angel, A., Koudas, N., Sarkas, N., Srivastava, D., Svendsen, M., Tirthapura, S.: Dense subgraph maintenance under streaming edge weight updates for real-time story identification. VLDB J. **23**, 175–199 (2013). https://doi.org/10.1007/s00778-013-0340-z
2. Balalau, O.D., Bonchi, F., Chan, T., Gullo, F., Sozio, M.: Finding subgraphs with maximum total density and limited overlap. In: WSDM, pp. 379–388 (2015)
3. Baños, R., Gil, C., Montoya, M.G., Ortega, J.: A new pareto-based algorithm for multi-objective graph partitioning. In: Aykanat, C., Dayar, T., Körpeoğlu, İ. (eds.) ISCIS 2004. LNCS, vol. 3280, pp. 779–788. Springer, Heidelberg (2004). https://doi.org/10.1007/978-3-540-30182-0_78
4. Bhaskara, A., Charikar, M., Chlamtac, E., Feige, U., Vijayaraghavan, A.: Detecting high log-densities: an $O(n^{1/4})$ approximation for densest $k$-subgraph. In: STOC, pp. 201–210 (2010)
5. Boden, B., Günnemann, S., Hoffmann, H., Seidl, T.: Mining coherent subgraphs in multi-layer graphs with edge labels. In: SIGKDD, pp. 1258–1266 (2012)
6. Carranza, A.G., Rossi, R.A., Rao, A., Koh, E.: Higher-order spectral clustering for heterogeneous graphs. arXiv preprint arXiv:1810.02959 (2018)
7. Chan, H., Han, S., Akoglu, L.: Where graph topology matters: the robust subgraph problem. In: SIAM, pp. 10–18 (2015)
8. Chandran, B.G., Hochbaum, D.S.: A computational study of the pseudoflow and push-relabel algorithms for the maximum flow problem. Oper. Res. **57**, 358–376 (2009)
9. Charikar, M.: Greedy approximation algorithms for finding dense components in a graph. In: Jansen, K., Khuller, S. (eds.) APPROX 2000. LNCS, vol. 1913, pp. 84–95. Springer, Heidelberg (2000). https://doi.org/10.1007/3-540-44436-X_10
10. Chen, J., Saad, Y.: Dense subgraph extraction with application to community detection. TKDE **24**, 1216–1230 (2012)
11. Combe, D., Largeron, C., Géry, M., Egyed-Zsigmond, E.: I-louvain: an attributed graph clustering method. In: Fromont, E., De Bie, T., van Leeuwen, M. (eds.) IDA 2015. LNCS, vol. 9385, pp. 181–192. Springer, Cham (2015). https://doi.org/10.1007/978-3-319-24465-5_16
12. Ehrgott, M.: Multicriteria Optimization. Springer, Heidelberg (2005). https://doi.org/10.1007/3-540-27659-9
13. Falih, I., Grozavu, N., Kanawati, R., Bennani, Y.: Community detection in attributed network. In: Companion Proceedings of WWW. pp. 1299–1306 (2018)
14. Fang, Y., Cheng, R., Luo, S., Hu, J.: Effective community search for large attributed graphs. VLDB **9**(12), 1233–1244 (2016)
15. Gallo, G., Grigoriadis, M.D., Tarjan, R.E.: A fast parametric maximum flow algorithm and applications. J. Comp. **18**, 30–55 (1989)
16. Goldberg, A.V.: Finding a maximum density subgraph. University of California Berkeley, CA (1984)
17. Hochbaum, D.S.: The pseudoflow algorithm: a new algorithm for the maximum-flow problem. Oper. Res. **56**, 992–1009 (2008)
18. Hooi, B., Song, H.A., Beutel, A., Shah, N., Shin, K., Faloutsos, C.: Fraudar: bounding graph fraud in the face of camouflage. In: SIGKDD, pp. 895–904 (2016)
19. Khuller, S., Saha, B.: On Finding dense Subgraphs. In: Albers, S., Marchetti-Spaccamela, A., Matias, Y., Nikoletseas, S., Thomas, W. (eds.) ICALP 2009. LNCS, vol. 5555, pp. 597–608. Springer, Heidelberg (2009). https://doi.org/10.1007/978-3-642-02927-1_50

20. Sánchez, P.I., et al.: Efficient algorithms for a robust modularity-driven clustering of attributed graphs. In: SIAM, pp. 100–108 (2015)
21. Shelokar, P., Quirin, A., Cordón, O.: Mosubdue: a pareto dominance-based multi-objective subdue algorithm for frequent subgraph mining. KAIS **34**, 75–108 (2013). https://doi.org/10.1007/s10115-011-0452-y
22. Tsourakakis, C.: The $k$-clique densest subgraph problem. In: WWW (2015)
23. Tsourakakis, C., Bonchi, F., Gionis, A., Gullo, F., Tsiarli, M.: Denser than the densest subgraph: extracting optimal quasi-cliques with quality guarantees. In: SIGKDD, pp. 104–112 (2013)
24. Valari, E., Kontaki, M., Papadopoulos, A.N.: Discovery of top-k dense subgraphs in dynamic graph collections. In: Ailamaki, A., Bowers, S. (eds.) SSDBM 2012. LNCS, vol. 7338, pp. 213–230. Springer, Heidelberg (2012). https://doi.org/10.1007/978-3-642-31235-9_14
25. Wu, Y., Jin, R., Zhu, X., Zhang, X.: Finding dense and connected subgraphs in dual networks. In: ICDE, pp. 915–926 (2015)
26. Yikun, B., Xin, L., Ling, H., Yitao, D., Xue, L., Wei, X.: No place to hide: catching fraudulent entities in tensors. In: The World Wide Web Conference (2019)

# Towards Description of Block Model on Graph

Zilong Bai[1(✉)], S. S. Ravi[2], and Ian Davidson[1]

[1] University of California, Davis, USA
zlbai@ucdavis.edu, davidson@cs.ucdavis.edu
[2] University of Virginia, Charlottesville, USA
ssravi0@gmail.com

**Abstract.** Existing block modeling methods can detect communities as blocks. However it remains a challenge to easily explain to a human why nodes belong to the same block. Such a description is very useful for answering *why people in the same community tend to interact cohesively.* In this paper we explore a novel problem: Given a block model already found, describe the blocks using an auxiliary set of information. We formulate a combinatorial optimization problem which finds a *unique* disjunction of the auxiliary information shared by the nodes either in the same block or between a pair of different blocks. The former terms intra-block description, the latter inter-block description. Given an undirected graph and its $k-$block model, our method generates $k + \frac{k(k-1)}{2}$ different descriptions. If the tags are descriptors of events occurring at the vertices, our descriptions can be interpreted as common events occurring within blocks and between blocks. We show that this problem is intractable even for simple cases, e.g., when the underlying graph is a tree with just two blocks. However, simple and efficient ILP formulations and algorithms exist for its relaxation and yield insights different from a state-of-the-art related work in unsupervised description. We empirically show the power of our work on multiple real-world large datasets.

**Keywords:** Explainable artificial intelligence · Unsupervised graph analysis · Block model

## 1 Introduction

Block modeling is an effective community detection method (cf. [1,30]). Conventionally, communities can be defined as subgraphs that are densely connected within but weakly connected to each other (cf. [24,43]). However, block models find communities based on different definitions of similarity/equivalence in terms of how they connect to the rest of the graph, such as structural equivalence (i.e., second-order proximity [38,43]) and stochastic equivalence (e.g., stochastic block model [1]). Equation 1 exemplifies a basic formulation of block modeling (cf. [15]),

ⓒ Springer Nature Switzerland AG 2021
F. Hutter et al. (Eds.): ECML PKDD 2020, LNAI 12459, pp. 37–53, 2021.
https://doi.org/10.1007/978-3-030-67664-3_3

which simultaneously discovers $k$ blocks stacked column-wise in an $n \times k$ block allocation matrix $\mathbf{F}$ and the block-level connectivity in a $k \times k$ image matrix $\mathbf{M}$.[1]

$$\underset{\mathbf{F}\geq 0,\mathbf{M}\geq 0}{Minimize} \|\mathbf{G} - \mathbf{FMF}^T\|_F \ s.t., \mathbf{F}^T\mathbf{F} = \mathbf{D} \qquad (1)$$

Block modeling produces a useful abstract view of the graph it simplifies by finding, for instance, structural equivalence [43]. However, existing block models are ineffective in explaining to a human domain expert *why* the members in the same block are equivalent which limits their use in some domains. Consider the sociological definition of a "community" which requires several properties[2] including i) interacting people, and ii) members who share common values, beliefs, or behaviors. A block model finds the former but not the latter. Despite existing work on posterior descriptions of detected communities (e.g., [29]) and efforts on descriptive community detection (e.g., [5,6,21]), it remains understudied in finding a complete explanation for a given block model. A complete explanation would describe both communities themselves *and* differences between the communities. Explaining differences between communities is important as in block models, in contrast to conventionally defined communities, strong inter-block connectivity can exist (e.g., [20,22]). This is particularly important if the blocks are discovered as *roles* [34], as each role is not defined individually but based on its interactions with others (e.g., [3,27]).

In this paper we consider the setting where we are given a block model $\mathcal{B} = \{\mathbf{F}^{n \times k}, \mathbf{M}^{k \times k}\}$ discovered from an adjacency matrix $\mathbf{A}$ of graph $G(V, E)$. We aim to explain the block model $\mathcal{B}$ using auxiliary information which are human-interpretable tags. We explore a novel problem of finding a description as a *disjunction* comprising the most frequently used tags for the edges within each block and between each pair of different blocks. The former is denoted as intra-block description, the latter inter-block description. If $G$ is an undirected simple graph, our method generates $k$ intra-block descriptions and $\frac{k(k-1)}{2}$ inter-block descriptions. We introduce two different disjointness constraints between the descriptions of different edge sets to address *how each edge set is unique or special to the other edge sets*. We illustrate our work with a toy example in Fig. 1 and a running example in Table 1 by comparing against DTDM [12]. Albeit closely related to our work in terms of Integer Linear Programming (ILP) formulations, DTDM considers *item* cluster description whereas ours focuses on *edge* subsets induced by the block model on the graph.

The major contributions of this work are as follows: 1. We propose a novel combinatorial optimization formulation, Valid Tag Assignment to Edges (VTAE), for describing the result of an existing block model based on edge covering (Sect. 3). 2. We show VTAE is intractable (Sect. 3), hence no exact efficient algorithm is possible. 3. We propose ILP formulations for VTAE and a relaxation with efficient algorithm that guarantees feasibility for this relaxation (Sect. 4). 4. We demonstrate our method on three real-world datasets (Sect. 5).

---

[1] Here $\mathbf{D}$ denotes a diagonal matrix. We use $\| \bullet \|_F$ to denote the Frobenius norm.
[2] https://www.oxfordbibliographies.com/view/document/obo-9780199756384/obo-9780199756384-0080.xml.

**Table 1.** Partial experimental results of descriptions learnt by our VTAE with *global* disjointness (see Sect. 3) and DTDM [12] applied to a Twitter dataset of 880 nodes given its precomputed 4-block block model.

| Method | Block 1 "Trump supporters" | Block 2 "Hillary supporters" | Between block 1 and 2 |
|---|---|---|---|
| Our VTAE-g | Hillary, TrumpRally, CNN, 2A, Ohio, USA, AmericaFirst, trump2016, TeamTrump, TRUMP2016, VoteTrump2016 | BernieSanders, Bernie, Hillary2016, p2, PAprimary, UniteBlue, MDPrimary, BernieOrBust, CT, ArizonaPrimary | NewYorkValues, ISIS, AZPrimary, MSM, NewYorkPrimary, pjnet, WI, trumptrain, 1A, maga, CCOT, CRUZ, SECPrimary |
| DTDM [12] | GOPdebate, DNCinPHL, Clinton, GOPDEBATE, BlackLivesMatter, MAGA, Sanders, CrookedHillary, Hannity, UniteBlue, CCOT | NeverTrump, DemsInPhilly, GOP, 1, SCPrimary, FeelTheBern, PrimaryDay, DemTownHall, FITN, NewYork, DumpTrump | N/A |

Our VTAE demonstrates advantages over DTDM [12] in: i) representing the similarity/difference between different block models (Fig. 6), ii) being robust against small perturbations in a given block allocation (Fig. 7).

## 2   Related Work

There has been a growing need for explainable machine learning and data mining approaches [16]. However many approaches focus on supervised learning [36], such as explaining per-instance prediction/classification result (e.g., [7,14,33]) of a *black-box* deep learning model [2]. Some other efforts are about describing unsupervised learning results. In line with the focus of this work, we briefly review former efforts on describing unsupervised learning (on graph).

**Conceptual Clustering** [18] attempts to use the *same features* for finding the clustering and explaining it. Extensions to find clusterings in one view and a description in another exist [10] but is limited to a single view of categorical variables. To find a clustering and explanation simultaneously in multiple views, recent work has explored conceptual clustering extensions with constraint programming [23], SAT [25] or ILP [10,28] but their scalability is limited due to the Pareto optimization formulation. Our work is different from this setting as we explicitly consider the edges in a graph.

**Clustering on Vertex-Labeled Graphs** explores finding vertex clusterings based on graph topology and vertex labels. Recent advances on vertex/edge-attributed graphs include novel extensions of community detection [4,17] or SBM [1]. Representative work of the latter includes SBM on graphs where the nodes are associated to binary attributes (e.g., [42]) or continuous values (e.g., [37]). A general underlying premise of such work is that the graph topology and tag/label information compensate for each other [13]. Although it seems relevant, this line of research does not create succinct descriptions for the *subgraphs* induced from a graph by a vertex set partitioning, hence is different from ours.

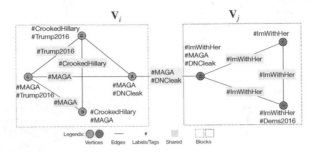

**Fig. 1.** An example using hashtags to describe two blocks/communities $V_i, V_j$, red and blue, on a Twitter network. The edge between vertices A and E is the only edge between $V_i$ and $V_j$ covered by {DNCleak}. The edges within $V_i$ are covered by {CrookedHillary, MAGA, Trump2016}, $V_j$ by {ImWithHer}. Previous work [12] finds overly simple distinct descriptions {MAGA, Trump2016} and {ImWithHer} to cover $V_i$ and $V_j$.

**Descriptive Community Detection** simultaneously *discovers* communities and their descriptions on vertex/edge-labeled graphs. Representative approaches include seminal work [6] and its later extension [5] that develop a search-based algorithm where each node on the search tree is a candidate community that can be described by a conjunction of labels. Another approach discusses three different 0−1 predicates to describe the edges within each community - i.e., conjunction, majority, and disjunction, and explores the first two in their algorithm development and experiments [21]. A framework alternates between two phases: i) the community detection phase and ii) inducing a disjunction of possible negated conjunctions to match the community and reshape the community if no perfect match is found [32]. Different from descriptive community detection approaches, our work aims to faithfully describe the effect of a given block model on a graph instead of discovering or modifying the community structure. We are post-processing results hence agnostic to how a block model is generated. Moreover, existing work is primarily concerned with finding a description for *each* community with assumption of community structure in a conventional sense - i.e., the nodes are well-connected to each other within the same community but are less connected to the outside nodes (cf. [19,31]). Whereas our work aims to find a succinct description for each set of intra-block or inter-block edges induced from a given block model, where significant inter-block/community edges can exist and need to be described.

**Post-processing Output of Existing Algorithms (i)** *Given* a community structure, [29] explores topic modeling; it *separately* builds topic models with Latent Dirichlet Allocation (LDA) (cf. [39,40]) and discovers communities based on the Girvan-Newman community detection algorithm (cf. [24]), and then combines the two by computing the probabilistic distribution over the topics for each community. However, such topic model only applies to each community without modeling the interactions between different communities. Our work considers a

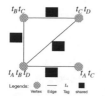 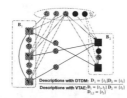 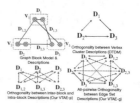

(a) Our VTAE vs DTDM [12] on a single community.

(b) Advantage of our VTAE over DTDM [12] in a two-block example.

(c) Comparison of disjointness constraints.

**Fig. 2.** Comparing our VTAE against DTDM [12].

fundamentally different type of description as a disjunction of tags and simultaneously discovers descriptions for both the inter-block edges and the intra-block edges induced by a block model on a graph. **(ii)** Most similar to ours, [12] formulates the novel cluster description problem as a combinatorial optimization problem to find concise but different descriptors of each cluster. However, that work is not specifically for graphs and ignores any graph structure. A technical limitation of [12] is that the description it learns for one community can actually cover a lot of nodes in other communities. Our method alleviates this issue as we simultaneously discover descriptions with disjointness constraints for both intra-block and inter-block edge sets, which can reduce the odds of using the attributes shared by nodes from two different blocks in describing the intra-block edges. See Fig. 2 for more illustrative examples. In this paper, we do not compare with [35] which proposes efficient algorithms based on random rounding as a follow-up work on [12], since we focus on formulation-wise contributions.

## 3   Problem Definition and Complexity Results

In this section we define the description problem for block models and show that the problem is intractable even for relaxed variations. We first introduce some necessary notation to formalize the problem. Each undirected graph $G(V, E)$ considered here has the following properties.

1. Each vertex $v \in V$ has an associated set $T(v)$ of tags. Some times we will refer to all tags for all instances as a matrix $L$.
2. For each edge $\{x, y\} \in E$, the associated tag sets $T(x)$ and $T(y)$ have *at least one common tag*; that is, $T(x) \cap T(y) \neq \emptyset$.[3]

**Input.** Given an undirected graph $G(V, E)$ and a subset $V_i \subseteq V$ of vertices found by the block model, we use $G_i(V_i, E_i)$ to denote the subgraph of $G$ *induced* on the vertex set $V_i$. (Thus, $E_i$ contains all and only those edges $\{x, y\}$ such that

---

[3] When this condition does not hold, our method can still be used but just with these edges removed as they are unexplainable.

both $x$ and $y$ are in $V_i$ and $\{x, y\} \in E$.) We refer to each set $E_i$ as an **intra-block edge set**. For each pair of subsets $V_i$ and $V_j$, with $i \neq j$, we use $E_{ij}$ to denote the set of edges that join a vertex $x \in V_i$ to a vertex $y \in V_j$. We refer to each such set as an **inter-block edge set**. Each partition of $V$ into $k$ subsets gives rise to $k$ sets of intra-block edges and $k(k-1)/2$ sets of inter-block edges. We denote the collection of these edge sets as $\mathcal{E}(G)$. Some of these edge sets may be empty and can be ignored if their corresponding image matrix entries are zero.

**Output.** Let $\Gamma = \cup_{v \in V} T(v)$ denote the collection/union of all the tags assigned to the vertices which is the universe of all tags. The goal of the problem considered below is to find a description for each edge set (both intra-block and inter-block). To do this it is required to first find a function $\tau : E \longrightarrow \Gamma$ that assigns a tag $\tau(e) \in \Gamma$ to each edge $e \in E$. Given a tag assignment function $\tau$ and an edge set $X \subseteq E$, we use $\tau(X)$ to denote the set of all the tags assigned to the edges in $X$. A tag assignment function $\tau$ is **valid** if and only if it satisfies *both* of the following conditions.

(a) [**Compatibility:**] For each edge $e = \{x, y\} \in E$, $\tau(e) \in T(x) \cap T(y)$; that is, the tag $\tau(e)$ assigned to $e$ appears in the tag sets of both the end points of $e$.

(b) [**Disjointness:**] For a pair of *different* edge sets $X$ and $Y$ in $\mathcal{E}(G)$, $\tau(X) \cap \tau(Y) = \emptyset$. That is, the sets of tags assigned to $X$ and $Y$ must be *disjoint*.

In our work and experiments we explored two types of disjointness: i) Partial Disjointness which requires $X$ to be an intra-block edge set and $Y$ to be an inter-block edge set and ii) Global Disjointness which requires disjointness between any two different edge sets from $\mathcal{E}(G)$.

We consider the following decision problem and show it is intractable. The optimization version of the problem adds the requirement of finding the most compact description (see Eq. 2).

**Valid Tag Assignment to Edges** (VTAE)

<u>Instance:</u>  An undirected graph $G(V, E)$, for each vertex $v \in V$ a nonempty set $T(v)$ of tags so that for any edge $\{x, y\} \in E$, $T(x) \cap T(y) \neq \emptyset$, a partition of $V$ into $k$ subsets $V_1, V_2, \ldots, V_k$.

<u>Question:</u>  Is there a valid tag assignment function $\tau : E \longrightarrow \Gamma$, where $\Gamma$ is the union of all the tag sets of vertices?

**The Complexity of VTAE**

We first observe that if the partition of the vertex set $V$ consists of just one block (namely, the vertex set $V$), the VTAE problem is trivial: for each edge $\{x, y\}$, one can choose any tag from the nonempty set $T(x) \cap T(y)$, since there is no inter-block edge set in this case. The following theorem shows that the VTAE problem becomes computationally intractable even when the underlying graph is a tree and the given partition of $V$ contains just two blocks/subsets.

**Theorem 1.** *The VTAE problem with either **partial** or **global** disjointness is **NP**-complete even when the graph $G(V, E)$ is a tree and the vertex set is partitioned into just two subsets.*

**Proof:** See supplementary material[4].

## 4   ILP Formulations and Algorithms

In this section we first formulate a basic instantiation of the VTAE problem and an extension with enhanced statistical significance as ILP formulations (Eq. 2 and 3). We then relax VTAE to ensure that feasible solutions exist and facilitate scalability. Particularly, our **divide and conquer** Algorithm 1 can describe the block models on a graph of $2 \times 10^6 s$ edges over $10^4$ vertices with marginally ignored edges in about 2 min on a machine with 4-core Intel Xeon CPU (Fig. 5).

**Notations.** We are given an unweighted & undirected graph $G(V, E)$ over $n$ vertices and a $k$-way vertex set partitioning $\{V_1, \dots, V_k\}$ of a block model. Each vertex is associated with a subset of the $|\Gamma|$ tags. These tags are in an $n \times |\Gamma|$ matrix L. The block model induces $k$ **intra**-block and $\frac{k(k-1)}{2}$ **inter**-block edge sets on the graph. The collection of these edge sets is $\mathcal{E}(G)$. For each edge set $Z \in \mathcal{E}(G)$ we solve for a $|\Gamma|$-dimensional binary vector $D_Z$ where $D_Z(t) = 1$ iff tag $t$ is chosen to describe this edge set. We define other frequently used notations in Table 2. Note they are precomputed to facilitate ILP formulations.

**Minimal VTAE and an Extension - ILP Formulations.** Here we present the ILP formulation for VTAE with an objective to find the most compact collection of descriptions in Eq. 2. To enhance the statistical significance of description, we propose an extension to VTAE which aims to find the most frequently used tags to build each description, additionally constrained by compactness (e.g., an integer upper bound $c_Z$) in Eq. 3.

**Table 2.** Frequently used notations.

| Notation | Definition and computation |
|---|---|
| $L_Z$ | **Tag allocation matrix** of edge set $Z$. To assist formulating **compatibility** constraint as a set coverage requirement. $|Z| \times |\Gamma|$ binary matrix. $L_Z(e, t) = 1$ **iff** $e \in Z, \{x, y\} = e, t \in T(x) \cap T(y)$ |
| $U_Z$ | **Tag universe** indicator where the most concise explanation for edge set $Z$ draws tags from. $|\Gamma|$-d binary vector. $U_Z(t) = 1$ **iff** $L_Z(e, t) = 1, \exists e \in Z$ |
| $w_Z$ | **Edge coverage** of tags in $Z$. $|\Gamma|$-d integer vector. $w_Z(t) = \sum_{e \in Z} L_Z(e, t)$ |

---

[4] Proofs for our theorems, our codes and other supplementary materials are available on https://github.com/ZilongBai/ECMLPKDD2020TDBMG. .

$$\underset{D_Z(t)\in\{0,1\}}{Minimize} \underset{Z\ t}{\Sigma\Sigma}\ D_Z(t)$$

$$s.t.\underset{t}{\Sigma}D_Z(t) \times L_Z(e,t) \geq 1, \forall e \in Z, \forall Z \in \mathcal{E}(G)\ (\textbf{cover}) \tag{2}$$

$$D_X(t) + D_Y(t) \leq 1, \forall t, \forall X \neq Y \in \mathcal{E}(G)\ (\textbf{disjointness})$$

$$\underset{D_Z(t)\in\{0,1\}}{Maximize} \underset{Z\ t}{\Sigma\Sigma}\ D_Z(t) \times w_Z(t)$$

$$s.t.\underset{t}{\Sigma}D_Z(t) \times L_Z(e,t) \geq 1, \forall e \in Z, \forall Z \in \mathcal{E}(G)\ (\textbf{cover})$$

$$D_X(t) + D_Y(t) \leq 1, \forall t, \forall X \neq Y \in \mathcal{E}(G)\ (\textbf{disjointness}) \tag{3}$$

$$\underset{t}{\Sigma}\ D_Z(t) \leq c_Z, \forall Z \in \mathcal{E}(G)$$

**Relaxations to Ensure Feasibility and Enhance Efficiency.** Due to the rigorous constraints (**cover**) and (**disjointness**), the solver on real-world datasets commonly finds Eq. 2 and 3 to be *infeasible*, and the problem for whether VTAE is feasible is computationally *intractable* (see Sect. 3).

Inspired by [12] we could have two intuitive solutions to side-step the infeasibility issue: (**1**) allow limited overlaps between descriptions and (**2**) "cover or forget" edges to guarantee disjointness. Note that neither of these changes makes the VTAE problem (with either disjointness) tractable. For example for limited overlapping descriptions Sect. 2 of the supplementary material (see Footnote 4) shows that problem is also intractable (Theorem 2). We discuss ILP formulations for these two relaxations in the supplementary material (see Footnote 4) due to their limited value of practical use. In particular, **minimizing overlap** yields significant overlap between the descriptions in our empirical study and **cover or forget** demands an auxiliary binary variable to identify whether *each* edge needs to be forgotten, rendering a prohibitively large search space for our edge covering.

We propose a **divide and conquer** Algorithm 1 to explore the idea of selectively removing tags from the tag universes such that the tag universes are *disjoint* and each edge set covering subproblem can be solved *independently*. Firstly, we use Eq. 4 and 5 to find binary vector $R_Z(t), \forall Z \in \mathcal{E}(G)$, where $R_Z(t) = 1$ iff $t$ needs to be removed from $U_Z$ to make the tag universes disjoint. Let $u^*$ be the minimal feasible function value of Eq. 4, we set $u = u^*$ in Eq. 5 to find tags to exclude with small impact on the coverage of edges. Then we set the $t$-th *column* in matrix $L_Z$ to all-zero if $R_Z(t) = 1$. The edges whose *rows* in $L_Z$ turn all-zero are removed from the following covering problem. Figure 3 shows how to preprocess the example in Fig. 1. Finally, on each preprocessed edge set $Z$ we learn description using the most frequently used tags with Eq. 7 constrained by the optimal succinctness discovered by Eq. 6. Our algorithm effectively reduces the search space and scales well in practice. The independency between subproblems can facilitate parallel implementation.

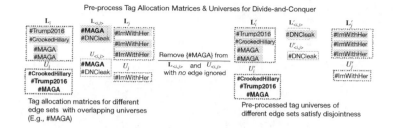

**Fig. 3.** Preprocess tag allocation matrices of the example in Fig. 1.

---

**Algorithm 1:** Solve VTAE with Divide-and-Conquer.

**Input:** Unweighted simple graph $G(V, E)$, vertex tags $L$, block structure
$$\mathcal{V} = \{V_1, V_2, \ldots, V_k\}$$

1 Precompute tag allocation matrices $L_Z, \forall Z \in \mathcal{E}(G)$ based on $G$ and $\mathcal{V}$ ;
2 Solve Eq. 4 for the minimal upper-bound of tags to exclude $u^*$ to achieve tag universe disjointness ;
3 Solve Eq. 5 with $u^*$ in **(upper-bound)** for the tags to exclude: $R_Z^*(t)$ ;
4 Exclude $R_Z^*(t)$ from each row of $L_Z$ respectively ;
5 Remove empty rows from $L_Z, \forall Z \subset \mathcal{E}(G)$ ;
6 **while** $Z \in \mathcal{E}(G)$ **do**
7     Solve Eq. 6 for $D_Z$, set $d \leftarrow \sum_t D_Z(t)$ ;
8     Solve Eq. 7 for $D_Z^*$ ;
9 **end**
10 **return** $D_Z^*, \forall Z \in \mathcal{E}(G)$

---

$$\underset{R_Z(t)\in\{0,1\}, Z\in\mathcal{E}(G)}{Minimize} \quad u$$

$$s.t. \sum_t R_Z(t) \leq u, \forall Z \in \mathcal{E}(G) \; (\textbf{upper}-\textbf{bound})$$

$$U_X(t) \times (1 - R_X(t)) + U_Y(t) \times (1 - R_Y(t)) \leq 1,$$
$$\forall t, \forall X \neq Y \in \mathcal{E}(G) \; (\textbf{disjointness}) \tag{4}$$

$$\underset{R_Z(t)\in\{0,1\}, Z\in\mathcal{E}(G)}{Minimize} \quad \underset{Z}{\Sigma}\underset{t}{\Sigma} w_Z(t) \times R_Z(t)$$

$$s.t. \sum_t R_Z(t) \leq u, \forall Z \in \mathcal{E}(G) \; (\textbf{upper}-\textbf{bound})$$

$$U_X(t) \times (1 - R_X(t)) + U_Y(t) \times (1 - R_Y(t)) \leq 1,$$
$$\forall t, \forall X \neq Y \in \mathcal{E}(G) \; (\textbf{disjointness}) \tag{5}$$

$$\underset{D_Z(t)\in\{0,1\}}{Minimize} \; \underset{t}{\Sigma} \; D_Z(t) \quad s.t. \quad \underset{t}{\Sigma} D_Z(t) \times L_Z(e,t) \geq 1, \quad \forall e \in Z \; (\textbf{cover}) \tag{6}$$

$$\underset{D_Z(t)\in\{0,1\}}{Maximize} \ \underset{t}{\Sigma} \ D_Z(t) \times w_Z(t)$$
$$s.t.\underset{t}{\Sigma}D_Z(t) \times L_Z(e,t) \geq 1, \quad \forall e \in Z \quad (\textbf{cover});\underset{t}{\Sigma}D_Z(t) \leq d \quad (\textbf{upper}-\textbf{bound})$$
$$(7)$$

## 5   Experiments

We evaluate our methods, VTAE when the disjointness is **global** (VTAE-g) or **partial** (VTAE-p), on three real-world datasets from different application domains: i) Twitter data with *two* graphs, ii) the BlogCatalog dataset[5] [26,41], and iii) graphs constructed from brain imaging data (fMRI of BOLD) from ADNI[6]. The details for the construction of the Twitter graphs and the brain imaging graphs are in the supplementary material (see Footnote 4). We summarize their basic statistics in Table 3. On these real-world large datasets, we aim to address the following questions:

**Table 3.** Statistics for the datasets.

| Dataset | Twitter (smaller) | Twitter (larger) | BlogCatalog | Brain imaging graphs |
|---|---|---|---|---|
| # of Nodes | 880 | 10,000 | 5,196 | 1,730 |
| # of Edges | 73,136 | 2,301,732 | 171,743 | 575,666.05 (mean) |
| # of Tags | 136 | 136 | 8,189 | 1,730 |

**Q1.** *Novelty in description:* Can our descriptions offer novel insights different from existing methods for cluster description (i.e., [12])?

**Q2.** *Scalability:* How well does our method scale to large datasets?

**Q3.** *Usefulness for model selection:* Can our descriptions better represent the similarity/difference between different block models on the same graph?

**Q4.** *Robustness to perturbations in block allocation:* Is our description sensitive to small perturbations in block allocation of nodes?

**Q5.** *Description for historically/scientifically known model:* Is our method useful in situations where the block structure is not a result of an algorithm but historically/scientifically known?

**Q6.** *Stability/diversity w.r.t. graph topology:* When the block structure is fixed but graph topology changes, can our descriptions demonstrate *stability/diversity* consistent with domain knowledge?

---

[5] http://people.tamu.edu/~xhuang/BlogCatalog.mat.zip.
[6] http://adni.loni.usc.edu/study-design/collaborative-studies/dod-adni.

We address **Q1** on the smaller Twitter graph, **Q2** and **Q3** on the larger Twitter graph. We explore **Q4** on the larger Twitter graph and the BlogCatalog dataset. To demonstrate the versatility of our method, we explore **Q5** and **Q6** on the brain imaging graphs where the labels are not explicitly given. The experiments are conducted on a machine with 4-core Intel Xeon CPU. The ILP formulations are solved by `Gurobi` (https://www.gurobi.com/) in `Python`.

**Precompute Block Model with NMtF.** For Twitter and BlogCatalog data we use Nonnegative Matrix tri-Factorization (NMtF) which is not jointly convex with multiplicative update rules in seminal work [15] to generate block models (block allocation **F**, image matrix **M**) on the adjacency matrix of a graph **G**.

**Scientifically Known Block Structure.** For the fMRI data we use the well-known default mode network (DMN) that divides the brain into multiple interacting regions. We consider two regions - i.e., foreground and background as explored in [8,9]. See supplementary material (see Footnote 4) for visualization of DMN and the two-region partitioning.

**Baseline.** We choose the **cover or forget** relaxation of DTDM [12] (DTDM-`cof`) to find a succinct and distinct description $D_i$, a $|\Gamma|$-dimensional binary vector for vertex cluster $V_i$, $i = 1, \ldots, k$. Follow [12] we define and precompute binary matrices $L_i$ for vertex cluster $V_i$ where $L_i(v, t) = 1$ `iff` vertex $v$ is assigned to cluster $V_i, 1 \leq i \leq k$ and is associated with tag $t$. We use $|V|$ binary variables $f_v$, each marks whether a vertex $v$ needs to be forgotten or not. The number of forgotten vertices (i.e., $\sum_{v \in V} f_v$) is part of the minimization objective (Eq. 8).

$$\underset{D_i(t) \in \{0,1\}, f_v \in \{0,1\}}{Minimize} \sum_i \sum_t D_i(t) + \sum_{v \in V} f_v$$
$$s.t. \ f_v + \sum_t D_i(t) \times L_i(v, t) \geq 1, \forall v \in V_i, \forall i; \ D_i(t) + D_j(t) \leq 1, \forall t, \forall i \neq j \tag{8}$$

**Answering Q1.** We simplified the smaller Twitter graph (Table 3) into 4 blocks. Upon examination of the members in each block (see supplementary material (see Footnote 4)) we observe a distinct political polarization as expected [11]. We then describe which hashtags are commonly used over the edges within a block $E_1, \ldots, E_4$ and between blocks $E_{1,2}, E_{2,3}, \ldots, E_{3,4}$ (see Fig. 4). Compared to Table 5 the edges demonstrate higher complexity than the vertices in the sense of associated content (i.e., can demand more tags in a description upon minimal length). Our VTAE-g generates description for each **intra**-block edge set that drastically differs from its corresponding vertex block description of baseline method DTDM [12]. See VTAE-p results in the supplementary material (see Footnote 4).

**Table 4.** The statistics of interacting blocks found using block modeling [15] for us to describe. See supplementary material (see Footnote 4) for full member list in each block.

| Block | Prominent members | Size | Intra. edges | Ext. edges |
|---|---|---|---|---|
| 1: Trump supporters | Trump Train, Italians4Trump | 184 | 7,332 | 20,646 |
| 2: Clinton and supporters | Hillary Clinton, WeNeedHillary | 227 | 5,288 | 21,566 |
| 3: Other-candidates | Ted Cruz, John Kasich | 309 | 10,756 | 23,307 |
| 4: Trump inner circle | DonaldTrump, DonaldTrumpJr | 160 | 5,880 | 20,931 |

**Table 5.** Vertex cluster descriptions of DTDM-cof [12] (Eq. 8) for Table 4.

| Block | Vertex cluster description |
|---|---|
| $V_1$ | GOPdebate, DNCinPHL, Clinton, GOPDEBATE, BlackLivesMatter, MAGA, Sanders, CrookedHillary, Hannity, UniteBlue, CCOT |
| $V_2$ | NeverTrump, DemsInPhilly, GOP, 1, SCPrimary, FeelTheBern, PrimaryDay, DemTownHall, FITN, NewYork, DumpTrump |
| $V_3$ | Trump, RNCinCLE, SuperTuesday, IowaCaucus, NYPrimary, DonaldTrump, Election2016, MakeAmericaGreatAgain, TrumpRally, NewYorkValues, DNCleak, USA, trump2016 |
| $V_4$ | DemDebate, Trump2016, NHPrimary, trump, ImWithHer, iacaucus, BernieSanders, TRUMP, CruzSexScandal |

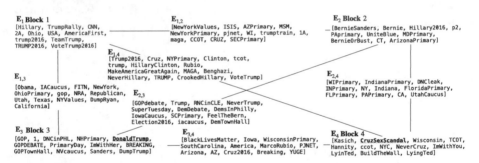

**Fig. 4.** Edge set explanations discovered by VTAE-g. Underlined are tags used by both our VTAE-g and DTDM-cof [12] for the same block (edges within and the vertices) in Table 5. Our result and the baseline use very few tags in common.

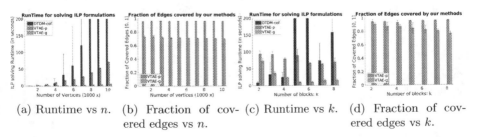

(a) Runtime vs $n$.    (b) Fraction of covered edges vs $n$.    (c) Runtime vs $k$.    (d) Fraction of covered edges vs $k$.

**Fig. 5.** Runtime of our methods (and DTDM-cof [12]) on Twitter (larger).

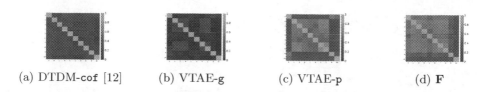

(a) DTDM-cof [12]    (b) VTAE-g    (c) VTAE-p    (d) **F**

**Fig. 6.** Similarity matrices at $k = 7$. Corresponds to $k = 7$ in Table 6. Visually our method VTAE-p better represents the similarity/difference between **F**'s than DTDM [12].

**Answering Q2.** We address this problem on the larger Twitter graph (Table 3). We explore both the effects of the number of sampled vertices given a fixed 5-way block model (Fig. 5a, 5b) and $k$ the number of blocks (Fig. 5c, 5d). For the former we randomly sample 10 times from the entire $10^4$ individuals at each given number of vertices. For the latter we generate 10 block models ($\{\mathbf{F}, \mathbf{M}\}$'s) with random initializations at each $k$ with the NMtF formulation [15]. We show that the combined runtime of ILP formulations in our Algorithm 1 are comparable to earlier work [12] despite finding $k + \frac{k(k-1)}{2}$ descriptions for *edge* sets versus $k$ descriptions for *vertex* clusters. Such runtime efficiency of our methods is achieved by marginally ignoring edges (Fig. 5d, 5b). Particularly, our method can explain the block modeling results on this large Twitter graph within 2 min, ignoring $\leq 5\%$ edges for VTAE-p given an 8-block model.

**Answering Q3.** To be useful for model selection, the descriptions should represent the similarity/difference between different block models for the same graph. We investigate the descriptions generated in answering **Q2** on the large Twitter graph. For the 10 block models precomputed at each $k$, we measure the cosine similarity between the descriptions of each pair of block models generated by our VTAE-p, VTAE-g, and the baseline method DTDM [12], respectively. Each yields a $10 \times 10$ similarity matrix $\mathbf{S}^k_{[method]}$. We compute $\mathbf{S}^k_g$ for **F**'s as ground-truth. Since a block allocation is unique up to permutation, all the similarities are measured after aligning the descriptions and **F**'s according to their corresponding image matrices **M**'s. We visualize the results at $k = 7$ in Fig. 6 and present $\|\mathbf{S}^k_{[method]} - \mathbf{S}^k_{\mathbf{F}}\|_F$ in Table 6 for each method at $k = 2, \dots, 8$. We do not use NMI or ACC to measure the similarity between **F**'s as neither provides useful information to align the **inter**-block descriptions for a fair comparison.

**Table 6.** Difference between similarity matrices of each method and **F**'s $\|\mathbf{S}^k_{[method]} - \mathbf{S}^k_{\mathbf{F}}\|_F$. Our VTAE-p outperforms DTDM [12] at $k = 5, 6, 7$.

| k | 2 | 3 | 4 | 5 | 6 | 7 | 8 |
|---|---|---|---|---|---|---|---|
| DTDM [12] | **0.7045** | **1.2834** | **1.6660** | 2.3027 | 2.1325 | 2.5608 | **1.9610** |
| Our VTAE-g | 1.1270 | 2.0602 | 2.8341 | 3.6892 | 3.3882 | 4.0660 | 3.4347 |
| Our VTAE-p | 1.2043 | 1.8109 | 1.6930 | **1.2069** | **1.9389** | **1.0781** | 2.5427 |

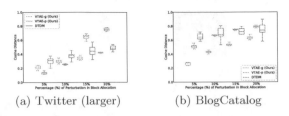

(a) Twitter (larger)          (b) BlogCatalog

**Fig. 7.** Changes in descriptions due to perturbations in block allocation.

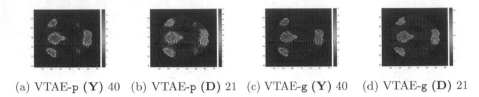

(a) VTAE-p **(Y)** 40    (b) VTAE-p **(D)** 21    (c) VTAE-g **(Y)** 40    (d) VTAE-g **(D)** 21

**Fig. 8.** Heat maps for subjects using voxels in their descriptions.

**Answering Q4.** We select one block model generated at $k = 4$ for the larger Twitter data and generate block model at $k = 6$ (i.e., the number of ground-truth classes) for the BlogCatalog data. We artificially introduce different levels of perturbations in its **F**: We change the block membership of randomly selected $p = 5\%, 10\%, 15\%, 20\%$ of nodes. We measure the *cosine distance* between the descriptions learnt for perturbed block models and the original block model (Fig. 7). Our method demonstrates stronger robustness than [12] against small perturbations in most cases.

**Answering Q5 and Q6.** We generate vertex tags with the 10 nearest neighbors of these vertices on the correlation graphs, each graph measures how much the BOLD signals of each two voxels correlate. Our method then explains the *strong* interactions of the default mode cognitive network in terms of the nearest neighbors to it on the correlation graph. We summarize the descriptions at the cohort level - i.e., control cohort **(Y)** and Alzheimer's affected cohort **(D)**, as *heat maps* in Fig. 8. The color of each pixel indicates the number of subjects using it in their descriptions in a cohort. Red means many subjects using this subject. Figure 8a and Fig. 8c show that our descriptions for the control cohort subjects are confined within the DMN region. Figure 8b and Fig. 8d show our descriptions for the Alzheimer's affected group with less well-defined regions. Importantly, our method generates *stable* descriptions for the control cohort despite having 40 subjects, while our descriptions for the demented cohort with 21 are highly *diverse*. Note our method does not explicitly distinguish subjects from different cohorts. This *stability/diversity* result is consistent with the neuroscience domain knowledge that similar intra-DMN interactions exist amongst the young and healthy subjects from the control cohort whilst the Alzheimer's affected subjects can have demented DMN interactions for *varied* causes.

# 6    Conclusion

In this paper we have researched a novel problem of describing a given block modeling result on a graph by using labels or tags associated to the nodes. These nodal tags are not utilized in the block modeling process. We have formally defined the problem as Valid Tag Assignment to Edges (VTAE) to find descriptions for both the intra-block edge sets and the inter-block edge sets. The descriptions of different edge sets are built with disjoint subsets of tags. We show that the decision version of VTAE is intractable but relaxations exist and can be efficiently solved by existing ILP solver `Gurobi`. Experiments on Twitter data show that even for the same block modeling result on the same graph, our formulations can generate descriptions significantly different from the baseline approach for cluster description [12]. Our numerical evaluations on Twitter and BlogCatalog datasets demonstrate advantages of our method over the baseline method [12]. In some settings, our descriptions can i) better represent the similarity/difference between different block models on the same graph and ii) be more robust against small perturbations in block allocation. On graphs built with brain imaging data, we show the versatility of our method as it can be useful even when the tags are not explicitly given but generated with nearest neighbors on a weighted graph. Experimental results on this dataset also show that the distribution of our descriptions is consistent with domain knowledge at the cohort-level. In the future, we aim to derive extensions of existing models to enhance the block modeling performance in more challenging settings, such as large noisy and incomplete real-world graphs.

**Acknowledgments.** This research is supported by ONR Grant N000141812485 and NSF Grants IIS-1910306, IIS-1908530, OAC-1916805, ACI-1443054 (DIBBS), IIS-1633028 (BIG DATA) and CMMI-1745207 (EAGER).

# References

1. Abbe, E.: Community detection and stochastic block models: recent developments. J. Mach. Learn. Res. **18**(1), 6446–6531 (2017)
2. Adadi, A., Berrada, M.: Peeking inside the black-box: a survey on explainable artificial intelligence (XAI). IEEE Access **6**, 52138–52160 (2018)
3. Akar, E., Mardikyan, S.: User roles and contribution patterns in online communities: a managerial perspective. Sage Open **8**(3), 2158244018794773 (2018)
4. Atzmueller, M.: Descriptive community detection. In: Missaoui, R., Kuznetsov, S.O., Obiedkov, S. (eds.) Formal Concept Analysis of Social Networks. LNSN, pp. 41–58. Springer, Cham (2017). https://doi.org/10.1007/978-3-319-64167-6_3
5. Atzmueller, M., Doerfel, S., Mitzlaff, F.: Description-oriented community detection using exhaustive subgroup discovery. Inf. Sci. **329**, 965–984 (2016)
6. Atzmueller, M., Mitzlaff, F.: Efficient descriptive community mining. In: FLAIRS (2011)
7. Bach, S., Binder, A., Montavon, G., Klauschen, F., Müller, K.R., Samek, W.: On pixel-wise explanations for non-linear classifier decisions by layer-wise relevance propagation. PLoS One **10**(7), e0130140 (2015)

8. Bai, Z., Qian, B., Davidson, I.: Discovering models from structural and behavioral brain imaging data. In: SIGKDD, pp. 1128–1137 (2018)
9. Bai, Z., Walker, P., Tschiffely, A., Wang, F., Davidson, I.: Unsupervised network discovery for brain imaging data. In: SIGKDD, pp. 55–64 (2017)
10. Chabert, M., Solnon, C.: Constraint programming for multi-criteria conceptual clustering. In: Beck, J.C. (ed.) CP 2017. LNCS, vol. 10416, pp. 460–476. Springer, Cham (2017). https://doi.org/10.1007/978-3-319-66158-2_30
11. Conover, M.D., Ratkiewicz, J., Francisco, M., Gonçalves, B., Menczer, F., Flammini, A.: Political polarization on Twitter. In: ICWSM (2011)
12. Davidson, I., Gourru, A., Ravi, S.: The cluster description problem-complexity results, formulations and approximations. In: NIPS, pp. 6190–6200 (2018)
13. Deshpande, Y., Sen, S., Montanari, A., Mossel, E.: Contextual stochastic block models. In: NIPS, pp. 8581–8593 (2018)
14. Dhurandhar, A., et al.: Explanations based on the missing: towards contrastive explanations with pertinent negatives. In: NIPS, pp. 592–603 (2018)
15. Ding, C., Li, T., Peng, W., Park, H.: Orthogonal nonnegative matrix t-factorizations for clustering. In: SIGKDD, pp. 126–135 (2006)
16. Došilović, F.K., Brčić, M., Hlupić, N.: Explainable artificial intelligence: a survey. In: 2018 41st MIPRO, pp. 0210–0215. IEEE (2018)
17. Falih, I., Grozavu, N., Kanawati, R., Bennani, Y.: Community detection in attributed network. In: WWW, pp. 1299–1306 (2018)
18. Fisher, D.H.: Knowledge acquisition via incremental conceptual clustering. Mach. Learn. 2(2), 139–172 (1987). https://doi.org/10.1007/BF00114265
19. Fortunato, S.: Community detection in graphs. Phy. Rep. 486(3–5), 75–174 (2010)
20. Funke, T., Becker, T.: Stochastic block models: a comparison of variants and inference methods. PLoS One 14(4), e0215296 (2019)
21. Galbrun, E., Gionis, A., Tatti, N.: Overlapping community detection in labeled graphs. Data Min. Knowl. Disc. 28(5), 1586–1610 (2014). https://doi.org/10.1007/s10618-014-0373-y
22. Ganji, M., et al.: Image constrained blockmodelling: a constraint programming approach. In: SDM, pp. 19–27 (2018)
23. Garey, M.R., Johnson, D.S.: Computers and Intractability; A Guide to the Theory of NP-Completeness. W. H. Freeman & Co., New York (1990)
24. Girvan, M., Newman, M.E.: Community structure in social and biological networks. PNAS 99(12), 7821–7826 (2002)
25. Guns, T., Nijssen, S., De Raedt, L.: k-Pattern set mining under constraints. IEEE TKDE 25(2), 402–418 (2011)
26. Huang, X., Li, J., Hu, X.: Label informed attributed network embedding. In: WSDM, pp. 731–739 (2017)
27. Kao, H.T., Yan, S., Huang, D., Bartley, N., Hosseinmardi, H., Ferrara, E.: Understanding cyberbullying on Instagram and Ask.fm via social role detection. In: WWW, pp. 183–188 (2019)
28. Kotthoff, L., O'Sullivan, B., Ravi, S., Davidson, I.: Complex clustering using constraint programming: Modelling electoral map (2015)
29. Li, D., et al.: Community-based topic modeling for social tagging. In: CIKM (2010)
30. Müller, B., Reinhardt, J., Strickland, M.T.: Neural Networks: An Introduction. Springer Science & Business Media, Heidelberg (2012)
31. Newman, M.E.: Modularity and community structure in networks. PNAS 103(23), 8577–8582 (2006)
32. Pool, S., Bonchi, F., Leeuwen, M.V.: Description-driven community detection. TIST 5(2), 1–28 (2014)

33. Ribeiro, M.T., Singh, S., Guestrin, C.: "Why should I trust you?" explaining the predictions of any classifier. In: SIGKDD, pp. 1135–1144 (2016)
34. Rossi, R.A., Ahmed, N.K.: Role discovery in networks. IEEE TKDE **27**(4), 1112–1131 (2014)
35. Sambaturu, P., Gupta, A., Davidson, I., Ravi, S., Vullikanti, A., Warren, A.: Efficient algorithms for generating provably near-optimal cluster descriptors for explainability. In: AAAI, pp. 1636–1643 (2020)
36. Samek, W., Montavon, G., Vedaldi, A., Hansen, L.K., Müller, K.-R. (eds.): Explainable AI: Interpreting, Explaining and Visualizing Deep Learning. LNCS (LNAI), vol. 11700. Springer, Cham (2019). https://doi.org/10.1007/978-3-030-28954-6
37. Stanley, N., Bonacci, T., Kwitt, R., Niethammer, M., Mucha, P.J.: Stochastic block models with multiple continuous attributes. Appl. Netw. Sci. **4**(1), 1–22 (2019)
38. Tang, J., Qu, M., Wang, M., Zhang, M., Yan, J., Mei, Q.: Line: large-scale information network embedding. In: WWW, pp. 1067–1077 (2015)
39. Tang, J., Jin, R., Zhang, J.: A topic modeling approach and its integration into the random walk framework for academic search. In: IEEE ICDM (2008)
40. Tang, J., Zhang, J., Yao, L., Li, J., Zhang, L., Su, Z.: Arnetminer: extraction and mining of academic social networks. In: SIGKDD, pp. 990–998 (2008)
41. Tang, L., Liu, H.: Relational learning via latent social dimensions. In: SIGKDD, pp. 817–826 (2009)
42. Yang, J., McAuley, J., Leskovec, J.: Community detection in networks with node attributes. In: IEEE ICDM, pp. 1151–1156 (2013)
43. Zhang, D., Yin, J., Zhu, X., Zhang, C.: Network representation learning: a survey. IEEE Trans. Big Data **6**, 3–28 (2018)

# Large-Scale Optimization and Differential Privacy

# Orthant Based Proximal Stochastic Gradient Method for $\ell_1$-Regularized Optimization

Tianyi Chen[1(✉)], Tianyu Ding[2], Bo Ji[3], Guanyi Wang[4], Yixin Shi[1],
Jing Tian[5], Sheng Yi[1], Xiao Tu[1], and Zhihui Zhu[6]

[1] Microsoft, Redmond, USA
Tianyi.Chen@microsoft.com
[2] Johns Hopkins University, Baltimore, USA
tding1@jhu.edu
[3] Zhejiang University, Hangzhou, China
jibo27@zju.edu.cn
[4] Georgia Institute of Technology, Atlanta, USA
gwang93@gatech.edu
[5] University of Washington, Seattle, USA
jingtc20@uw.edu
[6] University of Denver, Denver, USA
zhihui.zhu@du.edu

**Abstract.** Sparsity-inducing regularization problems are ubiquitous in machine learning applications, ranging from feature selection to model compression. In this paper, we present a novel stochastic method – Orthant Based Proximal Stochastic Gradient Method (OBProx-SG) – to solve perhaps the most popular instance, *i.e.*, the $\ell_1$-regularized problem. The OBProx-SG method contains two steps: (i) a proximal stochastic gradient step to predict a support cover of the solution; and (ii) an orthant step to aggressively enhance the sparsity level via orthant face projection. Compared to the state-of-the-art methods, *e.g.*, Prox-SG, RDA and Prox-SVRG, the OBProx-SG not only converges comparably in both convex and non-convex scenarios, but also promotes the sparsity of the solutions substantially. Particularly, on a large number of convex problems, OBProx-SG outperforms the existing methods comprehensively in the aspect of sparsity exploration and objective values. Moreover, the experiments on non-convex deep neural networks, *e.g.*, MobileNetV1 and ResNet18, further demonstrate its superiority by generating the solutions of much higher sparsity without sacrificing generalization accuracy, which further implies that OBProx-SG may achieve significant memory and energy savings. The source code is available at https://github.com/tianyic/obproxsg.

**Keywords:** Stochastic learning · Sparsity · Orthant prediction

© Springer Nature Switzerland AG 2021
F. Hutter et al. (Eds.): ECML PKDD 2020, LNAI 12459, pp. 57–73, 2021.
https://doi.org/10.1007/978-3-030-67664-3_4

## 1   Introduction

Plentiful tasks in machine learning and deep learning require formulating and solving particular optimization problems [3,9], of which the solutions may not be unique. From the perspective of the application, people are usually interested in a subset of the solutions with certain properties. A common practice to address the issue is to augment the objective function by adding a regularization term [23]. One of the best known examples is the sparsity-inducing regularization, which encourages highly sparse solutions (including many zero elements). Besides, such regularization typically has shrinkage effects to reduce the magnitude of the solutions [22]. Among the various ways of introducing sparsity, the $\ell_1$-regularization is perhaps the most popular choice. Its utility has been demonstrated ranging from improving the interpretation and accuracy of model estimation [20,21] to compressing heavy model for efficient inference [7,12].

In this paper, we propose and analyze a novel efficient stochastic method to solve the following large-scale $\ell_1$-regularization problem

$$\operatorname*{minimize}_{x\in\mathbb{R}^n} \left\{ F(x) \overset{\text{def}}{=} \underbrace{\frac{1}{N} \sum_{i=1}^{N} f_i(x)}_{f(x)} + \lambda\|x\|_1 \right\}, \tag{1}$$

where $\lambda > 0$ is a weighting term to control the level of sparsity in the solutions, and $f(x)$ is the raw objective function. We pay special interests to the $f(x)$ as the average of numerous $N$ continuously differentiable instance functions $f_i : \mathbb{R}^n \to \mathbb{R}$, such as the loss functions measuring the deviation from the observations in various data fitting problems. A larger $\lambda$ typically results in a higher sparsity while sacrifices more on the bias of model estimation, hence $\lambda$ needs to be carefully fine-tuned to achieve both low $f(x)$ and high sparse solutions. Above formulation is widely appeared in many contexts, including convex optimization, e.g., LASSO, logistic regression and elastic-net formulations [22,32], and non-convex problems such as deep neural networks [29,30].

Problem (1) has been well studied in deterministic optimization with various methods that capable of returning solutions with both low objective value and high sparsity under proper $\lambda$. Proximal methods are classical approaches to solve the structured non-smooth optimization problems with the formulation (1), including the popular proximal gradient method (Prox-FG) and its variants, e.g., ISTA and FISTA [2], in which only the first-order derivative information is used. They have been proved to be quite useful in practice because of their simplicity. Meanwhile, first-order methods are limited due to the local convergence rate and lack of robustness on ill-conditioned problems, which can often be overcome by employing the second-order derivative information as is used in proximal-Newton methods [18,28]. However, when $N$ is enormous, a straightforward computation of the full gradients or Hessians could be prohibitive because of the costly evaluations over all $N$ instances. Thus, in modern large-scale machine learning applications, it is inevitable to use stochastic methods that operate on a small subset of above summation to economize the computational cost at every iteration.

Nevertheless, in stochastic optimization, the studies of $\ell_1$-regularization (1) become somewhat limited. In particular, the existing state-of-the-art stochastic algorithms rarely achieve both fast convergence and highly sparse solutions simultaneously due to the stochastic nature [25]. Proximal stochastic gradient method (Prox-SG) [10] is a natural extension of Prox-FG by using a mini-batch to estimate the full gradient. However, there are two potential drawbacks of Prox-SG: (i) the lack of exploitation on the certain problem structure, *e.g.*, the $\ell_1$ regularization (1); (ii) the slower convergence rate than Prox-FG due to the variance introduced by random sampling. To exploit the regularization structure more effectively (produce sparser solutions), regularized dual-averaging method (RDA) [25] is proposed by extending the simple dual averaging scheme in [19]. The key advantages of RDA are to utilize the averaged accumulated gradients of $f(x)$ and an aggressive coefficient of the proximal function to achieve a more aggressive truncation mechanism than Prox-SG. As a result, in convex setting, RDA usually generates much sparser solutions than that by Prox-SG in solving (1) but typically has slower convergence. On the other hand, to reduce the variance brought by the stochastic approximation, proximal stochastic variance-reduced gradient method (Prox-SVRG) [26] is developed based on the well-known variance reduction technique SVRG developed in [15]. Prox-SVRG has both capabilities of decent convergence rate and sparsity exploitation in convex setting, while its per iteration cost is much higher than other approaches due to the calculation of full gradient for achieving the variance reduction.

The above mentioned Prox-SG, RDA and Prox-SVRG are valuable state-of-the-art stochastic algorithms with apparent strength and weakness. RDA and Prox-SVRG are derived from proximal gradient methods, and make use of different averaging techniques cross all instances to effectively exploit the problem structure. Although they explore sparsity well in convex setting, the mechanisms may not perform as well as desired in non-convex formulations [8]. Moreover, observing that the proximal mapping operator is applicable for any non-smooth penalty function, this generic operator may not be sufficiently insightful if the regularizer satisfies extra properties. In particular, the non-smooth $\ell_1$-regularized problems of the form (1) degenerate to a smooth reduced space problem if zero elements in the solution are correctly identified.

This observation has motivated the exploitation of orthant based methods, a class of deterministic second-order methods that utilizes the particular structure within the $\ell_1$-regularized problem (1). During the optimization, they predict a sequence of orthant faces, and minimize smooth quadratic approximations to (1) on those orthant faces until a solution is found [1,4,5,16]. Such a process normally equips with second-order techniques to yield superlinear convergence towards the optimum, and introduces sparsity by Euclidean projection onto the constructed orthant faces. Orthant based methods have been demonstrated competitiveness in deterministic optimization to proximal methods [5,6,16]. In contrast, related prior work in stochastic settings is very rare, perhaps caused by the expensive and non-reliable orthant face selection under randomness.

***Our Contributions.*** In this paper, we propose an Orthant Based Proximal Stochastic Gradient Method (OBProx-SG) by capitalizing on the advantages of orthant based methods and Prox-SG, while avoiding their disadvantages. Our OBProx-SG is efficient, promotes sparsity more productively than others, and converges well in both practice and theory. Specifically, we have the following contributions.

– We provide a novel stochastic algorithmic framework that utilizes Prox-SG Step and reduced space Orthant Step to effectively solve problem (1). Comparing with the existing stochastic algorithms, it exploits the sparsity significantly better by combining the moderate truncation mechanism of Prox-SG and an aggressive orthant face projection under the control of a switching mechanism. The switching mechanism is specifically established in the stochastic setting, which is simple but efficient, and performs quite well in practice. Moreover, we present the convergence characteristics under both convex and non-convex formulations, and provide analytic and empirical results to suggest the strategy of the inherent switching hyperparameter selection.
– We carefully design the Orthant Step for stochastic optimization in the following aspects: (i) it utilizes the sign of the previous iterate to select an orthant face, which is more efficient compared with other strategies that involve computations of (sub)-gradient in the deterministic orthant based algorithms [1,16]; (ii) instead of optimizing with second-order methods, only the first-order derivative information is used to exploit on the constructed orthant face.
– Experiments on both convex (logistic regression) and non-convex (deep neural networks) problems show that OBProx-SG usually outperforms the other state-of-the-art methods comprehensively in terms of the sparsity of the solution, final objective value, and runtime. Particularly, in the popular deep learning applications, without sacrificing generalization performance, the solutions computed by OBProx-SG usually possess multiple-times higher sparsity than those searched by the competitors.

## 2    The OBProx-SG Method

To begin, we summarize the proposed Orthant Based Proximal Stochastic Gradient Method (OBProx-SG) in Algorithm 1. In a very high level, it proceeds one of the two subroutines at each time, so called Prox-SG Step (Algorithm 2) and Orthant Step (Algorithm 3). There exist two switching parameters $N_{\mathcal{P}}$ and $N_{\mathcal{O}}$ that control how long we are sticking to each step and when to switch to the other. We will see that the switching mechanism (choices of $N_{\mathcal{P}}$ and $N_{\mathcal{O}}$) is closely related to the convergence of OBProx-SG and the sparsity promotions. But we defer the detailed discussion till the end of this section, while first focus our attention on the Prox-SG Step and Orthant Step.

**Prox-SG Step.** In Prox-SG step, the algorithm performs one iteration of standard proximal stochastic gradient step to approach a solution of (1). Particularly, at $k$-th iteration, we sample a mini-batch $\mathcal{B}_k$ to make an unbiased estimate of the full gradient of $f$ (line 2, Algorithm 2). Then we utilize the following proximal mapping to yield next iterate as

$$x_{k+1} = \text{Prox}_{\alpha_k \lambda \|\cdot\|_1}(x_k - \alpha_k \nabla f_{\mathcal{B}_k}(x_k))$$
$$= \underset{x \in \mathbb{R}^n}{\text{argmin}} \ \frac{1}{2\alpha_k}\|x - (x_k - \alpha_k \nabla f_{\mathcal{B}_k}(x_k))\|_2^2 + \lambda\|x\|_1. \tag{3}$$

---

**Algorithm 1.** Outline of OBProx-SG for solving (1).

---

1: **Input:** $x_0 \in \mathbb{R}^n$, $\alpha_0 \in (0,1)$, and $\{N_\mathcal{P}, N_\mathcal{O}\} \subset \mathbb{Z}^+$.
2: **for** $k = 0, 1, 2, \ldots$ **do**
3:     **Switch** Prox-SG Step or Orthant Step by Algorithm 4.
4:     **if** Prox-SG Step is selected **then**
5:         Compute the Prox-SG Step update:
            $x_{k+1} \leftarrow \text{Prox-SG}(x_k, \alpha_k)$ by Algorithm 2.
6:     **else if** Orthant Step is selected **then**
7:         Compute the Orthant Step update:
            $x_{k+1} \leftarrow \text{Orthant}(x_k, \alpha_k)$ by Algorithm 3.
8:     Update $\alpha_{k+1}$ given $\alpha_k$ according to some rule.

---

**Algorithm 2.** Prox-SG Step.

---

1: **Input:** Current iterate $x_k$, and step size $\alpha_k$.
2: Compute the stochastic gradient of $f$ on $\mathcal{B}_k$

$$\nabla f_{\mathcal{B}_k}(x_k) \leftarrow \frac{1}{|\mathcal{B}_k|} \sum_{i \in \mathcal{B}_k} \nabla f_i(x_k). \tag{2}$$

3: Compute $x_{k+1} \leftarrow \text{Prox}_{\alpha_k \lambda \|\cdot\|_1}(x_k - \alpha_k \nabla f_{\mathcal{B}_k}(x_k))$ .
4: **Return** $x_{k+1}$.

---

It is known that the above subproblem (3) has a closed form solution [2]. Denote the trial iterate $\widehat{x}_{k+1} := x_k - \alpha_k \nabla f_{\mathcal{B}_k}(x_k)$, then $x_{k+1}$ is computed efficiently as:

$$[x_{k+1}]_i = \begin{cases} [\widehat{x}_{k+1}]_i - \alpha_k\lambda, & \text{if } [\widehat{x}_{k+1}]_i > \alpha_k\lambda \\ [\widehat{x}_{k+1}]_i + \alpha_k\lambda, & \text{if } [\widehat{x}_{k+1}]_i < -\alpha_k\lambda \\ 0, & \text{otherwise} \end{cases} \tag{4}$$

In OBProx-SG, Prox-SG Step generally serves as a globalization mechanism to guarantee convergence and predict a cover of supports (non-zero entries) in the solution. But it alone is insufficient to exploit the sparsity structure because of the relatively moderate truncation mechanism effected in a small projection region, i.e., the trial iterate $\widehat{x}_{k+1}$ is projected to zero only if it falls into $[-\alpha_k\lambda, \alpha_k\lambda]$. Our remedy here is to combine it with our Orthant Step, which exhibits an aggressive sparsity promotion mechanism while still remains efficient.

**Orthant Step.** Since the fundamental to Orthant Step is the manner in which we handle the zero and non-zero elements, we define the following index sets for any $x \in \mathbb{R}^n$:

$$\mathcal{I}^0(x) := \{i : [x]_i = 0\}, \ \mathcal{I}^+(x) := \{i : [x]_i > 0\}, \ \mathcal{I}^-(x) := \{i : [x]_i < 0\} \quad (6)$$

---

**Algorithm 3.** Orthant Step.

1: **Input:** Current iterate $x_k$, and step size $\alpha_k$.
2: Compute the stochastic gradient of $\widetilde{F}$ on $\mathcal{B}_k$

$$\nabla\widetilde{F}_{\mathcal{B}_k}(x_k) \leftarrow \frac{1}{|\mathcal{B}_k|} \sum_{i\in\mathcal{B}_k} \nabla\widetilde{F}_i(x_k) \quad (5)$$

3: Compute $x_{k+1} \leftarrow \text{Proj}_{\mathcal{O}_k}(x_k - \alpha_k\nabla\widetilde{F}_{\mathcal{B}_k}(x_k))$.
4: **Return** $x_{k+1}$.

---

Furthermore, we denote the non-zero indices of $x$ by $\mathcal{I}^{\neq 0}(x) := \mathcal{I}^+(x) \cup \mathcal{I}^-(x)$. To proceed, we define the orthant face $\mathcal{O}_k$ that $x_k$ lies in to be

$$\mathcal{O}_k := \{x \in \mathbb{R}^n : \text{sign}([x]_i) = \text{sign}([x_k]_i) \text{ or } [x]_i = 0, 1 \le i \le n\} \quad (7)$$

such that $x \in \mathcal{O}_k$ satisfies: (i) $[x]_{\mathcal{I}^0(x_k)} = 0$; (ii) for $i \in \mathcal{I}^{\neq 0}(x_k)$, $[x]_i$ is either $0$ or has the same sign as $[x_k]_i$.

The key assumption for Orthant Step is that an optimal solution $x^*$ of problem (1) inhabits $\mathcal{O}_k$, i.e., $x^* \in \mathcal{O}_k$. In other words, the orthant face $\mathcal{O}_k$ already covers the support (non-zero entries) of $x^*$. Our goal becomes now minimizing $F(x)$ over $\mathcal{O}_k$, i.e., solving the following subproblem:

$$x_{k+1} = \underset{x\in\mathcal{O}_k}{\text{argmin}} \ F(x) = f(x) + \lambda\|x\|_1. \quad (8)$$

By the definition of $\mathcal{O}_k$, we know $[x]_{\mathcal{I}^0(x_k)} \equiv 0$ are fixed, and only the entries of $[x]_{\mathcal{I}^{\neq 0}(x_k)}$ are free to move. Hence, (8) is essentially a reduced space optimization problem. Observing that for any $x \in \mathcal{O}_k$, $F(x)$ can be written precisely as a smooth function $\widetilde{F}(x)$ in the form

$$F(x) \equiv \widetilde{F}(x) := f(x) + \lambda\,\text{sign}(x_k)^T x, \quad (9)$$

therefore (8) is equivalent to the following smooth problem

$$x_{k+1} = \underset{x \in \mathcal{O}_k}{\mathrm{argmin}} \ \widetilde{F}(x). \qquad (10)$$

A direct way for solving problem (10) is the projected stochastic gradient descent method, as stated in Algorithm 3. It performs one iteration of stochastic gradient descent (SGD) step combined with projections onto the orthant face $\mathcal{O}_k$. At $k$-th iteration, a mini-batch $\mathcal{B}_k$ is sampled, and is used to approximate the full gradient $\nabla\widetilde{F}(x_k)$ by the unbiased estimator $\nabla\widetilde{F}_{\mathcal{B}_k}(x_k)$ (line 2 , Algorithm 3). The standard SGD update computes a trial point $\widehat{x}_{k+1} = x_k - \alpha_k\nabla\widetilde{F}_{\mathcal{B}_k}(x_k)$, which is then passed into a projection operator $\mathrm{Proj}_{\mathcal{O}_k}(\cdot)$ defined as

$$\left[\mathrm{Proj}_{\mathcal{O}_k}(z)\right]_i := \begin{cases} [z]_i & \text{if } \mathrm{sign}([z]_i) = \mathrm{sign}([x_k]_i) \\ 0 & \text{otherwise} \end{cases}. \qquad (11)$$

Notice that $\mathrm{Proj}_{\mathcal{O}_k}(\cdot)$ is an Euclidean projector, and ensures that the trial point $\widehat{x}_{k+1}$ is projected back to the current orthant face $\mathcal{O}_k$ if it happens to be outside, as illustrated in Fig. 1. In the demonstrated example, the next iterate $x_{k+1} = \mathrm{Proj}_{\mathcal{O}_k}(\widehat{x}_{k+1})$ turns out to be not only a better approximated solution but also sparser compared with $x_k$ since $[x_{k+1}]_2 = 0$ after the projection, which suggests the power of Orthant Step in sparsity promotion. In fact, compared with Prox-SG, the orthant-face projection (11) is a more aggressive sparsity truncation mechanism. Particularly, Orthant Step enjoys a much larger projection region to map a trial iterate to zero comparing with other stochastic algorithms. Consider the 1D example in Fig. 2, where $x_k > 0$, it is clear that the projection region of Orthant Step $(-\infty, \alpha_k\lambda]$ is a superset of that of Prox-SG and Prox-SVRG $[-\alpha_k\lambda, \alpha_k\lambda]$, and it is apparently larger than that of RDA.

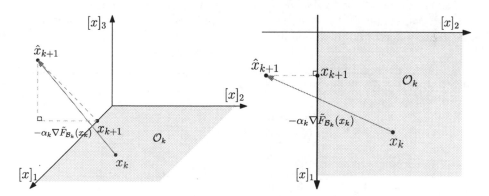

**Fig. 1.** Illustration of Orthant Step with projection in (11), where $\mathcal{O}_k = \{x \in \mathbb{R}^3 : [x]_1 \geq 0, [x]_2 \geq 0, [x]_3 = 0\}$. (L): 3D view. (R): top view.

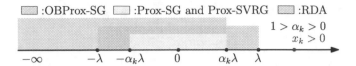

**Fig. 2.** Projection regions of different methods for 1D case at $x_k > 0$.

In practice, by taking advantage of the fact that (10) is a reduced space problem, *i.e.*, $[x_{k+1}]_{\mathcal{I}^0(x_k)} \equiv 0$, we only need to store a small part of stochastic gradient information $[\nabla \widetilde{F}(x_k)]_{\mathcal{I}^{\neq 0}(x_k)}$, and compute the projection of $[\widetilde{x}_{k+1}]_{\mathcal{I}^{\neq 0}(x_k)}$. This makes the whole procedure even more efficient when $|\mathcal{I}^{\neq 0}(x_k)| \ll n$.

We emphasize that Orthant Step is one of the keys to the success of our proposed OBProx-SG method in terms of sparsity exploration. It is originated from the orthant based methods in deterministic optimization, which normally utilize second-order information. When borrowing the idea, we make the selection of orthant face $\mathcal{O}_k$ more effective by looking at the sign of the previous iterate (see (7)). Then, we make use of a stochastic version of the projected gradient method in solving the subproblem (10) to introduce sparsity aggressively. As a result, Orthant Step always makes rapid progress to the optimality, while at the same time promotes sparse solutions dedicatedly.

***Switching Mechanism.*** To complete the discussion of the OBProx-SG framework, we now explain how we select Prox-SG or Orthant Step at each iteration, which is crucial in generating accurate solutions with high sparsity. A popular switching mechanism for deterministic multi-routine optimization algorithms utilizes the optimality metric of each routine (typically the norm of (sub)gradient) [5,6]. However, in stochastic learning, this approach does not work well in general due to the additional computation cost of optimality metric and the randomness that may deteriorate the progress of sparsity exploration as numerically illustrated in Appendix C.

---

**Algorithm 4.** Switching Mechanism.

---

1: **Input:** $k, N_\mathcal{P}, N_\mathcal{O}$.
2: **if** $\text{mod}(k, N_\mathcal{P} + N_\mathcal{O}) < N_\mathcal{P}$ **then**
3:     **Return** Prox-SG Step is selected.
4: **else**
5:     **Return** Orthant Step is selected.

---

To address this issue, we specifically establish a simple but efficient switching mechanism consisting of two hyperparameters $N_\mathcal{P}$ and $N_\mathcal{O}$, which performs quite well in both practice and theory. As stated in Algorithm 4, $N_\mathcal{P}$ ($N_\mathcal{O}$) controls how many consecutive iterations we would like to spend doing Prox-SG Step (Orthant Step), and then switch to the other. OBProx-SG is highly flexible to different choices of $N_\mathcal{P}$ and $N_\mathcal{O}$. For example, an alternating strategy between one Prox-SG Step and one Orthant Step corresponds to set $N_\mathcal{P} = N_\mathcal{O} = 1$, and a strategy of first doing several Prox-SG Steps then followed by Orthant Step all the time corresponds to set $N_\mathcal{P} < \infty, N_\mathcal{O} = \infty$. A larger $N_\mathcal{P}$ helps to predict a better orthant face $\mathcal{O}_k$ which hopefully covers the support of $x^*$, and a larger $N_\mathcal{O}$ helps to explore more sparsity within $\mathcal{O}_k$.

As we will see in Sect. 3, convergence of Algorithm 1 requires either doing Prox-SG Step infinitely many times ($N_\mathcal{P} \le \infty, N_\mathcal{O} < \infty$), or doing finitely many Prox-SG Steps followed by infinitely many Orthant Steps ($N_\mathcal{P} < \infty, N_\mathcal{O} = \infty$) given the support of $x^*$ has been covered by some $\mathcal{O}_k$. In practice, without knowing $x^*$ ahead of time, we can always start by employing Prox-SG Step $N_\mathcal{P}$ iterations, followed by running Orthant Step $N_\mathcal{O}$ iterations, then repeat until convergence. Meanwhile, experiments in Sect. 4 show that first performing Prox-SG Step sufficiently many times then followed by running Orthant Step all the time usually produces even slightly better solutions. Moreover, for the latter case, a bound for $N_\mathcal{P}$ is provided in Sect. 3. For simplicity, we refer the OBProx-SG under ($N_\mathcal{P} < \infty, N_\mathcal{O} = \infty$) as OBProx-SG+ throughout the remainder of this paper.

We end this section by giving empirical suggestions of setting $N_\mathcal{P}$ and $N_\mathcal{O}$. Overall, in order to obtain accurate solutions of high sparsity, we highly recommend to start OBProx-SG with Prox-SG Step and ends with Orthant Step. Practically, employing finitely many Prox-SG Steps followed by sticking on Orthant Steps ($N_\mathcal{P} < \infty, N_\mathcal{O} = \infty$) until the termination, is more preferable because of its attractive property regarding maintaining the progress of sparsity exploration. In this case, although the theoretical upper bound of $N_\mathcal{P}$ is difficult to be measured, we suggest to keep running Prox-SG Step until reaching some acceptable evaluation metrics e.g., objectives or validation accuracy, then switch Orthant Step to promote sparsity.

## 3   Convergence Analysis

In this section, we give a convergence analysis of our proposed OBProx-SG and OBProx-SG+, referred as OBProx-SG(+) for simplicity. Towards that end, we first make the following assumption.

**Assumption 1** *The function $f : \mathbb{R}^n \to \mathbb{R}$ is continuously differentiable, and bounded below on the compact level set $\mathcal{L} := \{x \in \mathbb{R}^n : F(x) \le F(x_0)\}$, where $x_0$ is the initialization of Algorithm 1. The stochastic gradient $\nabla f_{\mathcal{B}_k}$ and $\nabla \tilde{F}_{\mathcal{B}_k}$ evaluated on the mini-batch $\mathcal{B}_k$ are Lipschitz continuous on the level set $\mathcal{L}$ with a shared Lipschitz constant $L$ for all $\mathcal{B}_k$. The gradient $\nabla \tilde{F}_{\mathcal{B}_k}(x)$ is uniformly bounded over $\mathcal{L}$, i.e., there exists a $M < \infty$ such that $\|\nabla \tilde{F}_{\mathcal{B}_k}(x)\|_2 \le M$.*

Remark that many terms in Assumption 1 appear in numerical optimization literatures [5,26,27]. Let $x^*$ be an optimal solution of problem (1), $F^*$ be the minimum, and $\{x_k\}_{k=0}^{\infty}$ be the iterates generated by Algorithm 1. We then define the gradient mapping and its estimator on mini-batch $\mathcal{B}$ as follows

$$\mathcal{G}_\eta(x) = \frac{1}{\eta}\left(x - \mathrm{Prox}_{\eta\lambda\|\cdot\|_1}(x - \eta\nabla f(x))\right), \text{ and} \tag{12}$$

$$\mathcal{G}_{\eta,\mathcal{B}}(x) = \frac{1}{\eta}\left(x - \mathrm{Prox}_{\eta\lambda\|\cdot\|_1}(x - \eta\nabla f_\mathcal{B}(x))\right). \tag{13}$$

Here we define the noise $e(x)$ be the difference between $\mathcal{G}_\eta(x)$ and $\mathcal{G}_{\eta,\mathcal{B}}(x)$ with zero-mean due to the random sampling of $\mathcal{B}$, i.e., $\mathbb{E}_\mathcal{B}[e(x)] = 0$, of which variance is bounded by $\sigma^2 > 0$ for one-point mini-batch. $\tilde{x}$ is so-called a stationary point of $F(x)$ if $\mathcal{G}_\eta(\tilde{x}) = 0$. Additionally, establishing some convergence results require the below constants to measure the least and largest magnitude of non-zero entries in $x^*$:

$$0 < \delta_1 := \frac{1}{2}\min_{i\in\mathcal{I}^{\neq 0}(x^*)}|[x^*]_i|, \text{ and } 0 < \delta_2 := \frac{1}{2}\max_{i\in\mathcal{I}^{\neq 0}(x^*)}|[x^*]_i|, \tag{14}$$

Now we state the first main theorem of OBProx-SG.

**Theorem 1.** *Suppose* $N_\mathcal{P} < \infty$ *and* $N_\mathcal{O} < \infty$.

*(i) the step size* $\{\alpha_k\}$ *is* $\mathcal{O}(1/k)$, *then* $\liminf_{k\to\infty}\mathbb{E}\|\mathcal{G}_{\alpha_k}(x_k)\|_2^2 = 0$.
*(ii)* $f$ *is* $\mu$-*strongly convex, and* $\alpha_k \equiv \alpha$ *for any* $\alpha < \min\{\frac{1}{2\mu}, \frac{1}{L}\}$, *then*

$$\mathbb{E}[F(x_{k+1}) - F^*] \leq (1 - 2\alpha\mu)^{\kappa_\mathcal{P}}[F(x_0) - F^*] + \frac{LC^2}{2\mu}\alpha, \tag{15}$$

*where* $\kappa_\mathcal{P}$ *is the number of Prox-SG Steps employed until k-th iteration.*

Theorem 1 implies that if OBProx-SG employs Prox-SG Step and Orthant Step alternatively, then the gradient mapping converges to zero zero in expectation under decaying step size for general $f$ satisfying Assumption 1 even if $f$ is non-convex on $\mathbb{R}^n$. In other words, the iterate $\{x_k\}$ converges to some stationary point in the sense of vanishing gradient mapping. Furthermore, if $f$ is $\mu$-strongly convex and the step size $\alpha_k \equiv \alpha$ is constant, we obtain a linear convergence rate up to a solution level that is proportional to $\alpha$, which is mainly derived from the convergence properties of Prox-SG to optimality. However, in practice, we may hesitate to repeatedly switch back to Prox-SG Step since most likely it is going to ruin the sparsity from the previous iterates by Orthant Step due to the stochastic nature. It is worth asking that if the convergence is still guaranteed by doing only finitely many Prox-SG Steps and then keeping doing Orthant Steps, where the below Theorem 2 is drawn in line with this idea.

**Theorem 2.** *Suppose* $N_\mathcal{P} < \infty$, $N_\mathcal{O} = \infty$, $f$ *is convex on* $\{x : \|x - x^*\|_2 \leq \delta_1\}$ *and* $\|x_{N_\mathcal{P}} - x^*\|_2 \leq \frac{\delta_1}{2}$. *Set* $k := N_\mathcal{P} + t$, $(t \in \mathbb{Z}^+)$, *step size* $\alpha_k = \mathcal{O}(\frac{1}{\sqrt{N}t}) \in (0, \min\{\frac{1}{L}, \frac{\delta_1^2}{M(\delta_1 + 2\delta_2)}\})$, *and mini-batch size* $|\mathcal{B}_k| = \mathcal{O}(t) \leq N - \frac{N}{2M}$. *Then for any* $\tau \in (0, 1)$, *we have* $\{x_k\}$ *converges to some stationary point in expectation with probability at least* $1 - \tau$, *i.e.,* $\mathbb{P}(\liminf_{k\to\infty}\mathbb{E}\|\mathcal{G}_{\alpha_k}(x_k)\|_2^2 = 0) \geq 1 - \tau$.

Theorem 2 states the convergence is still ensured if the last iterate yielded by Prox-SG Step locates close enough to $x^*$, i.e., $\|x_{N_{\mathcal{P}}} - x^*\|_2 < \delta_1/2$. We will see in appendix that it further indicates $x^*$ inhabits the orthant faces $\{\mathcal{O}_k\}_{k \in \mathcal{S}_{\mathcal{O}}}$ of all subsequent iterates updated by Orthant Steps. Consequently, the convergence is then naturally followed by the property of Project Stochastic Gradient Method. Note that the local convexity-type assumption that $f$ is convex on $\{x : \|x - x^*\|_2 \leq \delta_1\}$ appears in many non-convex problem analysis, such as: tensor decomposition [11] and one-hidden-layer neural networks [31]. Although the assumption $\|x_{N_{\mathcal{P}}} - x^*\|_2 < \delta_1/2$ is hard to be verified in practice, setting $N_{\mathcal{P}}$ to be large enough and $N_{\mathcal{O}} = \infty$ usually performs quite well, as we will see in Sect. 4. To end this part, we present an upper bound of $N_{\mathcal{P}}$ via the probabilistic characterization to reveal that if the step size is sufficiently small, and the mini batch size is large enough, then after $N_{\mathcal{P}}$ Prox-SG Steps, OBProx-SG computes iterate $x_{N_{\mathcal{P}}}$ sufficiently close to $x^*$ with high probability.

**Theorem 3.** *Suppose $f$ is $\mu$-strongly convex on $\mathbb{R}^n$. There exists some constants $C > 0, \frac{1}{2L} > \gamma > 0$ such that for any constant $\tau \in (0,1)$, if $\alpha_k$ satisfies $\alpha_k \equiv \alpha < \min\left\{\frac{2\gamma\mu\tau\delta_1^2}{(2L\gamma-1)C}, \frac{1}{2\mu}, \frac{1}{L}\right\}$, and the mini-batch size $|\mathcal{B}_k|$ satisfies $|\mathcal{B}_k| > \frac{8\gamma\mu\sigma^2}{2\gamma\mu\tau\delta_1^2\,(2L\gamma-1)C\alpha}$, then the probability of $\|x_{N_{\mathcal{P}}} - x^*\|_2 \leq \delta_1/2$ is at least $1 - \tau$ for any $N_{\mathcal{P}} \geq K$ where $K := \left\lceil \frac{\log\,(poly(\tau\delta_1^2, 1/|\mathcal{B}_k|, \alpha)/(F(x_0) - F^*))}{\log\,(1 - 2\mu\alpha)} \right\rceil$ and $poly(\cdot)$ represents some polynomial of $\tau\delta_1^2, 1/|\mathcal{B}_k|$ and $\alpha$.*

In words, Theorem 3 implies that after sufficient number of iterations, with high probability Prox-SG produces an iterate $x_{N_{\mathcal{P}}}$ that is $\delta$-close to $x^*$. However, we note that it does not guarantee $x_{N_{\mathcal{P}}}$ as sparse as $x^*$; as we explained before, due to the limited projection region and randomness, $x_{N_{\mathcal{P}}}$ may still have a large number of non-zero elements, though many of them could be small. As will be demonstrated in Sect. 4, the following Orthant Steps will significantly promote the sparsity of the solution.

## 4 Numerical Experiments

In this section, we consider solving $\ell_1$-regularized classification tasks with both convex and non-convex approaches. In Sect. 4.1, we focus on logistic regression (convex), and compare OBProx-SG with other state-of-the-art methods including Prox-SG, RDA and Prox-SVRG on numerous datasets. Three evaluation metrics are used for comparison: (i) final objective function value, (ii) density of the solution (percentage of nonzero entries), and (iii) runtime. Next, in Sect. 4.2, we apply OBProx-SG to deep neural network (non-convex) with popular architectures designed for classification tasks to further demonstrate its effectiveness and superiority. For these extended non-convex experiments, we also evaluate the generalization performance on unseen test data.

## 4.1   Convex Setting: Logistic Regression

We first focus on the convex $\ell_1$-regularized logistic regression with the form

$$\underset{(x;b)\in\mathbb{R}^{n+1}}{\text{minimize}} \frac{1}{N} \sum_{i=1}^{N} \log(1 + e^{-l_i(x^T d_i + b)}) + \lambda\|x\|_1, \qquad (16)$$

for binary classification, where $N$ is the number of samples, $n$ is the feature size of each sample, $b$ is the bias, $d_i \in \mathbb{R}^n$ is the vector representation of the $i$-th sample, $l_i \in \{-1, 1\}$ is the label of the $i$-th sample, and $\lambda$ is the regularization parameter. We set $\lambda = 1/N$ throughout the convex experiments, and test problem (16) on 8 public large-scale datasets from LIBSVM repository[1], as summarized in Table 1.

We train the models with a maximum number of epochs as 30. Here "one epoch" means we partition $\{1, \cdots, N\}$ uniformly at random into a set of mini-batches. The mini-batch size $|\mathcal{B}|$ for all the convex experiments is set to be $\min\{256, \lceil 0.01N \rceil\}$ similarly to [27]. The step size $\alpha_k$ for Prox-SG, Prox-SVRG and OBProx-SG is initially set to be 1.0, and decays every epoch with a factor 0.995. For RDA, we fine tune its hyperparameter $\gamma$ per dataset to reach the best results. The switching between Prox-SG Step and Orthant Step plays a crucial role in OBProx-SG. Following Theorem 1(i), we set $N_{\mathcal{P}} = N_{\mathcal{O}} = 5N/|\mathcal{B}|$ in Algorithm 1, namely first train the models 5 epochs by Prox-SG Step, followed by performing Orthant Step 5 epochs, and repeat such routine until the maximum number of epochs is reached. Inspired by Theorem 1(ii), we also test OBProx-SG+ with $N_{\mathcal{P}} = 15N/|\mathcal{B}|, N_{\mathcal{O}} = \infty$ such that after 15 epochs of Prox-SG Steps we stick to Orthant Step till the end. Experiments are conducted on a 64-bit machine with an 3.70 GHz Intel Core i7 CPU and 32 GB of main memory.

We compare the performance of OBProx-SG(+) with other methods on the datasets in Table 1, and report the final objective value $F$ and $f$ (Table 2), density

**Table 1.** Summary of datasets

| Dataset | N | n | Attribute | Dataset | N | n | Attribute |
|---|---|---|---|---|---|---|---|
| a9a | 32561 | 123 | binary $\{0, 1\}$ | real-sim | 72309 | 20958 | real $[0, 1]$ |
| higgs | 11000000 | 28 | real $[-3, 41]$ | rcv1 | 20242 | 47236 | real $[0, 1]$ |
| kdda | 8407752 | 20216830 | real $[-1, 4]$ | url_combined | 2396130 | 3231961 | real $[-4, 9]$ |
| news20 | 19996 | 1355191 | unit-length | w8a | 49749 | 300 | binary $\{0, 1\}$ |

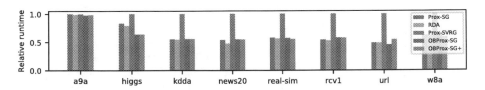

**Fig. 3.** Relative runtime for tested algorithms on convex problems

[1] https://www.csie.ntu.edu.tw/~cjlin/libsvmtools/datasets/.

(percentage of non-zero entries) in the solution (Table 3) and runtime (Fig. 3). For ease of comparison, we mark the best result as bold in the tables.

Our observations are summarized as follows. Table 2 shows that our OBProx-SG(+) performs significantly better than RDA, and is competitive to Prox-SG and Prox-SVRG in terms of the final $F$ and $f$ (round up to 3 decimals), which implies that OBProx-SG(+), Prox-SG and Prox-SVRG can reach comparable convergence results in practice. Besides the convergence, we have a special concern about the sparsity of the solutions. As is demonstrated in Table 3, OBProx-SG(+) is no doubt the best solver. In fact, OBProx-SG achieves the solutions of highest sparsity (lowest density) on 1 out of 8 datasets, while OBProx-SG+ performs even better, which computes all solutions with the highest sparsity. Apparently, OBProx-SG(+) has strong superiority in promoting sparse solutions while retains almost the same accuracy. Finally, for runtime comparison, we plot the relative runtime of these solvers, which is scaled by the maximum runtime consumed by a particular solver on that dataset. Figure 3 indicates that Prox-SG, RDA and OBProx-SG(+) are almost as efficient as each other, while Prox-SVRG takes much more time due to the computation of full gradient.

**Table 2.** Objective function values $F/f$ for tested algorithms on convex problems

| Dataset | Prox-SG | RDA | Prox-SVRG | OBProx-SG | OBProx-SG+ |
|---|---|---|---|---|---|
| a9a | 0.332/0.330 | 0.330/0.329 | 0.330/0.329 | **0.327/0.326** | 0.329/0.328 |
| higgs | **0.326/0.326** | **0.326/0.326** | **0.326/0.326** | **0.326/0.326** | **0.326/0.326** |
| kdda | **0.102/0.102** | 0.103/0.103 | 0.105/0.105 | **0.102/0.102** | **0.102/0.102** |
| news20 | **0.413/0.355** | 0.625/0.617 | **0.413/0.355** | **0.413/0.355** | **0.413/0.355** |
| real-sim | **0.164/0.125** | 0.428/0.421 | **0.164/0.125** | **0.164/0.125** | **0.164/0.125** |
| rcv1 | **0.242/0.179** | 0.521/0.508 | **0.242/0.179** | **0.242/0.179** | **0.242/0.179** |
| url_combined | 0.050/0.049 | 0.634/0.634 | 0.078/0.077 | 0.050/0.049 | **0.047/0.046** |
| w8a | **0.052/0.048** | 0.080/0.079 | **0.052/0.048** | **0.052/0.048** | **0.052/0.048** |

**Table 3.** Density (%) of solutions for tested algorithms on convex problems

| Dataset | Prox-SG | RDA | Prox-SVRG | OBProx-SG | OBProx-SG+ |
|---|---|---|---|---|---|
| a9a | 96.37 | 86.69 | 61.29 | 62.10 | **59.68** |
| higgs | 89.66 | 96.55 | 93.10 | **70.69** | **70.69** |
| kdda | 0.09 | 18.62 | 3.35 | 0.08 | **0.06** |
| news20 | 4.24 | 0.44 | 0.20 | 0.20 | **0.19** |
| real-sim | 53.93 | 52.71 | 22.44 | 22.44 | **22.15** |
| rcv1 | 16.95 | 9.61 | 4.36 | 4.36 | **4.33** |
| url_combined | 7.73 | 41.71 | 6.06 | 3.26 | **3.00** |
| w8a | 99.00 | 99.83 | 78.07 | 78.03 | **74.75** |

The above experiments in convex setting demonstrate that the proposed OBProx-SG(+) outperform the other state-of-the-art methods, and have apparent strengths in generating much sparser solutions efficiently and reliably.

## 4.2  Non-convex Setting: Deep Neural Network

We now apply OBProx-SG(+) to the non-convex setting that solves classification tasks by Deep Convolutional Neural Network (CNN) on the benchmark datasets CIFAR10 [17] and Fashion-MNIST [24]. Specifically, we are testing two popular CNN architectures, *i.e.*, MobileNetV1 [14] and ResNet18 [13], both of which have proven successful in many image classification applications. We add an $\ell_1$-regularization term to the raw problem, where $\lambda$ is set to be $10^{-4}$ throughout the non-convex experiments.

We conduct all non-convex experiments for 200 epochs with a mini-batch size of 128 on one GeForce GTX 1080 Ti GPU. The step size $\alpha_k$ in Prox-SG, Prox-SVRG and OBProx-SG(+) is initialized as 0.1, and decay by a factor 0.1 periodically. The $\gamma$ in RDA is fine-tuned to be 20 for CIFAR10 and 30 for Fashion-MNIST in order to achieve the best performance. Similar to convex experiments, we set $N_{\mathcal{P}} = N_{\mathcal{O}} = 5N/|\mathcal{B}|$ in OBProx-SG, and set

**Table 4.** Final objective values $F/f$ for tested algorithms on non-convex problems

| Backbone | Dataset | Prox-SG | RDA | Prox-SVRG | OBProx-SG | OBProx-SG+ |
|---|---|---|---|---|---|---|
| MobileNetV1 | CIFAR10 | 1.473/0.049 | 4.129/0.302 | 1.921/0.079 | 1.619/**0.048** | **1.453**/0.063 |
| | Fashion-MNIST | 1.314/**0.089** | 4.901/0.197 | 1.645/0.103 | 2.119/**0.089** | **1.310**/0.099 |
| ResNet18 | CIFAR10 | 0.781/0.034 | 1.494/0.051 | 0.815/0.031 | **0.746/0.021** | 0.755/0.044 |
| | Fashion-MNIST | 0.688/0.103 | 1.886/0.081 | 0.683/**0.074** | **0.682/0.074** | 0.689/0.116 |

**Table 5.** Density/testing accuracy (%) for tested algorithms on non-convex problems

| Backbone | Dataset | Prox-SG | RDA | Prox-SVRG | OBProx-SG | OBProx-SG+ |
|---|---|---|---|---|---|---|
| MobileNetV1 | CIFAR10 | 14.17/**90.98** | 74.05/81.48 | 92.26/87.85 | 9.15/90.54 | **2.90**/90.91 |
| | Fashion-MNIST | 5.28/94.23 | 74.67/92.12 | 75.40/93.66 | 4.15/94.28 | **1.23/94.39** |
| ResNet18 | CIFAR10 | 11.60/92.43 | 41.01/90.74 | 37.92/92.48 | 2.12/**92.81** | **0.88**/92.45 |
| | Fashion-MNIST | 6.34/94.28 | 42.46/93.66 | 35.07/94.24 | 5.44/**94.39** | **0.29**/93.97 |

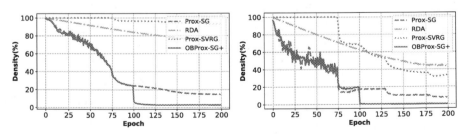

**Fig. 4.** Density. (L): MobileNetV1 on CIFAR10. (R): ResNet18 on Fashion-MNIST

$N_{\mathcal{P}} = 100N/|\mathcal{B}|$, $N_{\mathcal{O}} = \infty$ in OBProx-SG+ since running Prox-SG Step 100 epochs already achieves an acceptable validation accuracy.

Based on the experimental results, the conclusions that we made previously in convex setting still hold in the current non-convex case: (i) OBProx-SG(+) performs competitively among the methods with respect to the final objective function values, see Table 4; (ii) OBProx-SG(+) computes much sparser solutions which are significantly better than other methods as shown in Table 5. Particularly, OBProx-SG+ achieves the highest sparse (lowest dense) solutions on all non-convex tests, of which the solutions are 4.24 to 21.86 times sparser than those of Prox-SG, while note that RDA and Prox-SVRG perform not comparable on the sparsity exploration because of the ineffectiveness of variance reduction techniques for deep learning [8]. In addition, we evaluate how well the solutions generalize on unseen test data. Table 5 shows that all the methods reach a comparable testing accuracy except RDA.

Finally, we investigate the sparsity evolution of the iterates to reveal the superiority of Orthant Step on sparsity promotion, where we use OBProx-SG+ as the representative of OBProx-SG(+) for illustration. As shown in Fig. 4, OBProx-SG+ produces the highest sparse (lowest dense) solutions compared with other methods. Particularly, at the early $N_{\mathcal{P}}$ iterations, OBProx-SG+ performs merely the same as Prox-SG. However, after the switching to Orthant Step at the 100th epoch, OBProx-SG+ outperforms all the other methods dramatically. It is a strong evidence that because of the construction of orthant face subproblem and the larger projection region, our orthant based technique is more remarkable than the standard proximal gradient step and its variants in terms of the sparsity exploration. As a result, the solutions computed by OBProx-SG generally have a better interpretation under similar generalization performances. Furthermore, OBProx-SG may be further used to save memory and hard disk storage consumption drastically by constructing sparse network architectures.

## 5 Conclusions

We proposed an Orthant Based Proximal Stochastic Gradient Method (OBProx-SG) for solving $\ell_1$-regularized problem, which combines the advantages of deterministic orthant based methods and proximal stochastic gradient method. In theory, we proved that it converges to some global solution in expectation for convex problems and some stationary point for non-convex formulations. Experiments on both convex and non-convex problems demonstrated that OBProx-SG usually achieves competitive objective values and much sparser solutions compared with state-of-the-arts stochastic solvers.

**Acknowledgments.** We would like to thank the four anonymous reviewers for their constructive comments. T. Ding was partially supported by NSF grant 1704458. Z. Zhu was partially supported by NSF grant 2008460.

# References

1. Andrew, G., Gao, J.: Scalable training of $l_1$-regularized log-linear models. In: Proceedings of the 24th International Conference on Machine Learning, pp. 33–40. ACM (2007)
2. Beck, A., Teboulle, M.: A fast iterative shrinkage-thresholding algorithm for linear inverse problems. SIAM J. Imaging Sci. **2**(1), 183–202 (2009)
3. Bradley, S., Hax, A., Magnanti, T.: Applied Mathematical Programming. Addison-Wesley, Boston (1977)
4. Chen, T.: A Fast Reduced-Space Algorithmic Framework for Sparse Optimization. Ph.D. thesis, Johns Hopkins University (2018)
5. Chen, T., Curtis, F.E., Robinson, D.P.: A reduced-space algorithm for minimizing $\ell_1$-regularized convex functions. SIAM J. Optim. **27**(3), 1583–1610 (2017)
6. Chen, T., Curtis, F.E., Robinson, D.P.: Farsa for $\ell_1$-regularized convex optimization: local convergence and numerical experience. Optim. Methods Softw. **33**, 396–415 (2018)
7. Cheng, Y., Wang, D., Zhou, P., Zhang, T.: A survey of model compression and acceleration for deep neural networks. arXiv preprint arXiv:1710.09282 (2017)
8. Defazio, A., Bottou, L.: On the ineffectiveness of variance reduced optimization for deep learning. In: Advances in Neural Information Processing Systems (2019)
9. Dixit, A.K.: Optimization in Economic Theory. Oxford University Press on Demand, Oxford (1990)
10. Duchi, J., Singer, Y.: Efficient online and batch learning using forward backward splitting. J. Mach. Learn. Res. **10**, 2899–2934 (2009)
11. Ge, R., Huang, F., Jin, C., Yuan, Y.: Escaping from saddle points–online stochastic gradient for tensor decomposition. In: Conference on Learning Theory, pp. 797–842 (2015)
12. Han, S., Mao, H., Dally, W.J.: Deep compression: Compressing deep neural networks with pruning, trained quantization and Huffman coding. arXiv preprint arXiv:1510.00149 (2015)
13. He, K., Zhang, X., Ren, S., Sun, J.: Deep residual learning for image recognition. In: Proceedings of the IEEE Conference on Computer Vision and Pattern Recognition (2016)
14. Howard, A.G., et al.: MobileNets: Efficient convolutional neural networks for mobile vision applications. arXiv preprint arXiv:1704.04861 (2017)
15. Johnson, R., Zhang, T.: Accelerating stochastic gradient descent using predictive variance reduction. Adv. Neural Inf. Process. Syst. **26**, 315–323 (2013)
16. Keskar, N.S., Nocedal, J., Oztoprak, F., Waechter, A.: A second-order method for convex $\ell_1$-regularized optimization with active set prediction. arXiv preprint arXiv:1505.04315 (2015)
17. Krizhevsky, A., Hinton, G.: Learning multiple layers of features from tiny images. Master's thesis, Department of Computer Science, University of Toronto (2009)
18. Lee, J., Sun, Y., Saunders, M.: Proximal newton-type methods for convex optimization. In: Advances in Neural Information Processing Systems, pp. 836–844 (2012)
19. Nesterov, Y.: Primal-dual subgradient methods for convex problems. Math. Program. **120**, 221–259 (2009). https://doi.org/10.1007/s10107-007-0149-x
20. Riezler, S., Vasserman, A.: Incremental feature selection and l1 regularization for relaxed maximum-entropy modeling. In: Empirical Methods in Natural Language Processing (2004)

21. Sra, S.: Fast projections onto $\ell_{1,q}$-norm balls for grouped feature selection. In: Joint European Conference on Machine Learning and Knowledge Discovery in Databases (2011)
22. Tibshirani, R.: Regression shrinkage and selection via the lasso. J. Roy. Stat. Soc.: Ser. B (Methodol.) **58**(1), 267–288 (1996)
23. Tikhonov, N., Arsenin, Y.: Solution of Ill-Posed Problems. Winston and Sons, Washington, D.C. (1977)
24. Xiao, H., Rasul, K., Vollgraf, R.: Fashion-mnist: a novel image dataset for benchmarking machine learning algorithms (2017)
25. Xiao, L.: Dual averaging methods for regularized stochastic learning and online optimization. J. Mach. Learn. Res. **11**, 2543–2596 (2010)
26. Xiao, L., Zhang, T.: A proximal stochastic gradient method with progressive variance reduction. SIAM J. Optim. **24**(4), 2057–2075 (2014)
27. Yang, M., Milzarek, A., Wen, Z., Zhang, T.: A stochastic extra step quasi-newton method for nonsmooth nonconvex optimization. arXiv preprint arXiv:1910.09373 (2019)
28. Yuan, G.X., Ho, C.H., Lin, C.J.: An improved GLMNET for l1-regularized logistic regression. J. Mach. Learn. Res. **13**(1), 1999–2030 (2012)
29. Zaremba, W., Sutskever, I., Vinyals, O.: Recurrent neural network regularization. arXiv preprint arXiv:1409.2329 (2014)
30. Zeiler, M.D., Fergus, R.: Stochastic pooling for regularization of deep convolutional neural networks. arXiv preprint arXiv:1301.3557 (2013)
31. Zhong, K., Song, Z., Jain, P., Bartlett, P.L., Dhillon, I.S.: Recovery guarantees for one-hidden-layer neural networks. In: International Conference on Machine Learning (2017)
32. Zou, H., Hastie, T.: Regularization and variable selection via the elastic net. J. Roy. Stat. Soci. B (Stat. Methodol.) **67**, 301–320 (2005)

# Efficiency of Coordinate Descent Methods for Structured Nonconvex Optimization

Qi Deng[1(✉)] and Chenghao Lan[2]

[1] Shanghai University of Finance and Economics, Shanghai, China
qideng@sufe.edu.cn
[2] Cardinal Operations, Shanghai, China
lanchenghao@shanshu.ai

**Abstract.** We propose novel coordinate descent (CD) methods for minimizing nonconvex functions comprising three terms: (i) a continuously differentiable term, (ii) a simple convex term, and (iii) a concave and continuous term. First, by extending randomized CD to nonsmooth nonconvex settings, we develop a coordinate subgradient method that randomly updates block-coordinate variables by using block composite subgradient mapping. This method converges asymptotically to critical points with proven sublinear convergence rate for certain optimality measure. Second, we develop a randomly permuted CD method with two alternating steps: linearizing the concave part and cycling through variables. We prove asymptotic convergence to critical points and establish sublinear complexity rate for objectives with both smooth and concave components. Third, we develop a randomized proximal difference-of-convex (DC) algorithm whereby we solve the subproblem inexactly by accelerated coordinate descent (ACD). Convergence is guaranteed with at most a few number of ACD iterations for each DC subproblem, and convergence complexity is established for identifying certain approximate critical points. Fourth, we further develop the third method to minimize smooth and composite weakly convex functions, and show advantages of the proposed method over gradient methods for ill-conditioned nonconvex functions, namely the weakly convex functions with high Lipschitz constant to negative curvature ratios. Finally, an empirical study on sparsity-inducing learning models demonstrates that CD methods are superior to gradient-based methods for certain large-scale problems.

**Keywords:** Coordinate descent method · Nonconvex optimization · Nonsmooth optimization

## 1 Introduction

Coordinate descent (CD) methods update only a subset of coordinate variables in each iteration, keeping other variables fixed. Due to their scalability to the

Q. Deng acknowledges funding from National Natural Science Foundation of China (Grant 11831002).

C. Lan—Work done primarily while at SUFE.

F. Hutter et al. (Eds.): ECML PKDD 2020, LNAI 12459, pp. 74–89, 2021.
https://doi.org/10.1007/978-3-030-67664-3_5

so-called "big data" problems (see [5,15,20,23]), CD methods have attracted significant attention from machine learning and data science. In this paper, new CD methods are developed for large-scale structured nonconvex problems in the following form:

$$\min_{x \in \mathbb{R}^d} F(x) = f(x) + \phi(x) - h(x), \tag{1}$$

where $f(x)$ is continuously differentiable, $\phi(x)$ is convex lower-semicontinuous with a simple structure, and $h(x)$ is convex continuous. The nonconvex problem formed in (1) is sufficiently powerful to express a variety of machine learning applications, including sparse regression, low rank optimization, and clustering (see [13,16,24]).

Our primary contribution to the field is that we propose several novel CD methods with guaranteed convergence for a broad class of nonconvex problems described by (1). For the proposed algorithms, we not only provide guarantees to asymptotic convergence but also prove rates of convergence for properly defined optimality measures. To the best of our knowledge, this is the first study of coordinate descent methods for such nonsmooth and nonconvex optimization with complexity efficiency guarantee. We summarize the results as follows.

Our first result is a new randomized coordinate subgradient descent (RCSD) method for nonsmooth nonconvex and composite problems. While our algorithm recovers existing nonconvex CD methods [22] as a special case, it allows the nonsmooth part to be inseparable and concave, and coordinates to be sampled either uniformly or non-uniformly at random. We show the asymptotic convergence to critical points, and we establish the sublinear rate of convergence for a proposed optimality measure which naturally extends the proximal gradient mapping to nonsmooth and nonconvex settings.

Motivated by block coordinate gradient descent (BCGD) [5], we propose a new randomly permuted coordinate descent (RPCD) for nonsmooth and nonconvex optimization. Our primary innovation is to alternate RPCD between linearizing the concave part $h(x)$ and successively updating all the block-coordinate variables based on some cycling order. The cycling order can be deterministic or randomly shuffled, provided that each block of variables is updated once in each loop. We provide asymptotic convergence of RPCD method to critical points. For a certain case ($\phi(x) = 0$), we also establish a sublinear rate of convergence of the subgradient norm.

We next extend ACD methods to the nonsmooth and nonconvex setting by viewing (1) as a difference-of-convex function. We propose an ACD-based proximal DC (ACPDC) algorithm by transforming $F(x)$ into a sequence of strongly convex functions that are approximately minimized by the ACD method. We show that only a few rounds of ACD (linear in block numbers) are needed in each outer loop to guarantee the same $\mathcal{O}(1/\varepsilon)$ complexity of DCA, with difference up to some constant factor. Hence, ACPDC offers significant improvements compared to the classic DC algorithm which requires exact optimal solutions to the subproblems. ACPDC also offers advantages to the proximal DC algorithm, an extension of the DC algorithm that performs one proximal gradient descent step to minimize the majorized function. Taking advantages of the fast convergence of ACD, ACPDC is more efficient in optimizing the convex subproblem, offering a much better trade-off between iteration complexity and running time than proximal DC.

Finally, we draw attention to minimization of weakly convex functions, the nonconvex functions with bounded negative curvature, and propose a new ACD-based proximal point method (ACPP) for solving such problems. We establish complexity of convergence to some approximate stationary point solutions. We also show that ACPP can be significantly more efficient than classic CD method or gradient based method for optimizing smooth ill-conditioned problems; ill-conditioned problems are those that have a relatively high ratio between Lipschitz constant to negative curvature.

*Related Work.* There is a significant body of work on nonconvex coordinate descent methods. We refer to [15] for some general strategies to develop block update algorithms for nonconvex optimization. [27] proposed proximal cyclic CD methods for minimizing composite functions with nonconvex separable regularizers. However, their work didn't include the nonconvex and nonsmooth function described in (1). See such an example in Sect. 7. A recent work [14] extended the cyclic block mirror descent to the nonsmooth setting, and guaranteed asymptotic convergence to block-wise optimality for minimizing Problem (1). In contrast, our work directly guarantees convergence to the critical points by employing a different linearization technique to handle the nonsmooth part $h(x)$. The work [4] proposed efficient CD-type algorithms for achieving specific coordinate-wise optimality for problems with group sparsity constraints. Besides, nonconvex CD methods with complexity guarantees have also been proposed for smooth composite optimization [22] and stochastic composite optimization [7]. However, none of these proposals consider a nonsmooth concave part in the objective, nor do they improve on algorithm efficiency for the ill-conditioned nonconvex problem; we fully address both challenges for CD methods in this paper.

Another related research direction is the so-called DC optimization (see [13, 24, 25, 29]). Specified for minimizing the difference-of-convex (DC) functions, a DC algorithm alternates between linearizing the concave part and optimizing some convex surrogate of DC function by applying convex algorithms. Lately, much progress has been made towards more efficient DC algorithms and their applications to machine learning and statistics (see [1, 13, 16, 21, 26, 28]). We refer to the recent review [25] for DC algorithms and their applications. Notably, as a special case of DC functions, the weakly convex function has been increasingly popular due to its importance in machine learning and statistics [9, 11, 30]. We refer to [6, 8, 9, 17, 18] for recent advance in this area. However, for the weakly-convex and the more general DC problems, it remains to develop efficient CD-type methods that are scalable to high-dimensional and large-scale data.

*Outline of the Paper.* Section 2 introduces notations and preliminaries. Section 3 and Sect. 4 present RCSD and RPCD for nonsmooth nonconvex optimization, and then establish their convergence results, respectively. Section 5 presents a new DC algorithm based on ACD and demonstrates its asymptotic convergence to critical points and its convergence complexity of a proposed optimality measure. Section 6 considers the nonconvex problem with bounded negative curvature and presents an even faster proximal point method based on using ACD. Section 7 discusses the applications of proposed methods on sparsity-inducing machine learning models, and present experiments to demonstrate the advantages of our proposed CD methods over state-of-the-art full gradient methods. Section 8 draws conclusions. See Appendix sections for proofs of the theoretical results.

## 2  Notations and Preliminaries

We denote $[m] = \{1, 2, ..., m\}$. Let $d$ be a positive integer and $\mathbf{I}_d$ be the $d \times d$ identity matrix. Assume that matrix $\mathbf{U}_i \in \mathbb{R}^{d \times d_i}$ ($i \subset [m]$) satisfies: $[\mathbf{U}_1, \mathbf{U}_2, ..., \mathbf{U}_m] = \mathbf{I}_d$ where $\sum_{i=1}^m d_i = d$. Let $x_i = \mathbf{U}_i^T x$ be the restriction of $x$ to the $i$-th block, and we hereby express $x = \sum_{i=1}^m \mathbf{U}_i x_i$. Let $\|\cdot\|_i$ be the standard Euclidean norm on $\mathbb{R}^{d_i}$, and the norm $\|\cdot\|$ on $\mathbb{R}^d$ is denoted by $\|x\| = \sqrt{\sum_{i=1}^m \|x_i\|_i^2}$. We say that $f$ is block-wise Lipschitz smooth, if there are constants $L_1, L_2, ..., L_m > 0$ such that for any $x, x + \mathbf{U}_i t \in \mathbb{R}^{d_i}$ and $t \in \mathbb{R}^{d_i}$, $i \in [m]$, we have $\|\nabla_i f(x) - \nabla_i f(x + \mathbf{U}_i t)\|_i \leq L_i \|t\|_i$. For any $s \in [0, 1]$, we define $T_s = \sum_{i=1}^m L_i^s$ and $\|x\|_{[s]} = \sqrt{\sum_{i=1}^m L_i^s \|x_i\|_i^2}$. The dual norm is defined by $\|y\|_{[s],*} = \sqrt{\sum_{i=1}^m L_i^{-s} \|y_i\|_i^2}$. We denote $L_{\max} = \max_{1 \leq i \leq m} L_i$ and $L_{\min} = \min_{1 \leq i \leq m} L_i$.

We say that a function $f(x)$ is $\mu_s$-strongly convex with respect to $\|\cdot\|_{[s]}$ if $f(y) \geq f(x) + \langle \nabla f(x), y - x \rangle + \frac{\mu_s}{2} \|x - y\|_{[s]}^2$. It immediately follows that $\mu_s \in [0, 1]$.

Given a proper lower semi-continuous (lsc) function $f : \mathbb{R}^d \to \mathbb{R}$, for any $x \in \text{dom}(f)$, the limiting subdifferential of $f$ at $x$ is defined as

$$\partial f(x) = \Big\{ u : \exists x^k \to x, u^k \to u, \text{ with } f(x^k) \to f(x)$$

$$\text{and } \liminf_{y \neq x, y \to x^k} \frac{f(y) - f(x^k) - \langle u^k, y - x^k \rangle}{\|y - x^k\|} \geq 0 \text{ as } k \to \infty \Big\}.$$

Using the limiting subdifferential, we can define the optimality measure of the proposed algorithms. A point $x$ is known as a critical point of Problem (1) if $[\nabla f(x) + \partial \phi(x)] \cap \partial h(x) \neq \emptyset$, and it is known as a stationary point of Problem (1) if $\partial h(x) \subseteq \nabla f(x) + \partial \phi(x)$. While critical points are weaker than stationary points as an optimality measure, these two notions coincide when $h(x)$ is smooth: $\partial h(x) = \{\nabla h(x)\}$.

Throughout this paper, we make the following assumptions.

1. $f(x)$ is block-wise Lipschitz smooth with parameters $L_1, L_2, ..., L_m$.
2. $\phi(x)$ is block-wise separable: $\phi(x) = \sum_{i=1}^m \phi_i(x_i)$, where $\phi_i : \mathbb{R}^{d_i} \to \mathbb{R}$ ($i \in [m]$) is a proper convex lsc function.
3. $h(x)$ is convex and continuous, and its subgradient is easy to compute.
4. There exists an optimal solution $x^*$ such that $F(x^*) = \min_x F(x) > -\infty$.

## 3   Randomized Coordinate Subgradient Descent

Our goal in this section is to develop the proposed randomized coordinate subgradient descent (RCSD) for the nonconvex problem (1) in Algorithm 1. We can regard this method as a block-wise proximal-subgradient-type algorithm that updates some random coordinates iteratively while keeping the other coordinates fixed. Let $\gamma = [\gamma_1, \gamma_2, ..., \gamma_m]^T$ be a positive vector. For any $y \in \mathbb{R}^{d_i}$, the composite block proximal mapping is given by

$$\mathcal{P}_i\left(\bar{x}_i, y_i, \gamma_i\right) = \operatorname*{argmin}_{x \in \mathbb{R}^{d_i}}\left\{\langle y_i, x\rangle + \phi_i(x) + \tfrac{\gamma_i}{2}\|\bar{x}_i - x\|_i^2\right\}, \tag{2}$$

and we denote $\mathcal{P}\left(\bar{x}, y, \gamma\right) = \sum_{i=1}^{m} \mathbf{U}_i \mathcal{P}_i\left(\bar{x}_i, y_i, \gamma_i\right)$.

---

**Algorithm 1:** RCSD

---

**Input:** $x^0$;
**for** $k{=}0,1,2,...K$ **do**
  Sample $i_k \in [m]$ randomly with $\operatorname{prob}(i_k = i) = p_i$;
  Compute $\nabla_{i_k} f(x^k)$ and $v_{i_k}^k$ where $v^k \in \partial h(x^k)$;
  $x_{i_k}^{k+1} = \mathcal{P}_{i_k}\left(x_{i_k}^k, \nabla_{i_k} f(x^k) - v_{i_k}^k, \gamma_{i_k}\right)$;
  $x_j^{k+1} = x_j^k$ if $j \neq i_k$;
**end**

---

To analyze the convergence of RCSD, we must first establish some optimality measure. Let $v \in \partial h(x)$, we define the composite block subgradient as

$$g_i(x, \nabla_i f(x) - v_i, \gamma_i) = \gamma_i\left(x_i - \mathcal{P}_i(x_i, \nabla_i f(x) - v_i, \gamma_i)\right), \tag{3}$$

and we define the composite subgradient as

$$g(x, \nabla f(x) - v, \gamma) = \sum_{i=1}^{m} \mathbf{U}_i g_i(x, \nabla_i f(x) - v_i, \gamma_i). \tag{4}$$

We use notations $g(x, \nabla f(x) - v, \gamma)$ and $g(x)$ interchangeably when there is no ambiguity. It can be seen that $\|g(x)\|_{[s],*} \neq 0$ when $x$ is a non-critical point, and $\|g(x)\|_{[s],*} = 0$ for some $v \in \partial h(x)$ when $x$ is a critical point. This result is proven in the following proposition:

**Proposition 1.** $\bar{x}$ *is a critical point of* (1) *if and only if there exists* $v \in \partial h(\bar{x})$ *such that* $g(\bar{x}, \nabla f(\bar{x}) - v, \gamma) = 0$.

The following theorem presents the general convergence property of RCSD.

**Theorem 2.** *Let* $\{x^k\}$ *be sequence generated by Algorithm 1.*

1) Assume that $\gamma_i > \frac{L_i}{2}$. Then almost surely (a.s.), every limit point of the sequence is a critical point.

2) Assume that $\gamma_i = L_i$, $i \in [m]$, and that blocks are sampled with probability $p_i \propto L_i^{1-s}$ $(0 \leq s \leq 1)$. Then

$$\min_{0 \leq k \leq K} \mathbb{E} \left\| g(x^k, \nabla f(x^k) - v^k, \gamma) \right\|_{[s],*}^2 \leq \frac{2(\sum_{j=1}^{m} L_j^{1-s})[F(x^0) - F(x^*)]}{K+1},$$

where the expectation is regarding $i_0, i_1, ..., i_K$.

We note that the sublinear rate in Theorem 2 is typical for first order methods on nonconvex problems. If $h(x)$ is void and uniform sampling $(s = 1)$ is performed, we recover the rate for 1-RCD [22]. Another difference between our work and [22] is that our analysis also adapts to nonuniform sampling $(s < 1)$, whereby the composite subgradient is measured by the norm $\| \cdot \|_{[s],*}$. Suppose that our goal is to find some $\varepsilon$-accurate solution $(\min_k \mathbb{E}[\|g^k\|^2] \leq \varepsilon)$, the total number of iterations required by non-uniform sampling RCSD $(s = 0)$ is $\mathcal{O}((\sum_{i=1}^{m} L_i)/\varepsilon)$. In contrast, since $\|g^k\|_{[1],*}^2 \geq L_{\max}^{-1} \|g^k\|^2$, the bound provided by uniform sampling RCSD $(s = 1)$ is $\mathcal{O}((L_{\max}m)/\varepsilon)$.

# 4 Randomly Permuted Coordinate Descent

In this section, our goal is to develop a randomly permuted coordinate descent (RPCD) method in Algorithm 2. When analyzing the convergence of permuted CD, we normally require the triangle inequality to bound the gradient norm by the sum of point distances. However, this may be difficult to achieve in the nonsmooth setting since subgradient is not necessarily Lipschitz continuous. To avoid this problem, Algorithm 2 computes the subgradient of $h(x)$ only once after scanning all the blocks but always uses the newest block gradient of $f(x)$ to update the corresponding block variables. The update order can be either deterministic or randomly shuffled, provided that all the blocks are updated in each outer loop of the algorithm. The following theorem summarizes the main convergence property of RPCD.

**Theorem 3.** Let $\{x^k\}_{k=0,1,...}$ be the generated sequence in Algorithm 2.

1) If $\gamma_i \geq L_i$ $(1 \leq i \leq m)$, then every limit point of $\{x^k\}$ is a critical point.

2) If we assume $\phi(x) = 0$ and $\gamma_i = L_i$, then

$$\min_{0 \leq k \leq K} \|\nabla f(x^k) - v^k\|^2 \leq 4\left(L_{\max} + \frac{mL^2}{L_{\min}}\right) \frac{F(x^0) - F(x^*)}{(K+1)}.$$

We remark that the complexities of RPCD and RCSD are analyzed on quite different measures: the efficiency estimate of RPCD bounds the error in the worst case while the result for RCSD bounds the expected error. Nevertheless, in the experiments (presented in Sect. 7.1), we observe that RPCD and RCSD exhibit a very similar empirical performance. We note that similar observations have been made in [5] when comparing randomized and cyclic CD in the convex setting.

---

**Algorithm 2: RPCD**

---

  **Input:** $x^0$;
  **for** $k=0,1,2,...K\text{-}1$ **do**
    |  Compute $v^k \in \partial h(x^k)$ and set $\widetilde{x}^0 = x^k$;
    |  Generate permutation: $\pi_0, \pi_1, \pi_2, \dots \pi_{m-1}$;
    |  **for** $t = 0, 2, ..., m-1$ **do**
    |    |  Set $\widetilde{x}_{\pi_t}^{t+1} = \mathcal{P}_{\pi_t}\left(\widetilde{x}_{\pi_t}^t, \nabla_{\pi_t} f(\widetilde{x}^t) - v_{\pi_t}^k, \gamma_{\pi_t}\right)$;
    |    |  Set $\widetilde{x}_j^{t+1} = \widetilde{x}_j^t$ if $j \neq \pi_t$;
    |  **end**
    |  Set $x^{k+1} = \widetilde{x}^m$;
  **end**

---

## 5   Randomized Proximal DC Method for Nonconvex Optimization

In this section we will develop a new randomized proximal DC algorithm based on ACD, with the simple observation that $F(x)$ can be reformulated as a difference-of-convex function. Specifically, suppose that $f(x)$ has a bounded negative curvature ($\mu$-weakly convex): $f(x) \geq f(y) + \langle \nabla f(y), x - y \rangle - \frac{\mu}{2}\|x - y\|^2$. By this definition, $f(x) + \frac{\mu}{2}\|x\|^2$ is convex and hence $f(x)$ has a DC expression: $f(x) = \left(f(x) + \frac{\mu}{2}\|x\|^2\right) - \frac{\mu}{2}\|x\|^2$. Therefore, we express $F(x)$ in the following DC form:

$$F(x) = f(x) + \frac{\mu}{2}\|x\|^2 + \phi(x) - \left[\frac{\mu}{2}\|x\|^2 + h(x)\right].$$

Without loss of generality, we assume that $f(x)$ is convex for the remainder of this section. We begin with the most basic DC algorithm (DCA):

$$x^{k+1} = \underset{x}{\operatorname{argmin}} \widetilde{F}(x) = f(x) + \phi(x) - h(x^k) - \langle v^k, x - x^k \rangle, \quad v^k \in \partial h(x^k). \quad (5)$$

Despite its simplicity, the main drawback of DCA is that exactly minimizing $\widetilde{F}(x)$ can be potentially slow for many large-scale problems. To avoid this difficulty, recent works (such as [3,13,26]) propose proximal DC algorithms (pDCA) by performing one step of proximal gradient descent to solve (5) approximately. pDCA can be interpreted as an application of DCA via a different DC representation: $F(x) = \left[\frac{L}{2}\|x\|^2 + \phi(x)\right] - \left[h(x) + \frac{L}{2}\|x\|^2 - f(x)\right]$. It follows from the DC algorithm that we obtain $x^{k+1}$ by

$$x^{k+1} = \underset{x}{\operatorname{argmin}} \left\{\frac{L}{2}\|x\|^2 + \phi(x) - \langle v^k + Lx^k - \nabla f(x^k), x \rangle\right\}$$

$$= \underset{x}{\operatorname{argmin}} \left\{\langle \nabla f(x^k) - v^k, x \rangle + \phi(x) + \frac{L}{2}\|x - x^k\|^2\right\}.$$

While pDCA significantly improves the per-iteration running time, it does not efficiently exploit the convex structure of $f(x) + \phi(x)$. Consider the extreme case that $h(x) = 0$, then pDCA is reduced to the proximal gradient descent. Yet it is well known that proximal gradient descent has a suboptimal worst-case complexity and the optimal complexity is achieved by Nesterov's accelerated methods.

---

**Algorithm 3: ACPDC**

**Input:** $x_0$, $\mu$, $t$;
Compute $\widetilde{\mu}$;
**for** $k=0,1,2,...K$ **do**

 Compute $v^k \in \partial h(x^k)$ and define:
 $F_k(x) = f(x) + \phi(x) - h(x^k) - \langle v^k, x - x^k \rangle + \frac{\mu}{2}\|x - x^k\|_{[1]}^2$
 Obtain $x^{k+1}$ from running APCC with input $F_k(\cdot)$, initial point $x^k$,
 strong convexity parameter $\widetilde{\mu}$ and iteration number $t$;

**end**

---

To overcome the mentioned drawbacks, in Algorithm 3, we propose a new randomized proximal DC method (ACPDC) by using accelerated coordinate descent for the DC subproblem. The key ideas behind the algorithm are as follows. First, at the $k$-th iteration, we form a majorized function $F_k(x)$ by linearizing $h(x)$ and adding a strongly convex function $\frac{\mu}{2}\|x - x^k\|_{[1]}^2 = \frac{\mu}{2}\sum_{i=1}^m L_i\|x_i - x_i^k\|_i^2$. By setting the weights for each block separately, we are able to exploit the local curvature information more efficiently than the gradient-based DC algorithm. Second, to solve the subproblem fast, we use an accelerated proximal coordinate gradient (APCG, [19]) method because of its superior performance in convex optimization. When the objective is $\widetilde{\mu}$-strongly convex and smooth, APCG has an iteration complexity of $\mathcal{O}(m/\sqrt{\widetilde{\mu}}\log(1/\varepsilon))$ for attaining an $\varepsilon$-optimality gap.

To develop the complexity result of the overall procedure, we need to define some optimality criterion. Let us denote

$$F_\mu(y, x, v) = f(y) + \phi(y) - h(x) - \langle v, y - x \rangle + \frac{\mu}{2}\|y - x\|_{[1]}^2, \quad v \in \partial h(x), \quad (6)$$

and denote $\bar{x} = \underset{y}{\arg\min} F_\mu(y, x, v)$. Let $p(x, v, \mu) = \mu\sum_{i=1}^m [L_i U_i(x_i - \bar{x}_i)]$ be the prox-mapping. Based on this definition, $x$ is exactly a critical point when $\|p(x, v, \mu)\|_{[1]} = 0$, and the norm of $\|p(x, v, \mu)\|_{[1]}$ can be viewed as a measure of the optimality of $x$. It is worth to note that the optimality criterion based on $\|p(x, v, \mu)\|_{[1],*}$ is consistent with the earlier proposed criterion of using composite subgradient (4). Specifically, there exist constants $c_1, c_2 > 0$ such that

$$c_1\|p(x, v, \mu)\|_{[1],*} \leq \|g(x, \nabla f(x) - v)\|_{[1],*} \leq c_2\|p(x, v, \mu)\|_{[1],*}.$$

We establish the equivalence between different optimality criteria more rigorously in Appendix section.

The following theorem presents the main complexity result of Algorithm 3.

**Theorem 4.** *Assume that $f(x)$ is convex, and that there exists $M > 0$ such that $M = \sup_{v \in \partial h(x)} \|v\| < +\infty$. In Algorithm 3 if we set $t = t_0 = \left\lceil \ln 4 \frac{m}{\sqrt{\mu/(1+\mu)}} \right\rceil$, then every limit point of the sequence is a critical point, a.s. Moreover, we have*

$$\min_{1 \le k \le K} \mathbb{E}\left[\|p(x^k, v^k, \mu)\|_{[1],*}^2\right] \le \frac{2\mu\left[F(x^0) - F(x^*) + 4M\|x^0 - x^*\| + \mu\|x^0 - x^*\|_{[1]}^2\right]}{K}.$$

Theorem 4 implies that ACPDC only requires $\mathcal{O}(m)$ steps of ACD for each DC subproblem, matching the computation complexity of proximal gradient approach in many applications (c.f. Sect. 7). As a result, ACPDC obtains the same $\mathcal{O}(1/\varepsilon)$ complexity of RCSD, with difference up to some constant factors.

# 6 Randomized Proximal Point Method for Weakly Convex Problems

This section develops a new randomized proximal point algorithm based on ACD for minimizing weakly convex functions, and establishes new rates of convergence to approximate stationary point solutions. We assume that $h(x)$ is void and consider the following form

$$\min_x F(x) = f(x) + \phi(x). \tag{7}$$

Here, we assume that $f(x)$ is $\mu$-weakly convex: $f(x) \ge f(y) + \langle \nabla f(y), x - y \rangle - \frac{\mu}{2}\|x - y\|^2$. We present the ACD-based proximal point method (ACPP) in Algorithm 4. At the $k$-th iteration of ACPP, given the initial point $x^k$, we employ APCG to solve convex composite problem $\min_x F_k(x) = F(x) + \mu\|x - x^k\|^2$ to some appropriate accuracy.

We define some optimality criterion for problem (7). Following the setup in [18], we say that a point $x$ is an $(\varepsilon, \delta)$ *approximate stationary point* if there exists $\bar{x}$ such that $\|x - \bar{x}\|^2 \le \delta$ and $\left[\text{dist}(0, \partial F(\bar{x}))\right]^2 \le \varepsilon$. Moreover, $x$ is a stochastic $(\varepsilon, \delta)$ approximate stationary point if $\mathbb{E}\|x - \bar{x}\|^2 \le \delta$ and $\mathbb{E}\left[\text{dist}(0, \partial F(\bar{x}))\right]^2 \le \varepsilon$. Here we define $\text{dist}(y, X) = \inf_{x \in X} \|x - y\|$. Conceptually, an approximate stationary point is in proximity to some nearly-stationary point. Note that similar criteria have been proposed in [8,9] for minimizing nonsmooth and weakly convex functions.

The next theorem shows that ACPP obtains a stochastic $(O(1/K), O(1/K))$ approximate solution, matching the existing result for nonconvex proximal point methods. For simplicity, we skip the asymptotic part, as it is very similar to the analysis of ACPDC.

**Theorem 5.** *In Algorithm 4, let $\kappa = \widetilde{L}_{\max}/\widetilde{L}_{\min}$, $\eta = \sqrt{\widetilde{\mu}_1}/m$ and $t = \left\lceil -\eta^{-1} \ln \widetilde{\lambda} \right\rceil$ where $\widetilde{\lambda} = \min\left\{\frac{1}{4}, \frac{\mu^2}{L^2}, \frac{\mu}{2L}\right\}$, then*

$$\mathbb{E}\|x^{\hat{k}} - x^{\hat{k}^*}\|^2 \le \frac{8\kappa\mu}{KL^2}\left\{\mu\|x^0 - x^*\|^2 + \frac{\mu}{L}[F(x^0) - F(x^*)]\right\}, \tag{8}$$

$$\mathbb{E}\left[\text{dist}(0, \partial F(x^{\hat{k}^*}))\right]^2 \le \frac{16\kappa\mu}{K}\left\{\mu\|x^0 - x^*\|^2 + \frac{\mu}{L}[F(x^0) - F(x^*)]\right\}. \tag{9}$$

**Algorithm 4: ACPP**

**Input:** $x^0$, $\mu$, $t$;
Compute $\tilde{\mu}_1$;
**for** $k=0,1,2,...,K$ **do**
  Set $F_k(x) = F(x) + \mu\|x - x^k\|^2$;
  Obtain $x^{k+1}$ from running Algorithm ?? with input $F_k(x)$, $x^k$, $\tilde{\mu}_1$ and $t$;
**end**
**Output:** $x^{\hat{k}}$, where $\hat{k}$ is chosen from $\{2, 3, ..., K+1\}$ uniformly at random.

## 6.1 Smooth Nonconvex Optimization

We describe a version of ACPP for smooth nonconvex optimization: $\min_x F(x)$ where $F(x)$ is Lipschitz smooth with negative curvature $\mu$. This is a special case of the studied composite problem when $\phi(x)$ is void. Under this assumption, $F_k(\cdot)$ in the subproblem of ACPP is both smooth and strongly convex. Therefore, we can apply more efficient non-uniform accelerated coordinate descent (NUACDM, [2]) to further improve ACPP. We highlight in below the convergence result of ACPP with NUACDM as the subproblem solver. More details are in Appendix section.

**Theorem 6.** *Assume that $f(x)$ is $\mu$-weakly convex and that $\phi(x) = 0$ in Problem (7), and NUACDM is used to solved the subproblem of ACPP. To guarantee that $\mathbb{E}\|\nabla F(x^{\hat{k}})\|^2 \le \varepsilon$, the total number of block gradient computations is*

$$N_\varepsilon = \mathcal{O}\left(\left(\sum_{i=1}^m \sqrt{L_i}\right)\sqrt{\mu}\log\left(\frac{L}{\mu}\right)\frac{\mu\|x^0 - x^*\|^2}{\varepsilon}\right).$$

We remark that the $\mathcal{O}\left((\sum_{i=1}^m \sqrt{L_i}\sqrt{\mu})/\varepsilon\right)$ complexity of ACPP is much better then the $\mathcal{O}\left((\sum_{i=1}^m L_i)/\varepsilon\right)$ complexity of RCSD (c.f. Theorem 2) when the problem is ill-conditioned, namely, $L_i \gg \mu$ ($i \in [m]$). Moreover, it is interesting to compare efficiency of CD and AGD ([17,18]). Since computing full gradient can run $O(m)$ times slower than computing block-gradient, the total complexity of AGD is $\mathcal{O}\left((m\sqrt{L\mu})/\varepsilon\right)$. Therefore, if $\{L_i\}$ are non-uniform, ACPP is faster than AGD by a factor up to $m$.

## 7 Applications

As a notable application of Problem (1), we consider sparse machine learning that is described in the following form:

$$\min_{x\in\mathbb{R}^d} \quad f(x) \quad \text{s.t.} \quad \|x\|_0 \le k, \tag{10}$$

where $f(x)$ is the loss function and the $l_0$-norm constraint promotes sparsity. Due to the nonconvexity and discontinuity of $l_0$-norm, direct optimization of the above problem is generally intractable. An alternative way is to translate (10) into a regularized problem with some sparsity-inducing penalty $\psi(x)$.

$$\min_{x \in \mathbb{R}^d} f(x) + \psi(x) \tag{11}$$

While convex relaxation (e.g. $\ell_1$penalty: $\psi(x) = \lambda \|x\|_1$) was widely studied, nonconvex penalties have received increasing popularity (for example, see [12, 13,24]). Here, we consider a wide class of nonconvex regularizers expressed in DC functions. For example, SCAD [11,12,26] penalty has the separable form $\psi_{\lambda,\theta}(x) = \sum_{i=1}^{m} [\phi_\lambda(x_i) - h_{\lambda,\theta}(x_i)]$ where $\phi_\lambda(\cdot)$ and $h_{\lambda,\theta}(\cdot)$ are defined by

$$\phi_\lambda(x) = \lambda|x_i|, \quad \text{and} \quad h_{\lambda,\theta}(x) = \begin{cases} 0 & \text{if } |x| \leq \lambda \\ \frac{x^2 - 2\lambda|x| + \lambda^2}{2(\theta-1)} & \text{if } \lambda < |x| \leq \theta\lambda \\ \lambda|x| - \frac{1}{2}(\theta+1)\lambda^2 & \text{if } |x| > \theta\lambda \end{cases} \tag{12}$$

It is routine to check that $h_{\lambda,\theta}(x)$ is Lipschitz smooth with constant $1/(\theta-1)$.

Another interesting sparsity-inducing regularizer arises from direct reformulation of the $\ell_0$ term. For instance, the constraint $\{\|x\|_0 \leq k\}$ can be equivalently expressed as $\{\|x\|_1 - \|\|x\|\|_k = 0\}$, where the largest-$k$ norm $\|\|x\|\|_k$ is defined by $\|\|x\|\|_k = \sum_{i=1}^{k} |x_{j_i}|$, where $j_1, j_2, ...j_d$ are coordinate indices in descending order of absolute value. In this view, [13] studied the following formulation:

$$\min_{x \in \mathbb{R}^d} F(x) = f(x) + \lambda\|x\|_1 - \lambda\|\|x\|\|_k. \tag{13}$$

Notice that existing nonconvex coordinate descent [22,27] are not applicable to (13), since the norm $\|\|\cdot\|\|_k$ is neither smooth nor separable. In contrast, our methods can be applied to Problem (13) since the concave part of the objective is allowed to be nonsmooth and inseparable. Moreover, efficient implementations of our methods are viable: in RPCD and ACPDC, full subgradient of $\|\|\cdot\|\|_k$ is computed only once a while; in RCSD, block subgradient of $\|\|\cdot\|\|_k$ can be computed in logarithmic time by maintaining the coordinates in max heaps.

## 7.1   Experiments

We compare our algorithms with two gradient-based methods on the nonconvex problems described above. The first algorithm is the proximal DC algorithm (pDCA, [13,16]) which performs a single step of proximal gradient descent in each iteration. The second algorithm is an enhanced proximal DC algorithm using extrapolation technique (pDCA$_e$ [26]). In the paper [26], pDCA$_e$ has been reported to outperform pDCA on a variety of nonconvex learning problems. The test-cases, including both synthetic and real data, are presented in Table 1. We refer to Appendix for more details about data preparation and algorithm parameter-tuning.

*Logistic Loss + Largest-k Norm Penalty.* Our first experiment compares the proposed algorithms on logistic loss classification with largest-$k$ norm penalty:

$$\min_{x \in \mathbb{R}^d} \frac{1}{n} \sum_{i=1}^{n} \log(1 + \exp(-b_i(a_i^T x))) + \frac{\rho}{d}(\|x\|_1 - \|\|x\|\|_k).$$

**Table 1.** Dataset description. R stands for regression and C stands for classification.

| Datasets | $n$ | $d$ | $m$ | Nonzeros | Problem type |
|---|---|---|---|---|---|
| synthetic | 500 | 5000 | 1000 | 100.00% | R |
| E2006-tfidf | 16087 | 150360 | 1000 | 0.83% | R |
| E2006-log1p | 16087 | 4272227 | 10000 | 0.14% | R |
| UJIndoorLoc | 19937 | 554 | 554 | 94.22% | R |
| resl-sim | 72309 | 20959 | 1000 | 0.25% | C |
| news20.binary | 19996 | 1355191 | 10000 | 0.03% | C |
| mnist | 60000 | 718 | 718 | 21.02% | C |
| dexter | 600 | 20000 | 1000 | 0.85% | C |

We use the following real datasets: `resl-sim`, `news20.binary`, `dexter` and `mnist`. For `mnist`, we formulate a binary classification problem by labeling the digits 0, 4, 5, 6, 8 positive and the rest negative. We plot the function objectives after 100 passes to the dataset for various values of $\rho$; our results are shown in Fig. 1.

In Fig. 1, we find that pDCA$_e$ consistently outperforms pDCA in all experiments. This result suggests that, despite the unclear theoretical advantage of pDCA$_e$ over pDCA, extrapolation indeed has empirical an advantage in nonsmooth and nonconvex optimization. Moreover, RPCD exhibits at least the same (sometimes better) performance as RCSD, while both RPCD and RCSD have superior performance when compared with gradient methods. Finally, ACPDC achieves the best performance among all the tested algorithms, confirming the advantage of accelerated coordinate descent in the nonconvex setting.

*Smoothed $\ell_1$ Loss + SCAD Penalty.* Our next experiment targets the $\ell_1$ loss regression with SCAD penalty:

$$\min_{x\in\mathbb{R}^d} F(x) = \frac{1}{n}\sum_{i=1}^n \left|b_i - a_i^T x\right| + \frac{\rho}{d}\psi_{\lambda,\theta}(x). \tag{14}$$

Although Problem (14) does not exactly satisfy our assumption because $\ell_1$-loss is nonsmooth, by introducing a small term $\delta$, we can approximate $\ell_1$ loss by the following Huber loss: $H_\delta(a) = \frac{a^2}{2\delta}$ if $|a| \le \delta$, $H_\delta(a) = |a| - \frac{\delta}{2}$ if $|a| > \delta$. Our problem of interest, thus, follows:

$$\min_{x\in\mathbb{R}^d} F_\delta(x) = \frac{1}{n}\sum_{i=1}^n H_\delta(b_i - a_i^T x) + \frac{\rho}{d}\psi_{\lambda,\theta}(x), \tag{15}$$

It is easy to verify that $f_\delta(x) = \frac{1}{n}\sum_{i=1}^n H_\delta(b_i - a_i^T x)$ is block-wise Lipschitz smooth with $\frac{\|Ae_i\|^2}{n\delta}$. In view of the Lipschitz smoothness of $f_\delta(x)$ and the weak convexity of $\psi_{\lambda,\theta}(x)$, $F_\delta(x)$ has a condition number of $O(\frac{\rho(\theta-1)}{d\delta})$. To set the parameters, we choose $\delta$ from $\{10^{-2}, 10^{-3}\}$.

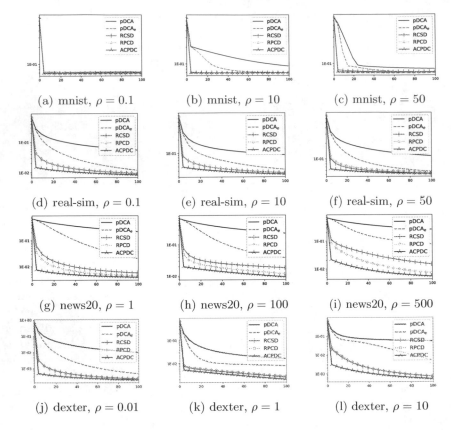

**Fig. 1.** Experimental results on logistic loss classification with largest-$k$ norm penalty. $y$-axis: objective value (log scaled). $x$-axis: number of passes to the dataset.

We conduct experiments on datasets `synthetic`, `E2006`, `UJIndoorLoc` and `log1p.E2006`. For `UJIndoorLoc`, we consider predicting the location (longitude) of users inside of buildings. Throughout this experiment we fix $\lambda = 1$ and $\theta = 2$. Convergence performance over 1000 passes to the datasets is shown in Fig. 2. When the value of $\delta$ decreases, the function $F_\delta$ becomes more ill-conditioned, thereby being increasingly difficult to optimize. Indeed, we find that the convergence of all the tested algorithms slows down when $\delta$ decreases from $10^{-2}$ to $10^{-3}$. Nonetheless, we observe that CD methods consistently outperform gradient-based methods, while pDCA$_e$ consistently outperforms pDCA. Among CD methods, we find that both ACPDC and ACPP exhibit fast convergence and they both outperform RCSD and RPCD, further confirming the advantage of using ACD in the nonconvex settings.

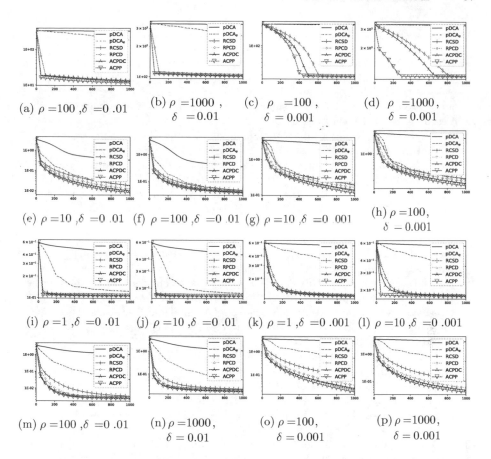

(a) $\rho = 100, \delta = 0.01$   (b) $\rho = 1000, \delta = 0.01$   (c) $\rho = 100, \delta = 0.001$   (d) $\rho = 1000, \delta = 0.001$

(e) $\rho = 10, \delta = 0.01$   (f) $\rho = 100, \delta = 0.01$   (g) $\rho = 10, \delta = 0.001$   (h) $\rho = 100, \delta = 0.001$

(i) $\rho = 1, \delta = 0.01$   (j) $\rho = 10, \delta = 0.01$   (k) $\rho = 1, \delta = 0.001$   (l) $\rho = 10, \delta = 0.001$

(m) $\rho = 100, \delta = 0.01$   (n) $\rho = 1000, \delta = 0.01$   (o) $\rho = 100, \delta = 0.001$   (p) $\rho = 1000, \delta = 0.001$

**Fig. 2.** Experimental results on smoothed $l_1$ regression with SCAD penalty. $y$-axis: objective value (log scaled). $x$-axis: number of passes to the dataset. Test datasets (from top to bottom): `synthetic`, `E2006-tfidf` and `UJIndoorLoc` and `E2006-log1p`.

## 8   Conclusion

In this paper, we developed novel CD methods for minimizing a class of non-smooth and nonconvex functions. The proposed randomized coordinate sub-gradient descent (RCSD) and randomly permuted coordinate descent (RPCD) methods naturally extend randomized coordinate descent and cyclic coordinate descent, respectively, to the realm of nonsmooth and nonconvex optimization. We also developed a new randomized proximal DC algorithm (ACPDC) for composite DC problems and a new randomized proximal point algorithm (ACPP) for weakly convex problems, based on the fast convergence of ACD for convex programming. We developed new optimality measures and established iteration complexities for the proposed algorithms. Both theoretical and experimental results demonstrate the advantage of our proposed CD approaches over state-of-the-art gradient-based methods.

# References

1. Ahn, M., Pang, J.S., Xin, J.: Difference-of-convex learning: directional stationarity, optimality, and sparsity. SIAM J. Optim. **27**(3), 1637–1665 (2017)
2. Allen-Zhu, Z., Qu, Z., Richtárik, P., Yuan, Y.: Even faster accelerated coordinate descent using non-uniform sampling. In: International Conference on Machine Learning, pp. 1110–1119 (2016)
3. An, N.T., Nam, N.M.: Convergence analysis of a proximal point algorithm for minimizing differences of functions. Optimization **66**(1), 129–147 (2017)
4. Beck, A., Hallak, N.: Optimization problems involving group sparsity terms. Math. Program. **178**(1), 39–67 (2018). https://doi.org/10.1007/s10107-018-1277-1
5. Beck, A., Tetruashvili, L.: On the convergence of block coordinate descent type methods. SIAM J. Optim. **23**(4), 2037–2060 (2013)
6. Carmon, Y., Duchi, J.C., Hinder, O., Sidford, A.: Accelerated methods for nonconvex optimization. arXiv preprint arXiv:1611.00756 (2016)
7. Dang, C.D., Lan, G.: Stochastic block mirror descent methods for nonsmooth and stochastic optimization. SIAM J. Optim. **25**(2), 856–881 (2015)
8. Davis, D., Grimmer, B.: Proximally guided stochastic subgradient method for nonsmooth, nonconvex problems. SIAM J. Optim. **29**(3), 1908–1930 (2019)
9. Drusvyatskiy, D., Paquette, C.: Efficiency of minimizing compositions of convex functions and smooth maps. Math. Program. **178**(1), 503–558 (2018). https://doi.org/10.1007/s10107-018-1311-3
10. Dua, D., Graff, C.: UCI machine learning repository (2017)
11. Fan, J., Li, R.: Variable selection via nonconcave penalized likelihood and its Oracle properties. J. Am. Stat. Assoc. **96**(456), 1348–1360 (2001)
12. Gong, P., Zhang, C., Lu, Z., Huang, J.Z., Ye, J.: A general iterative shrinkage and thresholding algorithm for non-convex regularized optimization problems. Int. Conf. Mach. Learn. **28**(2), 37–45 (2013)
13. Gotoh, J., Takeda, A., Tono, K.: DC formulations and algorithms for sparse optimization problems. Math. Program. **169**(1), 141–176 (2017). https://doi.org/10.1007/s10107-017-1181-0
14. Hien, L.T.K., Gillis, N., Patrinos, P.: Inertial block mirror descent method for non-convex non-smooth optimization. arXiv preprint arXiv:1903.01818 (2019)
15. Hong, M., Razaviyayn, M., Luo, Z.Q., Pang, J.S.: A unified algorithmic framework for block-structured optimization involving big data: with applications in machine learning and signal processing. IEEE Signal Process. Mag. **33**(1), 57–77 (2016)
16. Khamaru, K., Wainwright, M.J.: Convergence guarantees for a class of non-convex and non-smooth optimization problems. In: International Conference on Machine Learning, pp. 2606–2615 (2018)
17. Kong, W., Melo, J.G., Monteiro, R.D.: Complexity of a quadratic penalty accelerated inexact proximal point method for solving linearly constrained nonconvex composite programs. arXiv preprint arXiv:1802.03504 (2018)
18. Lan, G., Yang, Y.: Accelerated stochastic algorithms for nonconvex finite-sum and multi-block optimization. arXiv preprint arXiv:1805.05411 (2018)
19. Lin, Q., Lu, Z., Xiao, L.: An accelerated randomized proximal coordinate gradient method and its application to regularized empirical risk minimization. SIAM J. Optim. **25**(4), 2244–2273 (2015)
20. Nesterov, Y.: Efficiency of coordinate descent methods on huge-scale optimization problems. SIAM J. Optim. **22**(2), 341–362 (2012)

21. Nouiehed, M., Pang, J.-S., Razaviyayn, M.: On the pervasiveness of difference-convexity in optimization and statistics. Math. Program. **1**, 1–28 (2018). https://doi.org/10.1007/s10107-018-1286-0

22. Patrascu, A., Necoara, I.: Efficient random coordinate descent algorithms for large-scale structured nonconvex optimization. J. Global Optim. **61**(1), 19–46 (2014). https://doi.org/10.1007/s10898-014-0151-9

23. Richtárik, P., Takáč, M.: Iteration complexity of randomized block-coordinate descent methods for minimizing a composite function. Math. Program. **144**(1), 1–38 (2012). https://doi.org/10.1007/s10107-012-0614-z

24. Thi, H.L., Dinh, T.P., Le, H., Vo, X.: DC approximation approaches for sparse optimization. Eur. J. Oper. Res. **244**(1), 26–46 (2015)

25. Le Thi, H.A., Pham Dinh, T.: DC programming and DCA: thirty years of developments. Math. Program. **169**(1), 5–68 (2018). https://doi.org/10.1007/s10107-018-1235-y

26. Wen, B., Chen, X., Pong, T.K.: A proximal difference-of-convex algorithm with extrapolation. Comput. Optim. Appl. **69**(2), 297–324 (2017). https://doi.org/10.1007/s10589-017-9954-1

27. Xu, Y., Yin, W.: A globally convergent algorithm for nonconvex optimization based on block coordinate update. J. Sci. Comput. **72**(2), 700–734 (2017)

28. Xu, Y., Qi, Q., Lin, Q., Jin, R., Yang, T.: Stochastic optimization for DC functions and non-smooth non-convex regularizers with non-asymptotic convergence. In: International Conference on Machine Learning, pp. 6942–6951 (2019)

29. Yuille, A.L., Rangarajan, A.: The concave-convex procedure (CCCP). Adv. Neural Inf. Process. Syst. **14**, 1033–1040 (2002)

30. Zhang, C.H., Zhang, T.: A general theory of concave regularization for high-dimensional sparse estimation problems. Stat. Sci. **27**(4), 576–593 (2012)

# Escaping Saddle Points of Empirical Risk Privately and Scalably via DP-Trust Region Method

Di Wang[1,2(✉)] and Jinhui Xu[1]

[1] Department of Computer Science and Engineering, State University of New York at Buffalo, Buffalo, NY 14260, USA
{dwang45,jinhui}@buffalo.edu
[2] King Abdullah University of Science and Technology, Thuwal, Saudi Arabia

**Abstract.** It has been shown recently that many non-convex objective/loss functions in machine learning are known to be strict saddle. This means that finding a second-order stationary point (*i.e.,* approximate local minimum) and thus escaping saddle points are sufficient for such functions to obtain a classifier with good generalization performance. Existing algorithms for escaping saddle points, however, all fail to take into consideration a critical issue in their designs, that is, the protection of sensitive information in the training set. Models learned by such algorithms can often implicitly memorize the details of sensitive information, and thus offer opportunities for malicious parties to infer it from the learned models. In this paper, we investigate the problem of privately escaping saddle points and finding a second-order stationary point of the empirical risk of non-convex loss function. Previous result on this problem is mainly of theoretical importance and has several issues (*e.g.,* high sample complexity and non-scalable) which hinder its applicability, especially, in big data. To deal with these issues, we propose in this paper a new method called Differentially Private Trust Region, and show that it outputs a second-order stationary point with high probability and less sample complexity, compared to the existing one. Moreover, we also provide a stochastic version of our method (along with some theoretical guarantees) to make it faster and more scalable. Experiments on benchmark datasets suggest that our methods are indeed more efficient and practical than the previous one.

**Keywords:** Differential privacy · Empirical Risk Minimization · Private machine learning

## 1 Introduction

Learning from sensitive data is a frequently encountered challenging task in many data analytic applications. It requires the learning algorithm to not only

**Electronic supplementary material** The online version of this chapter (https://doi.org/10.1007/978-3-030-67664-3_6) contains supplementary material, which is available to authorized users.

F. Hutter et al. (Eds.): ECML PKDD 2020, LNAI 12459, pp. 90–106, 2021.
https://doi.org/10.1007/978-3-030-67664-3_6

learn effectively from the data but also provide a certain level of guarantee on privacy preserving. As a rigorous notion for statistical data privacy, differential privacy (DP) has received a great deal of attentions in the past decade [11,12]. DP works by injecting random noise into the statistical results obtained from sensitive data so that the distribution of the perturbed results is insensitive to any single-record change in the original dataset. A number of methods with DP guarantees have been discovered and recently adopted in industry [14].

As a fundamental supervised-learning problem in machine learning, Empirical Risk Minimization (ERM) has been extensively studied in recent years since it encompasses a large family of classical models such as linear regression, LASSO, SVM, logistic regression and ridge regression. Its Differentially Private (DP) version (DP-ERM) can be formally defined as follows.

**Definition 1 (DP-ERM** [34]**).** *Given a dataset $D = \{x_1, \cdots, x_n\}$ from a data universe $\mathcal{X}$, DP-ERM is to find an estimator $w^{priv} \in \mathbb{R}^p$ so as to minimize the empirical risk, i.e.*

$$L(w, D) = \frac{1}{n} \sum_{i=1}^{n} \ell(w, x_i), \tag{1}$$

*with the guarantee of being differentially private, where $\ell(\cdot, \cdot)$ is the loss function.*

*If the loss function is convex, the utility of the private estimator is measured by the expected excess empirical risk, i.e. $\mathbb{E}_{\mathcal{A}}[L(w^{priv}, D)] - \min_{x \in \mathbb{R}^p} L(w, D)$, where the expectation of $\mathcal{A}$ is taking over all the randomness of the algorithm.*

Previous research on DP-ERM has mainly focused on convex loss functions [7] (see the section of Related Work for more details). However, empirical studies have revealed that non-convex loss functions typically achieve better classification accuracy than the convex ones [25]. Furthermore, recent developments in deep learning [17] also suggest that loss functions are more likely to be non-convex in real world applications. Thus, there is an urgent need for the research community to shift our focus from convex to non-convex loss functions. So far, very few papers [28,31,34,37] have considered DP-ERM with non-convex loss functions. This is mainly due to the fact that finding the global minimum of a non-convex loss function is NP-hard.

Different from convex loss functions, non-convex functions have adopted a few different ways to measure the utility. The authors of [37] studied the problem with smooth loss function and proposed using the $\ell_2$ gradient-norm of a private estimator, *i.e.*, $\|\nabla L(w^{priv}, D)\|_2$, to measure the utility, which was then extended in [31,34] to the cases of non-smooth loss functions and high dimensional space. It is well known that $\ell_2$ gradient-norm can estimate only the first-order stationary point (or critical point)[1], and thus may lead to inferior generalization performance [10]. The authors of [28] are the first to show that the utility of general non-convex loss functions can also be measured in the same way as convex loss functions by the expected excess empirical risk. However, their upper bound

---

[1] A point $w$ of a function $F(\cdot)$ is called a first-order stationary point (critical point) if it satisfies the condition of $\|\nabla F(w)\| = 0$.

$O(\frac{p}{\log n\epsilon^2})$ is quite large compared with the convex case and needs to assume that $n \geq O(\exp(p))$, which may not be satisfied in some real-world datasets. They also showed that for some special loss functions such as sigmoid loss, the bound can be further improved. But such improvements are dependent on the special structures or some assumptions of the loss functions and thus cannot be extended to the general case.

Due to the intrinsic challenge of approximating global minimum and issues related to saddle points, recent research on deep neural network training [16,23] and many other machine learning problems [5,15] has shifted the attentions to obtaining local minima. It has been shown that fast convergence to a local minimum is actually sufficient for such tasks to have good generalization performance. This motivates us to investigate efficient techniques for finding local minima. However, as shown in [2], computing a local minimum could be quite challenging as it is actually NP-hard for general non-convex functions. Fortunately, many non-convex functions in machine learning are known to be strict saddle [15], meaning that a second-order stationary point (or approximate local minimum) is sufficient to obtain a close enough point to some local minimum. With this, the authors in [28] used a new way to measure the utility, based on the $\ell_2$-gradient norm and the minimal eigenvalue of a Hessian matrix (see Preliminaries section for details), where the goal is to design an algorithm with the ability of escaping saddle points and approximating some second-order stationary point. Specifically, they showed that when $n$ is large enough, the classical differentially private gradient descent method could escape saddle points and meanwhile output an $\alpha$-second-order stationary point ($\alpha$-SOSP). But their method has several issues, which hamper its applications in big data. Firstly, their sample complexity (or equivalently error bound) is relatively high. It is not clear whether it can be improved. Secondly, their method needs to calculate the gradient and Hessian matrix of the whole objective function in each iteration, which is prohibitive in large scale datasets. Finally, their result mainly focuses on theoretical development and does not provide any experimental study. Thus, it is not clear how practical it is.

**Our Contributions:** To address the aforementioned theoretical and practical issues, we propose in this paper a new method called Differentially Private Trust Region (DP-TR) which is capable of escaping saddle points privately. Particularly, we first show that our algorithm can output an $\alpha$-SOSP with high probability and less sample complexity (compared to the one in [28]). To make our method scalable, we then present a stochastic version of DP-TR called Differentially Private Stochastic Trust Region (DP-STR) with the same functionality. We show that DP-STR is much faster and has asymptotically the same sample complexity as DP-TR. Finally, we provide comprehensive experimental studies on the practical performance of our methods in escaping saddle point under differential privacy model.

Due to space limit, all proofs are left to the Supplementary Material.

## 2    Related Work

DP-ERM is a fundamental problem in both machine learning and differential privacy communities. There are quite a number of results on DP-ERM with convex loss functions, which investigate the problem from different perspectives. For example, [29,30,35] considered ERM in the non-interactive local model. [21,27] and [1] investigated the regret bound in online settings. [36] explored the problem from the perspective of learnability and stability. The problem has also been well-studied in the offline central model [4,7,8,20,34], as well as in high dimensional space [22,26].

For general non-convex loss functions, as mentioned earlier, existing results have used three different ways to measure the utility of the private estimator [28,31,34,37]. For $\ell_2$-gradient norm based utility, [31] provided a comprehensive study following the work of [37]. However, since the first order stationary points often have inferior performance to the second order stationary points in practice, which is the focus of this paper, their results are incomparable with ours. For the expected excess empirical (or population) risk based utility, it has been applied only to some special non-convex loss functions before the work of [28]. For example, [34] showed a near optimal bound for some special non-convex loss functions satisfying the Polyak-Lojasiewicz condition. [3] studied the problem of optimizing privately piecewise Lipschitz functions which satisfy the dispersion condition in online settings. Recently, the authors of [28] provided the first result for general non-convex functions. However, their bound is loose compared to the convex case and needs to assume that $n \geq O(\exp(p))$, which may not hold in practice.

For the third type of utility (based on the $\ell_2$-gradient norm and the minimal eigenvalue of a Hessian matrix), [28] was the first to use it to measure the closeness of the private estimator to some second order stationary points. Compared with theirs, our proposed methods improve considerably the sample complexity. More precisely, we show that to achieve an $\alpha$-SOSP (see Definition 12 for details), the sample complexity of DP-TR and DP-STR is $O(\frac{p\sqrt{\ln\frac{1}{\delta}}}{\alpha^{1.75}\epsilon})$ (omitting other terms), while it is $O(\frac{p\sqrt{\ln\frac{1}{\delta}}}{\alpha^2\epsilon})$ in [28]. Equivalently, with a fixed data size $n$, our algorithms yield an $O\left((\frac{p\sqrt{\ln\frac{1}{\delta}}}{n\epsilon})^{\frac{4}{7}}\right)$-SOSP, while [28] can output only an $O\left((\frac{p\sqrt{\ln\frac{1}{\delta}}}{n\epsilon})^{\frac{1}{2}}\right)$-SOSP (with high probability). Moreover, we also show in experiments that our methods are more efficient and scalable.

## 3    Preliminaries

In this section, we review some definitions related to differential privacy and some terminologies and lemmas in optimization.

## 3.1 Differential Privacy

Informally speaking, DP ensures that an adversary cannot infer whether or not a particular individual is participating in the database query, even with unbounded computational power and access to every entry in the database except for that particular individual's data. DP considers a centralized setting that includes a trusted data curator, who generates the perturbed statistical information (e.g., counts and histograms) by using some randomized mechanism. Formally, it can be defined as follows.

**Definition 2 (Differential Privacy [12]).** *Given a data universe $\mathcal{X}$, we say that two datasets $D, D' \subseteq \mathcal{X}$ are neighbors if they differ by only one entry, which is denoted as $D \sim D'$. A randomized algorithm $\mathcal{A}$ is $(\epsilon, \delta)$-differentially private (DP) if for all neighboring datasets $D, D'$ and for all events $S$ in the output space of $\mathcal{A}$, the following holds*

$$\mathbb{P}(\mathcal{A}(D) \in S) \leq e^{\epsilon} \mathbb{P}(\mathcal{A}(D') \in S) + \delta.$$

*When $\delta = 0$, $\mathcal{A}$ is $\epsilon$-differentially private.*

In this paper, we will study only $(\epsilon, \delta)$-DP and use the Gaussian Mechanism [12] to guarantee $(\epsilon, \delta)$-DP.

**Definition 3 (Gaussian Mechanism).** *Given any function $q : \mathcal{X}^n \to \mathbb{R}^p$, the Gaussian Mechanism is defined as:*

$$\mathcal{M}_G(D, q, \epsilon) = q(D) + Y,$$

*where $Y$ is drawn from Gaussian Distribution $\mathcal{N}(0, \sigma^2 I_p)$ with $\sigma \geq \frac{\sqrt{2 \ln(1.25/\delta)} \Delta_2(q)}{\epsilon}$, and $\Delta_2(q)$ is the $\ell_2$-sensitivity of the function $q$, i.e. $\Delta_2(q) = \sup_{D \sim D'} ||q(D) - q(D')||_2$. Gaussian Mechanism ensures $(\epsilon, \delta)$-differential privacy.*

We will use the sub-sampling property and the advanced composition theorem to ensure $(\epsilon, \delta)$-DP for the DP-STR algorithm.

**Lemma 4 (Advanced Composition Theorem [13]).** *Given target privacy parameters $0 < \epsilon, \delta \leq 1$, to ensure $(\epsilon, \delta)$-DP over $T$ mechanisms, it suffices that each mechanism is $(\epsilon', \frac{\delta}{2T})$-DP, where $\epsilon' = \frac{\epsilon}{2\sqrt{2T \ln(2/\delta)}}$.*

**Lemma 5 ([4]).** *Over a domain of datasets $\mathcal{X}^n$, if an algorithm $\mathcal{A}$ is $(\epsilon, \delta)$-DP, then for any $n$-size dataset $D$, executing $\mathcal{A}$ on uniformly random $\gamma n$ entries of $D$ ensures $(2\gamma\epsilon, \delta)$-DP.*

We will use a relaxation of DP called zero-Concentrated Differential Privacy (zCDP) [6] to guarantee $(\epsilon, \delta)$-DP for our DP-TR method. Due to its optimal composition proposition, zCDP is easier to analyze and can achieve a tighter bound, compared to those using the advanced composition theorem to ensure $(\epsilon, \delta)$-DP (Lemma 4) [31].

**Definition 6 (zCDP[6]).** *A randomized mechanism $\mathcal{A}$ is $\rho$-zero concentrated differentially private if for all $D \sim D'$ and all $\alpha \in (1, \infty)$,*

$$D_\alpha(\mathcal{A}(D)\|\mathcal{A}(D')) \leq \rho\alpha,$$

*where $D_\alpha(\mathcal{A}(D)\|\mathcal{A}(D'))$ is the $\alpha$-Rényi divergence[2] between the distribution of $\mathcal{A}(D)$ and $\mathcal{A}(D')$.*

The following lemma shows the connection between zCDP and $(\epsilon, \delta)$-DP.

**Lemma 7 ([6]).** *If $\mathcal{A}$ is $\rho$-zCDP, then $\mathcal{A}$ is $(\rho + 2\sqrt{\rho \ln \frac{1}{\delta}}, \delta)$-DP for any $\delta > 0$.*

The following lemma says that adding Gaussian noise could also achieve zCDP.

**Lemma 8 ([6]).** *Given any function $q : \mathcal{X}^n \mapsto \mathbb{R}^p$, the Gaussian Mechanism $\mathcal{M}_G(D, q, \epsilon) = q(D) + Y$, where $Y$ is drawn from Gaussian Distribution $\mathcal{N}(0, \sigma^2 I_p)$ with $\sigma \geq \frac{\Delta_2(q)}{\sqrt{2\rho}}$, is $\rho$-zCDP.*

Similar to DP, zCDP also has the composition property.

**Lemma 9 ([6]).** *Let $\mathcal{A}$ be $\rho$-zCDP and $\mathcal{A}'$ be $\rho'$-zCDP, then their composition $\mathcal{A}''(D) = (\mathcal{A}(D), \mathcal{A}'(D))$ is $(\rho + \rho')$-zCDP.*

## 3.2 Optimization

We first specify the necessary assumptions on our loss functions that are commonly used in other related work such as [24,28,38].

*Assumption 1.* We assume that $L(\cdot, D)$ is bounded from below and its global minimum is achieved at $w^*$. We let $\Delta$ denote

$$\Delta = L(w^0, D) - L(w^*, D),$$

where $w^0$ is the initial vector of our algorithms.

*Assumption 2.* We assume that for each $x \in \mathcal{X}$, the loss function $\ell(\cdot, x)$ is $G$-Lipschitz, that is, for all $w, w' \in \mathbb{R}^d$

$$|\ell(w, x) - \ell(w', x)| \leq G\|w - w'\|_2.$$

We also assume that $\ell(\cdot, x)$ is $M$-smooth, that is, for all $w, w' \in \mathbb{R}^p$

$$\|\nabla\ell(w, x) - \nabla\ell(w', x)\|_2 \leq M\|w - w'\|_2.$$

Finally, we assume that $\ell(\cdot, x)$ is twice differentiable and $\rho$-Hessian Lipschitz, that is, for all $w, w' \in \mathbb{R}^p$

$$\|\nabla^2\ell(w, x) - \nabla^2\ell(w', x)\|_2 \leq \rho\|w - w'\|_2,$$

where $\|A\|_2$ is the spectral norm of a matrix $A$.

---

[2] Generally, $D_\alpha(P\|Q)$ is the Rényi divergence between $P$ and $Q$ which is defined as

$$D_\alpha(P\|Q) = \frac{1}{\alpha - 1} \log \mathbb{E}_{x \sim Q}(\frac{P(x)}{Q(x)})^\alpha.$$

Note that the above assumption indicates that for any $w, h \in \mathbb{R}^p$

$$L(w + h, D) \leq L(w, D) + \langle \nabla L(w, D), h \rangle + \frac{1}{2} h^T \nabla^2 L(w, D) h + \frac{\rho}{6} \|h\|_2^3.$$

In this paper, we focus on approximating a second order stationary point, which is defined as follows.

**Definition 10.** *A point $w$ is called a second-order stationary point (SOSP) of a twice differentiable function $F$ if*

$$\|\nabla F(w)\|_2 = 0 \ and \ \lambda_{\min}(\nabla^2 F(w)) \geq 0 \ ,$$

*where $\lambda_{\min}$ denotes its minimal eigenvalue.*

Since it is extremely challenging to find an exact SOSP [15], we turn to its approximation. The following definition of $\alpha$-approximate SOSP relaxes the first- and second-order optimality conditions.

**Definition 11** ([15]). *$w$ is an $\alpha$-second-order stationary point ($\alpha$-SOSP) or $\alpha$-approximate local minimum[3] of a twice differentiable function $F$ which is $\rho$-Hessian Lipschitz, if*

$$\|\nabla F(w)\|_2 \leq \alpha \ and \ \lambda_{\min}(\nabla^2 F(w)) \geq -\sqrt{\rho \alpha}. \tag{2}$$

Based on this, we now formally define our problem of DP-SOSP.

**Definition 12 (DP-SOSP).** *Given $\alpha, \epsilon, \delta > 0$, DP-SOSP is to identify the smallest sample complexity $n(\alpha, p, \epsilon, \delta)$ such that when $n \geq n(\alpha, p, \epsilon, \delta)$, for any dataset $D$ of size $n$, there is an $(\epsilon, \delta)$-DP algorithm which outputs an $\alpha$-SOSP of the empirical risk (1) with high probability.*

Since our ideas are derived from the trust region method proposed in [9], we now briefly introduce the trust region method. In each step of the trust region method for a function $F(\cdot)$, it solves a Quadratic Constraint Quadratic Program (QCQP):

$$h^k = \arg \min_{h \in \mathbb{R}^d, \|h\|_2 \leq r} \langle \nabla F(w^k), h \rangle + \frac{1}{2} \langle \nabla^2 F(w^k) h, h \rangle, \tag{3}$$

where $r$ is called the *trust-region radius*. Then, it updates in the following way

$$w^{k+1} = w^k + h^k.$$

Since the function $F(w)$ is non-convex, this indicates that the sub-problem (3) is non-convex. However, its global minimum can be characterized by the following lemma.

---

[3] This is a special version of $(\epsilon, \gamma)$-SOSP [15]. Our results can be easily extended to the general definition. The same applies to the constrained case.

**Lemma 13 (Corollary 7.2.2 in [9]).** *Any global minimum of the problem (3) should satisfy*

$$(\nabla^2 F(w^k) + \lambda I)h^k = -\nabla F(w^k), \tag{4}$$

*where the dual variable $\lambda \geq 0$ should satisfies the conditions of $\nabla^2 F(x^k)+\lambda I \succ 0$ and $\lambda(\|h^k\|_2 - r) = 0$.*

It is worth noting that in practice sub-problem (3) can be solved by the Lanczos method efficiently (see [18] for details). For the dual variable $\lambda$ in Lemma 13, it can be solved by almost any QCQP solver such as CVX [19].

## 4   Methodology

In this section, we first introduce our main method, DP-TR, and then extend to its stochastic version, *i.e.*, DP-STR.

### 4.1   Differentially Private Trust Region Method

The key idea of our DP-TR is the following. In each iteration, instead of using the gradient and Hessian of the empirical risk (1) directly to the sub-problem (3), we use their perturbed versions to ensure DP. That is, we use $\tilde{\nabla}L(w^k, D) = \nabla L(w^k, D) + \epsilon_k$ and $\tilde{\nabla}^2 L(w^k, D) = \nabla^2 L(w^k, D) + H_k$, where $\epsilon_t$ is a Gaussian vector and $H_t$ is a randomized symmetric Gaussian matrix (since a Hessian matrix is symmetric, we need to add a symmetric random matrix). The main steps of DP-TR are given in Algorithm 1.

For the stopping criteria, we use the dual variable $\lambda^k$ and see whether the value is greater or less than some threshold. This criteria enable the last-term convergence analysis in Theorem 15.

The following theorem shows that Algorithm 1 is $(\epsilon, \delta)$-DP.

**Theorem 14.** *For any $\epsilon, \delta > 0$, Algorithm 1 is $(\epsilon, \delta)$-differentially private under Assumption 2.*

The following theorem shows that when the data size $n$ is large enough, then with high probability the output of Algorithm 1 will be an $\alpha$-SOSP.

**Theorem 15.** *Under Assumptions 1 and 2, for any given $\alpha$, if we take $r = \sqrt{\frac{\alpha}{\rho}}$, $T = \frac{6\sqrt{\rho}\Delta}{\alpha^{1.5}}$, then with probability at least $1 - \zeta - \frac{T}{p^c}$ for some universal constant $c > 0$ and $\zeta > 0$, the algorithm outputs a point which is an $O(\alpha)$-SOSP if $n$ satisfies*

$$n \geq \Omega(\frac{p\ln\frac{1}{\zeta}\sqrt{\ln\frac{1}{\delta}}}{\alpha^{1.75}\epsilon}), \tag{5}$$

*where the Big-$\Omega$ notation omits the terms of $G, M, \rho, \Delta, \ln\frac{1}{\alpha}$.*

---

**Algorithm 1.** DP-TR

---

**Input**: Privacy parameters $\epsilon, \delta$, trust-region radius $r$, iteration number $T$ (to be specified later), initial vector $w^0$ and error term $\alpha$

1: Let $\phi = (\sqrt{\epsilon + \ln \frac{1}{\delta}} - \sqrt{\ln \frac{1}{\delta}})^2$.
2: **for** $k = 0, \cdots, T-1$ **do**
3:     Denote $\tilde{\nabla}L(w^k, D) = \nabla L(w^k, D) + \epsilon_k$, where $\epsilon_t \sim \mathcal{N}(0, \sigma^2 I_d)$ with $\sigma^2 = \frac{4G^2 T}{n^2 \phi}$.
4:     Denote $\tilde{\nabla}^2 L(w^k, D) = \nabla^2 L(w^k, D) + H_k$, where $H_t$ is a symmetric matrix with its upper triangle (including the diagonal) being i.i.d samples from $\mathcal{N}(0, \sigma_2^2)$, $\sigma_2^2 = \frac{4pM^2 T}{n^2 \phi}$, and each lower triangle entry is copied from its upper triangle counterpart.
5:     Solve the following QCQP and get $h^k$ and dual variable $\lambda^k$,

$$h^k = \arg \min_{h \in \mathbb{R}^d, \|h\|_2 \leq r} \langle \tilde{\nabla}L(w^k, D), h \rangle + \frac{1}{2} \langle \tilde{\nabla}^2 L(w^k, D)h, h \rangle,$$

6:     Let $w^{k+1} = w^k + h^k$.
7:     **if** $\lambda^k \leq \sqrt{\alpha \rho}$ **then**
8:         Output $w_\alpha = w^{k+1}$.
9:     **end if**
10: **end for**

---

*Remark 16.* We note that in the previous work [28], to output an $O(\alpha)$-SOSP with high probability, the data size $n$ needs to satisfy $n \geq \Omega(\frac{p\sqrt{\ln \frac{1}{\delta}}}{\alpha^2 \epsilon})$, while the dependency on $\alpha$ in (5) is $\frac{1}{\alpha^{1.75}}$. Thus, we improve the sample size by a factor of $O(\frac{1}{\alpha^{0.25}})$. Equivalently, if we fix $n$, Theorem 15 ensures that Algorithm 1 outputs a point which is $O\left((\frac{p\sqrt{\ln \frac{1}{\delta}}}{n\epsilon})^{\frac{4}{7}}\right)$-SOSP, while the previous work in [28] outputs a point which is $O\left((\frac{p\sqrt{\ln \frac{1}{\delta}}}{n\epsilon})^{\frac{1}{2}}\right)$-SOSP. We can see that our algorithm yields better approximate SOSP than the previous one. We leave as open problems to determine whether the sample complexity in (5) can be further improved and what is the optimal bound of the sample complexity.

Also, in [28] the number of iterations is $T = \tilde{O}(\frac{1}{\alpha^2})$, while Algorithm 1 needs only $O(\frac{1}{\alpha^{1.5}})$ iterations. This means that the running time of Algorithm 1 is $O(\frac{n\text{Poly}(p)}{\alpha^{1.5}})$, while it is $O(\frac{n\text{Poly}(p)}{\alpha^2})$ in [28]. Thus, our algorithm has an improved time complexity for the term of $\frac{1}{\alpha}$ compared with the previous one. Moreover, as we will see in the experiment section, our algorithms is indeed faster than the previous one.

Theorem 15 shows the explicit step size control of the DP-TR method: Since the dual variable satisfies $\lambda^k > \sqrt{\alpha \rho}$ for all but the last iteration. Thus we can always find a solution to the trust-region sub-problem (3) in the boundary, *i.e.*, $\|h^k\|_2 = r$, according to Lemma 13.

## 4.2   Differentially Private Stochastic Trust Region Method

In the previous section we show that our method DP-TR needs less samples and is faster than DP-GD proposed in [28]. However, as mentioned in Remark 16, the time complexities of both algorithms are linearly dependent on the sample size $n$, which is prohibitive in large scale datasets. Thus, a natural question is to determine whether it is possible to design an algorithm that shares the advantages of DP-TR and meanwhile is scalable. In this section we give an affirmative answer to this question by providing a stochastic version of DP-TR called Differentially Private Stochastic Trust Region method (DP-STR)

The key idea of DP-STR is that, instead of evaluating the gradient and Hessian matrix of the whole function $L(w, D)$ in each iteration, we will uniformly sub-sample two sets of indices $\mathcal{S}, \mathcal{T} \subseteq [n]$ and calculate the gradients and Hessian matrix of the loss function with the samples corresponding to the set $\mathcal{S}$ and $\mathcal{T}$, respectively. That is

$$\nabla L(w^k, \mathcal{S}) = \frac{1}{|\mathcal{S}|} \sum_{i \in \mathcal{S}} \nabla \ell(w^k, x_i), \tag{6}$$

$$\nabla^2 L(w^k, \mathcal{T}) = \frac{1}{|\mathcal{T}|} \sum_{i \in \mathcal{T}} \nabla^2 \ell(w^k, x_i). \tag{7}$$

Then, similar to DP-TR, we add some Gaussian noise and random Gaussian matrix to $\nabla L(w^k, \mathcal{S})$ and $\nabla^2 L(w^k, \mathcal{T})$, respectively, to ensure $(\epsilon, \delta)$-DP. See Algorithm 2 for details. Note that since zCDP can not be guaranteed by sub-sampling, we use the traditional advanced composition theorem Lemma 4 and sub-sampling property Lemma 5 to guarantee $(\epsilon, \delta)$-DP.

**Theorem 17.** *For any $0 < \epsilon, \delta < 1$, Algorithm 2 is $(\epsilon, \delta)$-differentially private.*

**Theorem 18.** *Under Assumptions 1 and 2, for a given $\alpha$, if we take $r = \sqrt{\frac{\alpha}{\rho}}$, $T = \frac{6\sqrt{\rho}\Delta}{\alpha^{1.5}}$, $|\mathcal{S}| \geq \Omega(\frac{L^2 \ln \frac{p}{\zeta}}{\alpha^2})$ and $|\mathcal{T}| \geq \Omega(\frac{M^2 \ln \frac{p}{\zeta}}{\alpha\rho})$ in Algorithm 2, then with probability at least $1 - 3\zeta - \frac{T}{p^c}$ for some universal constant $c > 0$ and $\zeta > 0$, the algorithm outputs a point that is an $O(\alpha)$-SOSP if $n$ satisfies (5), which is the same as in Theorem 15.*

Comparing with Theorem 15, we can see that the sample complexity of Theorem 18 is the same while the time complexity of Algorithm 2 is $O\left(T(|\mathcal{S}| + |\mathcal{T}|)\text{Poly}(p)\right) = O(\frac{\text{Poly}(p)}{\alpha^{3.5}})$, which is independent of the sample size $n$. This means that DP-STR is faster and scalable to large scale datasets.

*Remark 19.* We note that it is unknown whether the algorithm in [28] can be extended to a stochastic version whose time complexity is independent of the

---

**Algorithm 2.** DP-STR

---

**Input**: Privacy parameters $\epsilon, \delta$, trust-region radius $r$, iteration number $T$, sub-sampling size $|\mathcal{S}|, |\mathcal{T}|$ (to be specified later), initial vector $w^0$ and error term $\alpha$.

1: **for** $k = 0, \cdots, T - 1$ **do**
2:    Uniformly sub-sample two independent indices sets $\mathcal{S}, \mathcal{T} \subseteq [n]$ with size $|\mathcal{S}|$ and $|\mathcal{T}|$, respectively.
3:    Denote $\tilde{\nabla}L(w^k, \mathcal{S}) = \nabla L(w^k, \mathcal{S}) + \epsilon_k$, where $\epsilon_t \sim \mathcal{N}(0, \sigma^2 I_d)$ with $\sigma^2 = \frac{256 G^2 \ln \frac{5T}{\delta} \ln \frac{2}{\delta}}{n^2 \epsilon^2}$ and $\nabla L(w^k, \mathcal{S})$ is given in (6)
4:    Denote $\tilde{\nabla}^2 L(w^k, \mathcal{T}) = \nabla^2 L(w^k, \mathcal{T}) + H_k$, where $H_t$ is a symmetric matrix with its upper triangle (including the diagonal) being i.i.d samples from $\mathcal{N}(0, \sigma_2^2)$, $\sigma_2^2 = \frac{256 p M^2 T \ln \frac{2}{\delta} \ln \frac{5T}{\delta}}{n^2 \epsilon^2}$, and each lower triangle entry is copied from its upper triangle counterpart. $\nabla^2 L(w^k, \mathcal{T})$ is given in (7).
5:    Solve the following QCQP and get $h^k$ and dual variable $\lambda^k$,

$$h^k = \arg \min_{h \in \mathbb{R}^d, \|h\|_2 \leq r} \langle \tilde{\nabla}L(w^k, \mathcal{S}), h \rangle + \frac{1}{2} \langle \tilde{\nabla}^2 L(w^k, \mathcal{T}) h, h \rangle,$$

6:    Let $w^{k+1} = w^k + h^k$.
7:    **if** $\lambda^k \leq \sqrt{\alpha \rho}$ **then**
8:        Output $w_\alpha = w^{k+1}$.
9:    **end if**
10: **end for**

---

size $n$. The algorithm in [28] consists of two routines, one is the Differentially Private Gradient Descent method and the other one is the procedure of selecting an $\alpha$-SOSP. The first one can be easily extend to a stochastic version, which is similar as the one in [37]. However, for the second one, it needs to calculate the whole Hessian matrix and verify some conditions as stopping criteria, but it is unknown whether we can extend it to a stochastic version. Compared with their algorithm, in Algorithm 1 we use the Hessian matrix for Trust-Region sub-problem and use the dual variable $\lambda^k$ as our stopping criteria. Thus, this is why we can extend Algorithm 1 to a stochastic version.

Note that in Algorithm 2 we use the basic subsampling technique for DP-STR to improve the time complexity. In [34], the authors proposed the Stochastic Variance Reduction Gradient method to improve the gradient complexity for DP-ERM with convex less functions and show it is superior to the DP-SGD method. Thus, it is unknown whether we can use the same idea to our problem to further improve the time complexity or gradient complexity. Moreover, in both of Algorithm 1 and 2, we assume that we can exactly solve the Trust-Region sub-problem (3). However, in most cases, exactly solving the problem is quite hard and costly. Thus whether we can relax this assumption is still an open problem. We leave these as further research.

# 5   Experiments

In this section, we present numerical experiments for different non-convex Empirical Risk Minimization problems on different datasets to demonstrate the advantage of our DP-TR and DP-STR algorithms in finding SOSP under differential privacy.

## 5.1   Experimental Settings

*Baselines* As mentioned in previous section, the only known method for this problem is DP-GD given in [28]. Thus, we compare it with our methods (DP-TR and DP-STR) after carefully tuning the algorithms for a fair comparison. For the QCQP sub-problem in Algorithm 1 and 2, we use the CVX package [19] to solve it.

*Datasets.* We evaluate the algorithms on real-world datasets with $n \gg p$. Specifically, we use the datasets, Covertype and IJCNN, which are commonly used in the study of DP-ERM such as [32–34]. More information about these datasets is listed in Table 1. We normalize each row of the datasets as preprocessing.

Table 1. Summary of datasets used in the experiments.

| Dataset | Sample size $n$ | Dimension $p$ |
|---------|-----------------|---------------|
| Covertype | 581,012 | 54 |
| IJCNN | 35,000 | 22 |

*Evaluated Problems.* For the loss functions we will follow the studies in [24,31,38]. The first non-convex problem that will be investigated is logistic regression with a non-convex regularizer $r(w) = \sum_{i=1}^{p} \frac{\lambda w_i^2}{1+w_i^2}$. Specifically, suppose that we are given training data $\{(x_i, y_i)\}_{i=1}^{n}$, where $x_i \in \mathbb{R}^p$ and $y \in \{-1, 1\}$ are, respectively, the feature vector and label of the $i$-th data record. The corresponding ERM is

$$\min_{w \in \mathbb{R}^p} \frac{1}{n} \sum_{i=1}^{n} \log(1 + \exp(-y_i \langle x_i, w \rangle)) + r(w).$$

In the experiment, we set $\lambda = 10^{-3}$.

The second problem that will be considered is the sigmoid regression with $\ell_2$ norm regularizer. Given training dataset $\{(x_i, y_i)\}_{i=1}^{n}$ where $x_i \in \mathbb{R}^p$ and $y \in \{-1, 1\}$ are, respectively, the feature vector and label of the $i$-th data record. Then, minimization problem is

$$\min_{w \in \mathbb{R}^p} \frac{1}{n} \sum_{i=1}^{n} \frac{1}{1 + \exp(-y_i \langle x_i, w \rangle)} + \frac{\lambda}{2} \|w\|_2^2.$$

In the experiment, we set $\lambda = 10^{-3}$.

*Measurements.* We first study how the optimality gap, *i.e.*, $L(w_\alpha, D) - \min_{w \in \mathbb{R}^p} L(w, D)$, changes w.r.t the privacy level $\epsilon$ or time (second). For the optimal solution of the problem $\min_{w \in \mathbb{R}^p} L(w, D)$, we obtain it through multiple runs of the classical trust region method and taking the best one. Besides the expected excess empirical risk, we also use the gradient norm, *i.e.*, $\|\nabla L(w_\alpha, D)\|_2$, to measure the utility. For logistic regression, we also consider its classification accuracy w.r.t privacy level, where the non-private case is obtained by running the trust region method and taking the best. For each experiment, we run 10 times and take the average as the final output. In all experiments, we set $\delta = \frac{1}{n}$ and $\alpha = 10^{-1}$.

## 5.2  Experimental Results

Figure 1 shows the classification accuracy of the private classifier given by the sigmoid regression on the Covertype and IJCNN datasets w.r.t different privacy levels. We can see that the accuracy increases when $\epsilon$ becomes larger, which means that the algorithm will be non-private. From Remark 16 we can see that this is due to the fact that when $\epsilon$ is larger, we can output an SOSP which is closer to the local minimum. Also, the accuracy of the non-private case is 86% and 95% for Covertype and IJCNN dataset, respectively. This indicates that the accuracy is comparable to the non-private case when $\epsilon \geq 1.5$.

The first and second subfigures of Fig. 2, 3, 4 and 5 depict the optimality gap and the gradient norm w.r.t different privacy level $\epsilon$ of the two non-convex problems on Covertype and IJCNN datasets. For Covertype, we set the batchsize as 50000, while for IJCNN we set it as 5000. From the figures, we can see that compared with DP-GD, our DP-TR method has better performance on both the optimality gap and the gradient norm. This is due to the fact that DP-TR has improved the bound of SOSP (see Remark 16). However, the results of DP-STR are worse than that of DP-GD and DP-TR. We attribute this to the fact that the noise level of DP-STR added in each iteration (steps 2 and 3) is higher than that of DP-TR and DP-GD. For example, in Step 2 of Algorithm 2 we add a Gaussian noise with variance $\sigma^2 = \frac{256L^2 \ln \frac{5T}{\delta} \ln \frac{2}{\delta}}{n^2 \epsilon^2}$ to each coordinate, while in step 3 of Algorithm 1 we only need to add a Gaussian noise with variance $\sigma^2 = \frac{4L^2 T}{n^2 \phi} \approx \frac{64L^2 T \log \frac{1}{\delta}}{n^2 \epsilon^2}$. Equivalently, the sub-optimality of DP-STR is due to the higher level of noise that needs to be added, which is required by the Advanced Composition Theorem to ensure $(\epsilon, \delta)$-DP. We leave it as an open problem to determine how to improve the practical performance of DP-STR.

The third subfigures of Fig. 2, 3, 4 and 5 show the results on the optimality gap w.r.t time of the two non-convex problems on the datasets of Covertype and IJCNN. Here we fix $\epsilon$ to be 1 in all the experiments. We can see that although the gap of DP-STR is worse than that of DP-GD and DP-TR, its running time is the least one. This is due to the fact that DP-STR needs only to evaluate a subset of the gradient and Hessian matrix, instead of the full ones as in DP-TR and DP-GD.

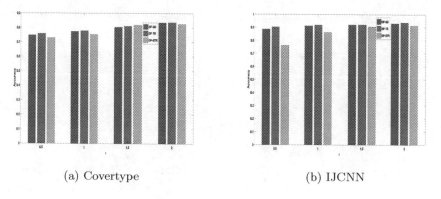

(a) Covertype                          (b) IJCNN

**Fig. 1.** Accuracy w.r.t privacy level on Covertype and IJCNN datasets

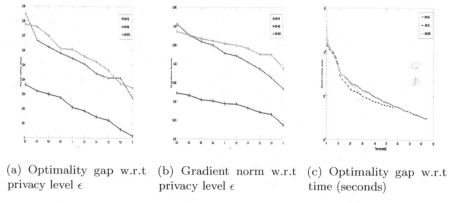

(a) Optimality gap w.r.t   (b) Gradient norm w.r.t   (c) Optimality gap w.r.t
privacy level $\epsilon$                privacy level $\epsilon$                time (seconds)

**Fig. 2.** Results of logistic regression with non-convex regularizer on Covertype dataset

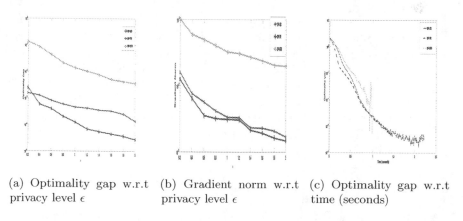

(a) Optimality gap w.r.t   (b) Gradient norm w.r.t   (c) Optimality gap w.r.t
privacy level $\epsilon$                privacy level $\epsilon$                time (seconds)

**Fig. 3.** Results of logistic regression with non-convex regularizer on IJCNN dataset

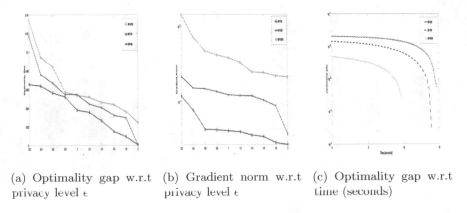

(a) Optimality gap w.r.t privacy level $\epsilon$    (b) Gradient norm w.r.t privacy level $\epsilon$    (c) Optimality gap w.r.t time (seconds)

**Fig. 4.** Results of sigmoid regression with $\ell_2$ norm regularizer on Covertype dataset

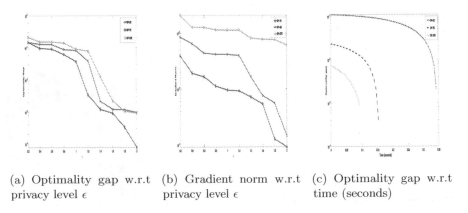

(a) Optimality gap w.r.t privacy level $\epsilon$    (b) Gradient norm w.r.t privacy level $\epsilon$    (c) Optimality gap w.r.t time (seconds)

**Fig. 5.** Results of sigmoid regression with $\ell_2$ norm regularizer on IJCNN dataset

## 6  Conclusion

In this paper we study the problem of escaping saddle points of empirical risk in the differential privacy model and propose a new method called DP-Trust Region (DP-TR) along with its stochastic version called DP-STR. Particularly, we show that to achieve an $\alpha$-SOSP with high probability, DP-TR and DP-STR have lower sample complexities compared with the existing algorithm DP-GD. We also show that DP-TR is faster than DP-GD; DP-STR is more scalable and much faster than DP-TR. Experimental results on benchmark datasets confirm our theoretical claims.

## References

1. Agarwal, N., Singh, K.: The price of differential privacy for online learning. In: Proceedings of the 34th International Conference on Machine Learning, ICML 2017, Sydney, NSW, Australia, 6–11 August 2017, pp. 32–40 (2017)

2. Anandkumar, A., Ge, R.: Efficient approaches for escaping higher order saddle points in non-convex optimization. In: Conference on Learning Theory, pp. 81–102 (2016)

3. Balcan, M.F., Dick, T., Vitercik, E.: Dispersion for data-driven algorithm design, online learning, and private optimization. In: 2018 IEEE 59th Annual Symposium on Foundations of Computer Science (FOCS), pp. 603–614. IEEE (2018)

4. Bassily, R., Smith, A., Thakurta, A.: Private empirical risk minimization: efficient algorithms and tight error bounds. In: 2014 IEEE 55th Annual Symposium on Foundations of Computer Science (FOCS), pp. 464–473. IEEE (2014)

5. Bhojanapalli, S., Neyshabur, B., Srebro, N.: Global optimality of local search for low rank matrix recovery. In: Advances in Neural Information Processing Systems, pp. 3873–3881 (2016)

6. Bun, M., Steinke, T.: Concentrated differential privacy: simplifications, extensions, and lower bounds. In: Hirt, M., Smith, A. (eds.) TCC 2016. LNCS, vol. 9985, pp. 635–658. Springer, Heidelberg (2016). https://doi.org/10.1007/978-3-662-53641-4_24

7. Chaudhuri, K., Monteleoni, C.: Privacy-preserving logistic regression. In: Advances in Neural Information Processing Systems, pp. 289–296 (2009)

8. Chaudhuri, K., Monteleoni, C., Sarwate, A.D.: Differentially private empirical risk minimization. J. Mach. Learn. Res. **12**, 1069–1109 (2011)

9. Conn, A.R., Gould, N.I., Toint, P.L.: Trust region methods. SIAM **1** (2000)

10. Dauphin, Y.N., Pascanu, R., Gulcehre, C., Cho, K., Ganguli, S., Bengio, Y.: Identifying and attacking the saddle point problem in high-dimensional non-convex optimization. In: Advances in Neural Information Processing Systems, pp. 2933–2941 (2014)

11. Dwork, C.: Differential privacy: a survey of results. In: Agrawal, M., Du, D., Duan, Z., Li, A. (eds.) TAMC 2008. LNCS, vol. 4978, pp. 1–19. Springer, Heidelberg (2008). https://doi.org/10.1007/978-3-540-79228-4_1

12. Dwork, C., McSherry, F., Nissim, K., Smith, A.: Calibrating noise to sensitivity in private data analysis. In: Halevi, S., Rabin, T. (eds.) TCC 2006. LNCS, vol. 3876, pp. 265–284. Springer, Heidelberg (2006). https://doi.org/10.1007/11681878_14

13. Dwork, C., Roth, A., et al.: The algorithmic foundations of differential privacy. Found. Trends® Theor. Comput. Sci. **9**(3–4), 211–407 (2014)

14. Erlingsson, Ú., Pihur, V., Korolova, A.: RAPPOR: randomized aggregatable privacy-preserving ordinal response. In: Proceedings of the 2014 ACM SIGSAC Conference on Computer and Communications Security, pp. 1054–1067. ACM (2014)

15. Ge, R., Huang, F., Jin, C., Yuan, Y.: Escaping from saddle points-online stochastic gradient for tensor decomposition. In: Conference on Learning Theory, pp. 797–842 (2015)

16. Ge, R., Lee, J.D., Ma, T.: Learning one-hidden-layer neural networks with landscape design. In: International Conference on Learning Representations (2018)

17. Goodfellow, I., Bengio, Y., Courville, A., Bengio, Y.: Deep Learning, vol. 1. MIT Press, Cambridge (2016)

18. Gould, N.I., Lucidi, S., Roma, M., Toint, P.L.: Solving the trust-region subproblem using the Lanczos method. SIAM J. Optim. **9**(2), 504–525 (1999)

19. Grant, M., Boyd, S.: CVX: Matlab software for disciplined convex programming, version 2.1, March 2014. http://cvxr.com/cvx

20. Huai, M., Wang, D., Miao, C., Xu, J., Zhang, A.: Pairwise learning with differential privacy guarantees. In: The Thirty-Fourth AAAI Conference on Artificial Intelligence, AAAI 2020, New York City, New York, USA, 7–12 February 2020 (2020)
21. Jain, P., Kothari, P., Thakurta, A.: Differentially private online learning. In: Conference on Learning Theory, pp. 24.1–24.34 (2012)
22. Kasiviswanathan, S.P., Jin, H.: Efficient private empirical risk minimization for high-dimensional learning. In: International Conference on Machine Learning, pp. 488–497 (2016)
23. Kawaguchi, K.: Deep learning without poor local minima. In: Advances in Neural Information Processing Systems, pp. 586–594 (2016)
24. Kohler, J.M., Lucchi, A.: Sub-sampled cubic regularization for non-convex optimization. In: Proceedings of the 34th International Conference on Machine Learning-Volume 70, pp. 1895–1904. JMLR. org (2017)
25. Mei, S., Bai, Y., Montanari, A., et al.: The landscape of empirical risk for nonconvex losses. Ann. Stat. **46**(6A), 2747–2774 (2018)
26. Talwar, K., Thakurta, A.G., Zhang, L.: Nearly optimal private LASSO. In: Advances in Neural Information Processing Systems, pp. 3025–3033 (2015)
27. Thakurta, A.G., Smith, A.: (nearly) optimal algorithms for private online learning in full-information and bandit settings. In: Advances in Neural Information Processing Systems, pp. 2733–2741 (2013)
28. Wang, D., Chen, C., Xu, J.: Differentially private empirical risk minimization with non-convex loss functions. In: International Conference on Machine Learning, pp. 6526–6535 (2019)
29. Wang, D., Gaboardi, M., Xu, J.: Empirical risk minimization in non-interactive local differential privacy revisited (2018)
30. Wang, D., Smith, A., Xu, J.: Noninteractive locally private learning of linear models via polynomial approximations. In: Algorithmic Learning Theory, pp. 897–902 (2019)
31. Wang, D., Xu, J.: Differentially private empirical risk minimization with smooth non-convex loss functions: a non-stationary view (2019)
32. Wang, D., Xu, J.: Differentially private empirical risk minimization with smooth non-convex loss functions: a non-stationary view. In: Proceedings of the AAAI Conference on Artificial Intelligence, vol. 33, pp. 1182–1189 (2019)
33. Wang, D., Xu, J.: On sparse linear regression in the local differential privacy model. In: International Conference on Machine Learning, pp. 6628–6637 (2019)
34. Wang, D., Ye, M., Xu, J.: Differentially private empirical risk minimization revisited: faster and more general. In: Advances in Neural Information Processing Systems, pp. 2722–2731 (2017)
35. Wang, D., Zhang, H., Gaboardi, M., Xu, J.: Estimating smooth GLM in non-interactive local differential privacy model with public unlabeled data. arXiv preprint arXiv:1910.00482 (2019)
36. Wang, Y.X., Lei, J., Fienberg, S.E.: Learning with differential privacy: stability, learnability and the sufficiency and necessity of ERM principle. J. Mach. Learn. Res. **17**(183), 1–40 (2016)
37. Zhang, J., Zheng, K., Mou, W., Wang, L.: Efficient private ERM for smooth objectives. In: Proceedings of the 26th International Joint Conference on Artificial Intelligence, pp. 3922–3928. AAAI Press (2017)
38. Zhou, D., Xu, P., Gu, Q.: Stochastic variance-reduced cubic regularized Newton method. In: International Conference on Machine Learning, pp. 5985–5994 (2018)

# Boosting and Ensemble Methods

# To Ensemble or Not Ensemble: When Does End-to-End Training Fail?

Andrew Webb[1], Charles Reynolds[1], Wenlin Chen[1], Henry Reeve[2],
Dan Iliescu[3], Mikel Luján[1], and Gavin Brown[1(✉)]

[1] University of Manchester, Manchester, UK
Gavin.Brown@manchester.ac.uk
[2] University of Bristol, Bristol, UK
[3] University of Cambridge, Cambridge, UK

**Abstract.** End-to-End training (E2E) is becoming more and more popular to train complex Deep Network architectures. An interesting question is whether this trend will continue—are there any clear failure cases for E2E training? We study this question in depth, for the specific case of E2E training an *ensemble* of networks. Our strategy is to blend the gradient smoothly in between two extremes: from independent training of the networks, up to to full E2E training. We find clear failure cases, where overparameterized models *cannot be trained E2E*. A surprising result is that the optimum can sometimes lie in between the two, neither an ensemble or an E2E system. The work also uncovers links to Dropout, and raises questions around the nature of ensemble diversity and multi-branch networks.

## 1 Introduction

Ensembles of neural networks are a common sight in the Deep Learning literature, often at the top of Kaggle leaderboards, and a key ingredient of now classic results, e.g. AlexNet (Krizhevsky et al. 2012). In recent literature, an equally common sight is *End-to-End* (E2E) training of a deep learning architecture, using a single loss to train all components simultaneously. An interesting scientific question is whether the E2E paradigm has limits, examined in some detail by Glasmachers (2017), who concluded that E2E can be inefficient, and *"does not make optimal use of the modular design of present neural networks"*.

In line with this question, some have explored training an *ensemble* with E2E, as if it was a modular, multi-branch architecture. Dutt et al. (2020) explore E2E ensemble training and demonstrate it allows one to *"significantly reduce the number of parameters"* while maintaining accuracy. Anastasopoulos and Chiang (2018) use the principle to improve accuracy in multi-source language translation. Furlanello et al. (2018) show good performance across a variety of standard

**Electronic supplementary material** The online version of this chapter (https://doi.org/10.1007/978-3-030-67664-3_7) contains supplementary material, which is available to authorized users.

F. Hutter et al. (Eds.): ECML PKDD 2020, LNAI 12459, pp. 109–123, 2021.
https://doi.org/10.1007/978-3-030-67664-3_7

benchmarks. This may seem promising, however, Lee et al. (2015) conclude the opposite, that such End-to-End training of an ensemble is *harmful* to generalization accuracy. The idea raises interesting scientific questions about learning in ensemble/modular architectures. Some authors have noted that some architectures (e.g. with multiple branches, or skip connections) may be viewed as an ensemble of jointly trained sub-networks, e.g. ResNets (Veit et al. 2016; Zhao et al. 2016). It is widely accepted that "diversity" of *independently* trained ensembles is beneficial (Dietterich 2000; Kuncheva and Whitaker 2003). and hard evidence lies in the success of the many published variants of Random Forests and Bagging (Breiman 1996). But if we train a set of networks *End-to-End*, they share a loss function—so their parameters are strongly correlated—seemingly the opposite of diversity. There is clearly a subtle relation here, raising hard questions, e.g., what is the meaning of 'diversity' in E2E ensembles?

Our aim is to understand when each scheme—*Independent* ensemble training, or *End-to-End* ensemble training—is appropriate, and to shed light on previously reported results (Fig. 1).

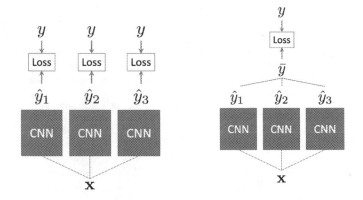

**Fig. 1.** Computational graphs representing the training of a CNN ensemble: Independent (Left) versus End-to-End training (Right). We study the dynamics of learning when interpolating smoothly in-between these two extremes.

Our strategy is to interpolate the gradient smoothly between the two—a convex combination of Independent and End-to-End training. This uncovers complex dynamics, highlighting a tension between individual model capacity and diversity, as well as generating interesting properties, including ensemble robustness with respect to faults.

**Organisation of the Paper**

In Sect. 2 we cover the necessary notation and probabilistic view we adopt for the paper. In Sect. 3 we ask and expand upon the reasoning behind the primary question of the paper: *is it an ensemble of models, or one big model?*. This leads us to define a hybrid training scheme, which turns out to generalise some previous work in ensemble diversity training schemes. In Sects. 4 and 5 we thoroughly investigate the question, uncovering a rich source of unexpectedly complex dynamics. Finally in Sects. 6 and 7 we outline how this connects to a wide range of work already published, and opens new doors for research.

## 2 Background and Design Choices

We assume a standard supervised learning scenario, with training set $S = \{(\mathbf{x}_i, y_i)\}_{i=1}^n$ drawn i.i.d. from $P(\mathcal{XY})$. For each observation $\mathbf{x}$, we assume the corresponding $y$ is a class label drawn from a true distribution $p(y|\mathbf{x})$. We approximate this distribution with a model $q(y|\mathbf{x})$, corresponding to some deep network architecture. We will explore a variety of architectures: including simple MLPs, medium/large scale CNNs, and finally DenseNets (Huang et al. 2018).

To have an ensemble, there is a choice to make in how to combine the network outputs. In our work we choose to average the network logits, and re-normalise with softmax. This was used by Hinton et al. (2015), and Dutt et al. (2020)— we replicate one of their experiments later. There are of course alternatives, e.g. majority voting or averaging probability estimates. Majority voting rules out the possibility of E2E ensemble training by gradient methods, since the vote operation is inherently non-differentiable. Averaging probability estimates is common, but we present a result below that encouraged us to consider the averaging logits approach instead, beyond simply replicating Dutt et al. (2020).

We approach the classification problem from a probabilistic viewpoint: minimising the KL-divergence from a model $q$ to the target distribution $p$. Given a set of such probability models, we could ask, what is the optimal combiner rule that preserves the contribution from each, in the sense of minimising KL divergence? In a formal statement we refer to this as the 'central' model, denoted $\bar{q}$, that lies at the centre of the set of probability estimates:

$$\bar{q} = \arg\min_z \left[ \frac{1}{M} \sum_{i=1}^M D(z \,||\, q_i) \right] = \arg\min_z \left[ \frac{1}{M} \sum_{i=1}^M \int z(y|\mathbf{x}) \log \frac{z(y|\mathbf{x})}{q_i(y|\mathbf{x})} \, dy \right] \quad (1)$$

It can easily be proved that the minimizer here is the normalized geometric mean, corresponding to a Product of Experts model, though modeling only the means (i.e. a PoE of Generalized Linear Models):

$$\bar{q}(y \mid \mathbf{x}) \overset{\text{def}}{=} Z^{-1} \prod_{j=1}^M q_i(y \mid \mathbf{x})^{1/M}. \quad (2)$$

This can be written in terms of the distribution's canonical link $f$ and its inverse $f^{-1}$. The inverse link for the Categorical distribution is the softmax $f^{-1}(\boldsymbol{\eta})$, where $\boldsymbol{\eta}$ is a vector of logits. Correspondingly, the logits are given by the link applied to class probabilities $\boldsymbol{\eta} = f(q(y|\mathbf{x}))$.

$$\bar{q}(y|\mathbf{x}) = f^{-1}\left( \frac{1}{M} \sum_i f(q_i(y|\mathbf{x})) \right) \quad (3)$$

i.e. a softmax operation on the averaged logits—this provides additional motivation for the ensemble combination rule used by Hinton et al. (2015) and Dutt et al. (2020), since it is the rule that preserves the most information from the individual models, in the sense of KL divergence.

## 3    Is it an Ensemble, or One Big Model?

Ensemble methods are a well-established part of the ML landscape. A traditional explanation for the success of ensembles is the 'diversity' of their predictions, and particularly their errors. When we reach the limit of what a single model can do, we create an ensemble of such models that exhibit 'diverse' errors, and combine their outputs. This *diversity* can lead to the individual errors being averaged out, and overall lower ensemble error obtained.

A common heuristic is to have the individual models intentionally lower capacity than they might be, and compensate via diversity. Traditional ensemble methods, such as Bagging and Random Forests, generate diversity via randomization heuristics such as feeding different bootstraps of training data to each model (Brown et al. 2005a). *Stacking* (Wolpert 1992) is similar in spirit to E2E ensemble training, in that it trains the combiner function, although only after individual models are fixed, using them as meta-features.

However, if the ensemble combining procedure was fully *differentiable*, we could in theory train all networks End-to-End (E2E), as if they were branches or components of one "big model". In this case, what role is there for *diversity*? Furthermore, with modern deep networks, it is easy to have extremely high capacity models, but regularised to avoid overfitting. With this in mind, is there much benefit to ensembling deep networks? Various empirical successes, e.g. Krizhevsky et al. (2012), suggest a tentative "yes", but understanding the overfitting behaviour of deep network architectures is one of the most complex open challenges in the field today. It is now common to heavily *over-parameterize* and regularize deep networks. These observations raise interesting questions on the benefits of such architectures, and for ensemble diversity.

To study the question of failure cases for E2E ensembles, we could simply train a system E2E and report outcomes. However, to get a more detailed picture, we define a hybrid loss function, *interpolating between* the likelihood for an independent ensemble and the E2E ensemble likelihood. We refer to this as the *Joint Training* loss, since it treats the networks jointly as components of a single (larger) network, or as members of an ensemble. We stress that we are not advocating this as a means to achieve SOTA results, but merely as a forensic tool to understand the behaviour of the E2E paradigm.

**Definition 1** (*Joint Training Loss*):

$$L_\lambda \stackrel{\text{def}}{=} \lambda \mathrm{D}(p \, \| \, \bar{q}) + (1 - \lambda)\frac{1}{M} \sum_{j=1}^{M} \mathrm{D}(p \, \| \, q_j) \,, \tag{4}$$

where $D$ is the KL-divergence as defined before. The loss is a convex combination of two extremes: $\lambda = 1$, where we train $\bar{q}$ as one system, and $\lambda = 0$, where we train the ensemble independently (with a learning rate scaled by $1/M$). When $\lambda$ lies between the two, it could be seen as $\bar{q}$ being 'regularized' by the individual networks partially fitting the data themselves. Alternatively, we can view the same architecture as an *ensemble*, but trained *interactively*. This is made clear by rearranging Eq. (4) as:

$$L_\lambda = \frac{1}{M} \sum_{j=1}^{M} D(p \,\|\, q_j) - \frac{\lambda}{M} \sum_{j=1}^{M} D(\bar{q} \,\|\, q_j). \tag{5}$$

A simple proof is available in supplementary material. This alternative view shows that $\lambda$ controls a balance between the average of the individual losses, and a term measuring the *diversity between the ensemble members*.

From this perspective, this is a diversity-forcing training scheme similar to previous work Liu and Yao (1999). In fact for the special case of a Gaussian $p, q$, Eq. (5) is *exactly* that presented by Liu and Yao (1999). However, the probabilistic view allows us to generalise, and for the case of classification problems, this can be seen as *managing the diversity in a classification ensemble*. However, seen as a 'hybrid' loss, (4), explicitly varies the gradient in-between the two extremes of an ensemble and a single multi-branch architecture, thus including architectures studied previously (Lee et al. 2015; Dutt et al. 2020) as special cases. In the following section we use this to study the learning dynamics in terms of tensions between model capacity and diversity.

## 4 Experiments

End-to-End training of ensembles raises several interesting questions: How does End-to-End training behave when varying the capacity/size of networks? Should we have a large number of simple networks? Or the opposite—a small number of complex networks? What effect does varying $\lambda$ have in this setting, smoothly varying from Independent to End-to-End training? Suppose we have a high capacity, well-tuned *single* network, performing close to SOTA. In this situation, presumably, independent training of an ensemble of such models will still *marginally* improve over a single small model, due to variance reduction. However, it is much less clear in this case whether there will be further gains as $\lambda \to 1$, toward End-to-End training. We investigate this first with simple MLPs, exploring trade-offs while the number of learnt parameters is held constant, and then with higher capacity CNNs, including DenseNets (Huang et al. 2017).

*Spending a Fixed Parameter Budget.* We first compare a single large network to an ensemble of very small networks: each with the *same* number of parameters. The study of such architectures may have implications for IoT/Edge compute scenarios, with memory/power constraints. We set a budget of memory or number of parameters, and ask how best to "spend" them in different architectures.

This type of trade-off is well recognised as highly relevant in the current energy-conscious research climate, e.g. very recently Zhu et al. (2019). We stress again however that we are not chasing state of the art performance, and instead observe the dynamics of Independent vs E2E training in a controlled manner.

We use single layer MLPs in four configurations with ~815K parameters: a single network with 1024 nodes (1 × 1024H), 16 networks each with 64 nodes (16×64H), 64 networks with 16 nodes (64×16H), and 256 networks with 4 nodes (256 × 4H). We evaluate these on the 10-class classification problem, Fashion-MNIST. Full experimental details are in the supplementary material.

Figure 2 summarizes results for the 1 × 1024H network versus the 256 × 4H ensemble. Again, these have the same number of configurable parameters, just deployed in a different manner. The 'monolithic' 1024 node network (horizontal dashed line, 10.2% error) significantly outperforms the ensemble trained independently (16.3%), as well as the 'classic' methods of Bagging and Stacking. However, *End-to-End training* of the small network ensemble almost matches performance, and Joint Training ($\lambda = 0.95$) gets closer with 10.3% error.

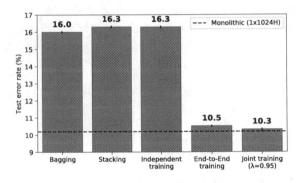

**Fig. 2.** Large single network vs. small net ensemble. Joint Training the ensemble ($\lambda = 0.95$) comes closest to match the large network accuracy, while classic ensemble methods significantly underperform.

Figure 3 shows detailed results, varying $\lambda$ on other configurations. The distinction between End-to-End training and independent training is most pronounced for the large ensemble of small networks, 256 × 4H, where *E2E training is sub-optimal.*

However, as the capacity of the networks increases, the difference between End-to-End and independent training vanishes—illustrated by the almost uniform response to varying $\lambda$ with larger networks. This is investigated further in the next section with state-of-the-art *DenseNet* networks (Huang et al. 2017) with millions of parameters.

*High Capacity Individual Models.* In deep learning, a recent effective strategy is to massively overparameterize a model and then rely on the implicit regularizing properties of stochastic gradient descent. We ask, is there anything to be

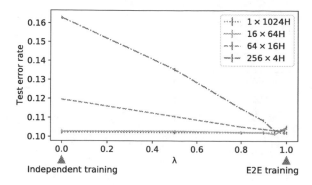

**Fig. 3.** Error rates for MLPs with 815K parameters, on Fashion-MNIST. Joint Training with higher $\lambda$ values improves the performance of small network ensembles, matching the performance of a single large network.

gained from End-to-End training of such large models? What does the concept of diversity even mean when the models have almost zero training error? We address this by examining ensembles of DenseNets (Huang et al. 2017), which achieve close to SOTA at the time of writing. We train a variety of DenseNet ensembles on the CIFAR-100 dataset, with the size of the ensemble increasing as the complexity of each ensemble member decreases, such that each configuration occupies approximately 12 GB of GPU RAM. These configurations are described in Table 1. Full experimental details can be found in the supplementary material.

**Table 1.** DenseNet ensembles. A proxy measure of capacity is $(d, k)$, the depth $d$ and growth rate $k$, which we decrease as the size $M$ of the ensemble increases. Each architecture occupies approximately 12 GB RAM.

| Name | Depth | $k$ | $M$ | Memory | Params. |
|---|---|---|---|---|---|
| DN-High | 100 | 12 | 4 | ~12 GB | 3.2M |
| DN-Mid | 82 | 8 | 8 | ~12 GB | 2.1M |
| DN-Low | 64 | 6 | 16 | ~12 GB | 1.7M |

Table 2 shows results for independent, End-to-End, and Joint training, vs. Bagging/Stacking. Each row contains results for a DenseNet ensemble, with minimum error rate in bold. Results for DN-High replicate the results of Dutt et al. (2020, Table 2).

In the previous section, we saw End-to-End training outperforming independent training for large ensembles of very small models. Here, *we find the opposite is true for small ensembles of large, SOTA models.* In every DenseNet configuration, *End-to-End training (i.e. $\lambda = 1$) is sub-optimal.* In all but DN-Low, which is the configuration with the largest ensemble of smallest models, independent training ($\lambda = 0$) achieves the lowest test error. **These results indicate there**

**Table 2.** Error rates (%) for DenseNet ensembles. Independent training is optimal (bold) for all but the smallest capacity networks.

|  | DN-Low | DN-Mid | DN-High |
|---|---|---|---|
| $\lambda = 0.0$ (Independent) | 25.0 | **19.9** | **17.8** |
| $\lambda = 0.5$ | **23.5** | 20.0 | 18.8 |
| $\lambda = 0.9$ | 25.9 | 22.4 | 20.3 |
| $\lambda = 1.0$ (End-to-End) | 29.6 | 25.7 | 22.1 |
| Bagging | 32.5 | 29.5 | 28.4 |
| Stacking | 28.3 | 23.4 | 21.4 |

**is little to no benefit in test error when E2E ensemble training SOTA deep neural networks**. However, the result for DN-Low suggests that there may be a relatively smooth transition, somewhere between E2E ensembles and independent training, as the complexity of the ensemble members increases.

*Intermediate Capacity.* In this section we further explore the relationship between ensemble member complexity and E2E ensemble training. We train ensembles of convolutional networks on CIFAR-100. Each ensemble has 16 members of varying complexity, from 70,000 parameters up to 1.2 million. Full details can be found in the supplementary material. Note that these experiments are not intended to be competitive with the SOTA, but to illustrate relative benefits of E2E ensembles as the complexity of individual members varies.

Figure 4 shows the test error rate and standard errors for each $\lambda$ value for each configuration. The 'dip' in each of the four lines in the figure is the optimal $\lambda$. We find that the optimal point smoothly decreases: E2E ensembles are the preferred option with simpler networks, but this changes as the networks become more complex. At 1.2M parameters, the optimal point lies clearly *between* the two extremes. This supports the trend observed in the previous sections, that the benefits of E2E training as a scheme depend critically on the complexity of individual networks.

*When Does End-to-End Ensemble Training Fail?* These results suggest a trend: ensembles of low capacity models benefit from E2E training, and high capacity models perform better if trained independently. We suggest a possible explanation for this trend, and a diagnostic for determining during training whether E2E training will perform well. We find that ensembles of high capacity models, when trained E2E, exhibit a 'model dominance' effect, shown in Fig. 6. There is a sharp transition in behaviour at $\lambda = 1$ (E2E), whereby a single ensemble member individually performs well—in both training and test error—while all other members perform poorly. This effect is not observed in ensembles of low capacity models.

We suggest that this dominance effect can explain the poor performance of E2E training seen in Table 2. We also note that model dominance occurs in the

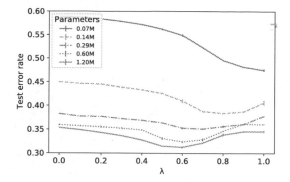

**Fig. 4.** Ensemble test error against $\lambda$ for ensembles of convolutional networks of varying complexity on CIFAR-100. The optimal $\lambda$ value shifts away from E2E ensemble training with greater individual network capacity.

E2E training experiments of Dutt et al. (2020, Table 2) (called 'coupled training (FC)' by the authors); the authors report the average and standard deviation (within a single trial) of the ensemble member error rates. In the case with 2 ensemble members, it can be inferred that one ensemble member achieves a much lower error rate than the other.

In our own experiments, the dominance effect manifests early in training (Fig. 6). In each trial, the ensemble member with the lowest error rate by the end of epoch 3 dominates by the end of training; model dominance can be used as an early diagnosis during training that the ensemble members are overcapacity for E2E training. Note that model dominance can occur in classification because the ensemble prediction is a normalized geometric mean; a network closely fitting the data can arbitrarily reduce the ensemble error—the well-known 'veto' effect in Products of Experts (Welling 2007)—and the parameters of other networks can be prevented from moving far from their initial values. This 'stagnation' of other ensemble members can be seen in Fig. 5, which shows some of the filters in the first layer of a ConvNet trained independently and E2E, showing a small number of strong filters in the latter case.

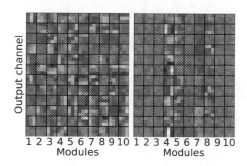

**Fig. 5.** Model dominance in filters in the first layer of 10 CNN modules. Trained independently (left) and E2E (right). In E2E training, only network 4 has strong filters.

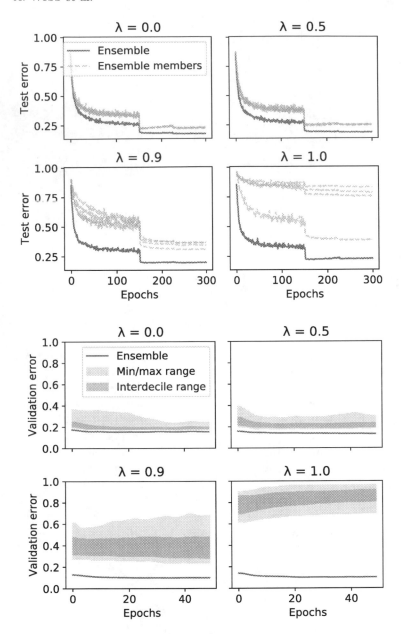

**Fig. 6.** Ensemble vs Member error: Over-capacity models (DenseNets, top sub-figure) experience a 'model dominance' effect when trained E2E, whereas there is no such effect with $256 \times 4H$ undercapacity models (bottom sub-figure).

# 5   What Happens *Between* E2E and Independent Training?

The previous section has demonstrated clearly that E2E training of an ensemble can be sub-optimal. More interestingly perhaps, the complex behaviour seems to lie between independent and E2E. This section goes beyond just ensemble test error rate, and looks into properties of the networks *inside* the ensemble.

*Robustness.* We find that Joint Training with $\lambda$ very slightly less than 1 can greatly increase the *robustness* of an ensemble, in the sense that dropping the response from a subset of the networks at test time does not significantly harm the accuracy. This *partial evaluation*, performing inference for only a subset of the networks, may be useful in resource limited scenarios, e.g. limited power budgets in Edge/IoT devices.

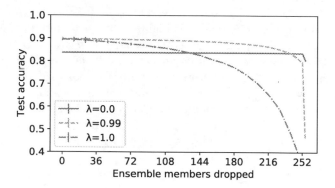

**Fig. 7.** Independent training ($\lambda = 0$) gives robust ensembles; however, many members are redundant. End-to-End training ($\lambda = 1$) gives brittle ensembles. Joint Training with $\lambda \approx 1$ (but strictly less than 1) retains the accuracy whilst adding robustness.

Figure 7 analyses the $256 \times 4$H ensemble, dropping a random subset of networks for each test example in Fashion-MNIST, averaged over 20 repeats. Independent training ($\lambda = 0$) builds a highly robust ensemble, but which could also be seen to be redundant—adding more networks does not increase accuracy. End-to-End training ($\lambda = 1$) achieves higher test accuracy, but the ensemble is not at all robust. Joint Training with $\lambda = 0.99$ achieves the same accuracy as End-to-End with all members, but can maintain accuracy even if more than 50% of the networks are dropped. A similar behaviour is observed with the $64 \times 16$H architecture, in Fig. 8.

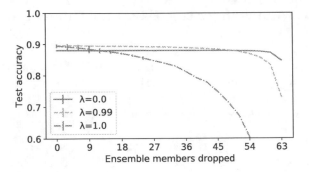

**Fig. 8.** Robustness with the 64 × 16H architecture. The same robustness appears, but relative benefit is less, as the independently trained ensemble can perform better.

***Robust Ensembles: Why Does this Occur?*** When training independently, the networks have no "idea" of each others' existence—so solve the problem as individuals—and hence the ensemble is highly robust to removal of networks.

When we *End-to-End* train the *exact same architecture*, all networks are effectively "slave" components in a larger network. They cannot directly target their own losses, so individual errors would not be expected to be very low. However, as a system, they work together, and achieve low ensemble error. The problem is that the networks *rely* on the others too much—so when removing networks, the performance degrades rapidly. This is illustrated in Fig. 9—we see individual network loss rapidly increasing as we approach End-to-End training (i.e. $\lambda = 1$).

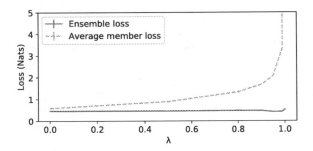

**Fig. 9.** 64 × 16H: ensemble vs. average loss. The average grows sharply as $\lambda \to 1$, a symptom of the individuals relying on others to correct mistakes.

The "sweet spot" is a small amount of *joint training* with $\lambda = 0.99$, which ensure the networks cannot rely on each other completely, but are still working together. This is highly reminiscent of *Dropout*, where one of Hinton's stated motivations was to make each *"hidden unit more robust and drive it towards creating useful features on its own without relying on other hidden units to cor-*

*rect its mistakes."* (Srivastava et al. 2014, Section 2). This similarity is not a coincidence, and is discussed in the next section.

## 6   Discussion and Future Work

The primary question for this paper was *"when does E2E ensemble training fail?"*, with the intention to explain conflicting results in the literature, e.g. Lee et al. (2015); Dutt et al. (2020). The answer turns out to be intimately tied with the over-parameterization (or not) of individual networks in the ensemble. Dutt et al. (2020); Lee et al. (2015) concluded that End-to-End training of ensembles leads to poor performance—however, only very large state-of-the-art models were studied. Our results agree with theirs *exactly*, in situations where ensemble members are over capacity—Table 2 reproduces the results of Dutt et al. (2020, Table 2)—but, the advantage of E2E emerges with lower capacity individuals, where there are also interesting and potentially useful dynamics in-between Independent/E2E training.

**Relations to Prior Work.** The investigation turned up links to literature going back over 20 years, in terms of ensemble diversity (Heskes 1998). As mentioned, we presented results primarily on classification problems—however, the framework presented in Sect. 2 applies generally for targets following any exponential family distribution. For the special case of a Gaussian, and when we re-arrange the loss as (5), the gradient is *exactly* equivalent to a previously proposed method, Negative Correlation Learning (NCL) (Liu and Yao 1999; Brown et al. 2005b). Thus, an alternative view of the loss is as a generalization of NCL to arbitrary exponential family distributions, where the Categorical distribution assumption here can be seen as *managing diversity in classification ensembles*, echoing the title of Brown et al. (2005b).

**The Dropout Connection?** We observed a qualitative similarity between ensemble robustness, and the motivations behind Dropout (Srivastava et al. 2014). It is interesting to note that Negative Correlation Learning Liu and Yao (1999) can be proven equivalent to a stochastic dropping of ensemble members during training (Reeve et al. 2018), i.e. Dropout at the network level. With some investigation we have determined that their result depends critically on the Gaussian assumption and hence holds only for NCL, not more generally for Joint Training loss with any exponential family. Despite this, the observations in Sect. 5 do suggest a connection, worthy of future work.

**Should We Always E2E Train Multi-branch Networks?** The End-to-End training methodology treats the ensemble as if it were a single multi-branch architecture. A similar view could be taken of *any* multi-branch architecture, e.g. ResNeXt (Xie et al. 2017) that it is either an ensemble of branches, or a single system. We have found that if ensemble members are over capacity, then one network can 'dominate' an ensemble, leading to poor performance overall. This raises the obvious open question, of whether it is also true for sub-branches in general branched architectures like ResNeXt.

**Ill-Conditioning?** As $\lambda \to 1$, the condition number of the Hessian for the JT Loss tends to infinity, as we show in the supplementary material. Therefore there may be cases where E2E training with first-order optimization methods (e.g. simple SGD) performs poorly, and where second-order methods behave markedly differently. We leave this question for future study.

## 7    Conclusions

We have presented a detailed analysis of the question *"When does End-to-End Ensemble training fail?"*. This is in response to the recent trend of end-to-end training being used more and more in deep learning, and specifically for an ensemble (Lee et al. 2015; Furlanello et al. 2018; Dutt et al. 2020). Our strategy for this was to study a convex combination of the likelihood-based losses for independent and E2E training, blending the gradient slowly from one to the other.

Our answer to the question is that End-to-End training tends to underperform when member networks are significantly over-parameterized, though it is possible to diagnose whether this will happen by examining the relative network training errors in the first few epochs. In this case we suggest alternative classical methods such as Bagging, or Independent training. Further, we conjecture that this may have implications for general multi-branch architectures—should we always train them E2E, or perhaps individually?

The space between independent and E2E turned out to be a rich source of quite unexpectedly complex dynamics, generating robust ensembles, with links to Dropout, general multi-branch deep learning, and early literature on ensemble diversity. We suggest the book is not yet closed and perhaps there is much more to learn in this space.

**Acknowledgements.** The authors gratefully acknowledge the support of the EPSRC for the LAMBDA project (EP/N035127/1).

## References

Anastasopoulos, A., Chiang, D.: Leveraging translations for speech transcription in low-resource settings. arXiv:1803.08991 (2018)

Breiman, L.: Bagging predictors. Mach. Learn. **24**(2), 123–140 (1996)

Brown, G., Wyatt, J., Harris, R., Yao, X.: Diversity creation methods: a survey and categorisation. Inf. Fusion **6**(1), 5–20 (2005a)

Brown, G., Wyatt, J.L., Tiňo, P.: Managing diversity in regression ensembles. J. Mach. Learn. Res. **6**, 1621–1650 (2005b). http://dl.acm.org/citation.cfm?id=1046920. 1194899

Dietterich, T.G.: Ensemble methods in machine learning. In: Kittler, J., Roli, F. (eds.) MCS 2000. LNCS, vol. 1857, pp. 1–15. Springer, Heidelberg (2000). https://doi.org/10.1007/3-540-45014-9_1

Dutt, A., Pellerin, D., Quénot, G.: Coupled ensembles of neural networks. Neurocomputing **396**, 346–357 (2020)

Furlanello, T., Lipton, Z.C., Tschannen, M., Itti, L., Anandkumar, A.: Born again neural networks (2018). arXiv:1805.04770

Glasmachers, T.: Limits of end-to-end learning (2017). arXiv preprint arXiv:1704.08305

Heskes, T.: Selecting weighting factors in logarithmic opinion pools. In: NIPS, pp. 266–272. The MIT Press (1998)

Hinton, G., Vinyals, O., Dean, J.: Distilling the knowledge in a neural network (2015). arXiv:1503.02531

Huang, G., Chen, D., Li, T., Wu, F., van der Maaten, L., Weinberger, K.: Multi-scale dense networks for resource efficient image classification. In: ICLR (2018). http://openreview.net/forum?id=Hk2aImxAb

Huang, G., Liu, Z., Maaten, L.v.d., Weinberger, K.Q.: Densely connected convolutional networks. In: CVPR, pp. 2261–2269 (2017)

Krizhevsky, A., Sutskever, I., Hinton, G.E.: ImageNet classification with deep convolutional neural networks. In: NIPS (2012)

Kuncheva, L.I., Whitaker, C.J.: Measures of diversity in classifier ensembles and their relationship with the ensemble accuracy. Mach. Learn. **51**(2), 181–207 (2003)

Lee, S., Purushwalkam, S., Cogswell, M., Crandall, D., Batra, D.: Why M heads are better than one: training a diverse ensemble of deep networks. arXiv preprint arXiv:1511.06314 (2015)

Liu, Y., Yao, X.: Ensemble learning via negative correlation. Neural Netw. **12**(10), 1399–1404 (1999)

Reeve, H.W.J., Mu, T., Brown, G.: Modular dimensionality reduction. In: Berlingerio, M., Bonchi, F., Gärtner, T., Hurley, N., Ifrim, G. (eds.) ECML PKDD 2018. LNCS (LNAI), vol. 11051, pp. 605–619. Springer, Cham (2019). https://doi.org/10.1007/978-3-030-10925-7_37

Srivastava, N., Hinton, G., Krizhevsky, A., Sutskever, I., Salakhutdinov, R.: Dropout: a simple way to prevent neural networks from overfitting. J. Mach. Learn. Res. 1929–1958 (2014). http://jmlr.org/papers/v15/srivastava14a.html

Veit, A., Wilber, M.J., Belongie, S.: Residual networks behave like ensembles of relatively shallow networks. In: Advances in Neural Information Processing Systems, pp. 550–558 (2016)

Welling, M.: Product of experts. Scholarpedia **2**(10), 3879 (2007). revision #137078

Wolpert, D.H.: Stacked generalization. Neural Netw. **5**(2), 241–259 (1992)

Xie, S., Girshick, R., Dollár, P., Tu, Z., He, K.: Aggregated residual transformations for deep neural networks. In: CVPR, pp. 5987–5995 (2017)

Zhao, L., Wang, J., Li, X., Tu, Z., Zeng, W.: On the connection of deep fusion to ensembling. Technical report MSR-TR-2016-1118 (2016)

Zhu, S., Dong, X., Su, H.: Binary ensemble neural network: more bits per network or more networks per bit? In: CVPR, pp. 4923–4932 (2019)

# Learning Gradient Boosted Multi-label Classification Rules

Michael Rapp[1](✉), Eneldo Loza Mencía[1], Johannes Fürnkranz[2],
Vu-Linh Nguyen[3], and Eyke Hüllermeier[3]

[1] Knowledge Engineering Group, TU Darmstadt, Darmstadt, Germany
mrapp@ke.tu-darmstadt.de, research@eneldo.net
[2] Computational Data Analysis Group, JKU Linz, Linz, Austria
juffi@faw.jku.at
[3] Heinz Nixdorf Institute, Paderborn University, Paderborn, Germany
vu.linh.nguyen@uni-paderborn.de, eyke@upb.de

**Abstract.** In multi-label classification, where the evaluation of predictions is less straightforward than in single-label classification, various meaningful, though different, loss functions have been proposed. Ideally, the learning algorithm should be customizable towards a specific choice of the performance measure. Modern implementations of boosting, most prominently gradient boosted decision trees, appear to be appealing from this point of view. However, they are mostly limited to single-label classification, and hence not amenable to multi-label losses unless these are label-wise decomposable. In this work, we develop a generalization of the gradient boosting framework to multi-output problems and propose an algorithm for learning multi-label classification rules that is able to minimize decomposable as well as non-decomposable loss functions. Using the well-known Hamming loss and subset 0/1 loss as representatives, we analyze the abilities and limitations of our approach on synthetic data and evaluate its predictive performance on multi-label benchmarks.

**Keywords:** Multi-label classification · Gradient boosting · Rule learning

## 1 Introduction

Multi-label classification (MLC) is concerned with the per-instance prediction of a subset of relevant labels out of a predefined set of available labels. Examples of MLC include real-world applications like the assignment of keywords to documents, the identification of objects in images, and many more (see, e.g., [23] or [24] for an overview). To evaluate the predictive performance of multi-label classifiers, one needs to compare the predicted label set to a ground-truth set of labels. As this can be done in many different ways, a large variety of loss functions have been proposed in the literature, many of which are commonly used to compare MLC in experimental studies. As these measures may conflict with each other, optimizing for one loss function often leads to deterioration

F. Hutter et al. (Eds.): ECML PKDD 2020, LNAI 12459, pp. 124–140, 2021.
https://doi.org/10.1007/978-3-030-67664-3_8

with respect to another loss. Consequently, an algorithm is usually not able to dominate its competitors on all measures [7].

For many MLC algorithms it is unclear what measure they optimize. On the other hand, there are also approaches that are specifically tailored to a certain loss function, such as the F1-measure [17], or the subset 0/1 loss [16]. *(Label-wise) decomposable* losses are particularly easy to minimize, because the prediction for individual labels can be optimized independently of each other [7]. For example, binary relevance (BR) learning transforms an MLC problem to a set of independent binary classification problems, one for each label. On the other hand, much of the research in MLC focuses on incorporating label dependencies into the prediction models, which is in general required for optimizing *non-decomposable* loss functions such as subset 0/1.

While BR is appropriate for optimizing the Hamming loss, another reduction technique, label powerset (LP), is a natural choice for optimizing the subset 0/1 loss [7]. The same applies to (probabilistic) classifier chains [4,18], which seek to model the joint distribution of labels. Among the approaches specifically designed for MLC problems, boosting algorithms that allow for minimizing multi-label losses are most relevant for this work. Most of these algorithms (e.g. [13,19,21,25] and variants thereof) require the loss function to be label-wise decomposable. This restriction also applies to methods that aim at minimizing ranking losses (e.g. [5,15], or [19]) or transform the problem space to capture relations between labels (e.g. [14] or [2]). To our knowledge, AdaBoost.LC [1] is the only attempt at directly minimizing non-decomposable loss functions.

The first contribution of this work is a framework that allows for optimizing multivariate loss functions (Sect. 3). It inherits the advantages of modern formulations of the gradient boosting framework, in particular the ability to flexibly use different loss functions and to incorporate regularization into the learning objective. The proposed framework allows the use of non-decomposable loss functions and enables the ensemble members to provide loss-minimizing predictions for several labels at the same time, hence taking label dependencies into account. This is in contrast to AdaBoost.LC, where the base classifiers may only predict for individual labels. Our experiments suggest that this ability is crucial to effectively minimize non-decomposable loss functions on real-world data sets.

Our second contribution is BOOMER, a concrete instantiation of this framework for learning multi-label classification rules (Sect. 4). While there are several rule-based boosting approaches for conventional single-label classification, e.g., the ENDER framework [6] which generalizes several rule-based boosting methods such as RuleFit [9], their use has not yet been systematically investigated for MLC. We believe that rules are a natural choice in our framework, because they define a more general concept class than the commonly used decision trees: While each tree can trivially be viewed as a set of rules, not every rule set may be encoded as a tree. Also, an ensemble of rules provides more flexibility in how the attribute space is covered. While in an ensemble of $T$ decision trees, each example is covered by exactly $T$ rules, an ensemble of rules can distribute the rules in a more flexible way, using more rules in regions where predictions are difficult and fewer rules in regions that are easy to predict.

## 2   Preliminaries

In this section, we introduce the notation used throughout the remainder of this work and present the type of models we use. Furthermore, we present relevant loss functions and recapitulate to what extent their optimization may benefit from the exploitation of label dependencies.

In contrast to binary and multi-class classification, in multi-label classification an example can be associated with several class labels $\lambda_k$ out of a predefined and finite label set $\mathcal{L} = \{\lambda_1, \dots, \lambda_K\}$. An example $\boldsymbol{x}$ is represented in attribute-value form, i.e., it consists of a vector $\boldsymbol{x} = (x_1, \dots, x_L) \in \mathcal{X} = A_1 \times \dots \times A_L$, where $x_l$ specifies the value associated with a numeric or nominal attribute $A_l$. In addition, each example is associated with a binary label vector $\boldsymbol{y} = (y_1, \dots, y_K) \in \mathcal{Y}$, where $y_k$ indicates the absence $(-1)$ or presence $(+1)$ of label $\lambda_k$. We denote the set of possible labelings by $\mathcal{Y} = \{-1, +1\}^K$.

We deal with MLC as a supervised learning problem in which the task is to learn a predictive model $f : \mathcal{X} \to \mathcal{Y}$ from a given set of labeled training examples $\mathcal{D} = \{(\boldsymbol{x}_1, \boldsymbol{y}_1), \dots, (\boldsymbol{x}_N, \boldsymbol{y}_N)\} \subset \mathcal{X} \times \mathcal{Y}$. A model of this kind maps a given example to a predicted label vector $f(\boldsymbol{x}) = (f_1(\boldsymbol{x}), \dots, f_K(\boldsymbol{x})) \in \mathcal{Y}$. It should generalize well beyond the given observations, i.e., it should yield predictions that minimize the expected risk with respect to a specific loss function. In the following, we denote the binary label vector that is predicted by a multi-label classifier as $\hat{\boldsymbol{y}} = (\hat{y}_1, \dots, \hat{y}_K) \in \mathcal{Y}$.

### 2.1   Ensembles of Additive Functions

We are concerned with ensembles $F = \{f_1, \dots, f_T\}$ that consist of $T$ additive classification functions $f_t \in \mathcal{F}$, referred to as *ensemble members*. By $\mathcal{F}$ we denote the set of potential classification functions. In this work, we focus on classification rules (cf. Sect. 2.2). Given an example $\boldsymbol{x}_n$, all of the ensemble members predict a vector of numerical confidence scores

$$\hat{\boldsymbol{p}}_n^t = f_t(\boldsymbol{x}_n) = (\hat{p}_{n1}^t, \dots, \hat{p}_{nK}^t) \in \mathbb{R}^K, \tag{1}$$

where each score expresses a preference for predicting the label $\lambda_k$ as absent if $\hat{p}_k < 0$ or as present if $\hat{p}_k > 0$. The scores provided by the individual members of an ensemble can be aggregated into a single vector of confidence scores by calculating the vector sum

$$\hat{\boldsymbol{p}}_n = F(\boldsymbol{x}_n) = \hat{\boldsymbol{p}}_n^1 + \dots + \hat{\boldsymbol{p}}_n^T \in \mathbb{R}^K, \tag{2}$$

which can subsequently be turned into the final prediction of the ensemble in the form of a binary label vector (cf. Sect. 4.3).

## 2.2    Multi-label Classification Rules

As ensemble members, we use conjunctive classification rules of the form

$$f : b \to \hat{p} \, ,$$

where $b$ is referred to as the *body* of the rule and $\hat{p}$ is called the *head*. The body $b : \mathcal{X} \to \{0,1\}$ consists of a conjunction of conditions, each being concerned with one of the attributes. It evaluates to 1 if a given example satisfies all of the conditions, in which case it is said to be *covered* by the rule, or to 0 if at least one condition is not met. An individual condition compares the value of the $l$-th attribute of an example to a constant by using a relational operator, such as $=$ and $\neq$ (if the attribute $A_l$ is nominal), or $\leq$ and $>$ (if $A_l$ is numerical).

In accordance with (1), the head of a rule $\hat{p} = (\hat{p}_1, \dots, \hat{p}_K) \in \mathbb{R}^K$ assigns a numerical score to each label. If a given example $x$ belongs to the axis-parallel region in the attribute space $\mathcal{X}$ that is covered by the rule, i.e., if it satisfies all conditions in the rule's body, the vector $\hat{p}$ is predicted. If the example is not covered, a null vector is predicted. Thus, a rule can be considered as a mathematical function $f : \mathcal{X} \to \mathbb{R}^K$ defined as

$$f\left(x\right) = b\left(x\right) \hat{p} \, . \tag{3}$$

This is similar to the notation used by Dembczyński et al. [6] in the context of single-label classification. However, in the multi-label setting, we consider the head as a vector, rather than a scalar, to enable rules to predict for several labels.

## 2.3    Multi-label Loss Functions

Various measures for evaluating the predictions provided by a multi-label classifier are commonly used in the literature (see, e.g., [23] for an overview). We focus on measures that assess the quality of predictions for $N$ examples and $K$ labels in terms of a single score $\mathcal{L}(Y, \hat{Y}) \in \mathbb{R}_+$, where $Y, \hat{Y} \in \{-1, +1\}^{N \times K}$ are matrices that represent the true labels according to the ground truth, respectively the predicted labels provided by a classifier.

**Selected Evaluation Measures.** The *Hamming loss* measures the fraction of incorrectly predicted labels among all labels and is defined as

$$\mathcal{L}_{\text{Hamm.}}\left(Y, \hat{Y}\right) := \frac{1}{NK} \sum_{n=1}^{N} \sum_{k=1}^{K} [\![ y_{nk} \neq \hat{y}_{nk} ]\!] \, , \tag{4}$$

where $[\![ P ]\!] = 1$ if the predicate $P$ is true $= 0$ otherwise.

In addition, we use the *subset 0/1 loss* to measure the fraction of examples for which at least one label is predicted incorrectly. Is it formally defined as

$$\mathcal{L}_{\text{subs.}}\left(Y, \hat{Y}\right) := \frac{1}{N} \sum_{n=1}^{N} [\![ y_n \neq \hat{y}_n ]\!] \, . \tag{5}$$

Both, the Hamming loss and the subset 0/1 loss, can be considered as generalizations of the 0/1 loss known from binary classification. Nevertheless, as will be discussed in the following Sect. 2.4, they have very different characteristics.

**Surrogate Loss Functions.** As seen in Sect. 2.1, the members of an ensemble predict vectors of numerical confidence scores, rather than binary label vectors. For this reason, discrete functions, such as the bipartition measures introduced above, are not suited to assess the quality of potential ensemble members during training. Instead, continuous loss functions that can be minimized in place of the actual target measure ought to be used as surrogates. For this purpose, we use multivariate (instead of univariate) loss functions $\ell : \{-1, +1\}^K \times \mathbb{R}^K \to \mathbb{R}_+$, which take two vectors $\boldsymbol{y}_n$ and $\hat{\boldsymbol{p}}_n$ as arguments. The former represents the true labeling of an example $\boldsymbol{x}_n$, whereas the latter corresponds to the predictions of the ensemble members according to (2).

As surrogates for the Hamming loss and the subset 0/1 loss, we use different variants of the *logistic loss*. This loss function, which is equivalent to *cross-entropy*, is the basis for logistic regression and is commonly used in boosting approaches to single-label classification (early uses go back to Friedman et al. [8]). To cater for the characteristics of the Hamming loss, the *label-wise logistic loss* applies the logistic loss function to each label individually:

$$\ell_{\text{l.w.-log}} (\boldsymbol{y}_n, \hat{\boldsymbol{p}}_n) := \sum_{k=1}^{K} \log \left(1 + \exp \left(-y_{nk}\hat{p}_{nk}\right)\right) . \tag{6}$$

Following the formulation of this objective, $\hat{p}_{nk}$ can be considered as log-odds, which estimates the probability of $y_{nk} = 1$ as logistic $(\hat{p}_{nk}) = \frac{1}{1+\exp(-\hat{p}_{nk})}$. Under the assumption of label independence, the logistic loss has been shown to be a consistent surrogate loss for the Hamming loss [5,11].

As the label-wise logistic loss can be calculated by aggregating the values that result from applying the loss function to each label individually, it is label-wise decomposable. In contrast, the *example-wise logistic loss*

$$\ell_{\text{ex.w.-log}} (\boldsymbol{y}_n, \hat{\boldsymbol{p}}_n) := \log \left(1 + \sum_{k=1}^{K} \exp \left(-y_{nk}\hat{p}_{nk}\right)\right) \tag{7}$$

is non-decomposable, as it cannot be computed via label-wise aggregation. This smooth and convex surrogate is proposed by Amit et al. [1], who show that it provides an upper bound of the subset 0/1 loss.

## 2.4    Label Dependence and (Non-)Decomposability

The idea of modeling correlations between labels to improve the predictive performance of multi-label classifiers has been a driving force for research in MLC for many years. However, Dembczyński et al. [7] brought up strong theoretical and empirical arguments that the type of loss function to be minimized, as well as the type of dependencies that occur in the data, strongly influence to what extent the exploitation of label dependencies may result in an improvement. The authors distinguish between two types of dependencies, namely *marginal (unconditional)* and *conditional dependence*. While the former refers to a lack of (stochastic) independence properties of the joint probability distribution $p(\boldsymbol{y}) = p(y_1, \ldots, y_K)$ on labelings $\boldsymbol{y}$, the latter concerns the conditional probabilities $p(\boldsymbol{y} \mid \boldsymbol{x})$, i.e., the distribution of labelings conditioned on an instance $\boldsymbol{x}$.

According to the notion given in Sect. 2.2, the rules we aim to learn contribute to the final prediction of labels for which they predict a non-zero confidence score and abstain otherwise. The head of a *multi-label rule* contains multiple non-zero scores, which enables one to express conditional dependencies between the corresponding labels, where the rule's body is tailored to cover a region of the attribute space where these dependencies hold. In contrast, *single label rules* are tailored to exactly one label and ignore the others, for which reason they are unable to explicitly express conditional dependencies.

As discussed in Sect. 2.3, we are interested in the Hamming loss and the subset 0/1 loss, which are representatives for decomposable and non-decomposable loss functions, respectively. In the case of decomposability, modeling dependencies between the labels cannot be expected to drastically improve predictive performance [7]. For this reason, we expect that single-label rules suffice for minimizing Hamming loss on most data sets. In contrast, given that the labels in a data set are not conditionally independent, the ability to model dependencies is required to effectively minimize non-decomposable losses. Hence, we expect that the ability to learn multi-label rules is crucial for minimizing the subset 0/1 loss.

## 3   Gradient Boosting Using Multivariate Loss Functions

As our first contribution, we formulate an extension of the gradient boosting framework to multivariate loss functions. This formulation, which should be flexible enough to use any decomposable or non-decomposable loss function, as long as it is differentiable, serves as the basis of the MLC method that is proposed in Sect. 4 as the main contribution of this paper.

### 3.1   Stagewise Additive Modeling

We aim at learning an ensemble of additive functions $F = \{f_1, \ldots, f_T\}$ as introduced in Sect. 2.1. It should be trained in a way such that the expected empirical risk with respect to a certain (surrogate) loss function $\ell$ is minimized. Thus, we are concerned with minimizing the regularized training objective

$$\mathcal{R}(F) = \sum_{n=1}^{N} \ell(\boldsymbol{y}_n, \hat{\boldsymbol{p}}_n) + \sum_{t=1}^{T} \Omega(f_t), \tag{8}$$

where $\Omega$ denotes an (optional) regularization term that may be used to penalize the complexity of the individual ensemble members to avoid overfitting and to ensure the convergence towards a global optimum if $\ell$ is not convex.

Unfortunately, constructing an ensemble of additive functions that minimizes the objective given above is a hard optimization problem. In gradient boosting, this problem is tackled by training the model in a stagewise procedure, where the individual ensemble members are added one after the other, as originally proposed by Friedman et al. [8]. At each iteration $t$, the vector of scores $F_t(\boldsymbol{x}_n)$ that is predicted by the existing ensemble members for an example $\boldsymbol{x}_n$ can be calculated based on the predictions of the previous iteration:

$$F_t(\boldsymbol{x}_n) = F_{t-1}(\boldsymbol{x}_n) + f_t(\boldsymbol{x}_n) = \left(\hat{\boldsymbol{p}}_n^1 + \cdots + \hat{\boldsymbol{p}}_n^{t-1}\right) + \hat{\boldsymbol{p}}_n^t . \tag{9}$$

Substituting the additive calculation of the predictions into the objective function given in (8) yields the following objective to be minimized by the ensemble member that is added in the $t$-th iteration:

$$\mathcal{R}(f_t) = \sum_{n=1}^{N} \ell\left(\boldsymbol{y}_n, F_{t-1}(\boldsymbol{x}_n) + \hat{\boldsymbol{p}}_n^t\right) + \Omega(f_t) . \tag{10}$$

## 3.2   Multivariate Taylor Approximation

To be able to efficiently minimize the training objective when adding a new ensemble member $f_t$, we rewrite (10) in terms of the second-order multivariate Taylor approximation

$$\mathcal{R}(f_t) \approx \sum_{n=1}^{N} \left(\ell\left(\boldsymbol{y}_n, F_{t-1}(\boldsymbol{x}_n)\right) + \boldsymbol{g}_n\hat{\boldsymbol{p}}_n^t + \frac{1}{2}\hat{\boldsymbol{p}}_n^t H_n \hat{\boldsymbol{p}}_n^t\right) + \Omega(f_t) , \tag{11}$$

where $\boldsymbol{g}_n = (g_{n1}, \ldots, g_{nK})$ denotes the vector of first-order partial derivatives of the loss function $\ell$ with respect to the existing ensemble members' predictions for a particular example $\boldsymbol{x}_n$ and individual labels $\lambda_k$. Accordingly, the Hessian matrix $H_n = ((h_{n11} \ldots h_{n1K}), \ldots, (h_{nK1} \ldots h_{nKK}))$ consists of all second-order partial derivatives:

$$g_{ni} = \frac{\partial\ell}{\partial\hat{p}_{ni}}(\boldsymbol{y}_n, F_{t-1}(\boldsymbol{x}_n)), \quad h_{nij} = \frac{\partial\ell}{\partial\hat{p}_{ni}\partial\hat{p}_{nj}}(\boldsymbol{y}_n, F_{t-1}(\boldsymbol{x}_n)) . \tag{12}$$

By removing constant terms, (11) can be further simplified, resulting in the approximated training objective

$$\widetilde{\mathcal{R}}(f_t) = \sum_{n=1}^{N} \left(\boldsymbol{g}_n\hat{\boldsymbol{p}}_n^t + \frac{1}{2}\hat{\boldsymbol{p}}_n^t H_n \hat{\boldsymbol{p}}_n^t\right) + \Omega(f_t) . \tag{13}$$

In each training iteration, the objective function $\widetilde{\mathcal{R}}$ can be used as a quality measure to decide which of the potential ensemble members improves the current model the most. This requires the predictions of the potential ensemble members for examples $\boldsymbol{x}_n$ to be known. How to find these predictions depends on the type of ensemble members used and the loss function at hand. In Sect. 4, we present solutions to this problem, using classification rules as ensemble members.

---

**Algorithm 1:** Learning an ensemble of boosted classification rules

**input**  : Training examples $\mathcal{D} = \{(\boldsymbol{x}_n, \boldsymbol{y}_n)\}_n^N$, first and second derivative $\ell'$ and $\ell''$ of the loss function, number of rules $T$, shrinkage parameter $\eta$

**output**: Ensemble of rules $F$

1  $\mathcal{G} = \{\boldsymbol{g}_n\}_n^N, \mathcal{H} = \{H_n\}_n^N$ = calculate gradients and Hessians w.r.t. $\ell'$ and $\ell''$

2  $f_1 : b_1 \to \hat{\boldsymbol{p}}_1$ with $b_1(\boldsymbol{x}) = 1, \forall \boldsymbol{x}$ and $\hat{\boldsymbol{p}}_1 = \text{FIND\_HEAD}(\mathcal{D}, \mathcal{G}, \mathcal{H}, b_1)$     ▷ Section 4.1

3  **for** $t = 2$ **to** $T$ **do**

4  $\quad$ $\mathcal{G}, \mathcal{H}$ = update gradients and Hessians of examples covered by $f_{t-1}$

5  $\quad$ $\mathcal{D}'$ = randomly draw $N$ examples from $\mathcal{D}$ (with replacement)

6  $\quad$ $f_t : b_t \to \hat{\boldsymbol{p}}_t - \text{REFINE\_RULE}(\mathcal{D}', \mathcal{G}, \mathcal{H})$     ▷ Section 4.2

7  $\quad$ $\hat{\boldsymbol{p}}_t = \text{FIND\_HEAD}(\mathcal{D}, \mathcal{G}, \mathcal{H}, b_t)$

8  $\quad$ $\hat{\boldsymbol{p}}_t = \eta \cdot \hat{\boldsymbol{p}}_t$

9  **return** ensemble of rules $F = \{f_1, \ldots, f_T\}$

---

# 4    Learning Boosted Multi-label Classification Rules

Based on the general framework defined in the previous section, we now present BOOMER, a concrete stagewise algorithm for learning an ensemble of gradient boosted single- or multi-label rules $F = \{f_1, \ldots, f_T\}$ that minimizes a given loss function in expectation[1]. The basic algorithm is outlined in Algorithm 1.

Because rules, unlike other classifiers like e.g. decision trees, only provide predictions for examples they cover, the first rule $f_1$ in the ensemble is a *default rule*, which covers all examples, i.e., $b_1(\boldsymbol{x}) = 1, \forall \boldsymbol{x} \in \mathcal{X}$. In subsequent iterations, more specific rules are added. All rules, including the default rule, contribute to the final predictions of the ensemble according to their confidence scores, which are chosen such that the objective function in (13) is minimized. In each iteration $t$, this requires the gradients and Hessians to be (re-)calculated based on the confidence scores $\hat{p}_{nk}$ that are predicted for each example $\boldsymbol{x}_n$ and label $\lambda_k$ by the current model ($\hat{p}_{nk} = 0, \forall n, k$ if $t = 1$), as well as the true labels $y_{nk} \in \{-1, +1\}$ (cf. Algorithm 1, lines 1 and 4). While the default rule always provides a confidence score for each label, all of the remaining rules may either predict for a single label or for all labels, depending on a hyper-parameter. We consider both variants in the experimental study presented in Sect. 5. The computations necessary to obtain loss-minimizing predictions for the default rule and each of the remaining rules are presented in Sect. 4.1.

To learn the rules $f_2, \ldots, f_T$, we use a greedy procedure where the body is iteratively refined by adding new conditions and the head is adjusted accordingly in each step. The algorithm used for rule refinement is discussed in detail in Sect. 4.2. To reduce the variance of the ensemble members, each rule is learned

---

[1] An implementation is available at https://www.github.com/mrapp-ke/Boomer.

on a different subsample of the training examples, randomly drawn with replacement as in *bagging*, which results in more diversified and less correlated rules. However, once a rule has been learned, we recompute its predictions on the entire training data (cf. Algorithm 1, line 7), which we have found to effectively prevent overfitting the sub-sample used for learning the rule. As an additional measure to reduce the risk of fitting noise in the data, the scores predicted by a rule may be multiplied by a *shrinkage* parameter $\eta \in (0, 1]$ (cf. Algorithm 1, line 8). Small values for $\eta$, which can be considered as the learning rate, reduce the impact of individual rules on the model [12].

## 4.1  Computation of Loss-Minimizing Scores

As illustrated in Algorithm 2, the function FIND_HEAD is used to find optimal confidence scores to be predicted by a particular rule $f$, i.e., scores that minimize the objective function $\widetilde{\mathcal{R}}$ introduced in (13). Because of the fact that rules provide the same predictions $\hat{\boldsymbol{p}}$ for all examples $\boldsymbol{x}_n$ they cover and abstain for the others, the objective function can be further simplified. As the addition is commutative, we can sum up the gradient vectors and Hessian matrices that correspond to the covered examples (cf. Algorithm 2, line 1), resulting in the objective

$$\widetilde{\mathcal{R}}\left(f_t\right) = \boldsymbol{g}\hat{\boldsymbol{p}} + \frac{1}{2}\hat{\boldsymbol{p}}H\hat{\boldsymbol{p}} + \Omega\left(f_t\right) , \tag{14}$$

where $\boldsymbol{g} = \sum_n b\left(\boldsymbol{x}_n\right)\boldsymbol{g}_n$ denotes the element-wise sum of the gradient vectors and $H = \sum_n b\left(\boldsymbol{x}_n\right)H_n$ corresponds to the sum of the Hessian matrices.

To penalize extreme predictions, we use the $L_2$ regularization term

$$\Omega_{\text{L2}}\left(f_t\right) = \frac{1}{2}\lambda\left\|\hat{\boldsymbol{p}}^t\right\|_2^2 , \tag{15}$$

where $\|\boldsymbol{x}\|_2$ denotes the Euclidean norm and $\lambda \geq 0$ is the regularization weight.

To ensure that the predictions $\hat{\boldsymbol{p}}$ minimize the regularized training objective $\widetilde{\mathcal{R}}$, we equate the first partial derivative of (14) with respect to $\hat{\boldsymbol{p}}$ with zero:

$$\frac{\partial\widetilde{\mathcal{R}}}{\partial\hat{\boldsymbol{p}}}\left(f_t\right) = \boldsymbol{g} + H\hat{\boldsymbol{p}} + \lambda\hat{\boldsymbol{p}} = \boldsymbol{g} + \left(H + \text{diag}\left(\lambda\right)\right)\hat{\boldsymbol{p}} = 0 \tag{16}$$

$$\Longleftrightarrow \left(H + \text{diag}\left(\lambda\right)\right)\hat{\boldsymbol{p}} = -\boldsymbol{g} ,$$

where $\text{diag}\left(\lambda\right)$ is a diagonal matrix with $\lambda$ on the diagonal.

(16) can be considered as a system of $K$ linear equations, where $H + \text{diag}\left(\lambda\right)$ is a matrix of coefficients, $-\boldsymbol{g}$ is a vector of ordinates and $\hat{\boldsymbol{p}}$ is the vector of unknowns to be determined. For commonly used loss functions, including the ones in Sect. 2.3, the sums of Hessians $h_{ij}$ and $h_{ji}$ are equal. Consequently, the matrix of coefficients is symmetrical.

In the general case, i.e., if the loss function is non-decomposable, the linear system in (16) must be solved to determine the optimal multi-label head $\hat{\boldsymbol{p}}$. However, when dealing with a decomposable loss function, the first and second

---

**Algorithm 2:** FIND_HEAD

**input** : (Sub-sample of) training examples $\mathcal{D} = \{(\boldsymbol{x}_n, \boldsymbol{y}_n)\}_n^N$,
          gradients $\mathcal{G} = \{\boldsymbol{g}_n\}_n^N$, Hessians $\mathcal{H} = \{H_n\}_n^N$, body $b$
**output**: Single- or multi-label head $\hat{\boldsymbol{p}}$
1 $\boldsymbol{g} = \sum_n b(\boldsymbol{x}_n)\,\boldsymbol{g}_n, H = \sum_n b(\boldsymbol{x}_n)\,H_n$
2 **if** loss function is decomposable **or** searching for a single-label head **then**
3     $\hat{\boldsymbol{p}}$ = obtain $\hat{p}_k$ w.r.t. $\boldsymbol{g}$ and $H$ for each label independently acc. to (17)
4     **if** searching for a single-label head **then**
5        $\hat{\boldsymbol{p}}$ = find best single-label prediction $\hat{p}_k \in \hat{\boldsymbol{p}}$ w.r.t. (13)

6 **else**
7     $\hat{\boldsymbol{p}}$ = obtain $(\hat{p}_1, \ldots, \hat{p}_K)$ w.r.t. $\boldsymbol{g}$ and $H$ by solving the linear system in (16)

8 **return** head $\hat{\boldsymbol{p}}$

---

derivative with respect to a particular element $\hat{p}_i \in \hat{\boldsymbol{p}}$ is independent of any other element $\hat{p}_j \in \hat{\boldsymbol{p}}$. This causes the sums of Hessians $h_{ij}$ that do not exclusively depend on $\hat{p}_i$, i.e., if $i \neq j$, to become zero. In such case, the linear system reduces to $K$ independent equations, one for each label. This enables to compute the optimal prediction $\hat{p}_i$ for the $i$-th label as

$$\hat{p}_i = -\frac{g_i}{h_{ii} + \lambda}. \tag{17}$$

Similarly, when dealing with single-label rules that predict for the $i$-th label, the predictions $\hat{p}_j$ with $j \neq i$ are known to be zero, because the rule will abstain for the corresponding labels. Consequently, (17) can be used to determine the predictions of single-label rules even if the loss function is non-decomposable.

## 4.2 Refinement of Rules

To learn a new rule, we use a top-down greedy search, commonly used in inductive rule learning (see, e.g., [10] for an overview). Algorithm 3 is meant to outline the general procedure and does not include any algorithmic optimizations that can drastically improve the computational efficiency in practice.

The search starts with an empty body that is successively refined by adding additional conditions. Adding conditions to its body causes the rule to become more specific and results in less examples being covered. The conditions, which may be used to refine an existing body, result from the values of the covered examples in case of nominal attributes or from averaging adjacent values in case of numerical attributes. In addition to bagging, we use *random feature selection*, as in random forests, to ensure that the rules in an ensemble are more diverse (cf. Algorithm 3, line 2). For each condition that may be added to the current body at a particular iteration, the head of the rule is updated via the

---

**Algorithm 3:** REFINE_RULE

    **input**  : (Sub-sample of) training examples $\mathcal{D} = \{(\boldsymbol{x}_n, \boldsymbol{y}_n)\}_n^N$,
                 gradients $\mathcal{G} = \{\boldsymbol{g}_n\}_n^N$, Hessians $\mathcal{H} = \{H_n\}_n^N$, current rule $f$
      (optional)
    **output**: Best rule $f^*$

1  $f^* = f$
2  $A' =$ randomly select $\lfloor \log_2 (L-1) + 1 \rfloor$ out of $L$ attributes from $\mathcal{D}$
3  **foreach** possible condition $c$ on attributes $A'$ and examples $\mathcal{D}$ **do**
4    |  $f' : b' \to \hat{\boldsymbol{p}}' =$ copy of current rule $f$
5    |  add condition $c$ to body $b'$
6    |  $\hat{\boldsymbol{p}}' =$ FIND_HEAD $(\mathcal{D}, \mathcal{G}, \mathcal{H}, b')$                    ▷ Section 4.1
7    |  **if** $\widetilde{\mathcal{R}}(f') < \widetilde{\mathcal{R}}(f^*)$ w.r.t. $\mathcal{G}$ and $\mathcal{H}$ **then**
8    |      ⌊ $f^* = f'$

9  **if** $f^* \neq f$ **then**
10  |  $\mathcal{D}' =$ subset of $\mathcal{D}$ covered by $f^*$
11  ⌊ **return** REFINE_RULE $(\mathcal{D}', \mathcal{G}, \mathcal{H}, f^*)$

12  **return** best rule $f^*$

---

function FIND_HEAD that has already been discussed in Sect. 4.1 (cf. Algorithm 3, line 6). As a restriction, in the case of single-label rules, each refinement of the current rule must predict for the same label (omitted in Algorithm 3 for brevity). Among all refinements, the one that minimizes the regularized objective in (14) is chosen. If no refinement results in an improvement according to said objective, the refinement process stops. We do not use any additional stopping criteria.

### 4.3 Prediction

Predicting for an example $\boldsymbol{x}_n$ involves two steps: First, the scores provided by individual rules are aggregated into a vector of confidence scores $\hat{\boldsymbol{p}}_n$ according to (2). Second, $\hat{\boldsymbol{p}}_n$ must be turned into a binary label vector $\hat{\boldsymbol{y}}_n \in \mathcal{Y}$. As it should minimize the expected risk with respect to the used loss function, i.e., $\hat{\boldsymbol{y}}_n = \arg\min_{\hat{\boldsymbol{y}}_n} \ell(\boldsymbol{y}_n, \hat{\boldsymbol{y}}_n)$, it should be chosen in a way that is tailored to the loss function at hand.

In case of the label-wise logistic loss in (6), we compute the prediction as

$$\hat{\boldsymbol{y}}_n = (\text{sgn}(\hat{p}_{n1}), \ldots, \text{sgn}(\hat{p}_{nK})) , \tag{18}$$

where $\text{sgn}(x) = +1$ if $x > 0$ and $= -1$ otherwise.

To predict the label vector that minimizes the example-wise logistic loss given in (7), we return the label vector among the vectors in the training data, which minimizes the loss, i.e.,

$$\hat{\boldsymbol{y}}_n = \arg\min_{\boldsymbol{y} \in \mathcal{D}} \ell_{\text{ex.w.-log}}(\boldsymbol{y}, \hat{\boldsymbol{p}}_n) . \tag{19}$$

The main reason for picking only a known label vector was to ensure a fair comparison to the main competitor for non-decomposable losses, LP, which is also restricted to returning only label vectors seen in the training data. Moreover, Senge et al. [20] argue that often only a small fraction of the $2^K$ possible label subsets are observed in practice, for which reason the correctness of an unseen label combination becomes increasingly unlikely for large data sets, and, in fact, our preliminary experiments confirmed this.

## 5   Evaluation

We evaluate the ability of our approach to minimize Hamming and subset 0/1 loss, using the label-wise and example-wise logistic loss as surrogates. In each case, we consider both, single- and multi-label rules, resulting in four different variants (*l.w.-log. single, l.w.-log. multi, ex.w.-log. single,* and *ex.w.-log. multi*). For each of them, we tune the shrinkage parameter $\eta \in \{0.1, 0.3, 0.5\}$ and the regularization weight $\lambda \in \{0.0, 0.25, 1.0, 4.0, 16.0, 64.0\}$, using *Bootstrap Bias Corrected Cross Validation (BBC-CV)* [22], which allows to incorporate parameter tuning and evaluation in a computationally efficient manner.

### 5.1   Synthetic Data

We use three synthetic data sets, each containing $10,000$ examples and six labels, as proposed by Dembczyński et al. [5] for analyzing the effects of using single- or multi-label rules with respect to different types of label dependencies. The attributes correspond to two-dimensional points drawn randomly from the unit circle and the labels are assigned according to linear decision boundaries through its origin. The results are shown in Fig. 1.

In the case of *marginal independence*, points are sampled for each label independently. Noise is introduced by randomly flipping 10% of the labels, such that the scores that are achievable by the Bayes-optimal classifier on an independent test set are 10% for Hamming loss and ≈47% for subset 0/1 loss. As it is not possible—by construction of the data set—to find a rule body that is suited to predict for more than one label, multi-label rules cannot be expected to provide any advantage. In fact, despite very similar trajectories, the single-label variants achieve slightly better losses in the limit. This indicates that, most probably due to the number of uninformative features, the approaches that aim at learning multi-label rules struggle to induce pure single-label rules.

For modeling a *strong marginal dependence* between the labels, a small angle between the linear boundaries of two labels is chosen, so that they mostly coincide for the respective examples. Starting from different default rules, depending on the loss function they minimize, the algorithms converge to the same performances, which indicates that all variants can similarly deal with marginal dependencies. However, when using multi-label rules, the final subset 0/1 score is approached in a faster and more steady way, as a single rule provides predictions for several labels at once. In fact, it is remarkable that the ex.w.-log. multi

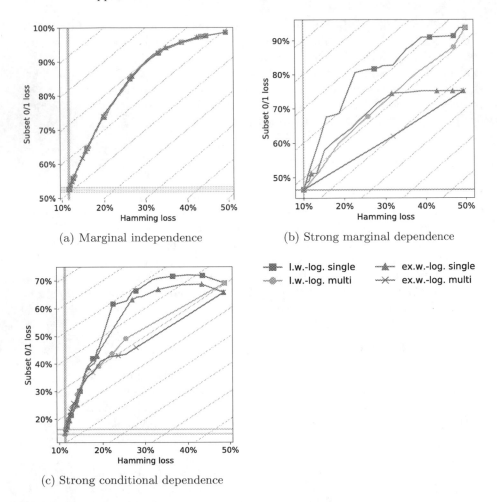

(a) Marginal independence    (b) Strong marginal dependence

(c) Strong conditional dependence

**Fig. 1.** Predictive performance of different approaches with respect to the Hamming loss and the subset 0/1 loss on three synthetic data sets. Starting at the top right, each curve shows the performance as more rules are added. The slope of the isometrics refers to an even improvement of Hamming and subset 0/1 loss. The symbols indicate $1, 2, 4, 8, \ldots, 512, 1000$ rules in the model.

variants already converge after two rules, whereas ex.w.-log. single needs six, one for each label, and optimizing for label-wise logarithmic loss takes much longer.

Finally, *strong conditional dependence* is obtained by randomly switching all labels for 10% of the examples. As a result, the score that is achievable by the Bayes-optimal classifier is 10% for both losses. While the trajectories are similar to before, the variants that optimize the non-decomposable loss achieve better results in the limit. Unlike the approaches that consider the labels independently, they seem to be less prone to the noise that is introduced at the example-level.

**Table 1.** Predictive performance (in percent) of different approaches with respect to Hamming loss, subset 0/1 loss and the example-based F1-measure. For each evaluation measure and data set, we report the ranks of the different approaches (small numbers) and highlight the best approach (bold text).

| | | l.w.-log. | | ex.w.-log. | | XGBoost | | |
| --- | --- | --- | --- | --- | --- | --- | --- | --- |
| | | Single | multi | Single | Multi | BR | LP | CC |
| Subset 0/1 loss | BIRDS | 38.72 2 | 40.43 6 | 39.57 4.5 | 39.15 3 | 39.57 4.5 | 40.85 7 | **38.30** 1 |
| | EMOTIONS | 73.33 6.5 | 73.33 6.5 | 70.48 3.5 | 65.24 2 | 70.48 3.5 | **60.00** 1 | 71.43 5 |
| | ENRON | 87.81 7 | 86.82 6 | 84.35 4 | 83.53 2 | 85.34 5 | **82.70** 1 | 83.86 3 |
| | LLOG | 78.85 5 | 78.85 5 | 78.27 3 | 76.35 2 | 80.96 7 | **70.38** 1 | 78.85 5 |
| | MEDICAL | 28.45 3 | 30.16 4 | 23.90 2 | **23.04** 1 | 56.90 7 | 41.11 5 | 44.95 6 |
| | SCENE | 39.33 7 | 33.93 5 | 25.17 3 | **23.26** 1 | 34.72 6 | 24.16 2 | 30.11 4 |
| | SLASHDOT | 65.29 5 | 62.89 4 | 53.30 3 | **49.75** 1 | 72.04 7 | 52.94 2 | 66.96 6 |
| | YEAST | 83.72 7 | 82.27 5 | 78.60 4 | 75.81 2 | 82.72 6 | **75.70** 1 | 76.59 3 |
| | Avg. rank | 5.31 | 5.19 | 3.38 | **1.75** | 5.75 | 2.50 | 4.13 |
| Hamming loss | BIRDS | 3.52 5 | **3.16** 1 | 3.58 6 | 3.23 2 | 3.43 3 | 4.03 7 | 3.45 4 |
| | EMOTIONS | 18.81 5 | 20.00 7 | 18.17 2 | 18.41 3 | 18.73 4 | **18.02** 1 | 19.29 6 |
| | ENRON | 4.56 3 | **4.49** 1 | 4.90 5 | 4.91 6 | 4.52 2 | 5.75 7 | 4.63 4 |
| | LLOG | 1.48 2.5 | 1.48 2.5 | 1.49 4.5 | **1.45** 1 | 1.49 4.5 | 2.13 7 | 1.60 6 |
| | MEDICAL | 0.84 2.5 | 0.86 4 | 0.84 2.5 | **0.79** 1 | 1.69 7 | 1.55 6 | 1.36 5 |
| | SCENE | 8.16 7 | 7.42 6 | 7.21 4 | **6.63** 1 | 7.19 3 | 7.27 5 | 6.85 2 |
| | SLASHDOT | **4.15** 1 | 4.17 2 | 5.03 6 | 4.64 5 | 4.61 4 | 5.16 7 | 4.54 3 |
| | YEAST | 19.27 2 | 19.84 5 | 19.44 4 | 19.29 3 | **19.13** 1 | 21.22 7 | 20.11 6 |
| | Avg. rank | 3.56 | 3.56 | 4.19 | **2.75** | 3.56 | 5.88 | 4.50 |
| Example-based F1 | BIRDS | 70.38 4 | **72.42** 1 | 68.00 7 | 71.18 2 | 69.68 6 | 70.06 5 | 70.96 3 |
| | EMOTIONS | 58.19 7 | 59.06 6 | 64.56 3 | 65.75 2 | 61.90 5 | **69.24** 1 | 62.59 4 |
| | ENRON | 52.86 6 | 53.87 4 | 53.73 5 | 53.91 3 | 54.76 2 | 46.61 7 | **56.46** 1 |
| | LLOG | 22.51 6 | 23.21 5 | 23.28 4 | 26.30 2 | 19.36 7 | **33.17** 1 | 24.16 3 |
| | MEDICAL | 81.68 3 | 80.07 4 | 85.51 2 | **85.97** 1 | 53.14 7 | 71.34 5 | 64.72 6 |
| | SCENE | 65.90 7 | 71.69 5 | 80.30 2 | **81.69** 1 | 70.30 6 | 79.72 3 | 74.23 4 |
| | SLASHDOT | 40.94 5 | 44.04 4 | 54.78 2 | **58.62** 1 | 32.57 7 | 54.01 3 | 39.15 6 |
| | YEAST | 61.56 6 | 61.03 7 | 62.78 3 | **63.41** 1 | 62.54 4 | 61.71 5 | 62.92 2 |
| | Avg. rank | 5.50 | 4.50 | 3.50 | **1.63** | 5.50 | 3.75 | 3.63 |

## 5.2 Real-World Benchmark Data

We also conducted experiments on eight benchmark data sets from the Mulan and MEKA projects.[2] As baselines we considered binary relevance (BR), label powerset (LP) and classifier chains (CC) [18], using XGBoost [3] as the base classifier. For BR and CC we used the logistic loss, for LP we used the softmax objective. While we tuned the number of rules $T \in \{50, 100, \ldots, 10000\}$ for our algorithm, XGBoost comes with an integrated method for determining the number of trees. The learning rate and the $L_2$ regularization weight were tuned in the same value ranges for both. Moreover, we configured XGBoost to sample 66% of the training examples at each iteration and to choose from $\lfloor \log_2 (L-1) + 1 \rfloor$ random attributes at each split. As classifier chains are sensitive to the order of labels, we chose the best order among ten random permutations. According to

---

[2] Data sets are available at http://mulan.sourceforge.net/datasets-mlc.html and https://sourceforge.net/projects/meka/files/Datasets.

their respective objectives, the BR baseline is tuned with respect to Hamming, LP and CC are tuned with respect to subset 0/1 loss.

In Table 1, we report the predictive performance of our methods and their competitors in terms of Hamming and subset 0/1 loss. For completeness, we also report the example-based F1 score (see, e.g., [23]). The Friedman test indicates significant differences for all but the Hamming loss. The Nemenyi post-hoc test yields critical distances between the average ranks of 2.91/3.19 for $\alpha = 0.1/0.05$.

On average, *ex.w.-log. multi* ranks best in terms of subset 0/1 loss. It is followed by LP, its counterpart *ex.w.-log. single* and CC. As all of them aim at minimizing subset 0/1 loss, it is expected that they rank better than their competitors which aim at the Hamming loss. Most notably, since example-wise optimized multi-label rules achieve better results than single-label rules on all data sets (statistically significant with $\alpha = 0.1$), we conclude that the ability to induce such rules, which is a novelty of the proposed method, is crucial for minimizing subset 0/1 loss.

On the other hand, in terms of Hamming loss, rules that minimize the label-wise logistic loss are competitive to the BR baseline, without a clear preference for single- or multi-label rules. Interestingly, although the example-wise logistic loss aims at minimizing subset 0/1 loss, when using multi-label rules, it also achieves remarkable results w.r.t. Hamming loss on some data sets and consequently even ranks best on average.

## 6    Conclusion

In this work, we presented an instantiation of the gradient boosting framework that supports the minimization of non-decomposable loss functions in multi-label classification. Building on this framework, we proposed an algorithm for learning ensembles of single- or multi-label classification rules. Our experiments confirm that it can successfully target different losses and is able to outperform conventional state-of-the-art boosting methods on data sets of moderate size.

While the use of multivariate loss functions in boosting has not received much attention so far, our framework could serve as a basis for developing algorithms specifically tailored to non-decomposable loss functions, such as F1 or Jaccard, and their surrogates. The main drawback is that the computation of predictions for a large number of labels $n$ is computationally demanding in the non-decomposable case—solving the linear system has complexity $\mathcal{O}\left(n^3\right)$. To compensate for this, we plan to investigate approximations that exploit the sparsity in label space. As the training complexity in MLC not only increases with the number of examples and attributes, but also with the number of labels, such optimizations are generally required for handling very large data sets.

**Acknowledgments.** This work was supported by the German Research Foundation (DFG) under grant number 400845550. Computations were conducted on the Lichtenberg high performance computer of the TU Darmstadt.

# References

1. Amit, Y., Dekel, O., Singer, Y.: A boosting algorithm for label covering in multilabel problems. In: In Proceedings of International Conference AI and Statistics (AISTATS), pp. 27–34 (2007)
2. Bhatia, K., Jain, H., Kar, P., Varma, M., Jain, P.: Sparse local embeddings for extreme multi-label classification. In: Advances in Neural Information Processing Systems 28, pp. 730–738. Curran Associates, Inc. (2015)
3. Chen, T., Guestrin, C.: XGBoost: A scalable tree boosting system. In: Proceedings 22nd International Conference on Knowledge Discovery and Data Mining (KDD), p. 785–794 (2016)
4. Cheng, W., Hüllermeier, E., Dembczyński, K.: Bayes optimal multilabel classification via probabilistic classifier chains. In: Proceedings of 27th International Conference on Machine Learning (ICML), pp. 279–286 (2010)
5. Dembczyński, K., Kotłowski, W., Hüllermeier, E.: Consistent multilabel ranking through univariate losses. In: Proceedings of 29th International Conference on Machine Learning (ICML), pp. 1319–1326. Omnipress (2012)
6. Dembczyński, K., Kotłowski, W., Słowiński, R.: ENDER: a statistical framework for boosting decision rules. Data Min. Knowl. Discov. **21**(1), 52–90 (2010)
7. Dembczyński, K., Waegeman, W., Cheng, W., Hüllermeier, E.: On label dependence and loss minimization in multi-label classification. Mach. Learn. **88**(1–2), 5–45 (2012)
8. Friedman, J.H., Hastie, T., Tibshirani, R.: Additive logistic regression: a statistical view of boosting. Ann. Stat. **28**(2), 337–407 (2000)
9. Friedman, J.H., Popescu, B.E.: Predictive learning via rule ensembles. Ann. Appl. Stat. **2**, 916–954 (2008)
10. Fürnkranz, J., Gamberger, D., Lavrač, N.: Foundations of Rule Learning. Springer, Berlin (2012)
11. Gao, W., Zhou, Z.H.: On the consistency of multi-label learning. Artif. Intell. **199–200**, 22–44 (2013)
12. Hastie, T., Tibshirani, R., Friedman, J.: The Elements of Statistical Learning: Data Mining, Inference, and Prediction. Springer, New York (2009)
13. Johnson, M., Cipolla, R.: Improved image annotation and labelling through multi-label boosting. In: Proceedings of British Machine Vision Conference (BMVC) (2005)
14. Joly, A., Wehenkel, L., Geurts, P.: Gradient tree boosting with random output projections for multi-label classification and multi-output regression. arXiv preprint arXiv:1905.07558 (2019)
15. Jung, Y.H., Tewari, A.: Online boosting algorithms for multi-label ranking. In: Proceedings of 21st International Conference on AI and Statistics (AISTATS), pp. 279–287 (2018)
16. Nam, J., Loza Mencía, E., Kim, H.J., Fürnkranz, J.: Maximizing subset accuracy with recurrent neural networks in multi-label classification. In: Advances in Neural Information Processing Systems 30 (NeurIPS), pp. 5419–5429 (2017)
17. Pillai, I., Fumera, G., Roli, F.: Designing multi-label classifiers that maximize F measures: state of the art. Pattern Recogn. **61**, 394–404 (2017)
18. Read, J., Pfahringer, B., Holmes, G., Frank, E.: Classifier chains for multi-label classification. In: Buntine, W., Grobelnik, M., Mladenić, D., Shawe-Taylor, J. (eds.) ECML PKDD 2009. LNCS (LNAI), vol. 5782, pp. 254–269. Springer, Heidelberg (2009). https://doi.org/10.1007/978-3-642-04174-7_17

19. Schapire, R.E., Singer, Y.: BoosTexter: a boosting-based system for text categorization. Mach. Learn. **39**(2), 135–168 (2000)
20. Senge, R., del Coz, J.J., Hüllermeier, E.: Rectifying classifier chains for multi-label classification. In: Proceedings Lernen, Wissen & Adaptivität, pp. 151–158 (2013)
21. Si, S., Zhang, H., Keerthi, S.S., Mahajan, D., Dhillon, I.S., Hsieh, C.J.: Gradient boosted decision trees for high dimensional sparse output. In: Proceedings of 34th International Conference on Machine Learning (ICML), pp. 3182–3190 (2017)
22. Tsamardinos, I., Greasidou, E., Borboudakis, G.: Bootstrapping the out-of-sample predictions for efficient and accurate cross-validation. Mach. Learn. **107**(12), 1895–1922 (2018). https://doi.org/10.1007/s10994-018-5714-4
23. Tsoumakas, G., Katakis, I., Vlahavas, I.: Mining multi-label data. In: Data Mining and Knowledge Discovery Handbook, pp. 667–685. Springer, Boston (2010) https://doi.org/10.1007/978-0-387-09823-4_34
24. Zhang, M.L., Zhou, Z.H.: A review on multi-label learning algorithms. IEEE Trans. Knowl. Data Eng. **26**(8), 1819–1837 (2013)
25. Zhang, Z., Jung, C.: GBDT-MO: Gradient boosted decision trees for multiple outputs. arXiv preprint arXiv:1909.04373 (2019)

# Landmark-Based Ensemble Learning with Random Fourier Features and Gradient Boosting

Léo Gautheron[1](✉), Pascal Germain[2], Amaury Habrard[1], Guillaume Metzler[1], Emilie Morvant[1], Marc Sebban[1], and Valentina Zantedeschi[3]

[1] Univ Lyon, UJM-Saint-Etienne, CNRS, Institut d Optique Graduate School, Laboratoire Hubert Curien UMR 5516, 42023 Saint-Etienne, France
{leo.gautheron,amaury.habrard,guillaume.metzler,emilie.morvant, marc.sebban}@univ-st-etienne.fr, leo_g_autheron@hotmail.fr
[2] Département d'informatique et de génie logiciel, Université Laval, Québec, Canada
pascal.germain@ift.ulaval.ca
[3] GE - Global Research, 1 Research Circle, Niskayuna, NY 12309, USA
vzantedeschi@gmail.com

**Abstract.** This paper jointly leverages two state-of-the-art learning stra-tegies—gradient boosting (GB) and kernel Random Fourier Features (RFF)—to address the problem of kernel learning. Our study builds on a recent result showing that one can learn a distribution over the RFF to produce a new kernel suited for the task at hand. For learning this distribution, we exploit a GB scheme expressed as ensembles of RFF weak learners, each of them being a kernel function designed to fit the residual. Unlike Multiple Kernel Learning techniques that make use of a pre-computed dictionary of kernel functions to select from, at each iteration we fit a kernel by approximating it from the training data as a weighted sum of RFF. This strategy allows one to build a classifier based on a small ensemble of learned kernel "landmarks" better suited for the underlying application. We conduct a thorough experimental analysis to highlight the advantages of our method compared to both boosting-based and kernel-learning state-of-the-art methods.

**Keywords:** Gradient boosting · Random Fourier features · Kernel learning

## 1 Introduction

Kernel methods are among the most popular approaches in machine learning due to their capability to address non-linear problems, their robustness and their

**Electronic supplementary material** The online version of this chapter (https://doi.org/10.1007/978-3-030-67664-3_9) contains supplementary material, which is available to authorized users.

F. Hutter et al. (Eds.): ECML PKDD 2020, LNAI 12459, pp. 141–157, 2021.
https://doi.org/10.1007/978-3-030-67664-3_9

simplicity. However, they exhibit two main flaws in terms of memory usage and time complexity. Landmark-based kernel approaches [2] can be used to drastically reduce the number of instances involved in the comparisons, but they heavily depend on the choice and the parameterization of the kernel. Multiple Kernel Learning [13] and Matching Pursuit methods [12] can provide alternative solutions to this problem but these require the use of a pre-defined dictionary of base functions. Another strategy to improve the scalability of kernel methods is to use approximation techniques such as the Nyström [3] or Random Fourier Features (RFF) [10]. The latter is probably the most used thanks to its simplicity and ease of computation. It allows the approximation of any shift-invariant kernel based on the Fourier transform of the kernel. Several works have extended this technique by allowing one to adapt the RFF approximation directly from the training data [1,6,11]. Among them, the recent work of Letarte *et al.* [6] introduces a method to obtain a weighting distribution over the random features by a single pass over them. This strategy is derived from a statistical learning analysis, starting from the observation that each random feature can be interpreted as a weak hypothesis in the form of trigonometric functions obtained by the Fourier decomposition. However, in practice, this method requires the use of a fixed set of landmarks selected beforehand and independently from the task before being able to learn the representation in a second step. This leads to three important limitations: *(i)* the need for a heuristic strategy for selecting relevant landmarks, *(ii)* these latter and the associated representation might not be adapted for the underlying task, and *(iii)* the number of landmarks might not be minimal *w.r.t.* that task, inducing higher computational and memory costs.

We propose in this paper to tackle these issues with a gradient boosting approach [4]. Our aim is to learn iteratively the classifier and a compact and efficient representation at the same time. Our greedy optimization method is similar to Oglic & Gärtner's one [8], which at each iteration of the functional gradient descent [7] refines the representation by adding the base function minimizing a residual-based loss function. But unlike our approach, their method does not allow to learn a classifier at the same time. Instead, we propose to jointly optimize the classifier and the base functions in the form of kernels by leveraging both gradient boosting and RFF. Interestingly, we further show that we can benefit from a significant performance boost by *(i)* considering each weak learner as a single trigonometric feature, and *(ii)* learning the random part of the RFF.

**Organization of the Paper.** Section 2 describes the notations and the necessary background knowledge. We present our method in Sect. 3 as well as two efficient refinements before presenting an extensive experimental study in Sect. 4, comparing our strategy with boosting-based and kernel learning methods.

## 2    Notations and Related Work

We consider binary classification tasks from a $d$-dimensional input space $\mathbb{R}^d$ to a label set $Y=\{-1,+1\}$. Let $S=\left\{(\mathbf{x}_i, y_i)\right\}_{i=1}^{n}$ be a training set of $n$ points.

We focus on kernel-based algorithms that rely on pre-defined kernel functions $k : \mathbb{R}^d \times \mathbb{R}^d \to \mathbb{R}$ assessing the similarity between any two points of the input space. These methods present a good performance when the parameters of the kernels are learned and the chosen kernels are able to fit the distribution of the data. However, selecting the right kernel and tuning its parameters is computationally expensive, in general. To reduce this overhead, one can resort to Multiple Kernel Learning techniques [13] which boil down to selecting the combination of kernels that fits the best the training data: a dictionary of $T$ base functions $\{k^t\}_{t=1}^T$ is composed of various kernels associated with some fixed parameters, and a combination is learned, defined as

$$H(\mathbf{x}, \mathbf{x}') = \sum_{t=1}^{T} \alpha^t \, k^t(\mathbf{x}, \mathbf{x}'), \tag{1}$$

with $\alpha^t \in \mathbb{R}$ the weight of the kernel $k^t(\mathbf{x}, \mathbf{x}')$. As shown in Sect. 3, our main contribution is to address this issue of optimizing a linear combination of kernels by leveraging RFF and gradient boosting (we recall basics on it in Sect. 3.1). To avoid the dictionary of kernel functions in Eq. (1) from being pre-computed, we propose a method inspired from Letarte et al. [6] to learn a set of approximations of kernels tailored to the underlying classification task. Unlike Letarte et al., we learn such functions so that the representation and the classifier are jointly optimized. We consider landmark-based shift-invariant kernels relying on the value $\boldsymbol{\delta} = \mathbf{x}^t - \mathbf{x} \in \mathbb{R}^d$ and usually denoted by abuse of notation by $k(\boldsymbol{\delta}) = k(\mathbf{x}^t - \mathbf{x}) = k(\mathbf{x}^t, \mathbf{x})$, where $\mathbf{x}^t \in \mathbb{R}^d$ is a point—called landmark—lying on the input space which all the instances are compared to, and that strongly characterizes the kernel. At each iteration of our gradient boosting procedure, we optimize the kernel function itself, exploiting the flexibility of the framework of Letarte et al., where a kernel is a weighted sum of RFF [10] defined as

$$k_{q^t}(\mathbf{x}^t - \mathbf{x}) = \sum_{j=1}^{K} q_j^t \cos\left(\boldsymbol{\omega}_j \cdot (\mathbf{x}^t - \mathbf{x})\right), \tag{2}$$

where the $\boldsymbol{\omega}_j$ are drawn from the Fourier transform of a shift-invariant kernel $k$ denoted by $p(\boldsymbol{\omega})$ and defined as

$$p(\boldsymbol{\omega}) = \frac{1}{(2\pi)^d} \int_{\mathbb{R}^d} k(\boldsymbol{\delta}) e^{-i\boldsymbol{\omega} \cdot \boldsymbol{\delta}} d\boldsymbol{\delta}. \tag{3}$$

When $q^t$ is uniform, we retrieve the setting of RFF and we have $k(\boldsymbol{\delta}) \simeq k_{q^t}(\boldsymbol{\delta})$ where larger number of random features $K$ give better approximations [10]. Letarte et al. [6] aim to learn the weights of the random Fourier features $q^t$. To do so, they consider a loss function $\ell$ that measures the quality of the similarities computed using the kernel $k_{q^t}$. Their theoretical study on $\ell$ leads to a closed-form solution for $q^t$ computed as

$$\forall j \in \{1, \ldots, K\}, \quad q_j^t = \frac{1}{Z^t} \exp\left(\frac{-\beta\sqrt{n}}{n} \sum_{i=1}^{n} \ell(h_{\omega^j}^t(\mathbf{x}_i))\right), \tag{4}$$

---

**Algorithm 1:** Gradient boosting [4]

**Inputs** : Training set $S = \left\{ (\mathbf{x}_i, y_i) \right\}_{i=1}^{n}$; Loss $\ell$; Number of iterations $T$

**Output**: sign $\left( H^0(\mathbf{x}) + \sum_{t=1}^{T} \alpha^t h_{a^t}(\mathbf{x}) \right)$

1: $\forall i = 1, \ldots, n, \quad H^0(\mathbf{x}_i) = \operatorname{argmin}_\rho \sum_{i=1}^{n} \ell(y_i, \rho)$

2: **for** $t = 1, \ldots, T$ **do**

3: $\quad \forall i = 1, \ldots, n, \quad \tilde{y}_i = -\dfrac{\partial \ell\left(y_i, H^{t-1}(\mathbf{x}_i)\right)}{\partial H^{t-1}(\mathbf{x}_i)}$

4: $\quad a^t = \operatorname{argmin}_a \sum_{i=1}^{n} \left( \tilde{y}_i - h_a(\mathbf{x}_i) \right)^2$

5: $\quad \alpha^t = \operatorname{argmin}_\alpha \sum_{i=1}^{n} \ell\left(y_i, H^{t-1}(\mathbf{x}_i) + \alpha h_{a^t}(\mathbf{x}_i)\right)$

6: $\quad \forall i = 1, \ldots, n, \quad H^t(\mathbf{x}_i) - H^{t-1}(\mathbf{x}_i) + \alpha^t h_{a^t}(\mathbf{x}_i)$

7: **end for**

---

with $\beta \geq 0$ a parameter to tune, $h_\omega^t(\mathbf{x}) = \cos\left(\boldsymbol{\omega} \cdot (\mathbf{x}^t - \mathbf{x})\right)$, and $Z^t$ a normalization constant such that $\sum_{j=1}^{K} q_j^t = 1$. They learn a representation of the input space of $n_L$ features where each of them is computed using $k_{q^t}$ with the landmark $(\mathbf{x}^t, y^t)$ selected randomly from the training set. Once the new representation is computed, a (linear) predictor is learned from it, in a second step.

It is worth noticing that this kind of procedure exhibits two limitations. First, the model can be optimized only after having learned the representation. Second, the landmarks have to be fixed before learning the representation. Thus, the constructed representation is not guaranteed to be compact and relevant for the learning algorithm considered. To tackle these issues, we propose in the following a strategy that performs both steps at the same time through a gradient boosting process that allows to jointly learn the set of landmarks and the final predictor.

# 3    Gradient Boosting Random Fourier Features

The approach we propose follows the widely used gradient boosting framework first introduced by Friedman [4]. We briefly recall it below.

## 3.1    Gradient Boosting in a Nutshell

Gradient boosting is an ensemble method that aims at learning a weighted majority vote over an ensemble of $T$ weak predictors in a greedy way by learning one classifier per iteration. The final majority vote is of the form

$$\forall \mathbf{x} \in \mathbb{R}^d, \; \operatorname{sign}\left( H^0(\mathbf{x}) + \sum_{t=1}^{T} \alpha^t h_{a^t}(\mathbf{x}) \right),$$

where $H^0$ is an initial classifier fixed before the iterative process (usually set such that it returns the same value for every sample), and $\alpha^t$ is the weight associated

---

**Algorithm 2: GBRFF1**

---

**Inputs** : Training set $S = \left\{(\mathbf{x}_i, y_i)\right\}_{i=1}^{n}$; Number of iterations $T$;
$\qquad K$ number of random features; Parameters $\gamma$ and $\beta$

**Output**: sign $\left( H^0(\mathbf{x}) + \sum_{t=1}^{T} \alpha^t \sum_{j=1}^{K} q_j^t \cos\left(\boldsymbol{\omega}_j^t \cdot (\mathbf{x}^t - \mathbf{x})\right) \right)$

1: $H^0 \leftarrow H^0(\mathbf{x}_i) = \frac{1}{2} \ln \frac{1+\frac{1}{n}\sum_{j=1}^{n} y_j}{1-\frac{1}{n}\sum_{j=1}^{n} y_j}$

2: **for** $t = 1, \ldots, T$ **do**

3: $\quad \forall i = 1, \ldots, n, \quad w_i = \exp(-y_i H^{t-1}(\mathbf{x}_i))$

4: $\quad \forall i = 1, \ldots, n, \quad \tilde{y}_i = y_i w_i$

5: $\quad \forall j = 1, \ldots, K,$ draw $\boldsymbol{\omega}_j^t \sim \mathcal{N}(0, 2\gamma)^d$

6: $\quad \mathbf{x}^t = \underset{\mathbf{x} \in \mathbb{R}^d}{\operatorname{argmin}} \ \frac{1}{n} \sum_{i=1}^{n} \exp\left( -\tilde{y}_i \frac{1}{K} \sum_{j=1}^{K} \cos(\boldsymbol{\omega}_j^t \cdot (\mathbf{x} - \mathbf{x}_i)) \right)$

7: $\quad \forall j = 1, \ldots, K, \ q_j^t = \frac{1}{Z^t} \exp\left( \frac{-\beta\sqrt{n}}{n} \sum_{i=1}^{n} \exp\left( -\tilde{y}_i \cos\left(\boldsymbol{\omega}_j^t \cdot (\mathbf{x}^t - \mathbf{x}_i)\right) \right) \right)$

8: $\quad \alpha^t = \frac{1}{2} \ln \dfrac{\sum_{i=1}^{n} \left(1 + y_i \sum_{j=1}^{K} q_j^t \cos\left(\boldsymbol{\omega}_j^t \cdot (\mathbf{x}^t - \mathbf{x}_i)\right)\right) w_i}{\sum_{i=1}^{n} \left(1 - y_i \sum_{j=1}^{K} q_j^t \cos\left(\boldsymbol{\omega}_j^t \cdot (\mathbf{x}^t - \mathbf{x}_i)\right)\right) w_i}$

9: $\quad \forall i = 1, \ldots, n, \ H^t(\mathbf{x}_i) = H^{t-1}(\mathbf{x}_i) + \alpha^t \sum_{j=1}^{K} q_j^t \cos\left(\boldsymbol{\omega}_j^t \cdot (\mathbf{x}^t - \mathbf{x}_i)\right)$

10: **end for**

---

to the predictor $h_{a^t}$ and is learned at the same time as the parameters $a^t$ of that classifier. Given a differentiable loss $\ell$, the objective of the gradient boosting algorithm is to perform a gradient descent where the variable to be optimized is the ensemble and the function to be minimized is the empirical loss. The pseudo-code of gradient boosting is reported in Algorithm 1. First, the ensemble is constituted by only one predictor: the one that outputs a constant value minimizing the loss over the whole training set (line **1**). Then at each iteration, the algorithm computes for each training example the negative gradient of the loss (line **3**), also called the residual and denoted by $\tilde{y}_i$. The next step consists in optimizing the parameters of the predictor $h_{a^t}$ that fits the best the residuals (line **4**), before learning the optimal step size $\alpha^t$ that minimizes the loss by adding $h_{a^t}$, weighted by $\alpha^t$, to the current vote (line **5**). Finally, the model is updated by adding $\alpha^t h_{a^t}(\cdot)$ (line **6**) to the vote.

### 3.2  Gradient Boosting with Random Fourier Features

Our main contribution takes the form of a learning algorithm which jointly optimizes a compact representation of the data and the model. Our method, called **GBRFF1**, leverages both Gradient Boosting and RFF. We describe its pseudo-code in Algorithm 2 which follows the steps of Algorithm 1. The loss function $\ell$ at the core of our algorithm is the exponential loss:

$$\ell\left(H^T\right) = \frac{1}{n} \sum_{i=1}^{n} \exp\left( -y_i H^T(\mathbf{x}_i) \right). \tag{5}$$

Given $\ell\left(H^T\right)$, line **1** of Algorithm 1 amounts to setting the initial learner as

$$\forall i \in \{1,\ldots,n\}, \quad H^0(\mathbf{x}_i) = \frac{1}{2}\ln\frac{1+\frac{1}{n}\sum_{j=1}^n y_j}{1-\frac{1}{n}\sum_{j=1}^n y_j}. \tag{6}$$

The residuals of line **3** are defined as $\tilde{y}_i = -\dfrac{\partial\ell\left(y_i,H^{t-1}(\mathbf{x}_i)\right)}{\partial H^{t-1}(\mathbf{x}_i)} = y_i\,e^{-y_i H^{t-1}(\mathbf{x}_i)}$.

Line 4 of Algorithm 1 tends to learn a weak learner that outputs exactly the residuals' values by minimizing the squared loss; but, this is not well suited in our setting with the exponential loss (Eq. (5)). To benefit from the exponential decrease of the loss, we are rather interested in weak learners that output predictions having a large absolute value and being of the same sign as the residuals. Thus, we aim at favoring parameter values minimizing the exponential loss between the residuals and the predictions of the weak learner as follows:

$$a^t = \underset{a}{\arg\min}\,\frac{1}{n}\sum_{i=1}^n \exp\left(-\tilde{y}_i h_a(\mathbf{x}_i)\right). \tag{7}$$

Following the RFF principle, we can now define our weak learner as

$$h_{a^t}(\mathbf{x}_i) = \sum_{j=1}^K q_j^t \cos\left(\boldsymbol{\omega}_j^t \cdot (\mathbf{x}^t - \mathbf{x}_i)\right), \tag{8}$$

where its parameters are given by $a^t=(\{\boldsymbol{\omega}_j^t\}_{j=1}^K,\mathbf{x}^t,q^t)$. Instead of using a pre-defined set of landmarks [6], we build this set iteratively, *i.e.*, we learn one landmark per iteration. To benefit from the closed form of Eq. (4), we propose the following greedy approach to learn the parameters $a^t$. At each iteration $t$, we draw $K$ vectors $\{\boldsymbol{\omega}_j^t\}_{j=1}^K\sim p^K$ with $p$ the Fourier transform of a given kernel (as defined in Eq. (3)); then we look for the optimal landmark $\mathbf{x}^t$. Plugging Eq. (8) into Eq. (7) and assuming a uniform prior distribution over the random features, $\mathbf{x}^t$ is learned to minimize

$$\mathbf{x}^t = \underset{\mathbf{x}\in\mathbb{R}^d}{\arg\min}\,f(\mathbf{x}) = \frac{1}{n}\sum_{i=1}^n\exp\left(-\tilde{y}_i\frac{1}{K}\sum_{j=1}^K\cos(\boldsymbol{\omega}_j^t\cdot(\mathbf{x}-\mathbf{x}_i))\right). \tag{9}$$

Even if this problem is non-convex due to the cosine function, we can still compute its derivative and perform a gradient descent to find a possible solution. The partial derivative of Eq. (9) with respect to $\mathbf{x}$ is given by

$$\frac{\partial f}{\partial\mathbf{x}}(\mathbf{x})=\frac{1}{Kn}\sum_{i=1}^n\left[\frac{\tilde{y}_i}{K}\sum_{j=1}^K\sin(\boldsymbol{\omega}_j^t\cdot(\mathbf{x}-\mathbf{x}_i))\right]\exp\left[-\frac{\tilde{y}_i}{K}\sum_{j=1}^K\cos(\boldsymbol{\omega}_j^t\cdot(\mathbf{x}-\mathbf{x}_i))\right]\sum_{j=1}^K\boldsymbol{\omega}_j^t.$$

According to Letarte *et al.* [6], given the landmark $\mathbf{x}^t$ found by gradient descent, we can now compute the weights of the random features $q^t$ as

$$\forall j \in \{1,\ldots,K\}, \quad q_j^t=\frac{1}{Z^t}\exp\left[\frac{-\beta\sqrt{n}}{n}\sum_{i=1}^n\exp\left(-\tilde{y}_i\cos\left(\boldsymbol{\omega}_j^t\cdot(\mathbf{x}^t-\mathbf{x}_i)\right)\right)\right], \tag{10}$$

with $\beta\geq 0$ a parameter to tune and $Z^t$ the normalization constant.

The last step concerns the step size $\alpha^t$. It is computed so as to minimize the combination of the current model $H^{t-1}$ with the weak learner $h^t$, i.e.,

$$\alpha^t = \underset{\alpha}{\operatorname{argmin}} \sum_{i=1}^{n} \exp\left[-y_i(H^{t-1}(\mathbf{x}_i)+\alpha h^t(\mathbf{x}_i))\right] = \underset{\alpha}{\operatorname{argmin}} \sum_{i=1}^{n} w_i \exp\left[-y_i\alpha h^t(\mathbf{x}_i)\right],$$

where $w_i = \exp(-y_i H^{t-1}(\mathbf{x}_i))$. In order to have a closed-form solution of $\alpha$, we use the convexity of the above quantity and the fact that $h^t(\mathbf{x}_i) \in [-1,1]$ to bound the loss function to optimize. Indeed, we get

$$\sum_{i=1}^{n} w_i e^{-y_i\alpha h^t(\mathbf{x}_i)} \leq \sum_{i=1}^{n} \left[\frac{1-y_i h^t(\mathbf{x}_i)}{2}\right] w_i e^{\alpha} + \sum_{i=1}^{n} \left[\frac{1+y_i h^t(\mathbf{x}_i)}{2}\right] w_i e^{-\alpha}.$$

This upper bound is strictly convex. Its minimum $\alpha^t$ can be found by setting to 0 the derivative w.r.t. $\alpha$ of the right-hand side of the previous equation. We get

$$\sum_{i=1}^{n} \left(\frac{1-y_i h^t(\mathbf{x}_i)}{2}\right) w_i\, e^{\alpha} = \sum_{i=1}^{n} \left(\frac{1+y_i h^t(\mathbf{x}_i)}{2}\right) w_i\, e^{-\alpha},$$

for which the solution is given by $\alpha^t = \frac{1}{2} \ln \left(\frac{\sum_{i=1}^{n}(1 - y_i h^t(\mathbf{x}_i))w_i}{\sum_{i=1}^{n}(1 + y_i h^t(\mathbf{x}_i))w_i}\right).$
The same derivation can be used to find the initial predictor $H^0$.

As usually done in the RFF literature [1,10,11] we use the RBF kernel $k_\gamma(\mathbf{x},\mathbf{x}')=e^{-\gamma\|\mathbf{x}-\mathbf{x}'\|^2}$ with as Fourier transform vectors of $d$ numbers each drawn from the normal law with zero mean and variance $2\gamma$ that we denote $\mathcal{N}(0,2\gamma)^d$.

### 3.3   Refining GBRFF1

In **GBRFF1**, the number of random features $K$ used at each iteration has a direct impact on the computation time of the algorithm. Moreover $\boldsymbol{\omega}^t$ is drawn according to the Fourier transform of the RBF kernel and thus is not learned. The second part of our contribution is to propose two refinements. First, we bring to light the fact that one can drastically reduce the complexity of **GBRFF1** by learning a rough approximation of the kernel, yet much simpler and still very effective, using $K=1$. In this scenario, we show that learning the landmarks boils down to finding a single real number in $[-\pi, \pi]$. Then, to speed up the convergence of the algorithm, we suggest to optimize $\boldsymbol{\omega}^t$ after a random initialization from the Fourier transform. We show that a simple gradient descent with respect to this parameter allows a faster convergence with better performance. These two improvements lead to a variant of our original algorithm, called **GBRFF2** and presented in Algorithm 3.

**Cheaper Landmark Learning Using the Periodicity of the Cosine.** As we set $K=1$, the weak learner $h_{a^t}(\mathbf{x})$ is now simply defined as

$$h_{a^t}(\mathbf{x}) = \cos\left(\boldsymbol{\omega}^t \cdot (\mathbf{x}^t - \mathbf{x}_i)\right),$$

---

**Algorithm 3:** GBRFF2

---

**Inputs** : Training set $S = \left\{(\mathbf{x}_i, y_i)\right\}_{i=1}^{n}$; Number of iterations $T$;
             Parameters $\gamma$ and $\lambda$

**Output:** $\mathrm{sign}\left(H^0(\mathbf{x}) + \sum_{t=1}^{T} \alpha^t \cos\left(\boldsymbol{\omega}^t \cdot \mathbf{x}_i - b^t\right)\right)$

1:   $H^0 \leftarrow H^0(\mathbf{x}_i) = \frac{1}{2}\ln \frac{\sum_{j=1}^{n}\left(1+y_j\right)}{\sum_{j=1}^{n}\left(1-y_j\right)}$

2:   **for** $t = 1, \ldots, T$ **do**

3:      $\forall i = 1, \ldots, n, \quad w_i = \exp(-y_i H^{t-1}(\mathbf{x}_i))$

4:      $\forall i = 1, \ldots, n, \quad \tilde{y}_i = y_i w_i$

5:      Draw $\boldsymbol{\omega} \sim \mathcal{N}(0, 2\gamma)^d$

6:      $b^t = \underset{b \in [-\pi, \pi]}{\mathrm{argmin}} \; \frac{1}{n}\sum_{i=1}^{n} \exp\left(-\tilde{y}_i \cos\left(\boldsymbol{\omega} \cdot \mathbf{x}_i - b\right)\right)$

7:      $\boldsymbol{\omega}^t = \underset{\boldsymbol{\omega} \in \mathbb{R}^d}{\mathrm{argmin}} \; \lambda\|\boldsymbol{\omega}\|_2^2 + \frac{1}{n}\sum_{i=1}^{n} \exp\left(-\tilde{y}_i \cos\left(\boldsymbol{\omega} \cdot \mathbf{x}_i - b^t\right)\right).$

8:      $\alpha^t = \frac{1}{2}\ln \dfrac{\sum_{i=1}^{n}\left(1+y_i \cos\left(\boldsymbol{\omega}^t \cdot \mathbf{x}_i - b^t\right)\right)w_i}{\sum_{i=1}^{n}\left(1-y_i \cos\left(\boldsymbol{\omega}^t \cdot \mathbf{x}_i - b^t\right)\right)w_i}$

9:      $\forall i = 1, \ldots, n, \; H^t(\mathbf{x}_i) = H^{t-1}(\mathbf{x}_i) + \alpha^t \cos\left(\boldsymbol{\omega}^t \cdot \mathbf{x}_i - b^t\right)$

10: **end for**

---

where its parameters are given by $a^t = (\boldsymbol{\omega}^t, \mathbf{x}^t)$. This formulation allows us to eliminate the dependence on the hyper-parameter $K$. Moreover, one can also get rid of $\beta$, because learning the weights $q_j^t$ (line **7** of Algorithm 2) is no more necessary. Instead, since $K=1$, we can see $\alpha^t$ learned at each iteration as a surrogate of these weights. As our weak learner is based on a single random feature, the objective function (line **6**) to learn the landmark at iteration $t$ becomes

$$\mathbf{x}^t = \underset{\mathbf{x} \in \mathbb{R}^d}{\mathrm{argmin}} \; f_{\boldsymbol{\omega}^t}(\mathbf{x}) = \frac{1}{n}\sum_{i=1}^{n} \exp\left(-\tilde{y}_i \cos(\boldsymbol{\omega}^t \cdot (\mathbf{x} - \mathbf{x}_i))\right).$$

Let $c \in [\![1, d]\!]$ be the index of the $c$-th coordinate of the landmark $\mathbf{x}^t$. We can rewrite the objective function as

$$f_{\boldsymbol{\omega}^t}(\mathbf{x}^t) = \frac{1}{n}\sum_{i=1}^{n} e^{-\tilde{y}_i \cos\left(\boldsymbol{\omega}^t \cdot \mathbf{x}^t - \boldsymbol{\omega}^t \cdot \mathbf{x}_i\right)} = \frac{1}{n}\sum_{i=1}^{n} e^{-\tilde{y}_i \cos\left(\omega_c^t x_c^t + \sum_{j \neq c} \omega_j^t x_j^t - \boldsymbol{\omega}^t \cdot \mathbf{x}_i\right)}.$$

We leverage the periodicity of the cosine function along each direction to find the optimal $c$-th coordinate of the landmark $\mathbf{x}_c^t \in [\frac{-\pi}{\omega_c^t}, \frac{\pi}{\omega_c^t}]$ that minimizes $f_{\boldsymbol{\omega}^t}(\mathbf{x}^t)$ by fixing all the other coordinates. Figure 1 illustrates this phenomenon on the two-moons dataset when applying **GBRFF1** with $K=1$. The plots in the first row show the periodicity of the loss represented as repeating diagonal green/yellow stripes (light yellow is associated to the smallest loss). There is an infinite number of landmarks giving such a minimal loss at the middle of the yellow stripes. Thus, by setting one coordinate of the landmark to an arbitrary value, the algorithm is still able at any iteration to find along the second coordinate a value that minimizes the loss (the resulting landmark at the current iteration is depicted by a white cross). The second row shows that such a strategy allows us to get an accuracy of 100% on this toy dataset after 10 iterations. By generalizing this, instead of learning a landmark vector $\mathbf{x}^t \in \mathbb{R}^d$, we fix all but one coordinate of the landmark to 0, and then learn a single scalar $b^t \in [-\pi, \pi]$ that minimizes

$$f_{\boldsymbol{\omega}^t}(b^t) = \frac{1}{n} \sum_{i=1}^n \exp\left(-\tilde{y}_i \cos\left(\boldsymbol{\omega}^t \cdot \mathbf{x}_i - b^t\right)\right).$$

**Learning $\boldsymbol{\omega}^t$ for Faster Convergence.** The second refinement concerns the randomness of the RFF due to vector $\boldsymbol{\omega}^t$. So far, the latter was drawn according $p$ and then used to learn $b^t$. We suggest instead to fine-tune $\boldsymbol{\omega}^t$ by doing a gradient descent with as initialization the vector drawn from $p$. Supported by the experiments performed in the following, we claim that such a strategy allows us to both speed up the convergence of the algorithm and boost the accuracy. This update requires to add a line of code, just after line **6** of Algorithm 2, expressed as a regularized optimization problem:

$$\boldsymbol{\omega}^t = \operatorname*{argmin}_{\boldsymbol{\omega} \in \mathbb{R}^d} \lambda \|\boldsymbol{\omega}\|_2^2 + \frac{1}{n} \sum_{i=1}^n \exp\left(-\tilde{y}_i \cos\left(\boldsymbol{\omega} \cdot \mathbf{x}_i - b^t\right)\right),$$

its derivative being   $\dfrac{\partial f_{\boldsymbol{\omega}}}{\partial \boldsymbol{\omega}}(\boldsymbol{\omega}) = 2\lambda\boldsymbol{\omega} + \dfrac{1}{n} \sum_{i=1}^n \mathbf{x}_i \tilde{y}_i \sin(\boldsymbol{\omega} \cdot \mathbf{x}_i - b^t) \, e^{-\tilde{y}_i \cos(\boldsymbol{\omega} \cdot \mathbf{x}_i - b^t)}.$

## 4   Experimental Evaluation

The objective of this section is three-fold: first, we aim to bring to light the interest of learning the landmarks rather than fixing them as done in Letarte et al. [6]; second we study the impact of the number $K$ of random features; lastly, we perform an extensive experimental comparison of our algorithms. The Python code of all experiments and the data used are publicly available[1].

### 4.1   Setting

For **GBRFF1** and **GBRFF2**, we select by cross-validation (CV) the hyper-parameter $\gamma \in \dfrac{2^{\{-2,\ldots,2\}}}{d}$. For **GBRFF2**, we also tune $\lambda \in \{0, 2^{\{-5,\ldots,-2\}}\}$. We compare our two methods with the following algorithms.

- **LGBM** [5] is a state-of-the-art gradient boosting method using trees as base predictors. We select by CV the maximum tree depth in $\{1, \ldots, 10\}$ and the L2 regularization parameter $\lambda \in \{0, 2^{\{-5,\ldots,-2\}}\}$.

---

[1] The code is available here: https://leogautheron.github.io.

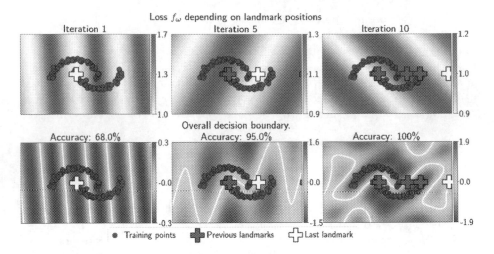

**Fig. 1. GBRFF1** with $K=1$ on the two-moons dataset at different iterations. Top row shows the periodicity of the loss (light yellow indicates the minimal loss). Bottom row shows the resulting decision boundaries between the classes (blue & red) by fixing arbitrarily one coordinate of the landmark and minimizing the loss along the other one. (Color figure online)

- **BMKR** [13] is a Multiple Kernel Learning method based on gradient boosting with least square loss. It selects at each iteration the best kernel plugged inside an SVR to fit the residuals among 10 RBF kernels with $\gamma \in 2^{\{-4,...,5\}}$ and the linear kernel $k(\mathbf{x}, \mathbf{x}') = \mathbf{x}^\top \mathbf{x}'$. We select by CV the SVR parameter $C \in 10^{\{-2,...,2\}}$.

- **GFC** [8] is a greedy feature construction method based on functional gradient descent. It iteratively refines the representation learned by adding a feature that matches the residual function defined for the least squared loss. We use the final representation to learn a linear SVM where $C \in 10^{\{-2,...,2\}}$ is selected by CV.

- **PBRFF** [6] that (1) draws with replacement $n_L$ landmarks from the training set; (2) learns a representation of $n_L$ features where each feature is computed using Eq. (2) based on $K=10$ vectors drawn like our methods from $\mathcal{N}(0, 2\gamma)^d$; (3) learns a linear SVM on the new representation. We select by CV its parameters $\gamma \in \frac{2^{\{-2,...,2\}}}{d}$, $\beta \in 10^{\{-2,...,2\}}$ and the SVM parameter $C \in 10^{\{-2,...,2\}}$.

We consider 16 datasets coming mainly from the UCI repository that we binarized as described in Table 1. We generate for each dataset 20 random 70%/30% train/test splits. Datasets are pre-processed such that each feature in the training set has 0 mean and unit variance; the factors computed on the training set are then used to scale each feature in the test set. All parameters are tuned by 5-fold CV on the training set by performing a grid search.

**Table 1.** Description of the datasets (n: number of examples, d: number of features, c: number of classes) and the classes chosen as negative (−1) and positive (+1).

| Name | N | D | C | Label −1 | Label +1 | Name | N | D | C | Label −1 | Label +1 |
|------|---|---|---|----------|----------|------|---|---|---|----------|----------|
| Wine | 178 | 13 | 3 | 2, 3 | 1 | Australian | 690 | 14 | 2 | 0 | 1 |
| Sonar | 208 | 60 | 2 | M | R | Pima | 768 | 8 | 2 | 0 | 1 |
| Newthyroid | 215 | 5 | 3 | 1 | 2, 3 | Vehicle | 846 | 18 | 4 | Van | Bus, opel, saab |
| Heart | 270 | 13 | 2 | 1 | 2 | German | 1000 | 23 | 2 | 1 | 2 |
| Bupa | 345 | 6 | 2 | 2 | 1 | Splice | 3175 | 60 | 2 | +1 | −1 |
| Iono | 351 | 34 | 2 | G | B | Spambase | 4597 | 57 | 2 | 0 | 1 |
| Wdbc | 569 | 30 | 2 | B | M | Occupancy | 20560 | 5 | 2 | 0 | 1 |
| Balance | 625 | 4 | 3 | B, R | L | Bankmarketing | 45211 | 51 | 2 | No | Yes |

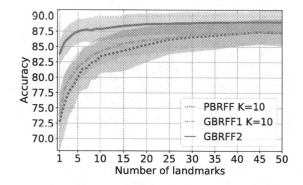

**Fig. 2.** Mean test accuracy over 20 train/test splits over the 16 datasets. We train the three methods using from 1 to 50 landmarks.

## 4.2 Influence of Learning the Landmarks

We present in Fig. 2 the behavior of the three methods that make use of landmarks and RFF, that is **PBRFF**, **GBRFF1** and **GBRFF2**. With more than 25 landmarks, **PBRFF** and **GBRFF1** show similar mean accuracy and reach about 87.5% after 50 iterations. However, for a small set of landmarks (in particular smaller than 25) **GBRFF1** is consistently superior by about 1 point higher than **PBRFF**, showing the interest of learning the landmarks. But the certainly most striking result comes from the performance of our variant **GBRFF2** which outperforms the two competing methods. This is particularly true for a small amount of landmarks. Notice that **GBRFF2** is able to reach its maximum with about 20 landmarks, while **GBRFF1** and **PBRFF** require more iterations without reaching the same performance. This definitely shows the benefit of learning the random features compared to drawing them randomly.

## 4.3 Influence of the Number of Random Features

A key parameter of **GBRFF1** is $K$, the number of random features used at each iteration. To highlight its impact, we report in Fig. 3 the mean test accuracy of

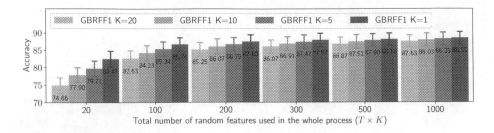

**Fig. 3.** Mean results over the 16 datasets *w.r.t.* the same total number of random features $T \times K$ for $K \in \{1, 5, 10, 20\}$, with $T$ the number of boosting iterations.

**GBRFF1** with $K \in \{1, 5, 10, 20\}$ across all datasets and over the 20 train/test splits. To have a fair study, the comparison is performed according to the same total number of random features after the whole boosting process, that is $T \times K$ with $T$ the number of iterations. First of all, we observe that with a total of $1,000$ random features, $K$ does not have a big impact on the performance. However, when decreasing the value of $T \times K$, it becomes much more interesting in terms of accuracy to set $K$ to a small value. This shows that the more we want a compact final representation, the more we need to refine the random features: it is better to weight each of the features greedily with $\alpha^t$ (line **8** of Algorithm 3) rather than using the closed-form solution of Eq. (10) (line **7** of Algorithm 2) to weight them all at once. Even if in the usual context of RFF it is desirable to have a large $K$ value to approximate a kernel, this series of experiments shows that a simple rough approximation with $K=1$ along with a sufficient number of iterations allows the final ensemble to mimic the approximation of a new kernel suited for the task at hand.

### 4.4    Influence of the Number of Samples on the Computation Time

The specificities of **GBRFF2** come from the number of random features $K$ set to 1 at each iteration and the learning of $\omega^t$. We have already shown in Fig. 2 that this allows us to get better results. We study in this section how **GBRFF2** scales compared to the other methods. To do so, we consider artificial datasets with an increasing number of samples (generated with scikit-learn [9] library's `make_classification` function). The initial size is set to 150 samples, and we successively generate datasets with a size equal to the previous dataset size multiplied by 1.5. Here, we do not split the datasets in train and test as we are not interested in the accuracy. We report the time in seconds necessary to train the models and to predict the labels on the whole datasets. The parameters are fixed as follows: $C = 1$ for the methods using SVM or SVR; the tree depth is set to 5 for **LGBM**; $K = 10$, $\gamma = \frac{1}{d}$, and $\beta = 1$ for **PBRFF** and **GBRFF1**; $\gamma = \frac{1}{d}$ and $\lambda = 0$ for **GBRFF2**. All the methods are run with 100 iterations (or landmarks) and are not run on datasets requiring more than 1000 s of execution time (because larger datasets requiring more than 1000 s by the fastest method

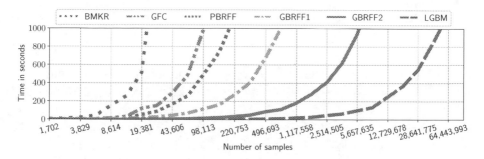

**Fig. 4.** Computation time in seconds required to train and test the six methods with fixed parameters on an artificial dataset having an increasing number of samples. The whole dataset is used for training and testing, and a method requiring more than 1000 seconds at a given step is not trained on the larger datasets.

do not fit in the RAM memory of the computer used for the experiments). We report the results in Fig. 4.

We first recall that **GBRFF2** learns at each iteration a random feature and a landmark while **GBRFF1** only learns the landmark and **PBRFF** draws them randomly. Thus, **GBRFF1** should present higher computation times compared to **PBRFF**. However, for datasets with a number of samples larger than 20,000, **GBRFF1** becomes cheaper than **PBRFF**. This is due to the fact that the SVM classifier learned by **PBRFF** does not scale as well as gradient boosting-based methods. The two-step method **GFC** is in addition also slower than **GBRFF1**. This shows the computational advantage of having a one-step procedure to learn both the representation and the final classifier. When looking at the time limit of 1000 seconds, both **GBRFF1** and **GBRFF2** are the fastest kernel-based methods compared to **BMKR**, **GFC** and **PBRFF**. This shows the efficiency of learning kernels in a greedy fashion. We also see that **GBRFF2** performs faster than **GBRFF1** for any number of samples. At the limit of 1000 seconds, it is able to deal with datasets that are 10 times larger than **GBRFF1**, due to the lower complexity of the learned weak learner used in **GBRFF2**. Finally, **GBRFF2** is globally the second-fastest method behind the gradient boosting method **LGBM** that uses trees as base classifiers.

### 4.5 Performance Comparison Between All Methods

Table 2 presents for each dataset the mean results over the 20 splits using 100 iterations/landmarks for each method. Due to the size of the dataset "bankmarketing", we do not report the results of the algorithms that do not converge in time for this dataset, and we compute the average ranks and mean results over the other 15 datasets. In terms of accuracy, **GBRFF2** shows very good results compared with the state-of-the-art as it obtains the best average rank among the six methods and on average the best mean accuracy leaving apart

**Table 2.** Mean test accuracy ± standard deviation over 20 random train/test splits. A '-' in the last row indicates that the algorithm did not converge in time on this dataset. Average ranks and mean results are computed over the first 15 datasets.

| Dataset | BMKR | GFC | PBRFF | GBRFF1 | LGBM | GBRFF2 |
|---|---|---|---|---|---|---|
| Wine | **99.5** ± 1.0 | 99.3 ± 1.1 | 98.1 ± 2.1 | 98.3 ± 1.5 | 96.6 ± 3.2 | 98.5 ± 1.6 |
| Sonar | 78.8 ± 7.2 | 76.6 ± 3.2 | 76.7 ± 5.2 | 81.8 ± 3.5 | 82.4 ± 4.3 | **83.0** ± 5.0 |
| Newthyroid | 96.5 ± 1.7 | 96.5 ± 2.1 | 96.5 ± 1.5 | 95.3 ± 2.2 | 94.8 ± 2.9 | **96.9** ± 2.1 |
| Heart | **85.6** ± 4.0 | 79.4 ± 4.5 | 85.4 ± 3.5 | 83.6 ± 4.0 | 83.0 ± 3.5 | 83.1 ± 4.0 |
| Bupa | 68.1 ± 4.9 | 64.7 ± 3.2 | 69.0 ± 4.2 | 70.3 ± 4.9 | **72.0** ± 3.3 | 71.2 ± 4.5 |
| Iono | **94.2** ± 1.4 | 91.5 ± 2.3 | 94.2 ± 1.8 | 88.2 ± 2.3 | 93.3 ± 2.5 | 89.2 ± 2.1 |
| Wdbc | 96.1 ± 1.2 | 95.8 ± 1.3 | 96.5 ± 1.1 | 96.8 ± 1.1 | 95.8 ± 1.5 | **97.3** ± 1.2 |
| Balance | 96.0 ± 1.2 | 95.1 ± 2.0 | **98.9** ± 1.1 | 97.7 ± 0.7 | 93.5 ± 2.6 | 97.7 ± 0.6 |
| Australian | 85.9 ± 2.0 | 80.9 ± 2.4 | 84.6 ± 2.3 | 86.7 ± 1.7 | 85.5 ± 1.9 | **86.9** ± 1.9 |
| Pima | 76.4 ± 2.0 | 68.7 ± 2.6 | 76.1 ± 2.5 | 76.5 ± 2.7 | 75.5 ± 2.7 | **77.1** ± 2.5 |
| Vehicle | 96.6 ± 1.3 | 95.9 ± 0.8 | 96.5 ± 1.4 | 96.3 ± 1.2 | 96.7 ± 1.0 | **97.1** ± 1.0 |
| German | 72.3 ± 1.8 | 64.3 ± 2.8 | 72.4 ± 1.4 | 73.7 ± 1.6 | 73.5 ± 1.7 | **74.0** ± 1.3 |
| Splice | 87.5 ± 1.0 | 87.0 ± 1.0 | 83.5 ± 0.7 | 83.9 ± 1.1 | **97.0** ± 0.5 | 92.4 ± 0.8 |
| Spambase | 93.5 ± 0.4 | 91.3 ± 0.6 | 91.6 ± 0.7 | 90.7 ± 0.7 | **95.6** ± 0.4 | 92.8 ± 0.6 |
| Occupancy | **99.3** ± 0.1 | 98.9 ± 0.7 | 98.9 ± 0.1 | 98.8 ± 0.1 | 99.3 ± 0.1 | 98.9 ± 0.1 |
| Mean | 88.4 ± 2.1 | 85.7 ± 2.0 | 87.9 ± 2.0 | 87.9 ± 2.0 | 89.0 ± 2.1 | **89.1** ± 2.0 |
| Average Rank | 2.88 | 4.94 | 3.75 | 3.81 | 3.44 | **2.19** |
| Bankmarketing | – | – | – | 89.7 ± 0.2 | **90.8** ± 0.2 | 90.0 ± 0.2 |

"bankmarketing". Interestingly, our method is the only kernel-based one that scales well enough to be applied to this latter dataset.

## 4.6    Comparison of LGBM and GBRFF2 on Toy Datasets

In this last experiment, we focus on **LGBM** and **GBRFF2** which have been shown to be the two best performing methods in terms of accuracy and execution time. Even if **BMKR** is among the three best methods in terms of accuracy, we do not consider it for this experiment due to its poor execution time. Learning a classifier based on non-linear kernels through **GBRFF2** has the advantage of being able to capture non-linear decision surfaces, whereas **LGBM** is not well suited for this because it uses trees as base learner. To illustrate this advantage, we consider three synthetics 2D datasets with non-linearly separable classes. The first one, called "swiss", represents two spirals of two classes side by side. The second one, namely "circles", consists of four circles with the same center and an increasing radius by alternating the class of each circle. The third dataset, called "board", consists of a four by four checkerboard with alternating classes in each cell. Here, both **LGBM** and **GBRFF2** are run for 1000 iterations to ensure their convergence and parameters are tuned by CV as previously.

Figure 5 gives evidence that **GBRFF2** is able to achieve better results than **LGBM** using only a small amount of training examples, *i.e.*, 500 or less. The performances are asymptotically similar for both methods on the board and circle datasets with a faster rate of convergence for **GBRFF2**. Furthermore, if we look at the decision boundaries and their associated performances at train and test time, we can see that **LGBM** is prone to overfit the training data compared to our approach, showing a drastic drop in performance between learning and testing. The learned decision boundaries are also smoother with **GBRRF2** than with **LGBM**. These experiments show the advantage of having a non-linear weak learner in a gradient boosting approach.

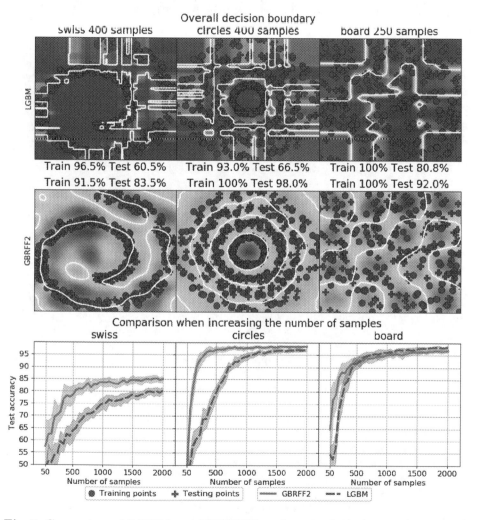

**Fig. 5.** Comparison of **LGBM** and **GBRFF2** on three synthetic datasets in terms of classification accuracy and decision boundaries (upper part of the figure) and in terms of performance *w.r.t.* the number of examples (last row of plots).

## 5   Conclusion and Perspectives

In this paper, we take advantages of two machine learning approaches, gradient boosting and random Fourier features, to derive a novel algorithm that jointly learns a compact representation and a model based on random features. Building on a recent work [6], we learn a kernel by approximating it as a weighted sum of RFF [10]. The originality is that we learn such kernels so that the representation and the classifier are jointly optimized. We show that we can benefit from a performance boost in terms of accuracy and computation time by considering each weak learner as a single trigonometric feature and learning the random part of the RFF. The experimental study shows the competitiveness of our method with state-of-the-art boosting and kernel learning methods.

The optimization of the random feature and of the landmark at each iteration can be computationally expensive when the number of iterations is large. A promising future line of research to speed-up the learning is to derive other kernel approximations where these two parameters can be computed with a closed-form solution. Other perspectives regarding the scalability include the use of standard gradient boosting tricks [5] such as sampling or learning the kernels in parallel.

**Acknowledgements.** Work supported in part by French projects APRIORI ANR-18-CE23-0015, LIVES ANR-15-CE23-0026 and IDEXLYON ACADEMICS ANR-16-IDEX-0005, and in part by the Canada CIFAR AI Chair Program.

## References

1. Agrawal, R., Campbell, T., Huggins, J., Broderick, T.: Data-dependent compression of random features for large-scale kernel approximation. In: the 22nd International Conference on Artificial Intelligence and Statistics, pp. 1822–1831 (2019)
2. Balcan, M., Blum, A., Srebro, N.: Improved guarantees for learning via similarity functions. In: the 21st Annual Conference on Learning Theory, pp. 287–298 (2008)
3. Drineas, P., Mahoney, M.W.: On the Nyström method for approximating a gram matrix for improved kernel-based learning. J. Mach. Learn. Res. **6**, 2153–2175 (2005)
4. Friedman, J.H.: Greedy function approximation: a gradient boosting machine. Ann. Stat. **29**, 1189–1232 (2001)
5. Ke, G., et al.: Lightgbm: a highly efficient gradient boosting decision tree. In: Advances in Neural Information Processing Systems, pp. 3146–3154 (2017)
6. Letarte, G., Morvant, E., Germain, P.: Pseudo-Bayesian learning with kernel fourier transform as prior. In: The 22nd International Conference on Artificial Intelligence and Statistics, pp. 768–776 (2019)
7. Mason, L., Baxter, J., Bartlett, P.L., Frean, M.: Functional gradient techniques for combining hypotheses. In: Advances in Neural Information Processing Systems, pp. 221–246 (1999)
8. Oglic, D., Gärtner, T.: Greedy feature construction. In: Advances in Neural Information Processing Systems, pp. 3945–3953 (2016)
9. Pedregosa, F., et al.: Scikit-learn: machine learning in python. J. Mach. Learn. Res. **12**, 2825–2830 (2011)

10. Rahimi, A., Recht, B.: Random features for large-scale kernel machines. In: Advances in Neural Information Processing Systems, pp. 1177–1184 (2008)
11. Sinha, A., Duchi, J.C.: Learning kernels with random features. In: Advances in Neural Information Processing Systems, pp. 1298–1306 (2016)
12. Vincent, P., Bengio, Y.: Kernel matching pursuit. Mach. Learn. **48**(1–3), 165–187 (2002)
13. Wu, D., Wang, B., Precup, D., Boulet, B.: Boosting based multiple kernel learning and transfer regression for electricity load forecasting. In: Altun, Y. (ed.) ECML PKDD 2017. LNCS (LNAI), vol. 10536, pp. 39–51. Springer, Cham (2017). https://doi.org/10.1007/978-3-319-71273-4_4

# A General Machine Learning Framework for Survival Analysis

Andreas Bender(✉)⬤, David Rügamer⬤, Fabian Scheipl⬤, and Bernd Bischl⬤

Department of Statistics, LMU Munich, Ludwigstr. 33, 80539 Munich, Germany
andreas.bender@stat.uni-muenchen.de

**Abstract.** The modeling of time-to-event data, also known as survival analysis, requires specialized methods that can deal with censoring and truncation, time-varying features and effects, and that extend to settings with multiple competing events. However, many machine learning methods for survival analysis only consider the standard setting with right-censored data and proportional hazards assumption. The methods that do provide extensions usually address at most a subset of these challenges and often require specialized software that can not be integrated into standard machine learning workflows directly. In this work, we present a very general machine learning framework for time-to-event analysis that uses a data augmentation strategy to reduce complex survival tasks to standard Poisson regression tasks. This reformulation is based on well developed statistical theory. With the proposed approach, any algorithm that can optimize a Poisson (log-)likelihood, such as gradient boosted trees, deep neural networks, model-based boosting and many more can be used in the context of time-to-event analysis. The proposed technique does not require any assumptions with respect to the distribution of event times or the functional shapes of feature and interaction effects. Based on the proposed framework we develop new methods that are competitive with specialized state of the art approaches in terms of accuracy, and versatility, but with comparatively small investments of programming effort or requirements for specialized methodological know-how.

**Keywords:** Survival analysis · Gradient boosting · Neural networks · Competing risks · Multi-state models

## 1 Introduction

Survival analysis is a branch of statistics that provides a framework for the analysis of time-to-event data, i.e., the outcome is defined by the time it takes until an event occurs. Analysis of such data requires specialized techniques because, in contrast to standard regression or classification tasks,

(a) the outcome can often not be observed fully (censoring, truncation),

© Springer Nature Switzerland AG 2021
F. Hutter et al. (Eds.): ECML PKDD 2020, LNAI 12459, pp. 158–173, 2021.
https://doi.org/10.1007/978-3-030-67664-3_10

(b) the features can change their value during the observation period (time-varying features (TVF),

(c) the association of the feature(s) with the outcome changes over time (time-varying effects (TVE)),

(d) one or more other events occur that make it impossible to observe the event of interest (competing risks (CR)),

(e) more generally, in a multi-state setting, observation units can move from and to different states (multi-state models (MSM)).

Failure to take these issues into consideration usually results in biased estimates, incorrect interpretation of feature effects on the outcome, loss of predictive accuracy or a combination thereof. In this work, we use a reformulation of the survival task to a standard regression task that provides a holistic approach to survival analysis. Within this framework, censoring, truncation and time-varying features (TVF) can be incorporated by specific data transformations and extensions to time-varying effects (TVE) as well as competing risks and multi-state models can be re-expressed in terms of interaction effects. This abstraction of the survival task away from specialized algorithms is illustrated schematically in Fig. 1. Task-appropriate pre-processing (leftmost subgraph) yields a standardized data format that allows the estimation of feature-conditional hazard rates using any learning algorithm that can minimize the negative Poisson log-likelihood, such as, GBT, deep neural networks (DNN), regularization based methods, and others (middle subgraph).

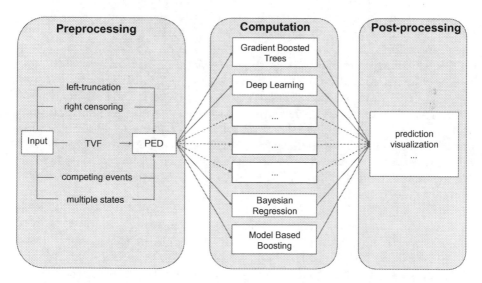

**Fig. 1.** An abstraction of survival analysis for different tasks. The structure of the piece-wise exponential data (PED) depends on the task requirements, e.g., left truncation or competing risks. Given the appropriate pre-processing, the estimation step is computationally independent of the survival task, except for an appropriate use of interaction terms.

*Our Contributions*

We define a general machine learning framework for survival analysis based on piece-wise exponential models (cf. Sect. 2). Within this framework, different concepts specific to time-to-event data analysis can be understood in terms of data augmentation and inclusion of interaction terms. By re-expressing the survival task as a Poisson regression task, a large variety of algorithms become available for survival analysis. Based on the proposed approach, we implement a gradient boosted trees algorithm with comparatively low development effort and show that it achieves state-of-the-art performance (cf. Sect. 3).

*Related Work*

The machine learning community has developed many highly efficient methods for high-dimensional settings in different domains, including survival analysis. The individual methods and implementations, however, often only support a subset of the cases relevant for time-to-event analysis mentioned above. For example, the random survival forest (RSF) proposed in [22] was later extended to the competing risks setting [21], but does not support left-truncation, TVF and TVE, or multistate models. Another popular implementation of random forests [35] only supports right censored data and proportional hazards models. An extension of RSF, the oblique RSF (ORSF, [23]) was shown to outperform other RSF based algorithms, but has the same limitations. With respect to TVF and TVE, a review of tree- and forest-based methods for survival analysis stated that "the modeling of time–varying features and time–varying effects deserves much more attention" [6]. Similarly, a more recent review of machine learning methods for survival analysis [34] only lists the time-dependent Cox model [25, Ch. 9.2], and L1- and L2-regularized extensions thereof, as a possibility for the inclusion of TVF.

Deep learning based methods for time-to-event data have also received much attention lately. An early use of neural networks for Cox type models was proposed by [10]. More recently, [31] presented a framework for deep single event survival analysis based on a joint latent process for features and survival times using deep exponential families. For competing risks data, a deep learning framework based on Gaussian processes was described in [1]. Another recent framework is DeepHit, which can handle competing risks using a custom loss function [28] and was extended to handle TVF [27], but did not discuss left-truncation, multistate models and TVE.

Boosting has also been a popular technique for high-dimensional survival analysis. For example, [5] propose a Cox-type boosting approach for the estimation of proportional sub-distribution hazards. A flexible multi-state model based on the stratified Cox partial likelihood in the context of model-based boosting [17] is presented in [32]. Furthermore, an implementation of gradient boosted trees (GBT) for the Cox PH model is also available for the popular XGBoost implementation [8], which was also shown to perform well compared with the ORSF [23]. Recently, [29] derived a custom algorithm for gradient boosted trees that support TVF and demonstrate that their inclusion improves predictive performance compared to boosting algorithms that don't take TVF into account.

Compared to methods based on Cox regression, few publications have developed methods based on the piece-wise exponential model, on which the framework proposed here is based. Among them is an early application of neural networks to survival analysis suggested in [30] and extended by [4]. The latter offers a general framework based on the representation of generalized linear models via feed forward neural networks, but does not discuss MSM. Piece-wise exponential trees with TVF and splits based on the piece-wise exponential survival function were suggested by [19]. A spline based estimation of the hazard function was discussed in [7], which could also be represented via neural networks (cf. [11]). A flexible estimation of piece-wise-exponential model based multi-state models with shared effects using structured fusion Lasso was developed in [33]. All of these methods can be viewed as special cases within the proposed framework. For example, [4] could be extended to different neural network architectures and MSMs, [19] could be extended to forests.

## 2   Survival Analysis as Poisson Regression

In the context of survival analysis, an observation usually consists of a tuple $(t_i, \delta_i, \mathbf{x}_i)$, where $t_i$ is the observed event time for observation unit $i = 1, \ldots, n$, $\delta_i \in \{0, 1\}$ is the event- or status-indicator (i.e. 1 if event occurred, 0 if the observation of censored) and $\mathbf{x}_i$ is the $p$-dimensional feature vector. The presence of censoring requires special estimation techniques, as the time-to-event can not be observed when censoring occurs before the event of interest. Thus $t_i = \min(T_i, C_i)$, where $T_i$ and $C_i$ random variables of the event time and censoring time, respectively. A classic example is the time until death when censoring occurs as patients drop out of the study (unrelated to the event of interest, $T_i \perp C_i$). Left-truncation occurs when the event of interest already occurred before the subject could be included into the sample and thus presents a form of sampling bias. In some settings, another event could preclude observation of the event of interest or change the probability of its occurrence. In this case we speak of competing risks (CR), thus the observation consists of $(t_i, \delta_i, k, \mathbf{x}_i)$, where $k = 1, \ldots, K$ indicates the type of event that occurred at, $t_i$ if $\delta_i = 1$. More generally, there might be multiple states that the observation units can transition from and to. We then speak of multi-state models (MSM) and $k$ is an indicator for different transitions (cf. Eq. (8)).

In general, the goal of survival analysis is to estimate the conditional distribution of event times defined by the survival probability $S(t|\mathbf{x}) = P(T > t|\mathbf{x})$. While some methods focus on the estimation of $S(t|\mathbf{x})$ directly, it is often more convenient to estimate the (log-)hazard

$$\lambda(t|\mathbf{x}) := \lim_{\Delta t \to 0} \frac{P(t \leq T < t + \Delta t | T \geq t, \mathbf{x})}{\Delta t} \tag{1}$$

from which $S(t|\mathbf{x})$ follows as

$$S(t|\mathbf{x}) = \exp\left(-\int_0^t \lambda(s|\mathbf{x})\mathrm{d}s\right). \tag{2}$$

Here we represent (1) via

$$\lambda(t|\mathbf{x}(t)) = \exp(g(\mathbf{x}(t), t)), \tag{3}$$

where $g$ is a general function of potentially TVF $\mathbf{x}(t)$, that can include high-order feature interactions, non-linearity and time-dependence of feature effects (TVE) via an interaction with $t$.

In this work, we approximate (3) using the piece-wise exponential model [13]. Let $t_i$ the observed event or censoring time and $\delta_i \in \{0, 1\}$ the respective censoring or event indicator for observation units $i = 1, \ldots, n$. The distribution of censoring times can depend on features but is assumed to be independent of the event time process $T$. By partitioning the follow-up, i.e., the time span under investigation, into $j = 1, \ldots, J$ intervals with cut-points $\kappa_0 = 0 < \cdots < \kappa_J$ and partitions $(\kappa_0, \kappa_1], \ldots, (\kappa_{j-1}, \kappa_j], \ldots (\kappa_{J-1}, \kappa_J]$, we can rewrite (3) using piece-wise constant hazard rates

$$\lambda(t|\mathbf{x}_i(t)) \equiv \exp(g(\mathbf{x}_{ij}, t_j)) := \lambda_{ij}, \quad \forall t \in (\kappa_{j-1}, \kappa_j], \tag{4}$$

with $t_j$ a representation of time in interval $j$, e.g., $t_j := \kappa_j$ and $\mathbf{x}_{ij}$ the value of the TVF in interval $j$. Depending on the desired resolution, additional cut-points can be introduced at each time point at which feature values are updated, otherwise multiple feature values have to be aggregated in one interval. This model assumes that only the current value of $\mathbf{x}_{ij}$ affects the hazard in interval $j$, but more sophisticated approaches have been suggested within this framework that take into account the entire history of TVF [3]. Piece-wise constant hazards imply piece-wise exponential log-likelihood contributions

$$\ell_i = \log(\lambda(t_i; \mathbf{x}_i)^{\delta_i} S(t_i; \mathbf{x}_i)) = \sum_{j=1}^{J_i} (\delta_{ij} \log \lambda_{ij} - \lambda_{ij} t_{ij}), \tag{5}$$

where $J_i$ is the last interval in which observation unit $i$ was observed, such that $t_i \in (\kappa_{J_i-1}, \kappa_{J_i}]$ and

$$\delta_{ij} = \begin{cases} 1 & t_i \in (\kappa_{j-1}, \kappa_j] \wedge \delta_i = 1 \\ 0 & \text{else} \end{cases}, \quad t_{ij} = \begin{cases} t_i - \kappa_{j-1} & \delta_{ij} = 1 \\ \kappa_j - \kappa_{j-1} & \text{else} \end{cases}. \tag{6}$$

Concrete examples for the type of data transformations required to obtain (6) for right-censored data (including TVF) are provided in [2] (cf. Tables 1 and 2.

Using the working assumption $\delta_{ij} \overset{iid}{\sim} Poisson(\mu_{ij} = \lambda_{ij} t_{ij})$ and with $f(\delta_{ij})$ the Poisson density function, [13] showed that the Poisson log-likelihood

$$\ell_i = \log \left( \prod_{j=1}^{J_i} f(\delta_{ij}) \right) = \sum_{j=1}^{J_i} (\delta_{ij} \log \lambda_{ij} + \delta_{ij} \log t_{ij} - \lambda_{ij} t_{ij}) \tag{7}$$

is proportional to (5) and therefore the former can be minimized using Poisson regression. Note that (7) can be directly extended to the setting with left-truncated event times [16] by replacing $j = 1$ with $j_i$, the first interval in which observation unit $i$ is in the risk set. The expectation is defined by $\mu_{ij} = \lambda_{ij} t_{ij} = \exp(g(\mathbf{x}_{ij}, t_j) + \log(t_{ij}))$. For estimation, $\log(t_{ij})$ is included as an offset, thus the hazard rate $\frac{\mu_{ij}}{t_{ij}} = \lambda_{ij} = g(\mathbf{x}_{ij}, t_j)$ is defined as the conditional expectation of having an event in interval $j$ divided by the time under risk. Note that the Poisson assumption is simply a computational vehicle for the estimation of the hazard (4) rather than an assumption about the distribution of the event times. Despite the partition of the follow-up into intervals, this is a method for continuous event times as the information about the time under risk in each interval is contained in the offset and thus used during estimation. The number and placement of cut points controls the approximation of the hazard and could thus be viewed as a potential tuning parameter. In our experience, however, setting cut points at the unique event times $\{t_i : \delta_i = 1, i = 1, \ldots, n\}$ in the training data always leads to a good approximation (at least with enough regularization) as the number of cut-points will increase in areas with many events. For larger data sets, however, we recommend to set these cut-points on a smaller representative sub-sample of the data set (cf. Sect. 4).

For the extension of (3) to MSMs, we define

$$\lambda(t|\mathbf{x}, k) = \exp\left(f(\mathbf{x}(t), t, k)\right), \quad k = 1, \ldots, K, \tag{8}$$

as the transition specific hazard for the transition indexed by $k$, i.e., $k$ is an index of transitions $m_k \to m'_k$ where $m_k$ is the initial state and $m'_k$ a transient or absorbing state. The set of possible transitions is given by $\{m_k \to m'_k : k = 1, \ldots, K\} \subseteq \{m \to m' : m, m' \in \{0, \ldots, M\}, m \neq m'\}$, where $M + 1$ denotes the total number of possible states. $f(\mathbf{x}(t), t, k)$ is a function of potentially time-varying features $\mathbf{x}(t)$, including multivariate and/or non-linear effects. The dependency of $f(\mathbf{x}(t), t, k)$ on time $t$ (TVE) and transition $k$ (MSM) is expressed in terms of interactions by defining $\tilde{\mathbf{x}} := (\mathbf{x}(t), t, k)$ and $f(\mathbf{x}(t), t, k) = f(\tilde{\mathbf{x}}(t))$. Let $t_{i,k}$ be the event or censoring time w.r.t. transition $m_k \to m'_k$ and $\delta_{i,k} \in \{0, 1\}$ the respective transition indicator. As extension of (6) we define

$$\delta_{ij,k} = \begin{cases} 1 & t_{i,k} \in (\kappa_{j-1}, \kappa_j] \wedge \delta_{i,k} = 1 \\ 0 & \text{else,} \end{cases}, t_{ij,k} = \begin{cases} t_{i,k} - \kappa_{j-1} & \delta_{ij,k} = 1 \\ \kappa_j - \kappa_{j-1} & \text{else} \end{cases}.$$

Table 1 shows how the data must be transformed in order to estimate (8) via PEMs for the competing risks setting, i.e., $k = 1, \ldots, K$ is an index of transitions $m_k = 0 \to m'_k, m'_k \in \{1, \ldots, M\}$; a concrete example is given in Table 2. For each $i = 1, \ldots, n$, there is one row for each interval the observation unit was under risk for a specific transition. Thus, one data set is created for each transition such that transitions to state $m'_k$ are encoded as 1 and everything else, i.e., censoring and transition to other states is encoded as 0. These transition-specific data sets, each containing a feature vector with the transition index $k$, are then concatenated. Note that we used the same interval split points $\kappa_j$ for all transitions in Table 1. However, it would also be possible to choose transition specific cut-points $\kappa_{j,k}$, or, more generally, even use multiple time-scales

[20]. In the general multi-state setting, the number of observation units under risk might depend on the transition and the intervals visited by $i$ are defined by $(t_{i,m}, \kappa_{j_{i,k}}], \ldots, (\kappa_{J_{i,k}-1}, \kappa_{J_{i,K}}]$, where $t_{i,m}$ is the time-point at which $i$ enters state $m$.

**Table 1.** Data structure after transformation to the piece-wise exponential data format in the competing risks setting. Horizontal lines indicate a new observation unit $i = 1, \ldots, n$. Double horizontal lines indicate a new transition indexed $k = 1, \ldots, K$. As before, $J_i$ is the interval in which $i$ was observed last, i.e., $t_i \in (\kappa_{J_i-1}, \kappa_{J_i}]$. Features $x$ depend on time via $x_{i,p}(t) = x_{ij,p} \; \forall t \in (\kappa_{j-1}, \kappa_j], p = 1, \ldots, P$. This is not a strict requirement, as additional split points could be set at each time point a feature value is updated. See Table 2 for a concrete example.

| $i$ | $j$ | $\delta_{ij,k}$ | $t_j$ | $t_{ij}$ | $k$ | $x_{ij,1}$ | $\ldots$ | $x_{ij,P}$ |
|---|---|---|---|---|---|---|---|---|
| 1 | 1 | $\delta_{11,1}$ | $t_1$ | $t_{11}$ | 1 | $x_{11,1}$ | $\ldots$ | $x_{11,P}$ |
| 1 | 2 | $\delta_{12,1}$ | $t_2$ | $t_{12}$ | 1 | $x_{12,1}$ | $\ldots$ | $x_{12,P}$ |
| $\vdots$ | $\vdots$ | $\vdots$ | $\vdots$ | $\vdots$ | $\vdots$ | $\vdots$ | $\vdots$ | $\vdots$ |
| 1 | $J_1$ | $\delta_{1J_1,1}$ | $t_{J_1}$ | $t_{1J_1}$ | 1 | $x_{1J_1,1}$ | $\ldots$ | $x_{1J_1,P}$ |
| 2 | 1 | $\delta_{21,1}$ | $t_1$ | $t_{21}$ | 1 | $x_{21,1}$ | $\ldots$ | $x_{21,P}$ |
| $\vdots$ | $\vdots$ | $\vdots$ | $\vdots$ | $\vdots$ | $\vdots$ | $\vdots$ | $\vdots$ | $\vdots$ |
| $n$ | 1 | $\delta_{n1,1}$ | $t_1$ | $t_{n1}$ | 1 | $x_{n1,1}$ | $\ldots$ | $x_{n1,P}$ |
| $\vdots$ | $\vdots$ | $\vdots$ | $\vdots$ | $\vdots$ | $\vdots$ | $\vdots$ | $\vdots$ | $\vdots$ |
| $n$ | $J_n$ | $\delta_{nJ_n,1}$ | $t_{J_n}$ | $t_{nJ_n}$ | 1 | $x_{nJ_n,1}$ | $\ldots$ | $x_{nJ_n,P}$ |
| 1 | 1 | $\delta_{11,2}$ | $t_1$ | $t_{11}$ | 2 | $x_{11,1}$ | $\ldots$ | $x_{11,P}$ |
| $\vdots$ | $\vdots$ | $\vdots$ | $\vdots$ | $\vdots$ | $\vdots$ | $\vdots$ | $\vdots$ | $\vdots$ |
| $\vdots$ | $\vdots$ | $\vdots$ | $\vdots$ | $\vdots$ | $\vdots$ | $\vdots$ | $\vdots$ | $\vdots$ |
| 1 | 1 | $\delta_{11,K}$ | $t_1$ | $t_{11}$ | $K$ | $x_{11,1}$ | $\ldots$ | $x_{11,P}$ |
| $\vdots$ | $\vdots$ | $\vdots$ | $\vdots$ | $\vdots$ | $\vdots$ | $\vdots$ | $\vdots$ | $\vdots$ |

In Sect. 3 we evaluate the suggested approach using an implementation based on GBT that we refer to as GBT (PEM). As a concrete computing engine we used the extreme gradient boosting (XGBoost) library [8] without any alterations to the algorithm. Therefore all features of the library can be used directly when estimating the hazard on the transformed data set. Note, however, that depending on the algorithm used, one must be able to specify an offset during estimation and potentially some other, algorithm or implementation specific settings. For example, when using XGBoost to estimate the GBT (PEM), the objective function needs to be set to the Poisson objective and the base score must be set to 1, because the default of 0.5 would imply a wrong offset, while $\log(1) = 0$. The offset $(\log(t_{ij}))$ must be attached to the data via the base margin

**Table 2.** Data transformation for a hypotethical competing risks example ($K = 2$) for 3 subjects, $i = 1, \ldots, 3$. Subject $i = 1$ experienced an event of type $k = 2$ at $t_1 = 1.3$, subject $i = 3$ experienced an event of type $k = 1$ at time $t_3 = 2.7$, subject 2 was censored at $t_2 = 0.5$. Tables present the transformed data with intervals $(0, 1], (1, 1.5], (1.5, 3]$ for causes $k = 1$ (left) and $k = 2 = K$ (right). These can be used to estimate cause specific hazards, by applying the algorithm to each of the tables separately or cause specific hazards with potentially shared effects, by stacking the tables and using $k$ as a feature.

| $i$ | $j$ | $\delta_{ij}$ | $t_j$ | $t_{ij}$ | $k$ |
|---|---|---|---|---|---|
| 1 | 1 | 0 | 1 | 1 | 1 |
| 1 | 2 | 0 | 1.5 | 0.3 | 1 |
| 2 | 1 | 0 | 1 | 0.5 | 1 |
| 3 | 1 | 0 | 1 | 1 | 1 |
| 3 | 2 | 0 | 1.5 | 0.5 | 1 |
| 3 | 3 | 1 | 3 | 1.2 | 1 |

| $i$ | $j$ | $\delta_{ij}$ | $t_j$ | $t_{ij}$ | $k$ |
|---|---|---|---|---|---|
| 1 | 1 | 0 | 1 | 1 | 2 |
| 1 | 2 | 1 | 1.5 | 0.3 | 2 |
| 2 | 1 | 0 | 1 | 0.5 | 2 |
| 3 | 1 | 0 | 1 | 1 | 2 |
| 3 | 2 | 0 | 1.5 | 0.5 | 2 |
| 3 | 3 | 0 | 3 | 1.2 | 2 |

argument during estimation. In contrast, for the prediction of the conditional hazard $\lambda(t|\mathbf{x}_i) - \lambda_{ij}$ based on new data points, the offset should be omitted, otherwise the algorithm will predict $\hat{\mu}_{ij} = \hat{\lambda}_{ij} \cdot t_{ij}$ (the expectation) instead of $\hat{\lambda}_j(\mathbf{x})$ (the hazard). When predicting the cumulative hazard or survival probability, however, the time under risk in each interval must be taken into account, such that $\hat{S}(t|\mathbf{x}_i) = \exp\left(-\int \hat{\lambda}(t|\mathbf{x}_i)dt\right) = \exp\left(-\sum_{j=1}^{j(t)} \hat{\lambda}_{ij}\tilde{t}_j\right)$, where $j(t)$ indicates the interval for which $t \in (\kappa_{j-1}, \kappa_j]$ and $\tilde{t}_j = \min(\kappa_j - \kappa_{j-1}, t - \kappa_{j-1})$ is the time spent in interval $j$. A prototype implementation of the GBT (PEM) algorithm that takes these issues into account and also provides the necessary helper functions for data transformation, estimation, tuning, prediction and evaluation is provided at https://github.com/adibender/pem.xgb.

# 3 Experiments

We perform a set of benchmark experiments with real world and synthetic data sets, exclusively using openly and directly available data, including a subset of data sets from recent publications on oblique random survival forests (ORSF, [23]) and DeepHit [28]. DeepHit and ORSF both have been shown to outperform other approaches such as RSF [22], conditional forests [18], regularized Cox regression [12] and DeepSurv [31]. We compare our approach in benchmarks against the two algorithms, which are evaluated separately based on evaluation measures used in the respective publications to ensure comparability. All code to perform respective analyses as well as additional supplementary files are provided in a GitHub repository: https://github.com/adibender/machine-learning-for-survival-ecml2020.

The data sets used for single event comparisons are listed in Table 3. The "synthetic (TVE)" data set is created based on an additive predictor $g(\mathbf{x}, t) = f_0(x_0, t) \cdot 6 - 0.1 \cdot x_1 + f_2(x_2, t) + f_3(x_3, t)$, where $f_0$, $f_2$ and $f_3$ are bivariate, non-linear functions of the inputs (see code repository for details) and $x_0, \ldots, x_3$ feature columns comprised in $\mathbf{x}$. Additionally, 20 noise variables are drawn from the uniform distribution $U(0, 1)$.

For the comparison with ORSF, we use the Integrated Brier Score $\text{IBS}(\tau) = \frac{1}{\tau} \int_0^\tau \widehat{\text{BS}}(u, \hat{S}) \mathrm{d}u$, where $\widehat{\text{BS}}(t, \hat{S})$ is the estimated Brier Score at time $t$ weighted by the inverse probability of censoring weights [15] and $\hat{S}$ the estimated survival probability function of the respective algorithm. In addition, [23] report the time-dependent C-Index [14]. We only consider the IBS here as it measures calibration as well as discrimination, while the C-index only measures the latter. Note that the IBS depends on the specific evaluation time $\tau$ and different methods might perform better at different evaluation times. Therefore, we calculate the IBS for three different time-points, the 25%, 50% and 75% quantiles of the event times in the test data, in the following referred to as Q25, Q50 and Q75, respectively.

**Table 3.** Data sets used in benchmark experiments for comparison with ORSF.

| | Name | N | P | Censoring |
|---|---|---|---|---|
| 1 | PBC | 412 | 14 | 61.90 |
| 2 | Breast | 614 | 1690 | 78.20 |
| 3 | GBSG 2 | 686 | 8 | 56.40 |
| 4 | Tumor | 776 | 7 | 51.70 |
| 5 | Synthetic (TVE) | 1000 | 24 | ~33% |

For comparison with DeepHit, we use the metabric data set (cf. [28]) for single event comparison as well as two CR data sets. The "MGUS 2" data is described in [26]. The "synthetic (TVE CR)" data set is simulated using an additive predictor identical to the one used for the "synthetic (TVE)" data simulation for the first cause. The predictor for the second cause, has a simpler structure $f_0(x_0, t) + 2 \cdot x_4 - .1 \cdot x_5$, however, with non-proportional baseline hazard with respect to $x_0 \in \{-1, 1\}$. The number of noise variables is limited to 10 for this setting. Here we report the weighted C-index alongside the weighted Brier Score as it was the main measure reported in [28]. The proposed GBT (PEM) approach for CR (cf. Sect. 2) is a cause specific hazards model, however, the parameters of both causes are estimated jointly and the hazards of both causes can have shared effects (see Fig. 2). The simulation setting, therefore, constitutes a difficult setup because there are no shared effects and optimization w.r.t. the first cause will favor parameters that allow flexible models while the optimization w.r.t. the second cause favors sparse models and thus parameters that would restrict flexibility.

**Table 4.** Data sets used for the comparison with DeepHit. MGUS2 and Synthetic (TVE CR) are data set with two competing risks and additional right-censoring.

| Name | N | P | Censoring (%) |
|------|------|----|------|
| METABRIC | 1981 | 79 | 55.20 |
| MGUS 2 | 1384 | 6 | 29.6 |
| Synthetic (TVE CR) | 500 | 14 | ~23% |

## 3.1 Evaluation

We compare four algorithms, the non-parametric Kaplan Meier estimate (Reference) as a minimal baseline, the Cox proportional hazards model [9] (baseline for linear, time-constant effects), the Oblique Random Survival Forest (ORSF) [23] and DeepHit [28]. For each experimental replication for a specific data set, 70% of the data is randomly assigned as training data and the remaining 30% is used to calculate the evaluation measures at three time points Q25, Q50 and Q75. Algorithms are tuned on the training data using random search with a fixed budget and 4-fold cross-validation. Each algorithm is then retrained on the entire training data set using the best set of parameters before making final predictions on the test set. The random search consists of 20 iteration for each algorithm. For the GBT (PEM), we define the search space as follows (possible range in brackets): maximum tree depth $\{1, \ldots, 20\}$, minimum loss reduction $[0, 5]$, minimum child weight $\{5, \ldots, 50\}$, subsample percentage (rows) $[0.5, 1]$, subsample percentage of features in each tree $[.5, 1]$, L2-regularization $[1, 3]$. The learning rate is set to 0.05 and number of rounds to 5000, with early stopping after 50 rounds without improvement. For the ORSF we tune the elastic net mixing parameter $(0, 1)$, the parameter that penalizes complexity of the linear predictor in each node $(0.25, 0.75)$, minimum number of events to split node $\{5, \ldots, 20\}$ and minimum observations to split node $\{10, 40\}$. For DeepHit, we use 50 random search iterations, where we search through $\{1,2,3,5\}$ shared layers with $\{50, 100, 200, 300\}$ dimensions, $\{1,2,3,5\}$ cause-specific network layers with $\{50,100,200,300\}$ dimensions, ReLU, eLU or Tanh as activation function in these layers, a batch size in $\{32,64,128\}$, a maximum of 50000 iterations, a dropout rate of 0.6 (taken from the original paper) and a learning rate of 0.0001. The network specific parameters $\alpha$ and $\gamma$ are also chosen in accordance with the original paper and set to 1 and 0, respectively, while the network specific parameter $\beta$ is varied in the random search with possible values in $\{0.1, 0.5, 1, 3, 5\}$.

## 3.2 Results

The results for the experiments based on single-event scenarios comparison with ORSF are summarized in Table 5. The proposed method performs well in many settings in comparison to ORSF. Notably, both algorithms are often not much better than the Cox PH models indicating that the PH assumption is not violated strongly in those data sets and the sample size might be too small to detect small deviations w.r.t. to non-linearity of feature effects, interaction effects and time-varying effects. The "synthetic (TVE)" setting illustrates that in the presence of strong, non-linear and non-linearly TVE our approach clearly outperforms the other methods. For the PBC data we additionally ran an analysis including TVF with GBT (PEM). In this case, the inclusion of TVF resulted in a worse performance (IBS of 4.3 (Q25), 6.4 (Q50) and 9.2 (Q75)), which indicates that the inclusion of TVF lead to overfitting or that simple inclusion of the last observed value and carrying the last value forward is not appropriate in this setting.

**Table 5.** Results of benchmark experiments for single event data comparing GBT (PEM) with ORSF. Bold numbers indicate the best performance for each setting.

| Data | | Kaplan-Meier | Cox-PH | ORSF | GBT (PEM) |
|---|---|---|---|---|---|
| Breast | Q25 | **1.9** | – | 2.0 | 2.0 |
| | Q50 | 4.1 | – | **4.0** | **4.0** |
| | Q75 | 7.2 | – | **6.7** | **6.7** |
| GBSG 2 | Q25 | 3.1 | 3.1 | **2.9** | 3.0 |
| | Q50 | 6.8 | 6.5 | **6.2** | 6.4 |
| | Q75 | 12.5 | 11.4 | **11.1** | 11.3 |
| PBC | Q25 | 5.4 | **3.7** | 4.0 | 3.8 |
| | Q50 | 9.1 | **5.3** | 6.1 | 5.5 |
| | Q75 | 14.0 | 8.1 | 8.6 | **7.8** |
| Synthetic (TVE) | Q25 | 9.8 | 7.3 | 7.0 | **4.6** |
| | Q50 | 19.2 | 10.3 | 9.9 | **6.7** |
| | Q75 | 23.7 | 11.1 | 11.7 | **8.6** |
| Tumor | Q25 | 6.7 | 6.0 | **5.5** | 5.8 |
| | Q50 | 12.3 | 11.2 | **10.8** | 10.9 |
| | Q75 | 17.6 | 16.3 | **16.2** | **16.2** |

Table 6 summarizes the results of comparisons with DeepHit. The GBT (PEM) again shows good overall performance. For the synthetic data set our method clearly outperforms the other approaches because it is capable of estimating non-linearity as well as time-variation. On the MGUS 2 data set, Deep-Hit shows the best performance for cause 1, while GBT (PEM) outperforms the other approaches for cause 2. On the synthetic data, the cause-specific Cox-PH model shows good discrimination (C-Index) for the second cause, but is worse than GBT (PEM) and DeepHit w.r.t. to the Brier Score.

**Table 6.** Results of benchmark experiments comparing GBT (PEM) with DeepHit for single event and CR data. Bold numbers indicate the best performance for each setting.

| Data | Index | Method | Cause 1 | | | Cause 2 | | |
|------|-------|--------|-----|-----|-----|-----|-----|-----|
| | | | Q25 | Q50 | Q75 | Q25 | Q50 | Q75 |
| METABRIC | Brier Score | Cox-PH | 13.3 | 22.1 | 26.4 | – | – | – |
| | | DeepHit | 14.3 | 23.5 | 27.0 | – | – | – |
| | | GBT (PEM) | **12.8** | **21.3** | **25.9** | – | – | – |
| | C-Index | Cox-PH | 63.7 | 65.1 | 64.7 | – | – | – |
| | | DeepHit | 68.6 | 63.3 | 54.9 | – | – | – |
| | | GBT (PEM) | **71.9** | **71.5** | **67.7** | – | – | – |
| MGUS 2 | Brier Score | Cox-PH (CS) | 23.6 | 43.7 | 64.3 | 13.4 | 20.5 | 22.3 |
| | | DeepHit | **22.8** | **41.0** | **57.8** | 14.9 | 27.0 | 41.5 |
| | | GBT (PEM) | 22.9 | 41.6 | 60.5 | **13.0** | **20.1** | **22.1** |
| | C-Index | Cox-PH (CS) | 66.7 | **65.9** | **62.4** | 68.8 | 69.4 | 70.1 |
| | | DeepHit | 59.6 | 57.0 | 52.3 | 65.5 | 67.2 | 68.6 |
| | | GBT (PEM) | **68.4** | 62.9 | 60.5 | **72.6** | **70.9** | **70.8** |
| Synthetic (TVE, CR) | Brier Score | Cox-PH | 9.4 | 13.1 | 25.1 | 35.5 | 44.3 | 50.6 |
| | | DeepHit | 9.5 | 16.0 | 28.9 | 33.0 | 38.8 | **41.0** |
| | | GBT (PEM) | **7.2** | **11.6** | **20.6** | **30.1** | **38.0** | 43.6 |
| | C-Index | Cox-PH | 90.2 | 89.5 | 85.4 | **86.5** | **83.9** | **81.6** |
| | | DeepHit | 92.3 | 90.8 | 84.6 | 82.0 | 80.1 | 79.8 |
| | | GBT (PEM) | **93.9** | **92.2** | **87.5** | 80.9 | 80.8 | 81.0 |

# 4  Algorithmic Details and Complexity Analysis

We now briefly describe algorithmic details and discuss the complexity of the resulting algorithms when using the proposed framework.

*Algorithmic Details*

The proposed framework is general in the sense that it transforms a survival task into a regression task. Nevertheless, different methods (and algorithms) have different strengths and weaknesses and different strategies can be applied to specify various alternative models within this framework. For example, in tree based methods, time-variation of feature effects could be controlled by allowing interactions of the time variable only with a subset of features, e.g. based on prior information, and similarly in order to control shared vs. transition specific effects in the multi-state setting. Tree-based methods are particularly intuitive when it comes to understanding the integration of TVE and extension to multi-state models via interaction terms into the model. This is illustrated in Fig. 2. For example, in panel (A) of Fig. 2, features and split points before the split w.r.t. time indicate feature effects common to all time-points. Once the data in panel (A) is split w.r.t. time $t$, the predicted hazard will be different for

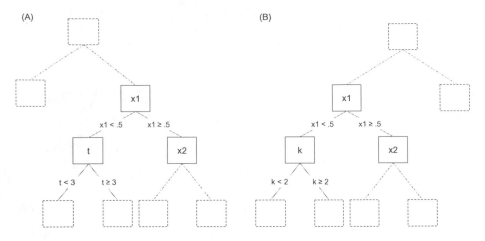

**Fig. 2.** Illustration of how TVE and shared vs. transitions specific effects can be understood in terms of feature interactions in tree based models.

intervals with $\kappa_j < 3$ and $\kappa_j \geq 3$ for observations with $x_1 < .5$. Similarly, in a multi-state setting (panel (B) in Fig. 2), splits above the split w.r.t. $k$ indicate shared effects for all transitions, while splits below indicate different effects for transitions $k < 2$ vs. $k \geq 2$. Forcing a split w.r.t. to $k$ at the root node would be equivalent to an estimation of cause specific hazards on each subset and no shared effects.

Neural networks are particularly flexible when it comes to the specification of different PEMs. For example, the network could be split in two subnetworks, one for the temporal component, one for features, which is equivalent to the specification of a proportional hazards model, while allowing for non-linearity and high-dimensional interactions in feature effects. Similarly, defining subnetworks of the time variable for each category of a categorical feature would imply a stratified proportional hazards model.

*Complexity Analysis*
As described in the literature review, various approaches exist that account for special survival characteristics like TVF, CR or continuous time-scale prediction by altering the underlying method. While adapting the structure of the algorithm itself potentially increases the complexity of the method, our approach leaves the algorithm of choice unchanged as different time points and transitions are simply included as features. This allows to employ commonly used prediction methods without introducing further algorithmic complexity. We note, however, that our approach might be improved upon in terms of scaling with respect to the number of intervals $J$ relative to the number of observations $n$. In the worst case, the number of total data points is quadratic in $n$ (or more precisely $\mathcal{O}(n(n+1)/2)$) when one interval cut-point is introduced for each observed event or censoring time. We therefore propose a refinement of the presented method that improves run-times without forfeiting performance. Instead of setting cut-points at all

unique event times, we suggest to define cut points more sparsely, for example, based on a sub-sample of the original data.

To investigate this strategy we conduct a scaling experiment where the sample size was consecutively doubled starting from $n = 400$ up to $n = 3200$. For each sample size, ten replications of one experiment as described for the "synthetic TVE" setting in Sect. 3 were performed and the elapsed time (hours) as well as performance (IBS) for two different strategies of cut-point selection was measured. The first strategy (full) uses all event times ($t_i$ where $\delta_i = 1$) as cut-points. The second strategy (sub-sample) is equivalent to the first strategy, but event times were chosen based on a sub-sample of $n' = 200$, selected randomly from the training data in each iteration. Results in Table 7 show that the "sub-sample" strategy leads to an approximately linear increase in computation time while the performance remains virtually unchanged. Potentially, a sparser choice of cut-points could also lead to a more robust and thus improved hazard estimation, as more events are available in each interval, but we did not conduct a formal investigation in that regard.

**Table 7.** Results of the scaling experiments with $n$ the number of observation in the simulated data. "strategy" refers to the way interval cut-points were selected with "full" splits at all unique event times ($t_i$ where $\delta_i = 1$) and "sub-sample" refers to the selection of cut-points based on unique event times based on a random sub-sample of size $n' = 200$ (but all observations were used for estimation). Mean time (in hours) and IBS over 10 replications are reported for each setting.

|  | Strategy | N | | | |
|---|---|---|---|---|---|
|  |  | 400 | 800 | 1600 | 3200 |
| Time (hours) | Full | 0.10 | 0.48 | 2.49 | 8.94 |
|  | Sub-sample | **0.09** | **0.20** | **0.51** | **1.04** |
| IBS | Full | 8.10 | 6.50 | 6.40 | **5.90** |
|  | Sub-sample | **8.00** | **6.40** | **6.20** | **5.90** |

## 5   Conclusion

We have presented a general machine learning framework for time-to-event analysis based on a data augmentation strategy that reduces a large variety of survival analysis tasks to the optimization of a Poisson likelihood. We demonstrated its versatility and state-of-the-art performance. The availability of Poisson regression for most machine learning frameworks provides additional practical advantages. For example, photon-ML [37] is a scalable machine learning library for Apache Spark [36] that has no native support for survival analysis, but implements generalized linear mixed models. Therefore, survival modeling with high cardinality random effects (frailty) is directly available using our framework. Similarly, lightGBM [24], a high-performance implementation of GBT, currently has no implementation of survival methods, but could be also used for high-dimensional survival tasks based on PEMs, including reliability analysis or churn analysis with intermediate states.

**Acknowledgements.** This work has been funded by the German Federal Ministry of Education and Research (BMBF) under Grant No. 01IS18036A. The authors of this work take full responsibilities for its content.

# References

1. Alaa, A.M., van der Schaar, M.: Deep multi-task gaussian processes for survival analysis with competing risks. In: Proceedings of the 31st International Conference on Neural Information Processing Systems, pp. 2326–2334 (2017)
2. Bender, A., Groll, A., Scheipl, F.: A generalized additive model approach to time-to-event analysis. Statistical Modelling p. 1471082X17748083 (2018)
3. Bender, A., Scheipl, F., Hartl, W., Day, A.G., Küchenhoff, H.: Penalized estimation of complex, non-linear exposure-lag-response associations. Biostatistics **20**(2), 315–331 (2018)
4. Biganzoli, E., Boracchi, P., Marubini, E.: A general framework for neural network models on censored survival data. Neural Netw. **15**(2), 209–218 (2002)
5. Binder, H., Allignol, A., Schumacher, M., Beyersmann, J.: Boosting for high-dimensional time-to-event data with competing risks. Bioinformatics **25**(7), 890–896 (2009)
6. Bou-Hamad, I., Larocque, D., Ben-Ameur, H.: A review of survival trees. Stat. Surv. **5**, 44–71 (2011)
7. Cai, T., Hyndman, R.J., Wand, M.P.: Mixed model-based hazard estimation. J. Comput. Graph. Stat. **11**(4), 784–798 (2002)
8. Chen, T., Guestrin, C.: XGBoost: a scalable tree boosting system. In: Proceedings of the 22nd ACM SIGKDD International Conference on Knowledge Discovery and Data Mining - KDD 2016, pp. 785–794 (2016). arXiv: 1603.02754
9. Cox, D.R.: Regression models and life-tables. J. Royal Stat. Soc. Series B (Methodological) **34**(2), 187–220 (1972)
10. Faraggi, D., Simon, R.: A neural network model for survival data. Stat. Med. **14**(1), 73–82 (1995)
11. Fornili, M., Ambrogi, F., Boracchi, P., Biganzoli, E.: Piecewise exponential artificial neural networks (PEANN) for modeling hazard function with right censored data. In: Formenti, E., Tagliaferri, R., Wit, E. (eds.) CIBB 2013 2013. LNCS, vol. 8452, pp. 125–136. Springer, Cham (2014). https://doi.org/10.1007/978-3-319-09042-9_9
12. Friedman, J.H., Hastie, T., Tibshirani, R.: Regularization paths for generalized linear models via coordinate descent. J. Stat. Softw. **33**(1), 1–22 (2010). number: 1
13. Friedman, M.: Piecewise exponential models for survival data with covariates. Ann. Stat. **10**(1), 101–113 (1982)
14. Gerds, T.A., Kattan, M.W., Schumacher, M., Yu, C.: Estimating a time-dependent concordance index for survival prediction models with covariate dependent censoring. Stat. Med. **32**(13), 2173–2184 (2013)
15. Gerds, T.A., Schumacher, M.: Consistent estimation of the expected brier score in general survival models with right-censored event times. Biometrical J. **48**(6), 1029–1040 (2006)

16. Guo, G.: Event-history analysis for left-truncated data. Sociol. Methodol. **23**, 217–243 (1993)
17. Hothorn, T., Bühlmann, P.: Model-based boosting in high dimensions. Bioinformatics **22**(22), 2828–2829 (2006)
18. Hothorn, T., Hornik, K., Zeileis, A.: Unbiased recursive partitioning: a conditional inference framework. J. Comput. Graph. Stat. **15**(3), 651–674 (2006)
19. Huang, X., Chen, S., Soong, S.j.: Piecewise exponential survival trees with time-dependent covariates. Biometrics **54**(4), 1420–1433 (1998)
20. Iacobelli, S., Carstensen, B.: Multiple time scales in multi-state models. Stat. Med. **32**(30), 5315–5327 (2013)
21. Ishwaran, H., et al.: Random survival forests for competing risks. Biostatistics **15**(4), 757–773 (2014)
22. Ishwaran, H., Kogalur, U.B., Blackstone, E.H., Lauer, M.S.: Random survival forests. Ann. Appl. Stat. **2**(3), 841–860 (2008)
23. Jaeger, B.C., et al.: Oblique random survival forests. Ann. Appl. Stat. **13**(3), 1847–1883 (2019)
24. Ke, G., et al.: LightGBM: a highly efficient gradient boosting decision tree. In: Guyon, I., Luxburg, U.V., Bengio, S., Wallach, H., Fergus, R., Vishwanathan, S., Garnett, R. (eds.) Advances in Neural Information Processing Systems, vol. 30, pp. 3146–3154. Curran Associates, Inc. (2017)
25. Klein, J.P., Moeschberger, M.L.: Survival Analysis: Techniques for Censored and Truncated Data. Springer, New York (2006)
26. Kyle, R.A., et al.: A long-term study of prognosis in monoclonal gammopathy of undetermined significance. N. Engl. J. Med. **346**(8), 564–569 (2002)
27. Lee, C., Yoon, J., Schaar, M.V.D.: Dynamic-DeepHit: a deep learning approach for dynamic survival analysis with competing risks based on longitudinal data. IEEE Trans. Bio-Med. Eng. **67**(1), 122–133 (2020)
28. Lee, C., Zame, W.R., Yoon, J., Schaar, M.V.d.: DeepHit: a deep learning approach to survival analysis with competing risks. In: Thirty-Second AAAI Conference on Artificial Intelligence (April 2018)
29. Lee, D.K.K., Chen, N., Ishwaran, H.: Boosted nonparametric hazards with time-dependent covariates. arXiv:1701.07926 [stat] (November 2019)
30. Liestbl, K., Andersen, P.K., Andersen, U.: Survival analysis and neural nets. Stat. Med. **13**(12), 1189–1200 (1994)
31. Ranganath, R., Perotte, A., Elhadad, N., Blei, D.: Deep Survival Analysis. arXiv:1608.02158 (August 2016)
32. Reulen, H., Kneib, T.: Boosting multi-state models. Lifetime Data Anal. **22**(2), 241–262 (2015). https://doi.org/10.1007/s10985-015-9329-9
33. Sennhenn-Reulen, H., Kneib, T.: Structured fusion lasso penalized multi-state models. Stat. Med. **35**(25), 4637–4659 (2016)
34. Wang, P., Li, Y., Reddy, C.K.: Machine learning for survival analysis: a survey. ACM Comput. Surv. (CSUR) **51**(6), 110:1–110:36 (2019)
35. Wright, M.N., Ziegler, A.: Ranger: a fast implementation of random forests for high dimensional data in C++ and r. J. Stat. Softw. **77**(1), 1–17 (2017)
36. Zaharia, M., et al.: Apache spark: a unified engine for big data processing. Commun. ACM **59**(11), 56–65 (2016)
37. Zhang, X., Zhou, Y., Ma, Y., Chen, B.C., Zhang, L., Agarwal, D.: Glmix: generalized linear mixed models for large-scale response prediction. In: Proceedings of the 22nd ACM SIGKDD International Conference on Knowledge Discovery and Data Mining, pp. 363–372 (2016)

# Fairness by Explicability and Adversarial SHAP Learning

James M. Hickey$^{(\boxtimes)}$, Pietro G. Di Stefano, and Vlasios Vasileiou

Experian UK&I and EMEA DataLabs, London, UK
james.hickey@experian.com

**Abstract.** The ability to understand and trust the fairness of model predictions, particularly when considering the outcomes of unprivileged groups, is critical to the deployment and adoption of machine learning systems. SHAP values provide a unified framework for interpreting model predictions and feature attribution but do not address the problem of fairness directly. In this work, we propose a new definition of fairness that emphasises the role of an external auditor and model explicability. To satisfy this definition, we develop a framework for mitigating model bias using regularizations constructed from the SHAP values of an adversarial surrogate model. We focus on the binary classification task with a single unprivileged group and link our fairness explicability constraints to classical statistical fairness metrics. We demonstrate our approaches using gradient and adaptive boosting on: a synthetic dataset, the UCI Adult (Census) dataset and a real-world credit scoring dataset. The models produced were fairer and performant.

**Keywords:** Algorithmic fairness · SHAP values · Adversarial learning · Machine learning interpretability

## 1 Introduction

The last few decades have seen machine learning algorithms become even more performant and leverage larger varieties of data. These advances have led to wide-spread adoption of machine learning in nearly every industry. The potential damage and wider societal harm that could be caused by large-scale automated decisioning systems is palpable amongst regulators, industry practitioners and consumers [10,25,34]. Two specific concerns that have emerged center on the interpretability and fairness of the decisions resulting from these algorithms. These are not unjustified with cases of unfair decisioning systems manifesting in multiple domains from criminal recidivism [10] to credit worthiness

Supported by Experian Ltd.

**Electronic supplementary material** The online version of this chapter (https://doi.org/10.1007/978-3-030-67664-3_11) contains supplementary material, which is available to authorized users.

© Springer Nature Switzerland AG 2021
F. Hutter et al. (Eds.): ECML PKDD 2020, LNAI 12459, pp. 174–190, 2021.
https://doi.org/10.1007/978-3-030-67664-3_11

assessment. In the European Union, these concerns have manifest in the General Data Protection Regulation [14, 19] that enshrines each individual's right to fair and transparent processing. This combined societal and legislative scrutiny has resulted in model interpretability and algorithmic fairness coming to the fore in research [13, 31].

At the broadest level, the concept of algorithmic fairness tackles whether members of specific unprivileged groups are more likely to receive unfavourable decisions from the predictions of a machine learning system. Recent advances have enabled modellers to incorporate fairness at every point of the model building process [11, 16, 31, 40]. One embodiment incorporates fairness constraints into the training procedure [1, 6, 12, 17, 20, 29, 29, 33, 47], typically these constraints rely on statistical measures of fairness and are subject to drawbacks [22] and trade-offs. These measures rely on *a priori* worldviews and do not incorporate the role of external model auditing or decision explicability in their fairness criteria. This is poorly aligned with how these issues are dealt within industry, where external actors often question the model fairness through building surrogate explanatory models, even if mentally, using the information available to them.

To address these issues, we propose a new definition of fairness we dub "Fairness by Explicability". Under this definition, if an external actor's surrogate model cannot produce a *narrative* (i.e., a set of explanations) against the fairness of a particular model, then that particular model can be considered *explicably fair*. This definition explicitly frames the perception of an algorithm's fairness as one determined by a combination of an auditor's worldview, data availability, model interpretability framework and measurement/modelling approach. It can be considered complementary to the existing ways of evaluating a model's fairness, since while those may capture risk arising from non-adherence to regulatory requirements, our new "fairness by explicability" viewpoint captures the additional and independent risk that may arise from analyses performed by one's own clients [30].

To enforce our "Fairness by Explicability" definition, we leverage model interpretability methodologies [27, 35, 49] to incorporate fairness constraints through adversarial learning. More explicitly, we utilize the SHAP [26, 27] values of a surrogate adversary model in two ways. The first works by constructing a differentiable fairness regularization term. The second is a modification to the classic AdaBoost algorithm [15] to include adversarial attribution values in the weight updates.

We link our fairness approach to statistical fairness [41] via the construction of an appropriate surrogate model. Our approaches are illustrated using a synthetic dataset, the UCI Adult Census Dataset [5], and a commercial credit scoring dataset. These datasets present a diverse evaluation set, with the real-world dataset providing assurance that these approaches are viable in industrial applications. The structure of the papers is as follows: in Sect. 2 we introduce our notation; in Sect. 3 we provide a brief account of SHAP values and Sect. 4 discusses statistical fairness measures. Section 5 introduces the "Fairness by

Explicability" worldview and in Sect. 6 we present our SHAP-regularized algorithms before discussing the results of the experiments in Sect. 7. We then state our conclusions and highlight areas of further research in Sect. 8.

## 2   Notation

To measure the fairness of any algorithm output one needs to define the task objective, the un-/privileged groups to measure fairness against and the favourable outcomes. For the remainder of this paper, we focus on binary classification tasks with a single privileged group indicator $Z$. We denote the other covariates present with $\mathbb{X}$ and the combination of $Z$ with those covariates by $\tilde{\mathbb{X}}$. Furthermore, and without loss of generality, we define the value of 1 for the target $Y$ and the corresponding model outcomes $\hat{Y}$ as the favourable label. Model outcomes are constructed by applying a threshold to the scores $\bar{Y}$. For each instance $i$, we denote the corresponding values with the appropriate lowercase symbol and subscript, i.e. $y_i$, $z_i$, $x_i$, etc. In this case, $x_i$ and $\tilde{x}_i$ denote vectors and the value of the $j$th covariate is given by $x_{ij}$ and $\tilde{x}_{ij}$.

## 3   SHapley Additive Explanations (SHAP)

SHapley Additive Explanations, or SHAP values [26,27], provide a unified framework for interpreting model predictions. This approach was built off the insight that many other modern explanatory frameworks such as LIME [35] and DeepLIFT [38] could be recast as variants of a generic additive feature attribution paradigm. In this paradigm, a simplified explanatory model $\sigma$ is built to explain the original prediction $f$ using simplified binary input vectors $\tilde{x}'_i \in \{0,1\}^M$, where $M$ is the number of features and $i$ is the instance label. These simplified inputs are related to the original feature vectors $\tilde{x}_i$ through the mapping $\tilde{x}_i = h_{\tilde{x}_i}(\tilde{x}'_i)$ and the local explanatory model is given by:

$$\sigma(\tilde{x}'_i) = \phi_0^{i,f} + \sum_{j=1}^{M} \phi_j^{i,f} \tilde{x}'_{ij}. \tag{1}$$

The local feature effect of feature $j$ for model $f$ is $\phi_j^{i,f}$ and global explanations are calculated via the statistics of these values across a dataset. The different explanatory frameworks, e.g. LIME, emerge from specific choices of the mapping function $h_{\tilde{x}_i}$, the kernel weighting of instances in the objective ($\pi_{\tilde{x}}$) and any additional regularization terms $\Omega(\sigma)$ used to fit $\sigma$. These choices influence the properties of the surrogate model. In Ref. [27], they showed that only one $\sigma$ satisfies these 3 desirable properties:

1. Local Accuracy: $f(\tilde{x}_i) = \sigma(\tilde{x}'_i) = \sum_{j=0}^{M} \phi_j^{i,f}$, when $\tilde{x}_i = h_{\tilde{x}_i}(\tilde{x}'_i)$.
2. Missingness: $\tilde{x}'_{ij} = 0 \implies \phi_j^{i,f} = 0$.

3. Attribution Consistency: for any two models $f, f'$, the ordering of the differences of the model output when a feature is present vs missing is reflected in
their respective attributions of that feature.

Its attributions $\phi_j$ are the same Shapley Values first identified in cooperative
game theory [24, 37, 39, 44]:

$$\phi_j^{\bullet, f} = \sum_{z' \subseteq \tilde{x}'} \frac{|z'|!(M - |z'| - 1)!}{M!} [f_{\tilde{x}}(z') - f_{\tilde{x}}(z' \setminus j)]. \tag{2}$$

Here, $|z'|$ is the number of non-zero entries in $z'$, $z' \setminus j$ denotes setting the $j$th
element of $z'$ to 0 and the summation is over all $z'$ where the non-zero entries are
a subset of the non-zero entries of $\tilde{x}'$. These SHAP values can be estimated for
a generic model using KernelSHAP [27] while for specific model families there
are efficient computational methods and analytic approximations [26, 39].

## 4  Metrics and Statistical Fairness

To estimate fairness metrics one requires a dataset of $N$ instances with $Y$ and
$Z$ as well as the outcomes. Given this data, the appropriate fairness metric is
often defined by the worldview(s) [43] of those auditing the outcomes. These
worldviews tend to fall into three broad categories: "We're all equal" [2], "What
you see is what you get" [13, 36] and causal [9, 12, 21, 23, 42, 48]. The first two
categories are statistical in nature and we now discuss their application to the
binary task domain.

   Statistical fairness metrics relate to the conditional probabilities involving $Y$,
$\hat{Y}$ and $Z$. The "We're all equal" worldview has numerous group fairness metrics
associated with it. These metrics measure any differences in outcome given group
membership and seek to balance said outcomes. Contrastingly, "What you see is
what you get" asserts that the observed data captures the underlying "truth" and
typically prefers to offer individuals similar outcomes conditional on $Y$. In this
work, we consider two of the most common statistical fairness metrics from these
categories: "statistical parity" difference (SPD) and "equality of opportunity"
difference (EOD). More formally, these are defined as:

$$\text{SPD} = |P(\hat{Y} = 1|Z = 1) - P(\hat{Y} = 1|Z = 0)|, \tag{3}$$

$$\text{EOD} = |P(\hat{Y} = 1|Y = 1, Z = 1)$$
$$- P(\hat{Y} = 1|Y = 1, Z = 0)|. \tag{4}$$

   Note that a target SPD value can also be calculated by replacing $\hat{Y}$ with
$Y$ respectively in Eq. 3. Both of these measures are estimated from a specified
dataset, their value of zero denotes a maximally fair model, and both have trade-
offs [22] and limitations. For example, SPD can be minimized through randomly
modifying outcomes while ignoring all other covariates $\mathbb{X}$ and so can be viewed
as a lazy penalization. Contrastingly, minimizing EOD may not reduce any gap
in the rate of favourable outcomes between the groups.

## 5 Fairness by Explicability

The traditional statistical fairness metrics presented in Sect. 4 are not explicitly linked to the domain of model interpretability. Recent work [7] demonstrated empirically that the SHAP values of $Z$ could capture statistical unfairness provided $Z$ was used as a feature of the model. To formalize an explicit link between model fairness and explicability, we first recall that statistical fairness measures emerge from the worldviews of individuals auditing the model outcomes for fairness. Typically, when trying to understand observations, a human agent (an external actor/auditor) will construct a surrogate model to obtain explanations for their observations. The role of $Z$ in these explanations determines whether the outcomes constructed are perceived as fair or not. Building on this idea, we propose a new worldview to capture the mechanism by which model decisions are evaluated by external actors.

**Definition 1.** *Consider a model trained by an auditor to predict $\bar{Y}$ using $Z$ and, optionally, a combination from $\{Y, \mathbb{X}\}$. If this model does not detect any difference in the $Z$ attribution between the $Z = 0, 1$ groups, then the predictor model is explicably fair with respect to the auditor.*

We dub this worldview "Fairness by Explicability". The precise measure of fairness one attains is determined by: the population examined by the auditor, the interpretability framework used, how attributions are calculated and aggregated, and the auditor model developed. This definition can be specialized into a *strong* "Fairness by Explicability" form by further requiring that total attribution for $Z$ is also reduced to zero.

Auditors are usually interested in the average attribution of the two groups given a population of data. This informs the metrics used to quantify how "explicably fair" a model is. These are:

$$\text{FE} = |\frac{\sum_{i, s.t. Z=1} \phi_Z^{i,l}}{N_1} - \frac{\sum_{i, s.t. Z=0} \phi_Z^{i,l}}{N_0}|, \tag{5}$$

$$\text{SFE} = \frac{\sum_i |\phi_Z^{i,l}|}{N}, \tag{6}$$

where $\phi_Z^{i,l}$ is the SHAP value of $Z$ for instance $i$ for auditor model $l$, $N$ is the total number of instances in the dataset and $N_{1(0)}$ is number of examples when $Z = 1(0)$. FE measures the difference in mean attribution between the two groups. When it is minimized the model is considered fair according to our "Fairness by Explicability" definition. The second metric (SFE) measures the total attribution of $Z$ across the population, when minimized the auditor model concludes that the model satisfies the strong version of "Fairness by Explicability" and, by definition, the first metric is also zero. These metrics are equivalent to those of Ref. [7] but in this instance are applied to an external auditor model and are informed by how a typical auditor would aggregate their explicability scores.

From this discussion, "Fairness by Explicability" may appear intuitive but difficult to implement and, in general, being "explicably fair" does not provide any guarantees of statistical fairness. However, an initial informal connection to the prior fairness worldviews can be made through consideration of specific forms of the auditor models. Intuitively, removing the dependency on $Z$ as measured by an external $l$ will tend to reduce $\bar{Y}$'s dependency on $Z$. This will generally lead to improved SPD and EOD although the decision policy plays a large role in how these two connect.

# 6    Achieving Fairness by Explicability

We now present two different approaches for imposing "Fairness by Explicability" directly into the training process of gradient-based and adaptive boosting (specifically AdaBoost) algorithms. These approaches rely on inserting a surrogate model $g$ directly into the iterative training procedures. The form of $g$ is then chosen to account for the examination of an anticipated external auditor whose model is $l$. Both approaches require $Z$ during the training phase only, hence any sensitive attributes defining $Z$ do not need to be supplied at prediction time. In addition to this presentation, we also discuss how the approaches can be linked to the SPD and EOD.

## 6.1    SHAPSqueeze

The first approach to imposing "Fairness by Explicability" uses a series of differentiable regularizations to penalize unfair attributions. We consider a differentiable loss function of the form:

$$\mathcal{L}_{\text{fair}} = (1 - \lambda) * \mathcal{L}_o + \lambda * \mathcal{R}, \tag{7}$$

which we can optimize through gradient-based methods, e.g. stochastic gradient descent. At each iteration, a surrogate model $g$ is fit to the $\bar{Y}$ values. From $g$, the SHAP values of $Z$, and optionally $Y$, are used to calculate the appropriate regularization term ($\mathcal{R}$). In this work, $\mathcal{L}_o$ is the binary cross-entropy. Considering the case where $l$ and $g$ are identical, when the associated $\mathcal{R}$ is minimized then the attributions to $Z$ will be zero and *strong* "Fairness by Explicability" is satisfied by the model scores $\bar{Y}$.

The specific form of $g$ we examine is a linear regression model, see the first row of Table 1. The SHAP values of interest are given by:

$$\phi_Z^{i,g} = \beta(z_i - \mathbb{E}[Z]). \tag{8}$$

Equation (8) directly relates the SHAP values of $Z$ to its model coefficient, $\beta$, and the specific realisation of $Z$ for instance $i$. The regularization $\mathcal{R}$ is then simply the sum of the squares of these SHAP values scaled by a constant $C$, see Table 1. This constant is used to make the size of the gradients coming from $\mathcal{R}$ and $\mathcal{L}_o$ comparable, while $\lambda$ is used to adjust the balance between these two quantities. Moreover, we note that the explicability fairness metrics in Eq. 5 are proportional to $\beta$ in this instance. Therefore, these specific $g$ and $\mathcal{R}$ will seek to eliminate the linear dependence of the model predictions $\bar{Y}$ on $Z$. Consequently, we expect reductions in the SPD as the model becomes explicably fairer.

To conclude, we note that the use of linear regression makes both the model fitting and SHAP value derivative calculations computationally efficient to perform. However, the approach described is applicable to any $g$ whose SHAP values are differentiable with respect to $\bar{Y}$ and so parametric/kernel regression models could also be employed. In combination with adding more features, this can allow for the consideration of more complex auditors with different worldviews.

Table 1. The surrogate models and regularizations considered in this work.

| Algorithm | Surrogate - $g$ | Regularization |
|---|---|---|
| SHAPSqueeze | $\bar{Y} = \beta Z + \alpha$ | $\mathcal{R} = C \sum_i (\phi_Z^{i,g})^2$ |
| SHAPEnforce | | $\mathcal{P} = \begin{cases} -\phi_Z^{i,g}, \text{if} y_i = 1 \\ 0, \text{otherwise} \end{cases}$ |

## 6.2   SHAPEnforce

The classic AdaBoost algorithm [15] trains a model that is a weighted linear combination of weak classifiers. The training process is iterative, with each weak learner ($k_m$) being fitted to a reweighted version of the training data. After $R$ iterations, the outputted model is given by $C_R = \sum_{m=1}^{R} \alpha_m k_m$. We consider learners that output a score and whose classification output, $\{0, 1\}$, is obtained by thresholding. Traditionally, AdaBoost generates the instance weights for the $m^{\text{th}}$ training round, $\omega_i^m$, by scaling the previous iteration's weights $\omega_i^{(m-1)}$. Instances $k_m$ that are incorrectly classified have their weights enhanced by $e^{\alpha_m}$, while correctly classified instances are downweighted by $e^{-\alpha_m}$. As training proceeds, the algorithm increasingly focuses on erroneous examples to improve its predictive performance. To incorporate "Fairness by Explicability" into AdaBoost, we adjust its reweighting process to consider the SHAP values $\{\phi_j^{i,g}\}, i = 1, \ldots, N$, of the features $\{j\}$ of a surrogate $g$. This SHAP weighting is introduced through a penalty function ($\mathcal{P}(\{\phi_j^{i,g}\})$) and fairness regularization weight ($\lambda$) which trades off the original weight update with the new penalty.

In effect, this forces weak learners to not only focus on erroneous examples but also those with specific SHAP values as determined by $g$ and $\mathcal{P}$. This pushes the algorithm to improve its predictions on instances with specific SHAP values and is dubbed "SHAPEnforce". Furthermore, in contrast to SHAPSqueeze, it is fully non-parametric and only requires that the SHAP values of $g$ can be computed.

Algorithm 1 presents the pseudo-code for "SHAPEnforce". The learning approach can be qualitatively interpreted as a two-player game. Expanding on this view, at each stage the predictive learner makes a move by constructing a weak learner and attempts to reweight the training data as-if the surrogate had not

---

**Algorithm 1:** SHAPEnforce

---

**Input**: training examples $\{(x_i, y_i, z_i)\}_{i=1}^N$, specification of favourable outcome, a surrogate model $g$, a SHAP penalty function $\mathcal{P}$, and the number of boosting rounds $R$.

**Output**: A classifier $C_R(x) = \sum_{m=1}^R \alpha_m k_m(x)$.

1  Initialize weights $\omega_i^1 = 1/N, \forall i$.

2  **for** $m = 1$ ***to*** $R$ **do**

3      Fit a weak learner $k_m(x)$ using the training data with weights $\omega_i^m$.

4      Compute the probability of favourable outcome, $\bar{y}_i^m$, and the predicted label $\hat{y}_i^m$ from $k_m$.

5      Fit $g$ - taking features and the target from $\{(x_i, y_i, z_i, \bar{y}_i^m)\}$.

6      Compute the $\phi_j^{i,g}$, and the corresponding weight adjustment $\mathcal{P}(\{\phi_j^{i,g}\})$.

7      Compute $e_m \leftarrow \mathbb{E}_{\omega^m}[\mathbb{1}_{(y \neq k_m(x))}]$.

8      Compute $\alpha_m = \log((1 - e_m)/e_m)$.

9      Update the instance weights:

$$\omega_i^{(m+1)} \leftarrow \omega_i^m [(1 - \lambda) * e^{\alpha_m(\mathbb{1}_{(y \neq k_m(x))} - \mathbb{1}_{(y = k_m(x))})} + \lambda e^{\mathcal{P}(\{\phi_j^{i,g}\})}].$$

10      Set $\omega_i^{m+1} \leftarrow \frac{\omega_i^{m+1}}{\sum_i \omega_i^{m+1}}$.

11  **end**

---

acted up to that point. Similarly, once the learner is constructed the surrogate acts to reweight the dataset in its own best interest. The regularization weight $\lambda$ then controls the resulting outcome between these two competing actions.

In this work, we consider a linear surrogate model trained on data where $Y = 1$ whose form and associated $\mathcal{P}$ is shown in Table 1. We again compute the SHAP values using Eq. 8. The $\mathcal{P}$ considered is local in nature and, conditioned on $Y = 1$, will downweight any examples with positive SHAP values while upweighting those with negative values. This forces the predictor model to focus on instances where the $Z$ attributions have a negative impact on the favourable outcome and where the weak learner has made mistakes when the target is favourable, i.e. $Y = 1$. By focusing on the examples with negative $Z$ attribution, their $Z$ attribution will be increased at the next round, hence the explicability fairness, as determined by an equivalent $l$, will tend to increase. This choice of $\mathcal{P}$ further reflects the intuition that unprivileged groups are likely to have unfavourable predictions from weak learners and hence negative $Z$ attribution. Furthermore, with the focus on examples where $Y = 1$ we expect this modification to reduce the EOD. Finally, $\mathcal{P}$ is related to a fairness regularization term previously applied to neural networks [4]. Our work formalizes this previously ad-hoc loss as a "Fairness by Explicability" regularizer with an appropriate auditor.

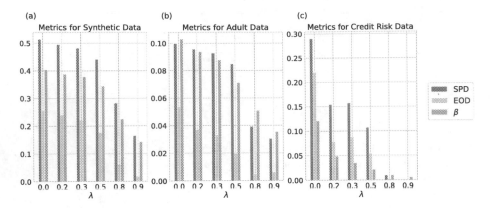

**Fig. 1.** Fairness metrics for SHAPSqueeze plotted with varying regularization strength for the (a) synthetic, (b) Adult and (c) Credit Risk test datasets. We set $C = 1$ for the synthetic data, $C = 10$ for Adult and $C = 100$ for the Credit Risk evaluations.

# 7    Computational Experiments

To evaluate our algorithms we consider three binary classification datasets: a synthetic dataset, the UCI Adult dataset [5], and a commercial Credit Risk dataset. The train/test splits are shown in Table 2. The datasets were preprocessed so categorical variables were one-hot encoded and numeric variables were converted to their standard score.

We exemplify the SHAPSqueeze objectives using XGBoost [8]. In each experiment, we evaluate the algorithms predictive performance, as measured by accuracy/precision and ROC AUC, as well as measuring the SPD and EOD. To determine these quantities, we use a fixed threshold policy. For SHAPSqueeze, in the case of the synthetic and UCI Adult dataset, this threshold is 0.5 while a more risk-averse threshold of 0.85 is set for the commercial Credit Risk dataset. This higher threshold better reflects real-world business practices in this domain. SHAPEnforce, being a modification to AdaBoost, is less calibrated than the SHAPSqueeze implementation and so a threshold of 0.5 is used in all cases. Additionally, we build linear regression auditor models on the test set to measure the explicability fairness. The equations defining $l$ are the same as the $g$ employed, and so the explicability fairness is given by the coefficient $\beta$ of the fitted $l$, see Table 1. Note for SHAPEnforce, $l$ is built on the data subset where $Y = 1$.

## 7.1    Datasets

**Synthetic Data.** The synthetic dataset was generated to exhibit a very large SPD. To construct this, the distribution of $\mathbb{X}$ is conditional on $Z$ and $Y$ is determined by $Z$ and $\mathbb{X}$. Specifically, $Z$ was sampled from a Bernoulli distribution and $\mathbb{X}$ contains three sets of covariates: "safe" covariates $\mathbb{X}_s \sim \mathcal{N}(0, 1)$, "proxy"

**Table 2.** Datasets used for the algorithm evaluation.

| Dataset | Train size | Test size |
|---|---|---|
| Synthetic | 75000 | 25000 |
| Adult | 32561 | 16281 |
| Credit Risk | 48112 | 23697 |

covariates ($\mathbb{X}_p$) and "indirect effect" covariates ($\mathbb{X}_i$). The latter two are sampled from $\mathcal{N}(Z, 1)$. From this, the log-odds of the binary target ($S_Y$) are given by $0.25\mathbf{w} \cdot (\mathbb{X}_i + \mathbb{X}_s) + 1.25Z$, $\mathbf{w}$ is a vector of ones. The target $Y$ is then sampled from Bern $\left(\frac{1}{1+e^{-S_Y}}\right)$. Using this approach we sampled a dataset with 10 safe, 4 indirect effect and 2 proxy variables. Furthermore, the sampled dataset was such that approximately 90% of the favourable outcomes were obtained by the privileged group.

**Adult Census.** The goal is to predict whether a person will have an income below or above \$50k. In this dataset, we consider the variable sex as our protected attribute and removed race, marital status, native country and relationship from our models. The other covariates measure financial information, occupation and education.

**Private Credit Risk Dataset.** In this dataset, we are trying to infer a customer's default probability given curated information on their current account transactions. We are interested in removing bias related to age. We binarize the age variable dividing our examples in two groups, an "older" (unprivileged) group of people over 50 and a "younger" group of people under 50 years old.

## 7.2   Results

Results for SHAPSqueeze on the test datasets are shown in Fig. 1. We observe that across all 3 datasets increasing $\lambda$ induces fairness as observed by reductions in SPD, EOD and $\beta$. We set $C = 1$ for the synthetic dataset, $C = 10$ for Adult and for the Credit Risk dataset we set $C = 100$. These values were chosen to ensure the mean gradients from $\mathcal{L}_o$ and $\mathcal{R}$ in the intermediate stages of training, i.e. $\sim 100$ iterations, when $\lambda = 0.5$ were on the same order of magnitude and effective. For the synthetic data, we observe $\beta$ drops from 0.404 at $\lambda = 0$ to 0.142 at $\lambda = 0.9$. It is accompanied by a tolerable drop in the AUC and accuracy of 0.04 in both cases. Similarly, the SPD is reduced by roughly 0.35 while the EOD is almost eliminated, taking a value of 0.018 at $\lambda = 0.9$. Increasing $\lambda$ further, $\beta$ approaches zero and is faithful to our *strong* "Fairness by Explicability" definition.

We observe the same patterns for the fairness metrics when SHAPSqueeze is applied to the Adult and Credit Risk datasets. In the former, we observe a

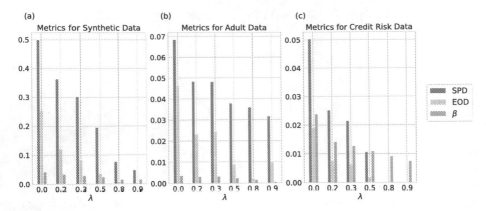

**Fig. 2.** Fairness metrics for SHAPEnforce, using the penalty $\mathcal{P}$ in Table 1, plotted with varying regularization strength. Results for the synthetic, Adult and Credit Risk test datasets are shown in (a), (b) and (c) respectively.

reduction of roughly 0.04 in accuracy and AUC with increasing $\lambda$, at $\lambda = 0.9$ these take values of 0.80 and 0.83 respectively. In the latter case, the precision is reduced by $\approx 0.12$ and the AUC drops by $\approx 0.07$ as we change $\lambda$ from 0 to 0.9. Contrastingly, for Adult, we observe an increase in precision (from 0.76 to 0.98) as the fairness regularization increases the scores beyond the classification threshold. A similar effect is seen in the Credit Risk dataset where we observed an increase in the accuracy from 0.744 to 0.837 as $\lambda$ was increased to 0.9. This increased accuracy is attributed to the conservative threshold of 0.85 employed. This threshold also results in the SPD and EOD being eliminated at $\lambda = 0.9$ as the regularization pushes all of the scores above 0.85. At this point $\beta$ is roughly 0.006 demonstrating that even when the SPD and EOD are zero a model may not be 100% explicably fair. This highlights the differences in fairness definition and, in particular, the use of $\bar{Y}$ and not $\hat{Y}$ when measuring explicable fairness. To avoid this scenario one would either reduce $C$ or select a different $\lambda$ value. At $\lambda = 0.7$, the model has SPD, EOD and $\beta$ values of 0.035, 0.013 and 0.014 respectively. It is also performant with tolerable drops in the AUC (0.06) and precision (0.1) observed.

The results for SHAPEnforce are presented in Fig. 2. In all cases, we observe the EOD, SPD, AUC and accuracy decrease with increasing $\lambda$. For the synthetic data, the accuracy drops by approximately 0.08 from 0.828 to 0.75 as we increase $\lambda$. This is accompanied by a drop of $\approx 0.03$ in the AUC from 0.868 to 0.836 as we change $\lambda$ from 0 to 0.9. Compared to the statistical fairness metrics, we observe smaller improvements in the explicable fairness. Furthermore, the decreasing trend of $\beta$ is less pronounced and consistent compared to SHAP-Squeeze. This was expected for two reasons. Firstly, the unregularized AdaBoost model is explicably fairer than XGBoost and so there is less explicable unfairness to remove. Secondly, we expected the heuristic nature of the modification provides no guarantees on explicable fairness and so the magnitude of the reduction is not guaranteed. For the synthetic dataset we observe a decrease in $\beta$ of $\approx 63\%$ as we increase $\lambda$ from 0 to 0.9. Moving to the Adult results, we observe $\beta$ decreases by approximately $84\%$ on changing $\lambda$ from 0 to 0.9. The SPD and EOD are reduced to 0.03 and 0.01 respectively with tolerable drops in accuracy (0.02) and AUC (0.01) observed. For the Credit Risk data, we again observe explicable fairness improvements, on the order of $69\%$ as we increase $\lambda$. This is accompanied with the SPD and EOD being eliminated for $\lambda > 0.6$. Similar to SHAPSqueeze, this elimination is due to the regularization pushing all scores below the threshold for $\lambda > 0.6$. In practice one would use a model from another $\lambda$, such as $\lambda = 0.2$, where the SPD and EOD are reduced by roughly $50\%$ and $61\%$ respectively while the precision and AUC take values of 0.85 and 0.84 respectively. This represents a drop of $\approx 0.01$ for the former while the latter is consistent with the unregularized model. However, at this point, $\beta$ is only reduced by approximately $40.5\%$ compared to $\lambda = 0$.

## 8    Conclusions

In this work, we developed a novel fairness definition, "Fairness by Explicability", that gives the explanations of an auditor's surrogate model primacy when determining model fairness. We demonstrated how to incorporate this definition into model training using adversarial learning with surrogate models and SHAP values. This approach was implemented through appropriate regularization terms and a bespoke adaptation of AdaBoost. We exemplified these approaches on 3 datasets, using XGBoost in combination with our regularizations, and connected our choices of surrogate model to "statistical parity" and "equality of opportunity" difference. In all cases, the models trained were explicably and statistically fairer, yet still performant. This methodology can be readily extended to other interpretability frameworks, such as LIME [35], with the only constraint being that $\mathcal{R}$ must be appropriately differentiable. Future work will explore more complex surrogate models and different explicability scores in the proposed framework.

## 9    Related Work

In recent years, there has been significant work done in both model interpretability, adversarial learning and fairness constrained machine learning model training.

*Interpretability:* Ref. [27] provided a unified framework for interpreting model predictions. This framework unified several existing frameworks, e.g. LIME [35] and DeepLift [38], and it can be argued to be the "gold standard" for model interpretability. It provides both local and global measures of feature attribution and through the KernelSHAP algorithm, is model agnostic. Further work has introduced computationally efficient approximations to the SHAP values of [27] for tree-based models [26]. Other works in interpretability have focussed on causality for model interpretability. These approaches provide insight into *why* the decision was made, rather than an explanation of the model predictive accuracy and are frequently qualitative in nature. Ref. [32] is a recent exception, where the counterfactual examples generated obey realistic constraints to ensure practical use and are examined quantitatively through bespoke metrics.

*Adversarial Training:* Ref. [3] used adversarial training to remove EOD while a framework for learning adversarially fair representations was developed in Ref. [28]. Similar, in Ref. [47] an adversarial network [18] was used to debias a predictor network, their specific approach compared favourably to the approach of [3].

*Training Fair Models:* Typically, fair learning methodologies have tended to focus on incorporating statistical fairness constraints directly into the model objective function. Ref. [29] combined neural networks with statistical fairness regularizations but their form restricts their applicability to neural networks. Similarly, Ref. [17] trained a fair logistic regression using convex regularizations and addresses proportionally fair classification. Other works have viewed fair model training as one of constrained optimization [45, 46] or have created meta-algorithms for fair classification [6].

In these works, the approaches to fair learning have tended to focus on fairness metrics associated with more traditional worldviews and less focus on model explicability. Similarly, the role of model explicability in fairness, to the authors' knowledge, has not been used directly in fair model training but instead research has focussed on the consistency and transparency of explanations. Our work is novel as it places the role of model explicability at the core of a new fairness definition and develops an adversarial learning methodology that is applicable to adaptive boosting and any model trained via gradient-based optimization. In the former case, our proposed algorithm is fully non-parametric where the adversary can come from any model family provided the corresponding explicability scores, in this case SHAP values, can be computed.

# References

1. Agarwal, A., Beygelzimer, A., Dudik, M., Langford, J., Wallach, H.: A reductions approach to fair classification. In: Dy, J., Krause, A. (eds.) Proceedings of the 35th International Conference on Machine Learning. PMLR, 10–15 July 2018. http://proceedings.mlr.press/v80/agarwal18a.html. Proceedings of Machine Learning Research, vol. 80, pp. 60–69
2. Barocas, S., Selbst, A.D.: Big data's disparate impact. Calif. Law Rev. **104**, 671 (2016)
3. Beutel, A., Chen, J., Zhao, Z., Chi, E.H.: Data decisions and theoretical implications when adversarially learning fair representations. arXiv preprint arXiv:1707.00075 (2017)

4. Beutel, A., et al.: Putting fairness principles into practice: challenges, metrics, and improvements. In: Proceedings of the 2019 AAAI/ACM Conference on AI, Ethics, and Society, AIES 2019, pp. 453–459. Association for Computing Machinery, New York (2019). https://doi.org/10.1145/3306618.3314234

5. University of California, I.: Census income dataset (1996). https://archive.ics.uci.edu/ml/datasets/census+income

6. Celis, L.E., Huang, L., Keswani, V., Vishnoi, N.K.: Classification with fairness constraints: a meta-algorithm with provable guarantees. In: Proceedings of the Conference on Fairness, Accountability, and Transparency, pp. 319–328. Association for Computing Machinery, New York (2019). https://doi.org/10.1145/3287560.3287586

7. Cesaro, J., Gagliardi Cozman, F.: Measuring unfairness through game-theoretic interpretability. In: Cellier, P., Driessens, K. (eds.) ECML PKDD 2019. CCIS, vol 1167, pp. 253–264. Springer, Cham (2020). https://doi.org/10.1007/978-3-030-43823-4_22

8. Chen, T., Guestrin, C.: XGBoost: a scalable tree boosting system. In: Proceedings of the 22nd ACM SIGKDD International Conference on Knowledge Discovery and Data Mining, KDD 2016, pp. 785–794. Association for Computing Machinery, New York (2016). https://doi.org/10.1145/2939672.2939785

9. Chiappa, S., Gillam, T.: Path-specific counterfactual fairness. In: Proceedings of the AAAI Conference on Artificial Intelligence, vol. 33, February 2018. https://doi.org/10.1609/aaai.v33i01.33017801

10. Chouldechova, A.: Fair prediction with disparate impact: a study of bias in recidivism prediction instruments. Big Data 5(2), 153–163 (2017). https://doi.org/10.1089/big.2016.0047. pMID: 28632438

11. Corbett-Davies, S., Goel, S., Morgenstern, J., Cummings, R.: Defining and designing fair algorithms. In: Proceedings of the 2018 ACM Conference on Economics and Computation, p. 705. Association for Computing Machinery, New York (2018). https://doi.org/10.1145/3219166.3277556

12. Di Stefano, P., Hickey, J., Vasileiou, V.: Counterfactual fairness: removing direct effects through regularization. arXiv preprint arXiv:2002.10774 (2020)

13. Dwork, C., Hardt, M., Pitassi, T., Reingold, O., Zemel, R.: Fairness through awareness. In: Proceedings of the 3rd Innovations in Theoretical Computer Science Conference, ITCS 2012, pp. 214–226. Association for Computing Machinery, New York (2012). https://doi.org/10.1145/2090236.2090255

14. European Parliament and Council of the European Union: Regulation (EU) 2016/679 regulation on the protection of natural persons with regard to the processing of personal data and on the free movement of such data, and repealing directive 95/46/EC (data protection directive). OJ L 119, 1–88 (2016). https://gdpr-info.eu/

15. Freund, Y., Schapire, R.E.: A decision-theoretic generalization of on-line learning and an application to boosting. J. Comput. Syst. Sci. 55(1), 119–139 (1997). https://doi.org/10.1006/jcss.1997.1504. http://www.sciencedirect.com/science/article/pii/S002200009791504X

16. Friedler, S., Choudhary, S., Scheidegger, C., Hamilton, E., Venkatasubramanian, S., Roth, D.: A comparative study of fairness-enhancing interventions in machine learning. In: FAT* 2019 - Proceedings of the 2019 Conference on Fairness, Accountability, and Transparency, pp. 329–338. Association for Computing Machinery, Inc, January 2019. https://doi.org/10.1145/3287560.3287589

17. Goel, N., Yaghini, M., Faltings, B.: Non-discriminatory machine learning through convex fairness criteria (2018). https://www.aaai.org/ocs/index.php/AAAI/AAAI18/paper/view/16476
18. Goodfellow, I., et al.: Generative adversarial nets. In: Ghahramani, Z., Welling, M., Cortes, C., Lawrence, N.D., Weinberger, K.Q. (eds.) Advances in Neural Information Processing Systems 27, pp. 2672–2680. Curran Associates, Inc. (2014). http://papers.nips.cc/paper/5423-generative-adversarial-nets.pdf
19. Goodman, B., Flaxman, S.: European union regulations on algorithmic decision-making and a "right to explanation". AI Mag. **38**(3), 50–57 (2017). https://doi.org/10.1609/aimag.v38i3.2741. https://www.aaai.org/ojs/index.php/aimagazine/article/view/2741
20. Kamishima, T., Akaho, S., Asoh, H., Sakuma, J.: Fairness-aware classifier with prejudice remover regularizer. In: Flach, P.A., De Bie, T., Cristianini, N. (eds.) ECML PKDD 2012. LNCS (LNAI), vol. 7524, pp. 35–50. Springer, Heidelberg (2012). https://doi.org/10.1007/978-3-642-33486-3_3
21. Kilbertus, N., Rojas-Carulla, M., Parascandolo, G., Hardt, M., Janzing, D., Schölkopf, B.: Avoiding discrimination through causal reasoning, June 2017
22. Kleinberg, J.: Inherent trade-offs in algorithmic fairness. In: Abstracts of the 2018 ACM International Conference on Measurement and Modeling of Computer Systems, SIGMETRICS 2018, p. 40. Association for Computing Machinery, New York (2018). https://doi.org/10.1145/3219617.3219634
23. Kusner, M.J., Loftus, J., Russell, C., Silva, R.: Counterfactual fairness. In: Guyon, I., et al. (eds.) Advances in Neural Information Processing Systems 30, pp. 4066–4076. Curran Associates, Inc. (2017). http://papers.nips.cc/paper/6995-counterfactual-fairness.pdf
24. Lipovetsky, S., Conklin, M.: Analysis of regression in game theory approach. Appl. Stoch. Models Bus. Ind. **17**(4), 319–330 (2001). https://doi.org/10.1002/asmb.446. https://onlinelibrary.wiley.com/doi/abs/10.1002/asmb.446
25. Lum, K., Isaac, W.: To predict and serve? Significance **13**(5), 14–19 (2016). https://doi.org/10.1111/j.1740-9713.2016.00960.x
26. Lundberg, S.M., et al.: Explainable AI for trees: from local explanations to global understanding. arXiv preprint arXiv:1905.04610 (2019)
27. Lundberg, S.M., Lee, S.I.: A unified approach to interpreting model predictions. In: Guyon, I., et al. (eds.) Advances in Neural Information Processing Systems 30, pp. 4765–4774. Curran Associates, Inc. (2017)
28. Madras, D., Creager, E., Pitassi, T., Zemel, R.: Learning adversarially fair and transferable representations. In: Dy, J., Krause, A. (eds.) Proceedings of the 35th International Conference on Machine Learning. Proceedings of Machine Learning Research, vol. 80, pp. 3384–3393. PMLR, Stockholmsmässan, Stockholm Sweden, 10–15 July 2018. http://proceedings.mlr.press/v80/madras18a.html
29. Manisha, P., Gujar, S.: A neural network framework for fair classifier. arXiv preprint arXiv:1811.00247 (2018)
30. Mansoor, S.: A viral Tweet accused apple's new credit card of being 'sexist'. Now New York state regulators are investigating. TIME Mag. (2019). https://time.com/5724098/new-york-investigating-goldman-sachs-apple-card/
31. Mehrabi, N., Morstatter, F., Saxena, N., Lerman, K., Galstyan, A.: A survey on bias and fairness in machine learning. arXiv preprint arXiv:1908.09635 (2019)
32. Mothilal, R.K., Sharma, A., Tan, C.: Explaining machine learning classifiers through diverse counterfactual explanations. In: Proceedings of the 2020 Conference on Fairness, Accountability, and Transparency, FAT* 2020, pp. 607–617. Association for Computing Machinery, New York (2020). https://doi.org/10.1145/3351095.3372850

33. Nabi, R., Shpitser, I.: Fair inference on outcomes. In: Proceedings of the AAAI Conference on Artificial Intelligence 2018, pp. 1931–1940, February 2018. https://www.ncbi.nlm.nih.gov/pubmed/29796336. 29796336[pmid]

34. O'Neil, C.: Weapons of Math Destruction: How Big Data Increases Inequality and Threatens Democracy. Crown Publishing Group, New York (2016)

35. Ribeiro, M.T., Singh, S., Guestrin, C.: "Why should I trust you?": explaining the predictions of any classifier. In: Proceedings of the 22nd ACM SIGKDD International Conference on Knowledge Discovery and Data Mining, pp. 1135–1144. Association for Computing Machinery, New York (2016). https://doi.org/10.1145/2939672.2939778

36. Roemer, J.E., Trannoy, A.: Equality of opportunity. In: Handbook of Income Distribution, vol. 2, pp. 217–300. Elsevier (2015)

37. Shapley, L.S.: A value for n-person games. In: Contributions to the Theory of Games, vol. 2, no. 28, pp. 307–317 (1953) https://doi.org/10.1515/9781400881970-018

38. Shrikumar, A., Greenside, P., Kundaje, A.: Learning important features through propagating activation differences. In: Proceedings of the 34th International Conference on Machine Learning - Volume 70, pp. 3145–3153. JMLR.org (2017)

39. Štrumbelj, E., Kononenko, I.: Explaining prediction models and individual predictions with feature contributions. Knowl. Inf. Syst. **41**(3), 647–665 (2013). https://doi.org/10.1007/s10115-013-0679-x

40. Suresh, H., Guttag, J.V.: A framework for understanding unintended consequences of machine learning. arXiv preprint arXiv:1901.10002 (2019)

41. Verma, S., Rubin, J.: Fairness definitions explained. In: Proceedings of the International Workshop on Software Fairness, pp. 1–7. Association for Computing Machinery, New York (2018). https://doi.org/10.1145/3194770.3194776

42. Wachter, S., Mittelstadt, B.D., Russell, C.: Counterfactual explanations without opening the black box: automated decisions and the GDPR. Harvard J. Law Technol. **31**, 842–854 (2018)

43. Yeom, S., Tschantz, M.C.: Discriminative but not discriminatory: a comparison of fairness definitions under different worldviews. arXiv preprint arXiv:1808.08619v4 (2019)

44. Young, H.P.: Monotonic solutions of cooperative games. Int. J. Game Theory **14**, 65–72 (1985). https://doi.org/10.1007/BF01769885

45. Zafar, M.B., Valera, I., Rodriguez, M.G., Gummadi, K.P., Weller, A.: From parity to preference-based notions of fairness in classification. In: Proceedings of the 31st International Conference on Neural Information Processing Systems, NIPS 2017, pp. 228–238. Curran Associates Inc., Red Hook (2017)

46. Zafar, M.B., Valera, I., Rogriguez, M.G., Gummadi, K.P.: Fairness constraints: mechanisms for fair classification. In: Singh, A., Zhu, J. (eds.) Proceedings of the 20th International Conference on Artificial Intelligence and Statistics. PMLR, Fort Lauderdale, 20–22 April 2017. http://proceedings.mlr.press/v54/zafar17a.html. Proceedings of Machine Learning Research, vol. 54, pp. 962–970

47. Zhang, B.H., Lemoine, B., Mitchell, M.: Mitigating unwanted biases with adversarial learning. In: Proceedings of the 2018 AAAI/ACM Conference on AI, Ethics, and Society, AIES 2018, pp. 335–340. Association for Computing Machinery, New York (2018). https://doi.org/10.1145/3278721.3278779

48. Zhang, L., Wu, Y., Wu, X.: A causal framework for discovering and removing direct and indirect discrimination. In: Proceedings of the Twenty-Sixth International Joint Conference on Artificial Intelligence, IJCAI 2017, pp. 3929–3935 (2017). https://doi.org/10.24963/ijcai.2017/549
49. Zhao, Q., Hastie, T.: Causal interpretations of black-box models. J. Bus. Econ. Stat. 1–10 (2019). https://doi.org/10.1080/07350015.2019.1624293

# End-to-End Learning for Prediction and Optimization with Gradient Boosting

Takuya Konishi[1(✉)] and Takuro Fukunaga[2,3,4]

[1] National Institute of Informatics, Chiyoda City, Japan
`takuya-ko@nii.ac.jp`
[2] Chuo University, Hachioji, Japan
`fukunaga.07s@g.chuo-u.ac.jp`
[3] JST PRESTO, Kawaguchi, Japan
[4] RIKEN Center for Advanced Intelligence Project, Chuo City, Japan

**Abstract.** Mathematical optimization is a fundamental tool in decision making. However, it is often difficult to obtain an accurate formulation of an optimization problem due to uncertain parameters. Machine learning frameworks are attractive to address this issue: we predict the uncertain parameters and then optimize the problem based on the prediction. Recently, end-to-end learning approaches to predict and optimize the successive problems have received attention in the field of both optimization and machine learning. In this paper, we focus on gradient boosting which is known as a powerful ensemble method, and develop the end-to-end learning algorithm with maximizing the performance on the optimization problems directly. Our algorithm extends the existing gradient-based optimization through implicit differentiation to the second-order optimization for efficiently learning gradient boosting. We also conduct computational experiments to analyze how the end-to-end approaches work well and show the effectiveness of our end-to-end approach.

**Keywords:** Combinatorial optimization · Boosting/ensemble methods

## 1 Introduction

Mathematical optimization is a fundamental tool in decision making. Nowadays many efficient solvers are developed for solving optimization problems such as linear programs and integer programs. Once a decision making problem is formulated as an optimization problem, we find a reasonable solution in many cases. However, it is often difficult to obtain an accurate formulation of an optimization problem because the problem contains uncertain parameters; they are simply unknown, are difficult to observe, or fluctuate frequently due to noisy measurement and dynamical environments.

T. Konishi—Supported by JSPS KAKENHI Grant Number 17K12743 and JP18H05291, Japan.
T. Fukunaga—Supported by JST PRESTO grant JPMJPR1759, Japan.

© Springer Nature Switzerland AG 2021
F. Hutter et al. (Eds.): ECML PKDD 2020, LNAI 12459, pp. 191–207, 2021.
https://doi.org/10.1007/978-3-030-67664-3_12

For example, consider finding the fastest route from one place to another. This problem is formulated as the shortest path problem, which is one of the most well-studied optimization problems. While there are many scalable solvers for the shortest path problem, it is usually hard to access precise parameters of the objective function: the parameters, in this case, are edge weights, which represent how much time or cost it takes to traverse roads and thus constantly change by several factors, e.g., weather conditions and ongoing events.

Machine learning frameworks are useful in such situations. In this paper, we formulate the situations as a supervised learning problem. We first prepare a machine learning model trained by data about the uncertain parameters of an optimization problem. The machine learning model predicts the uncertain parameters, and then a solver computes a solution for the optimization problem defined from the predicted parameters.

A naive approach for this task is to solve the prediction and the optimization phases separately: the machine learning model ignores the optimization problem and is trained simply by minimizing a popular loss function, e.g., the squared loss and the cross-entropy loss. We call this approach *two-stage approach*. Although the two-stage approach is natural, it could be suboptimal because the machine learning model trained in this approach is agnostic to how the prediction affects the results of the optimization problems. As we demonstrate in Sect. 5, this issue is critical unless the model predicts uncertain parameters perfectly.

Recently, optimization-aware approaches for the above-mentioned task have been studied actively. These approaches incorporate information on target optimization problems into the prediction phase. Several previous studies report that these approaches give better solutions than the two-stage approach [4,5,9,17]. Among these previous studies, we focus on the *end-to-end learning approach* proposed by Wilder, Dilkina, and Tambe [17]. This work considers learning neural networks (NNs), defines a loss function that is derived from a target optimization problem, and directly minimizes this optimization-aware loss. The technical challenge in this approach is the discontinuity of the loss function. Wilder et al. address this challenge by leveraging implicit differentiation, which enables us to minimize the loss function by gradient-based optimization.

Motivated by the work of Wilder et al., we propose introducing the end-to-end learning approach with gradient boosting. Gradient boosting is known as a powerful class of machine learning models, and thus it would benefit a wide range of optimization problems. The standard learning algorithm of gradient boosting often adopts second-order optimization, which requires the Hessian matrix in addition to the gradient of the loss function. However, it is highly nontrivial to compute the Hessian matrix of the loss function when the loss function is derived from optimization problems. We derive the Hessian matrix with implicit differentiation and develop an end-to-end learning algorithm based on second-order optimization. This is our main technical contribution. By this contribution, the model of the gradient boosting is fitted so as to output a good solution for the target optimization problem.

We also conduct computational experiments to verify the effectiveness of our end-to-end learning algorithm. In the experiments, we first compare end-to-end and two-stage approaches on a toy example, which clarifies the essential differences between these two approaches. The result illustrates in which case the end-to-end learning approach is useful. Previous studies on the end-to-end learning approach lack this kind of discussion, and we believe that our toy example is of independent interest. We then evaluate our end-to-end learning approach to several baselines on the shortest path and bipartite matching problems.

## 2 Related Work

Optimization-aware approaches have been recently considered by several previous works. Elmachtoub and Grigas [6] proposed the Smart Predict and Optimize (SPO) framework that introduces a loss function that represents the error of the solution of considered optimization problems. The SPO framework leverages a convex surrogate loss to avoid the computational issue of the original loss. Mandi et al. [14] also considered the SPO framework for the combinatorial optimization problems. Ito et al. [9] clarified that the estimation by the two-stage approach has an optimistic bias that causes the overestimate of the objective function. To address this issue, they proposed a resampling technique that achieves asymptotically unbiased estimation. Demirovic et al. [4] proposed a semi-direct approach that indirectly utilizes the information of optimization problems. Demirovic et al. [5] considered a optimization-aware framework for combinatorial optimization problems that have a ranking objective; the solution of the problems depends on the relative order of the parameters of the objective function.

Our work is closely related to the recent progress on the end-to-end learning approaches through implicit differentiation. Amos and Kolter [1] proposed Opt-Net for solving quadratic programming. OptNet is a layer of NNs that receives input parameters and outputs the solution of a quadratic program. To learn this layer with gradient-based optimization, they considered implicit differentiation for the KKT condition; it enables us to compute the gradient of the optimal solution of the quadratic program. The closest work to ours is the end-to-end learning approach by Wilder et al. [17]: it employed OptNet for solving three combinatorial optimization problems formulated as linear programs and submodular maximization. While the motivation of [17] is the same as our work, the difference is that we consider gradient boosting which is known as yet another powerful model. However, popular implementations of gradient boosting use second-order optimization which additionally requires the Hessian matrix of the loss function for faster convergence. A technical contribution of this paper is to adapt the framework of OptNet to gradient boosting by extending it to second-order optimization.

Gradient boosting is a widely used class of machine learning models [7]. It is an ensemble method for making a strong prediction model by sequentially adding weak models such as decision trees. Recently, several efficient algorithms and their implementations have been developed. XGBoost is a scalable implementation of gradient boosting decision tree (GBDT) and has popularized the effectiveness of gradient boosting algorithms through many machine learning competitions [2]. LightGBM is also an efficient implementation that achieves faster training and higher accuracy [10]. These seminal works and their implementations consider the case when the loss function is convex. However, it is not suitable for our problem: as presented in Sect. 4, our proposed framework requires addressing the non-convex loss function where the Hessian matrix is not always positive definite. We address this issue by introducing the modified Hessian, which provides the efficient learning framework of GBDT even if the loss is non-convex.

## 3   Setting

Given a convex objective function $f$, consider the convex optimization problem

$$
\begin{aligned}
\text{minimize} \quad & f(x; \theta) \\
\text{subject to} \quad & Ax - b \leq 0,
\end{aligned}
\tag{1}
$$

where $x \in \mathbb{R}^q$ are optimization variables, and $\theta \in \mathbb{R}^p$ are parameters of $f$. A solution $x$ needs to satisfy $Ax - b \leq 0$, which consists of $r$ linear constraints determined by the matrix $A \in \mathbb{R}^{r \times q}$ and the vector $b \in \mathbb{R}^r$.

We address a setting where values of parameters $\theta$ are not given, and thus we have to estimate $\theta$ from other information. Specifically, we assume that each element $\theta_i$ of $\theta$ is associated with a feature vector $y_i \in \mathbb{R}^D$. We let $Y \in \mathbb{R}^{p \times D}$ denote the matrix its $i$th row of which is $y_i$. The task for a machine learning algorithm is to make a prediction model that takes a feature vector $y_i$ as an input and a parameter $\theta_i$ as the output. For this, we are given $N$ training instances $\{(Y^{(n)}, \theta^{(n)})\}_{n=1}^N$ where $Y^{(n)}$ and $\theta^{(n)}$ are the $n$th training feature matrix and parameter.

In the following, we present examples of concrete settings with two popular optimization problems.

*Shortest Path Problem:* We are given an edge-weighted directed graph $G = (V, E)$, source vertex $s$, and target vertex $t$. The problem is to find a shortest path $P$ from $s$ to $t$, i.e., a subset of the edge set $E$ where the sum of the weights is the smallest among all paths from $s$ to $t$. The uncertain parameters $\theta$ are edge weights whose $i$th element $\theta_i$ corresponds to the weight of the $i$th edge. The feature vector $y_i$ indicates some information explaining the $i$th edge.

*Bipartite Matching Problem:* We are given an edge-weighted undirected bipartite graph $G = (V, E)$, and seek a maximum weight matching $M$ i.e., a subset of the edge set $E$ no two edges of which are incident to the same vertex. As in the shortest path problem, the uncertain parameters $\theta$ are edge weights, and the $i$th edge has a weight $\theta_i$ and a feature vector $y_i$.

# 4    End-to-End Learning of Gradient Boosting

## 4.1    GBDT

We adopt GBDT as our gradient boosting model in the same way as popular implementations. A GBDT model $m$ takes the form of a tree ensemble $\theta = m(Y) = \sum_{t=1}^{T} m_t(Y)$, where $m_t$ is the $t$th tree in the ensemble.

GBDT is trained through a greedy procedure. For each $t = 2, ..., T$, the procedure constructs $m_t$ from $m_1, \ldots, m_{t-1}$ so as to achieve a small value of the following loss function:

$$\mathcal{L}(m_t) = \sum_{n-1}^{N} \ell\left(\theta^{(n)}, \hat{\theta}_{t-1}^{(n)} + m_t(Y^{(n)})\right) + \Omega(m_t), \tag{2}$$

where $\hat{\theta}_{t-1}^{(n)}$ is an estimate of $\theta^{(n)}$ predicted by the ensemble of $m_1, ..., m_{t-1}$, $\ell(a, b)$ is a per-example loss function that measures the discrepancy between $a$ and $b$, and $\Omega(m_t)$ is a regularization term for $m_t$.

To minimize the loss function (2), we adopt the second-order optimization method for gradient boosting [7], which is commonly employed in popular implementations [2,10]. In this method, we first compute

$$g^{(n)} = \frac{\partial \ell\left(\theta^{(n)}, \hat{\theta}_{t-1}^{(n)}\right)}{\partial \hat{\theta}_{t-1}^{(n)}} \in \mathbb{R}^p, \quad H^{(n)} = \frac{\partial^2 \ell\left(\theta^{(n)}, \hat{\theta}_{t-1}^{(n)}\right)}{\partial \hat{\theta}_{t-1}^{(n)2}} \in \mathbb{R}^{p \times p}, \tag{3}$$

which are the gradient and Hessian matrix of $\ell$ at the $n$th training instance, respectively. Then, the function (2) is approximated as

$$\sum_{n=1}^{N} \left[ \ell\left(\theta^{(n)}, \hat{\theta}_{t-1}^{(n)}\right) + g^{(n)\top} m_t(Y^{(n)}) + \frac{1}{2} m_t(Y^{(n)})^{\top} H^{(n)} m_t(Y^{(n)}) \right] + \Omega(m_t). \tag{4}$$

This approximation is analytically minimized, and the optimal value of the function can be used as a score of evaluating the decision tree $m_t$. The decision tree $m_t$ is constructed by greedily adding branches based on this score.

## 4.2    End-to-End Learning Framework

Let us discuss what $\ell$ is suitable for our problem. Below, we consider the $n$th training instance for making $m_t$ and omit the superscript $(n)$ and subscript $t$ from the variables for notational convenience. If we address the problem by the two-stage approach which separately optimizes the optimization problem (1) and the learning problem (4), one can apply the squared loss when $\theta$ is continuous, and the cross-entropy loss when $\theta$ is binary, as in standard supervised learning. However, such standard losses ignore the optimization problem (1); the final goal should be to obtain a nice solution of the problem (1).

To take into account the problem (1) during the training procedure, our end-to-end approach takes the following loss function for a true parameter $\theta$ and its estimate $\hat{\theta}$:

$$\ell(\theta, \hat{\theta}) = f(x(\hat{\theta}); \theta). \qquad (5)$$

Here $x(\hat{\theta})$ denotes the optimal solution of the problem (1) when the parameter is $\hat{\theta}$. We can view this loss as a measure of how the estimated solution $x(\hat{\theta})$ performs well in the problem (1) with the true parameter $\theta$.

While the loss function (5) can be computed given $\theta$ and $\hat{\theta}$, the difficulty is to compute its gradients and Hessians. Even if the objective function $f$ is differentiable, the computation requires the derivatives of $x(\hat{\theta})$ with respect to $\hat{\theta}$. Since $x(\hat{\theta})$ is a solution of the optimization problem, it needs argmin differentiation, which is intractable in general. While OptNet has been recently developed for computing the gradients of the solution of the optimization problems through implicit differentiation, it is not enough for GBDT, that additionally requires Hessians. Thus we extend their method to computation of Hessians.

We note that the gradient $g$ and the Hessian $H$ of (5) are written as

$$g = \frac{\partial \ell(\theta, \hat{\theta})}{\partial \hat{\theta}} = \frac{\partial f(x(\hat{\theta}); \theta)}{\partial x(\hat{\theta})} \frac{\partial x(\hat{\theta})}{\partial \hat{\theta}}, \qquad (6)$$

$$H = \frac{\partial^2 \ell(\theta, \hat{\theta})}{\partial \hat{\theta}^2}$$
$$= \left[ \left( \frac{\partial x(\hat{\theta})}{\partial \hat{\theta}} \right)^{\top} \frac{\partial^2 f(x(\hat{\theta}); \theta)}{\partial x(\hat{\theta})^2} \frac{\partial x(\hat{\theta})}{\partial \hat{\theta}} + \left( \frac{\partial f(x(\hat{\theta}); \theta)}{\partial x(\hat{\theta})} \otimes I_p \right) \frac{\partial^2 x(\hat{\theta})}{\partial \hat{\theta}^2} \right], \qquad (7)$$

where $\otimes$ is the Kronecker product, and $I_p$ is the identity matrix of size $p$ [13]. The technical challenge is to compute $\frac{\partial x(\hat{\theta})}{\partial \hat{\theta}} \in \mathbb{R}^{q \times p}$ and $\frac{\partial^2 x(\hat{\theta})}{\partial \hat{\theta}^2} \in \mathbb{R}^{qp \times p}$ which are the first and second derivatives of $x$ with respect to $\theta$. The other terms can be easily derived if the objective function $f$ is at least twice differentiable.

## 4.3  Computing $\frac{\partial x(\hat{\theta})}{\partial \hat{\theta}}$ and $\frac{\partial^2 x(\hat{\theta})}{\partial \hat{\theta}^2}$

We derive $\frac{\partial x(\hat{\theta})}{\partial \hat{\theta}}$ and $\frac{\partial^2 x(\hat{\theta})}{\partial \hat{\theta}^2}$ from the KKT condition of problem (1). The Lagrangian relaxation of (1) is formulated as

$$\max_{\lambda \geq 0} \min_{x} f(x; \theta) + \lambda^{\top}(Ax - b), \qquad (8)$$

where $\lambda \in \mathbb{R}^r$ denotes the Lagrange multipliers. It is known that the optimal objective value of (1) is equal to that of (8), and an optimal solution $x$ for (1) also forms an optimal solution for (8). Note that both of (1) and (8) are parameterized by $\theta$. To clarify this fact, We denote an optimal solution for (8) by $x(\theta)$ and $\lambda(\theta)$. $x(\theta)$ and $\lambda(\theta)$ satisfy the KKT condition

$$
\begin{cases}
\dfrac{\partial f(x(\theta); \theta)}{\partial x(\theta)} + A^\top \lambda(\theta) = 0 \\[2mm]
\mathrm{diag}(\lambda(\theta))(Ax(\theta) - b) = 0,
\end{cases}
\tag{9}
$$

where $\mathrm{diag}(a)$ is the diagonal matrix whose diagonal elements are the vector $a$. Applying the implicit function theorem, we differentiate (9) and obtain the following set of linear equations [1, 17]:

$$
\begin{pmatrix}
\dfrac{\partial^2 f(x(\theta); \theta)}{\partial x(\theta)^2} & A^\top \\[3mm]
\mathrm{diag}(\lambda(\theta))A & \mathrm{diag}(Ax(\theta) - b)
\end{pmatrix}
\begin{pmatrix}
\dfrac{\partial x(\theta)}{\partial \theta} \\[3mm]
\dfrac{\partial \lambda(\theta)}{\partial \theta}
\end{pmatrix}
= -
\begin{pmatrix}
\dfrac{\partial^2 f(x(\theta); \theta)}{\partial x(\theta)\partial \theta} \\[3mm]
0
\end{pmatrix}
\tag{10}
$$

Solving (10) gives $\frac{\partial x(\theta)}{\partial \theta}$. Further differentiating (10), we also obtain another set of linear equations

$$
\begin{pmatrix}
I_p \otimes \dfrac{\partial^2 f(x(\theta); \theta)}{\partial x(\theta)^2} & I_p \otimes A^\top \\[3mm]
I_p \otimes \mathrm{diag}(\lambda(\theta))A & I_p \otimes \mathrm{diag}(Ax(\theta) - b)
\end{pmatrix}
\begin{pmatrix}
\dfrac{\partial^2 x(\theta)}{\partial \theta^2} \\[3mm]
\dfrac{\partial^2 \lambda(\theta)}{\partial \theta^2}
\end{pmatrix}
$$
$$
= -
\begin{pmatrix}
\dfrac{\partial}{\partial \theta}\left( \dfrac{\partial^2 f(x(\theta); \theta)}{\partial x(\theta)^2} \dfrac{\partial x(\theta)}{\partial \theta} + \dfrac{\partial^2 f(x(\theta); \theta)}{\partial x(\theta)\partial \theta}\right) \\[3mm]
\dfrac{\partial\mathrm{diag}(\lambda(\theta))A}{\partial \theta}\dfrac{\partial x(\theta)}{\partial \theta} + \dfrac{\partial\mathrm{diag}(Ax(\theta) - b)}{\partial \theta}\dfrac{\partial \lambda(\theta)}{\partial \theta}
\end{pmatrix},
\tag{11}
$$

which is derived from the product rule of matrix derivatives [12]. By substituting the solutions of (10) into (11) and then solving it, $\frac{\partial^2 x(\theta)}{\partial \theta^2}$ is also obtained.

### 4.4   Diagonal Approximation of Hessian

In practice, solving (11) is computationally expensive since it consists of $p(q + r)$ equations. To reduce the computation, we consider approximating the Hessian $H$ by its diagonal

$$
H \approx \mathrm{diag}(h),
\tag{12}
$$

where $h \in \mathbb{R}^p$ is the vector that consists of the diagonal elements of $H$. Such approximation is often used in the gradient boosting [7]. Here $x_i(\theta)$ denotes the $i$th element of $x(\theta)$. To obtain $h$, it is sufficient to compute $\frac{\partial^2 x_i(\theta)}{\partial \theta_j^2}$ for $i = 1, ...q$ and $j = 1, ..., p$ in the second derivative $\frac{\partial^2 x(\theta)}{\partial \theta^2}$.

Suppose that $\widehat{\frac{\partial^2 x(\theta)}{\partial \theta^2}}$ is the $q \times p$ matrix whose $(i, j)$ element is $\frac{\partial^2 x_i(\theta)}{\partial \theta_j^2}$, and $\widehat{\frac{\partial^2 \lambda(\theta)}{\partial \theta^2}}$ is the corresponding $r \times p$ matrix whose $(i, j)$ element is $\frac{\partial^2 \lambda_i(\theta)}{\partial \theta_j^2}$. We can reduce (11) to the following set of equations:

$$
\begin{pmatrix}
\dfrac{\partial^2 f(x(\theta); \theta)}{\partial x(\theta)^2} & A^\top \\
\mathrm{diag}(\lambda(\theta)) A \ \mathrm{diag}(Ax(\theta) - b)
\end{pmatrix}
\begin{pmatrix}
\dfrac{\widehat{\partial^2 x(\theta)}}{\partial \theta^2} \\
\dfrac{\widehat{\partial^2 \lambda(\theta)}}{\partial \theta^2}
\end{pmatrix}
= -
\begin{pmatrix}
P \\
Q
\end{pmatrix}.
\tag{13}
$$

Here $P \in \mathbb{R}^{q \times p}$ and $Q \in \mathbb{R}^{r \times p}$ are the elements of the right-hand side of (11) that correspond to $\frac{\widehat{\partial^2 x(\theta)}}{\partial \theta^2}$ and $\frac{\widehat{\partial^2 \lambda(\theta)}}{\partial \theta^2}$. When $f$ is linear or quadratic, $P$ is a zero matrix. $Q$ is $2\frac{\partial \lambda(\theta)}{\partial \theta} \odot \left(A\frac{\partial x(\theta)}{\partial \theta}\right)$ where $\odot$ is the Hadamard product. Since (13) consists of $q + r$ linear equations, the computational complexity for solving it is the same as that of (10). Although this diagonal approximation could lose important information on the original Hessian matrix, the reduction of the computational complexity yields a significant benefit for practical use.

Another practical merit of the diagonal approximation is the efficient compatibility with popular implementations. Since $\theta$ is a vector in general, we require multiple output GBDT. Current XGBoost and LightGBM support multi-output predictions by combining independent models into one model. The diagonal approximation can adopt these implementations by separately passing diagonal elements to the independent models.

When all elements of $\theta$ are the same type of parameters, e.g., edge weights of the shortest path problem, the diagonal approximation further allows GBDT to take a more efficient form. In such a case, multiple outputs can be modeled by one single model $m_{\mathrm{single}}$: $\theta_i = m_{\mathrm{single}}(y_i)$ for $i = 1, ..., p$. Since this model outputs a scalar, the gradient and Hessian are also scalar. We can see each of multiple outputs as a distinct training sample and use the $i$th element of (6) and the $i$th diagonal element of (12) for learning the $i$th output. This form changes the original problem of learning a $p$-dimensional output model with $N$ samples to the problem of learning a one-dimensional output model with $Np$ samples. The form is more sample-efficient when $N$ is small and $p$ is large.

## 4.5   Enhancing Second-Order Optimization

In practice, the above simple second-order optimization often fails due to the non-convexity of the per-example loss function (5). The Hessian matrix of the non-convex function is generally not positive definite, and hence the update of the model does not guarantee to decrease the training loss. To address this issue, we first apply the simple Tikhonov damping [15] to the diagonal approximation (12), which means replacing (12) by

$$
H_{\mathrm{tikhonov}} = \mathrm{diag}(h) + \alpha I_p,
\tag{14}
$$

where $\alpha > 0$ is a damping factor. We also leverage the saddle-free Newton (SFN) method for the second-order optimization [3]. The SFN method was proposed for avoiding saddle-points over the loss function and takes the absolute values of the eigenvalues of the Hessian matrix. For our diagonal approximation, the SFN method simply replaces the diagonal elements with their absolute values. We adopt two variants of the SFN method:

$$H_{\mathrm{abs}} = |\mathrm{diag}(h)|, \tag{15}$$

$$H_{\mathrm{sfn}} = |\mathrm{diag}(h)| + \alpha I_p, \tag{16}$$

where $|\mathrm{diag}(h)|$ denotes the diagonal matrix whose diagonal elements are their absolute values. (15) simply takes the absolute values while (16) is its damped version. To avoid increasing free hyperparameters, $\alpha$ is heuristically set to the minimum of the elements of $h$ if it is negative, otherwise set to 0.0.

## 5  Experiments

We conducted three computational experiments to evaluate the effectiveness of the proposed algorithm. The first two tasks are the shortest path problem, and the last task is the bipartite matching problem. Both tasks are formulated as linear programs which can be included in our setting. However, when the objective function is linear, the coefficient matrices in the left-hand sides of (10) and (11) are singular. To deal with this issue, we follow the approach of [17]: we use the regularized objective $f(x; \theta) = \theta^\top x + \frac{1}{2}\gamma\|x\|_2^2$ for computing (10) and (11). Here $\gamma$ is a regularized parameter. Note that this objective is only used for the training procedure of end-to-end learning approaches: we optimized the original linear programs when solving the test instances. As mentioned in Sect. 3, the parameters of the both tasks denote edge weights on a graph. Since we consider only two types of edge weights in all the experiments, the both tasks are formulated as binary classification at the prediction phase.

For the both tasks, we prepared a set of instances of the optimization problem. Each instance has a parameter $\theta$ of the objective function and its corresponding feature matrix $Y$. The set of instances were divided into 80% for training instances and 20% for test instances. The quality of a solution for test instances was evaluated by the objective function with ground-truth parameter $\theta$. For each experiment, we repeated this process 10 times while changing the train-test split, and report the average performance on them. We measure the performances of the models by the average objective values of the solution (Objective) obtained at the optimization phase. Also, to check the performance at the prediction phase, we also measure the average area under the ROC curve (AUC) which is used in evaluating the performance of binary classification.

We compared the proposed method with related algorithms. We first prepared four algorithms compared in [17]: the two-stage approach for a one-layer NN (nn1_twostage), the two-stage approach for a two-layer NN (nn2_twostage), the end-to-end approach for a one-layer NN (nn1_e2e), and the end-to-end approach for a two-layer NN (nn2_e2e). We also prepared four gradient boosting algorithms: the two-stage approach (gbm_twostage), the end-to-end approach

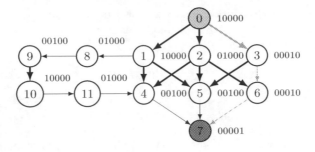

**Fig. 1.** A toy example of the shortest path problem. (Color figure online)

with Tikhonov damping (14) (gbm_e2e_tikhonov), the end-to-end approach with the SFN method (15) (gbm_e2e_abs) and the end-to-end approach with the damped SFN method (16) (gbm_e2e_sfn). To check the validity of the above-compared methods, we also prepared two algorithms without learning: the optimization with randomly initialized parameters (random) and the optimization with ground-truth parameters (optimal).

For all approaches, we used Gurobi Optimizer for solving linear programs. For all end-to-end learning approaches, $\gamma$ was set to 0.1. For all two-stage approaches, we used the cross-entropy loss and a sigmoid function as the output function. For NNs, we used the implementation of [17] for linear programs. For two-layer NNs, we used fully connected hidden layers where the number of hidden units is 50 and ReLU as the activation function. The optimization algorithm was Adam [11] where the number of epochs is 50, and the learning rate is 1e−4. For two-stage NNs, the batch size was set to 100. For gradient boosting, we employed the sample-efficient single output model presented in Sect. 4.4. We utilized the implementation of LightGBM [10] and set most of the hyperparameters to the default values. For the two-stage approach, we set the number of ensembles to 100, which is a default setting of LightGBM. For the end-to-end approach, we set the number of ensembles to 20. The reason of the fewer number of ensembles of the end-to-end approach than the two-stage approach is to simply reduce the computational cost: the gradients and Hessian matrices require more computation due to solving linear equations. All other parameters were commonly set to the default values. Our code is available at https://github.com/tconishi/gbm-e2e.

## 5.1  Toy Example

We first illustrate an advantage of the end-to-end learning approach by comparing its behavior with that of the two-stage approach on a toy example. Figure 1 illustrates an edge-weighted directed graph with 12 vertices. We consider solving the shortest path problem from vertex 0 (colored with red) to vertex 7 (colored with blue) on this graph. The edge thickness represents the weights of edges: the weights of thick and thin edges are 1.0 and 0.1, respectively. In this case,

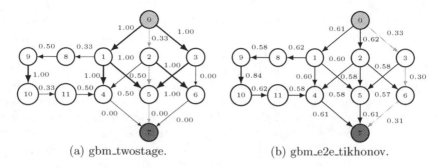

(a) gbm_twostage.    (b) gbm_e2e_tikhonov.

**Fig. 2.** Probabilities predicted by gbm_twostage (a) and gbm_e2e_tikhonov (b). The thick edges represent that they have a large probability. (Color figure online)

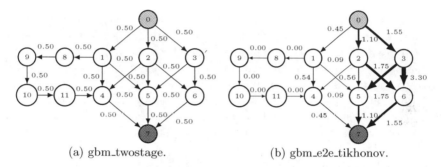

(a) gbm_twostage.    (b) gbm_e2e_tikhonov.

**Fig. 3.** Absolute values of gradients when making the first tree by gbm_twostage (a) and gbm_e2e_tikhonov (b). The thick edges represent that they have a large absolute gradient.

the unique shortest path is $(0, 3, 6, 7)$ colored with blue (here a path is represented by a sequence of vertices visited by it), and the sum of the weights is 1.2 $(= 1.0 + 0.1 + 0.1)$.

We consider a situation where the edge weights are unknown. In this case, the uncertain parameters are the edge weights, and the task at the learning phase is to predict the edge weights and we find a shortest path based on the predicted weights at the optimization phase. Assume every vertex has the feature of a five-dimensional one-hot vector (described beside the vertex in Fig. 1). The feature of an edge is simply represented as the concatenation of the features of its end vertices. For example, the feature of the edge $\{0, 2\}$ is a 10-dimensional binary vector 1000001000. We assume that 20 training instances are given for this graph and we learn a classifier with the training instances. To point out our findings clearly, we did not add any noise to the weights and features; i.e., we prepared the same 20 graphs as the training instances.

Figure 2 shows the results of prediction by gbm_twostage and gbm_e2e_tikhonov. The values of edges represent the predicted probabilities that the edges have large weights. The path colored with red denotes the solution based on the predicted weights. As shown in Fig. 2a, the two-stage approach fails to select the correct path. The reason is that the two-stage approach underestimates the weights of the path $(0, 2, 5)$ although the two edges $\{0, 2\}$ and $\{2, 5\}$ have the large weight in the ground-truth. This error is caused by the path $(0, 2, 5)$ has the same feature as $(1, 8, 9)$ and $(10, 11, 4)$ all of whose edges have a small weight. Since the prediction phase in the two-stage approach tries to minimize the total loss of all the edges in the graph, and thus it misclassified the weights of edges $\{0, 2\}$ and $\{2, 5\}$. Note that this issue could be avoided by replacing the sample-efficient single output model with the multi-output model. However, the multi-output model suffers from the small sample size if $N$ is small.

Contrarily, the end-to-end approach is successful in selecting the correct path as shown in Fig. 2b. To clarify the reason, we focus on the gradients in learning the models. Figure 3 illustrates the absolute values of the gradients when the approaches make the first tree. Since our implementation always returns 0.5 as an initially predicted probability, all the absolute gradients of the two-stage approach are also 0.5. On the other hand, the absolute gradients of the end-to-end approach take different values for each edge. In particular, they have large values around the correct shortest path and small values on unrelated paths. This means that, at the prediction phase of the end-to-end approach, edges around the correct shortest path contribute to the loss function more than other edges, and thus the end-to-end approach succeeded in ignoring the bad information for selecting the correct path.

Note that the total accuracy of prediction by the end-to-end approach is worse than the two-stage approach. In particular, the end-to-end approach misclassified the edge $\{0, 3\}$ although the edge is included in the shortest path. This is because the misclassification of edge $\{0, 3\}$ does not affect the total loss of the end-to-end approach: in this case, the end-to-end approach could select the shortest path regardless of the misclassification of $\{0, 3\}$.

## 5.2   Shortest Path Problem

In this experiment, we consider solving the shortest path problem on a real-world graph. We prepared the Facebook ego network dataset [16] for simulating real-world graphs. The Facebook dataset consists of 10 ego networks which include 4,039 vertices (i.e., users) and 88,234 edges (i.e., user friendships). We selected an ego network of a user whose id is 3980. This network contains 59 vertices and 292 edges. To select the source vertex and target vertex, we first computed the longest shortest paths on the network where we assume all the edges have the same weight. We then picked one path from them and selected the end vertices as the source and target. For this network, we made 25 instances where we fixed the source and the target but we assigned different edge weights for each instance. We are given training instances where the edge weights are observed, and the task is to find the shortest path from the source to the target for test instances where the edge weights are not known.

For each instance, we assigned feature vectors to edges according to the following procedure. In the Facebook dataset, every user has a binary feature vector that represents user profiles. We first made a feature vector of an edge between two users by concatenating the feature vectors of the users. We concatenated this vector with an instance-wise vector that is 20-dimensional and randomly generated for each instance. As a result, an edge has different feature vectors for each instance.

To assign edge weights, we prepared a fully connected and randomly initialized three-layer NN which consists of the ReLU activation and the sigmoid output. This network takes an input as a feature vector of the $i$th edge and generated an output $w_i \in [0, 1]$. The $i$th edge weight $\theta_i$ was determined by $\theta_i = 0.1 + 0.9 \cdot \text{round}(w_i)$, where $\text{round}(w_i)$ outputs a rounded value of $w_i$. $\theta_i$ takes 0.1 if $w_i < 0.5$ and 1.0 if $w_i \geq 0.5$. To simulate noisy settings, we flipped $\theta_i$ from 0.1 to 1.0 and vice versa with a probability $p = 0.0$, 0.1 and 0.2, respectively.

Table 1 shows the experimental results. Compared to the objective values of the gradient boosting algorithms, our end-to-end learning algorithms mostly achieved better performances than the two-stage algorithm. The two-stage algorithm consistently had better test AUC scores than end-to-end algorithms. Also, in this experiment, the gradient boosting algorithms showed better performances than the NN algorithms. Note that several methods have achieved the same objective values even though the corresponding AUC scores were different in a single setting. Such results could occur when these methods select the paths with the same sum of weights.

**Table 1.** Results on the shortest path problem of the Facebook dataset. The best algorithms except for random and optimal are shown in bold.

|  | $p=0.0$ | | $p=0.1$ | | $p=0.2$ | |
|---|---|---|---|---|---|---|
|  | Objective | AUC | Objective | AUC | Objective | AUC |
| nn1_twostage | 1.3820 | 0.7414 | 1.4900 | 0.6613 | 2.3000 | 0.6025 |
| nn2_twostage | 1.3820 | 0.8482 | 1.4900 | 0.7249 | 2.3000 | **0.6268** |
| nn1_e2e | 1.3820 | 0.6225 | 1.4900 | 0.5456 | 2.3000 | 0.5100 |
| nn2_e2e | 1.3820 | 0.5745 | 1.4900 | 0.5345 | 2.3000 | 0.4933 |
| gbm_twostage | 1.3940 | **0.9022** | 1.5060 | **0.7255** | 2.3580 | 0.6162 |
| gbm_e2e_tikhonov | 1.3820 | 0.5443 | 1.4540 | 0.5423 | **2.2820** | 0.5828 |
| gbm_e2e_abs | **1.3460** | 0.5962 | **1.4180** | 0.5570 | 2.5700 | 0.5000 |
| gbm_e2e_sfn | 1.3820 | 0.5420 | 1.4540 | 0.5443 | 2.3000 | 0.5624 |
| random | 2.1100 | – | 2.0640 | – | 2.8180 | – |
| optimal | 1.2960 | – | 1.2020 | – | 1.6840 | – |

**Table 2.** Results on the matching problem of the MovieLens dataset.

|  | $p = 0.0$ | | $p = 0.1$ | | $p = 0.2$ | |
|---|---|---|---|---|---|---|
|  | Objective | AUC | Objective | AUC | Objective | AUC |
| nn1_twostage | 10.3500 | 0.9758 | 12.3750 | 0.7873 | 12.9000 | 0.6721 |
| nn2_twostage | 13.6000 | 0.9952 | 14.2500 | **0.7953** | 14.4000 | 0.6755 |
| nn1_e2e | 8.7000 | 0.7585 | 10.2000 | 0.3042 | 11.3500 | 0.3834 |
| nn2_e2e | 12.7000 | 0.8502 | 13.8250 | 0.3797 | 13.7250 | 0.4062 |
| gbm_twostage | **15.2000** | **0.9986** | **15.3750** | 0.7950 | **15.7500** | **0.6766** |
| gbm_diffopt_tikhonov | 13.4750 | 0.7012 | 15.0000 | 0.6441 | 14.1250 | 0.6015 |
| gbm_diffopt_abs | 13.4500 | 0.6714 | 14.8000 | 0.6339 | 15.7250 | 0.5881 |
| gbm_diffopt_sfn | 13.0500 | 0.7196 | 15.2500 | 0.6505 | 14.7250 | 0.5998 |
| random | 5.5500 | – | 8.8000 | – | 12.0250 | – |
| optimal | 15.7000 | – | 39.0750 | – | 40.0000 | – |

## 5.3    Bipartite Matching Problem

In this experiment, we consider solving the maximum bipartite matching problem. We performed this problem on the MovieLens 100K dataset [8], which includes 1,000,000 ratings from 943 users on 1,682 movies. We considered complete bipartite graphs where the vertices on one side denote the set of users and the vertices on the other side denote the set of movies. The edge weights represent the degree of user-movie interactions, e.g., user preferences of movies. Suppose that we have several bipartite graphs as training instances where the edge weights are given. The task is to find a better matching of movies to users for test instances where the edge weights are not known.

To make instances, we randomly selected 40 users and 800 movies from the dataset. We fixed the 40 users for all instances and assigned different 40 movies of each instance. As a result, we obtained 20 instances to be considered.

We obtained feature vectors for users and movies from the dataset. For users, we used age, gender, occupation, and the first digit of the ZIP code, and transformed them into one vector; gender, occupation, and the first digit were encoded as a one-hot vector. For movies, we used genre information and transformed it into a binary vector. The feature vector of the edge between a user and a movie was constructed by concatenating the vectors of the user and the movie.

Since the MovieLens dataset contains the user ratings on movies, we generated the edge weights of bipartite graphs in a synthetic way. We prepared a fully connected three-layers NN that has the ReLU activation and the sigmoid output and was randomly initialized. For the $i$th edge, this network takes an input as a feature vector and generates an output $w_i \in [0, 1]$. The $i$th edge weight $\theta_i \in \{0, 1\}$ was determined by $\theta_i = \text{round}(w_i)$. As in Sect. 5.2, we flipped the $\theta_i$ from 0 to 1 and vice versa with a probability $p = 0.0, 0.1$ and $0.2$, respectively.

Table 2 shows the results of the experiment. In contrast to the experiment of the shortest path problem, the objective values of the end-to-end learning algorithms of gradient boosting are worse than the two-stage algorithm. As the noise level increases, the performance of end-to-end learning algorithms tends to be improved compared to the two-stage algorithm. Regarding the AUC scores, the two-stage algorithms still remain superior to the end-to-end algorithms.

## 5.4  Discussion

Through the above three experiments, we confirmed the effectiveness of our end-to-end learning approach. In Sect. 5.1, we first showed a toy example of the shortest path problem. While this example may be somewhat simple and artificial, it highlights the key difference between the two approaches. The two-stage approach tries to minimize the loss of all uncertain parameters. It is successful in predicting individual parameters but sometimes could fail to find the optimal solution due to the noise or incompleteness of the features unrelated to the solution. The end-to-end approach tries to minimize the loss of finding the optimal solution. It can capture the structure of the problem and pay attention to predicting uncertain parameters around the solution. Whereas, the prediction of individual parameters could be worse than the two-stage approach: the loss function is less accurate of the misspecification of all the parameters.

The results of the subsequent sections could be explained by this difference. For the shortest path problem reported in Sect. 5.2, our end-to-end learning algorithms of gradient boosting outperformed the two-stage algorithm in the optimization phase. We guess that this is because it is important for obtaining a good solution in the shortest path problem to predicting the parameters of edges around optimal solutions. As in the toy problem, the end-to-end approaches would be successful in this point for the real-world graph. On the contrary, for the bipartite matching problem in Sect. 5.3, the two-stage approach showed better performances than the end-to-end approaches. Compared to the shortest path problem, it would be more important for the bipartite matching problem to precisely predict individual edges rather than to focus on the edges around optimal solutions. One possible reason is that the bipartite matching problem could have other suboptimal solutions of good objective value apart from optimal solutions. Even when the models fail to obtain the optimal solutions, the two-stage approach could find good suboptimal solutions.

We note that a limitation of the end-to-end approaches is the large computational cost. The gradient and Hessian computations of all end-to-end approaches require solving linear systems and were significantly slower than two-stage ones. Although we focused on the objective values of the optimization problem, which are the performance measure to be primarily evaluated, developing more scalable algorithms would be an important research direction.

The performance of NN algorithms was not so good as opposed to the results of [17]. Although we also experimented with the bipartite matching problem in [17], we did not observe the superior performance of the end-to-end approach of NNs. Since the hyperparameter setting described in [17] was somewhat unclear, we referred to the code of [17]. However, it was still not evident how the setting was selected though the settings of two-stage and end-to-end approaches were different. Hence, we set the hyperparameters of the NNs (and also gradient boosting) in our experiments to mostly use the common settings to two-stage and end-to-end algorithms. We did not know the exact reasons for the discrepancy between our results and [17]. Future work could be devoted to investigating the performance of the models and learning algorithms through more systematic and comprehensive experiments.

## 6 Conclusion

We considered the problem of predicting uncertain parameters in mathematical optimization. We focused on the end-to-end approach for gradient boosting and derived the learning algorithms by extending gradient-based optimization to second-order optimization. In the experiments, we first clarified the property of the end-to-end approach by a toy example and then showed the effectiveness in the shortest path problem.

## References

1. Amos, B., Kolter, Z.: OptNet: differentiable optimization as a layer in neural networks. In: Proceedings of the 34th International Conference on Machine Learning, pp. 136–145 (2017)
2. Chen, T., Guestrin, C.: XGBoost: a scalable tree boosting system. In: Proceedings of the 22nd ACM SIGKDD International Conference on Knowledge Discovery and Data Mining, pp. 785–794 (2016)
3. Dauphin, Y.N., Pascanu, R., Gülçehre, Ç., Cho, K., Ganguli, S., Bengio, Y.: Identifying and attacking the saddle point problem in high-dimensional non-convex optimization. Advances in Neural Information Processing Systems **27**, 2933–2941 (2014)
4. Demirovic, E., et al.: An investigation into prediction + optimisation for the knapsack problem. In: Integration of Constraint Programming, Artificial Intelligence, and Operations Research, pp. 241–257 (2019)
5. Demirovic, E., Stuckey, P.J., Bailey, J., Chan, J., Leckie, C., Ramamohanarao, K., Guns, T.: Predict+optimise with ranking objectives: Exhaustively learning linear functions. In: Proceedings of the 28th International Joint Conference on Artificial Intelligence, pp. 1078–1085 (2019)
6. Elmachtoub, A.N., Grigas, P.: Smart "predict, then optimize". CoRR abs/1710.08005 (2017)
7. Friedman, J., Hastie, T., Tibshirani, R.: Additive logistic regression: a statistical view of boosting. Ann. Stat. **28**(2), 337–407 (2000)
8. Harper, F.M., Konstan, J.A.: The movielens datasets: History and context. ACM Trans. Interact. Intell. Syst. (TiiS) **5**(4), 19:1–19:19 (2016)
9. Ito, S., Yabe, A., Fujimaki, R.: Unbiased objective estimation in predictive optimization. In: Proceedings of the 35th International Conference on Machine Learning, pp. 2181–2190 (2018)
10. Ke, G., et al.: LightGBM: a highly efficient gradient boosting decision tree. In: Advances in Neural Information Processing Systems, vol. 30, pp. 3146–3154 (2017)
11. Kingma, D.P., Ba, J.: Adam: a method for stochastic optimization. In: 3rd International Conference on Learning Representations (2015)

12. Magnus, J.R.: On the concept of matrix derivative. J. Multivar. Anal. **101**(9), 2200–2206 (2010)
13. Magnus, J.R., Neudecker, H.: Matrix Differential Calculus with Applications in Statistics and Econometrics. Wiley (1988)
14. Mandi, J., Demirovic, E., Stuckey, P.J., Guns, T.: Smart predict-and-optimize for hard combinatorial optimization problems. CoRR abs/1911.10092 (2019). (to appear in proceedings of AAAI 2020)
15. Martens, J., Sutskever, I.: Training deep and recurrent networks with hessian-free optimization. In: Montavon, G., Orr, G.B., Müller, K.-R. (eds.) Neural Networks: Tricks of the Trade. LNCS, vol. 7700, pp. 479–535. Springer, Heidelberg (2012). https://doi.org/10.1007/978-3-642-35289-8_27
16. McAuley, J.J., Leskovec, J.: Learning to discover social circles in ego networks. In: Advances in Neural Information Processing Systems, vol. **25**, pp. 548–556 (2012)
17. Wilder, B., Dilkina, B., Tambe, M.: Melding the data-decisions pipeline: decision-focused learning for combinatorial optimization. In: The Thirty-Third AAAI Conference on Artificial Intelligence, pp. 1658–1665 (2019)

# Bayesian Methods

# Probabilistic Reconciliation of Hierarchical Forecast via Bayes' Rule

Giorgio Corani[✉], Dario Azzimonti, João P. S. C. Augusto, and Marco Zaffalon

Istituto Dalle Molle di Studi sull'Intelligenza Artificiale (IDSIA), USI - SUPSI, Manno, Switzerland
{giorgio,dario.azzimonti,zaffalon}@idsia.ch

**Abstract.** We present a novel approach for reconciling hierarchical forecasts, based on Bayes' rule. We define a prior distribution for the bottom time series of the hierarchy, based on the bottom base forecasts. Then we update their distribution via Bayes' rule, based on the base forecasts for the upper time series. Under the Gaussian assumption, we derive the updating in closed-form. We derive two algorithms, which differ as for the assumed independencies. We discuss their relation with the MinT reconciliation algorithm and with the Kalman filter, and we compare them experimentally.

**Keywords:** Hierarchical forecasting · Bayes' rule · Reconciliation · Linear-Gaussian model

## 1 Introduction

Often time series are organized into a hierarchy. For example, the total visitors of a country can be divided into regions and the visitors of each region can be further divided into sub-regions. The most disaggregated time series of the hierarchy are referred to as *bottom time series*, while the remaining time series are referred to as *upper time series*.

Forecasts of hierarchical time series should be *coherent*; for instance, the sum of the forecasts of the different regions should equal the forecast for the total. The forecasts are *incoherent* if they do not satisfy such constraints. A simple way for generating coherent forecasts is *bottom-up*: one takes the forecasts for the bottom time series and sums them up according to the summing constraints in order to produce forecasts for the entire hierarchy. Yet this approach does not consider the forecasts produced for the upper time series, which contain useful information. For instance, upper time series are smoother and allow to better estimate of the seasonal patterns and the effect of the covariates.

**Electronic supplementary material** The online version of this chapter (https://doi.org/10.1007/978-3-030-67664-3_13) contains supplementary material, which is available to authorized users.

Thus, modern reconciliation methods [9,18] proceed in two steps. First, *base forecasts* are computed by fitting an independent model to each time series. Then, the base forecasts are adjusted to become coherent; this step is called *reconciliation*. The forecasts for the entire hierarchy are then obtained by summing up the reconciled bottom time series. Reconciled forecasts are generally more accurate than the base forecasts, as they benefit from information coming from multiple time series. The state-of-the art reconciliation algorithm is MinT [18], which minimizes the mean squared error of the reconciled forecasts by solving a generalized least squares problem; its point forecast, besides being coherent, are generally more accurate than the base forecast.

Hierarchical probabilistic forecasting is however still an open area of research. The algorithm by [17] constructs a coherent forecast in a bottom-up fashion, modelling via copulas the joint distribution of the bottom time series, while [13] proposes a top-down approach, where the top time series is forecasted and then disaggregated. Both algorithms are based on numerical procedures which have no closed-form solution; hence they are not easily interpretable. In [6], a geometric interpretation of the reconciliation process is provided. It is moreover shown that the log score is not proper with respect to incoherent probabilistic forecasts. As an alternative, the energy score can be used for comparing reconciled to unreconciled probabilistic hierarchical forecasts. In [1] multivariate Gaussian predictive densities and bootstrap densities are experimentally compared for hierarchical probabilistic forecasting.

We address probabilistic reconciliation using Bayes' rule. We define the prior beliefs about the bottom time series, based on the base forecasts for the bottom time series. We then update them incorporating the information contained in the forecasts for the upper time series. Under the Gaussian assumption, we compute the update in closed form, obtaining the posterior distribution about the bottom time series and then about the entire hierarchy. Our reconciled forecasts minimize the mean squared error; indeed, we prove that they match the point predictions of MinT, whose optimality has been proven in a frequentist way. Our algorithm provides the joint predictive distribution for the hierarchy; thus we call it pMinT, which stands for probabilistic MinT. We also provide a variant of pMinT, obtained by making an additional independence assumption; we call it LG, as it is related to the linear-Gaussian model [3, Chap. 8.1.4].

We show a link between the reconciliation problem and the Kalman filter, opening the possibility of borrowing from the literature of the Kalman filter for future research. We then compare the algorithms on synthetic and real data sets, eventually concluding that pMinT yields more accurate probabilistic forecasts than both bottom-up and LG.

The paper is organized as follows: we introduce the reconciliation problem in Sect. 2, we discuss the algorithms in Sect. 3, we discuss the reconciliation of a simple hierarchy in Sect. 4 and we present the experiments in Sect. 5.

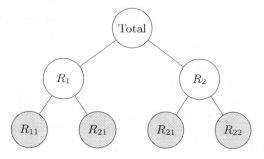

**Fig. 1.** A hierarchical time series which disaggregates the visitors into regions and sub-regions.

## 2   Time Series Reconciliation

Figure 1 shows a hierarchy. We could interpret it as the visitors of a country, which are disaggregated first by region ($R_1$, $R_2$) and then by sub-regions ($R_{11}$, $R_{12}$, $R_{21}$, $R_{22}$). The most disaggregated time series (*bottom time series*) are shaded. The hierarchy contains $m$ time series, of which $n$ are bottom time series. We denote by uppercase letters the random variables and by lowercase letters their observations. The vector of observations available at time $t$ for the entire hierarchy is $\mathbf{y}_t \in \mathbb{R}^m$; they are observations from the set of random variables $\mathbf{Y}_t$. Vector $\mathbf{y}_t$ can be broken down in two parts, namely $\mathbf{y}_t = [\mathbf{u}_t^T, \mathbf{b}_t^T]^T$; $\mathbf{b}_t \in \mathbb{R}^n$ contains the observations of the bottom time series while $\mathbf{u}_t \in \mathbb{R}^{m-n}$ contains the observations of the upper time series. At time $t$, the observations available for the hierarchy of Fig. 1 are thus:

$$\mathbf{y}_t = [y_{\text{Total}}, y_{R_1}, y_{R_2}, y_{R_{11}}, y_{R_{12}}, y_{R_{21}}, y_{R_{22}}]^T = [\mathbf{u}_t^T, \mathbf{b}_t^T]^T,$$

where:

$$\mathbf{u}_t = [y_{\text{Total}}, y_{R_1}, y_{R_2}]^T$$
$$\mathbf{b}_t = [y_{R_{11}}, y_{R_{12}}, y_{R_{21}}, y_{R_{22}}]^T.$$

The structure of the hierarchy is represented by the summing matrix $\mathbf{S} \in \mathbb{R}^{m \times n}$ such that:

$$\mathbf{y}_t = \mathbf{S}\mathbf{b}_t. \tag{1}$$

The $\mathbf{S}$ matrix of hierarchy in Fig.1 is:

$$\mathbf{S} = \begin{bmatrix} 1 & 1 & 1 & 1 \\ 1 & 1 & 0 & 0 \\ 0 & 0 & 1 & 1 \\ 1 & 0 & 0 & 0 \\ 0 & 1 & 0 & 0 \\ 0 & 0 & 1 & 0 \\ 0 & 0 & 0 & 1 \end{bmatrix} = \begin{bmatrix} \mathbf{A} \\ \hdashline \mathbf{I} \end{bmatrix}, \tag{2}$$

where the sub-matrix $\mathbf{A} \in \mathbb{R}^{(m-n)\times n}$ encodes which bottom time series should be summed up in order to obtain each upper time series.

We denote by $\widehat{\mathbf{y}}_{t+h} \in \mathbb{R}^m$ the base forecasts issued at time $t$ about of $y$ and referring to $h$ steps ahead. We separate base forecasts for bottom time series ($\widehat{\mathbf{b}}_{t+h} \in \mathbb{R}^n$) and upper time series ($\widehat{\mathbf{u}}_{t+h} \in \mathbb{R}^{m-n}$), namely $\widehat{\mathbf{y}}_{t+h} = [\widehat{\mathbf{u}}_{t+h}^T, \widehat{\mathbf{b}}_{t+h}^T]^T$. The variances of the error of the base forecasts will be used later. If forecasts for different time horizons are needed (e.g., $h = 1, 2, 3, ..$), the reconciliation is performed independently for each $h$. In the following we generically assume to reconcile the forecasts for $h$ steps ahead.

*The MinT Reconciliation.* Most reconciliation algorithms [9], including MinT [18], assume the reconciled bottom forecasts ($\widetilde{\mathbf{b}}_{t+h}$) to be a linear combination of the base forecasts ($\widehat{\mathbf{y}}_{t+h}$) available for the whole hierarchy, i.e. their objective is to find a matrix $\mathbf{P}_h \in \mathbb{R}^{n\times m}$ such that:

$$\widetilde{\mathbf{b}}_{t+h} = \mathbf{P}_h \widehat{\mathbf{y}}_{t+h}. \tag{3}$$

Let us denote by $\widehat{\mathbf{E}}_{t+h} = \mathbf{Y}_{t+h} - \widehat{\mathbf{y}}_{t+h} \in \mathbb{R}^m$ the vector of the errors of the base forecast $h$-steps ahead and by $\mathbf{W}_h = \mathbb{E}[\widehat{\mathbf{E}}_{t+h}\,\widehat{\mathbf{E}}_{t+h}^T \mid \mathcal{I}_t]$ their covariance matrix, where $\mathcal{I}_t$ denotes all the information available up to time $t$. In [18] it is proven that the reconciliation matrix given by:

$$\mathbf{P}_h = (\mathbf{S}^T \mathbf{W}_h^{-1} \mathbf{S})^{-1} \mathbf{S}^T \mathbf{W}_h^{-1} \tag{4}$$

is optimal, in the sense that it minimizes the trace of the reconciliation errors' covariance matrix. The reconciled forecasts for the whole hierarchy are obtained by summing the reconciled bottom forecasts, and they are proven to minimize the mean squared error over the entire hierarchy.

**Estimation of $\mathbf{W}_h$.** Estimating $\mathbf{W}_h$ differently for each $h$ is an open problem. For the case $h = 1$ the estimation is simpler. The variance of the forecasts equals the variance of the residuals (i.e., the errors on 1-step predictions made on the training data) and cross-covariances are estimated as the covariance of the residuals. The best estimates are obtained [18] by shrinking the full covariance matrix towards a diagonal matrix, using the method of [15].

The case $h > 1$ is instead problematic. The variance of the forecasts are obtained by increasing the 1-step variance through analytical formulas, which differ from the variance of the $h$-steps ahead residuals. Morevoer, the covariances in $\mathbf{W}_h$ have to be numerically estimated by looking at the $h$-steps residuals. However, the number of $h$-steps residuals decreases with $h$, making the estimate more noisy.

As a workaround, [18] assumes $\mathbf{W}_h = k_h \mathbf{W}_1$, where $k_h > 0$ is an unknown constant which depends on $h$ while $\mathbf{W}_1$ is the covariance matrix of the one-step ahead errors. The underlying assumption is thus that all terms within the variance/covariance matrix of the errors grow in the same way with $h$. The advantage of this approach is that $k_h$ cancels out when computing the reconciled forecasts, as it can be seen by setting $\mathbf{W}_h = k_h \mathbf{W}_1$ in Eq. (4). Yet, $k_h$ appears in the expression of the variance of the reconciled forecasts. In the following we refer to the assumption $\mathbf{W}_h = k_h \mathbf{W}_1$ as "the $k_h$ assumption".

# 3   Probabilistic Reconciliation

We address the reconciliation problem by merging the probabilistic information contained in the base forecasts for the bottom and the upper time series. We perform the fusion using Bayes' rule.

We first define the prior about the *bottom* time series. We have observed the time series up to time $t$ and we are interested in the reconciled forecasts for time $t + h$. We denote by $\mathbf{B}_{t+h}$ the vector of the bottom time series at time $t + h$; this is thus a vector of random variables and $B^i_{t+h}$ represents its $i$th element. We moreover denote by $\widehat{\mathbf{b}}_{t+h}$ the vector of base forecasts for the bottom time series for time $t + h$, and by $\mathbf{b}_{t+h}$ the actual observation of the bottom random variables at time $t + h$. Finally, $\mathcal{I}_{t,b}$ is the information available up to time $t$ regarding the bottom time series, i.e. the past values of the bottom time series: $\mathcal{I}_{t,b} = \{\mathbf{b}_1, \ldots, \mathbf{b}_t\}$.

In the following we adopt the $k_h$ assumption for all the covariance matrices assuming moreover that, for a given $h$, the value of $k_h$ is shared among all the involved covariance matrices. As we will show later, this is equivalent to the $k_h$ assumption made by MinT. Let us hence denote the covariance matrix of the forecast $h$-steps ahead by $\widehat{\boldsymbol{\Sigma}}_{B,h} = k_h \widehat{\boldsymbol{\Sigma}}_{B,1}$. Assuming the bottom time series to be jointly Gaussian we have:

$$p(\mathbf{B}_{t+h} \mid \mathcal{I}_{t,b}) = N\left(\widehat{\mathbf{b}}_{t+h}, k_h \widehat{\boldsymbol{\Sigma}}_{B,1}\right). \tag{5}$$

*Probabilistic Bottom-Up.* If we have no information about the upper time series, we can build a joint predictive distribution for the entire hierarchy by summing the bottom forecast via matrix $\mathbf{S}$:

$$p(\mathbf{Y}_{t+h} \mid \mathcal{I}_{t,b}) = N\left(\mathbf{S}\widehat{\mathbf{b}}_{t+h}, \mathbf{S}k_h\widehat{\boldsymbol{\Sigma}}_{B,1}\mathbf{S}^T\right), \tag{6}$$

which is a *probabilistic bottom-up* reconciliation. Note that, in this case, $k_h$ appears only in the expression of the variance.

*Updating.* If the forecasts $\widehat{\mathbf{U}}_{t+h}$ about the upper time series are available, then we can use them in order to update our prior. We assume:

$$\widehat{\mathbf{U}}_{t+h} = \mathbf{A}\mathbf{B}_{t+h} + \varepsilon^u_{t+h}, \tag{7}$$
$$\varepsilon^u_{t+h} \sim N\left(\mathbf{0}, \widehat{\boldsymbol{\Sigma}}_{U,h}\right),$$

where $\widehat{\boldsymbol{\Sigma}}_{U,h} = k_h \widehat{\boldsymbol{\Sigma}}_{U,1}$ is the covariance of the noise. We thus treat $\widehat{\mathbf{U}}_{t+h}$ as a set of different sums of the future values of the bottom time series, corrupted by noise. Hence:

$$p(\widehat{\mathbf{U}}_{t+h} \mid \mathbf{B}_{t+h}) = N\left(\mathbf{A}\mathbf{B}_{t+h}, k_h \widehat{\boldsymbol{\Sigma}}_{U,1}\right). \tag{8}$$

The posterior distribution of the bottom time series is given by Bayes' rule:

$$p(\mathbf{B}_{t+h} \mid \mathcal{I}_{t,b}, \widehat{\mathbf{U}}_{t+h}) \frac{p(\mathbf{B}_{t+h} \mid \mathcal{I}_{t,b})p(\widehat{\mathbf{U}}_{t+h} \mid \mathcal{I}_{t,b}, \mathbf{B}_{t+h})}{p(\widehat{\mathbf{U}}_{t+h} \mid \mathcal{I}_{t,b})}$$

$$= \frac{p(\mathbf{B}_{t+h} \mid \mathcal{I}_{t,b})p(\widehat{\mathbf{U}}_{t+h} \mid \mathbf{B}_{t+h})}{p(\widehat{\mathbf{U}}_{t+h} \mid \mathcal{I}_{t,b})}$$

$$\propto p(\mathbf{B}_{t+h} \mid \mathcal{I}_{t,b})p(\widehat{\mathbf{U}}_{t+h} \mid \mathbf{B}_{t+h})$$

$$= p(\mathbf{B}_{t+h} \mid \mathcal{I}_{t,b})p(\mathbf{A}\mathbf{B}_{t+h} + \varepsilon_{t+h}^u \mid \mathbf{B}_{t+h}) \tag{9}$$

## 3.1   Computing Bayes' Rule

The posterior of Eq. (9) can be computed in closed form by assuming the vector $(\mathbf{B}_{t+h}, \widehat{\mathbf{U}}_{t+h})$ to be jointly Gaussian distributed. The linear-Gaussian (LG) model [3, Chap. 8.1.4]. computes analytically the updating by further assuming $\varepsilon_{t+h}^u$ to be independent from $\mathbf{B}_{t+h}$. Yet this independence might not always hold in our case. Consider for instance a special event driving upwards most time series. As a result we would observe both high values of $\mathbf{B}_{t+h}$ and negative values of $\varepsilon_{t+h}^u$, due to the underestimation of the upper time series. This would result in a correlation between $\mathbf{B}_{t+h}$ and $\varepsilon_{t+h}^u$.

We thus generalize the LG model by accounting for such correlation. We will later compare experimentally the results obtained adopting the LG model and its generalized version. We denote $\mathrm{Cov}(\mathbf{B}_{t+1}, \varepsilon_{t+1}^u \mid \mathcal{I}_{t,b}) = \mathbf{M}_1 \in \mathbb{R}^{n \times (m-n)}$ and we assume $\mathrm{Cov}(\mathbf{B}_{t+h}, \varepsilon_{t+h}^u \mid \mathcal{I}_{t,b}) = k_h \mathbf{M}_1$.

Our first step for computing Bayes' rule is to express the joint distribution $p(\mathbf{B}_{t+h}, \widehat{\mathbf{U}}_{t+h} \mid \mathcal{I}_{t,b})$. Since $\widehat{\mathbf{U}}_{t+h} = \mathbf{A}\mathbf{B}_{t+h} + \varepsilon_{t+h}^u$, the expected values are:

$$\mathbb{E}[\mathbf{B}_{t+h} \mid \mathcal{I}_{t,b}] = \widehat{\mathbf{b}}_{t+h},$$

$$\mathbb{E}[\widehat{\mathbf{U}}_{t+h} \mid \mathcal{I}_{t,b}] = \mathbf{A}\widehat{\mathbf{b}}_{t+h}.$$

We now derive the different blocks of the covariance matrix. The cross-covariance between $\mathbf{B}_{t+h}$ and $\widehat{\mathbf{U}}_{t+h}$ is:

$$\mathrm{Cov}(\mathbf{B}_{t+h}, \widehat{\mathbf{U}}_{t+h} \mid \mathcal{I}_{t,b}) = \mathrm{Cov}(\mathbf{B}_{t+h}, \mathbf{A}\mathbf{B}_{t+h} + \varepsilon_{t+h}^u \mid \mathcal{I}_{t,b})$$

$$= \mathrm{Cov}(\mathbf{B}_{t+h}, \mathbf{B}_{t+h} \mid \mathcal{I}_{t,b})\mathbf{A}^T + \mathrm{Cov}(\mathbf{B}_{t+h}, \varepsilon_{t+h}^u \mid \mathcal{I}_{t,b})$$

$$= k_h(\widehat{\boldsymbol{\Sigma}}_{B,1}\mathbf{A}^T + \mathbf{M}_1) \in \mathbb{R}^{n \times (m-n)}$$

where $k_h > 0$ is the multiplicative constant (Sect. 2) that yields the variance of the forecasts $h$-steps ahead, given the covariances of the forecasts 1-step ahead.

The covariance of the upper forecasts is:

$$
\begin{aligned}
\mathrm{Cov}(\widehat{\mathbf{U}}_{t+h} \mid \mathcal{I}_{t,b}) &= \mathrm{Cov}(\mathbf{AB}_{t+h} + \varepsilon_{t+h}^{u}, \mathbf{AB}_{t+h} + \varepsilon_{t+h}^{u} \mid \mathcal{I}_{t,b}) \\
&= k_h \mathbf{A}\widehat{\boldsymbol{\Sigma}}_{B,1}\mathbf{A}^T + k_h \widehat{\boldsymbol{\Sigma}}_{U,1} + \mathrm{Cov}(\mathbf{AB}_{t+h}, \varepsilon_{t+h}^{u}) + \mathrm{Cov}(\varepsilon_{t+h}^{u}, \mathbf{AB}_{t+h}) \\
&= k_h(\mathbf{A}\widehat{\boldsymbol{\Sigma}}_{B,1}\mathbf{A}^T + \widehat{\boldsymbol{\Sigma}}_{U,1} + \mathbf{AM}_1 + \mathbf{M}_1^T\mathbf{A}^T).
\end{aligned}
$$

Hence the joint prior (i.e., before observing $\widehat{\mathbf{U}}_{t+h}$) is:

$$
\begin{pmatrix} \mathbf{B}_{t+h} \\ \widehat{\mathbf{U}}_{t+h} \end{pmatrix} \mid \mathcal{I}_{t,b} \sim N\left[ \begin{pmatrix} \widehat{\mathbf{b}}_{t+h} \\ \mathbf{A}\widehat{\mathbf{b}}_{t+h} \end{pmatrix}, \begin{pmatrix} k_h\widehat{\boldsymbol{\Sigma}}_{B,1} & k_h(\widehat{\boldsymbol{\Sigma}}_{B,1}\mathbf{A}^T + \mathbf{M}_1) \\ k_h(\mathbf{A}\widehat{\boldsymbol{\Sigma}}_{B,1} + \mathbf{M}_1^T) & k_h(\mathbf{A}\widehat{\boldsymbol{\Sigma}}_{B,1}\mathbf{A}^T + \widehat{\boldsymbol{\Sigma}}_{U,1} + \mathbf{AM}_1 + \mathbf{M}_1^T\mathbf{A}^T) \end{pmatrix} \right].
$$

Now we receive the forecast $\widehat{\mathbf{u}}_{t+h}$ for the upper time series (recall that $\widehat{\mathbf{U}}_{t+h}$ denotes the random variables while $\widehat{\mathbf{u}}_{t+h}$ denotes observations). We obtain the posterior distribution for the bottom time series $P(\mathbf{B}_{t+h} \mid \widehat{\mathbf{u}}_{t+h})$ by applying the standard formulas for the conditional distribution of a MVN distribution [12, Sec.4.3.1]. To have a shorter notation, let us define:

$$
\begin{aligned}
\mathbf{G} &= \left( k_h(\widehat{\boldsymbol{\Sigma}}_{B,1}\mathbf{A}^T + \mathbf{M}_1) \right) \left( k_h(\mathbf{A}\widehat{\boldsymbol{\Sigma}}_{B,1}\mathbf{A}^T + \widehat{\boldsymbol{\Sigma}}_{U,1} + \mathbf{AM}_1 + \mathbf{M}_1^T\mathbf{A}^T) \right)^{-1} \\
&= \left( \widehat{\boldsymbol{\Sigma}}_{B,1}\mathbf{A}^T + \mathbf{M}_1 \right) \left( \mathbf{A}\widehat{\boldsymbol{\Sigma}}_{B,1}\mathbf{A}^T + \widehat{\boldsymbol{\Sigma}}_{U,1} + \mathbf{AM}_1 + \mathbf{M}_1^T\mathbf{A}^T \right)^{-1},
\end{aligned} \tag{10}
$$

where $k_h$ disappears from the expression of $\mathbf{G}$.

The reconciled bottom time series have then the following mean and variance:

$$
\widetilde{\mathbf{b}}_{t+h} = \mathbb{E}\left[ \mathbf{B}_{t+h} \mid \mathcal{I}_{t,b}, \widehat{\mathbf{u}}_{t+h} \right] = \widehat{\mathbf{b}}_{t+h} + \mathbf{G}(\widehat{\mathbf{u}}_{t+h} - \mathbf{A}\widehat{\mathbf{b}}_{t+h}) \tag{11}
$$

$$
\mathrm{Var}\left[ \mathbf{B}_{t+h} \mid \mathcal{I}_{t,b}, \widehat{\mathbf{u}}_{t+h} \right] = k_h \left( \widehat{\boldsymbol{\Sigma}}_{B,1} - \mathbf{G}(\mathbf{A}\widehat{\boldsymbol{\Sigma}}_{B,1} + \mathbf{M}_1^T) \right) \tag{12}
$$

Thus the adjustment applied to the base forecasts is proportional to $(\widehat{\mathbf{u}}_{t+h} - \mathbf{A}\widehat{\mathbf{b}}_{t+h})$, i.e. the difference between the prior mean and the uncertain observation (i.e., the forecasts) of the upper time series. The term $(\widehat{\mathbf{u}}_{t+h} - \mathbf{A}\widehat{\mathbf{b}}_{t+h})$ is called the *incoherence* of the base forecasts in [18]. The mean of the reconciled bottom time series does not depend on $k_h$, while the variance does.

The reconciled point forecast and the covariance for the entire hierarchy are:

$$
\mathbb{E}[\mathbf{Y}_{t+h} \mid \mathcal{I}_{t,b}, \widehat{\mathbf{u}}_{t+h}] = \widetilde{\mathbf{y}}_{t+h} = \mathbf{S}\widetilde{\mathbf{b}}_{t+h} \tag{13}
$$

$$
\mathrm{Var}\left[ \mathbf{Y}_{t+h} \mid \mathcal{I}_{t,b}, \widehat{\mathbf{u}}_{t+h} \right] = \mathbf{S}\,\mathrm{Var}\left[ \mathbf{B}_{t+h} \mid \mathcal{I}_{t,b}, \widehat{\mathbf{u}}_{t+h} \right]\mathbf{S}^T. \tag{14}
$$

## 3.2   Related Works and Optimality of the Reconciliation

Bayes' rule is a well-known tool for information fusion [5, Sec. 2], and we apply it for the first time for forecast reconciliation. We will later prove that the posterior mean of our approach yields the same point predictions of MinT.

Yet, our algorithm additionally provides the predictive distribution for the entire hierarchy; we thus call it pMinT, where p stands for *probabilistic*. We also

218     G. Corani et al.

contribute a novel reconciliation approach based on the linear-Gaussian (LG) model, which is obtained by setting $\mathbf{M} = \mathbf{0}$ in the definition of $\mathbf{G}$. Both pMinT and LG are thus probabilistic reconciliation algorithms.

We point out for the first time a link between the reconciliation problem and the Kalman filter, whose state-update equation can be derived from the linear-Gaussian model [14]. In particular, Eq. (11) has the same structure of the state-update of a Kalman filter. According to the definition of $\mathbf{G}$ of Eq. (10), the LG reconciliation corresponds to the standard Kalman filter [16, Chap. 5], while the pMinT reconciliation corresponds to a generalized Kalman filter which assumes correlation between the noise of the state and the noise of the output [16, Chap. 7.1]. Thus future research could explore the literature of the Kalman filter in order to borrow ideas for the reconciliation problem.

The optimality of our approach can be informally proven by considering that it yields the posterior mean of the reconciled forecasts, which is the minimizer of the quadratic loss under the Gaussian assumption [12, Chap. 5.7]. Moreover, it has the same equation of the state-update step of the Kalman filter, which provably minimizes the mean squared error of the estimates without any distributional assumption [16, Chap. 5]. In Sect. (4.1) we will moreover show that our point predictions correspond to those of MinT, which have been proven to be the minimizer of the mean-squared error.

### 3.3 The Covariance Matrices $\widehat{\boldsymbol{\Sigma}}_{B,1}$ and $\widehat{\boldsymbol{\Sigma}}_{U,1}$

The element $(i,j)$ of $\widehat{\boldsymbol{\Sigma}}_{B,1}$ is the covariance $\mathrm{Cov}(B_{t+1}^i, B_{t+1}^j \mid \mathcal{I}_{t,b}) = \mathrm{Cov}(B_{t+1}^i, B_{t+1}^j \mid \mathbf{B}_{1:t} = \mathbf{b}_{1:t})$, where $\mathbf{B}_{1:t} = \mathbf{b}_{1:t}$ denotes a realization $\mathbf{b}_{1:t}$ of $\mathbf{B}_{1:t}$. Yet we only have one observation of $B_{t+1}^i, B_{t+1}^j$ *conditional* on $\mathcal{I}_{t,b}$, which prevents estimating the covariance. We can overcome the problem with the following result, which shows that we can approximate $\widehat{\boldsymbol{\Sigma}}_{B,1}$ by computing the covariance of the residuals.

Let us consider the vectors of bottom time series $\mathbf{B}_1, \dots, \mathbf{B}_t$ and the conditional expectation $\widehat{\mathbf{B}}_{t+1} = \mathbb{E}[\mathbf{B}_{t+1} \mid \mathbf{B}_1, \dots, \mathbf{B}_t] = \mathbb{E}[\mathbf{B}_{t+1} \mid \mathbf{B}_{1:t}]$. Note that $\widehat{\mathbf{B}}_{t+1}$ is a random vector as we have not yet observed $\mathbf{B}_i$, $i = 1, \dots, t$. In this section we show

$$\mathbb{E}[\mathrm{Cov}(B_{t+1}^i, B_{t+1}^j \mid \mathbf{B}_{1:t})] = \mathrm{Cov}(E_{t+1}^i, E_{t+1}^j) \qquad i,j = 1, \dots, n, \tag{15}$$

where $E_{t+1}^i := B_{t+1}^i - \widehat{B}_{t+1}^i$, $i = 1, \dots, n$ denotes the residual of the model fitted on the $i$-th time series, for the forecast horizon $t+1$. If we observe $\mathcal{I}_{t,b}$, then we can approximate $\mathrm{Cov}(B_{t+1}^i, B_{t+1}^j \mid \mathcal{I}_{t,b})$ with $\mathrm{Cov}(B_{t+1}^i - \widehat{b}_{t+1}^i, B_{t+1}^j - \widehat{b}_{t+1}^j)$, the covariance of the residuals of the models fitted on the bottom time series.

Consider now the conditional covariance on the left side of Eq. (15), we have

$$\mathrm{Cov}(B_{t+1}^i, B_{t+1}^j \mid \mathbf{B}_{1:t}) = \mathbb{E}\left[\left(B_{t+1}^i - \mathbb{E}[B_{t+1}^i \mid \mathbf{B}_{1:t}]\right)\left(B_{t+1}^j - \mathbb{E}[B_{t+1}^j \mid \mathbf{B}_{1:t}]\right) \mid \mathbf{B}_{1:t}\right]$$
$$= \mathbb{E}\left[\left(B_{t+1}^i - \widehat{B}_{t+1}^i\right)\left(B_{t+1}^j - \widehat{B}_{t+1}^j\right) \mid \mathbf{B}_{1:t}\right]$$

By taking the expectation on both sides we obtain

$$\mathbb{E}[\text{Cov}(B^i_{t+1}, B^j_{t+1} \mid \mathbf{B}_{1:t})] = \mathbb{E}\left[\mathbb{E}\left[\left(B^i_{t+1} - \widehat{B}^i_{t+1}\right)\left(B^j_{t+1} - \widehat{B}^j_{t+1}\right) \mid \mathbf{B}_{1:t}\right]\right]$$

$$= \mathbb{E}\left[\left(B^i_{t+1} - \widehat{B}^i_{t+1}\right)\left(B^j_{t+1} - \widehat{B}^j_{t+1}\right)\right]$$

$$= \text{Cov}\left(B^i_{t+1} - \widehat{B}^i_{t+1}, B^j_{t+1} - \widehat{B}^j_{t+1}\right) = \text{Cov}\left(E^i_{t+1}, E^j_{t+1}\right)$$

We thus estimate the covariance matrix $\widehat{\boldsymbol{\Sigma}}_{B,1}$ using the covariance of the residuals of the models fitted on the bottom time series.

**Computation of $\widehat{\boldsymbol{\Sigma}}_{U,1}$.** According to Eq. (7),

$$c^u_{t\mid 1} = \mathbf{AB}_{t\mid 1} - \widehat{\mathbf{U}}_{t\mid 1},$$

whose variances and covariances can be readily computed from the residuals.

## 4    Reconciliation of a Simple Hierarchy

We now illustrate how the base forecasts interact during the reconciliation of a simple hierarchy. We consider a hierarchy constituted by two bottom time series ($B_1$ and $B_2$) and an upper time series $U$.

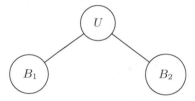

The base forecast for the bottom time series are the point forecasts $\widehat{b}_1$ and $\widehat{b}_2$ with variances $\sigma^2_1$ and $\sigma^2_2$. The prior beliefs about $B_1$ and $B_2$ are:

$$\binom{B_1}{B_2} \sim N\left[\binom{\widehat{b}_1}{\widehat{b}_2}, k_h \begin{pmatrix} \sigma^2_1 & \sigma_{1,2} \\ \sigma_{1,2} & \sigma^2_2, \end{pmatrix}\right]$$

where for simplicity we remove the forecast horizon $(t+h)$ from the notation.
The summing matrix is:

$$\mathbf{S} = \begin{bmatrix} 1 & 1 \\ 1 & 0 \\ 0 & 1 \end{bmatrix} = \begin{bmatrix} \mathbf{A} \\ 1 & 0 \\ 0 & 1 \end{bmatrix}.$$

We start considering the simpler case of reconciliation via the LG algorithm. The matrix $\mathbf{G}$ is:

$$\mathbf{G} = \widehat{\boldsymbol{\Sigma}}_{B,1}\mathbf{A}^T(\widehat{\boldsymbol{\Sigma}}_{U,1} + \mathbf{A}\widehat{\boldsymbol{\Sigma}}_{B,1}\mathbf{A}^T)^{-1} = \frac{1}{\sigma^2_u + \sigma^2_1 + \sigma^2_2 + 2\sigma_{1,2}}\begin{bmatrix} \sigma^2_1 + \sigma_{1,2} \\ \sigma^2_2 + \sigma_{1,2}, \end{bmatrix}$$

since $\mathbf{A} = [1\ 1]$, $\widehat{\boldsymbol{\Sigma}}_{U,1} = \sigma_u^2$ and moreover:

$$\widehat{\boldsymbol{\Sigma}}_{B,1}\mathbf{A}^T = [\sigma_1^2 + \sigma_{1,2}\ \ \sigma_{1,2} + \sigma_2^2]^T,$$
$$\mathbf{A}^T\widehat{\boldsymbol{\Sigma}}_{B,1}\mathbf{A} = \sigma_1^2 + \sigma_2^2 + 2\sigma_{1,2},$$
$$\widehat{\boldsymbol{\Sigma}}_{U,1} + \mathbf{A}\widehat{\boldsymbol{\Sigma}}_{B,1}\mathbf{A}^T = \sigma_u^2 + \sigma_1^2 + \sigma_2^2 + 2\sigma_{1,2}.$$

Note that $\mathbf{G}$ does not depend on $h$, as also shown in Eq. (10).

The reconciled bottom forecasts are:

$$\widetilde{\mathbf{b}} = \widehat{\mathbf{b}} + \mathbf{G}(\widehat{u} - \mathbf{A}\widehat{\mathbf{b}}), \tag{16}$$

where $\widehat{u}$ is the base forecast for $U$.

The reconciled bottom forecast can be written as:

$$\widetilde{\mathbf{b}} = \begin{bmatrix} \widehat{b}_1 \\ \widehat{b}_2 \end{bmatrix} + \begin{bmatrix} \sigma_1^2 + \sigma_{1,2} \\ \sigma_2^2 + \sigma_{1,2} \end{bmatrix} \frac{\widehat{u} - \mathbf{A}^T\widehat{\mathbf{b}}}{\sigma_u^2 + \sigma_1^2 + \sigma_2^2 + 2\sigma_{1,2}} \tag{17}$$

Equation (17) shows that the adjustment applied to the base forecasts depends on $\sigma_u^2$. If $\sigma_u^2$ is large the adjustment is small, since the upper forecast is not informative. If on the contrary $\sigma_u^2 = 0$, the sum of the reconciled bottom forecasts is forced to match $\widehat{u}$, i.e., $\widetilde{b}_1 + \widetilde{b}_2 = \widehat{u}$ (this can be shown by re-working Eq. (17)).

We now show that the reconciled bottom forecast are a linear combination of the base forecasts. Let us define:

$$g_1 = \frac{\sigma_1^2 + \sigma_{1,2}}{\sigma_u^2 + \sigma_1^2 + \sigma_2^2 + 2\sigma_{1,2}} \qquad g_2 = \frac{\sigma_2^2 + \sigma_{1,2}}{\sigma_u^2 + \sigma_1^2 + \sigma_2^2 + 2\sigma_{1,2}} \tag{18}$$

After some algebra we obtain:

$$\widetilde{\mathbf{b}} = \begin{bmatrix} \widehat{b}_1\left(1 - \frac{\sigma_1^2+\sigma_{1,2}}{\sigma_u^2+\sigma_1^2+\sigma_2^2+2\sigma_{1,2}}\right) + (\widehat{u} - \widehat{b}_2)\frac{\sigma_1^2+\sigma_{1,2}}{\sigma_u^2+\sigma_1^2+\sigma_2^2+2\sigma_{1,2}} \\ \widehat{b}_2\left(1 - \frac{\sigma_2^2+\sigma_{1,2}}{\sigma_u^2+\sigma_1^2+\sigma_2^2+2\sigma_{1,2}}\right) + (\widehat{u} - \widehat{b}_1)\frac{\sigma_2^2+\sigma_{1,2}}{\sigma_u^2+\sigma_1^2+\sigma_2^2+2\sigma_{1,2}} \end{bmatrix} = \begin{bmatrix} \widehat{b}_1\,(1 - g_1) + (\widehat{u} - \widehat{b}_2)g_1 \\ \widehat{b}_2\,(1 - g_2) + (\widehat{u} - \widehat{b}_1)g_2, \end{bmatrix} \tag{19}$$

Thus $\widetilde{b}_1$ is a weighted average of two estimates: $\widehat{b}_1$ and $(\widehat{u} - \widehat{b}_2)$; the weight of $\widehat{b}_1$ decreases with $\sigma_1^2$ and increases with $(\sigma_2^2 + \sigma_u^2)$.

The reconciliation carried out by pMinT is similar to what already discussed, once we adopt $g_1^*$ and $g_2^*$ in place of $g_1$ and $g_2$:

$$g_1^* = \frac{\sigma_1^2 + \sigma_{1,2} - \sigma_{u,1}}{\sigma_u^2 + \sigma_1^2 + \sigma_2^2 + 2\sigma_{1,2} - 2\sigma_{u,1} - 2\sigma_{u,2}} \tag{20}$$

$$g_2^* = \frac{\sigma_2^2 + \sigma_{1,2} - \sigma_{u,2}}{\sigma_u^2 + \sigma_1^2 + \sigma_2^2 + 2\sigma_{1,2} - 2\sigma_{u,1} - 2\sigma_{u,2}} \tag{21}$$

Thus pMinT accounts also for the cross-covariances $\sigma_{u,1}, \sigma_{u,2}$ between the bottom time series and of noise affecting the forecasts for the upper time series.

## 4.1   Relationship with MinT

Our reconciled bottom time series can be written as:

$$\tilde{\mathbf{b}} = \hat{\mathbf{b}} + \mathbf{G}(\hat{\mathbf{u}} - \mathbf{A}\hat{\mathbf{b}}) = (\mathbf{I} - \mathbf{G}\mathbf{A})\hat{\mathbf{b}} + \mathbf{G}\hat{\mathbf{u}}$$

$$= \begin{bmatrix} \mathbf{G} & (\mathbf{I} - \mathbf{G}\mathbf{A}) \end{bmatrix} \begin{bmatrix} \hat{\mathbf{u}} \\ \hat{\mathbf{b}} \end{bmatrix} = \mathbf{P}_h \hat{\mathbf{y}}. \tag{22}$$

The matrix $\mathbf{P}_h$ of pMinT is thus:

$$\mathbf{P}_{pMinT,h} = \begin{bmatrix} \mathbf{G} & (\mathbf{I} - \mathbf{G}\mathbf{A}) \end{bmatrix} \tag{23}$$

**Proposition 1.** *The matrices $\mathbf{P}_h$ of MinT and pMinT are equivalent. The proof is given in the supplementary material.*

## 5   Experiments

In this paper we take a probabilistic point of view; we thus assess the reconciled predictive distributions rather than the point forecasts. Our metric is the *energy score* (ES), a scoring rule for multivariate distributions [7]. The ES is the multivariate generalization of the continuous ranked probability score (CRPS), which is obtained by integrating the Brier score over the predictive distribution of the forecast [7]. Let $\mathbf{y}$ be the actual multivariate observation, and let us assume that we have $k$ samples $\mathbf{x}_1, \mathbf{x}_2, \ldots, \mathbf{x}_k$, from the multivariate predictive distribution $F$. The energy score is:

$$\text{ES}(\mathbf{y}, F) = \frac{1}{k}\sum_{i=1}^{k} \|\mathbf{x}_i - \mathbf{y}\| - \frac{1}{2k^2}\sum_{i=1}^{k}\sum_{j=1}^{k} \|\mathbf{x}_i - \mathbf{x}_j\| \tag{24}$$

The energy score is a loss function: the lower, the better. We consider three methods for probabilistic reconciliation: probabilistic bottom-up (BU), LG and pMinT. We did not find any package implementing the algorithms of [13,17]; thus we did not include them in our comparison. We estimate all the covariance matrices via the shrinkage estimator [15].

*Base Forecasts.* We consider two time series models to compute the base forecasts. The first is *ets*, which fits different exponential smoothing variants and eventually performs model selection via AICc. The second method is auto.arima. It first decides how to differentiate the time series to make it stationary; then, it looks for the best arma model which fits the stationary time series, performing model selection via AICc. Both approaches performed well in forecasting competitions; they are available from the *forecast* package [10] for R. In all simulations, we compute forecasts up to $h = 4$. As no particular pattern exists in the relative performance of the methods as $h$ varies, we present the results averaged over $h = 1, 2, 3, 4$.

*Setting $k_h$.* We are unaware of previous studies on how to set $k_h$. In this paper we compare two heuristics, acknowledging that this remains an open problem. The two options are $k_h = h$ and $k_h = 1$. The choice $k_h = h$ is based on the following approximation. The variance of $\hat{y}_{t+1}$ around $y_{t+1}$ is $\sigma^2$; assuming the independence of the errors, the variance of $\hat{y}_{t+1}$ around $y_{t+h}$ is $h\sigma^2$. The approximation lies in the fact that we are modeling the variance of $\hat{y}_{t+1}$ (not $\hat{y}_{t+h}$) around $y_{t+h}$.

Instead, the option $k_h = 1$ keeps the variance fixed with $h$. This represents the short-term behavior of models which contain only seasonal terms and no autoregressive terms. For instance, when dealing with a monthly time series, the variance of such models is constant up to $h = 12$.

*Code.* The code of our experiments is available at: https://github.com/iamthejao/BayesianReconciliation.

## 5.1 Synthetic Data

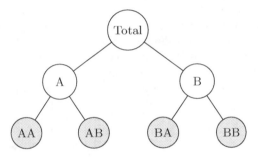

We generate synthetic data sets using the hierarchy above, previously considered in the experiments of [18]. We simulate the four bottom time series as AR(1) processes, drawing their parameters uniformly from the stationary region. The noises of the bottom time series at each time instant are correlated, multivariate Gaussian distributed, with mean $\mu = [0, 0, 0, 0]^T$ and covariance:

$$\Sigma = \begin{bmatrix} 5 & 3 & 2 & 1 \\ 3 & 5 & 2 & 1 \\ 2 & 2 & 5 & 3 \\ 1 & 1 & 3 & 5 \end{bmatrix}.$$

Thus $\Sigma$ enforces a stronger correlation between time series which have the same parents. At each time instant $t$ we add the noise $\eta_t \sim N(0, 10)$ to the time series AA and BA and the noise $(-\eta_t)$ to the time series AB and BB. In this way we simulate noisy bottom time series ($\eta_t$ and $-\eta_t$ cancel out when dealing with the upper time series) which can be encountered in real cases when several disaggregations are applied to the total time series. We consider the following length $T$ of the time series: $\{50; 100; 1000\}$. For each value of $T$ we perform 1000

**Table 1.** Mean energy score, averaged over 1000 simulations and over $h = 1, 2, 3, 4$. In each row we highlight the lower result.

| $T$ | Method | BU | pMinT | LG |
|---|---|---|---|---|
| 50 | arima | 9.7 | 9.5 | **9.4** |
| 50 | ets | 9.9 | 9.8 | **9.7** |
| 100 | arima | 9.1 | **9.0** | 9.0 |
| 100 | ets | 9.5 | 9.4 | **9.3** |
| 1000 | arima | 8.8 | **8.7** | 8.8 |
| 1000 | ets | 9.5 | **9.3** | 9.4 |

simulations. The averaged energy scores are given in Table 1; in each cell we report the lower energy score between the case $k_h = h$ and $k_h = 1$.

Since the time series are stationary, they basically fluctuate around their mean. In this case the magnitude of the incoherence is generally limited, allowing also the bottom-up reconciliation to be competitive. The ES of both pMinT and LG is on average 1.5% smaller than that of BU. We also note an advantage of LG over pMinT for small $T$, and instead the reverse for large $T$; this might be the effect of the additional covariances estimated by pMinT (see $\sigma_{u,1}$ and $\sigma_{u,2}$ in Sect. 4).

## 5.2   Experiments with Real Data Sets

We consider two hierarchical time series: *infantgts* and *tourism*. Both are *grouped* time series, which is a generalization of hierarchical time series. In particular the time series of a given level are always sums of some bottom time series, but they are not necessarily sums of time series of the adjacent lower level.

*Infantgts.* The *infantgts* is available within the **hts** [8] package for R. It contains infant mortality counts in Australia, disaggregated by sex and by eight different states. Each time series contains 71 yearly observations, covering the period 1933–2003. The bottom level contains 16 time series (8 states × 2 genders). The second level contains 2 time series: the counts of males and females, aggregated over the states. The third level sums males and females in each state, yielding 8 time series (one for each state). The fourth level is the total.

*Tourism.* The *tourism* data set regards the number of nights spent by Australians away from home. It is available in raw format from https:// robjhyndman.com/publications/MinT/. The time series cover the period 1998–2016 with monthly frequency. There are 304 bottom time series, referring to 76 regions and 4 purposes. The first level sums over the purposes, yielding 76 time series (one for each region); such values are further aggregated into macro-zones (27 time series) and states (7 time series). Other levels of the hierarchy aggregate the bottom time series of the same zone (yielding 108 time series: 27 zones × 4 purposes), which are then further aggregated into 28 time series (7 states × 4 purposes) and then 4 time series (4 purposes). The last level is the total. Overall the hierarchy contains 555 time series.

**Table 2.** Averaged energy scores. Each cell is the average over 200 reconciliations (50 different training/test splits × $h = 1, 2, 3, 4$). The rows corresponding to the best-performing values of $k_h$ are highlighted.

| dset | $k_h$ | method | BU | pMinT | LG |
|------|-------|--------|------|-------|------|
| infantgts | 1 | arima | **334.1** | 346.9 | 348.5 |
| | 1 | ets | 334.0 | **320.0** | 334.7 |
| | $h$ | arima | **327.2** | 335.1 | 331.0 |
| | $h$ | ets | 328.2 | **313.7** | 318.7 |
| tourism | 1 | arima | 2,737.6 | **2,412.0** | 2,547.4 |
| | 1 | ets | 2,496.0 | **2,403.7** | 2,520.1 |
| | $h$ | arima | 2,785.3 | **2,380.3** | 2,448.2 |
| | $h$ | ets | 2,527.1 | **2,353.6** | 2,410.3 |

We repeat 50 times the following procedure: split the time series into training and test, using a different split point; compute the base forecasts up to $h = 4$; reconcile the forecasts. The reconciliation is independently computed for each $h$. Each value of Table 2 is thus the average over 200 experiments (50 training/test splits × $h = 1, 2, 3, 4$). On *infantgts*, all the three reconciliation methods perform better with $k_h = h$, probably because the variance of the fitted time series models steadily increases with $h$. On the contrary, on *tourism* all reconciliation algorithms perform better with $k_h = 1$; in this case most models only contain the seasonal part. The rows referring to the best values of $k_h$ are highlighted in Table 2.

We call *setup* the combination of a data set and a forecasting method, such as < infangts,arima >. The pMinT algorithm yields the lowest energy score in most setups; in the next section we check whether the differences between methods are significant.

*Statistical Analysis.* For each setup we perform a significance tests for each pair of algorithms (pMinT vs BU, LG vs BU and pMinT vs LG), using the Bayesian signed-rank test [2], which returns the posterior probability of a method having lower median energy score than another (Table 3). Such posterior probabilities are numerically equivalent to $(1 - p\text{-}value)$, where $p\text{-}value$ is the p-value of the one-sided frequentist signed-rank test.

In most setups (Table 3) high posterior probabilities (implying low p-values) support the hypothesis of the pMinT having lower energy score than both BU and LG; moreover they also support the hypothesis of LG having lower energy score than BU.

**Table 3.** Posterior probability of the Bayesian signed rank test.

| dset | $k_h$ | method | Posterior probabilities | | |
|---|---|---|---|---|---|
| | | | pMinT < BU | pMinT < LG | LG < BU |
| infantgts | 1 | arima | 0.24 | 0.02 | 0.47 |
| | | ets | 1.00 | 0.85 | 1 |
| | h | arima | 0.89 | 0.00 | 1.00 |
| | | ets | 1.00 | 0.00 | 1 |
| tourism | 1 | arima | 1.00 | 1.00 | 1.00 |
| | | ets | 1.00 | 1.00 | 0.25 |
| | h | arima | 1.00 | 1.00 | 1.00 |
| | | ets | 1.00 | 1.00 | 1.00 |

*Meta-analysis.* We now perform a meta-analysis for each pair of algorithms across the different setups, adopting the Poisson-binomial approach [4,11]. Consider for instance pMinT vs BU. We model each setup as a Bernoulli trial, whose possible outcomes are the victory of pMinT or BU. The probability of pMinT winning is taken for each setup from Table 3 (the probability of BU winning is just its complement to 1). We then repeat 10,000 simulations, in which we draw the outcome of each setup according to the probabilities of Table 3.

We now report the probability of each method outperforming another method in more than half the setups, based on out of 10,000 simulations. Both pMinT and LG wins in more than half the setup with probability 1 against BU. Moreover, there is 0.85 probability of pMinT winning in more than half of the setups against LG. We thus recommend pMinT as a general default method for probabilistic reconciliation.

## 6    Conclusions

We have derived two algorithms (pMinT and LG) based on Bayes' rule for probabilistic reconciliation. We have also shown a didactic example which clarifies how base forecast and their variances interact during the reconciliation, In general pMinT yields better predictive distributions and thus we recommend it as a default. The LG method can be anyway an interesting alternative when dealing with small sample sizes. Future research could borrow ideas from the extensive literature of the Kalman filter, based on the link we pointed out between reconciliation and Kalman filter

**Acknowledgements.** We acknowledge support from grant n. 407540_167199 / 1 from Swiss NSF (NRP 75 Big Data).

# References

1. Athanasopoulos, G., Gamakumara, P., Panagiotelis, A., Hyndman, R., M., A.: Hierarchical forecasting. In: Macroeconomic Forecasting in the Era of Big Data pp. 689–719 (2020)
2. Benavoli, A., Corani, G., Demšar, J., Zaffalon, M.: Time for a change: a tutorial for comparing multiple classifiers through Bayesian analysis. J. Mach. Learn. Res. **18**(1), 2653–2688 (2017)
3. Bishop, C.: Pattern Recognition and Machine Learning. Springer, New York (2006)
4. Corani, G., Benavoli, A., Demšar, J., Mangili, F., Zaffalon, M.: Statistical comparison of classifiers through Bayesian hierarchical modelling. Mach. Learn. **106**(11), 1817–1837 (2017). https://doi.org/10.1007/s10994-017-5641-9
5. Durrant-Whyte, H., Henderson, T.C.: Multisensor data fusion. Springer Handbook of Robotics, pp. 585–610 (2008)
6. Gamakumara, P., Panagiotelis, A., Athanasopoulos, G., Hyndman, R.: Probabilisitic forecasts in hierarchical time series. Monash University, Department of Econometrics and Business Statistics, Technical report (2018)
7. Gneiting, T., Raftery, A.E.: Strictly proper scoring rules, prediction, and estimation. J. Am. Stat. Assoc. **102**(477), 359–378 (2007)
8. Hyndman, R., Lee, A., Wang, E., Wickramasuriya, S.: HTS: hierarchical and grouped time series (2018). https://CRAN.R-project.org/package=hts, R package version 5.1.5
9. Hyndman, R.J., Ahmed, R.A., Athanasopoulos, G., Shang, H.L.: Optimal combination forecasts for hierarchical time series. Comput. Stat. Data Anal. **55**(9), 2579–2589 (2011)
10. Hyndman, R.J., Khandakar, Y.: Automatic time series forecasting: the forecast package for R. J. Stat. Softw. **3**(27), 1–22 (2008)
11. Lacoste, A., Laviolette, F., Marchand, M.: Bayesian comparison of machine learning algorithms on single and multiple datasets. In: Proceedings of the 15th International Conference on Artificial Intelligence and Statistics, pp. 665–675 (2012)
12. Murphy, K.P.: Machine Learning: A Probabilistic Perspective. MIT Press, Cambridge (2012)
13. Park, M., Nassar, M.: Variational Bayesian inference for forecasting hierarchical time series. In: ICML 2014 Workshop on Divergence Methods for Probabilistic Inference, pp. 1–6 (2014)
14. Roweis, S., Ghahramani, Z.: A unifying review of linear Gaussian models. Neural Comput. **11**(2), 305–345 (1999)
15. Schäfer, J., Strimmer, K.: A shrinkage approach to large-scale covariance matrix estimation and implications for functional genomics. Stati. Appl. Genet. Mol. Biol. **4**(1), (2005)
16. Simon, D.: Optimal state estimation: Kalman, H infinity and nonlinear approaches. John Wiley & Sons (2006)
17. Taieb, S.B., Taylor, J.W., Hyndman, R.J.: Coherent probabilistic forecasts for hierarchical time series. In: Precup, D., Teh, Y.W. (eds.) Proceedings of the 34th International Conference on Machine Learning, vol. 70, pp. 3348–3357 (2017)
18. Wickramasuriya, S.L., Athanasopoulos, G., Hyndman, R.J.: Optimal forecast reconciliation for hierarchical and grouped time series through trace minimization. J. Am. Stat. Assoc. **114**(526), 804–819 (2019)

# Quantifying the Confidence of Anomaly Detectors in Their Example-Wise Predictions

Lorenzo Perini$^{(\boxtimes)}$ (iD), Vincent Vercruyssen$^{(\boxtimes)}$ (iD), and Jesse Davis$^{(\boxtimes)}$ (iD)

DTAI Research Group & Leuven.AI, KU Leuven, Leuven, Belgium
{lorenzo.perini,vincent.vercruyssen,jesse.davis}@kuleuven.be

**Abstract.** Anomaly detection focuses on identifying examples in the data that somehow deviate from what is expected or typical. Algorithms for this task usually assign a score to each example that represents how anomalous the example is. Then, a threshold on the scores turns them into concrete predictions. However, each algorithm uses a different approach to assign the scores, which makes them difficult to interpret and can quickly erode a user's trust in the predictions. This paper introduces an approach for assessing the reliability of any anomaly detector's example-wise predictions. To do so, we propose a Bayesian approach for converting anomaly scores to probability estimates. This enables the anomaly detector to assign a confidence score to each prediction which captures its uncertainty in that prediction. We theoretically analyze the convergence behaviour of our confidence estimate. Empirically, we demonstrate the effectiveness of the framework in quantifying a detector's confidence in its predictions on a large benchmark of datasets.

**Keywords:** Anomaly detection · Interpretability · Confidence scores

## 1 Introduction

Anomaly detection is a central task in data mining. It involves identifying portions of the data that do not correspond to expected normal behaviours. From a practical point of view, anomaly detection is important as anomalies often have significant costs in the real world. For example, fraudulent credit card transactions [2], retail store water leaks [18], or abnormal web traffic [15].

Typically, anomaly detection is tackled from an unsupervised perspective due to the costs and difficulties associated with acquiring labels for the anomaly class (e.g., you will not allow expensive equipment to breakdown simply to observe how it behaves in an anomalous state). The underlying assumption to these approaches is that anomalies are both (i) rare and (ii) somehow different from normal examples. Hence, anomalies may lie in low-density regions of the instance space or be far away from most other examples. The algorithms use these intuitions to assign a real-valued score to each example that denotes how anomalous

© Springer Nature Switzerland AG 2021
F. Hutter et al. (Eds.): ECML PKDD 2020, LNAI 12459, pp. 227–243, 2021.
https://doi.org/10.1007/978-3-030-67664-3_14

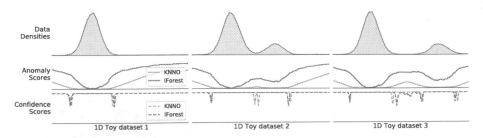

**Fig. 1.** Illustration of why confidence scores are important on three 1D toy datasets. The top plots show the data distributions under small perturbations. The middle plots show the anomaly scores assigned by KNNO and IFOREST. These two models produce non-standard scores, which are difficult to interpret and compare. The bottom plots show the corresponding confidence scores computed using our method (see Sect. 3). Small changes in the data distribution affect anomaly scores and predictions. The confidence scores capture clearly where the models (dis)agree. The dips in the confidence scores correspond to a transition in the predicted label of the underlying model.

an example is. Usually, these anomaly scores are converted to binary predictions (an example is normal or anomalous) by setting a threshold on the scores. However, the scores for many prominent approaches, such as KNNO, IFOREST, and SSDO are difficult for a human to interpret and compare. A natural way to address this issue is to transform the anomaly scores into a probability estimate. The standard approach is to *calibrate* the transformation such that for all examples that are predicted to have a $c\%$ probability of belonging to the anomaly class, $c\%$ of them should actually be anomalies. A user can now ignore any predictions with a low chance of being an anomaly. However, calibration requires labels which are generally not available in an anomaly detection setting.

The fact that anomalies are rare and unpredictable causes another issue. Hypothetically, even if we could collect multiple datasets, each one would contain distinct anomalies to which the anomaly detectors would assign different scores. Additionally, small perturbations in the training data might cause (large) differences in an example's anomaly score and, consequently, a different prediction. Consider the three one-dimensional toy datasets in Fig. 1. The middle row plots show the continuous anomaly scores that KNNO and IFOREST assign to each example in the distributions. These scores change as a result of small perturbations in the dataset, ultimately resulting in different predictions.

This paper tackles this challenge by providing a measure of how *uncertain* an anomaly detector's predictions are on an example-wise basis. The measure will allow a user to assess the reliability of anomaly detectors in different scenarios. We make the following four contributions. First, we propose a notion of a confidence measure that captures how consistent a model's prediction would be for that example if the training data were perturbed. Second, we propose EXCEED (*EXample-wise ConfidEncE of anomaly Detectors*), an approach that is able to compute our confidence measure for any anomaly detector that produces a real-valued anomaly score. The method begins by transforming the anomaly scores to *outlier probabilities* using a Bayesian approach. Then, it uses these probabilities to derive the example-wise confidence scores. This is illustrated in the bottom plots of Fig. 1, which show the computed confidence scores for KNNO's and IFOREST's prediction that each example is anomalous. The scores show in an interpretable way where the algorithms disagree and where they are uncertain about their prediction. Third, we perform a theoretical analysis of the convergence behaviour of our confidence estimates. Fourth, we perform an extensive empirical evaluation on 21 benchmark datasets.

## 2    Related Work

### 2.1    Assigning Anomaly Scores

Several different assumptions underpin anomaly detectors, but all exploit the fact that anomalies are rare and different than normal examples. From a geometric perspective, this means that anomalies are far away from normal examples. From a statistical perspective, this means that anomalies will fall in a low-density region of the instance space. Although any model producing scores can be used, we briefly describe three canonical unsupervised anomaly detection algorithms.

**kNNO** assigns an anomaly score based on the $k$-distance, which is the distance between an example and its $k$'th nearest neighbor [14]. Examples far away from other examples get high scores indicating they are more anomalous.

**iForest** assigns an anomaly score based on how difficult it is to isolate an example from the rest of the data by iteratively splitting the data along random dimensions [9]. Examples in low-density regions get higher scores.

**OCSVM** assigns an anomaly score based on the signed distance to the surface of the hypersphere encapsulating the normal examples. Examples outside the sphere get high scores indicating their anomalousness [16].

Because each algorithm produces a score in a completely different way, cross-comparisons between the algorithms are difficult. Consider the example of Fig. 1, where sometimes KNNO predicts anomalies while IFOREST does not, even though the KNNO scores are consistently lower.

### 2.2    From Anomaly Scores to Outlier Probabilities

A challenge with the anomaly scores produced by the aforementioned methods (as well as many others) is that it is difficult to interpret them. For example, understanding the $k$-distance requires context or domain knowledge (e.g., the number features, what constitutes a big distance, etc.). Therefore, a possible solution is to convert the anomaly score of an example to a probability estimate. The standard approach is to employ Platt scaling [13]:

$$\mathbb{P}(Y = 1S = s) = \frac{1}{1 + exp(\alpha \times s + \beta)}$$

where $Y$ is the true class for an example with anomaly score $s$, and $\alpha$ and $\beta$ are parameters that should be learned from the labeled data. Ideally, such a transformation should produce calibrated probability estimates. Intuitively, a probability $\mathbb{P}(Y = 1|S = s) = c$ is calibrated if out of all examples with this probability, about $c$ percent of these examples are member of the positive class.

However, in most anomaly detection applications we lack labeled data with which to train such a calibration model. We typically only know the contamination factor $\gamma$ which is the proportion of anomalies in the data. Hence, the standard approach is to calibrate the transformation such that $\gamma$ percent of the examples have a probability $> 0.5$. Beyond the logistic calibration approach, there is a long literature of approaches [5,7,8,10,11,17] for ensuring that this property is obtained and we now briefly describe some prominent approaches. Isotonic Calibration [20] is a non-parametric form of regression in which the transformation function is chosen from the class of all non-decreasing functions. Beta Calibration [8] is based on the assumption that scores are Beta distributed class-wise and transforms them according to the likelihood rate.

In anomaly detection, three methods are widely used to get calibrated outlier probabilities. The linear and squashing methods map the scores to probabilities using respectively a linear and a sigmoid transformation [4]. The unify method assumes that scores are normally distributed and estimates the outlier probability through the Gaussian cumulative distribution function [6].

## 3   A Theoretical Framework for Assessing an Anomaly Detector's Example-Wise Confidence

Using an anomaly detector in practice requires converting its returned anomaly score into a hard prediction. Typically, this is done by setting a threshold $\lambda$ on the scores. Then, any example $x$ with a score $s > \lambda$ will be classified by the model as an anomaly. Standard approaches [9,14,16] set a threshold by analyzing the data used to train the model. Hence, perturbing the training data would lead to a different threshold being picked, which in turn would affect an example's predicted class.

To capture this potential uncertainty, we propose a notion of a detector's example-wise confidence in its predictions, which works with any anomaly detector producing a real-valued scores. Intuitively, we can think of the example-wise confidence as the probability that a detector's prediction would change if a different dataset was observed. More formally, we define it as follows.

**Definition 1 (Example-Wise Confidence).** *Let* $f\colon \mathbb{R}^d \to \mathbb{R}$ *be a function that maps examples to anomaly scores. Given a dataset $D$ such that $|D| = n$, the model's confidence in its prediction for an example $x$ with anomaly score $s = f(x)$ can be defined as*

$$\mathscr{C}(\hat{Y})_x = \begin{cases} \mathbb{P}(\hat{Y} = 1 \mid s, n, \gamma, \hat{p}_s) & \text{if } \hat{Y} = 1 \\ 1 - \mathbb{P}(\hat{Y} = 1 \mid s, n, \gamma, \hat{p}_s) & \text{if } \hat{Y} = 0 \end{cases} \tag{1}$$

*where $\hat{Y}$ is the class label predicted by the anomaly detector, $\hat{p}_s$ is the estimated outlier probability (i.e., the probability that the example belongs to the anomaly class), and $\gamma$ is the expected proportion of positive examples.*

From now on, when we use the term *confidence* we will refer to $\mathbb{P}(\hat{Y} = 1 \,|\, s, n, \gamma, \hat{p}_s)$, as the case when $\hat{Y} = 0$ is directly computable from the previous one. Hence, when $\hat{Y} = 1$, high values of $\mathbb{P}(\hat{Y} = 1 \,|\, s, n, \gamma, \hat{p}_s)$ indicate that model is confident in its prediction that the example is an anomaly. One would expect confidence values around 0.5 when an example is near the decision boundary, that is on the border between normal and abnormal behaviors. We estimate the confidence in two steps. First, we employ a Bayesian approach to estimate the distribution of anomaly scores. This allows us to derive an example's outlier probability $\hat{p}_s$. Second, we use the outlier probability to estimate the confidence of the anomaly detector by considering how the combination of the observed training set and contamination factor $\gamma$ would be used to select the threshold $\lambda$ for converting anomaly scores into predictions.

## 3.1   Notation

Let $(\Omega, \Im, \mathbb{P})$ be a probability space, where $\Omega$ is the sample space, $\Im$ represents a $\sigma$-algebra over $\Omega$ and $\mathbb{P}$ is a probability measure. Let $X \colon \Omega \to \mathbb{R}^d$ be a multivariate real random variable with values in the feature space $\mathbb{R}^d$, and $Y \colon \Omega \to \{0, 1\}$ be a random variable identifying the class label. Assume that $D$ is an available dataset, which can be seen as an i.i.d. sample of size $n$ drawn from the joint distribution of $X$ and $Y$. An anomaly detection problem is the setting where there exists a function $f \colon \mathbb{R}^d \to \mathbb{R}$ that maps the feature points from the dataset $D$ to a single real value called anomaly score. A common assumption is that the function $f$ is measurable, so that $S = f(X)$ is a real random variable with anomaly scores as values. From now on, we will use the notation $D_n$ referring to the dataset of scores $D_n = \{s_1, \ldots, s_n\} = \{f(x_1), \ldots, f(x_n)\}$, with $x_1, \ldots, x_n \in D$. Given an Anomaly Detector $\Gamma$, for any example $x$ we define the *outlier probability of $x$* as the probability that $x$ is anomalous according to its anomaly score $s = f(x)$

$$\mathbb{P}(Y = 1 | f(X) = f(x)) = \mathbb{P}(Y = 1 | S = s) := \mathbb{P}(S \le s), \tag{2}$$

where $f$ is the function provided by $\Gamma$. Subsequently, the probability that one example is normal can be computed as

$$\mathbb{P}(Y = 0 | f(X) = f(x)) = 1 - \mathbb{P}(Y = 1 | S = s) = \mathbb{P}(S > s).$$

## 3.2   A Bayesian Approach for Assigning an Outlier Probability

Our goal is to infer the true class label based on the anomaly score. Formally, for any score $s \in \mathbb{R}$, we can model the example's true class given the score as the conditional random variable $Y | S = s$. Based on this framework, we can estimate an example's *outlier probability* (i.e., the probability it belongs to the anomaly class) as follows:

$$P_s := \mathbb{P}((Y|S) = 1|s) = \mathbb{P}(Y = 1|S = s) = \mathbb{P}(S \leq s).$$

Because $Y|S = s$ takes values in the set $\{0, 1\}$, we model its outcome using a Bernoulli distribution:

$$Y|S = s \sim Bernoulli(P_s)$$

where $P_s$ is the probability of success. If we knew $S$'s distribution, we could compute $\mathbb{P}(S \leq s)$–the probability that an example belongs to the anomaly class– using the cumulative distribution function of $S$. Unfortunately, the distribution of $S$ is usually unknown which makes it infeasible to directly approximate $P_s$.

Our solution is to take a Bayesian approach to this problem. The key insight is to measure the area of $\{S < s\}$ by drawing samples from the real distribution. We will view $P_s$ as a random variable and assume a uniform prior. Theoretically, we can derive the probability that an example belongs to the anomaly class as follows. First, we draw one example $a$ from the distribution of $S$, which simply entails drawing an example $x$ from $X$ and computing its anomaly score $a = f(x)$. Second, we record the event as a success (i.e., $b = 1$) if $a \leq s$ and as failure (i.e., $b = 0$) otherwise. We repeat the process $n$ times and record the total number of successes as $t$ and failures as $n - t$. In fact, the rate between successes and trials, corrected with other factors, will approximate the outlier probability as defined in Formula 2. Thanks to Bayes's rule we can use the following theorem.

**Theorem 2.** *Assume that a random variable $P_s$ follows a Beta distribution $Beta(\alpha, \beta)$ as prior. Given the events $b_1, \ldots, b_n$, which are i.i.d. examples drawn from a Bernoulli random variable $Bernoulli(P_s)$, then the posterior distribution of $P_s$ is still a Beta distribution with new parameters $Beta(\alpha + t, \beta + n - t)$, where $t = \sum_{i=1}^{n} b_i$ is the number of successes.*

*Proof.* According to the hypotheses, the prior distribution of $P_s$ is

$$\pi(q) = \frac{q^{\alpha-1}(1-q)^{\beta-1}}{\mathcal{B}(\alpha, \beta)},$$

where $\mathcal{B}(\alpha, \beta)$ is the Euler beta function. So, by using the Bayes's rule

$$\pi(q|b_1, \ldots, b_t) = \frac{\pi(q) \cdot \mathbb{P}(b_1, \ldots, b_t|q)}{\int_0^1 \pi(r) \cdot \mathbb{P}(b_1, \ldots, b_t|r) \, dr} = \frac{\frac{q^{\alpha-1}(1-q)^{\beta-1}}{\mathcal{B}(\alpha,\beta)} \cdot q^t (1-q)^{n-t}}{\int_0^1 \frac{r^{\alpha-1}(1-r)^{\beta-1}}{\mathcal{B}(\alpha,\beta)} \cdot r^t (1-r)^{n-t} \, dr}$$

$$= \frac{q^{\alpha+t-1}(1-q)^{\beta+n-t-1}}{\int_0^1 r^{\alpha+t-1}(1-r)^{\beta+n-t-1} \, dr} = Beta(\alpha + t, \beta + n - t)$$

where $t = \sum_{i=1}^{n} b_i$, $\pi(q|b_1, \ldots, b_n)$ is the posterior distribution of $P_s$ after i.i.d. sampling $n$ $Bernoulli(P_s)$ examples, and $\mathbb{P}(b_1, \ldots, b_n|q)$ is the likelihood.    $\square$

In our setting we assume that $P_s \sim Beta(1,1) = Unif(0,1)$. As a result, the posterior distribution of $P_s$ is still a Beta distribution

$$P_s|b_1, \ldots, b_n \sim Beta\left(1 + t, 1 + n - t\right). \tag{3}$$

In order derive an estimate of the outlier probability from $P_s$, we take the expectation of $P_s$. Since the posterior distribution is known from (3), $\mathbb{E}[P_s]$ can be obtained as a function of the parameters:

$$\hat{p}_s := \mathbb{E}[P_s] = \frac{1+t}{2+n}. \tag{4}$$

In practice we cannot sample from the true distribution and instead need to use the dataset $D_n$ to infer the posterior distribution. Thus, when drawing an example, we are restricted to sampling from the dataset. This limits us to drawing $n$ examples, that is, the total number of examples in the dataset. An additional consideration concerns the value of $t$. It represents the number of successes when sampling from $Y|S = s \sim Bernoulli(P_s)$. As a result, $t$ is a practical approximation of the real percentage $\theta$ of successes times the number of trials $n$

$$t \approx \theta \cdot n. \tag{5}$$

The reason why we use $t$ is to obtain a corrected estimate of the real parameter $\theta$, which would be the exact probability value of $Y|S = s$ if it were known.

### 3.3    Deriving a Detector's Confidence in Its Predictions

Although the second step of our framework works with any approach that converts anomaly scores into outlier probabilities, here $\hat{p}_s$ refers to the definition in Eq. 4. Deriving the confidence value requires estimating the proportion of times that an example will be predicted as being anomalous by the chosen anomaly detection algorithm. This requires analyzing how to set the threshold $\lambda$ for converting anomaly scores to predictions. Typically, anomaly detectors exploit the contamination factor $\gamma$ to pick the threshold $\lambda$. Note that $\gamma$ may be known from domain knowledge (e.g., historical anomaly rates) or it can be estimated from partially labeled data (c.f. [12]). There are two different scenarios:

$\gamma \in (0,1)$: **The training set contains some anomalies.** Here, the standard approach is to compute the expected number of anomalies in the training set as $k = \gamma \times n$.[1] Then, it ranks the training examples by their anomaly scores and sets the threshold to be the value in position $k$.

$\gamma = 0$: **The training set contains only normal examples.** In this case, the threshold has to be equal to the maximum anomaly score in the training set. Picking a lower value would result in a false positive on the training data.

---

[1] We assume that $k \in \mathbb{N}$, taking the floor function when needed.

In both cases the chosen threshold depends on the distribution of anomaly scores in the training set. In turn, these scores depend on the available data sample. That is, if we drew another training set from the population, the chosen threshold may change. This leads to our key insight: the task of measuring the model's confidence can be formulated as estimating the probability that an example with score $s$ will be classified as an anomaly based on a theoretical sample $D_n$ drawn from the population. Formally, given the training set size $n$ and the contamination factor $\gamma$, we want to compute the probability that an example $x$ with score $s = f(x)$ and outlier probability $\hat{p}_s$ will be classified as an anomaly when randomly drawing a training set of $n$ examples from the population of scores. In practice, the confidence can be seen as the probability that the chosen threshold $\lambda$ value will be less than or equal to the score $s$. This probability depends on our two cases for picking the threshold:

*Contamination Factor* $\gamma \in (0, 1)$. In this case by drawing theoretically from the Bernoulli distribution of $Y|S = s$ with parameter $P_s$, we should get at least $n - k + 1$ successes to classify $s$ as anomaly, where $k = \gamma \times n$ and "success" means that the drawn value is lower than $s$. As a result, our confidence is defined as:

$$\mathbb{P}(\hat{Y} = 1 \mid s, n, \gamma, \hat{p}_s) = \sum_{i=n(1-\gamma)+1}^{n} \binom{n}{i} \hat{p}_s^i (1 - \hat{p}_s)^{n-i} \qquad (6)$$

where $\hat{p}_s = \mathbb{E}[P_s]$, which is estimated using our Bayesian approach for computing an example's outlier probability (see Eq. 4). Hence, our confidence estimate explicitly relies on our outlier probability.

*Contamination Factor* $\gamma = 0$. In this case, the threshold is the maximum score in the training set, $\lambda = \max\{s_i\}_{i=1}^n$. We need to compute the probability that an example with score $s$ and outlier probability $\hat{p}_s$ will be classified as an anomaly when randomly drawing $n$ examples from the normal population. It is quite similar to the previous case, with the only difference being that no failures[2] are allowed. So, when $\gamma = 0$ we need to denote the confidence as

$$\mathbb{P}(\hat{Y} = 1 \mid s, n, 0, \hat{p}_s) = (\hat{p}_s)^n \qquad (7)$$

where again $\hat{p}_s = \mathbb{E}[P_s]$ comes from our Bayesian estimate of the example's outlier probability (see Eq. 4.)

## 4    Convergence Analysis of Our Confidence Estimate

This section analyzes the behaviour of our confidence estimate. In particular, given a fixed anomaly score $s$ for a test example, we want investigate how our confidence in the model's prediction changes as the number of training examples tends towards infinity. We would expect that as the size of the training set increases, our confidence estimate should converge. Again, we analyze the two cases based on whether or not the training set contains any anomalies.

---

[2] A failure would correspond to an training example having a higher anomaly score than the chosen threshold. Given the assumption that all training examples are normal, this would indicate a false positive.

## 4.1   Convergence Analysis when $\gamma \in (0,1)$

In this case, when we set the threshold based on a fixed dataset $D_n$, we can derive our confidence for a test example with score $s$ by merging Eq. 4 and 6:

$$\mathbb{P}(\hat{Y} = 1 \mid s, n, \gamma, \hat{p}_s) = \sum_{i=n(1-\gamma)+1}^{n} \binom{n}{i} \left(\frac{1+t}{2+n}\right)^i \left(\frac{1+n-t}{2+n}\right)^{n-i} \qquad (8)$$

where $t$ represents the successes in the Bayesian learning phase (Sect. 3.2).

This leads to the question: how does the confidence about the class prediction for a score $s$ behave as the number of training examples $n$ goes towards $+\infty$? In order to formally analyze this, we rewrite Formula 8 as

$$\mathbb{P}(\hat{Y} - 1 \mid s, n, \gamma, \hat{p}_s) = F_T(n) \quad F_T(n - \gamma n)$$

where the sum in Eq. 8 is the cumulative distribution of a binomial random variable $T \sim \mathcal{B}(n, \gamma n, \hat{p}_s)$ with $n$ trials, $\gamma n$ successes, and probability $p = \hat{p}_s$. When $n$ increases, the central limit theorem yields:

$$\mathbb{P}\left(\frac{T - n\hat{p}_s}{\sqrt{n\hat{p}_s(1-\hat{p}_s)}} \le c\right) \to \Phi(c) \qquad \text{for } n \to +\infty, \ \forall c \in \mathbb{R}.$$

Consequently, assuming that $n$ is large enough, we assert that, $\forall c \in \mathbb{R}$,

$$\mathbb{P}(T \le c) = F_T(c) \approx \Phi\left(\frac{c - n\hat{p}_s + 0.5}{\sqrt{n\hat{p}_s(1-\hat{p}_s)}}\right),$$

where $+0.5$ is a correction due to the continuity of the Gaussian variable. Thus, the confidence can be approximated by the cumulative distribution function of a Gaussian variable $T^*$ with mean $\mu = n \cdot \hat{p}_s + 0.5$ and variance $\sigma^2 = n \cdot \hat{p}_s(1-\hat{p}_s)$,

$$T^* \sim \mathcal{N}(n\hat{p}_s + 0.5, \ n\hat{p}_s(1-\hat{p}_s)).$$

Therefore:

$$\mathbb{P}(\hat{Y} = 1 \mid s, n, \gamma, \hat{p}_s) = F_T(n) - F_T(n - \gamma n)$$

$$\approx \Phi\left(\frac{n - n\hat{p}_s + 0.5}{\sqrt{n\hat{p}_s(1-\hat{p}_s)}}\right) - \Phi\left(\frac{n(1-\gamma) - n\hat{p}_s + 0.5}{\sqrt{n\hat{p}_s(1-\hat{p}_s)}}\right)$$

$$= \mathbb{P}(n(1-\gamma) \le T^* \le n) = \mathbb{P}\left((1-\gamma) \le \frac{T^*}{n} \le 1\right).$$

Next, we analyze the behaviour of $\frac{T^*}{n}$ as $n \to \infty$ in order to interpret the final result. Since $n \in \mathbb{N}$, it still follows a normal distribution with new parameters:

$$\mathbb{E}\left[\frac{T^*}{n}\right] = \frac{1}{n}\mathbb{E}[T^*] = \hat{p}_s - \frac{1}{2n} = \frac{1+t}{2+n} - \frac{1}{2n};$$

$$\text{Var}\left[\frac{T^*}{n}\right] = \frac{1}{n^2}\text{Var}[T^*] = \frac{\hat{p}_s(1-\hat{p}_s)}{n} = \frac{1+n-t}{n(2+n)}.$$

Since the mean and the variance of $\frac{T^*}{n}$ are bounded (they both are decreasing sequences when $n$ increases), the sequence of Gaussian random variables $\frac{T^*}{n}$ converges in distribution to a Gaussian random variable parameterized by the limit of the mean and the limit of the variance, respectively:

$$\lim_{n\to\infty} \mathbb{E}\left[\frac{T^*}{n}\right] = \theta, \quad \lim_{n\to\infty} \text{Var}\left[\frac{T^*}{n}\right] = 0,$$

where $\theta$ is defined in Eq. 5. Hence, when $n \to +\infty$, the limit random variable is normally distributed with mean $\theta$ and variance 0, which means the only value it assumes is $\theta$, the true outlier probability (see the end of Sect. 3.2). Formally, calling the limit degenerate random variable $\theta^*$,

$$\lim_{n\to\infty} \mathbb{P}(\hat{Y} = 1 \mid s, n, \gamma, \hat{p}_s) = \mathbb{P}\left((1 - \gamma) \le \theta^* \le 1\right).$$

Intuitively, this means that as the number of training examples goes to infinity the population is perfectly estimated and represented by the sample, which yields two cases. In the first case, the true outlier probability $\theta$ is greater than $1 - \gamma$. Roughly speaking, the expected proportion of normal examples $(1 - \gamma)$ is not high enough to yield a threshold value less than the considered score $s$. Hence, the confidence will be 1, because $\mathbb{P}\left((1 - \gamma) \le \theta^* \le 1\right) = 1$ since $\theta^*$ takes the constant value $\theta$ and the inequalities are satisfied. In contrast, in the second case the inequalities are not respected, meaning that $\theta$ does not fall inside the interval $[1 - \gamma, 1]$. Hence, in this scenario the proportion of normal examples is such that the value of the threshold must be greater than $s$ and, as a result, the confidence is 0 when predicting class 1.

At the end, the limit of the confidence that $s$ is predicted to be an anomaly is

$$\lim_{n\to\infty} \mathbb{P}(\hat{Y} = 1 \mid s, n, \gamma, \hat{p}_s) = \begin{cases} 1 & \text{if } \theta \ge 1 - \gamma; \\ 0 & \text{if } \theta < 1 - \gamma. \end{cases}$$

This corresponds to our intuition of what should occur when given an infinite number of training examples.

## 4.2   Convergence Analysis when $\gamma = 0$

In this analysis, the main hypothesis is that the training set only contains normal examples, which corresponds to learning a one-class model. Hence, only a representative sample of the normal class can be used to train the model.

The problem we tackle is: Given a true anomaly with score $s^*$, how confident will the model be in predicting that it belongs to the anomaly class? Our intuitions might be misleading in this case. In fact, since the contamination factor is 0, the threshold will be set as the highest observed score in the training set and the definition of confidence slightly changes. In this case no failures are allowed (i.e., all training examples must have a score less than the chosen threshold), once one training example has a score greater than $s^*$, this implies that the chosen threshold will be greater than $s^*$ as well. In practice, if $s^*$ is always greater than or equal to the anomaly score for each training example, the model will always predict that the test example is anomalous but its confidence might not be so high. The reason is simple: since the sample $D_n$ represents the normal class, the model cannot learn from the anomalies. So, drawing a normal example with a high anomaly score is theoretically possible. Hence, for a fixed anomalous test example, we need to analyze how the model's confidence in its prediction for the example changes as the number of normal examples in the training increases.

**Theorem 3.** *Given a dataset $D_n$ with $n$ scores and given a score $s^*$ such that $s < s^*$ $\forall s \in D_n$, fixed $\gamma = 0$, the expected rate of anomalies in $D_n$, then*

$$\mathbb{P}(\hat{Y} = 1 \mid s^*, n, \gamma = 0, \hat{p}_{s^*}) \longrightarrow \frac{1}{e} \approx 0.368 \quad \text{for } n \to +\infty.$$

*Proof.* Assuming that $D_n$ contains no anomalies, we get $t = n$ successes. This yields an outlier probability of:

$$\hat{p}_{s^*} = \mathbb{E}[P_{s^*}] = \frac{1+t}{2+n} = \frac{1+n}{2+n}.$$

Then, using the estimated probability that one score is less than or equal to $s^*$, we can compute the confidence using the hypothesis that the contamination factor in the training set is 0, meaning that no failures are allowed:

$$\mathbb{P}(\hat{Y} = 1 \mid s^*, n, \gamma = 0, \hat{p}_{s^*}) = (\hat{p}_{s^*})^n.$$

In fact, if we drew a score greater than $s^*$ from the training set, then the threshold would be greater than $s^*$ (predicted class equal to 0). Let's now analyze the limit:

$$\lim_{n \to +\infty} \mathbb{P}(\hat{Y} = 1 \mid s^*, n, \gamma = 0, \hat{p}_{s^*}) = \lim_{n \to +\infty} (\hat{p}_{s^*})^n = \lim_{n \to +\infty} \left(\frac{1+n}{2+n}\right)^n$$

$$= \lim_{n \to +\infty} \left(\frac{2+n-1}{2+n}\right)^n = \lim_{n \to +\infty} \left[\left(1 + \frac{-1}{2+n}\right)^{2+n} \cdot \left(\frac{1+n}{2+n}\right)^{-2}\right] = \frac{1}{e},$$

where the first factor is a notable limit and converges to $\frac{1}{e}$, whereas the second term converges to 1 because of the rate of polynomials of degree 1.                □

This can be understood as follows. While the outlier probability for a true anomaly goes to 1 when $n \to +\infty$, the number of normal examples in the training data also increases. Thus, we are more likely to observe unlikely events, i.e., the training set containing a normal example with a high anomaly score.

## 5    Experiments

The goal of our empirical evaluation is to: (1) intuitively illustrate how our confidence score works; (2) evaluate the quality of our confidence scores; and (3) assess the effect of using our Bayesian approach for converting anomalies scores to outlier probabilities on the quality of the confidence scores.

**Table 1.** The 21 benchmark anomaly detection datasets from [1].

| Dataset | # Examples | # Vars | $\gamma$ | Dataset | # Examples | # Vars | $\gamma$ |
|---|---|---|---|---|---|---|---|
| ALOI | 12384 | 27 | 0.030 | PenDigits | 9868 | 16 | 0.002 |
| Annthyroid | 7129 | 21 | 0.075 | Pima | 625 | 8 | 0.200 |
| Arrhythmia | 450 | 259 | 0.457 | Shuttle | 1013 | 9 | 0.013 |
| Cardiotocography | 1734 | 21 | 0.050 | Spambase | 3160 | 57 | 0.200 |
| Glass | 214 | 7 | 0.042 | Stamps | 340 | 9 | 0.091 |
| HeartDisease | 270 | 13 | 0.444 | Waveform | 3443 | 21 | 0.029 |
| Hepatitis | 80 | 19 | 0.163 | WBC | 454 | 9 | 0.022 |
| Ionosphere | 351 | 32 | 0.359 | WDBC | 367 | 30 | 0.027 |
| Lymphography | 148 | 19 | 0.040 | Wilt | 4655 | 5 | 0.020 |
| PageBlocks | 5473 | 10 | 0.102 | WPBC | 198 | 33 | 0.237 |
| Parkinson | 60 | 22 | 0.200 | | | | |

## 5.1   Experimental Setup

Our experimental goal is to evaluate ExCEED's ability to recover the example-wise confidences of an anomaly detector as opposed to evaluating predictive performance or quality of calibration. Hence, metrics like the AUC or Brier score are not suitable for assessing how small perturbations in the training data affect the example-wise predictions of the detector. For example, AUC would treat models that flipped the positions of two anomalies (normals) in the ranking produced by each model as being equivalently performant. In contrast, we are explicitly interested in understanding the number and magnitude of such flips. Moreover, our approach for converting the anomaly score to an outlier probability is a monotone function and hence does not affect the AUC of the model. We consider a confidence score to be good if it accurately captures the consistency with which a detector predicts the same label for an example. Hence, we expect a detector to predict for all examples subject to a confidence value of $\mathscr{C}(\hat{Y})_x$ the same label $\mathscr{C}(\hat{Y})_x$-percent of the time when retraining the detector multiple times with slightly perturbed training datasets. Therefore, we propose a novel method for evaluating an anomaly detector's example-wise confidence as an indication of how consistently it predicts the same label for that example. The method (1) draws 1000 sub-samples from the training data with the size of each sub-sample randomly selected in $[0.2 \cdot n, n]$, (2) trains an anomaly detector on each sub-sample, (3) uses each detector to predict the class labels of every example in the test set, and finally (4) computes for each test set example $x$ the frequency $F_x$ with which the detector predicted the same class.

We carryout our study on a benchmark consisting of 21 standard anomaly detection datasets from [1]. The datasets vary in size, number of features, and proportion of anomalies (Table 1). Given a benchmark dataset, we can now evaluate our confidence scores as follows. First, we split the dataset into training and test sets with stratified 5-fold cross-validation. Then, we take the class-weighted average of the $L^2$ differences between the confidence score $\mathscr{C}(\hat{Y})_x$ and the earlier computed frequency $F_x$ of each test set example $x$, yielding:

$$error(\mathscr{C}, F) = \frac{1}{2|T_N|} \sum_{x \in T_N} \left(\mathscr{C}(\hat{Y})_x - F_x\right)^2 + \frac{1}{2|T_A|} \sum_{x \in T_A} \left(\mathscr{C}(\hat{Y})_x - F_x\right)^2$$

where $T_N$ are the true test set normals and $T_A$ the true test set anomalies. We take a class-weighted average because the large class-imbalances that characterize anomaly detection datasets, would otherwise skew the final error. We report averages over the folds. The underlying anomaly detector is either KNNO, IFOREST, or OCSVM. This results in $21 \times 3 = 63$ experiments for each method. We compare 9 approaches, which can be divided into three categories:

*Our Method.* **ExCeeD** as introduced in Sect. 3.[3]

*Naive Baselines.* **ExCeeD-m**, this is ExCEED with a different prior distribution $Beta(\gamma \cdot m, m \cdot (1 - \gamma))$ where $\gamma$ is the contamination factor and $m$ is such that $\gamma \cdot m = 10$ (suggested in [19]). A **Baseline** approach which assumes the confidence scores to be equal to the model's predictions.

*Outlier Probability Methods.* The unify [6], linear, and squash methods for estimating the outlier probabilities $\hat{p}_s$ from anomaly scores. These probabilities are *not* confidence scores. To obtain true confidence scores, we have to combine each of these methods with the second step of our framework, yielding **ExCeeD-Unify**, **ExCeeD-Linear**, and **ExCeeD-Squash**.

*Calibration Methods.* The **Logistic** [13], **Isotonic** [20], and **Beta** [8] methods to empirically estimate calibration frequencies. Although these methods do not compute confidence scores, probabilistic predictions are often (incorrectly) interpreted as such and therefore included for completeness.

## 5.2 Experimental Results

To illustrate the intuition beyond our approach, Fig. 2 compares how the confidence scores computed by the different methods evolve as we gradually move an example from the large normal central cluster to the small anomaly cluster in the bottom right (using KNNO). For ExCEED, the confidence scores start out high when the example is close to the cluster of normal points and the prediction is 0 (i.e., normal). The confidence gradually decreases as the example moves away from the normal cluster, eventually reaching about 50% when it is halfway between the normal and abnormal clusters. Once the example is far enough away from the normal cluster, the confidence increases again as the model

---

[3] Implementation available at: https://github.com/Lorenzo-Perini/Confidence_AD.

**Table 2.** Comparison of EXCEED with the baselines. The table shows: the weighted average $error(\mathscr{C}, F)$ rank $\pm$ standard deviation (SD) of each method; the weighted average $error(\mathscr{C}, F) \pm$ SD of each method (computed as in [3]); and the number of times EXCEED wins (lower error), draws, and loses (higher error) against each baseline.

| Method | Weighted avg. $error(\mathscr{C}, F)$ rank $\pm$ SD of each method | # times EXCEED | | | Weighted avg. $error(\mathscr{C}, F)$ $\pm$ SD $\times 10^2$ of each method |
|---|---|---|---|---|---|
| | | Wins | Loses | Draws | |
| ExCeeD | **1.429 ± 0.844** | – | – | – | **1.972 ± 2.637** |
| Baseline | 2.270 ± 0.641 | 52 | 3 | 8 | 2.679 ± 2.931 |
| ExCeeD-Unify | 4.103 ± 2.040 | 55 | 2 | 6 | 15.13 ± 16.29 |
| Isotonic | 5.151 ± 1.482 | 59 | 2 | 2 | 17.92 ± 13.68 |
| Beta | 5.421 ± 1.285 | 62 | 0 | 1 | 23.13 ± 29.65 |
| Logistic | 5.532 ± 1.525 | 62 | 0 | 1 | 23.89 ± 29.69 |
| ExCeeD-m | 5.937 ± 2.709 | 59 | 1 | 3 | 24.95 ± 37.38 |
| ExCeeD-Linear | 6.802 ± 1.350 | 61 | 2 | 0 | 31.47 ± 23.15 |
| ExCeeD-Squash | 8.357 ± 1.271 | 61 | 2 | 0 | 45.97 ± 36.67 |

changes its prediction and becomes more certain that the example is anomalous. Using EXCEED with its Bayesian outlier probability clearly captures the gradual change in confidence we would intuitively expect in this scenario.

**Question 1: Does ExCeeD produce good confidence scores?** Table 2 summarizes the comparison between EXCEED and the baselines in terms of $error(\mathscr{C}, F)$. Our method outperforms all baselines and has the lowest average error rank over the 63 experiments. It also achieves lower errors in at least 54 of the 63 experiments compared to every other method and achieves the lowest weighted average error. When the results are split out per underlying anomaly detector (Table 3), EXCEED still outperforms all baselines, winning against each baseline at least 20, 18 and 13 out of 21 times when the detectors are, respectively, KNNO, IFOREST and OCSVM.

**Question 2: Does the Bayesian approach to estimate outlier probabilities contribute to better confidence scores?** Our confidence score can be computed from any outlier probability measure. To evaluate specifically how our proposed Bayesian approach for estimating the outlier probabilities contributes to the confidence scores, we simply compare EXCEED with the EXCEED-Unify, EXCEED-Linear, and EXCEED-Squash baselines. The results are summarized in Tables 2 and 3. The Bayesian subroutine of EXCEED to compute the outlier probabilities outperforms the three baselines by a substantial margin, obtaining lower errors in respectively 55, 61, and 61 out of 63 experiments, indicating its effectiveness.

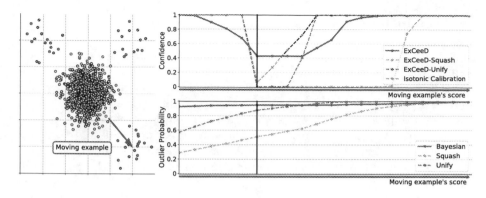

**Fig. 2.** Illustration of how moving an example along the purple arrow in the dataset (left plot) affects its outlier probability (bottom-right plot) and its confidence score derived from this probability (top-right plot). The underlying anomaly detector is KNNO. Left of the vertical black line, the detector predicts that the example belongs to the normal class. Because at first the example is embedded in the cluster of normal examples (the green points in the dataset), the initial confidence score is high. However, the detector's confidence in its prediction decreases as the example moves away from the normal points. Finally, it increases again when the example nears the anomalies (the red points) and the example is predicted to be an anomaly. ExCEED's confidence score captures our intuitions that the prediction should be confident (i.e., confidence score near 1.0) when the example is very obviously either normal or anomalous and uncertain (i.e., confidence score is near 0.5) when the example is equidistant from the normal and anomalous examples.

**Table 3.** Comparison of ExCEED with the baselines, split out per anomaly detector (KNNO, IFOREST, and OCSVM). The table presents the number of times ExCEED wins (lower error), draws, and loses (higher error) vs. each baseline.

| Method | KNNO: # of | | | IFOREST: # of | | | OCSVM: # of | | |
|---|---|---|---|---|---|---|---|---|---|
| | Wins | Draws | Losses | Wins | Draws | Losses | Wins | Draws | Losses |
| **ExCeeD** | – | – | – | – | – | – | – | – | – |
| **Baseline** | 21 | 0 | 0 | 18 | 0 | 3 | 13 | 3 | 5 |
| **ExCeeD-Unify** | 21 | 0 | 0 | 18 | 0 | 3 | 16 | 2 | 3 |
| **Isotonic** | 20 | 0 | 1 | 21 | 0 | 0 | 18 | 2 | 1 |
| **Beta** | 21 | 0 | 0 | 21 | 0 | 0 | 20 | 0 | 1 |
| **Logistic** | 21 | 0 | 0 | 21 | 0 | 0 | 20 | 0 | 1 |
| **ExCeeD-m** | 20 | 0 | 1 | 19 | 0 | 2 | 20 | 1 | 0 |
| **ExCeeD-Linear** | 21 | 0 | 0 | 21 | 0 | 0 | 19 | 2 | 0 |
| **ExCeeD-Squash** | 21 | 0 | 0 | 21 | 0 | 0 | 19 | 2 | 0 |

## 6    Conclusions

We proposed a method to estimate the confidence of anomaly detectors in their example-wise class predictions. We first formally defined the confidence as the probability that example-wise predictions change due to perturbations in the training set. Then, we introduced a method that estimates the confidence using a two step approach. First, we estimate smooth outlier probabilities using a Bayesian approach. Second, we use the estimated outlier probabilities to derive a confidence score on an example-by-example basis. A large experimental comparison shows that our approach can recover confidence scores matching empirical frequencies.

**Acknowledgements.** This work is supported by the Flemish government under the *"Onderzoeksprogramma Artificiële Intelligentie (AI) Vlaanderen"* programme (JD, LP, VV), FWO (G0D8819N to JD), and KU Leuven Research Fund (C14/17/07 to JD).

## References

1. Campos, G.O., et al.: On the evaluation of unsupervised outlier detection: measures, datasets, and an empirical study. Data Min. Knowl. Disc. **30**(4), 891–927 (2016). https://doi.org/10.1007/s10618-015-0444-8
2. Chandola, V., Banerjee, A., Kumar, V.: Anomaly detection: a survey. ACM Comput. Surv. (CSUR) **41**(3), 1–58 (2009)
3. Demšar, J.: Statistical comparisons of classifiers over multiple datasets. J. Mach. Learn. Res. **7**, 1–30 (2006)
4. Gao, J., Tan, P.N.: Converting output scores from outlier detection algorithms into probability estimates. In: Proceedings of Sixth IEEE International Conference on Data Mining, pp. 212–221. IEEE (2006)
5. Guo, C., Pleiss, G., Sun, Y., Weinberger, K.Q.: On calibration of modern neural networks. In: Proceedings of the 34th International Conference on Machine Learning, pp. 1321–1330 (2017)
6. Kriegel, H.P., Kroger, P., Schubert, E., Zimek, A.: Interpreting and unifying outlier scores. In: Proceedings of the 2011 SIAM International Conference on Data Mining, pp. 13–24. SIAM (2011)
7. Kull, M., Nieto, M.P., Kängsepp, M., Filho, T.S., Song, H., Flach, P.: Beyond temperature scaling: obtaining well-calibrated multi-class probabilities with Dirichlet calibration. In: Advances in Neural Information Processing Systems (2019)
8. Kull, M., Silva Filho, T.M., Flach, P., et al.: Beyond sigmoids: How to obtain well-calibrated probabilities from binary classifiers with beta calibration. Electron. J. Stat. **11**(2), 5052–5080 (2017)
9. Liu, F.T., Ting, K.M., Zhou, Z.H.: Isolation forest. In: Proceeding of 2008 Eighth IEEE International Conference on Data Mining, pp. 413–422. IEEE (2008)
10. Naeini, M.P., Cooper, G., Hauskrecht, M.: Obtaining well calibrated probabilities using Bayesian binning. In: Twenty-Ninth AAAI Conference on Artificial Intelligence (2015)
11. Perello-Nieto, M., De Menezes Filho, E.S.T., Kull, M., Flach, P.: Background check: a general technique to build more reliable and versatile classifiers. In: Proceedings of 16th IEEE International Conference on Data Mining. IEEE (2016)

12. Perini, L., Vercruyssen, V., Davis, J.: Class prior estimation in active positive and unlabeled learning. In: Proceedings of the 29th International Joint Conference on Artificial Intelligence and the 17th Pacific Rim International Conference on Artificial Intelligence (IJCAI-PRICAI) (2020)
13. Platt, J., et al.: Probabilistic outputs for support vector machines and comparisons to regularized likelihood methods. Adv. large Margin Classifiers **10**, 61–74 (1999)
14. Ramaswamy, S., Rastogi, R., Shim, K.: Efficient algorithms for mining outliers from large datasets. In: Proceedings of the 2000 ACM SIGMOD International Conference on Management of Data, pp. 427–438 (2000)
15. Robberechts, P., Bosteels, M., Davis, J., Meert, W.: Query log analysis: detecting anomalies in DNS traffic at a TLD resolver. In: Monreale, A., et al. (eds.) ECML PKDD 2018. CCIS, vol. 967, pp. 55–67. Springer, Cham (2019). https://doi.org/10.1007/978-3-030-14880-5_5
16. Schölkopf, B., Platt, J.C., Shawe-Taylor, J., Smola, A.J., Williamson, R.C.: Esti mating the support of a high-dimensional distribution. Neural Comput. **13**(7), 1443–1471 (2001)
17. Vaicenavicius, J., Widmann, D., Andersson, C., Lindsten, F., Roll, J., Schön, T.B.: Evaluating model calibration in classification. arXiv:1902.06977 (2019)
18. Vercruyssen, V., Wannes, M., Gust, V., Koen, M., Ruben, B., Jesse, D.: Semi-supervised anomaly detection with an application to water analytics. In: Proceedings of 18th IEEE International Conference on Data Mining, pp. 527–536. IEEE (2018)
19. Zadrozny, B., Elkan, C.: Obtaining calibrated probability estimates from decision trees and Naive Bayesian classifiers. In: Proceedings of ICML, pp. 609–616 (2001)
20. Zadrozny, B., Elkan, C.: Transforming classifier scores into accurate multiclass probability estimates. In: Proceedings of the Eighth ACM SIGKDD International Conference on Knowledge Discovery and Data Mining, pp. 694–699 (2002)

# Architecture of Neural Networks

# XferNAS: Transfer Neural Architecture Search

Martin Wistuba$^{(\boxtimes)}$

IBM Research, Dublin, Ireland
`martin.wistuba@ibm.com`

**Abstract.** The term Neural Architecture Search (NAS) refers to the automatic optimization of network architectures for a new, previously unknown task. Since testing an architecture is computationally very expensive, many optimizers need days or even weeks to find suitable architectures. However, this search time can be significantly reduced if knowledge from previous searches on different tasks is reused. In this work, we propose a generally applicable framework that introduces only minor changes to existing optimizers to leverage this feature. As an example, we select an existing optimizer and demonstrate the complexity of the integration of the framework as well as its impact. In experiments on CIFAR-10 and CIFAR-100, we observe a reduction in the search time from 200 to only 6 GPU days, a speed up by a factor of 33. In addition, we observe new records of 1.99 and 14.06 for NAS optimizers on the CIFAR benchmarks, respectively. In a separate study, we analyze the impact of the amount of source and target data. Empirically, we demonstrate that the proposed framework generally gives better results and, in the worst case, is just as good as the unmodified optimizer.

**Keywords:** Neural Architecture Search · AutoML · Metalearning

## 1 Introduction

For most recent advances in machine learning ranging across a wide variety of applications (image recognition, natural language processing, autonomous driving), deep learning has been one of the key contributing technologies. The search for an optimal deep learning architecture is of great practical relevance and is a tedious process that is often left to manual configuration. Neural Architecture Search (NAS) is the umbrella term describing all methods that automate this search process. Common optimization methods use techniques from reinforcement learning [1,3,4,34–36] or evolutionary algorithms [15,22,23], or are based on surrogate models [14,19]. The search is a computationally expensive task since it requires training hundreds or thousands of models, each of which requires few hours of training on a GPU [14,15,19,22,35,36]. All of these optimization methods have in common that they consider every new problem independently without considering previous experiences. However, it is a common knowledge that

© Springer Nature Switzerland AG 2021
F. Hutter et al. (Eds.): ECML PKDD 2020, LNAI 12459, pp. 247–262, 2021.
https://doi.org/10.1007/978-3-030-67664-3_15

well-performing architectures for one task can be transferred to other tasks and
even achieve good performance. Architectures discovered for CIFAR-10 have not
only been transferred to CIFAR-100 and ImageNet [14,19,22,36] but have also
been transferred from object recognition to object detection tasks [17,24]. This
suggests that the response function for different tasks, i.e. an architecture-score
mapping, shares commonalities.

The central idea in this work lies in the development of a search method
that uses knowledge acquired across previously explored tasks to speed up the
search for a new task. For this we assume that the response functions can be
decomposed into two parts: a universal part, which is shared across all tasks, and
a task-specific part (Fig. 1). We model these two functions with neural networks
and determine their parameters with the help of the collected knowledge of
architectures on different tasks. This allows the search for a new task to start
with the universal representation and only later learn and benefit from the task-
specific representation. This reduces the search time without negatively affecting
the final solution.

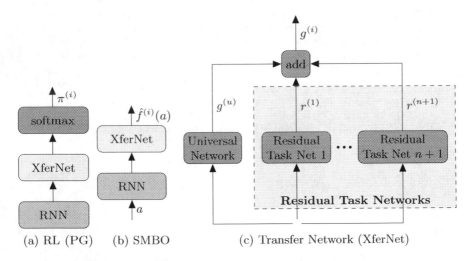

(a) RL (PG)        (b) SMBO            (c) Transfer Network (XferNet)

**Fig. 1.** An example of the integration of the transfer network into RL-based (a) and
surrogate model-based (b) optimizers. The transfer network (c) unravels independent
and task-specific influences.

The contributions in this paper are threefold:

- First, we propose a general, minimally invasive framework that allows existing
  NAS optimizers to leverage knowledge from other data sets, e.g. obtained
  during previous searches.
- Second, as an example, we apply the framework to NAO [19], a recent NAS
  optimizer, and derive XferNAS. This exemplifies the simplicity and elegance
  of extending existing NAS methods within our framework.

- Finally, we demonstrate the utility of the adapted optimizer XferNAS by searching for an architecture on CIFAR-10 [12]. In only 6 GPU days (NAO needed 200 GPU days) we discover a new architecture with improved performance compared to the state-of-the-art methods. We confirm the transferability of the discovered architecture by applying it, unchanged, to CIFAR-100.

## 2   Transfer Neural Architecture Search

In this section, we introduce our general, minimally invasive framework for NAS optimizers to leverage knowledge from other data sets, e.g. obtained during previous searches. First, we formally define the NAS problem and introduce our notation. Then we motivate our approach and introduce the framework. Finally, using the example of NAO [19], we show what steps are required to integrate the framework with existing optimizers. In this step, we derive XferNAS, which we examine in more detail in Sect. 4.

### 2.1   Problem Definition

We define a general deep learning algorithm $\mathbb{L}$ as a mapping from the space of data sets $D$ and architectures $A$ to the space of models $M$,

$$\mathbb{L} \; : \; D \times A \to M \; . \tag{1}$$

For any given data set $d \in D$ and architecture $a \in A$, this mapping returns the solution to the standard machine learning problem that consists in minimizing a regularized loss function $\mathcal{L}$ with respect to the model parameters $\theta$ of architecture $a$ using the data $d$,

$$\mathbb{L}\left(a, d\right) = \underset{m^{(a,\theta)} \in M^{(a)}}{\arg\min} \; \mathcal{L}\left(m^{(a,\theta)}, d^{(\mathrm{train})}\right) + \mathcal{R}\left(\theta\right) \; . \tag{2}$$

Neural Architecture Search solves the following nested optimization problem: given a data set $d$ and the search space $A$, find the optimal architecture $a^\star \in A$ which maximizes the objective function $\mathcal{O}$ (defined by classification accuracy in the scope of this work) on the validation data,

$$a^\star = \underset{a \in A}{\arg\max}\, \mathcal{O}\left(\mathbb{L}\left(a, d^{(\mathrm{train})}\right), d^{(\mathrm{valid})}\right) = \underset{a \in A}{\arg\max} f\left(a\right) \; . \tag{3}$$

Thus, Neural Architecture Search can be considered a global black-box optimization problem where the aim is to maximize the *response function* $f$. It is worth noting that the evaluation of $f$ at any point $a$ is computationally expensive since it involves training and evaluating a deep learning model.

In this work we assume that we have access to knowledge about the response functions on some *source tasks* $1, \ldots, n$, referred to as *source knowledge*. The idea is to leverage the source knowledge to address the NAS problem defined in Eq. (3) for a new task, the *target task* $n+1$. By no means, the source knowledge is sufficient to yield an optimal architecture for the new task. Sample architectures must be evaluated on the target task in order to gain knowledge about the target response function. We call the knowledge accumulated in this process the *target knowledge* and refer to the combined source and target knowledge as *observation history*. In the context of this work, we refrain from transferring model weights.

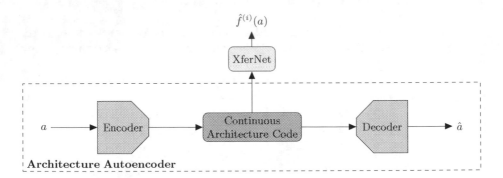

**Fig. 2.** XferNAS: Integration of the transfer network in NAO.

## 2.2 Transfer Network

The most common NAS optimizers are based on reinforcement learning (RL) [1, 3,4,34–36] or surrogate model-based optimization (SMBO) [14,19]. Many RL-based methods are based on policy gradient methods and use a controller (a neural network) to provide a distribution over architectures. The controller is optimized in order to learn a policy $\pi$ which maximizes the response function of the target task. Alternatively, SMBO methods use a surrogate model $\hat{f}^{(n+1)}$ to approximate the target response function $f^{(n+1)}$. Both of these approaches rely on the feedback gathered by evaluating the target response function for several architectures.

We propose a general framework to transfer the knowledge gathered in previous experiments in order to speed up the search on the target task. Current state-of-the-art NAS optimizers directly learn a task-dependent policy (RL) or a task-dependent surrogate model (SMBO) $g^{(i)}$. The core idea in this work is to disentangle the contribution of the *universal function* $g^{(u)}$ from the *task-dependent function* $g^{(i)}$ by assuming

$$g^{(i)} = g^{(u)} + r^{(i)} , \qquad (4)$$

where $r^{(i)}$ is a *task-dependent residual*. This disentanglement is achieved by learning all parameters jointly on the observation history, where the universal function is included for all tasks while the task-dependent residual included only for its corresponding task. The universal function can be interpreted as a function which models a good average solution across problems, whereas the task-dependent function is optimal for a particular task $i$. The advantage is that we can warm-start any NAS optimizer for the target task where $r^{(n+1)}$ is unknown by using only $g^{(u)}$. As soon as target knowledge is obtained, we can learn $r^{(n+1)}$ which enables us to benefit from the warmstart in the initial phase of the search and subsequently from the original NAS optimizer. We sketch the idea of this transfer network in Fig. 1 and provide an example for both, reinforcement learning and surrogate model-based optimizers. The functions $g^{(u)}$ and $r^{(i)}$ are modeled by a neural network, referred to as the universal network and the residual task networks, respectively. In the following, we exemplify this integration with the case of NAO [19] to provide a deeper understanding.

## 2.3   XferNAS

In principle any of the existing network-based optimizers (RL or SMBO) can be easily extended with our proposed transfer network in order to leverage the source knowledge. We demonstrate the use of the transfer network using the example of NAO [19], one of the state-of-the-art optimizers. NAO is based on two components, an auto-encoder and a performance predictor. The auto-encoder first transforms the architecture encoding into a continuous architecture code by an encoder and then reconstructs the original encoding using a decoder. The performance predictor predicts the accuracy on the validation split for a given architecture code. XferNAS extends this architecture by integrating the transfer network into the performance predictor, which not only predicts accuracy for the target task, but also for all source tasks (Fig. 2). Here, the different prediction functions for each task $i$ are divided into a universal prediction function and a task-specific residual,

$$\hat{f}^{(i)} = \hat{f}^{(u)} + r^{(i)} . \tag{5}$$

In this case, the universal prediction function can be interpreted as a prediction function that models the general architecture bias, regardless of the task. This is suitable for cases where there is no knowledge about the target task, as is the case at the beginning of a new search. The task-specific residual models the task-specific peculiarities. If knowledge about the task exists, it can be used to correct the prediction function for certain architecture codes.

The loss function to be optimized is identical to the original version of NAO and is a combination of the prediction loss function $L_{pred}$ and the reconstruction loss function $L_{rec}$,

$$L = \alpha L_{pred} + (1 - \alpha) L_{rec} . \tag{6}$$

However, the prediction loss function considers all observations to train the model for all tasks,

$$L_{pred} = \sum_{i=1}^{n+1} \sum_{a \in H^{(i)}} \left( f^{(i)}(a) - \hat{f}^{(i)}(a) \right)^2 . \tag{7}$$

$H^{(i)}$ is the set of all architectures which are evaluated for a task $i$ and for which the value of the response function is known. The joint optimization of both loss functions guarantees that architectures which are close in the architecture code space exhibit similar behavior across tasks.

Once the model has been trained by minimizing the loss function, potentially better architectures can be determined for the target task. The architecture codes of models with satisfactory performance serve as starting points for the optimization process. Gradient-based optimization is used to modify the current architecture code to maximize the prediction function of the target task

$$z \leftarrow z + \eta \frac{\partial \hat{f}^{(n+1)}}{\partial z} \ , \tag{8}$$

where $z$ is the current architecture code and $\eta$ is the step size. The architecture encoding is reconstructed by applying the decoder on the final architecture code. The step size is chosen large enough to get a new architecture, which is then evaluated on the target task.

XferNAS has two different phases. In the first phase, the system lacks target knowledge and relies solely on the source knowledge. The architectures with the highest accuracy on the source tasks serve as starting points for the determination of new candidates. This is achieved by means of the process described in Eq. (8) with $\eta = 10$. Having accumulated some target knowledge, the second phase selects as starting points the models with high accuracy on the target task. To keep the search time low, we only examine 33 architectures.

*Implementation Details.* The transfer network has been integrated into the publicly available code of NAO[1], allowing a fair comparison to Luo et al. [19]. In our experiments we retain the prescribed architecture and its hyperparameters. However, for the sake of completeness, we repeat this description. LSTM models are used to model the encoder and decoder. The encoder uses an embedding size of 32 and a hidden state size of 96, whereas the decoder uses a hidden state size of 96 in combination with an attention mechanism. Mean pooling is applied to the encoder LSTM's output to obtain a vector which serves as the input to the transfer network. The universal and the residual task networks are modelled with feed forward neural networks. Adam [11] is used to minimize Eq. (6) with learning rate set to $10^{-3}$, trade-off parameter $\alpha$ to 0.8, and weight decay to $10^{-4}$.

## 3   Related Work

Neural Architecture Search (NAS), the structural optimization of neural networks, is solved with a variety of optimization techniques. These include reinforcement learning [1,3,4,27,34–36], evolutionary algorithms [15,22,23,26], and surrogate model-based optimization [14,19] These techniques have made great advancements with the idea of sharing weights across different architectures

---

[1] https://github.com/renqianluo/NAO.

which are sampled during the search process [2,5,16,21,33] instead of training them from scratch. However, recent work shows that this idea is not working better than a random search [25]. For a detailed overview we refer to a recent survey [29].

A new but promising idea is to transfer knowledge from previous search processes to new ones [10,28,30], which is analogous to the behavior of human experts. We briefly discuss the current work in NAS for convolutional neural networks (CNNs) in this context. TAPAS [10] is an algorithm that starts with a simple architecture and extends it based on a prediction model. For the predictions, first a very simple network is trained on the target data set. Subsequently, the validation error is used to determine the similarity to previously examined data sets. Based on this similarity, predictions of the validation error of different architectures on the target data set are obtained. By means of these, a set of promising architectures are determined and evaluated on the target data set. However, the prediction model is not able to leverage the additional information collected on the target data set. T-NAML [30] seeks to achieve the same effect without searching for new architectures. Instead, it chooses a network which has been pre-trained on ImageNet and makes various decisions to adapt it to the target data set. For this purpose, it uses a reinforcement learning method, which learns to optimize neural architectures across several data sets simultaneously.

## 4   Experiments

In this section we empirically evaluate XferNAS and compare the discovered architectures to the state-of-the-art. Furthermore, we investigate the transferability of the discovered architectures by training it on a different data set without introducing any further changes. In our final ablation study, we investigate the impact of amount of source and target data on the surrogate model's predictions.

### 4.1   Architecture Search Space

In our experiments, we use the widely adopted NASNet search space [36], which is also used by most optimizers that we compare with. Architectures in this search space are based on two types of cells, normal cells and reduction cells. These cells are combined to form the network architecture, with repeating units comprising of N normal cells followed by a reduction cell. The sequence is repeated several times, doubling the number of filters each time. The difference of the reduction cell from the normal cell is that it halves the dimension of feature maps. Each cell consists of B blocks and the first sequence of N normal cells uses F filters. The output of each block is the sum of the result of two operations, where the choice of operation and its input is the task of the optimizer. There are #op different operations, the input can be the output of each of the previous blocks in the same cell or the output of the previous two cells. The output of a cell is defined by concatenating all block outputs that do not serve as input to at least one block. The considered 19 operations are: identity, convolution ($1 \times 1$, $3 \times 3$, $1 \times 3 + 3 \times 1$, $1 \times 7 + 7 \times 1$), max/average pooling ($2 \times 2$, $3 \times 3$, $5 \times 5$), max pooling ($7 \times 7$), min pooling ($2 \times 2$), (dilated) separable convolution ($3 \times 3$, $5 \times 5$, $7 \times 7$).

## 4.2    Training Details for the Convolutional Neural Networks

During the search process smaller architectures are trained (B = 5, N = 3, F = 32) for 100 epochs. The final architecture is trained for 600 epochs according to the specified settings. SGD with momentum set to 0.9 and cosine schedule [18] with $l_{max} = 0.024$ and without warmstart is used for training. Models are regularized by means of weight decay of $5 \cdot 10^{-4}$ and drop-path [13] probability of 0.3. We use a batch size of 128 but decrease it to 64 for computational reasons for architectures with $F \geq 64$. All experiments use a single V100 graphics card. The only exception are networks with $F = 128$ where we use two V100s to speed up the training process.

Standard preprocessing (whitening) and data augmentation are used. Images are padded to a size of $40 \times 40$ and then randomly cropped to $32 \times 32$. Additionally, they are flipped horizontally at random during training. Whenever cutout [6] is applied, a size of 16 is used.

## 4.3    Source Tasks

Image recognition on Fashion-MNIST [31], Quickdraw [8], CIFAR-100 [12] and SVHN [20] forms our four source tasks. For each of these data sets, we evaluated 200 random architectures, giving us a total of 800 different architectures. Every architecture is trained for 100 epochs with the settings described in Sect. 4.2. The default train/test splits of CIFAR-10, CIFAR-100, Fashion-MNIST and SVHN are used and the train split is further divided into 5,000 images for validation and the remaining for training. For computational reasons we refrain from using the entire Quickdraw data set (50 million drawings and 345 classes). We select 100 classes at random and select 450 drawings for training and 50 for validation per class at random. Each of these architectures was trained on exactly one data set and none of these architectures were evaluated on the target task before or during the search. Therefore, it is valid to conclude that the architecture found is new.

## 4.4    Image Recognition on CIFAR-10

We evaluate the proposed transfer framework using the CIFAR-10 benchmark data set and present the results in Table 1. The table is divided into four parts. In the first part we list the results that some manually created architectures achieve. In the second and third part we tabulate the results achieved by traditional NAS methods as well as those which transfer knowledge across tasks. In the last part we list our results. In contrast to some of the other search methods, we

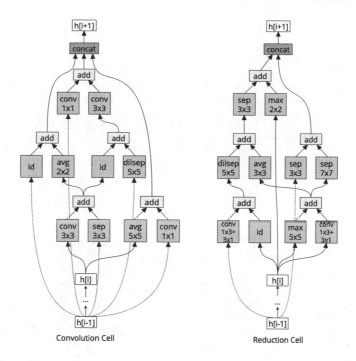

**Fig. 3.** Convolution and reduction cell of XferNASNet.

refrained from additional hyperparameter optimization of our final architecture (XferNASNet) (Fig. 3).

XferNAS is the extended version of NAO which additionally uses the transfer network; so this comparison is of particular interest. We not only observe a significant drop in the search effort (number of evaluated models reduced from 1,000 to 33, search time reduced from 200 GPU days to 6), but also on the error obtained on the test set. The smallest version of NAONet performs slightly better than XferNASNet (3.18 vs. 3.37), but also uses twice as many parameters. If the data augmentation technique cutout [6] is used, this minimal improvement turns around (2.11 vs. 1.99).

The transfer method TAPAS achieves significantly worse results (6.33 vs. 3.92 of the next better method). The other transfer method T-NAML achieves an error rate of 3.5 which is not better than XferNASNet (3.37). It should also be noted that T-NAML finetunes architectures which have been pre-trained on ImageNet. Thus, not only the number of parameters is probably much higher, more data is used and no new architectures are found. Therefore, it arguably solves a different task. A very simple baseline is to select the best architecture on the most similar source task (CIFAR-100). Objectively this baseline performs quite well (4.14) but compares poorly to XferNASNet.

Furthermore, we compare to other search methods that consider the NASNet search space. XferNASNet performs very well compared to current

**Table 1.** Classification error of discovered CNN models on CIFAR-10. We denote the total number of models trained during the search by M. B is the number of blocks, N the number of cells and F the number of filters. #op is the number of different operations considered in the cell which is an indicator of the search space complexity. For more details on the hyperparameters #op, B, N and F we refer to Sect. 4.1. We expand the results collected by [19] by the most recent works.

| Model | B | N | F | #op | Error (%) | #params | M | GPU Days |
|---|---|---|---|---|---|---|---|---|
| DenseNet-BC [9] | / | 100 | 40 | / | 3.46 | 25.6M | / | / |
| ResNeXt-29 [32] | / | / | / | / | 3.58 | 68.1M | / | / |
| NASNet-A [36] | 5 | 6 | 32 | 13 | 3.41 | 3.3M | 20000 | 2000 |
| AmoebaNet-A [22] | 5 | 6 | 36 | 10 | 3.34 | 3.2M | 20000 | 3150 |
| AmoebaNet-B [22] | 5 | 6 | 36 | 19 | 3.37 | 2.8M | 27000 | 3150 |
| AmoebaNet-B [22] | 5 | 6 | 128 | 19 | 2.98 | 34.9M | 27000 | 3150 |
| AmoebaNet-B + Cutout [22] | 5 | 6 | 128 | 19 | 2.13 | 34.9M | 27000 | 3150 |
| PNAS [14] | 5 | 3 | 48 | 8 | 3.41 | 3.2M | 1280 | 225 |
| ENAS [21] | 5 | 5 | 36 | 5 | 3.54 | 4.6M | / | 0.45 |
| DARTS + Cutout [16] | 5 | 6 | 36 | 7 | 2.83 | 4.6M | / | 4 |
| SNAS + Cutout [33] | 5 | 6 | 36 | 7 | 2.85 | 2.8M | / | 1.5 |
| NAONet [19] | 5 | 6 | 36 | 11 | 3.18 | 10.6M | 1000 | 200 |
| NAONet [19] | 5 | 6 | 64 | 11 | 2.98 | 28.6M | 1000 | 200 |
| NAONet + Cutout [19] | 5 | 6 | 128 | 11 | 2.11 | 128M | 1000 | 200 |
| TAPAS [10] | / | / | / | / | 6.33 | 2.7M | 1 | 0 |
| T-NAML [30] | / | / | / | / | 3.5 | N/A | 150 | N/A |
| Best on source task CIFAR-100 | 5 | 6 | 32 | 19 | 4.14 | 6.1M | 200 | / |
| XferNASNet | 5 | 6 | 32 | 19 | 3.37 | 4.5M | 33 | 6 |
| XferNASNet + Cutout | 5 | 6 | 32 | 19 | 2.70 | 4.5M | 33 | 6 |
| XferNASNet | 5 | 6 | 64 | 19 | 3.11 | 17.5M | 33 | 6 |
| XferNASNet + Cutout | 5 | 6 | 64 | 19 | 2.19 | 17.5M | 33 | 6 |
| XferNASNet | 5 | 6 | 128 | 19 | 2.88 | 69.5M | 33 | 6 |
| XferNASNet + Cutout | 5 | 6 | 128 | 19 | 1.99 | 69.5M | 33 | 6 |

gradient-based optimization methods such as DARTS and SNAS (2.70 versus 2.83 and 2.85, respectively). It also provides good results compared to architectures such as NASNet or AmoebaNet which were discovered by time-consuming optimization methods (3.37 in 2,000 GPU days or 3.34 in 3,150 GPU days versus 3.41 in 6 GPU days). We want to highlight that the results in part 2 of Table 1 are supposed to give the reader an idea how well XferNASNet performs on the NASNet search space compared to other methods operating on the same search space. However, a direct comparison to all methods but NAO is less relevant since this work is orthogonal to ours and can benefit of our transfer learning approach as well.

**Table 2.** Various architectures discovered on CIFAR-10 applied to CIFAR-100. Although the hyperparameters have not been optimized, XferNASNet achieves the best result.

| Model | B | N | F | #op | Error (%) | #params |
|---|---|---|---|---|---|---|
| DenseNet-BC [9] | / | 100 | 40 | / | 17.18 | 25.6M |
| Shake-shake [7] | / | / | / | / | 15.85 | 34.4M |
| Shake-shake + Cutout [6] | / | / | / | / | 15.20 | 34.4M |
| NASNet-A [36] | 5 | 6 | 32 | 13 | 19.70 | 3.3M |
| NASNet-A + Cutout [36] | 5 | 6 | 32 | 13 | 16.58 | 3.3M |
| NASNet-A + Cutout [36] | 5 | 6 | 128 | 13 | 16.03 | 50.9M |
| PNAS [14] | 5 | 3 | 48 | 8 | 19.53 | 3.2M |
| PNAS + Cutout [14] | 5 | 3 | 48 | 8 | 17.63 | 3.2M |
| PNAS + Cutout [14] | 5 | 6 | 128 | 8 | 16.70 | 53.0M |
| ENAS [21] | 5 | 5 | 36 | 5 | 19.43 | 4.6M |
| ENAS + Cutout [21] | 5 | 5 | 36 | 5 | 17.27 | 4.6M |
| ENAS + Cutout [21] | 5 | 5 | 128 | 5 | 16.44 | 52.7M |
| AmoebaNet-B [22] | 5 | 6 | 128 | 19 | 17.66 | 34.9M |
| AmoebaNet-B + Cutout [22] | 5 | 6 | 128 | 19 | 15.80 | 34.9M |
| NAONet + Cutout [19] | 5 | 6 | 36 | 11 | 15.67 | 10.8M |
| NAONet + Cutout [19] | 5 | 6 | 128 | 11 | 14.75 | 128M |
| TAPAS [10] | / | / | / | / | 18.99 | 15.2M |
| Best on source task CIFAR-100 | 5 | 6 | 32 | 19 | 19.96 | 6.3M |
| XferNASNet | 5 | 6 | 32 | 19 | 18.88 | 4.5M |
| XferNASNet + Cutout | 5 | 6 | 32 | 19 | 16.29 | 4.5M |
| XferNASNet | 5 | 6 | 64 | 19 | 16.35 | 17.6M |
| XferNASNet + Cutout | 5 | 6 | 64 | 19 | 14.72 | 17.6M |
| XferNASNet | 5 | 6 | 128 | 19 | 15.95 | 69.6M |
| XferNASNet + Cutout | 5 | 6 | 128 | 19 | 14.06 | 69.6M |

## 4.5 Architecture Transfer to CIFAR-100

A standard procedure to test the transferability is to apply the architectures discovered on CIFAR-10 to another data set, e.g. CIFAR-100 [19]. Although, some of the popular architectures have been adapted to the new data set through additional hyperparameter optimization, we refrain from this for XferNASNet. We compare the transferability of XferNASNet to other automatically discovered architectures and list the results in Table 2. For these results, we rely on the numbers reported by Luo et al. [19]. The XferNASNet architecture achieves an error of 18.88 without and an error of 16.29 with cutout. Thus, we achieve significantly better results than all other architectures except NAONet. However, when we increase the number of filters from 32 to 64, we achieve comparable

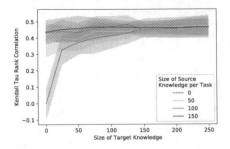

**Fig. 4.** Correlation coefficient between predicted and true validation accuracy with varying amount of source and target knowledge. The source knowledge significantly improves predictions when there is little target knowledge available.

results as NAO with 128 filters and much more parameters. Furthermore, when we increase the number of filters to 128, the error drops to about 14.06, which is significantly lower than that of NAONet (14.75), and notably with only about half the number of parameters. We also report the results obtained for the best architecture found during the random search (19.96) in order to reconfirm that XferNAS is discovering better architectures than the ones available for the source tasks.

## 4.6    Ablation Study of XferNet

At this point we would like to closely examine the benefits of knowledge transfer, especially the circumstances under which we observe a positive effect. For this we conduct an experiment where we evaluate 600 random architectures for CIFAR-10 according to Sect. 4.3 and hold out 50 to compute the correlation between predicted and true validation accuracy. The remaining 550 are candidates for the target knowledge available during the training of the surrogate model. In this experiment the amount of source and target knowledge is varied. For every amount of source and target knowledge, ten random splits are created which are used throughout the experiment. We train the surrogate model as described in Sect. 2.3 on all ten splits and test it on our held-out set. While the architectures within the observation history are evaluated on exactly one data set, the architectures for evaluation are unknown to the model. In Fig. 4, we visualize the mean and standard deviation of the correlation between the surrogate model prediction and the actual validation accuracy over the ten repetitions. The x-axis indicates the size of the target knowledge, and the four curves represent experiments corresponding to different sizes of source knowledge. The source knowledge size ranges from 0 architectures per source task (no knowledge transferred, equivalent to NAO) to 150. We elaborate on four scenarios in this context.

*How significant is the benefit of knowledge transfer for a new search (zero target knowledge)?* This is the scenario in which any method that does not transfer knowledge can not be better than random guessing (correlation of 0). If our hypothesis is correct and knowledge can be transferred, this should be the scenario in which our method achieves the best results. And indeed, the correlation is quite high and increases with the amount of source knowledge.

*Does the transfer model benefit from the target knowledge?* For any amount of source knowledge, additional target knowledge increases the correlation and, accordingly, improves the predictions. This effect depends inversely on the amount of source knowledge.

*What amount of target knowledge is sufficient, so that source knowledge no longer yields a positive effect?* For target knowledge comprising of 150 architectures (about 30 GPU days), the effect of source knowledge seems to fade away. Therefore, one can conclude that knowledge transfer does not contribute to any further improvement.

*When this threshold is reached, does the knowledge transfer harm the model?* We continue to experiment with larger sizes of target knowledge (up to 550 architectures, not shown in the plot) and empirically confirm that the additional source knowledge does not deteriorate the model performance. However, the correlation keeps improving with increasing amount of target knowledge for both cases, with and without knowledge transfer.

## 5  Conclusions and Future Work

In this paper, we present the idea of accelerating NAS by transferring the knowledge about architectures across different tasks. We develop a transfer framework that extends existing NAS optimizers with this capability and requires minimal modification. By integrating this framework into NAO, we demonstrate how simple and yet effective these changes are. We evaluate the resulting new XferNAS optimizer (NAO + Transfer Network) on CIFAR-10 and CIFAR-100. In just six GPU days, we discover XferNASNet which reaches a new record low for NAS optimizers on CIFAR-10 (2.11→1.99) and CIFAR-100 (14.75→14.06) with significantly fewer parameters. Thus, the addition of this component to NAO does not only reduce the search time from 200 GPU hours to only 6, it even improves the discovered architecture.

For the future we want to combine the Transfer Network with various other optimizers to evaluate its limitations. We are particularly interested in combining it with the recent optimizer methods that use weight-sharing which promise to be even faster. Furthermore, we want to evaluate the discovered XferNASNet on ImageNet, which is an experiment we could not conduct so far due to resource limitations.

# References

1. Baker, B., Gupta, O., Naik, N., Raskar, R.: Designing neural network architectures using reinforcement learning. In: 5th International Conference on Learning Representations, ICLR 2017, 24–26 April 2017, Toulon, France, Conference Track Proceedings (2017). https://openreview.net/forum?id=S1c2cvqee
2. Bender, G., Kindermans, P.J., Zoph, B., Vasudevan, V., Le, Q.: Understanding and simplifying one-shot architecture search. In: Dy, J., Krause, A. (eds.) Proceedings of the 35th International Conference on Machine Learning. Proceedings of Machine Learning Research. PMLR, 10–15 July 2018, Stockholmsmässan, Stockholm Sweden, vol. 80, pp. 550–559 (2018). http://proceedings.mlr.press/v80/bender18a.html
3. Cai, H., Chen, T., Zhang, W., Yu, Y., Wang, J.: Efficient architecture search by network transformation. In: Proceedings of the Thirty-Second AAAI Conference on Artificial Intelligence, (AAAI-18), the 30th Innovative Applications of Artificial Intelligence (IAAI-18), and the 8th AAAI Symposium on Educational Advances in Artificial Intelligence (EAAI-18), 2–7 February 2018, New Orleans, Louisiana, USA, pp. 2787–2794 (2018). https://www.aaai.org/ocs/index.php/AAAI/AAAI18/paper/view/16755
4. Cai, H., Yang, J., Zhang, W., Han, S., Yu, Y.: Path-level network transformation for efficient architecture search. In: Proceedings of the 35th International Conference on Machine Learning, ICML 2018, 10–15 July 2018, Stockholmsmässan, Stockholm, Sweden, pp. 677–686 (2018). http://proceedings.mlr.press/v80/cai18a.html
5. Cai, H., Zhu, L., Han, S.: ProxylessNAS: direct neural architecture search on target task and hardware. In: Proceedings of the International Conference on Learning Representations, ICLR 2019, New Orleans, Louisiana, USA (2019). https://openreview.net/forum?id=HylVB3AqYm
6. DeVries, T., Taylor, G.W.: Improved regularization of convolutional neural networks with cutout. CoRR abs/1708.04552 (2017). http://arxiv.org/abs/1708.04552
7. Gastaldi, X.: Shake-shake regularization. CoRR abs/1705.07485 (2017). http://arxiv.org/abs/1705.07485
8. Ha, D., Eck, D.: A neural representation of sketch drawings. In: 6th International Conference on Learning Representations, ICLR 2018, Vancouver, BC, Canada, 30 April–3 May 2018, Conference Track Proceedings (2018). https://openreview.net/forum?id=Hy6GHpkCW
9. Huang, G., Liu, Z., van der Maaten, L., Weinberger, K.Q.: Densely connected convolutional networks. In: 2017 IEEE Conference on Computer Vision and Pattern Recognition, CVPR 2017, 21–26 July 2017, Honolulu, HI, USA, pp. 2261–2269. IEEE Computer Society (2017). https://doi.org/10.1109/CVPR.2017.243
10. Istrate, R., Scheidegger, F., Mariani, G., Nikolopoulos, D.S., Bekas, C., Malossi, A.C.I.: TAPAS: train-less accuracy predictor for architecture search. In: Proceedings of the Thirty-Third AAAI Conference on Artificial Intelligence, (AAAI-19), Honolulu, Hawaii, USA (2019)
11. Kingma, D.P., Ba, J.: Adam: a method for stochastic optimization. In: 3rd International Conference on Learning Representations, ICLR 2015, 7–9 May 2015, San Diego, CA, USA, Conference Track Proceedings (2015), http://arxiv.org/abs/1412.6980
12. Krizhevsky, A.: Learning multiple layers of features from tiny images. Technical report (2009)

13. Larsson, G., Maire, M., Shakhnarovich, G.: FractalNet: ultra-deep neural networks without residuals. In: 5th International Conference on Learning Representations, ICLR 2017, 24–26 April 2017, Toulon, France, Conference Track Proceedings (2017). https://openreview.net/forum?id=S1VaB4cex

14. Liu, C., et al.: Progressive neural architecture search. In: Proceedings of the European Conference on Computer Vision (ECCV), pp. 19–34 (2018)

15. Liu, H., Simonyan, K., Vinyals, O., Fernando, C., Kavukcuoglu, K.: Hierarchical representations for efficient architecture search. In: 6th International Conference on Learning Representations, ICLR 2018, 30 April–3 May 2018, Vancouver, BC, Canada, Conference Track Proceedings (2018). https://openreview.net/forum?id=BJQRKzbA-

16. Liu, H., Simonyan, K., Yang, Y.: DARTS: differentiable architecture search. In: Proceedings of the International Conference on Learning Representations, ICLR 2019, New Orleans, Louisiana, USA (2019)

17. Liu, W., et al.: SSD: single shot multibox detector. In: Computer Vision - ECCV 2016–14th European Conference, 11–14 October 2016, Amsterdam, The Netherlands, Proceedings, Part I, pp. 21–37 (2016). https://doi.org/10.1007/978-3-319-46448-0_2

18. Loshchilov, I., Hutter, F.: SGDR: stochastic gradient descent with warm restarts. In: 5th International Conference on Learning Representations, ICLR 2017, 24–26 April 2017, Toulon, France, Conference Track Proceedings (2017). https://openreview.net/forum?id=Skq89Scxx

19. Luo, R., Tian, F., Qin, T., Chen, E., Liu, T.: Neural architecture optimization. In: Advances in Neural Information Processing Systems 31: Annual Conference on Neural Information Processing Systems 2018, NeurIPS 2018, 3–8 December 2018, Montréal, Canada, pp. 7827–7838 (2018). http://papers.nips.cc/paper/8007-neural-architecture-optimization

20. Netzer, Y., Wang, T., Coates, A., Bissacco, A., Wu, B., Ng, A.Y.: Reading digits in natural images with unsupervised feature learning. In: NIPS Workshop on Deep Learning and Unsupervised Feature Learning 2011 (2011). http://ufldl.stanford.edu/housenumbers/nips2011_housenumbers.pdf

21. Pham, H., Guan, M., Zoph, B., Le, Q., Dean, J.: Efficient neural architecture search via parameters sharing. In: Dy, J., Krause, A. (eds.) Proceedings of the 35th International Conference on Machine Learning. Proceedings of Machine Learning Research, vol. 80, pp. 4095–4104. PMLR, 10–15 July 2018, Stockholmsmässan, Stockholm Sweden (2018). http://proceedings.mlr.press/v80/pham18a.html

22. Real, E., Aggarwal, A., Huang, Y., Le, Q.V.: Regularized evolution for image classifier architecture search. In: The Thirty-Third AAAI Conference on Artificial Intelligence, AAAI 2019, The Thirty-First Innovative Applications of Artificial Intelligence Conference, IAAI 2019, The Ninth AAAI Symposium on Educational Advances in Artificial Intelligence, EAAI 2019, 27 January– 1 February 2019, Honolulu, Hawaii, USA, pp. 4780–4789 (2019)

23. Real, E., et al.: Large-scale evolution of image classifiers. In: Precup, D., Teh, Y.W. (eds.) Proceedings of the 34th International Conference on Machine Learning. Proceedings of Machine Learning Research. PMLR, International Convention Centre, 6–11 August 2017, Sydney, Australia, vol. 70, pp. 2902–2911 (2017)

24. Redmon, J., Divvala, S.K., Girshick, R.B., Farhadi, A.: You only look once: unified, real-time object detection. In: 2016 IEEE Conference on Computer Vision and Pattern Recognition, CVPR 2016, 27–30 June 2016, Las Vegas, NV, USA, pp. 779–788 (2016)

25. Sciuto, C., Yu, K., Jaggi, M., Musat, C., Salzmann, M.: Evaluating the search phase of neural architecture search. CoRR abs/1902.08142 (2019)
26. Wistuba, M.: Deep learning architecture search by neuro-cell-based evolution with function-preserving mutations. In: Berlingerio, M., Bonchi, F., Gärtner, T., Hurley, N., Ifrim, G. (eds.) ECML PKDD 2018. LNCS (LNAI), vol. 11052, pp. 243–258. Springer, Cham (2019). https://doi.org/10.1007/978-3-030-10928-8_15
27. Wistuba, M.: Practical deep learning architecture optimization. In: 5th IEEE International Conference on Data Science and Advanced Analytics, DSAA 2018, 1–3 October 2018, Turin, Italy, pp. 263–272 (2018). https://doi.org/10.1109/DSAA.2018.00037
28. Wistuba, M., Pedapati, T.: Inductive transfer for neural architecture optimization. CoRR abs/1903.03536 (2019). http://arxiv.org/abs/1903.03536
29. Wistuba, M., Rawat, A., Pedapati, T.: A survey on neural architecture search. CoRR abs/1905.01392 (2019). http://arxiv.org/abs/1905.01392
30. Wong, C., Houlsby, N., Lu, Y., Gesmundo, A.: Transfer learning with neural AutoML. In: Advances in Neural Information Processing Systems 31: Annual Conference on Neural Information Processing Systems 2018, NeurIPS 2018, 3–8 December 2018, Montréal, Canada, pp. 8366–8375 (2018)
31. Xiao, H., Rasul, K., Vollgraf, R.: Fashion-MNIST: a novel image dataset for benchmarking machine learning algorithms. CoRR abs/1708.07747 (2017)
32. Xie, S., Girshick, R.B., Dollár, P., Tu, Z., He, K.: Aggregated residual transformations for deep neural networks. In: 2017 IEEE Conference on Computer Vision and Pattern Recognition, CVPR 2017, 21–26 July 2017, Honolulu, HI, USA, pp. 5987–5995 (2017)
33. Xie, S., Zheng, H., Liu, C., Lin, L.: SNAS: stochastic neural architecture search. In: Proceedings of the International Conference on Learning Representations, ICLR 2019, New Orleans, Louisiana, USA (2019)
34. Zhong, Z., Yan, J., Wu, W., Shao, J., Liu, C.: Practical block-wise neural network architecture generation. In: 2018 IEEE Conference on Computer Vision and Pattern Recognition, CVPR 2018, 18–22 June 2018, Salt Lake City, UT, USA, pp. 2423–2432 (2018)
35. Zoph, B., Le, Q.V.: Neural architecture search with reinforcement learning. In: 5th International Conference on Learning Representations, ICLR 2017, 24–26 April 2017, Toulon, France, Conference Track Proceedings (2017)
36. Zoph, B., Vasudevan, V., Shlens, J., Le, Q.V.: Learning transferable architectures for scalable image recognition. In: 2018 IEEE Conference on Computer Vision and Pattern Recognition, CVPR 2018, 18–22 June 2018, Salt Lake City, UT, USA, pp. 8697–8710 (2018). https://doi.org/10.1109/CVPR.2018.00907

# Finding the Optimal Network Depth in Classification Tasks

Bartosz Wójcik, Maciej Wołczyk$^{(\boxtimes)}$, Klaudia Bałazy, and Jacek Tabor

GMUM, Jagiellonian University, Kraków, Poland
maciej.wolczyk@doctoral.uj.edu.pl

**Abstract.** We develop a fast end-to-end method for training lightweight neural networks using multiple classifier heads. By allowing the model to determine the importance of each head and rewarding the choice of a single shallow classifier, we are able to detect and remove unneeded components of the network. This operation, which can be seen as finding the optimal depth of the model, significantly reduces the number of parameters and accelerates inference across different processing units, which is not the case for many standard pruning methods. We show the performance of our method on multiple network architectures and datasets, analyze its optimization properties, and conduct ablation studies.

**Keywords:** Model compression and acceleration · Multi-head networks

## 1 Introduction

Although deep learning methods excel at various tasks in computer vision, natural language processing, and reinforcement learning, running neural networks used for those purposes is often computationally expensive [19, 22]. At the same time, many real-life applications require the models to run in real-time on hardware not specialized for deep learning techniques, which is infeasible for networks of this scale. Therefore, it is crucial to find methods for reducing both memory requirements and inference time without losing performance. In response to this problem, the field of model compression emerged, with pruning being one of the most important research directions [3, 4].

Pruning methods, which rely on removing parameters from a large trained network based on some metric (e.g. magnitude of the weights or approximated loss on removal) are very effective at reducing the size of the model. Unstructured pruning methods excel at creating sparse representations of overparameterized networks without losing much performance by removing single connections [7]. However, special hardware is required to take advantage of sparsity and decrease the inference time [17]. Structured pruning aims to solve this problem by removing whole channels instead of single connections, which accelerates the network even when running on general-purpose hardware. Nevertheless, recent research suggests that the acceleration-performance trade-off in networks obtained this

F. Hutter et al. (Eds.): ECML PKDD 2020, LNAI 12459, pp. 263–278, 2021.
https://doi.org/10.1007/978-3-030-67664-3_16

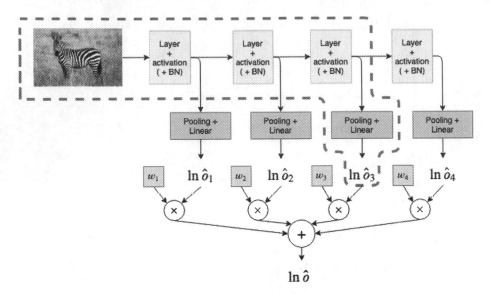

**Fig. 1.** We attach classifier heads (green blocks) to the base feed-forward network (yellow blocks) and then use their combined outputs as the final prediction of the model. The output of each classifier is weighted by its weight $w_k$, with $\sum_k w_k = 1$. During training, we force the model to converge to a single $w_l \approx 1$, and thus we can remove all layers after $l$ and all heads other than the $l$-th one to obtain a compressed network (red dashed line). (Color figure online)

way is worse than in shallower networks trained from scratch [5]. As such, pruning methods perform poorly at the task of reducing inference time of neural networks and an alternative approach is needed.

The approach of minimizing the depth of the network can be also motivated from a biological perspective. Recent research of functional networks in the human brain shows that learning a task leads to a reduction of communication distances [9,15]. Thus, it may be beneficial to try to simulate the same phenomenon in the case of artificial neural networks by decreasing the length of the paths in artificial neural networks (i.e. the depth) instead of minimizing the width (i.e. the number of connections in each layer).

Inspired by those findings, we introduce NETCUT, a quick end-to-end method for reducing the depth of the network in classification tasks. We add a classifier head on top of each hidden layer and use their combined predictions as the final answer of the model, where the influence of each head is a parameter updated using gradient descent. The basic scheme of the model is presented in Fig. 1.

Our main theoretical contribution is a method of aggregating outputs from individual classifier heads, which enforces choosing only one head during the training and simplifies the model in the process. We show that combining the logarithm of probabilities instead of probabilities themselves encourages the network to choose a model with a single classifier head. This method also allows us to avoid the numerical instability linked to using a standard softmax layer. Finally, we introduce a regularization component to the loss function, which approximates the time it takes the network to process the input. After the network converges to a single classifier head, we cut out the resulting network without a noticeable performance drop.

We test NETCUT in multiple settings, showing stable performance across different models, including experiments on highly non-linear network architectures obtained by generating random graphs. Our findings show that the resulting cut-out shallow networks are much faster than the original one, both on CPU and GPU while maintaining similar performance. Finally, we perform an extensive analysis of the optimization properties of our method. We examine the directions of gradients of each classifier, different initialization schemes for the weight layers and the effects of poor starting conditions.

## 2  Related Work

The importance of the depth of artificial neural networks has been shown both theoretically and empirically [2, 21]. Thus, carelessly reducing the number of layers may prove catastrophic in results, drastically reducing the expressive power of the neural network [14]. In NETCUT we explicitly model the trade-off of expressive power (measured by the negative classification loss) and the depth of the network by introducing both of those terms in the cost function.

Our method can be seen as the pruning of whole layers instead of single connections or neurons. Similar ideas have been investigated in the past. Wen et al. [18] introduced depth-regularization in ResNets by applying group Lasso on the whole layers. Huang & Wang [11] explored a similar idea from the perspective of forcing the output of the entire layer to fall to 0 instead of individual weights in the layer. However similar, those approaches are only applicable for architectures with residual connections, where NETCUT works for any chosen feed-forward network.

Attaching multiple output heads to a base neural network, which is a central idea in our approach, has been previously exhibited in many works. Considering the publications most strongly connected to our work, Huang et al. [10] used a multi-scale dense network in order to allow for the dynamical response time of the model, depending on the difficulty of the input sample. Zhang et al. [23] extended this idea by introducing knowledge distillation between classifiers in the model to improve the performance of earlier heads and using custom attention modules in order to avoid negative interference in the base of the network. However, since those methods focus on dynamical resource management, it is not clear how to use them to obtain smaller, quicker networks, which is the goal of our work.

## 3  Theoretical Model

In this section we describe our model. Subsection 3.1 presents a basic approach for using a multi-head network for classification. Subsection 3.2 introduces our

novel way of combining probabilities from classifiers. In Subsect. 3.3 we describe
our regularization scheme and show how compression is achieved.

## 3.1 Basic Multi-head Classification Model

NETCUT considers an overparameterized feed-forward neural network architecture, which we want to train for a classification task with the final goal of reducing the computational effort during inference.

We modify the original network by adding classifier heads on top of hidden layers, as it is presented in Fig. 1. Thus, for every input, we get multiple vectors $\hat{o}_1, \ldots, \hat{o}_n$, where $n$ is the number of classifiers, and by $\hat{o}_k^{(i)}$ we denote the probability of the $i$-th class given by the $k$-th classifier. This modification is non-invasive as the architecture of the core of the network remains unchanged, and the classifier heads have very few parameters compared to the original network.

The classifiers are then trained together by combining the probabilities $\hat{o}_k$ in order to produce the final prediction of the network $\hat{o}$, which is then used as an input to the cross-entropy loss

$$L_{\text{class}}(y, \hat{o}) = - \sum_i y^{(i)} \log \hat{o}^{(i)},$$

where $y$ is the true label represented as a one-hot vector.

In order to allow the model to choose which classifiers are useful for the task at hand, we introduce classifier head importance weights $w_k$, with $w_k \geq 0$ and $\sum_k w_k = 1$. The weights are parameters of the network, and the model updates them during the gradient descent step.

Then, the basic way of aggregating the outputs of each classifier would be to use a weighted average of the probabilities

$$\hat{o}^{(i)} = \sum_k w_k \hat{o}_k^{(i)}.$$

Therefore, the resulting model may be seen as an ensemble of classifiers of different depth.

## 3.2 NetCut Aggregation Scheme

The presented way of aggregating the outputs of the classifiers does not encourage the network in any way to choose just one classifier. This reduction is crucial for the proper compression of the model. In order to enforce it, we propose a novel way of combining the outputs of multiple classifiers, which heavily rewards choosing just one layer:

$$\hat{o}^{(i)} = \exp \sum_k w_k \ln \hat{o}_k^{(i)}.$$

The resulting vector $\hat{o}$ does not necessarily represent a probability distribution as its elements $\hat{o}^{(i)}$ do not always sum up to 1:

$$\sum_i \hat{o}^{(i)} = \sum_i \exp \sum_k w_k \ln \hat{o}_k^{(i)} \leq \sum_i \exp \ln \sum_k w_k \hat{o}_k^{(i)} = 1,$$

where the second transition comes from Jensen's inequality.

One can see that the cross-entropy loss is minimized when $\hat{o}^{(i)} = 1$ for the correct class $i$, which is possible only when equality in the above formula holds. This is the case when $w_l = 1$ for some $l$ and $w_k = 0$ for all $k \neq l$, since

$$\sum_i \hat{o}^{(i)} = \sum_i \exp \ln \hat{o}_l^{(i)} = 1.$$

Other equality conditions for Jensen's inequality cannot be satisfied in our setting, since the logarithm is not an affine function and it is impossible for every $\hat{o}_k$ to be equal because of the inherent noise of the training procedure. Thus, in order to reduce the value of the cross-entropy loss function in our approach, the method will strive towards leaving only one nonzero weight $w_l = 1$. This in turn is equivalent to choosing a single-head subnetwork which can be easily cut out from the starting network without losing performance.

Concluding, we arrive at the following theorem:

**Theorem 1.** *Suppose that the l-th classifier head of* NETCUT *network obtains a perfect classification rate, i.e. it always assigns probability 1 to the correct class. By minimizing the loss function of the model with our aggregation scheme, we will arrive at $w_l = 1$, and $w_k = 0$ for all $k \neq l$.*

Observe that the same result is not valid for the basic aggregation scheme.

Indeed in practice, as we show in Sect. 4, combining the log probabilities of each layer causes the network to quickly focus on only one of the classifiers. The same is not the case for weighting the probabilities directly, where the network spreads the mass across multiple head importance weights, as we show in Sect. 5.3. Such a subnetwork cannot be then easily extracted.

We emphasize the positive numerical properties of this approach. In order to combine the predictions by directly weighting the probabilities outputted by each layer, we would have to calculate the softmax function for each classifier, which could potentially introduce numerical instabilities. Meanwhile, in our approach we can avoid computing the softmax function by directly computing the log probabilities. This technique, known in the computer science community as the log-sum-exp trick is less prone to numerical under- and overflows [1].

### 3.3   Time-Regularization and Model Compression

Our goal is to train a network which will be significantly faster during inference, i.e. we want to minimize the time of processing a sample. However, the time the network takes to process a sample is not a differentiable object and thus cannot be used directly to update the network with standard gradient descent methods.

In order to approximate the inference time, we introduce a surrogate differentiable penalty based on the number of layers in the network (i.e. the depth):

$$L_{\text{reg}} = \sum_k w_k k.$$

For scaling the regularization cost, we introduce the $\beta$ hyperparameter, giving the final loss function of the model

$$L = L_{\text{class}} + \beta L_{\text{reg}}.$$

After training, we can reduce the size of the network by using only the layer chosen by the model, as given by $l = \arg \max_k w_k$. The trained network can be then easily compressed by cutting out all the layers with indices larger than $l$ and removing all the classifier heads other than $l$-th one, as presented in Fig. 1.

In theory, the "cut-out" network may perform worse than the original one, because of the lack of influence of the removed classifiers. However, in practice, we show that the final weight $w_l \approx 1$ at the end of the training, and the influence of the rest of the layers is negligible. This is not necessarily the case for weighting schemes other than our log softmax approach.

## 4   Experiments

In this section we show the performance of NETCUT on multiple models, highlighting its applicability in various settings. The following subsections describe experiments on standard vision architectures (traditional CNNs, ResNets, and fully connected networks). Then, Subsect. 4.4 presents experiments in which we apply NETCUT to randomly generated graph-based networks.

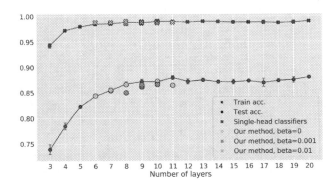

**Fig. 2.** Accuracy of NETCUT on a standard convolutional architecture trained on CIFAR-10. Results for the single-head baseline networks, representing all the sub-network our network can pick, are averaged over 3 runs and plotted with error bars. As each of the 5 runs of our method can converge to a different layer, we plot them separately. Observe that NETCUT shrinks the starting 20-layer network without a significant accuracy drop.

The head importance weights $w_k$ are initialized uniformly in each experiment, and the time-regularization coefficient $\beta$ is set to 0 unless specifically noted otherwise. We published the code used for conducting the experiments on GitHub: https://github.com/gmum/classification-optimal-network-depth.

## 4.1   Standard CNNs

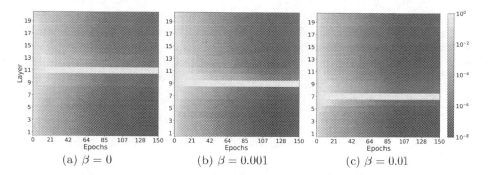

(a) $\beta = 0$          (b) $\beta = 0.001$          (c) $\beta = 0.01$

**Fig. 3.** Classifier head importance weights $w_k$ training progress for different $\beta$ values for a convolutional architecture trained on CIFAR-10. Colors in each row represent a change of a single weight through time. We use a logarithmic color map (as shown on the scale on the right) in order to show the differences in smaller values. The results show that the method decides on a layer very quickly and that by increasing the $\beta$ hyperparameter we can encourage the model to select a smaller subnetwork.

We begin testing our approach with a standard convolutional neural network with 20 base convolutional layers trained on the CIFAR-10 dataset. Each convolutional layer consists of 50 filters of size $5 \times 5$. We use batch normalization and the ReLU activation function. Each classifier head consists of a global max pooling layer and a fully-connected classification layer, giving in total $510 \cdot 20$ parameters which is very light compared to around 1.2M parameters in the base network. We train the network for 150 epochs using the Adam optimizer, batch size set to 128, with basic data augmentation (random crop, horizontal flip), and without explicit regularization other than the optional time-regularization described in Subsect. 3.3.

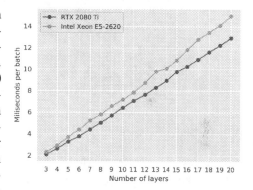

**Fig. 4.** Processing time of a single batch. The batch size is 1 for CPU and 64 for GPU. Decreasing the number of layers linearly reduces the inference time.

To evaluate the performance of NETCUT, we run the experiment 5 times for each hyperparameter setting. As a baseline, we compare ourselves to all the possible subnetworks our method could pick during the training. To do so, we train $n$-layer single-head subnetworks of our starting network, with $n$ ranging from 3 to 20. In Fig. 2 we report the train and test accuracy scores of those baseline single-head networks along with the results of our method with different $\beta$ hyperparameter values.

The results show that our approach is able to find a working subnetwork of much lower complexity, but similar accuracy. We can also control the shallowness of the network by changing the $\beta$ hyperparameter value, which determines the magnitude of the model complexity penalty in the loss function. In Fig. 3 we show how head importance weights $w_k$ change for a randomly chosen run from Fig. 2 for each $\beta$ value. The regularization effect is evident from the very start of the learning process.

To show that the networks obtained by NETCUT are significantly faster, we check the inference speed of single-head networks of varying depth and report the results in Fig. 4 for CPU and GPU. As expected, the time needed to process a single batch increases linearly with the number of layers. This means that networks obtained by our method can increase the inference speed up to 2.85 times compared to the starting network, with little to no performance drop.

## 4.2    ResNets

**Table 1.** ResNet-110 results for different $\beta$ values. The chosen head (block) and final accuracy are shown and compared to the original ResNet-110 architecture results. By changing $\beta$ we can directly balance the compression-performance trade-off.

| Reg. coef. $\beta$ | 0.000 | 0.002 | 0.004 | 0.008 | 0.010 | Original |
|---|---|---|---|---|---|---|
| Chosen block | 41 | 28 | 24 | 10 | 7 | 54 |
| Accuracy | 89.84% | 88.20% | 87.50% | 80.33% | 78.45% | 91.6% |

Subsequently, we test NETCUT on ResNet-110 [8] for CIFAR-10 by attaching a classification head after every block. Each head consists of a global average pooling layer and a fully-connected layer. We do not change any other architecture hyperparameters. We present the results in Table 1, which show the compression-performance trade-off for different values of the $\beta$ hyperparameter.

We have observed that for $\beta = 0$ our model always chooses one of the final classifier heads, even if we modify the network and use a larger number of blocks. This is not the case for the standard CNN architectures studied in the previous subsection. We hypothesize that this effect is caused by the presence of residual connections, which encourage iterative refinement of features and thus make adding more layers preferable [13].

## 4.3    Fully Connected Networks

To cover a wider range of model types, we also test NETCUT on simple fully-connected (FC) networks trained on MNIST and CIFAR-10 datasets. The width of every layer is 200 for MNIST and 1000 for CIFAR-10. Every head has the same architecture, consisting of a single fully-connected classification layer. We do not use batch normalization for this experiment.

Results presented in Fig. 5 show that the method achieves better test accuracy scores than the base 20-layer network and similar scores to the best network tested, while significantly reducing the computational complexity. We determine the nature of low accuracy in the deeper single-head networks as an effect of gradient instability and explore this problem further in Sect. 5.2.

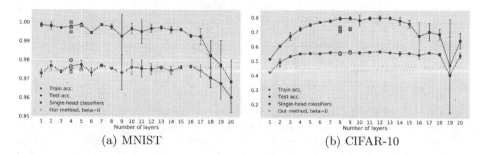

(a) MNIST                                      (b) CIFAR-10

**Fig. 5.** Accuracy of networks resulting from applying our method to FC architectures. For both datasets, our method finds shallower networks with almost equal test accuracy.

### 4.4   Graph-Based Networks

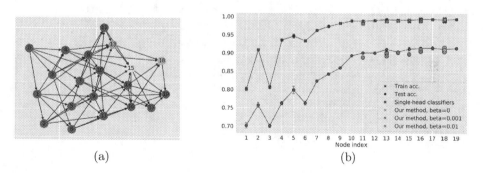

(a)                                             (b)

**Fig. 6.** (a) The graph used for generating the network used in the experiment. Nodes most commonly selected by the model for each $\beta$ are marked with colors corresponding to the legend on the plot on the right. (b) Accuracy of the network with NETCUT on CIFAR-10.

And lastly, we test NETCUT on a more general class of architectures with more complex connection patterns. We generate a random graph and then transform it into a neural network architecture, by treating each node as a ResNet-like block and using outputs of all its predecessors as inputs. This subsection is based on

Janik & Nowak [12], and we refer the reader to their work for details about the network generation process.[1]

For the first experiment, we use a randomly generated graph shown in Fig. 6. We add a classifier head to each ResNet-like block (represented by a node in the graph) and run our method with the expectation that we will be able to find the optimal subgraph of the original graph. In order to adapt our method to graphs, we redefine the time-regularization component of the loss function:

$$L_{reg} = \sum_k w_k e(k),$$

where $e(k)$ denotes the number of edges in a subgraph consisting of predecessors of node $k$ and the node itself. The form of this regularization component is again chosen to approximate the time it takes the network to process an example.

Similarly as before, we repeat each run of our method 5 times for each $\beta$ and compare the results to the baseline situation, where each possible subgraph has one classifier head on the final node. The results presented in Fig. 6 are consistent with the ones from previous experiments, further confirming the stability of NETCUT, even for non-standard architectures with complex connection patterns.

## 5    Analysis

### 5.1    Optimization Properties of the Network

The dynamics of training multi-head networks are non-trivial and still not well understood. In order to explore the optimization properties of NETCUT, we take inspiration from the multi-task learning literature. To estimate whether the interference between different tasks is beneficial or detrimental, several works compare the directions of gradients of each task [6,20]. If the gradients of two tasks point in the same direction, then positive reinforcement occurs. If they point in opposite directions, then tasks are harmful to each other, and for orthogonal directions they are independent.

For our analysis, we use the cosine similarity metric between two vectors:

$$\rho(v, u) = \frac{\langle v, u \rangle}{\sqrt{\langle v, v \rangle \langle u, u \rangle}},$$

where $\rho \in [-1, 1]$, with $-1$ appearing for opposite vectors, 0 for orthogonal vectors, and 1 for vectors pointing in the same direction.

We would like to measure the impact of each classifier in the network on the parameters. In order to do that, we introduce the notion of a partial gradient with respect to parameters $\theta$, defined as

$$\hat{g}_\theta^l = \frac{\partial L(\hat{o}_1, \ldots, \hat{o}_n)}{\partial \hat{o}_l} \frac{\partial \hat{o}_l}{\partial \theta},$$

[1] The implementation we are using for this experiment can be found in this GitHub repository: https://github.com/rmldj/random-graph-nn-paper.

where $n$ is the number of classifiers.

The partial gradient can be seen as the gradient of the loss function obtained by pushing the gradient only through the $\hat{o}_l$ and stopping it for all $\hat{o}_k$ where $k \neq l$. In practice, $\hat{g}_\theta^k$ represents the change of the parameters $\theta$ expected by the $k$-th classifier.

We can see that the sum of all the partial gradients gives us the "full" gradient:

$$\sum_k \frac{\partial L(\hat{o}_1, \ldots, \hat{o}_n)}{\partial \hat{o}_k} \frac{\partial \hat{o}_k}{\partial \theta} = \nabla_\theta L(\hat{o}_1, \ldots, \hat{o}_n) =: \hat{g}_\theta.$$

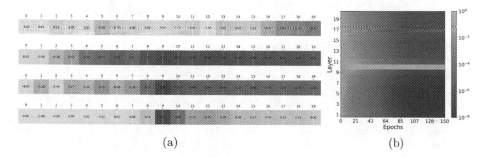

(a)                                                        (b)

**Fig. 7.** (a) Cosine similarities of the full gradient and partial gradients, calculated with respect to the first layer of the network. The subsequent vectors, starting from the top, were recorded at the beginning of the training, the 10th, the 45th and the 140th epoch, respectively. We see that after the training stabilizes, the partial gradient of the chosen layer is most directly correlated with the true gradient. (b) The head importance weights of the network throughout the training process.

To better understand the impact of different classifier heads on the gradient, we use this framework to analyze learning dynamics of our standard CNN architecture trained on CIFAR-10 in the setting described in Subsect. 4.1. The results presented in Fig. 7 show cosine similarities between partial gradients and the full gradient and the head importance weights throughout the training. The cosine similarity values for the 9th layer, which is the final choice of the model, are consistently increasing. This is because when the head's importance weight increases, so does the magnitude of its gradients and its impact on the full gradient.

Another interesting phenomenon appears in the earliest stages of training. At the very first step, most of the partial gradients are almost orthogonal to the full gradient, which suggests high levels of noise at the beginning of training. However, during the first few epochs, the cosine similarities of the later layers increase significantly, signaling that they agree with each other in large part. This is not the case for the earliest layers, as their partial gradients stay orthogonal to or even point in the opposite direction of the full gradient.

This observation shows why the importance weights of early classifier heads drop so quickly. Since the partial gradient of the early layers is usually orthogonal

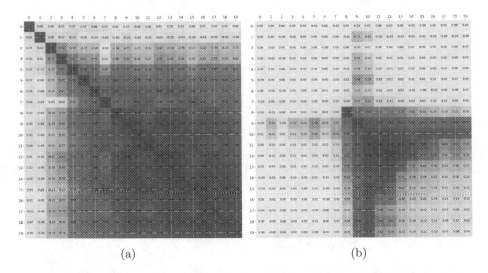

(a)                                                                (b)

**Fig. 8.** Cosine similarities of the partial gradients with respect to the first layer after (a) 10 epochs and (b) 45 epochs. The cell with coordinates $(x, y)$ shows the value of cosine similarity between respective partial gradients $\rho(\hat{g}_\theta^x, \hat{g}_\theta^y)$. We can see that the gradients of the later classifier heads are significantly more correlated than gradients of the earlier ones.

to the full gradient used for updating the parameters, those classifiers do not have an opportunity to learn any features directly relevant for classification, and thus their influence is quickly reduced.

To further investigate this issue, we check the cosine similarities between each pair of partial gradients $\hat{g}^k$ and $\hat{g}^l$ during training and present the results in Fig. 8. We observe that the cosine similarity in the later parts of the network is much higher than in the earlier parts of the network. Our explanation for this phenomenon is that the earlier heads strive to detect complex features that would enable proper classification even in a shallow network, while all of the later heads expect the base of the network to learn the same set of simple patterns. This finding seems to be consistent with the well-known phenomenon of early layers functioning as basic filters for detecting small local patterns such as edges, with only the deeper layers encoding more complex structures.

## 5.2   Poor Starting Conditions

An interesting question is how NETCUT performs in situations where the starting network fails at the given task, but contains a subnetwork that performs well. To investigate this point, we construct a setting where gradient vanishing or exploding phenomena occur. When the network is too deep and proper optimization techniques are not used, the noise in the gradients accumulating during the backward step will make it impossible to learn. However, if we were

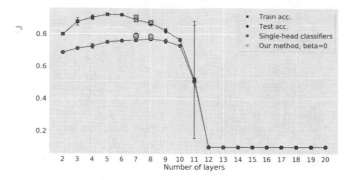

**Fig. 9.** Accuracy of NETCUT for models that exhibit vanishing or exploding gradient problems. We see that even if the starting network with 20 layers collapses, our method is able to find a well-performing subnetwork.

to use only a subset of layers of this network, such a model would be then able to achieve satisfactory performance.

To test this, we use our standard CNN architecture with 20 layers as described in Subsect. 4.1, but without the batch normalization layers, which are known for their effect on reducing the gradient instability [16]. The results presented in Fig. 9 show that networks with more than 10 layers are highly unstable, with their final performance being no better than random.

At the same time, our network is able to pick a subnetwork consisting of 7 layers, remaining in the region where the training is still stable and thus achieving good performance. This result suggests that NETCUT is able to counteract the effect of choosing a poor starting network for the given task.

### 5.3 Analysis of Head Importance Weights Behaviour

Head importance weights $w_k$ used for weighting the output of each classifier decide the shape of the final obtained network and as such are crucial for our method. To better understand the function they perform in NETCUT, we check their properties in different settings.

We examine the properties of the log softmax trick in NETCUT by investigating the baseline situation where for the final prediction we combine the probabilities themselves instead of the logarithm of the probabilities. We compare these two approaches on a fully-connected network with 20 layers trained on MNIST, following the setting described in Subsect. 4.3 and present the results in Fig. 10. As we can observe, combining the probabilities themselves leads to divergence of the training due to overflow happening in the softmax function, which is required for obtaining the probabilities. Similar issues do not appear in our approach since we can directly calculate the logarithm of the softmax operation, which is much more stable numerically.

Another important difference to note is that the baseline approach of combining probabilities does not lead to the choice of a single classifier, i.e. there are

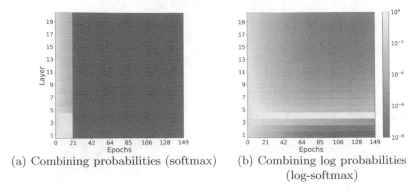

(a) Combining probabilities (softmax)    (b) Combining log probabilities (log-softmax)

**Fig. 10.** Two different ways of producing the final output for the network. The dark green values after the 21st epoch on the first plot represent NaN values appearing after overflow occurrence. Directly combining the probabilities does not guarantee that the model will converge on a single layer and can lead to numerical overflows. (Color figure online)

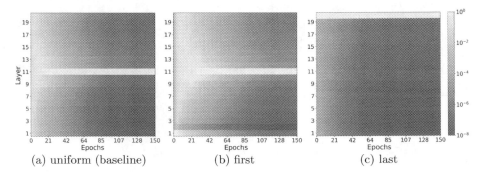

(a) uniform (baseline)    (b) first    (c) last

**Fig. 11.** Head importance weights progress with alternative init schemes with $\beta = 0$. Although the network behaves the same for uniform and first-head initializations, it stays at the last layer when initialized so. The achieved accuracy scores are 86.47%, 85.94%, and 87.98% for uniform, first, and last schemes, respectively.

multiple head importance weights $w_k$ with values significantly larger than zero. For such a model, cutting out a single-head classifier would lead to significant performance drop, because of the lack of influence of the removed heads.

To further understand the optimization properties of NETCUT, we examine the behavior of our method for different initialization schemes for importance head weights $w_k$. In the standard approach we initialize the weights uniformly. Here, we test two additional variants that put almost the entire weight onto the first or the last classification head, such that at initialization respectively $w_1 = 0.999$ or $w_n = 0.999$ with the rest of the probability mass spread uniformly. Results presented in Fig. 11 show that the first-head initialization does not differ significantly from our standard approach. However, starting with most of the probability mass on the last head, the network does not move the mass back to the earlier layers.

# 6    Conclusion

We presented NETCUT, a simple end-to-end method for compressing and accelerating neural networks, which improves the inference speed both on GPU and CPU. As part of our method, we introduced a novel way of combining outputs of multiple classifiers. Combining log probabilities instead of probabilities themselves allows us to converge to a single layer and avoid numerical instabilities connected with computing the softmax function. We show extensive analysis of the optimization properties of the network, including the investigation of gradient directions inspired by multi-task learning methods.

**Acknowledgements.** We thank Christopher Galias, Marek Śmieja and Przemysław Spurek for useful comments and discussion.

The work of Bartosz Wójcik was supported by the National Science Centre (Poland) grant no. 2018/31/B/ST6/00993. The work of Jacek Tabor was supported by the National Science Centre (Poland) the grant no. 2017/25/B/ST6/01271. Maciej Wołczyk carried out the work within the research project "Bio-inspired artificial neural networks" (grant no. POIR.04.04.00-00-14DE/18-00) within the Team-Net program of the Foundation for Polish Science co-financed by the European Union under the European Regional Development Fund.

# References

1. Blanchard, P., Higham, D.J., Higham, N.J.: Accurate computation of the Log-Sum-Exp and Softmax functions (2019)
2. Chatziafratis, V., Nagarajan, S.G., Panageas, I., Wang, X.: Depth-width trade-offs for reLU networks via Sharkovsky's theorem. arXiv preprint arXiv:1912.04378 (2019)
3. Cheng, Y., Wang, D., Zhou, P., Zhang, T.: A survey of model compression and acceleration for deep neural networks. arXiv preprint arXiv:1710.09282 (2017)
4. Choudhary, T., Mishra, V., Goswami, A., Sarangapani, J.: A comprehensive survey on model compression and acceleration. Artif. Intell. Rev. 1–43 (2020). https://doi.org/10.1007/s10462-020-09816-7
5. Crowley, E.J., Turner, J., Storkey, A., O'Boyle, M.: A closer look at structured pruning for neural network compression. arXiv preprint arXiv:1810.04622 (2018)
6. Du, Y., Czarnecki, W.M., Jayakumar, S.M., Pascanu, R., Lakshminarayanan, B.: Adapting auxiliary losses using gradient similarity. arXiv preprint arXiv:1812.02224 (2018)
7. Gale, T., Elsen, E., Hooker, S.: The state of Sparsity in deep neural networks. arXiv preprint arXiv:1902.09574 (2019)
8. He, K., Zhang, X., Ren, S., Sun, J.: Deep residual learning for image recognition. In: Proceedings of the IEEE Conference on Computer Vision and Pattern Recognition, pp. 770–778 (2016)
9. Heitger, M.H., Ronsse, R., Dhollander, T., Dupont, P., Caeyenberghs, K., Swinnen, S.P.: Motor learning-induced changes in functional brain connectivity as revealed by means of graph-theoretical network analysis. Neuroimage **61**(3), 633–650 (2012)
10. Huang, G., Chen, D., Li, T., Wu, F., van der Maaten, L., Weinberger, K.Q.: Multi-scale dense networks for resource efficient image classification. arXiv preprint arXiv:1703.09844 (2017)

11. Huang, Z., Wang, N.: Data-driven sparse structure selection for deep neural networks. In: Proceedings of the European Conference on Computer Vision (ECCV), pp. 304–320 (2018)
12. Janik, R.A., Nowak, A.: Neural networks on random graphs (2020)
13. Jastrzebski, S., Arpit, D., Ballas, N., Verma, V., Che, T., Bengio, Y.: Residual connections encourage iterative inference. arXiv preprint arXiv:1710.04773 (2017)
14. Nakkiran, P., Kaplun, G., Bansal, Y., Yang, T., Barak, B., Sutskever, I.: Deep double descent: where bigger models and more data hurt. In: International Conference on Learning Representations (2019)
15. Sami, S., Miall, R.C.: Graph network analysis of immediate motor-learning induced changes in resting state bold. Front. Hum. Neurosci. **7**, 166 (2013)
16. Santurkar, S., Tsipras, D., Ilyas, A., Madry, A.: How does batch normalization help optimization? (2018)
17. Turner, J., Cano, J., Radu, V., Crowley, E.J., O'Boyle, M., Storkey, A.: Characterizing across-stack optimizations for deep convolutional neural networks. In: 2018 IEEE International Symposium on Workload Characterization (IISWC), pp. 101–110. IEEE (2018)
18. Wen, W., Wu, C., Wang, Y., Chen, Y., Li, H.: Learning structured sparsity in deep neural networks. In: Advances in Neural Information Processing Systems, pp. 2074–2082 (2016)
19. Yang, Z., Dai, Z., Yang, Y., Carbonell, J., Salakhutdinov, R., Le, Q.V.: XLNet: leneralized autoregressive pretraining for language understanding (2019)
20. Yu, T., Kumar, S., Gupta, A., Levine, S., Hausman, K., Finn, C.: Gradient surgery for multi-task learning. arXiv preprint arXiv:2001.06782 (2020)
21. Zagoruyko, S., Komodakis, N.: Wide residual networks. CoRR abs/1605.07146 (2016). http://arxiv.org/abs/1605.07146
22. Zhang, C., Bengio, S., Hardt, M., Recht, B., Vinyals, O.: Understanding deep learning requires rethinking generalization (2016)
23. Zhang, L., Tan, Z., Song, J., Chen, J., Bao, C., Ma, K.: SCAN: a scalable neural networks framework towards compact and efficient models. In: Advances in Neural Information Processing Systems, pp. 4029–4038 (2019)

# Topological Insights into Sparse Neural Networks

Shiwei Liu[1]([⊠]), Tim Van der Lee[1], Anil Yaman[1,4], Zahra Atashgahi[1,2],
Davide Ferraro[3], Ghada Sokar[1], Mykola Pechenizkiy[1],
and Decebal Constantin Mocanu[1,2]

[1] Department of Mathematics and Computer Science,
Eindhoven University of Technology, 5600 MB Eindhoven, The Netherlands
{s.liu3,t.lee,a.yaman,z.atashgahi,g.a.z.n.sokar,m.pechenizkiy}@tue.nl
[2] Faculty of Electrical Engineering, Mathematics and Computer Science,
University of Twente, 7522NB Enschede, The Netherlands
d.c.mocanu@utwente.nl
[3] The Biorobotics Institute, Scuola Superiore Sant'Anna, Pisa, Italy
davide.ferraro@santannapisa.it
[4] Korea Advanced Institute of Science and Technology,
Daejeon 34141, Republic of Korea

**Abstract.** Sparse neural networks are effective approaches to reduce the resource requirements for the deployment of deep neural networks. Recently, the concept of adaptive sparse connectivity, has emerged to allow training sparse neural networks from scratch by optimizing the sparse structure during training. However, comparing different sparse topologies and determining how sparse topologies evolve during training, especially for the situation in which the sparse structure optimization is involved, remain as challenging open questions. This comparison becomes increasingly complex as the number of possible topological comparisons increases exponentially with the size of networks. In this work, we introduce an approach to understand and compare sparse neural network topologies from the perspective of graph theory. We first propose Neural Network Sparse Topology Distance (NNSTD) to measure the distance between different sparse neural networks. Further, we demonstrate that sparse neural networks can outperform over-parameterized models in terms of performance, even without any further structure optimization. To the end, we also show that adaptive sparse connectivity can always unveil a plenitude of sparse sub-networks with very different topologies which outperform the dense model, by quantifying and comparing their topological evolutionary processes. The latter findings complement the *Lottery Ticket Hypothesis* by showing that there is a much more efficient and robust way to find "winning tickets". Altogether, our results start enabling a better theoretical understanding of sparse neural networks, and demonstrate the utility of using graph theory to analyze them.

**Keywords:** Sparse neural networks · Neural Network Sparse Topology Distance · Adaptive sparse connectivity · Graph Edit Distance

© Springer Nature Switzerland AG 2021
F. Hutter et al. (Eds.): ECML PKDD 2020, LNAI 12459, pp. 279–294, 2021.
https://doi.org/10.1007/978-3-030-67664-3_17

# 1    Introduction

Deep neural networks have led to promising breakthroughs in various applications. While the performance of deep neural networks improving, the size of these usually over-parameterized models has been tremendously increasing. The training and deploying cost of the state-of-art models, especially pre-trained models like BERT [4], is very large.

Sparse neural networks are an effective approach to address these challenges. Discovering a small sparse and well-performing sub-network of a dense network can significantly reduce the parameters count (e.g. memory efficiency), along with the floating-point operations. Over the past decade, many works have been proposed to obtain sparse neural networks, including but not limited to magnitude pruning [9,10], Bayesian statistics [22,27], $l_0$ and $l_1$ regularization [23], reinforcement learning [19]. Given a pre-trained model, these methods can efficiently discover a sparse sub-network with competitive performance. While some works aim to provide analysis of sparse neural networks [6,7,21,34], they mainly focus on how to empirically improve training performance or to what extent the initialization and the final sparse structure contribute to the performance. Sparsity (the proportion of neural network weights that are zero-valued) inducing techniques essentially uncover the optimal sparse topologies (sub-networks) that, once initialized in a right way, can reach a similar predictive performance with dense networks as shown by the *Lottery Ticket Hypothesis* [6]. Such sub-networks are named "winning lottery tickets" and can be obtained from pre-trained dense models, which makes them inefficient during the training phase.

Recently, many works have emerged to achieve both, training efficiency and inference efficiency, based on adaptive sparse connectivity [3,5,20,26,28]. Such networks are initialized with a sparse topology and can maintain a fixed sparsity level throughout training. Instead of only optimizing model parameters - weight values (continuous optimization problem), in this case, the sparse topology is also optimized (combinatorial optimization problem) during training according to some criteria in order to fit the data distribution. In [5], it is shown that such metaheuristics approaches always lead to very-well performing sparse topologies, even if they are based on a random process, without the need of a pre-trained model and a *lucky* initialization as done in [6]. While it has been shown empirically that both approaches, i.e. winning lottery tickets and adaptive sparse connectivity, find very well-performing sparse topologies, we are generally lacking their understanding. Questions such as: *How different are these well-performing sparse topologies?, Can very different sparse topologies lead to the same performance?, Are there many local sparse topological optima which can offer sufficient performance (similar in a way with the local optima of the weights continuous optimization problem)?*, are still unanswered.

In this paper, we are studying these questions in order to start enabling a better theoretical understanding of sparse neural networks and to unveil high gain future research directions. Concretely, our contributions are:

- We propose the first metric which can measure the distance between two sparse neural networks topologies[1], and we name it Neural Network Sparse Topology Distance (NNSTD). For this, we treat the sparse network as a large *neural graph*. In NNTSD, we take inspiration from graph theory and Graph Edit Distance (GED) [31] which cannot be applied directly due to the fact that two different neural graphs may represent very similar networks since hidden neurons are interchangeable [18].
- Using NNSTD, we demonstrate that there exist many very different well-performing sparse topologies which can achieve the same performance.
- In addition, with the help of our proposed distance metric, we confirm and complement the findings from [5] by being able to quantify *how different* are the sparse and, at the same time, similarly performing topologies obtained with adaptive sparse connectivity. This implicitly implies that there exist many local well-performing sparse topological optima.

## 2    Related Work

### 2.1    Sparse Neural Networks

**Sparse Neural Networks for Inference Efficiency.** Since being proposed, the motivation of sparse neural networks is to reduce the cost associated with the deployment of deep neural networks (inference efficiency) and to gain better generalization [1,11,16]. Up to now, a variety of methods have been proposed to obtain inference efficiency by compressing a dense network to a sparse one. Out of them, pruning is certainly the most effective one. A method which iteratively alternates pruning and retraining was introduced by Han et al. [10]. This method can reduce the number of connections of AlexNet and VGG-16 on ImageNet by $9\times$ to $13\times$ without loss of accuracy. Further, Narang et al. [29] applied pruning to recurrent neural networks while getting rid of the retraining process. At the same time, it is shown in [35] that, with the same number of parameters, the pruned models (large-sparse) have better generalization ability than the small-dense models. A grow-and-prune (GP) training was proposed in [2]. The network growth phase slightly improves the performance. While unstructured sparse neural networks achieve better performance, it is difficult to be applied into parallel processors, since the limited support for sparse operations. Compared with fine-grained pruning, coarse-grained (filter/channel) pruning is more desirable to the practical application as it is more amenable for hardware acceleration [12,13].

**Sparse Neural Networks for Training Efficiency.** Recently, more and more works attempt to get memory and computational efficiency for the training phase. This can be naturally achieved by training sparse neural networks directly. However, while training them with a fixed sparse topology can lead to good performance [25], it is hard to find an optimal sparse topology to fit the data distribution before training. This problems was addressed by introducing the

---

[1] Our code is available at https://github.com/Shiweiliuiiiiiiii/Sparse_Topology_Distance.

adaptive sparse connectivity concept through its first instantiation, the Sparse Evolutionary Training (SET) algorithm [24,26]. SET is a straightforward strategy that starts from random sparse networks and can achieve good performance based on magnitude weights pruning and regrowing after each training epoch. Further, Dynamic Sparse Reparameterization (DSR) [28] introduced across-layer weights redistribution to allocate more weights to the layer that contributes more to the loss decrease. By utilizing the momentum information to guide the weights regrowth and across-layer redistribution, Sparse Momentum [3] can improve the classification accuracy for various models. However, the performance improvement is at the cost of updating and storing the momentum of every individual weight of the model. Very recently, instead of using the momentum, The Rigged Lottery [5] grows the zero-weights with the highest magnitude gradients to eliminate the extra floating point operations required by Sparse Momentum. Liu et al. [20] trained intrinsically sparse recurrent neural networks (RNNs) that can achieve usually better performance than dense models. Lee et al. [17] introduced single-shot network pruning (SNIP) that can discover a sparse network before training based on a connection sensitivity criterion. Trained in the standard way, the sparse pruned network can have good performance. Instead of using connection sensitivity, GraSP [32] prunes connections whose removal causes the least decrease in the gradient norm, resulting in better performance than SNIP in the extreme sparsity situation.

**Interpretation and Analysis of Sparse Neural Networks.** Some works are aiming to interpret and analyze sparse neural networks. Frankle & Carbin [6] proposed the *Lottery Ticket Hypothesis* and shown that the dense structure contains sparse sub-networks that are able to reach the same accuracy when they are trained with the same initialization. Zhou et al. [34] further claimed that the sign of the "lucky" initialization is the key to guarantee the good performance of "winning lottery tickets". Liu et al. [21] reevaluated the value of network pruning techniques. They showed that training a small pruned model from scratch can reach the same or even better performance than conventional network pruning and for small pruned models, the pruned architecture itself is more crucial to the learned weights. Moreover, magnitude pruning [35] can achieve better performance than $l_0$ regularization [23] and variational dropout [27] in terms of large-scale tasks [7].

## 2.2   Sparse Evolutionary Training

Sparse Evolutionary Training (SET) [26] is an effective algorithm that allows training sparse neural networks from scratch with a fixed number of parameters. Instead of starting from a highly over-parameterized dense network, the network topology is initialized as a sparse Erdős-Rényi graph [8], a graph where each edge is chosen randomly with a fixed probability, independently from every other edge. Given that the random initialization may not always guarantee good performance, adaptive sparse connectivity is utilized to optimize the sparse topology during training. Concretely, a fraction $\zeta$ of the connections with the smallest magnitude are pruned and an equal number of novel connections are re-grown after each training epoch. This adaptive sparse connectivity (pruning-and-regrowing) technique is capable of guaranteeing a constant sparsity level during the whole learning process and also improving the generalization ability. More precisely, at the beginning of the training, the connection $(W_{ij}^k)$ between neuron $h_j^{k-1}$ and $h_i^k$ exists with the probability:

$$p(W_{ij}^k) = \min(\frac{\epsilon(n^k + n^{k-1})}{n^k n^{k-1}}, 1) \qquad (1)$$

where $n^k, n^{k-1}$ are the number of neurons of layer $h^k$ and $h^{k-1}$, respectively; $\epsilon$ is a parameter determining the sparsity level. The smaller $\epsilon$ is, the more sparse the network is. By doing this, the sparsity level of layer $k$ is given by $1 - \frac{\epsilon(n^k + n^{k-1})}{n^k n^{k-1}}$. The connections between the two layers are collected in a sparse weight matrix $\mathbf{W}^k \in \mathbf{R}^{n^{k-1} \times n^k}$. Compared with fully-connected layers whose number of connections is $n^k n^{k-1}$, the SET sparse layers only have $n^W = \|\mathbf{W}^k\|_0 = \epsilon(n^k + n^{k-1})$ connections which can significantly alleviate the pressure of the expensive memory footprint. Among all possible adaptive sparse connectivity techniques, in this paper, we make use of SET due to two reasons: (1) its natural simplicity and computational efficiency, and (2) the fact that the re-grown process of new connections is purely random favoring in this way an unbiased study of the evolved sparse topologies.

## 3    Neural Network Sparse Topology Distance

In this section, we introduce our proposed method, NNSTD, to measure the topological distance between two sparse neural networks. The sparse topology locution used in this paper refers to the graph underlying a sparsely connected neural network in which each neuron represents a vertex in this graph and each existing connection (weight) represents an edge in the graph. Existing metrics to measure the distance between two graphs are not always applicable to artificial neural network topologies. The main difficulty is that two different graph topologies may represent similar neural networks since hidden neurons are interchangeable. All graph similarity metrics consider either labeled or unlabeled nodes to compute the similarity. With neural networks, input and output layers are labeled (each of their neurons corresponds to a concrete data feature or class, respectively), whereas hidden layers are unlabelled. In particular, we take multilayer perceptron networks (MLP) as the default.

The inspiration comes from Graph Edit Distance (GED) [31], a well-known graph distance metric. Considering two graphs $g_1$ and $g_2$, it measures the minimum cost required to transform $g_1$ into a graph isomorphic to $g_2$. Formally the graph edit distance is calculated as follows.

$$GED(g_1, g_2) = \min_{p \in P(g_1, g_2)} c(p) \qquad (2)$$

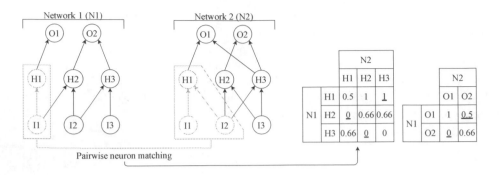

**Fig. 1.** NNSTD metric illustration.

where $p$ represents a sequence of transformation from $g_1$ into a graph isomorphic to $g_2$, and $c$ represents the total cost of such transformation. $P$ represents all possible transformations. This large panel of possibilities makes computing the GED a NP-hard problem when a subset of the nodes in the graphs are unlabeled (e.g. hidden neurons are interchangeable).

The proposed NNSTD metric is presented in Algorithm 1 and discussed next. A graphical example is also provided in Fig. 1. As an example, two neural networks $(N1, N2)$ are considered. For each hidden neuron $n$, a tree graph is constructed based on all direct inputs to this neuron, and these input neurons are collected in a set $g$. Per layer, for all possible pairs of neurons between the two networks, the Normalized Edit Distance (NED) is calculated between their input neurons, as defined in the second line of Algorithm 1. NED takes the value 1 if the two compared neurons have no input neurons in common, and 0 if they have the exact same neurons as input. To reduce the complexity of the search space, we take a greedy approach, and for any current layer we consider that the neurons of the previous layer are labeled (as they have been matched already by the proposed distance metric when the previous layer was under scrutiny), and that adding or deleting inputs have the same cost. For instance, for the neurons compared in Fig. 1, one input neuron is shared out of two different inputs considered, thus the distance between them is $NED(N1 : H1, N2 : H1) = 0.5$. The NNSTD matrix is solved using the Hungarian method to find the neuron (credit) assignment problem which minimizes the total cost, presented in underlined Fig. 1. The aggregated costs divided by the size of $L2$ gives the distance between the first layer of $N1$ and $N2$. To compare the next layer using the same method, the current layer must be fixed. Therefore the assignment solving the NNSTD matrix is saved to reorder the first layer of $N2$. To the end, an NNSTD value of 0 between two sparse layers (or two sparse networks) shows that the two layers are exactly the same, while a value of 1 (maximum possible) shows that the two layers are completely different.

---

**Algorithm 1:** Neural Network Sparse Topology Distance

---

**Function** NED($G1, G2$):

$\quad$ **return** $\dfrac{|(G1 \setminus G2) \cup (G2 \setminus G1)|}{|(G1 \setminus G2) \cup (G2 \setminus G1) \cup (G1 \cap G2)|}$;

**Function** CompareLayers($L1, L2$):

$\quad$ NNSTDmatrix;

$\quad$ **for** *neuron n1 in L1* **do**

$\qquad$ **for** *neuron n2 in L2* **do**

$\qquad\quad$ $G1 = input\_neurons\_set\_of(n1)$;

$\qquad\quad$ $G2 = input\_neurons\_set\_of(n2)$;

$\qquad\quad$ NNSTDmatrix$[(n1, n2)] = NED(G1, G2)$;

$\qquad$ **end**

$\quad$ **end**

$\quad$ neuron_assignment, normalized_cost $= solve($NNSTDmatrix$)$;

$\quad$ **return** neuron_assignment, normalized_cost$/$size(L2);

**Function** CompareNetworks($N1, N2$):

$\quad$ neuron_assignment $=$ Identity;

$\quad$ costs $= 0$;

$\quad$ **for** *layer l in* $[1, L]$ **do**

$\qquad$ neuron_assignment, normalized_cost $=$

$\qquad$ CompareLayers($N1[l], N2[l]$);

$\qquad$ reorder($N2[l]$, neuron_assignment);

$\qquad$ costs$+ =$ normalized_cost;

$\quad$ **end**

$\quad$ **return** costs$/L$;

---

## 4  Experimental Results

In this section, we study the performance of the proposed NNSTD metric and the sparse neural network properties on two datasets, Fashion-MNIST [33] and CIFAR-10 [15], in a step-wise fashion. We begin in Sect. 4.2 by showing that sparse neural networks can match the performance of the fully-connected counterpart, even without topology optimization. Next, in Sect. 4.3 we first validate NNSTD and then we apply it to show that adaptive sparse connectivity can find many well-performing very different sub-networks. Finally, we verify that adaptive sparse connectivity indeed optimizes the sparse topology during training in Sect. 4.4.

### 4.1  Experimental Setup

For the sake of simplicity, the models we use are MLPs with SReLU activation function [14] as it has been shown to provide better performance for SET-MLP [26]. For both datasets, we use 20% of the training data as the validation set and the test accuracy is computed with the model that achieves the highest validation accuracy during training.

For Fashion-MNIST, we choose a three-layer MLP as our basic model, containing 784 hidden neurons in each layer. We set the batch size to 128. The optimizer is stochastic gradient descent (SGD) with Nesterov momentum. We train these sparse models for 200 epochs with a learning rate of 0.01, Nesterov momentum of 0.9. And the weight decay is 1e−6.

The network used for CIFAR-10 consists of two hidden layers with 1000 hidden neurons. We use standard data augmentations (horizontal flip, random rotate, and random crop with reflective padding). We set the batch size to 128. We train the sparse models for 1000 epochs using a learning rate of 0.01, stochastic gradient descent with Nesterov momentum of $\alpha = 0.9$. And we use a weight decay of 1e-6.

### 4.2   The Performance of Sparse Neural Networks

We first verify that random initialized sparse neural networks are able to reach a competitive performance with the dense networks, even without any further topology optimization.

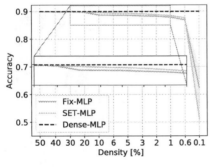 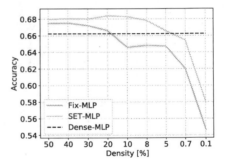

(a) Test accuracy with three-layer MLPs on Fashion-MNIST.

(b) Test accuracy with two-layer MLPs on CIFAR-10.

**Fig. 2.** Test accuracy of MLPs with various density levels. SET-MLP refers to the networks trained with adaptive sparse connectivity associated with SET. Fix-MLP refers to the networks trained without sparse topology optimization. The dashed lines represent the dense MLPs. Note that each line is the average of 8 trials and the standard deviation is very small.

For Fashion-MNIST, we train a group of sparse networks with density levels $(1 - sparsity)$ in the space $\{0.1\%, 0.6\%, 1\%, 2\%, 3\%, 5\%, 6\%, 10\%, 20\%, 30\%, 40\%, 50\%\}$. For each density level, we initialize two sparse networks with two different random seeds as root networks. For each root network, we generate a new network by randomly changing 1% connections. We perform this generating operation 3 times to have 4 networks in total including the root network for each random seed. Every new network is generated from the previous generation. Thus, the number of networks for each density level is 8 and the total number of sparse networks of Fashion-MNIST is 96. We train these sparse networks without any sparse topology optimization for 200 epochs, named as Fix-MLP. To evaluate the effectiveness of sparse connectivity optimization, we also train the same networks with sparse connectivity optimization proposed in SET [26] for 200 epochs, named as SET-MLP. The hyper-parameter of SET, pruning rate, is set to be 0.2. Besides this, we choose two fully-connected MLPs as the baseline.

The experimental results are given in Fig. 2a. We can see that, as long as the density level is bigger than 20%, both Fix-MLP and SET-MLP can reach a similar accuracy with the dense MLP. While decreasing the density level decreases the performance of sparse networks gradually, sparse MLPs still reach the dense accuracy with only 0.6% parameters. Compared with Fix-MLP, the networks trained with SET are able to achieve slightly better performance.

For CIFAR-10, we train two-layer MLPs with various density levels located in the range $\{0.1\%, 0.7\%, 5\%, 8\%, 10\%, 20\%, 30\%, 40\%, 50\%\}$. We use the same strategy with Fashion-MNIST to generate 72 networks in total, 8 for each density level. All networks are trained with and without adaptive sparse connectivity for 900 epochs. The two-layer dense MLP is chosen as the baseline.

The results are illustrated in Fig. 2b. We can observe that Fix-MLP consistently reaches the performance of the fully-connected counterpart when the percentage of parameters is larger than 20%. It is more surprising that SET-MLP can significantly improve the accuracy with the help of adaptive sparse connectivity. With only 5% parameters, SET-MLP can outperform the dense counterpart.

### 4.3    Topological Distance Between Sparse Neural Networks

**Evaluation of Neural Network Sparse Topology Distance.** In this part, we evaluate our proposed NNSTD metric by measuring the initial topological distance between three-layer MLPs on Fashion-MNIST before training. We first measure the topology distance between networks with the same density. We initialize one sparse network with a density level of 0.6%. Then, we generate 9 networks by iteratively changing 1% of the connections from the previous generation step. By doing this, the density of these networks is the same, whereas the topologies vary a bit. Therefore, we expect that the topological distance of each generation from the root network should increase gradually as the generation adds up, but still to have a small upper bound. The distance measured by our method is illustrated in Fig. 3a. We can see that the result is consistently in line with our hypothesis. Starting with the value close to zero, the distance increases as the topological difference adds up, but the maximum distance is still very small, around 0.2.

Further, we also evaluate NNSTD on sparse networks with different density levels. We use the same 96 sparse and two dense networks generated in Sect. 4.2. Their performance is given in Fig. 2a. Concretely, for each density level, we choose 8 networks generated by two different random seeds. For each density level in

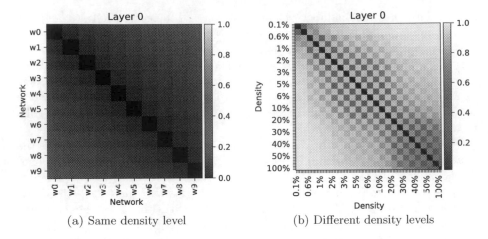

(a) Same density level    (b) Different density levels

**Fig. 3.** Evaluation of the proposed NNSTD metric. (a) refers to the sparse topology distance among 10 networks generated by randomly changing 1% connections with the same density level of 0.6%. $w_i(i = 1, 2, ..., 9)$ represents these gradually changed networks. (b) represents the sparse topology distance among networks generated with different density levels.

the plot, the first four networks are generated with one random seed and the latter four networks are generated with another one. We hypothesize that distance among the networks with the same density should be different from the networks with different density. The distance among networks with different density can be very large, since the density varies over a large range, from 0.1% to 100%. Furthermore, the topological distance increases as the density difference increases, since more cost is required to match the difference between the number of connections. We show the initial topology distance in Fig. 3b. We can see that the distance among different density levels can be much larger than among the ones with the same density, up to 1. The more similar the density levels are, the more similar the topologies are. As expected, the distance between networks with the same density generated with different random seeds is very big. This makes sense as initialized with different random seeds, two sparse connectivities between two layers can be totally different. We only plot the distance of the first layer, as all layers are initialized in the same way.

**Evolutionary Optimization Process Visualization.** Herein, we visualize the evolutionary optimization process of the sparse topology learned by adaptive sparse connectivity associated with SET on Fashion-MNIST and CIFAR-10.

First, we want to study that, initialized with very similar structures, how the topologies of these networks change when they are optimized by adaptive sparse connectivity. We choose the same 10 networks generated for Fashion-MNIST in Fig. 3a and train them with SET for 200 epochs. All the hyper-parameters are

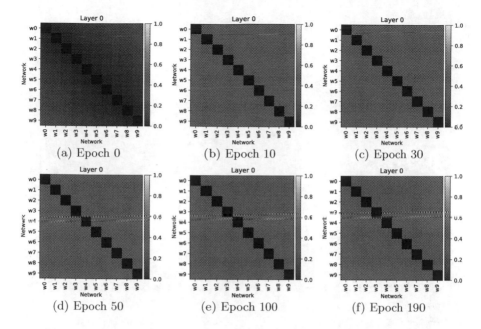

**Fig. 4.** Topological distance dynamics of 10 networks optimized by adaptive sparse connectivity with three-layer SET-MLP on Fashion-MNIST. The initial networks (epoch 0) have the same density level, with a tiny percentage (1%) of topological difference with each other. $w_i (i = 1, 2, ..., 9)$ represents different networks.

the same as in Sect. 4.2. We apply NNSTD to measure the pairwise topological distance among these 10 networks at the $10^{th}$, the $30^{th}$, the $100^{th}$ and the $190^{th}$ epoch. It can be observed in Fig. 4 that, while initialized similarly, the topological distance between networks gradually increases from 0 to 0.6. This means that similar initial topologies gradually evolve to very different topologies while training with adaptive sparse connectivity. It is worth noting that while these networks end up with very different topologies, they achieve very similar test accuracy, as shown in Table 1. This phenomenon shows that there are many sparse topologies obtained by adaptive sparse connectivity that can achieve good performance. This result can be treated as the complement of *Lottery Ticket Hypothesis*, which claims that, with "lucky" initialization, there are subnetworks yielding an equal or even better test accuracy than the original network. We empirically demonstrate that many sub-networks having good performance can be found by adaptive sparse connectivity, even without the "lucky" initialization. Besides, Fig. 5 depicts the comparison between the initial and the final topological distance among the 96 networks used in Fig. 2a. We can see that the distance among different networks also increases after the training process in varying degrees.

**Table 1.** The test accuracy of networks used for the evolutionary optimization process of adaptive sparse connectivity in Sect. 4.3, in percentage.

|  | W0 | W1 | W2 | W3 | W4 | W5 | W6 | W7 | W8 | W9 |
|---|---|---|---|---|---|---|---|---|---|---|
| Fashion-MNIST | 87.48 | 87.53 | 87.41 | 87.54 | 88.01 | 87.58 | 87.34 | 87.70 | 87.77 | 88.02 |
| CIFAR-10 | 65.46 | 65.62 | 65.26 | 65.46 | 65.00 | 65.57 | 65.61 | 64.92 | 64.86 | 65.58 |

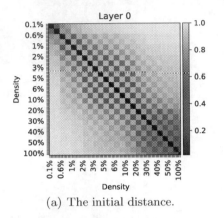
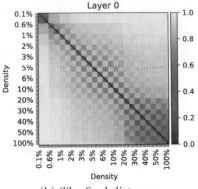

(a) The initial distance.    (b) The final distance.

**Fig. 5.** Heatmap representing the topological distance between the first layer of the 96 three-layers SET-MLP networks on Fashion-MNIST.

Second, we conduct a controlled experiment to study the evolutionary trajectory of networks with very different topologies. We train 10 two-layer SET-MLPs on CIFAR-10 for 900 epochs. All the hyperparameters of these 10 networks are the same except for random seeds. The density level that we choose for this experiment is 0.7%. With this setup, all the networks have very different topologies even with the same density level. The topologies are optimized by adaptive sparse connectivity (prune-and-regrow strategy) during training with a pruning rate of 20% and the weights are optimized by momentum SGD with a learning rate of 0.01.

The distance between different networks before training is very big as they are generated with different random seeds (Fig. 6a), while the expectation is that these networks will end up after the training process also with very different topologies. This is clearly reflected in Fig. 6b.

We are also interested in how the topology evolves within one network trained with SET. Are the difference between the final topology and the original topology big or small? To answer this question, we visualize the optimization process of the sparse topology during training within one network. We save the topologies obtained every 100 epochs and we use the proposed method to compare them with each other. The result is illustrated in Fig. 6c. We can see that the topological distance gradually increases from 0 to a big value, around 0.8. This means that, initialized with a random sparse topology, the network evolves towards a totally different topology during training.

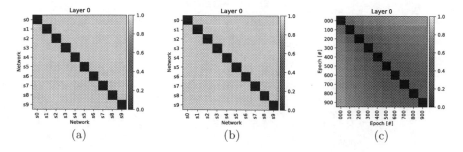

**Fig. 6.** Heatmap representing the topological distance between the first layer of the two-layer SET-MLP networks on CIFAR-10. (a) refers to distance before training. (b) refers to distance after training. (c) represents the topological distance evolution during training for the first network. $\partial_i (i = 1, 2, ..., 9)$ represents different networks.

In all cases, after training, the topologies end up with very different sparse configurations, while at the same time all of them have very similar performance as shown in Table 1. We highlight that this phenomenon is in line with Fashion-MNIST, which confirms our observation that there is a plenitude of sparse topologies obtained by adaptive sparse connectivity which achieve very good performance.

## 4.4  Combinatorial Optimization of Sparse Neural Networks

Although the sparse networks with fixed topology are able to reach similar performance with dense models, randomly initialized sparse networks can not always guarantee good performance, especially when the sparsity is very high as shown in Fig. 2. One effective way to optimize the sparse topology is adaptive sparse connectivity, a technique based on connection pruning followed by connection regrowing, which has shown good performance in the previous works [3,5,26,28]. Essentially, the learning process of the above-mentioned techniques based on adaptive sparse connectivity is a combinatorial optimization problem (model parameters and sparse topologies). The good performance achieved by these techniques can not be solely achieved by the sparse topologies, nor by their initialization [28].

Here, we want to further analyze if the topologies optimized by adaptive sparse connectivity contribute to better test accuracy or not. We hypothesize that, the test accuracy of the optimized topologies should continuously be improving until they converge. To test our hypothesis, we first initialize 10 two-layer MLPs with an extremely low density level (0.5%) under different random seeds and then train them using SET with a pruning rate of 0.2 for 900 epochs on CIFAR-10. We save the sparse networks per 100 epochs and retrain these networks for another 1000 epochs with randomly re-initialized weights. Besides this, to sanity check the effectiveness of the combinatorial optimization, we also retrain the saved networks for 1000 epochs starting from the learned weights by SET.

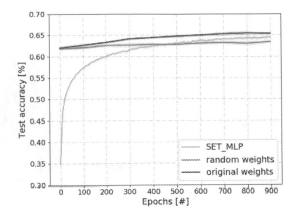

**Fig. 7.** Average test accuracy convergence of the SET network (yellow) and the average test accuracy of the retrained networks: starting from SET weights values (brown) and starting from random weights values (vivid cyan). Each line is the average of 10 trials. (Color figure online)

Figure 7 plots the learning curves of SET and the averaged test accuracy of the retrained networks. We can observe that, the test accuracy of random initialized networks consistently increases as the training epoch increases. This behavior highlights the fact that the adaptive sparse connectivity indeed helps the sparse topology to evolve towards an optimal one. Besides this, it seems that the topology learns faster at the beginning. However, the retrained networks which start from random initialized weights no longer match the performance of SET after about 400 epochs, which indicates that both, the weight optimization and the topology optimization, are crucial to the performance of sparse neural networks. Compared with random re-initialization, training further with the original weights is able to significantly improve the performance. This phenomenon provides a good indication on the behavior of sparse neural networks. It may also pinpoint directions for future research on sparse neural connectivity optimization, which, however, is out of the scope of this paper.

## 5   Conclusion

In this work, we propose the first method which can compare different sparse neural network topologies, namely NNSTD, based on graph theory. Using this method, we obtain novel insights into sparse neural networks by visualizing the topological optimization process of Sparse Evolutionary Training (SET). We demonstrate that random initialized sparse neural networks can be a good choice to substitute over-parameterized dense networks when there are no particularly high requirements for accuracy. Additionally, we show that there are many low-dimensional structures (sparse neural networks) that always achieve very good accuracy (better than dense networks) and adaptive sparse connectivity is an effective technique to find them.

In the light of these new insights, we suggest that, instead of exploring all resources to train over-parameterized models, intrinsically sparse networks with topological optimizers can be an alternative approach, as our results demonstrate that randomly initialized sparse neural networks with adaptive sparse connectivity offer benefits not just in terms of computational and memory costs, but also in terms of the principal performance criteria for neural networks, e.g. accuracy for classification tasks.

In the future, we intend to investigate larger datasets, like Imagenet [30], while considering also other types of sparse neural networks and other techniques to train sparse networks from scratch. We intend to invest more in developing hardware-friendly methods to induce sparsity.

Acknowledgements. This research has been partly funded by the NWO EDIC project.

# References

1. Chauvin, Y.: A back-propagation algorithm with optimal use of hidden units. In: Advances in Neural Information Processing Systems, pp. 519–526 (1989)
2. Dai, X., Yin, H., Jha, N.K.: Grow and prune compact, fast, and accurate LSTMs. arXiv preprint arXiv:1805.11797 (2018)
3. Dettmers, T., Zettlemoyer, L.: Sparse networks from scratch: faster training without losing performance. arXiv preprint arXiv:1907.04840 (2019)
4. Devlin, J., Chang, M.W., Lee, K., Toutanova, K.: BERT: pre-training of deep bidirectional transformers for language understanding. arXiv preprint arXiv:1810.04805 (2018)
5. Evci, U., Gale, T., Menick, J., Castro, P.S., Elsen, E.: Rigging the lottery: making all tickets winners. arXiv preprint arXiv:1911.11134 (2019)
6. Frankle, J., Carbin, M.: The lottery ticket hypothesis: finding sparse, trainable neural networks. arXiv preprint arXiv:1803.03635 (2018)
7. Gale, T., Elsen, E., Hooker, S.: The state of sparsity in deep neural networks. arXiv preprint arXiv:1902.09574 (2019)
8. Gilbert, E.N.: Random graphs. Ann. Math. Stat. **30**(4), 1141–1144 (1959)
9. Guo, Y., Yao, A., Chen, Y.: Dynamic network surgery for efficient DNNs. In: Advances in Neural Information Processing Systems, pp. 1379–1387 (2016)
10. Han, S., Pool, J., Tran, J., Dally, W.: Learning both weights and connections for efficient neural network. In: Advances in Neural Information Processing Systems, pp. 1135–1143 (2015)
11. Hassibi, B., Stork, D.G.: Second order derivatives for network pruning: optimal brain surgeon. In: Advances in Neural Information Processing Systems, pp. 164–171 (1993)
12. He, Y., Kang, G., Dong, X., Fu, Y., Yang, Y.: Soft filter pruning for accelerating deep convolutional neural networks. arXiv preprint arXiv:1808.06866 (2018)
13. He, Y., Liu, P., Wang, Z., Hu, Z., Yang, Y.: Filter pruning via geometric median for deep convolutional neural networks acceleration. In: Proceedings of the IEEE Conference on Computer Vision and Pattern Recognition, pp. 4340–4349 (2019)
14. Jin, X., Xu, C., Feng, J., Wei, Y., Xiong, J., Yan, S.: Deep learning with s-shaped rectified linear activation units. In: Thirtieth AAAI Conference on Artificial Intelligence (2016)

15. Krizhevsky, A., Hinton, G., et al.: Learning multiple layers of features from tiny images. Technical report, Citeseer (2009)
16. LeCun, Y., Denker, J.S., Solla, S.A.: Optimal brain damage. In: Advances in Neural Information Processing Systems, pp. 598–605 (1990)
17. Lee, N., Ajanthan, T., Torr, P.H.: SNIP: single-shot network pruning based on connection sensitivity. arXiv preprint arXiv:1810.02340 (2018)
18. Li, Y., Yosinski, J., Clune, J., Lipson, H., Hopcroft, J.E.: Convergent learning: do different neural networks learn the same representations? In: FE@ NIPS, pp. 196–212 (2015)
19. Lin, J., Rao, Y., Lu, J., Zhou, J.: Runtime neural pruning. In: Advances in Neural Information Processing Systems, pp. 2181–2191 (2017)
20. Liu, S., Mocanu, D.C., Pechenizkiy, M.: Intrinsically sparse long short-term memory networks. arXiv preprint arXiv:1901.09208 (2019)
21. Liu, Z., Sun, M., Zhou, T., Huang, G., Darrell, T.: Rethinking the value of network pruning. arXiv preprint arXiv:1810.05270 (2018)
22. Louizos, C., Ullrich, K., Welling, M.: Bayesian compression for deep learning. In: Advances in Neural Information Processing Systems, pp. 3288–3298 (2017)
23. Louizos, C., Welling, M., Kingma, D.P.: Learning sparse neural networks through $l\_0$ regularization. arXiv preprint arXiv:1712.01312 (2017)
24. Mocanu, D.C.: Network computations in artificial intelligence. Ph.D. thesis (2017)
25. Mocanu, D.C., Mocanu, E., Nguyen, P.H., Gibescu, M., Liotta, A.: A topological insight into restricted Boltzmann machines. Mach. Learn. **104**(2), 243–270 (2016). https://doi.org/10.1007/s10994-016-5570-z
26. Mocanu, D.C., Mocanu, E., Stone, P., Nguyen, P.H., Gibescu, M., Liotta, A.: Scalable training of artificial neural networks with adaptive sparse connectivity inspired by network science. Nat. Commun. **9**(1), 2383 (2018)
27. Molchanov, D., Ashukha, A., Vetrov, D.: Variational dropout sparsifies deep neural networks. In: Proceedings of the 34th International Conference on Machine Learning, vol. 70, pp. 2498–2507. JMLR. org (2017)
28. Mostafa, H., Wang, X.: Parameter efficient training of deep convolutional neural networks by dynamic sparse reparameterization. arXiv preprint arXiv:1902.05967 (2019)
29. Narang, S., Elsen, E., Diamos, G., Sengupta, S.: Exploring sparsity in recurrent neural networks. arXiv preprint arXiv:1704.05119 (2017)
30. Russakovsky, O., et al.: ImageNet large scale visual recognition challenge. Int. J. Comput. Vision **115**(3), 211–252 (2015)
31. Sanfeliu, A., Fu, K.: A distance measure between attributed relational graphs for pattern recognition. IEEE Trans. Syst. Man Cybern. **SMC-13**(3), 353–362 (1983). https://doi.org/10.1109/TSMC.1983.6313167
32. Wang, C., Zhang, G., Grosse, R.: Picking winning tickets before training by preserving gradient flow. arXiv preprint arXiv:2002.07376 (2020)
33. Xiao, H., Rasul, K., Vollgraf, R.: Fashion-MNIST: a novel image dataset for benchmarking machine learning algorithms. arXiv preprint arXiv:1708.07747 (2017)
34. Zhou, H., Lan, J., Liu, R., Yosinski, J.: Deconstructing lottery tickets: zeros, signs, and the supermask. arXiv preprint arXiv:1905.01067 (2019)
35. Zhu, M., Gupta, S.: To prune, or not to prune: exploring the efficacy of pruning for model compression. arXiv preprint arXiv:1710.01878 (2017)

# Graph Neural Networks

# GRAM-SMOT: Top-N Personalized Bundle Recommendation via Graph Attention Mechanism and Submodular Optimization

M. Vijaikumar[(✉)], Shirish Shevade, and M. N. Murty

Department of Computer Science and Automation, Indian Institute of Science, Bangalore, India
{vijaikumar,shirish,mnm}@iisc.ac.in

**Abstract.** Bundle recommendation – recommending a group of products in place of individual products to customers is gaining attention day by day. It presents two interesting challenges – (1) how to personalize and recommend existing bundles to users, and (2) how to generate personalized novel bundles targeting specific users. Recently, few models have been proposed for modeling the bundle recommendation problem. However, they have the following shortcomings. First, they do not consider the higher-order relationships amongst the entities (users, items and bundles). Second, they do not model the relative influence of items present in the bundles, which is crucial in defining such bundles. In this work, we propose GRAM-SMOT – a graph attention-based framework to address the above challenges. Further, we define a loss function based on the metric-learning approach to learn the embeddings of entities efficiently. To generate novel bundles, we propose a strategy that leverages submodular function maximization. To analyze the performance of the proposed model, we conduct comprehensive experiments on two real-world datasets. The experimental results demonstrate the superior performance of the proposed model over the existing state-of-the-art models.

**Keywords:** Personalized bundle recommendation · Graph attention mechanism · Submodular optimization

## 1 Introduction

In recent years, instead of individual products, recommending a group of products to customers is gaining widespread attention in e-commerce platforms. For example, Steam[1] – an online video gaming storefront sells collections of games to its users at a discounted price. Online music platforms such as NetEase-Music[2]

---

[1] https://store.steampowered.com.
[2] http://music.163.com.

© Springer Nature Switzerland AG 2021
F. Hutter et al. (Eds.): ECML PKDD 2020, LNAI 12459, pp. 297–313, 2021.
https://doi.org/10.1007/978-3-030-67664-3_18

and Spotify[3] recommend music playlists so that users need not go through the daunting process of hand-picking the songs for their playlist. These groups of products or items are called *bundles*, and the task of recommending such bundles is called *bundle recommendation*.

In addition to providing benefits to customers, recommending bundles in place of individual items have the following benefits: (i) Shipping the products as a group to the same destination is cost-effective, (ii) Exposes the unpopular products along with the popular products to potential customers, and (iii) Avoids the prolonged storage of products in large inventories.

There are primarily two challenges in bundle recommendations:

1. How to personalize and efficiently recommend existing bundles to users.
2. How to generate personalized novel bundles targeting specific users.

Here, personalizing bundles to customers' interests is essential. That is, though bundle recommendation is beneficial for customers as well as sellers, coming up with the right set of bundles for specific users is important [2]. Otherwise, for instance, when only half of the items in a bundle are relevant, customers may not show interest in those bundles. Besides, all the items in the bundle may not be of equal interest to the users. In addition, when constructing a bundle such as a music playlist for users, it may be a good idea to include songs related to the ones which the user has been exposed to earlier.

Generating personalized bundles for users is challenging due to the following reasons: (i) The search space or the full candidate set of possible bundles is doubly exponential, and (ii) "Ground truth" bundles, that is, the existing user-curated bundles are typically a few in number, compared to the number of possible bundles.

There are some recently proposed works that model the bundle recommendation problem in a personalized way [2–4,19]. In particular, the models – EFM [3] and DAM [4] – are proposed for recommending the existing bundles, and BR [19] and BGN [2] are proposed for generating new bundles along with recommending the existing bundles. Although they provide solutions to the bundle recommendation problem, they have the following shortcomings. First, when learning the representations of the entities, the model should take into account higher-order relationships (multi-hop neighborhood information) amongst entities. This is essential, because, for example, users not only get influenced by directly connected entities – bundles and items, but also by the other users who interacted with these entities. Second, it is essential to learn the influence values of items in defining bundles, and this should also be extended to the interactions between users and items, and users and bundles. However, the above arguments are seldom considered by the existing models.

**Contributions.** To address the above challenges, we propose a GRaph Attention Mechanism and SubModular OpTimization based framework – GRAM-SMOT for personalized bundle recommendation tasks. GRAM-SMOT learns

---

[3] https://www.spotify.com.

higher-order relationships between users, items and bundles by exploiting the inherent graph structure present among these entities. Further, it leverages graph attention mechanisms [21] to learn influence values amongst the entities. In addition, we define a loss function based on the metric-learning approach to learn the embeddings of entities efficiently. To generate novel personalized bundles, we design a greedy strategy via monotone submodular function maximization. Using this, we add items iteratively to the bundles that are personalized for individual users. We conduct extensive experiments on two real-world datasets – YouShu and NetEase. The experimental results show the superiority of the proposed model over state-of-the-art baselines. Our implementation is available at https://github.com/mvijaikumar/GRAM-SMOT.

## 2    Problem Formulation

Let $\mathcal{G}(\mathcal{V}, \mathcal{E})$ be a graph with nodes $(\mathcal{V})$ representing bundles, users and items and edges $(\mathcal{E})$ representing interactions between them. Let $\mathcal{B}$, $\mathcal{U}$ and $\mathcal{I}$ be the sets of bundles, users and items, respectively. Here, $\mathcal{V} = \mathcal{B} \cup \mathcal{U} \cup \mathcal{I}$. And, there exist interactions among users, items and bundles. However, there does not exist interaction amongst nodes of the same type. Let $\Omega_{UB} = \{(u, k) :$ user $u$ has interaction with bundle $k\}$, we define edge between user $u$ and bundle $k$ as

$$r_{uk} = \begin{cases} 1, \text{ if } (u, k) \in \Omega_{UB} \\ 0, \text{ otherwise} \end{cases}$$

Similarly, we denote by $\Omega_{UI}$ for interactions between users and items, and $\Omega_{BI}$ for interactions between bundles and items.

**Definition 1.** *Top-N existing bundle recommendation. Given $\mathcal{G}(\mathcal{V}, \mathcal{E})$, the objective of the top-N existing bundle recommendation is to enumerate a list of N existing bundles that a user is most likely to interact with in the future.*

**Definition 2.** *Bundle generation. Given $\mathcal{G}(\mathcal{V}, \mathcal{E})$, the objective of bundle generation is to generate personalized new bundles for each user such that users will most likely construct them in the future.*

## 3    The Proposed Approach

In this section, we discuss our proposed model for (i) recommending existing bundles, and (ii) generating new bundles in detail. The overall architecture of the proposed model is illustrated in Fig. 1.

### 3.1    GRAM-SMOT

**Embeddings for Users, Items and Bundles.** Let $P \in \mathbb{R}^{d \times |\mathcal{U}|}, Q \in \mathbb{R}^{d \times |\mathcal{I}|}$ and $B \in \mathbb{R}^{d \times |\mathcal{B}|}$ be embedding matrices containing $d$-dimensional representations of users, items and bundles, respectively. Let $x_u \in \mathbb{R}^{|\mathcal{U}| \times 1}, y_i \in \mathbb{R}^{|\mathcal{I}| \times 1}$ and $z_k \in$

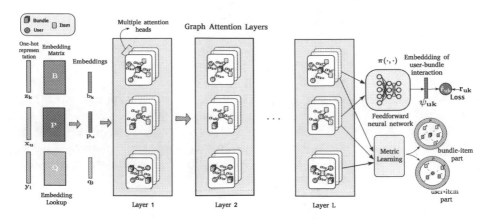

**Fig. 1.** The illustration of GRAM-SMOT architecture.

$\mathbb{R}^{|\mathcal{B}| \times 1}$ be one-hot representations for user $u$, item $i$ and bundle $k$, respectively. The initial embeddings of users, items and bundles are obtained as follows:

$$p_u = Px_u, q_i = Qy_i, \text{ and } b_k = Bz_k \text{ where } p_u, q_i, b_k \in \mathbb{R}^d. \tag{1}$$

**Graph Attention Mechanism.** All items present in a bundle may not be influential in characterizing the bundle. For example, in the Steam video game platform, though gamers often purchase games in bundles, their playing hours for most of the games in such bundles are very minimal[4] [19]. The same can be said of music playlists, where very few popular songs such as the ones that received, say, Grammy award, maybe the reason for the user's interaction with a playlist. Besides, even with isolated items, users may not be influenced by all the items they have interacted with equally. The above observations indicate that it is essential to learn the influence between entities (users, items and bundles).

We employ a graph attention mechanism [21] to tackle the above challenges. The influence value (attention value) between any pair of entities are obtained as

$$\bar{\alpha}_{ui} = f(p_u, q_i), \ \bar{\alpha}_{uk} = f(p_u, b_k) \text{ and } \bar{\alpha}_{ki} = f(b_k, q_i), \tag{2}$$

where $\bar{\alpha}_{ui}, \bar{\alpha}_{uk}$ and $\bar{\alpha}_{ki}$ are unnormalized attention values between user $u$ and item $i$, user $u$ and bundle $k$, and bundle $k$ and item $i$, respectively. Further,

$$f(p_u, q_i) = a(c \cdot [W_U p_u \| W_I q_i]), \tag{3}$$

where $\|$ denotes concatenation operation, $a(\cdot)$ denotes activation function and $x \cdot y$ denotes dot product between $x$ and $y$. Similarly,

$$\begin{aligned} f(p_u, b_k) &= a(c \cdot [W_U p_u \| W_B b_k]), \\ f(b_k, q_i) &= a(c \cdot [W_B b_k \| W_I q_i]) \end{aligned} \tag{4}$$

---

[4] https://cseweb.ucsd.edu/~jmcauley/datasets.html#steam_data.

where $c \in \mathbb{R}^{2d'}$ is a trainable parameter, $W_U, W_I$ and $W_B$ ($\in \mathbb{R}^{d' \times d}$) denote weight matrices for users, items and bundles, respectively. These weight matrices also act as projection matrices which handle the heterogeneity that exist amongst the entities.

Further, the normalized attention value between item $i$ and bundle $k$ is obtained as follows:

$$\alpha_{ki} = \text{softmax}\,(\bar{\alpha}_{ki})$$
$$= \frac{\exp(\bar{\alpha}_{ki})}{\sum_{i' \in \mathcal{N}_U(k)} \exp(\bar{\alpha}_{ki'}) + \sum_{u' \in \mathcal{N}_I(k)} \exp(\bar{\alpha}_{ku'}) + \exp(\bar{\alpha}_{kk})}, \tag{5}$$

where $\mathcal{N}_I(k)$ and $\mathcal{N}_U(k)$ denote the set of items that are part of bundle $k$ and the set of users who interacted with bundle $k$, respectively, and $\alpha_{kk}$ is a self-influence value of bundle $k$. Further, using the attention values, and the embeddings from the items that are part of the bundle $k$ and the interacted users, we obtain the bundle representation ($b'_k$) as follows:

$$b'_k = a(\alpha_{kk} W_B b_k + \sum_{u \in \mathcal{N}_U(k)} \alpha_{ku} W_U p_u + \sum_{i \in \mathcal{N}_I(k)} \alpha_{ki} W_I q_i). \tag{6}$$

In addition, it has been empirically shown that using multiple attention heads captures different aspects of the influence between entities [21]. Thus, with multiple attention heads ($H$), we obtain the bundle representation $b'_k$ as

$$b'_k = \Big\|_{h=1}^{H} a\Big(\alpha_{kk}^h W_B^h b_k + \sum_{u \in \mathcal{N}_U(k)} \alpha_{ku}^h W_U^h p_u + \sum_{i \in \mathcal{N}_I(k)} \alpha_{ki}^h W_I^h q_i\Big). \tag{7}$$

Similarly, we obtain representations for users ($p'_u$) and items ($q'_i$) by aggregating the embeddings from the interacted entities. To take into account higher-order relationships among the entities, this procedure is followed in multiple layers by having $b'_k$, $p'_u$ and $q'_i$ as input to the consecutive layers as shown in Fig. 1. Let $b_k^L$, $p_u^L$ and $q_i^L$ denote the embeddings of user $u$, item $i$ and bundle $k$ at layer $L$, respectively. Let $\pi(\cdot, \cdot)$ be a feedforward neural network, the probability that user $u$ will interact with bundle $k$ (denoted by $\hat{r}_{uk}^{UB}$) is obtained as follows:

$$\hat{r}_{uk}^{UB} = \sigma(w \cdot \psi_{uk}) \quad \text{where} \quad \psi_{uk} = \pi(p_u^L, b_k^L). \tag{8}$$

Here, $w$ is a weight vector, $\sigma(\cdot)$ denotes the sigmoid function and $\psi_{uk}$ denotes the resultant embedding of user-bundle interaction $(u, k)$ (See Fig. 1).

Further, we calculate the similarity values between bundle $k$ and item $i$, and user $u$ and item $i$ as

$$y_{ki}^{BI} = \phi(b_k^L, q_i^L) \quad \text{and} \quad y_{ui}^{UI} = \phi(p_u^L, q_i^L), \tag{9}$$

where $\phi(\cdot, \cdot)$ denotes a similarity function. For simplicity, we define $\phi(p_u^L, q_i^L) = p_u^L \cdot q_i^L$ and $\phi(b_k^L, q_i^L) = b_k^L \cdot q_i^L$, and $y_{ki}$ and $y_{ui}$ denote the respective similarity values.

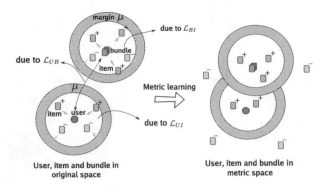

**Fig. 2.** The illustration of metric learning process.

**Loss Function via Metric-Learning Approach.** We leverage metric learning approach in constructing the loss function. The overall loss function is given as:

$$\min_{\mathcal{W}} \ \mathcal{L}(\mathcal{W}) = \mathcal{L}_{UI} + \mathcal{L}_{UB} + \mathcal{L}_{BI} + \lambda \, \mathcal{R}(\mathcal{W})$$

$$\text{where } \mathcal{L}_{UB} = - \sum_{(u,k) \in \mathcal{D}_{UB}} r_{uk} \ln \hat{r}_{uk} + (1 - r_{uk}) \ln(1 - \hat{r}_{uk})$$

$$\mathcal{L}_{UI} = \sum_{(u,i,i') \in \mathcal{D}_{UI}} s_{ui}[\mu - y_{ui}^{UI} + y_{ui'}^{UI}]_{+} \text{ and } \mathcal{L}_{BI} = \sum_{(k,i,i') \in \mathcal{D}_{BI}} s_{ki}[\mu - y_{ki}^{BI} + y_{ki'}^{BI}]_{+}$$

$$(10)$$

The overall loss function $\mathcal{L}(\cdot)$ consists of three individual loss functions – $\mathcal{L}_{UB}, \mathcal{L}_{UI}, \mathcal{L}_{BI}$, and a regularizer $\mathcal{R}(\cdot)$. Here, $\mathcal{W}$ consists of all the model parameters including embedding matrices, $\ln(\cdot)$ denotes natural logarithm, $[x]_{+}$ is defined as $max(0, x)$, $\lambda$ denotes non-negative hyperparameter, $\mu$ denotes margin, $s_{ki} = \ln(rank_m(k, i) + 1)$ where $rank_m(k, i)$ denotes the rank of item $i$ in bundle $k$ for a given metric $m$ [8]. Further, $\mathcal{L}_{UB}$ represents the loss between the user and bundle interactions. Besides, inspired by [8], leveraging the metric learning approach in defining $\mathcal{L}_{UI}$ and $\mathcal{L}_{BI}$, we pull the items that interacted with the users and bundles close to each other and push irrelevant items far apart with respect to given margin $\mu$. This process is illustrated in Fig. 2. We adopt negative sampling strategy [16] and Adam optimizer, and follow alternating minimization procedure in minimizing the losses $\mathcal{L}_{UB}, \mathcal{L}_{UI}$ and $\mathcal{L}_{BI}$. Here, $\mathcal{D}_{UB} = \mathcal{D}_{UB}^{+} \cup \mathcal{D}_{UB}^{-}$ where $\mathcal{D}_{UB}^{+} := \{(u, k) \in \Omega\}$ and $\mathcal{D}_{UB}^{-} \subset \{(u, k') \notin \Omega\}$, $\mathcal{D}_{BI} := \{(k, i, i') : (k, i) \in \Omega_{BI} \text{ and } (k, i') \notin \Omega_{BI}\}$ and $\mathcal{D}_{UI} := \{(u, i, i') : (u, i) \in \Omega_{UI} \text{ and } (u, i') \notin \Omega_{UI}\}$.

### 3.2 Personalized Bundle Generation

**Definition 3. *Submodular Function* [11].** *Let $V$ be a set, $\mathcal{F} : 2^V \to \mathbb{R}$ be a set function, $\Delta_{\mathcal{F}}(v/S) := \mathcal{F}(S \cup \{v\}) - \mathcal{F}(S)$ be discrete derivative of $\mathcal{F}$ with respect to $v$ where $S \subseteq V$ and $v \in V$.*

*A function $\mathcal{F} : 2^V \to \mathbb{R}$ is submodular, if $\forall A, B \subseteq V$ with $A \subseteq B, \forall v \in V$ when $v \in V \setminus B$, the following holds*

$$\Delta_{\mathcal{F}}(v/A) \geq \Delta_{\mathcal{F}}(v/B)$$

*Equivalently,* $\quad \mathcal{F}(A \cap B) + \mathcal{F}(A \cup B) \leq \mathcal{F}(A) + \mathcal{F}(B)$ $\qquad$ (11)

**Definition 4. Monotonicity.** *A set function $\mathcal{F}(\cdot)$ is called monotone non-decreasing if $\forall A \subseteq B, \mathcal{F}(A) \leq \mathcal{F}(B)$.*

For the problem of maximizing a monotone submodular function subject to a cardinality constraint, greedy algorithm gives $(1 - 1/e)$ approximation solution to the optimality [11].

In our setting, we construct $M$ bundles for each user with $N$ items in each bundle. We use $\bar{\mathcal{B}}_u$ to denote a set of bundles connected to user $u$. Further, due to the nature of the bundle generation task, we want to incorporate the following intuitions – new items should be added to bundles based on previously added items, and bundle characteristics should be defined according to items that it consists of. Therefore, our objective function is designed as follows:

$$
\begin{aligned}
\mathcal{I}^* &= \underset{\mathcal{I}'_{uk} \subseteq \mathcal{I} \text{ and } |\mathcal{I}'_{uk}| \leq N, \forall k \in \bar{\mathcal{B}}_u, \forall u \in \mathcal{U}}{\arg\max} \; \mathcal{F}(\{\mathcal{I}'_{uk}\}_{u,k}) \\
&= \underset{\mathcal{I}'_{uk} \subseteq \mathcal{I} \text{ and } |\mathcal{I}'_{uk}| \leq N, \forall k \in \bar{\mathcal{B}}_u, \forall u \in \mathcal{U}}{\arg\max} \sum_{u \in \bar{\mathcal{U}}} \sum_{k \in \bar{\mathcal{B}}_u} F(\mathcal{I}'_{uk}) \\
&= \underset{\mathcal{I}'_{uk} \subseteq \mathcal{I} \text{ and } |\mathcal{I}'_{uk}| \leq N, \forall k \in \bar{\mathcal{B}}_u, \forall u \in \mathcal{U}}{\arg\max} \sum_{u \in \bar{\mathcal{U}}} \sum_{k \in \bar{\mathcal{B}}_u} \sqrt{G(u, k, \mathcal{I}'_{uk})} \\
&= \underset{\mathcal{I}'_{uk} \subseteq \mathcal{I} \text{ and } |\mathcal{I}'_{uk}| \leq N, \forall k \in \bar{\mathcal{B}}_u, \forall u \in \mathcal{U}}{\arg\max} \sum_{u \in \bar{\mathcal{U}}} \sum_{k \in \bar{\mathcal{B}}_u} \sqrt{\sum_{i \in \mathcal{I}'_{uk}} g(u, k, i)}
\end{aligned}
$$

(12)

where $g(u, k, i) = \sigma(p_u^L \cdot q_i^L) + \sigma(b_k^L \cdot q_i^L)$ is the scoring function and $\mathcal{I}^*$ denotes the collection of the sets of items that are assigned to bundles $\{\bar{\mathcal{B}}_u\}_{u=1}^{|\mathcal{U}|}$.

**Greedy Bundle Generation Algorithm.** The bundle generation task is a subset selection problem, hence, it is NP-hard. However, the above objective is a non-decreasing non-negative concave function. As a consequence, it satisfies the monotone and submodularity criteria. Therefore, we devise a greedy strategy to maximize the above objective and it is given in Algorithms 1 and 2. We first create $M$ bundle nodes for each user and connect them with the corresponding users. We do warm-start training where we learn model parameters for $T_{ws}$ epochs. Then based on the learned parameters $\mathcal{W}$ and scoring function $g(\cdot, \cdot, \cdot)$, we select $M$ items, one for each bundle node, independently from the set of items $\mathcal{I}$ for each user and connect them with the newly added bundle nodes, accordingly. This process is continued after every $\tau$ epochs until we add $N$ items for each bundle. From the updated graph, one can recommend newly constructed bundles for each user using $\bar{\mathcal{B}}_u$.

---

**Algorithm 1:** BUNDLEGENERATION

---

**Input:** graph $\mathcal{G}(\mathcal{V}, \mathcal{E})$, warm-start epochs $T_{ws}$, updation step $\tau$, items
for each bundle to be added $N$

1  //warm-start
2  Initialize $\mathcal{W}$, set bundle generation epochs $T_{bg} = \tau * N$
3  **for** $t = 1 \rightarrow T_{ws}$ **do**
4       $\mathcal{W}^{new} \leftarrow \mathcal{W} - \eta_{ws} \frac{\partial \mathcal{L}(\mathcal{W}, \mathcal{G})}{\partial \mathcal{W}}$     $//\eta_{ws}$ is a learning rate
5       $\mathcal{W} \leftarrow \mathcal{W}^{new}$

6  //bundle generation
7  Update $\mathcal{G}(\mathcal{V}, \mathcal{E})$ with $M$ bundle nodes (denoted by set $\bar{\mathcal{B}}_u$) for each user
and connect them accordingly
8  $\mathcal{V} \leftarrow \cup_{u \in \mathcal{U}} \bar{\mathcal{B}}_u \cup \mathcal{V}$
9  **for** $t = 1 \rightarrow T_{bg}$ **do**
10       $\mathcal{W}^{new} \leftarrow \mathcal{W} - \eta_{bg} \frac{\partial \mathcal{L}(\mathcal{W}, \mathcal{G})}{\partial \mathcal{W}}$     $//\eta_{bg}$ is a learning rate
11       $\mathcal{W} \leftarrow \mathcal{W}^{new}$
12       **if** $(t+1) \% \tau == 0$ **then**
13           $\mathcal{G} \leftarrow$ UpdateGraph$(\mathcal{G}, \{\bar{\mathcal{B}}_u\}_u)$

14  **return** $\mathcal{G}$

---

# 4  Experiments

In this section, we describe the several experiments conducted, in view of addressing the following research questions:

**RQ1** How does our proposed model – GRAM-SMOT perform as compared to other state-of-the-art bundle recommendation models on real-world datasets? (Sect. 4.2.1)

**RQ2** Does the use of the metric learning strategy for bundle recommendation help in improving the performance? (Sect. 4.2.2)

**RQ3** Does learning user, item, and bundle representations through graph attention mechanism help in improving performance? (Sect. 4.2.3)

**RQ4** How does our proposed bundle generation approach via sub-modular optimization work in comparison with baseline approaches? (Sect. 4.4)

First, we discuss the experimental setting and then the experimental results.

## 4.1  Experimental Settings

**Datasets.** To evaluate the performance of our model, we conduct experiments on two real-world bundle recommendation datasets – YouShu[5] and NetEase[6]. In both the datasets, users have interactions with individual items as well as

---

[5] https://github.com/yliusysu/dam/tree/master/data/Youshu.
[6] This dataset was shared by the authors of the paper [3].

---

**Algorithm 2:** UPDATEGRAPH

**Input:** graph $\mathcal{G}(\mathcal{V}, \mathcal{E}), \{\bar{\mathcal{B}}_u\}_u$

1  **for** $u \in \mathcal{U}$ **do**
2      **for** $k \in \bar{\mathcal{B}}_u$ **do**
3          $\rho' = 0$
4          **for** $i \in \mathcal{I}$ **do**
5              $\rho = g(u, k, i)$  $//g(\cdot, \cdot, \cdot)$ is a scoring function
6              **if** $\rho > \rho'$ **then**
7                  $(\rho', i') \leftarrow (\rho, i)$
8      $\mathcal{E} \leftarrow \mathcal{E} \cup (k, i')$
9  **return** $\mathcal{G}$

---

**Table 1.** The statistics of the datasets.

| Dataset | YouShu | NetEase |
|---|---|---|
| # Users | 8,039 | 18,528 |
| # Bundles | 4,771 | 22,864 |
| # Items | 32,770 | 123,628 |
| # User-Bundle interactions | 51,377 | 302,303 |
| # User-Item interactions | 138,515 | 1,128,065 |
| # Bundle-Item interactions | 176,667 | 1,778,838 |
| Avg. Bundle interactions | 6.39 | 16.32 |
| Avg. Item interactions | 17.23 | 60.88 |
| Avg. Bundle size | 37.03 | 77.80 |
| User-Item density | 0.05% | 0.05% |
| User-Bundle density | 0.13% | 0.07% |

bundles. Further, bundles are constructed from any subset of existing items. The statistics of the datasets are given in Table 1.

***Youshu.*** This dataset was crawled from a Chinese book review website[7] by Cao *et al.* [4]. Here, users create a list of books they like under different genres. Each user has interacted with at least 10 bundles and 10 items. Further, each bundle consists of at least 10 items and each item appears in at least 5 bundles.

***NetEase.*** This dataset was crawled from NetEase Cloud Music website[8] by Chen *et al.* [3]. Here, each user in the dataset has interacted with at least 10 bundles and 10 items. Further, each bundle consists of at least 5 items and each item appears in at least 5 bundles.

---

[7] http://www.yousuu.com/.
[8] https://music.163.com.

**Evaluation Procedure.** We keep the experimental settings same as those followed in [4]. That is, we adopt *leave-one-out* strategy – we split the user-bundle part of the dataset into training, validation and test sets where one randomly selected bundle for each user are held-out for validation and test set. Here, user-item and bundle-item interactions are included as a part of the training set. Note that these user-item interactions do not include the interactions of users and items through bundles. Since it is time-consuming to rank all the bundles for each user, we randomly select 99 bundles that are not interacted with the users as negative samples to compare with the hold-out bundle. This procedure is repeated five times to get five different splits. We report the mean and standard deviation across these splits as the final results.

To compare the performance of the proposed model against the other models, following [4], we employ two well-known top-N recommendation metrics – Recall (Recall@N) and Mean Average Precision (MAP@N).

**Comparison Models.** We compare our model with the following models.

- **DAM** [4] is a deep attention-based multi-task model that aggregates the representations of items in a bundle using a factorized attention network. Further, it models user-item and user-bundle interactions jointly by a multi-task approach to get better user and bundle representations.
- **EFM** [3] (embedding factorization model) jointly learns the representations for users, items and bundles from the user-item, user-bundle and bundle-item interactions. This is an extension to the matrix factorization (MF) model, and devises an embedding based approach to incorporates item-item co-occurrence for improving bundle recommendation.
- **BR** [19] is a two-stage approach. In the first stage, it learns the user and item representations from the user-item matrix using pairwise loss function. In the second stage, it learns a model to appropriately combine the item representations using user-bundle and bundle-item interactions.
- **NCF** [7] is a state-of-the-art model for top-N recommendation tasks. It learns the user and bundle representations using MF and multi-layer neural networks combined to exploit the advantages of both the worlds.
- **BPR** [20] is a standard baseline for top-N recommendation tasks. It leverages pairwise loss function during optimization.

**Hyperparameter Setting and Reproducibility.** We implement our model using PyTorch 1.4. We tune the hyperparameters using a grid search on the validation set, and report the corresponding test set performance. Here, two-layer graph attention network is used with the number of attention head in the first and second layers is set to 4 and 2, and the number of nodes in the attention heads is set to 64 and 32 for YouShu, and 16 and 16 for NetEase, respectively. From the performance on the validation set, we set the number of training mini-batch size to 2048 for YouShu and 256 for NetEase, dropout on interaction layer to 0.3 and dropout on attention head to 0.6. Further, we use Adam optimizer with the learning rate set to 0.0007, Xavier initialization, and negative sampling strategy for the loss function corresponding to user-item interactions ($\mathcal{L}_{UI}$) with the number of negative samples set to 7. We borrow the results for the comparison models from [4].

## 4.2 Results and Discussion

### 4.2.1 Performance Comparison

We compare the performance of the proposed model – GRAM-SMOT over the other models in Table 2. As we can see from Table 2, GRAM-SMOT outperforms the other models significantly. We believe that, this is possibly due to the following reasons. When learning the representations, our model exploits the higher-order relationships between the entities (users, items and bundles) efficiently using a graph neural network-based technique. In addition, the model leverages graph attention mechanisms to learn the influence values of items that define the bundles. This is important because not all the items in bundles may have equal influence when it comes to the construction of the bundles. The same argument can be extended to the interactions between users and items and users and bundles. This shows the effectiveness of our model against other models for the top-N bundle recommendation task.

**Table 2.** Overall performance of different models on two real-world datasets – NetEase and YouShu (given in **Recall@5** and **MAP@5**) for top-N bundle recommendation. Here, the values in boldface indicate the best performance and 'Improvement' indicates the performance improvement over the next best model (shown in italics).

| Model | YouShu | | NetEase | |
|---|---|---|---|---|
| | Recall@5 | MAP@5 | Recall@5 | MAP@5 |
| BPR [20] | 0.5362 | 0.3608 | 0.3392 | 0.2018 |
| NCF [7] | 0.5715 | 0.3797 | 0.3534 | 0.2076 |
| BR [19] | 0.5752 | 0.3748 | 0.3318 | 0.1966 |
| EFM [3] | 0.5930 | 0.3903 | 0.3654 | 0.2140 |
| DAM [4] | *0.6109* | *0.4121* | *0.4016* | *0.2412* |
| GRAM-SMOT (ours) | **0.6704** | **0.4659** | **0.5493** | **0.3484** |
| | ± 0.0089 | ± 0.0071 | ± 0.0115 | ± 0.0098 |
| Improvement | 9.74 % | 13.06 % | 36.78 % | 44.44 % |

### 4.2.2 Ablation Study

In this section, we analyze the performance of various ablated versions of GRAM-SMOT. Here, we start with the graph attention framework devised for only user-bundle interaction data. Then, we incorporate user-item and bundle-item interaction data via a metric-learning approach. The results are illustrated in Fig. 3. From the experiments, we observe that including the knowledge obtained from

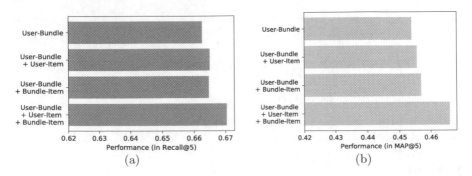

**Fig. 3.** Performance comparison of different ablated versions of GRAM-SMOT on YouShu dataset. Here, results obtained when different information incorporated via metric-learning approach are given in (a) Recall@5 and (b) MAP@5, respectively. We conduct paired $t$-test and improvement here are statistically significant with $p < .01$. Refer to Sect. 4.2.2 for more details.

user-item and bundle-item interactions via a metric-learning approach indeed improves the performance of our model. Further, we also observe that user-item and bundle-item interactions play a complementary role and removing either or both of them degrades performance drastically.

### 4.2.3   Graph Attention Mechanism

In this section, we conduct experiments to test whether devising the model based on graph attention mechanism helps in improving the performance. For this, we design a model based on graph convolution network (GCN) [9] by keeping all the other settings the same. Note that, the GCN framework aggregates information from connected neighbors equally and it does not learn the influence values. The comparison results are illustrated in Fig. 4(a). We observe from the figure that learning influence values helps in improving performance. From this, we can conclude that, for the top-N bundle recommendation task, it is essential to learn the influence values between the entities.

Figure 4(b) shows the performance of GRAM-SMOT with respect to the number of activation functions used. It demonstrates the improvement in the performance as the number of activation functions increased in the attention heads. Further, values of the different loss functions – $\mathcal{L}_{UB}, \mathcal{L}_{UI}$ and $\mathcal{L}_{BI}$, and validation error (in Recall@5 and MAP@5) with respect to the number of epochs are shown in Fig. 4(c) and (d).

### 4.3   Bundle Generation

Here, we present the experimental setting for the bundle generation task and then we will analyze the performance of our model.

**Evaluation Procedure for Bundle Generation.** Evaluating generated bundles offline is challenging due to the following reasons: (1) Human evaluation is

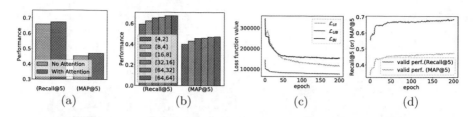

**Fig. 4.** (a) Performance of GRAM-SMOT via graph neural network without and with attention mechanism on YouShu dataset given in Recall@5 and MAP@5. (b) Performance of GRAM-SMOT with respect to different number of hidden units used in attention heads on YouShu dataset. (c) Loss function values – $\mathcal{L}_{UB}, \mathcal{L}_{UI}$ and $\mathcal{L}_{BI}$, and (d) performance of GRAM-SMOT on validation set in Recall@5 and MAP@5 against the number of epochs, respectively on YouShu dataset. Refer Sect. 4.2.3 for more details.

difficult because it is not possible to objectively determine whether generated bundles are valid or not and subjective evaluation is subject to bias. (2) As the number of available ground truth bundles is significantly less in number when compared to the number of possible bundles (the search space is doubly exponential), it is less likely that the generated bundles may intersect with the existing ground truth bundles. In order to tackle the above challenges and evaluate the generated bundles efficiently, we propose the following procedure.

First, we remove $\gamma_+$ (for example, 10) items from each bundle as positive samples and keep them aside for evaluation. Second, for each user, we select $\zeta$ (for example, 5) bundles for evaluation. Here, along with the positive items from each bundle, we sample and add $\gamma_-$ (for example, 90) negative items that are not part of the respective bundles. In this way, we construct candidate sets of items for bundles to be evaluated. Further, we evaluate the performance of the bundle generation algorithms based on how many positive items (from $\gamma_+$) that are included in the top-N list of items for each constructed bundle. And, this is repeated for all the users. We use the same top-N recommendation metrics – Recall@N and MAP@N for reporting performance.

**Comparison Baselines.** We compare our bundle generation approach with the following baselines:

- **Random.** This approach selects items randomly for the bundles from the candidate set of items.
- **BR** is a greedy approach proposed in [19]. Based on a defined scoring function, it randomly selects a set of items and finds the overall score. It repeats the above process for a certain number of epochs. In the end, it keeps the bundle with items having the best cumulative score as a newly constructed bundle.

## 4.4    Results and Discussion on Bundle Generation

We present the performance results of the proposed algorithm along with the baselines on bundle generation in Table 3. As we can see from the table, the

**Table 3.** Performance of GRAM-SMOT along with baselines on real-world datasets – YouShu and NetEase (given in **Recall@5** and **MAP@5**) for bundle generation task. Here, the values in boldface indicate the best performance. We conduct paired $t$-test and improvements here are statistically significant with $p < .01$.

| Model | YouShu | | NetEase | |
|---|---|---|---|---|
| | Recall@5 | MAP@5 | Recall@5 | MAP@5 |
| Random | 0.0507 | 0.0590 | 0.0498 | 0.0537 |
| | ± 0.0216 | ± 0.0249 | ± 0.0277 | ± 0.0298 |
| BR [19] | 0.1241 | 0.1297 | 0.1410 | 0.1462 |
| | ± 0.0188 | ± 0.0197 | ± 0.0169 | ± 0.0186 |
| GRAM-SMOT (ours) | **0.2086** | **0.2178** | **0.2314** | **0.2391** |
| | ± 0.0138 | ± 0.0153 | ± 0.0211 | ± 0.0240 |
| Improvement | 68.09% | 67.63% | 64.11% | 63.54% |

proposed approach outperforms the baselines. The improvement in performance could be explained by the following intuition – when a bundle is constructed, the initial items that are included are important as they determine the other items to be included. This, along with the use of submodular scoring function and graph attention mechanism for bundle generations, helps in generating appropriate bundles.

## 5    Related Work

**Deep Neural Networks for Recommendation Systems.** In the past decade, models in recommendation systems were mostly dominated by matrix factorization-based approaches and extensions thereof [10,20]. Followed by this, due to its ability to model the non-linear relationship between users and items efficiently, a surge of models based on deep neural networks have been proposed for recommendation systems [27]. However, they are not intentionally modeled for graph-structured data. As a consequence, they do not incorporate the information obtained from multi-hop neighborhoods which is essential for bundle recommendation tasks.

**Graph Neural Networks for Recommendation Systems.** Owing to its ability to model graph-structured data efficiently, several graph neural network frameworks have been adopted to recommendation systems [9,21]. For instance, graph convolutional networks [9] have been adopted for modeling item recommendation [24], social recommendation [6] and session-based recommendation [25]. Further, [23] leverages graph attention networks to incorporate useful information from the knowledge graph into user and item embeddings. However, to the best of our knowledge, there does not exist any work that models the bundle recommendation problem using the graph neural network framework.

**Next-Basket Recommendation.** Another line of research that is related to bundle recommendation is next-basket recommendations [14,15,26]. For instance, Li *et al.* [15] design a basket-sensitive random walk strategy on user-product bipartite graph to perform next-basket recommendations. In [26], Yu *et al.* propose a recurrent neural network-based model called the dynamic recurrent basket model for next basket recommendations. Further, a basket sensitive correlation network was proposed in [14]. However, the above approaches are mainly based on co-purchased item history and they are not personalized. Hence, the recommendations may not align with the individual interests of users.

**Personalized Bundle Recommendation.** Due to its industrial importance, personalized bundle recommendation is receiving increasing attention in recent years. In particular, Pathak *et al.* propose Bundle Ranking (BR) – a two-stage matrix factorization based model in [19]. First, it learns the embeddings for users and items through the user-item interaction matrix. Further, it utilizes this information along with user-bundle and bundle-item information to perform bundle recommendation. Cao *et al.* [3] propose a joint learning framework for the item and bundle recommendation using embedding factorization models. They extend traditional matrix factorization to incorporate knowledge coming from item-item co-occurrence into the embedding-based approach. Chen *et al.* propose a deep attentive multi-task model in [4]. Nevertheless, as discussed earlier, these models do not utilize higher-order relationships amongst entities.

**Bundle Generation.** Only minimal attention has been given on the aspect of generating novel personalized bundles for users. In particular, an attempt has been made in [19] to generate new bundles. They use a simple greedy approach to select good bundles based on a set of randomly selected bundles. Notably, Bai *et al.* [2] propose bundle generation networks (BGN), which use a structured determinantal point process (SDPP) based approach to generate new bundles. They attempt to integrate SDPP with a sequence-to-sequence neural network model to construct new bundles. However, since BGN requires side information to generate new bundles, it does not apply to our problem setting. Besides, the above approaches neither consider the higher-order relationships among entities nor learn the influence of items in defining bundles which are crucial for bundle recommendation tasks.

**Submodular Optimization.** Due to its theoretical guarantees and the existence of a greedy algorithm that provides better approximation solutions [1, 11,17], submodular functions are adopted in many real-world machine learning problems [12]. In the past, submodular functions have been used in the dictionary selection problem [5], natural language processing [13], robust graph representation learning [22], and group recommendations [18]. For instance, Das *et al.* study submodular functions in the context of feature selection and sparse approximation via dictionary selection. Further, Kumar *et al.* [13] formulate the paraphrase generation problem in terms of monotone submodular function maximization. However, to the best of our knowledge, GRAM-SMOT is the first approach that proposes a solution to personalized bundle recommendation via submodular function maximization.

## 6    Conclusion

In this work, we presented graph neural network-based framework – GRAM-SMOT for top-N bundle recommendation. We demonstrated the importance of leveraging graph attention mechanism and metric learning approach empirically using real-world datasets – YouShu and NetEase. Besides, we also introduced a submodular function maximization strategy to bundle generation tasks. The experimental results on various bundle recommendation tasks demonstrated that our proposed model outperforms existing state-of-the-art models.

In the future, we plan to devise an efficient strategy to generate potential candidate items for generating bundles. Further, we want to extend GRAM-SMOT for package-to-group recommendation tasks where the problem setting treats the subsets of users as different groups.

## References

1. Bach, F., et al.: Learning with submodular functions: a convex optimization perspective. Found. Trends® Mach. Learn. **6**(2–3), 145–373 (2013)
2. Bai, J., et al.: Personalized bundle list recommendation. In: WWW, pp. 60–71. ACM (2019)
3. Cao, D., Nie, L., He, X., Wei, X., Zhu, S., Chua, T.S.: Embedding factorization models for jointly recommending items and user generated lists. In: SIGIR (2017)
4. Chen, L., Liu, Y., He, X., Gao, L., Zheng, Z.: Matching user with item set: collaborative bundle recommendation with deep attention network. In: IJCAI, pp. 2095–2101. AAAI Press (2019)
5. Das, A., Kempe, D.: Submodular meets spectral: greedy algorithms for subset selection, sparse approximation and dictionary selection. In: ICML (2011)
6. Fan, W., et al.: Graph neural networks for social recommendation. In: WWW, pp. 417–426 (2019)
7. He, X., Liao, L., Zhang, H., Nie, L., Hu, X., Chua, T.S.: Neural collaborative filtering. In: WWW, pp. 173–182 (2017)
8. Hsieh, C.K., Yang, L., Cui, Y., Lin, T.Y., Belongie, S., Estrin, D.: Collaborative metric learning. In: WWW, pp. 193–201. WWW (2017)
9. Kipf, T.N., Welling, M.: Semi-supervised classification with graph convolutional networks. In: ICLR (2017)
10. Koren, Y., Bell, R.: Advances in collaborative filtering. In: Ricci, F., Rokach, L., Shapira, B. (eds.) Recommender Syst. Handb., pp. 77–118. Springer, Boston, MA (2015). https://doi.org/10.1007/978-1-4899-7637-6_3
11. Krause, A., Golovin, D.: Submodular function maximization. Tractability **3**, 71–104 (2014)
12. Krause, A., Guestrin, C.: Submodularity and its applications in optimized information gathering. TIST **2**(4), 1–20 (2011)
13. Kumar, A., Bhattamishra, S., Bhandari, M., Talukdar, P.: Submodular optimization-based diverse paraphrasing and its effectiveness in data augmentation. In: NAACL, pp. 3609–3619 (2019)

14. Le, D.T., Lauw, H.W., Fang, Y.: Correlation-sensitive next-basket recommendation. In: IJCAI (2019)

15. Li, M., Dias, B.M., Jarman, I., El-Deredy, W., Lisboa, P.J.: Grocery shopping recommendations based on basket-sensitive random walk. In: SIGKDD (2009)

16. Mikolov, T., Sutskever, I., Chen, K., Corrado, G.S., Dean, J.: Distributed representations of words and phrases and their compositionality. In: NeurIPS (2013)

17. Nemhauser, G.L., Wolsey, L.A., Fisher, M.L.: An analysis of approximations for maximizing submodular set functions—i. Math. Prog. **14**(1), 265–294 (1978) https://doi.org/10.1007/BF01588971

18. Parambath, S.A.P., Vijayakumar, N., Chawla, S.: SAGA: a submodular greedy algorithm for group recommendation. In: AAAI (2018)

19. Pathak, A., Gupta, K., McAuley, J.: Generating and personalizing bundle recommendations on steam. In: SIGIR, pp. 1073–1076. ACM (2017)

20. Rendle, S., Freudenthaler, C., Gantner, Z., Schmidt-Thieme, L.: BPR: Bayesian personalized ranking from implicit feedback. In: UAI, pp. 452–461. AUAI (2009)

21. Veličković, P., Cucurull, G., Casanova, A., Romero, A., Liò, P., Bengio, Y.: Graph attention networks. In: ICLR (2018)

22. Wang, L., et al.: Learning robust representations with graph denoising policy network. In: ICDM, pp. 1378–1383. IEEE (2019)

23. Wang, X., He, X., Cao, Y., Liu, M., Chua, T.S.: KGAT: knowledge graph attention network for recommendation. In: SIGKDD, pp. 950–958 (2019)

24. Wang, X., He, X., Wang, M., Feng, F., Chua, T.S.: Neural graph collaborative filtering. In: SIGIR, pp. 165–174 (2019)

25. Wu, S., Tang, Y., Zhu, Y., Wang, L., Xie, X., Tan, T.: Session-based recommendation with graph neural networks. AAAI **33**, 346–353 (2019)

26. Yu, F., Liu, Q., Wu, S., Wang, L., Tan, T.: A dynamic recurrent model for next basket recommendation. In: SIGIR, pp. 729–732 (2016)

27. Zhang, S., Yao, L., Sun, A., Tay, Y.: Deep learning based recommender system: a survey and new perspectives. CSUR **52**(1), 5 (2019)

# Temporal Heterogeneous Interaction Graph Embedding for Next-Item Recommendation

Yugang Ji[1], MingYang Yin[2], Yuan Fang[3], Hongxia Yang[2], Xiangwei Wang[2], Tianrui Jia[1], and Chuan Shi[1,4(✉)]

[1] Beijing University of Posts and Telecommunications, Beijing, China
{jiyugang,jiatianrui,shichuan}@bupt.edu.cn
[2] Alibaba Group, Hangzhou, China
{hengyang.yin,yang.yhx,florian.wxw}@alibaba-inc.com
[3] Singapore Management University, Singapore, Singapore
yfang@smu.edu.sg
[4] Peng Cheng Laboratory, Shenzhen, China

**Abstract.** In the scenario of next-item recommendation, previous methods attempt to model user preferences by capturing the evolution of sequential interactions. However, their sequential expression is often limited, without modeling complex dynamics that short-term demands can often be influenced by long-term habits. Moreover, few of them take into account the heterogeneous types of interaction between users and items. In this paper, we model such complex data as a *Temporal Heterogeneous Interaction Graph* (THIG) and learn both user and item embeddings on THIGs to address next-item recommendation. The main challenges involve two aspects: the *complex dynamics* and *rich heterogeneity* of interactions. We propose *THIG Embedding* (THIGE) which models the complex dynamics so that evolving short-term demands are guided by long-term historical habits, and leverages the rich heterogeneity to express the latent relevance of different-typed preferences. Extensive experiments on real-world datasets demonstrate that THIGE consistently outperforms the state-of-the-art methods.

**Keywords:** Temporal heterogeneous interaction graph · Next-item recommendation · Short-term demands · Long-term habits

## 1 Introduction

With the prevalence of e-commerce, our shopping styles are revolutionized in recent years. By modeling historical user-item interactions, recommender systems play a fundamental role in e-commerce [9,18]. In particular, the task of next-item recommendation—to predict the item that a user will interact with at the next time instance—not only caters to the business requirement of e-commerce platforms, but also enhances the user experience.

© Springer Nature Switzerland AG 2021
F. Hutter et al. (Eds.): ECML PKDD 2020, LNAI 12459, pp. 314–329, 2021.
https://doi.org/10.1007/978-3-030-67664-3_19

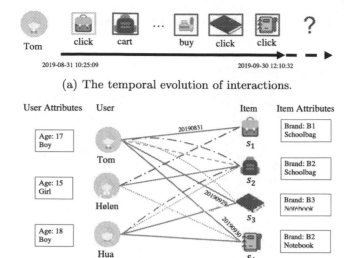

(a) The temporal evolution of interactions.

(b) An example of THIG.

**Fig. 1.** Toy example of next-item recommendation, from (a) a temporal sequence of interactions, to (b) a Temporal Heterogeneous Interaction Graph (THIG).

Earlier methods mainly exploit collaborative filtering (CF) [4,14], which models interactions without any temporal dynamics. However, temporal evolution often contributes significantly to the next-item recommendation. As shown in Fig. 1(a), Tom's current demand is more likely to be notebooks, rather than the general preference of bags which one would have concluded by analyzing his entire interaction history without temporal consideration. To capture such evolving demands of users, recurrent neural networks (RNN) [8,25] have been widely used by considering a sequence of interactions. While RNNs are only capable of modelling short-term preferences (e.g., demands of notebooks) from relatively recent interactions, capturing long-term preferences (e.g., preferred brands) from historical habits is also an important element of temporal dynamics [18]. However, existing methods usually model short- and long-term preferences independently, ignoring the role of habits in driving the current, evolving demands. Taking Fig. 1(a) as an example, when browsing similar items (e.g., two schoolbags), users prefer to click those with attributes they habitually care (e.g., the brands).

This presents the first research challenge: *How to effectively model the complex temporal dynamics, coupling both historical habits and evolving demands?* In this paper, we model historical habits as the long-term preferences, and the current, evolving demands as the short-term preferences. More importantly, we propose to guide demand learning with historical habits, and develop a habit-guided attention to tightly couple both long- and short-term preferences.

Another dimension overlooked by existing sequential models is the abundant heterogeneous structural information. Taking Fig. 1(b) as an example, there exists large-scale inter-linking between various users and items, e.g., Hua bought the carted $s_2$, which cannot be explicitly modeled by separate interaction sequences. More importantly, there are also rich user-item interactions of heterogeneous types, such as "click", "favorite", "cart" and "buy". As shown in Fig. 1(b), Hua prefers items of brand B2 for the interactions of "buy" and "cart", while $s_2$ is more popular to boys for the various interactions rather than a single click of Helen. While heterogeneous graphs [2,13,15,16] have been a *de facto* standard to model rich structural information and their representation learning have been studied extensively in heterogeneous network embedding and graph neural networks (GNN) [1,21,24], they ignore the complex temporal dynamics, treating the graph as a static snapshot. Furthermore, most of them treat the heterogeneous types of interaction independently, but in reality different types (e.g., clicks and buys) often express varying latent relevance w.r.t. each other. On the other hand, temporal graphs have been studied in some recent works in homogeneous settings [10,19] without modeling the rich heterogeneity.

This leads to the second research challenge: *How to make full use of the temporal heterogeneous interactions to model the preferences of different types?* Here we propose a Temporal Heterogeneous Interaction Graph (THIG) to model the heterogeneous interactions and the temporal dimension jointly. Compared with static graphs, THIGs can express the evolving preference of users and the changing popularity of items; compared with temporal graphs, THIGs can exploit rich heterogeneous factors that simultaneously contribute to preference learning. Particularly, we design a novel heterogeneous self-attention mechanism to distill the latent relevance and multifaceted preferences from multiple interactions.

Hinged on the above insights, we propose THIGE, a novel model for **T**emporal **H**eterogeneous **I**nteraction **G**raph **E**mbedding, to effectively learn user and item embeddings on THIGs for next-item recommendation. In THIGE, we first encode heterogeneous interactions with temporal information. Building upon the temporal encoding, we take into account the influence of long-term habits on short-term demands, and design a habit-guided attention mechanism to couple short- and long-term preferences. To fully exploit the rich heterogeneous interactions to enhance multifaceted preferences, we further capture the latent relevance of varying types of interaction via heterogeneous self-attention mechanisms.

We summarize the main contributions of this paper as follows.

- To our best knowledge, this work formulates and illustrates the first use of temporal heterogeneous interaction graphs for the problem of next-item recommendation. Different from previous sequential models which mainly focus on homogeneous sequences, here we fully utilize the structure information of multiple behaviors for item recommendation.
- We propose a novel model THIGE to couple long- and short-term preferences of heterogeneous nature, which fully exploits the temporal and heterogeneous interactions via habit-guided attention and heterogeneous self-

attention. Both the dependence of heterogeneous preferences and impact from historical habits to recent demands are effectively modeled.

- We perform extensive experiments on the public datasets Yelp and CloudTheme, and the industrial dataset UserBehavior. We compare THIGE against various state of the arts and obtain promising results.

## 2   Related Work

We introduce the related work in two main domains, one of which is graph embedding and the other is next-item recommendation.

**Graph Neural Networks.** Graph neural networks (GNN) are widely used for node representation on real world graphs including GCN [7], GraphSAGE [6] and GAT [20] to construct node embedding via neighborhood aggregation. Recently, taking the dynamic of links into considerations, Dyrep [19] and $M^2$DNE [10] split the graph into several snapshots to capture global evolution of local interests. Inspired by position embedding proposed in [13], it is promising to design the continuous-time function to generate time span-based temporal embedding. However, all these algorithms are proposed for homogeneous networks, which is not suitable to deal with graphs like THIGs in which nodes/edges are of multiple types. HetGCN [22] aggregates neighborhood information with meta-path guide random walks. HAN [21] proposes the hierarchical attention to guide aggregating heterogeneous neighborhood information. Furthermore, GATNE [1] combines heterogeneous information aggregated from different-typed neighborhoods via heterogeneous attention mechanisms. Furthermore, to fuse sequential information, MEIRec [5] captures evolution of users' same-typed interactions by LSTM. Unfortunately, this model is limited to the sequence length. How to model temporal heterogeneous graphs is challenging and meaningful [1].

**Next-Item Recommendation.** For dealing with next-item recommendation, sequence-based recommender systems are to understand the temporal dynamics between users and items [17]. The related works mainly focus on short-term interest learning. STAMP [9] captures the general and current interests by an attentive memory priority model. DIEN [25] respectively utilizes the classical and attentional GRUs on the interest extractor and evolving layers to capture short-term interest. Taking long-term preference into consideration, SHAN [23] designs the hierarchical attention to combine habits and demands modeling on hierarchical layers. M3R [18] mixes long-term, short-term and current interest modeling together to obtain the final interest. However, all of these sequential models cannot model the heterogeneous interactions. Le et al. [8] learn user preference from different-typed interactions by respectively modeling different-typed interactions with GRUs. Lv et al. [11] focus on capturing long-term preferences in different latent aspects like brands and categories. However, few of these models pay attention to the types of interactions within THIGs while different-typed interactions indicates various semantics.

# 3    Problem Formulation

Here we introduce the definition of THIGs and the problem of next-item recommendation on THIGs.

**Definition 1. *Temporal Heterogeneous Interaction Graph (THIG)*.** *A THIG is $\mathcal{G} = (\mathcal{V}, \mathcal{E}, \mathcal{T}, \mathcal{A}, \mathcal{O}, \mathcal{R}, \phi, \psi)$, where $\mathcal{V}$ is a set of nodes with types $\mathcal{O}$, $\mathcal{E}$ is the set of edges with types $\mathcal{R}$, $\mathcal{T}$ is the set of timestamps on edges, and $\mathcal{A}$ is the set of attributes on nodes. Moreover, $\phi : \mathcal{V} \to \mathcal{O}$ is the node type mapping function and $\psi : \mathcal{E} \to \mathcal{R}$ is the edge type mapping function. In THIGs, $|\mathcal{O}| \geq 2$ and $|\mathcal{R}| \geq 2$.*

As shown in Fig. 1(b), there are four types of interactions (edges), i.e., $\mathcal{R} = \{$click, favorite, cart, buy$\}$, between two types of nodes $\mathcal{O} = \{$user, item$\}$. A user may interact with the same item under multiple interactions like "click" and "buy" at different timestamps. Moreover, users and items contain their own features like age or brand. By modeling such heterogeneous interaction data with THIGs, richer semantics of dynamic interactions can be preserved for effective next-item recommendation.

**Definition 2. *Next-item recommendation on THIGs*.** *On a THIG, a user $u$ is associated with his/her historical interactions $\{(v_i, t_i, r_i) \mid 1 \leq i \leq n\}$ where the triple $(v_i, t_i, r_i)$ denotes that item $v_i$ is interacted under type $r_i$ at time $t_i$, for some $t_n < T$ such that $T$ is the current time. Similarly, an item $v$ is associated with its historical interactions $\{(u_j, t_j, r_j) \mid 1 \leq j \leq n\}$ where the triple $(u_j, t_j, r_j)$ denotes that item $u_j$ interact with item $v$ under type $r_j$ at time $t_j$ for some $t_n < T$. This task is to predict whether $u$ will interact with $v$ at the next time instance.*

For instance, as shown in Fig. 1(b), given that Tom has clicked or bought $S_1$, $S_2$, $S_3$ and $S_4$ before, our goal is to predict the next item he will interact with, based on each candidate item's interaction history. This problem is fundamental and meaningful in e-commerce platforms to understand both user and items simultaneously.

# 4    Proposed Approach: THIGE

In this section, we propose our model called THIGE. We begin with an overview, before zooming into the details.

## 4.1    Overview

The overall framework is shown in Fig. 2. Specifically, we divide the historical interactions of a user into long and short term based on their timestamps. For short-term preferences, we model users' sequences of recent interactions with gated recurrent units (GRU), to embed users' current demands $\boldsymbol{h}_u^{(S)}$. For long-term preference, we model users' long-term interactions with a heterogeneous

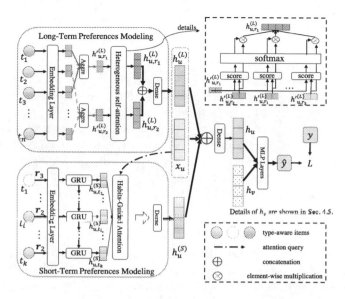

**Fig. 2.** Overall framework of user embedding in THIGE for next-item recommendation.

self-attention mechanism, to embed users' historical habits $\boldsymbol{h}_u^{(L)}$. Different from the decoupled combination (e.g., simple concatenation) of long- and short-term embeddings in previous methods [8,23], we propose to exploit the long-term historical habits to guide the learning of short-term demands using the habit-guided attention, which effectively captures the impact of habits on recent behaviors.

Note that Fig. 2 only shows the learning of user representations. For items, we do not distinguish their long- and short-term interactions, and only adopt a long-term model similar to that of users. The reason lies in the fact that there may be numerous users interacting with an item around a short period of time, and these users have no significant short-term sequential dependency.

## 4.2 Embedding Layer with Temporal Information

Each interacted item of a user is associated with not only attributes but also a timestamp. As shown in Fig. 2, the timestamps is in the form of $[t_1, t_2, \cdots, t_n]$.

Thus, the temporal embedding of an item $v$ consists of both a static and a temporal component. The static component $\boldsymbol{x}_v = \boldsymbol{W}_{\phi(v)} \boldsymbol{a}_v$, where the input vector $\boldsymbol{a}_v \in \mathbb{R}^{d_\phi(v)}$ encodes the attributes of $v$, $\boldsymbol{W}_{\phi(v)} \in \mathbb{R}^{d \times d_\phi(v)}$ denotes the latent projection, $d_\phi(v)$ and $d$ are the dimension of attributes and latent representation of $v$. Moreover, at time $t$, denoting $\Delta t$ as the time span before the current time $T$ and dividing the overall time span into $B$ buckets, the temporal component of $v$ is defined as $\boldsymbol{W}\xi(\Delta t)$, where $\xi(\Delta t) \in \mathbb{R}^B$ denotes the one-hot bucket representation of $\Delta t$, and $\boldsymbol{W} \in \mathbb{R}^{d_\mathcal{T} \times B}$ denotes the projection matrix and $d_\mathcal{T}$ is the output dimension. Thus, the temporal embedding of an item $v$ at time $t$ is

$$\boldsymbol{x}_{v,t} = [\boldsymbol{W}\xi(\Delta t) \oplus \boldsymbol{x}_v], \tag{1}$$

where $\oplus$ denotes concatenation. Similarly, we generate the static representation of a user $u$ as $\boldsymbol{x}_u \in \mathbb{R}^d$, and temporal representation of $u$ at time $t$ as $\boldsymbol{x}_{u,t}$.

To further consider the sequential evolution of heterogeneous interactions, we generate the $i^{th}$ interacted item embedding $\boldsymbol{x}_{v_i,t_i,r_i} = [\boldsymbol{x}_{v_i,t_i} \oplus \boldsymbol{r}_i]$ as the combination of the temporal embedding $\boldsymbol{x}_{v_i,t_i}$ and the corresponding type embedding $\boldsymbol{r}_i = \boldsymbol{W}_{\mathcal{R}} \boldsymbol{I}(r_i)$ where $\boldsymbol{I}(r_i)$ denotes the one-hot vector of $r_i$ with dimension $|\mathcal{R}|$, $\boldsymbol{W}_{\mathcal{R}} \in \mathbb{R}^{d_{\mathcal{R}} \times |\mathcal{R}|}$ is the projection matrix and $d_{\mathcal{R}}$ is the latent dimension. For long-term preference modeling, we input the temporal embedding into type-aware aggregators to distinguish preferences of different types.

### 4.3   Short-Term Preference with Habit-Guidance

Recent interactions of users usually indicate the evolving current demands. For instance, as shown in Fig. 1(a), Tom's current demand has been evolved from bags to notebooks. In order to model the short-term and evolving preferences, we adopt gated recurrent units (GRU) [3], which can capture the dependency of recent interactions. Consider a user $u$ here. Let his/her $k$ recent interactions be $\{(v_i, t_i, r_i) \mid 1 \le i \le k\}$, where $t_k$ is the most recent timestamp before the current time $T$. Subsequently, we encode the user preference at time $t_i$ as $\boldsymbol{h}_{u,t_i}^{(S)}$, using a GRU based on the embedding of interaction $(v_i, t_i, r_i)$, namely, $\boldsymbol{x}_{v_i,t_i,r_i}$, and his/her preference at $t_{i-1}$, as follows.

$$\boldsymbol{h}_{u,t_i}^{(S)} = \mathrm{GRU}(\boldsymbol{x}_{v_i,t_i,r_i}, \boldsymbol{h}_{u,t_{i-1}}^{(S)}), \quad \forall 1 < i \le k, \tag{2}$$

where $\boldsymbol{h}_{u,t_i}^{(S)} \in \mathbb{R}^d$. The time-dependent user embeddings $\{\boldsymbol{h}_{u,t_i}^{(S)} \mid 1 \le i \le k\}$ can be further aggregated to encode the current demand of user $u$.

However, the current and evolving demands of user are not only influenced by their recent transactions. Their long-term preferences, i.e., historical habits such as brands and lifestyle inclinations, often play a subtle but important role. Thus, we enhance the encoding of short-term preferences under the guidance of historical habits, in order to discover more fine-grained and personalized preferences. Specially, we propose a habit-guided attention mechanism to aggregate short-term user preferences, as follows.

$$\boldsymbol{h}_u^{(S)} = \sigma\left(W^{(S)} \cdot \sum_i a_{u,i} \boldsymbol{h}_{u,t_i}^{(S)} + b_s\right), \quad \forall 1 \le i \le k, \tag{3}$$

where $\boldsymbol{h}_u^{(S)} \in \mathbb{R}^d$ denotes the overall short-term preference of $u$, $W^{(S)} \in \mathbb{R}^{d \times d}$ denotes the projection matrix, $\sigma$ is the activation function and we adopt RELU here to ensure the non-linearity, $b_s$ is the bias, and $a_{u,i}$ is the habit-guided weight:

$$a_{u,i} = \frac{\exp\left([\boldsymbol{h}_u^{(L)} \oplus \boldsymbol{x}_u]^T \boldsymbol{W}_a \boldsymbol{h}_{u,t_i}^{(S)}\right)}{\sum_{j=1}^k \exp\left([\boldsymbol{h}_u^{(L)} \oplus \boldsymbol{x}_u]^T \boldsymbol{W}_a \boldsymbol{h}_{u,t_j}^{(S)}\right)}, \tag{4}$$

where $h_u^{(L)} \in \mathbb{R}^d$ is the long-term preference of $u$ which would have encoded the habits of $u$, and $W_a \in \mathbb{R}^{2d \times d}$ is a mapping to quantify the fine-grained relevance between the short-term preference of $u$ at different times, and the long-term habits of $u$. Therefore, how to encode the long-term habits $h_u^{(L)}$ in the context of heterogeneous interactions is the second key thesis of this work, as we will introduce next.

### 4.4  Long-Term Preference with Heterogeneous Interactions

Besides short-term preferences to encode current and evolving demands, users also exhibit long-term preferences to express personal and historical habits. In particular, there exist multiple types of heterogeneous interactions which have different relevance w.r.t. each other. For example, a "click" is more relevant to a "cart" or "buy" on the same item or similar items; "favorite" could be less relevant to "cart" or "buy", but is closely tied to the user's brand or lifestyle choices in the long run. Thus, different types of interactions entail both latent relevance and multifaceted preferences.

Thus, our goal is to fully encode the latent, fine-grained relevance of multi-faceted long-term preferences.

Consider a user $u$, and his/her long-term interactions $\{(v_i, t_i, r_i) \mid 1 \leq i \leq n\}$ where $n \gg k$ ($k$ is the count of recent interactions in short-term modeling). To differentiate the explicit interaction types, we first aggregate the embeddings of items which the user have interacted with under a specific type $r$:

$$h_{u,r}'^{(L)} = \sigma \left( W_r \cdot \text{aggre}(\{ x_{v_i, t_i} \mid 1 \leq i \leq n, r_i = r \}) \right), \tag{5}$$

where $h_{u,r}'^{(L)} \in \mathbb{R}^d$ is the type-$r$ long-term preferences of user $u$, $W_r \in \mathbb{R}^{d \times (d_T + d)}$ is the type-$r$ learnable mapping, $\text{aggre}(\cdot)$ is an aggregator, and we utilize mean-pooling here.

While we can simply sum or concatenate the type-specific long-term preferences into an overall representation, there exists latent relevance among the types (e.g., "click" and "buy"), and latent multifaceted preferences (e.g., brands and lifestyles). In this paper, we design a heterogeneous self-attention mechanism to express the latent relevance of different-typed interactions and long-term multi-faceted preferences. By concatenating all long-term preferences of different types as $H_u^{(L)} = \oplus_{r \in \mathcal{R}} h_{u,r}'^{(L)}$ with size $d$-by-$|\mathcal{R}|$, we first formulate the self-attention to capture the latent relevant of heterogeneous types in $\mathcal{R}$ w.r.t. each other:

$$h_{u,r}^{(L)} = \sum_{r' \in \mathcal{R}} \left( \frac{\exp \left( Q_{u,r}^T K_{u,r'} / \sqrt{d_a} \right)}{\sum_{r'' \in \mathcal{R}} \exp \left( Q_{u,r}^T K_{u,r''} / \sqrt{d_a} \right)} V_{u,r'} \right), \tag{6}$$

where $Q_u = W_Q H_u^{(L)}$, $K_u = W_K H_u^{(L)}$, $V_u = W_V H_u^{(L)}$, $W_Q, W_K \in \mathbb{R}^{d_a \times d}$ and $W_V \in \mathbb{R}^{d \times d}$ are the projection matrices, and $d_a$ is the dimension of keys and queries.

Next, to express multifaceted preferences, we adopt a multi-head approach to model latent, fine-grained facets. Specifically, the original embeddings of preferences are split into multi-heads and we adopt the self-attention for each head. The type-$r$ long-term preference is concatenated from the $h$ heads:

$$h_{u,r}^{(L)} = \oplus_{m=1}^{m=h} h_{u,r,m}^{(L)}, \tag{7}$$

where $h_{u,r,m}^{(L)}$ denotes the $m^{th}$ head based preference and there are $h$ heads. The overall long-term preference can also be derived by fusing different types in $\mathcal{R}$:

$$h_u^{(L)} = \sigma\left(W^{(L)}(\oplus_{r\in\mathcal{R}} h_{u,r}^{(L)}) + b_l\right), \tag{8}$$

where $W^{(L)} \in \mathbb{R}^{d\times|\mathcal{R}|d}$ and $b_l$ are the projection parameters. By now, both short- and long-term preferences haven been modeled. Taking the inherent attributes of users into consideration, the final representation of user $u$ is calculated by

$$h_u = \sigma(W_u[x_u \oplus h_u^{(S)} \oplus h_u^{(L)}] + b_u), \tag{9}$$

where $h_u \in \mathbb{R}^d$ will be used for next-item prediction, and $W_u \in \mathbb{R}^{d\times 3d}$ and $b_u$ are learnable parameters.

### 4.5   Preference Modeling of Items

The temporal interactions of an item is significantly different from those of a user. In practice, on a mass e-commerce platform, it is typical that many users interact with the same item around the same time constantly, without a meaningful sequential effect among different users. In other words, it is more reasonable to only model the general, long-term popularity of items. Thus, we model item representation $h_v^{(L)}$ similar to the long-term preference modeling of users in Eq. (8) with heterogeneous multi-head self-attention, and encode the item representation as follows:

$$h_v = \sigma(W_v[x_v \oplus h_v^{(L)}] + b_v), \tag{10}$$

where $h_v \in \mathbb{R}^d$ is the final representation of item $v$ for next-item prediction, and $W_v$ and $b_v$ are learnable parameters and $x_v$ is the attribute vector of item $v$.

### 4.6   Optimization Objective

To deal with next-item recommendation, we predict $\hat{y}_{u,v}$ between user $u$ and item $v$, indicating whether $u$ will interact with $v$ (under a given type) at the next time. Here we utilize a Multi-Layer Perception (MLP) [12]:

$$\hat{y}_{u,v} = \text{sigmoid}(\text{MLP}(h_u \oplus h_v)), \tag{11}$$

where $h_u$ and $h_v$ are the final representation of user $u$ and item $v$, respectively. Model parameters can be optimized with the following cross-entropy loss:

$$L = -\sum_{\langle u,v\rangle} (1 - y_{u,v})\log(1 - \hat{y}_{u,v}) + y_{u,v}\log(\hat{y}_{u,v}), \tag{12}$$

**Table 1.** Description of datasets.

| Dataset | Yelp | CloudTheme | UserBehavior |
|---|---|---|---|
| # User | 103,569 | 144,197 | 533,974 |
| # Item/Business | 133,502 | 272,334 | 4,152,242 |
| # Interaction | 1,889,132 | 1,143,567 | 122,451,055 |
| # Interaction type | 2 | 2 | 4 |
| # Training instance | 611,568 | 865,182 | 3,203,844 |
| (Training time span) | 5 years | 2 weeks | 1 weeks |
| # Test instance | 108,408 | 216,295 | 800,961 |
| (Test time span) | Next one quarter | Next day | Next day |

where $\langle u, v \rangle$ is a sample of user $u$ and item $v$, and $y_{u,v} \in \{0, 1\}$ is the ground truth of the sample. We also optimize the L2 regularization of latent parameters to ensure the robustness.

# 5   Experiments

In this section, we showcase the performances of our proposed THIGE[1] for next item-recommendation, and discuss the effectiveness of our design choices and key factors.

## 5.1   Datasets

We evaluate the empirical performance of THIGE for next-item recommendation on three real-world datasets including Yelp, CloudTheme and UserBehavior. The statistics are summarized in Table 1 and the details are introduced as follows.

- **Yelp**[2]: A public business dataset with two types of temporal interactions, namely, "review" and "tip" between users and businesses. Both users and businesses contain continuous and discrete features. We select interactions that happened before 14 Aug. 2019 as training data and the remaining as test data. For both training and test data, the last interacted business is labeled as positive instance, while five never interacted businesses of the same category are randomly sampled as negative instances.
- **CloudTheme**[3]: A public e-commerce dataset that records the click and purchase logs. Users and items are associated with embedding vectors given by the dataset. We select interactions happened before the last day as training data and the remaining as test data. For both training and test data, we treat the last interacted item as the positive instance and randomly sample five other items of the same theme as negative instances.

---

[1] The source is available at https://github.com/yuduo93/THIGE.
[2] https://www.yelp.com/dataset/documentation/main.
[3] https://tianchi.aliyun.com/dataset/dataDetail?dataId=9716.

– **UserBehavior**: An industrial dataset extracted from Taobao website, consisting of "click", "favorite", "cart" and "buy" interactions between users and items. For both training and test data, we utilize the actual feedback of users as labels—among the candidates displayed to users, we take the clicked item as the positive instance, and sample five other items from the remaining candidates as negative instances.

## 5.2   Baselines and Experimental Settings

We compare THIGE with six representative models and showcase the effectiveness of next-item recommendation and evaluate the effectiveness of our design choices. The baselines are listed as follows:

– **DIEN** [25] and **STAMP** [9] are two sequential models where the former is a hierarchical GRU to encode evolving interests and the latter is a short-term memory priority model to extract session-based interests;
– **SHAN** [23] and **M3R** [18] focus on modeling long-term interactions to enhance preference learning. SHAN adopts hierarchical attention mechanisms to fuse historical and recent interactions, while M3R models long- and short-term interests with GRUs and attention mechanisms respectively. The two methods treat short- and long-term preferences independently and combine their embeddings naïvely via concatenation or addition.
– **MEIRec** [5] and **GATNE** [1] are two heterogeneous GNN-based models. MEIRec focuses on aggregating information based on different meta-paths without paying attention to the relevance of meta-paths, while GATNE integrates multiple types of interactions with the attention mechanism but fails to model the dynamic.

For all baselines and our method, we set embedding size $d = 128$, $d_a = 128$, $d_\mathcal{T} = 16$, heads $h = 8$, the maximum iterations as 100, batch size as 128, learning rate as 0.001 and weight of regularization as 0.001 on all three datasets. The number of temporal buckets $B$ is set as 60, 14, and 7 on the three datasets, respectively. For DIEN, MEIRec and our THIGE, we set three-layers MLP with dimensions 64, 32 and 1. For our THIGE and all baselines learning long- or short-term preferences, we consider the last 10, 10 and 50 interactions as the short term, and sample up to 50, 50 and 200 historical interactions as the long term for Yelp, CloudTheme and UserBehavior respectively. We will further analyze the impact of the length of the short term in Sect. 5.5.

We evaluate the performance of next-item recommendation with the metrics of F1, PR-AUC and ROC-AUC.

## 5.3   Comparison with Baselines

We report the results of different methods for next-item recommendation in Table 2. In general, THIGE achieves the best performance on the three datasets, outperform the second best method by 4.04% on Yelp, 5.84% on CloudTheme and 0.51% on UserBehavior.

**Table 2.** Performance of next-item recommendation (with standard deviation). The best result is in bold while the second best is underlined. PR. denotes PR-AUC and ROC. denotes ROC-AUC.

| Dataset | Metric | DIEN | STAMP | SHAN | M3R | MEIRec | GATNE | THIGE |
|---------|--------|------|-------|------|-----|--------|-------|-------|
| Yelp | F1 | 39.52 (1.31) | 40.37 (0.94) | 40.17 (1.10) | 33.49 (1.04) | 42.86 (0.44) | 42.21 (0.96) | **43.77** (0.66) |
|  | PR. | 30.04 (0.37) | 31.36 (1.23) | 32.35 (1.10) | 26.40 (0.92) | 32.69 (0.54) | 33.39 (1.42) | **36.45** (1.66) |
|  | ROC. | 74.69 (0.57) | 73.74 (1.15) | 70.91 (1.14) | 72.03 (1.33) | 74.65 (0.23) | 76.15 (0.64) | **79.23** (0.80) |
| CT. | F1 | 25.70 (1.25) | 21.42 (0.91) | 26.25 (1.09) | 33.54 (1.67) | 25.02 (0.98) | 27.33 (0.50) | **37.17** (1.36) |
|  | PR. | 41.16 (0.22) | 25.65 (0.44) | 40.92 (1.09) | 34.23 (0.95) | 43.86 (0.42) | 44.74 (0.20) | **51.94** (0.43) |
|  | ROC. | 68.41 (0.34) | 52.97 (0.52) | 67.48 (1.06) | 62.92 (0.89) | 69.98 (0.35) | 71.22 (0.11) | **75.38** (0.33) |
| UB. | F1 | 67.32 (3.45) | 63.06 (1.51) | 58.84 (7.83) | 61.37 (2.20) | 66.48 (1.16) | **67.81** (1.14) | 67.19 (0.98) |
|  | PR. | 63.38 (0.19) | 59.09 (0.22) | 63.86 (4.76) | 57.68 (0.03) | 64.94 (0.15) | 65.42 (0.05) | **65.71** (0.09) |
|  | ROC. | 62.90 (0.23) | 58.29 (0.40) | 55.45 (3.98) | 57.82 (0.09) | 64.82 (0.16) | 65.06 (0.08) | **65.39** (0.06) |

Compared with sequential models (DIEN, STAMP, SHAN and M3R), the reason that THIGE is superior is twofold. First, THIGE designs a more effective way to integrate long- and short-term preferences, such that the current demands are explicitly guided by historical habits. Second, it also considers different types of interactions between users and items, leading to better performance.

Compared with GNN-based models (MEIRec and GATNE), the main improvement of THIGE comes from jointly modeling historical habits and evolving demands. Moreover, MEIRec models heterogeneous interactions in an entirely decoupled manner, whereas GATNE and THIGE achieves better performance by modeling their latent relevance. It is also not surprising that heterogeneous GNN-based methods typically outperform sequential models, as the former accounts for multi-typed interactions whereas the latter only models single-typed interactions.

## 5.4   Comparison of Model Variants

In this section, we analyze three categories of THIGE variants to evaluate the effectiveness of our design choices, as follows.

- Attention effect: **THIGE(hm)** only uses the heterogeneous multi-head self-attention, without the habit-guided attention; **THIGE(hb)** is the opposite, using only the habit-guided attention.
- Range of preferences: **THIGE(L)** models long range only while **THIGE(S)** models short rane only;
- Temporal effect: **HIGE** removes the temporal dimension from THIGE, treating the graph as static.

## 5.5   Analysis of Key Factors in THIGE

As shown in Fig. 3, our THIGE outperforms all three categories of variants. We make the following observations. (1) Compared with THIGE(hb) and THIGE(hm), THIGE models the latent relevance of heterogeneous preferences

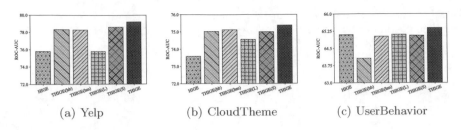

**Fig. 3.** Performance comparison of THIGE and its variants.

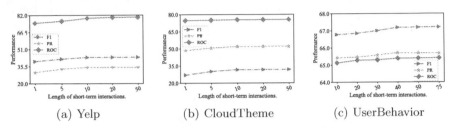

**Fig. 4.** Analyzing the length of short-term interactions in THIGE.

in a fine-grained manner, and model the impact of historical habits on current demands, leading to better performance. (2) Compared with THIGE(S) and THIGE(L), the joint modeling of short- and long-term preferences can improve performance, which also validates the assumption that the immediate decision of users is guided by their historical habits. (3) Compared with HIGE, the improvement in THIGE demonstrates the effectiveness of temporal embedding.

In THIGE, there are four key factors that may significantly affect the model performance: the length of short-term interactions, the samples of long-term interactions, the types of interactions and the number of latent preferences (i.e., the number of heads). In Fig. 4, we investigate how the length of the short term would impact the model. We respectively fix the samples of long-term interactions as 50, 50 and 200 for the three datasets, and then adjust the length of short-term interactions. Taking Fig. 4(c) as an example, we vary the length between 10 and 75 (i.e., treat the last 10–75 actions as the short term). When the length initially increases, the performance of THIGE is continuously improved, until reaching a saturation at about 40 or 50. Further treating more interactions as short-term has no additional benefit, which is expected as they can no longer be considered as current demands. This also justifies that the long-term, historical actions must be modeled differently to capture the user habit; simply extending the length of the short term does not work.

Next, in Fig. 5, we focus on detecting the influence of the length of long-term interactions. We respectively fix the length of short-term interactions as 10, 10 and 50 for the three datasets, and then vary the corresponding samples of long-term interactions. It is obvious that all improvements in performance are continuous but gradually weakened. There are two main reasons resulting

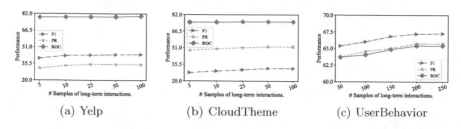

**Fig. 5.** Analyzing the samples of long-term interactions in THIGE.

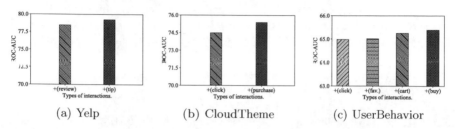

**Fig. 6.** Analyzing the types of interactions in THIGE.

in such phenomenons. On the one hand, with the length increases, the whole historical interactions of more and more users are captured and modeled. On the other hand, users who contain too many interactions may be abnormal and introduce noise that limits performance.

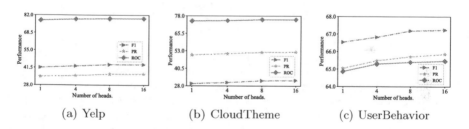

**Fig. 7.** Analyzing the number of heads in THIGE.

Furthermore, in Fig. 6, we demonstrate the contribution from different types of interactions. Taking UserBehavior as an example, we progressively include the interactions of "click", "favorite", "cart" and "buy", one at a time. As an example, the performance in Fig. 6(c) gradually improves, implying the effectiveness to integrate heterogeneous interactions. Moreover, comparing "favorite" and "cart", the "favorite" action has a smaller marginal return than "cart", which is not surprising given that "favorite" only have a weak tendency to induce future "cart", whereas "cart" actions are more likely to lead to purchases. That also means different types of interactions cannot be treated independently. Thus, modeling the relevance of different-typed interactions plays an important role.

Moreover, since the number of heads $h$ in heterogeneous multi-head self-attention mechanism reflects the number of latent preferences like categories and brands, we also vary the number of heads from 1 to 16 on the three datasets to analyze the influence of $h$ in THIGE. The experimental results in Fig. 7 indicate that $h = 8$ is a generally suitable and robust choice.

## 6  Conclusion

In this paper, we study the problem of representation learning on THIGs for next-item recommendation. To make full use of dynamic and heterogeneous information, we propose the THIGE to model short- and long-term preferences through habit-guided and heterogeneous self-attention mechanisms. The extensive experimental results on three real-world datasets demonstrate the effectiveness of our proposed model.

**Acknowledgements.** This work is supported in part by the National Natural Science Foundation of China (No. 61772082, 61806020, 61702296), the National Key Research and Development Program of China (2018YFB1402600), and Alibaba Group under its Alibaba Innovative Research (AIR) programme. This research is also supported by the National Research Foundation, Singapore under its International Research Centres in Singapore Funding Initiative. Any opinions, findings and conclusions or recommendations expressed in this material are those of the author(s) and do not reflect the views of National Research Foundation, Singapore. This work was done when Yugang Ji was visiting Alibaba as a research intern.

## References

1. Cen, Y., Zou, X., Zhang, J., et al.: Representation learning for attributed multiplex heterogeneous network. In: Proceedings of SIGKDD 2019, pp. 1358–1368 (2019)
2. Chang, S., Han, W., Tang, J., et al.: Heterogeneous network embedding via deep architectures. In: Proceedings of SIGKDD 2015, pp. 119–128 (2015)
3. Cho, K., Van Merriënboer, B., Gulcehre, C., et al.: Learning phrase representations using RNN encoder-decoder for statistical machine translation. In: Proceedings of EMNLP 2014, pp. 1724–1734 (2014)
4. Ekstrand, M.D., Riedl, J.T., Konstan, J.A., et al.: Collaborative Filtering Recommender Systems, vol. 4, no. 2, pp. 81–173 . Now Publishers Inc. (2011)
5. Fan, S., Zhu, J., Han, X., et al.: Metapath-guided heterogeneous graph neural network for intent recommendation. In: Proceedings of SIGKDD 2019, pp. 2478–2486 (2019)
6. Hamilton, W., Ying, Z., Leskovec, J.: Inductive representation learning on large graphs. In: Proceedings of NeurIPS 2017, pp. 1024–1034 (2017)
7. Kipf, T.N., Welling, M.: Semi-supervised classification with graph convolutional networks. In: Proceedings of ICLR 2017 (2017)
8. Le, D.T., Lauw, H.W., Fang, Y.: Modeling contemporaneous basket sequences with twin networks for next-item recommendation. In: IJCAI 2018, pp. 3414–3420 (2018)

9. Liu, Q., Zeng, Y., Mokhosi, R., Zhang, H.: STAMP: short-term attention/memory priority model for session-based recommendation. In: Proceedings of SIGKDD 2018, pp. 1831–1839 (2018)
10. Lu, Y., Wang, X., Shi, C., et al.: Temporal network embedding with micro-and macro-dynamics. In: Proceedings of CIKM 2019, pp. 469–478 (2019)
11. Lv, F., Jin, T., Yu, C., et al.: SDM: sequential deep matching model for online large-scale recommender system. In: Proceedings of CIKM 2019, pp. 2635–2643 (2019)
12. Pal, S.K., Mitra, S.: Multilayer perceptron, fuzzy sets, and classification. IEEE Trans. Neural Netw. **3**(5), 683–697 (1992)
13. Qu, M., Tang, J., Shang, J., et al.: An attention-based collaboration framework for multi-view network representation learning. In: Proceedings of CIKM 2017, pp. 1767–1776 (2017)
14. Sarwar, B., Karypis, G., Konstan, J., Riedl, J.: Item-based collaborative filtering recommendation algorithms. In: Proceedings of WWW 2001, pp. 285–295 (2001)
15. Shi, C., Li, Y., Zhang, J., Sun, Y., Philip, S.Y.: A survey of heterogeneous information network analysis. IEEE TKDE **29**(1), 17–37 (2016)
16. Shi, Y., Han, F., He, X., et al.: MVN2VEC: preservation and collaboration in multi-view network embedding. arXiv preprint arXiv:1801.06597 (2018)
17. Song, Y., Elkahky, A.M., He, X.: Multi-rate deep learning for temporal recommendation. In: Proceedings of SIGIR 2016, pp. 909–912 (2016)
18. Tang, J., Belletti, F., Jain, S., et al.: Towards neural mixture recommender for long range dependent user sequences. In: Proceedings of WWW 2019, pp. 1782–1793 (2019)
19. Trivedi, R., Farajtabar, M., Biswal, P., Zha, H.: Representation learning over dynamic graphs. In: Proceedings of ICLR 2019 (2019)
20. Veličković, P., Cucurull, G., Casanova, A., Romero, A., Lio, P., Bengio, Y.: Graph attention networks. In: Proceedings of ICLR (2018)
21. Wang, X., Ji, H., Shi, C., et al.: Heterogeneous graph attention network. In: Proceedings of WWW 2019, pp. 2022–2032 (2019)
22. Wang, Y., Duan, Z., Liao, B., et al.: Heterogeneous attributed network embedding with graph convolutional networks. Methods **25**(50), 75 (2019)
23. Ying, H., Zhuang, F., Zhang, F., et al.: Sequential recommender system based on hierarchical attention networks. In: Proceedings of AAAI 2018, pp. 3926–3932 (2018)
24. Zheng, V.W., Sha, M., Li, Y., et al.: Heterogeneous embedding propagation for large-scale e-commerce user alignment. In: Proceedings of ICDM 2018, pp. 1434–1439 (2018)
25. Zhou, G., Mou, N., Fan, Y., et al.: Deep interest evolution network for click-through rate prediction. In: Proceedings of AAAI 2019, pp. 5941–5948 (2019)

# Node Classification in Temporal Graphs Through Stochastic Sparsification and Temporal Structural Convolution

Cheng Zheng[1(✉)], Bo Zong[2], Wei Cheng[2], Dongjin Song[2], Jingchao Ni[2], Wenchao Yu[2], Haifeng Chen[2], and Wei Wang[1]

[1] Department of Computer Science, University of California, Los Angeles, CA, USA
{chengzheng,weiwang}@cs.ucla.edu
[2] NEC Laboratories America, Princeton, NJ, USA
{bzong,weicheng,dsong,jni,wyu,haifeng}@nec-labs.com

**Abstract.** Node classification in temporal graphs aims to predict node labels based on historical observations. In real-world applications, temporal graphs are complex with both graph topology and node attributes evolving rapidly, which poses a high overfitting risk to existing graph learning approaches. In this paper, we propose a novel $\underline{T}$emporal $\underline{S}$tructural $\underline{Net}$work (TSNet) model, which jointly learns temporal and structural features for node classification from the sparsified temporal graphs. We show that the proposed TSNet learns how to sparsify temporal graphs to favor the subsequent classification tasks and prevent overfitting from complex neighborhood structures. The effective local features are then extracted by simultaneous convolutions in temporal and spatial domains. Using the standard stochastic gradient descent and backpropagation techniques, TSNet iteratively optimizes sparsification and node representations for subsequent classification tasks. Experimental study on public benchmark datasets demonstrates the competitive performance of the proposed model in node classification. Besides, TSNet has the potential to help domain experts to interpret and visualize the learned models.

**Keywords:** Temporal graphs · Node classification · Graph sparsification · Temporal structural convolution

## 1 Introduction

Temporal graphs, as a data structure that carries both temporal and structural information from real-world data, has been widely adopted in applications from various domains, such as online social media [33], biology [27], action recognition [28], and so on. In this paper, we study the problem of node classification in temporal graphs [27]: Given a set of nodes with rich features and a temporal graph that records historical activities between nodes, the goal is to predict the label of every node. Consider the following application scenario.

© Springer Nature Switzerland AG 2021
F. Hutter et al. (Eds.): ECML PKDD 2020, LNAI 12459, pp. 330–346, 2021.
https://doi.org/10.1007/978-3-030-67664-3_20

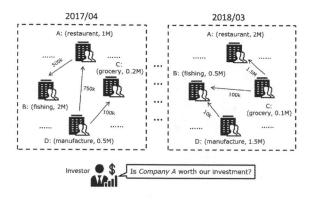

Fig. 1. An example of node classification in a temporal graph from the financial domain. Nodes are companies, and edges indicate monthly transactions. The goal is to predict which companies are promising for investment in the near future.

**Example.** In the financial domain, investors are eager to know which companies are promising for investment. As shown in Fig. 1, companies and their historical transactions naturally form a temporal graph, shown as a sequence of graph snapshots. Each snapshot encodes companies as nodes and transactions as edges within a month. The side information of companies (*e.g.*, industry sector and cash reserve) and transactions (*e.g.*, transaction amount) is represented by the node and edge attributes, respectively. In this task, we aim to predict each company's label: *promising* or *others* for future investment, with interpretable evidence for domain experts.

While node representation lies at the core of this problem, we face two main challenges from temporal graphs.

**Temporal Graph Sparsification.** Temporal graphs from real-life applications are large with high complexity. For example, the social graph on Facebook [5] and the financial transaction graph on Venmo [32] are densely connected with average node degrees of 500 and 111, respectively. Such complexity poses a high overfitting risk to existing machine learning techniques [17,20], and makes it difficult for domain experts to interpret and visualize learned models. While graph sparsification [16] suggests a promising direction to reduce graph complexity, existing methods perform sparsification by sampling subgraphs from predefined distributions [4,11,14,30]. The sparsified graphs may miss important information for classification because the predefined distributions could be irrelevant to subsequent tasks. Several recent efforts [8,22,34] strive to utilize supervision signals to remove noise edges and regularize the graph model training. However, the proposed methods are either transductive with difficulty to scale or of high gradient variance bringing increased training difficulty.

**Temporal-Structural Convolution.** Local features in the temporal-structural domain are the key to node classification in temporal graphs. Although existing techniques have investigated how to build convolutional operators to automatically learn and extract local features from either temporal domain [2] or structural domain [25], a naïve method that simply stacks temporal and structural operators could lead to suboptimal performance. An effective method that learns and extracts local features from joint temporal-structural space is still missing.

**Our Contribution.** We propose Temporal Structural Network (TSNet), a deep learning framework that performs supervised node classification in sparsified temporal graphs. TSNet consists of two major sub-networks: *sparsification network* and *temporal-structural convolutional network*.

1. The *sparsification network* aims to sparsify input temporal graphs by sampling edges from the one-hop neighborhood following a distribution that is learned from the subsequent supervised classification tasks.
2. The *temporal-structural convolutional network* takes sparsified temporal graphs as input and extracts local features by performing convolution in nodes' neighborhood defined in joint temporal-structural space.

As both sub-networks are differentiable, we can leverage standard stochastic gradient descent and backpropagation techniques to iteratively learn better parameters to sparsify temporal graphs and extract node representations. Experimental results on both public and private datasets show that TSNet can offer competitive performance on node classification tasks. Using a case study, we demonstrate the potential of TSNet to improve model interpretation and visualization of temporal graphs.

## 2   Problem Definition

In the following presentation, we use bold uppercase letters (*e.g.*, **A**) to represent tensors, uppercase letters (*e.g.*, $A$) to represent matrices, lowercase letters (*e.g.*, $a$) to denote scalars, and blackboard bold uppercase letters (*e.g.*, $\mathbb{A}$) to denote the concept of sets. We start with the definition of temporal graphs.

**Temporal Graphs.** In this work, we employ graph snapshot sequences to represent temporal graphs. Given $t$ discrete time points and $n$ nodes, a temporal graph is denoted as $G = (\mathbb{V}, \mathbf{V}, \mathbf{E}, \mathbf{A})$, where $\mathbb{V}$ is a set of nodes that appear in a temporal graph and $\mathbf{V} \in \mathbb{R}^{t \times n \times d_n}$ is a tensor that encodes $d_n$-dimensional node attributes across $t$ time points. $\mathbf{E} \in \mathbb{R}^{t \times n \times n}$ is a binary tensor where $\mathbf{E}(i, u, v) = 1$ if there is an edge between node $u$ and node $v$ at time $i$. $\mathbf{A} \in \mathbb{R}^{t \times n \times n \times d_e}$ is a tensor that encodes $d_e$-dimensional edge attributes across $t$ time points.

**Node Classification in Temporal Graphs.** Given a temporal graph $G = (\mathbb{V}, \mathbf{V}, \mathbf{E}, \mathbf{A})$ representing historical transactions and $\mathbb{Y}$ as a set of possible node labels, the goal is to predict labels $Y(u)$ for each node $u \in \mathbb{V}$.

In the following discussion, we focus on the cases where node labels are static for the ease of presentation. Note that with minor modification, our technique can easily be adapted for the cases where node labels dynamically evolve. We approach this problem by inductive supervised learning. In the training phase, we are given a temporal graph $G_{train}$ with training labels $Y_{train}$. In the testing phase, we use the trained model to infer node labels in testing graph $G_{test}$.

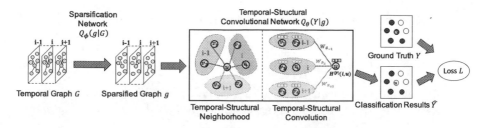

**Fig. 2.** The frameworks of TSNet. We utilize the two-step formulation of node classification problem. The sparsification network takes the temporal graph as input and generates sparsified subgraphs drawn from a learned distribution. The temporal-structural network extracts temporal and structural features simultaneously with the sparsified subgraph as input.

# 3   TSNet Overview

In this section, we start with a theoretical overview of the proposed TSNet.

## 3.1   A Two-Step Framework

Given input temporal graph $G$ and node label matrix $Y$, our objective is to learn $P(Y \mid G)$. Current Graph Neural Networks (GNNs) [11,13,25] learn node representation by aggregating node neighborhood features. However, in large and complex temporal graphs, node neighborhood tends to be dense with much noise which introduces high overfitting risk to existing approaches. To tackle the challenge, we leverage the two-step framework proposed in [34] to break node classification problem down into two steps: graph sparsification step and representation learning step.

$$P(Y \mid G) \approx \sum_{g \in \mathbb{S}_G} P(Y \mid g) P(g \mid G) \approx \sum_{g \in \mathbb{S}_G} Q_\theta(Y \mid g) Q_\phi(g \mid G) \qquad (1)$$

where $g$ is a sparsified subgraph, and $\mathbb{S}_G$ is a class of sparsified subgraphs of $G$. We approximate the distributions by tractable functions $Q_\theta$ and $Q_\phi$. With reparameterization tricks [10], we could differentiate the graph sparsification step to make efficient backpropagation. In the following, we will introduce our framework to find approximation function $Q_\phi(g \mid G)$ and $Q_\theta(Y \mid g)$.

## 3.2   Architecture

As shown in Fig. 2, the proposed TSNet consists of two major sub-networks: **sparsification network** and **temporal-structural convolutional network**.

- The **sparsification network** is a multi-layer neural network that implements $Q_\phi(g \mid G)$: Taking temporal graph $G$ as input, it generates a random sparsified subgraph of $G$ drawn from a learned distribution.

- The **temporal-structural convolutional network** implements $Q_\theta(Y \mid g)$ that takes a sparsified subgraph as input, extracts node representations by convolutional filtering on the temporal-structural neighborhood of each node, and makes predictions on node labels.

With differentiable operations in both sub-networks, our TSNet is an end-to-end supervised framework, which is trainable using gradient-based optimization.

## 4    Sparsification Network

In this section, we present the sparsification network, which optimizes temporal graph sparsification for subsequent node classification tasks.

### 4.1    Design Goals

The goal of sparsification network is to generate sparsified subgraphs for temporal graphs, serving as the approximation function $Q_\phi(g \mid G)$. Therefore, we need to answer the following three questions in the sparsification network.

1. As the essence of sparsification is to sample a subset of edges, how should we represent each edge so that we can differentiate edges for edge sampling?
2. What is the class of sparsified subgraphs $\mathbb{S}_G$? How to sample such sparsified subgraphs?
3. How to make sparsified subgraphs differentiable for end-to-end training?

### 4.2    Edge Representations

Given a temporal graph $G = (\mathbb{V}, \mathbf{V}, \mathbf{E}, \mathbf{A})$, an expected edge representation could consist of its edge attributes and certain information from the two connected nodes. Let $\mathbb{N}_{u,i}$ be the set of one-hop neighbors with respect to node $u$'s incoming edges at time $i$. The expected edge representation $\mathbf{X}(i, u, v)$ for the edge from $v$ to $u$ at time $i$ is calculated as follows.

$$\mathbf{X}(i, u, v) = \mathbf{V}'(i, u)||\mathbf{V}'(i, v)||\mathbf{A}(i, u, v) \tag{2}$$

where $||$ indicates vector concatenation and $\mathbf{A}(i, u, v)$ denotes edge attributes. $\mathbf{V}'(i, u)$ ($\mathbf{V}'(i, v)$) is the representation of node $u$ ($v$), which we calculate with mean aggregation [11] to capture both attribute and structural information,

$$\mathbf{V}'(i, u) = \sigma(W_{\phi_1} \cdot \mathbf{V}(i, u)||\mathrm{MEAN}(\mathbf{V}(i, u'), \forall u' \in \mathbb{N}_{u,i})) \tag{3}$$

where $W_{\phi_1}$ is the weights to be learned and $\sigma$ is a nonlinear activation function.

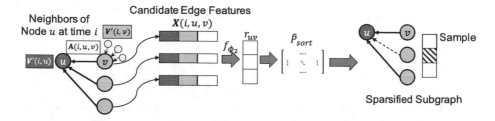

**Fig. 3.** An illustration of the proposed sparsification network. In this example, we focus on the node $u$ at time $i$ with 3 neighbor nodes and set $k$ as 2. Edge representations consist of both edge attributes and node representations. We implement the sparsification by a continuous relaxation of sorting and top-$k$ important incoming edge sampling.

### 4.3  Sampling Sparsified Subgraphs

We focus on $k$-neighbor subgraphs for $\mathbb{S}_G$. The concept of $k$-neighbor subgraph is originally proposed in the context of spectral sparsification for static graphs [23]: Given an input graph, each node of a $k$-neighbor subgraph can select no more than $k$ edges from its one-hop neighborhood. In this work, we extend the concept of $k$-neighbor subgraph to temporal graphs: Given a temporal graph $G$, each node of a $k$-neighbor subgraph can select no more than $k$ incoming edges *from its one-hop neighborhood in each graph snapshot of $G$*. Without loss of generality, we sketch this sampling process by focusing on a specific node $u$ in graph snapshot at time $i$. Let $\mathbb{N}_{u,i}$ be the set of one-hop neighbors with respect to $u$'s incoming edges at time $i$ and the cardinality of $\mathbb{N}_{u,i}$ is $d$.

1. For $v \in \mathbb{N}_{u,i}$, $r_{uv} = f_{\phi_2}(\mathbf{X}(i,u,v))$, where $r_{uv}$ is a scalar denoting the ranking score of the edge from node $v$ to $u$ at time $i$, and $f_{\phi_2}$ is a feedforward neural network (parameterized by $\phi_2$) that generates the score based on the edge representation $\mathbf{X}(i,u,v)$.
2. We sort the incoming edges based on their ranking scores, and select the top-$k$ edges with the largest ranking scores.
3. The above two steps are repeated for each node in each graph snapshot.

The parameters in $f_{\phi_2}$ are shared among all nodes in all graph snapshots; therefore, the number of parameters is independent to the size of temporal graphs.

### 4.4  Making Samples Differentiable

The conventional sorting operators are not differentiable such that it is difficult to utilize them for parameter optimization. To make sorting differentiable, we propose to implement the subgraph sampling based on the continuous relaxation of sorting operator [10]. Without loss of generality, we focus on a specific node $u$ at time $i$ in a temporal graph $G$. We implement the subgraph sampling in Sect. 4.3 as follows.

1. Let $\mathbb{N}_{u,i}$ be the set of one-hop neighbors with respect to $u$'s incoming edges at time $i$. We apply the reparameterization trick and introduce a fixed source of randomness [10] to the ranking score $r_{uv}$, $\forall v \in \mathbb{N}_{u,i}$,

$$\pi_{uv} = \log r_{uv} + g_{uv} \tag{4}$$

where $\pi_{uv}$ is a reparameterized scalar indicating the importance of the edge from node $v$ to $u$. $g_{uv}$ is a sample drawn from Gumbel$(0,1)$ and $g_{uv} = -\log(-\log(u))$ with $u \sim$ Uniform$(0,1)$. The reparameterization trick refines the stochastic computational graph for smooth gradient backward pass.

2. We relax the permutation matrix of the edge sorting operator $\hat{P}_{sort} \in \mathbb{R}^{d \times d}$ for node $u$ at time $i$, and its $j$-th row is

$$\hat{P}_{sort}(j,:)(\tau) = \text{softmax}[((d+1-2j)\pi_{u:} - A_\pi \mathbf{1})/\tau] \tag{5}$$

where $d$ is the cardinality of $\mathbb{N}_{u,i}$ and $\mathbf{1}$ denotes the column vector of all ones. $A_\pi$ is the matrix of absolute pairwise differences of the elements in $\{v \in \mathbb{N}_{u,i} \mid \pi_{uv}\}$, and the element at $x$-row and $y$-column is $A_\pi(x,y) = |\pi_{ux} - \pi_{uy}|$. $\tau$ is a hyper-parameter called *temperature* which controls the interpolation between discrete distribution and continuous categorical densities.

3. Before sparsification, the feature tensor of $\mathbb{N}_{u,i}$ is $\mathbf{V}_U(i,u) = \{\mathbf{V}(i,u,u'_1), \ldots, \mathbf{V}(i,u,u'_d)\}$, where $\mathbf{E}(i,u,u'_j) = 1$ and $\mathbf{V}_U(i,u) \in \mathbb{R}^{d \times d_n}$. By applying the relaxed sort operator $\hat{P}_{sort}$ to the unsparsified node features $\mathbf{V}_U(i,u)$, we then select first $k$ rows as the output

$$\mathbf{V}_S(i,u) = [\hat{P}_{sort}\mathbf{V}_U(i,u)](:k,:) \tag{6}$$

If $|\mathbb{N}_{u,i}| \leq k$ for node $u$, we will skip its sparsification and take all in $\mathbf{V}_U(i,u)$.

---

**Algorithm 1.** Sampling subgraphs by sparsification network

---

**Input:** Temporal graph $G = (\mathbb{V}, \mathbf{V}, \mathbf{E}, \mathbf{A})$ and integer $k$.
1: **for** $i = 1, \cdots, t$ **do**
2:     **for** $u \in \mathbb{V}$ **do**
3:         **if** $|\mathbb{N}_{u,i}| > k$ **then**
4:             **for** $v \in \mathbb{N}_{u,i}$ **do**
5:                 compute $\mathbf{X}(i,u,v)$ by Equation (2)
6:                 compute $\pi_{uv}$ by Equation (4)
7:             **end for**
8:             compute $\hat{P}_{sort}$ by Equation (5)
9:             compute $\mathbf{V}_S(i,u)$ by Equation (6)
10:         **end if**
11:     **end for**
12: **end for**

---

We sketch the full algorithm of sparsification network in a combinatorial manner in Algorithm 1. Let $\bar{d}$ be the average degree, $n$ be the total number of nodes in a temporal graph, and $t$ be the number of snapshots. The sparsification network visits each node's one-hop neighborhood and makes $\bar{d}^2$ calculations. The complexity of sampling subgraphs by the sparsification network is $O(\bar{d}^2nt)$.

# 5  Temporal-Structural Convolutional Network

As discussed in Sect. 3.1, the goal of the temporal-structural convolutional network (TSCN) is to serve as $Q_\theta(Y \mid g)$: it extracts node representations from the sparsified subgraphs generated by the sparsification network and leverages the vector representations to perform node classification. Inspired by the success of convolutional aggregation in the graph domain [4,11,13,25], the core idea behind the temporal-structural convolutional network is to simultaneously extract local temporal and structural features for node representations by convolutional aggregation in individual nodes' *temporal-structural neighborhood*.

## 5.1  Temporal-Structural Neighborhood

Unlike the neighborhood defined in static graphs that only tells "who are close to me", temporal-structural neighborhood stores information about "who and when are close to me". To accomplish this, we extend the notion of neighborhood to the temporal domain by aggregating the structural neighborhood across several preceding and/or subsequent snapshots of any given snapshot. Given a node $u$ at time $i$, its temporal-structural neighborhood can be represented by a matrix $F_{u,i} \in \mathbb{R}^{t \times n}$, where $F_{u,i}(j,v) = 1$ if node $v$ is in $u$'s (structural) neighborhood at time $j$; otherwise, $F_{u,i}(j,v) = 0$. In this work, we focus on the *first-order* temporal-structural neighborhood in the sparsified subgraphs. In other words, we have $F_{u,i}(j,v) = 1$ if the following two conditions hold: (1) $|i - j| = 1$, and (2) at time $j$, there is an incoming edge from node $v$ to $u$ in the sparsified temporal graph. Note that node $u$ at time $i$ is also in its own temporal-structural neighborhood, that is, $F_{u,i}(i,u) = 1$. With the notion of the temporal-structural neighborhood, we are ready to introduce the design of a temporal-structural convolutional layer.

## 5.2  Temporal-Structural Convolutional Layer

A temporal-structural convolutional layer performs feature aggregation in individual nodes' temporal-structural neighborhood. One could stack multiple convolutional layers to extract higher-order temporal-structural features.

Without loss of generality, we discuss the technical details of temporal-structural convolutional layer by focusing on a specific node $u$ at time $i$ in the $p$-th convolutional layer. The input is a temporal graph $G = (\mathbb{V}, \mathbf{V}, \mathbf{E}, \mathbf{A})$, node representations $\mathbf{H}^{(p-1)} \in \mathbb{R}^{t \times n \times d_n^{(p-1)}}$ and a relaxed sort operator $\hat{P}_{sort}$ from Sect. 4.4. With the same sort operator in Eq. 6 that sparsifies the node features of the first convolution layer, we obtain the sparsified node features of the $p$-th convolutional layer as

$$\mathbf{V}_U^{(p)}(i,u) = \{\mathbf{H}^{(p-1)}(i,u,u_1'),\ldots,\mathbf{H}^{(p-1)}(i,u,u_d')\} \tag{7}$$

$$\mathbf{V}_S^{(p)}(i,u) = [\hat{P}_{sort}\mathbf{V}_U^{(p)}(i,u)](:k,:) \tag{8}$$

The temporal-structural convolution performs as follows.

$$\mathbf{H}^{(p)}(i,u) = \sigma\left(\sum_{\{(j,v)|F_{u,i}(j,v)=1\}} \mathbf{V}_S^{(p)}(j,u,v)W_{i,u,j,v}^{(p)}\right) \tag{9}$$

where $\sigma(\cdot)$ is a non-linear activation function, $\mathbf{H}^{(p)} \in \mathbb{R}^{t\times n\times d_n^{(p)}}$ is the output node representations, and $W_{i,u,j,v}^{(p)} \in \mathbb{R}^{d_n^{(p-1)}\times d_n^{(p)}}$ is a customized convolution filter generated by

$$W_{i,u,j,v}^{(p)} = \mathrm{MLP}_{\theta_{i-j}^{(p)}}(\mathbf{V}_S^{(p)}(i,u),\mathbf{V}_S^{(p)}(j,v),\mathbf{A}(j,u,v)) \tag{10}$$

where $\mathrm{MLP}_{\theta_{i-j}^{(p)}}(\cdot)$ is a multi-layer neural network with parameters $\theta_{i-j}^{(p)}$ that generates customized convolutional filters based on node and edge features. In other words, in the case of first-order temporal-structural neighborhood, we utilize three networks $\mathrm{MLP}_{\theta_{-1}^{(p)}}$, $\mathrm{MLP}_{\theta_0^{(p)}}$, and $\mathrm{MLP}_{\theta_1^{(p)}}$ to model the temporal impacts from the temporal-structural neighborhood. Note that $\{\theta_{-1}^{(p)},\theta_0^{(p)},\theta_1^{(p)}\}$ are the only parameters in this convolutional layer and are shared by all nodes; therefore, the number of parameters in a temporal-structural convolutional layer is independent of the number of nodes, edges, or time points in a temporal graph.

As described in Eq. 9 and 10, for a single node, the computational cost is determined by the MLP structure as a fixed number ($c$). Therefore, the complexity of convolution layer is $O(cdtn)$ and is generally proportional to the number of nodes ($n$).

## 5.3   Network Architecture

Now we present the full temporal-structural convolutional network.

- **Convolutional layer.** As discussed in Sect. 5.2, one could stack multiple convolutional layers to hierarchically explore high-order temporal-structural neighborhood.
- **Pooling layer.** Let $\mathbf{H}^{(p)} \in \mathbb{R}^{t\times n\times d_n^{(p)}}$ be the output node representations from the $p$-th convolutional layers. The pooling layer performs another round of aggregation in temporal domain by $H = \mathrm{Pooling}(\mathbf{H}^{(p)})$, where $H \in \mathbb{R}^{n\times d_n^{(p)}}$. Possible pooling operations include *max*, *average*, and *sum* [11].
- **Output layer.** This layer employs a multi-layer neural network and the final output of TSNet is $\hat{Y} = \mathrm{MLP}_{\theta_o}(H)$, where $\theta_o$ denotes the parameters.

– **Objective function.** To handle the estimation variance brought by random sampling, the sparsification network generates $l$ sparsified subgraphs and we optimize the parameters in TSNet by minimizing the average loss from the $l$ samples. In particular, the objective function is formulated as follows.

$$J = \frac{1}{l} \sum_{i=1}^{l} L(Y, \hat{Y}_i) \tag{11}$$

The function $L$ is defined by cross entropy loss.

# 6    Experimental Study

In this section, we evaluate the performance of TSNet using real-life temporal graph datasets from multiple domains. In particular, we compare TSNet with state-of-the-art techniques in terms of classification accuracy and analyze its sensitivity to the hyper-parameters. Moreover, we provide a case study to demonstrate how sparsified subgraphs generated by TSNet could improve visualization. The supplementary material contains more detailed information.

## 6.1    Datasets

We employ four temporal graph datasets from different domains, including collaboration networks, online social media, and financial marketing. The dataset statistics are summarized in Table 1.

**Table 1.** Dataset statistics

|  | DBLP-3 | DBLP-5 | Reddit | Finance |
|---|---|---|---|---|
| # nodes | 1,662 | 5,994 | 128,858 | 45,542 |
| # edges | 33,808 | 113,062 | 29,009,401 | 661,586 |
| # snapshots / # in training | 10/5 | 10/5 | 31/16 | 36/18 |
| time granularity | 1 year | 1 year | 1 d | 1 month |
| # classes | 3 | 5 | 10 | 2 |

**DBLP-3/5.** The temporal graphs record co-author relationships in the DBLP computer science bibliography[1] from 2001 to 2010, where nodes represent authors and edges denote co-author relationships. There are 10 graph snapshots and each snapshot stores co-author relationships within one year. To generate the node attributes, we aggregate titles and abstracts of the corresponding author's papers published in that year into one document, represent this document by

---

[1] https://www.aminer.cn/citation.

the bag-of-words model. We aim to classify authors in DBLP-3 and DBLP-5 into three and five research areas respectively. The first 5 snapshots are in $G_{train}$.

**Reddit.** Reddit is a large online forum, where users contribute original posts or make comments/upvotes to existing posts. We extract posts and comments in 10 mid-sized subreddits in May 2015. Following the procedure in [11], we build a post-to-post temporal graph, where nodes are posts, and two posts become connected if they are both commented by at least one identical user. For node attributes, we aggregate the post and comment texts into one document, represent it by the bag-of-words model, and reduce the dimensionality to 20 by PCA. On this dataset, our goal is to classify each post into one of the 10 subreddit categories. The first 16 snapshots are in $G_{train}$.

**Finance.** This private dataset contains temporal graphs that record transaction history between companies from April 2014 to March 2017. Each node represents a company and each edge indicates a transaction between two companies. Node attributes are side information about the companies such as account balance, cash reserve, etc., which may change from year to year. In this dataset, we aim to classify companies into two categories: *promising* or *unpromising* for investment in the near future. We put the first 18 snapshots in $G_{train}$.

### 6.2   Baseline Methods

We implement four categories of baselines: (1) node classification methods for static graphs, including **GCN** [13], **GraphSAGE** [11], **GAT** [25], and **LDS** [8] that simultaneously learns the graph structure and GCN parameters; (2) stacking structural and temporal feature learning models, including **TempCNN-GCN** that extracts temporal features with temporal CNN [2] and then trains GCN model on static graphs, **GCN-TempCNN** in reverse order, and **Deepwalk-LSTM** that extracts structure features by DeepWalk [21] and then leverages LSTM to learn temporal features; (3) temporal graph learning models, including **DynamicTriad** [35], and **STAR** [27]; (4) variants of the proposed TSNet models, including **TSCN** that only uses temporal-structural convolutional network, **SS-TSCN** that samples subgraphs with spectral sparsification [23], and **DE-TSCN** that employs DropEdge [22] as graph sampling method.

### 6.3   Experimental Settings

**TSNet.** In our experiments, the hyper-parameter $k$ is searched between 3 and 15 for the optimal performance. For the temporal-structural convolutional network, it starts with two temporal-structural convolutional layers, with an internal single-layer feed forward network to generate convolutional filters. Then the output features pass a max-pooling layer over time and a non-linear layer which produces the logit for label prediction.

**Dataset Split and Accuracy Metrics.** We prepare the training and testing temporal graphs following the similar setting in [11]. We split the snapshots of

**Table 2.** Node classification performance

|  | DBLP-3 | | DBLP-5 | | Reddit | | Finance | |
|---|---|---|---|---|---|---|---|---|
|  | Macro-F1 | Micro-F1 | Macro-F1 | Micro-F1 | Macro-F1 | Micro-F1 | Macro-F1 | Micro-F1 |
| GCN | 0.862 | 0.859 | 0.679 | 0.678 | 0.357 | 0.290 | 0.480 | 0.464 |
| GraphSAGE | 0.850 | 0.847 | 0.814 | 0.814 | 0.399 | 0.336 | 0.496 | 0.448 |
| GAT | 0.875 | 0.863 | 0.821 | 0.830 | - | - | 0.509 | 0.495 |
| LDS | 0.876 | 0.869 | 0.797 | 0.794 | - | - | 0.499 | 0.474 |
| TempCNN-GCN | 0.851 | 0.857 | 0.720 | 0.710 | 0.411 | 0.340 | 0.532 | 0.548 |
| GCN-TempCNN | 0.676 | 0.691 | 0.720 | 0.711 | 0.384 | 0.313 | 0.440 | 0.397 |
| DeepWalk-LSTM | 0.913 | 0.913 | 0.772 | 0.777 | 0.370 | 0.303 | 0.493 | 0.446 |
| DynamicTriad | 0.753 | 0.745 | 0.713 | 0.717 | 0.393 | 0.324 | 0.419 | 0.430 |
| STAR | 0.908 | 0.908 | 0.811 | 0.815 | 0.439 | 0.367 | 0.541 | 0.502 |
| TSCN | 0.942 | 0.929 | 0.850 | 0.845 | 0.466 | 0.406 | 0.559 | 0.537 |
| SS-TSCN | 0.839 | 0.835 | 0.807 | 0.805 | 0.418 | 0.351 | 0.465 | 0.445 |
| DE-TSCN | 0.875 | 0.888 | 0.801 | 0.792 | 0.391 | 0.343 | 0.538 | 0.487 |
| TSNet | **0.954** | **0.955** | **0.859** | **0.860** | **0.475** | **0.416** | **0.630** | **0.610** |

temporal graphs into $G_{train}$ and $G_{test}$: test graphs remain unseen during training. We randomly sample 80% nodes in $G_{train}$ as the training nodes and provide their labels to the models. We test the performance of the models with all nodes in $G_{test}$. We evaluate the classification accuracy using macro-F1 and micro-F1 scores . The results show the average of 10 runs with random initializations.

## 6.4   Classification Accuracy

Table 2 summarizes the classification performance of TSNet and the baseline methods on all datasets. For DBLP-3, DBLP-5, Reddit, and Finance, the hyperparameter $k$ is set as 8, 9, 10, and 5, respectively. The hyper-parameter $l$ is set as 1 in this experiment. Note some results on Reddit is missing due to the out-of-memory error.

Overall, TSNet consistently outperforms all of the baseline methods in terms of macro-F1 and micro-F1 over all of the datasets. We make the following observations. (1) Compared with the deep learning techniques for static graphs, including GCN, GraphSAGE and GAT, TSNet achieves better performance by effectively utilizing temporal features in graphs. (2) In GCN-TempCNN and Deepwalk-LSTM, structural and temporal features are extracted from separate components. Because of the inter-component dependency, it becomes harder to adjust the parameters in structural feature learning than in temporal feature learning. Similarly, in TempCNN-GCN, parameters in temporal feature learning are harder to get trained. With temporal-structural convolution, TSNet can simultaneously adjust parameters for temporal and structural feature learning, and generate more effective features for better classification accuracy. (3) In comparison with LDS, SS-TSCN and DE-TSCN, TSNet outperforms because of the explicit supervision from downstream tasks, which shows that the sparsification network learns to sparsify temporal graphs to favor the subsequent

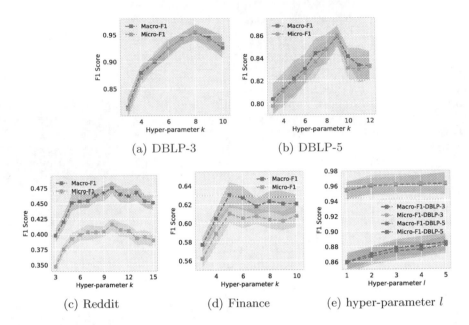

**Fig. 4.** Accuracy vs hyper-parameter $k$ and $l$

classification. (4) The comparison with TSCN is interesting: using the sparsified subgraphs from the sparsification network, it is easier to make TSCN generalized to unseen testing data with improved classification performance. (5) Compared with DynamicTriad and STAR, TSNet is more effective in node representations customized by the simultaneous learning of temporal and structural features.

### 6.5   Sensitivity to Hyper-parameters

Figure 4(a)-(d) demonstrates how classification accuracy responds when $k$ increases from 3. Over the four datasets, we observe a common phenomenon: there exists an optimal $k$ that delivers the best classification accuracy. When $k$ is small, TSNet can only make use of little structural information, which leads to suboptimal performance. When $k$ gets larger, the temporal-structural convolution involves more complex neighborhood aggregation with higher overfitting risk, which negatively impacts the classification performance for unseen testing data. By comparing across datasets, we observe that the optimal $k$ is associated with the average node degrees of the temporal graphs: higher $k$ in dense Reddit graph and lower $k$ in sparse Finance graph. Figure 4(e) shows how hyper-parameter $l$ impacts classification accuracy on DBLP-3 and DBLP-5. When $l$ increases from 1 to 5, we observe a relatively small improvement in classification accuracy. As the parameters in the sparsification network are shared by all edges in the temporal graphs, the estimation variance from random sampling could already be mitigated to some extent by a number of sampled edges in a sparsified subgraph.

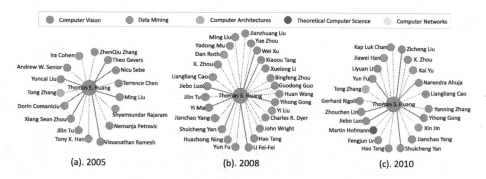

**Fig. 5.** One-hop neighborhood of *Thomas S. Huang* in DBLP-5. The visualization presents snapshots from (a) 2005, (b) 2008, and (c) 2010. Node colors indicate node labels from ground truth. Sparsification network selects edges with solid red lines. The sparsified graph supports downstream classification as well as model interpretation.

## 6.6   Case Study

In this section, we present a case study to demonstrate the potential of TSNet in enhancing model interpretation and visualization. Figure 5 visualizes the one-hop neighborhood of *Thomas S. Huang* with $k$ set as 7 in DBLP-5 temporal graph, and we only present snapshots from 2005, 2008, and 2010 due to space limitation. We make the following observations from Fig. 5. First, the central author is from the area of *computer vision*, and all selected edges also connect to authors from *computer vision*. When neighbors share identical labels consistently over time, temporal-structural features could boost the confidence of making classification decision. Second, instead of exploring all the neighbors, we can only focus on a subset of selected neighbors, which could make it easier for human experts to conduct the effort on model interpretation and result visualization.

# 7   Related Work

Our research is related to three lines of studies: graph sparsification, deep learning on graphs, and temporal graph modeling.

**Graph Sparsification.** The goal of graph sparsification is to find small subgraphs from input large graphs that best preserve desired properties. Existing techniques are mainly unsupervised and deal with simple graphs without node/edge attributes for preserving predefined graph metrics [12], information propagation traces [20], graph spectrum [1, 23], node degree distribution [7], node distance distribution [14], or clustering coefficient [19]. Importance based edge sampling has also been studied in a scenario where we could predefine the importance of edges [4]. Unlike existing sparsification methods, our method aims to optimize temporal graph sparsification with rich node and edge attributes by supervision signals from subsequent classification tasks.

**Deep Graph Learning.** Deep graph learning has made steady progress on automated node representation learning. Early studies [3,6] investigate convolutional filters in graph spectral domain under transductive settings. To enable inductive learning, convolutional filters in graph domain are proposed [13,25]. The graph feature learning models based on message passing mechanisms are also referred to as GNNs. Multiple latest studies [18,24] extend the GNNs into temporal graphs by introducing the recurrent operations into the GNN layers. Recent efforts also attempt to sample subgraphs from predefined distributions [8,31], and regularize graph learning by random edge dropping [22]. Our research investigates node representation learning in the temporal structural domain for general temporal graphs, where both node/edge attributes and graph topology could evolve over time. More importantly, our work answers a unique question: Given a limited budget for edge selection, how to optimize graph sparsification so that we could maximize performance in subsequent classification tasks.

**Temporal Graph Modeling.** It is an important yet challenging task to model dynamics and evolution in temporal graphs [9,27,28]. Recent studies attempt to model dynamics in temporal graphs using generative methods [9], matrix factorization [15], and deep learning approaches [28,29]. The model in [29] fuses the sequential and spatial graph convolution in a "sandwich" structure, which is yet dependent on the convolution order of sub-structure [26]. The ST-GCN model in [28] learns both spatial and temporal patterns by adding all temporally connected nodes into temporal kernels and conduct similar convolution as GCN [13] on skeleton graph with static topology. Comparing with these approaches, our model TSNet is novel as it can extract the temporal and structural features simultaneously and separate the temporal impact of different convolution filters.

## 8 Conclusion

In this paper, we propose Temporal Structural Network (TSNet) for node classification in temporal graphs. TSNet consists of two major sub-networks: (1) the sparsification network sparsifies input temporal graphs by sorting and sampling edges following a learned distribution; (2) the temporal-structural convolutional network performs convolution on the sparsified graphs to extract local features from the joint temporal-structural space. As an end-to-end model, the two sub-networks in TSNet are trained jointly and iteratively with supervised loss, gradient descent, and backpropagation techniques. In the experimental study, TSNet demonstrates superior performance over four categories of baseline models on public and private benchmark datasets. The qualitative case study suggests a promising direction for the interpretability of temporal graph learning.

**Acknowledgement.** We thank the anonymous reviewers for their careful reading and insightful comments on our manuscript. The work was partially supported by NSF (DGE-1829071, IIS-2031187).

# References

1. Adhikari, B., Zhang, Y., Amiri, S.E., Bharadwaj, A., Prakash, B.A.: Propagation-based temporal network summarization. IEEE Trans. Knowl. Data Eng. **30**(4), 729–742 (2018)
2. Bai, S., Kolter, J.Z., Koltun, V.: An empirical evaluation of generic convolutional and recurrent networks for sequence modeling. arXiv:1803.01271 (2018)
3. Bruna, J., Zaremba, W., Szlam, A., LeCun, Y.: Spectral networks and locally connected networks on graphs. In: ICLR (2014)
4. Chen, J., Ma, T., Xiao, C.: Fastgcn: fast learning with graph convolutional networks via importance sampling. In: ICLR (2018)
5. Ching, A., Edunov, S., Kabiljo, M., Logothetis, D., Muthukrishnan, S.: One trillion edges: graph processing at facebook-scale. Proc. VLDB Endowment **8**(12), 1804–1815 (2015)
6. Defferrard, M., Bresson, X., Vandergheynst, P.: Convolutional neural networks on graphs with fast localized spectral filtering. In: NIPS (2016)
7. Eden, T., Jain, S., Pinar, A., Ron, D., Seshadhri, C.: Provable and practical approximations for the degree distribution using sublinear graph samples. In: WWW (2018)
8. Franceschi, L., Niepert, M., Pontil, M., He, X.: Learning discrete structures for graph neural networks. In: ICML (2019)
9. Ghalebi, E., Mirzasoleiman, B., Grosu, R., Leskovec, J.: Dynamic network model from partial observations. In: NIPS (2018)
10. Grover, A., Wang, E., Zweig, A., Ermon, S.: Stochastic optimization of sorting networks via continuous relaxations. In: ICLR (2019)
11. Hamilton, W., Ying, Z., Leskovec, J.: Inductive representation learning on large graphs. In: NIPS (2017)
12. Hübler, C., Kriegel, H.P., Borgwardt, K., Ghahramani, Z.: Metropolis algorithms for representative subgraph sampling. In: ICDM (2008)
13. Kipf, T.N., Welling, M.: Semi-supervised classification with graph convolutional networks. In: ICLR (2017)
14. Leskovec, J., Faloutsos, C.: Sampling from large graphs. In: KDD (2006)
15. Li, J., Cheng, K., Wu, L., Liu, H.: Streaming link prediction on dynamic attributed networks. In: WSDM (2018)
16. Liu, Y., Safavi, T., Dighe, A., Koutra, D.: Graph summarization methods andapplications: a survey. ACM Comput. Surv. **51**(3), 1–34 (2018)
17. Loukas, A., Vandergheynst, P.: Spectrally approximating large graphs with smaller graphs. In: ICML (2018)
18. Ma, Y., Guo, Z., Ren, Z., Zhao, E., Tang, J., Yin, D.: Streaming graph neural networks. arXiv:1810.10627v2 (2019)
19. Maiya, A.S., Berger-Wolf, T.Y.: Sampling community structure. In: WWW (2010)
20. Mathioudakis, M., Bonchi, F., Castillo, C., Gionis, A., Ukkonen, A.: Sparsification of influence networks. In: KDD (2011)
21. Perozzi, B., Al-Rfou, R., Skiena, S.: Deepwalk: online learning of social representations. In: KDD (2014)
22. Rong, Y., Huang, W., Xu, T., Huang, J.: Dropedge: towards deep graph convolutional networks on node classification. In: ICLR (2020)
23. Sadhanala, V., Wang, Y.X., Tibshirani, R.: Graph sparsification approaches for laplacian smoothing. In: AISTATS (2016)

24. Trivedi, R., Farajtabar, M., Biswal, P., Zha, H.: Dyrep: learning representations over dynamic graphs. In: ICLR (2019)
25. Veličković, P., Cucurull, G., Casanova, A., Romero, A., Liò, P., Bengio, Y.: Graph attention networks. In: ICLR (2018)
26. Wu, Z., Pan, S., Long, G., Jiang, J., Zhang, C.: Graph wavenet for deep spatial-temporal graph modeling. In: IJCAI (2019)
27. Xu, D., Cheng, W., Luo, D., Liu, X., Zhang, X.: Spatio-temporal attentive RNN for node classification in temporal attributed graphs. In: IJCAI, pp. 3947–3953 (2019)
28. Yan, S., Xiong, Y., Lin, D.: Spatial temporal graph convolutional networks for skeleton-based action recognition. In: AAAI (2018)
29. Yu, B., Yin, H., Zhu, Z.: Spatio-temporal graph convolutional networks: a deep learning framework for traffic forecasting. In: IJCAI (2018)
30. Yu, W., et al.: Learning deep network representations with adversarially regularized autoencoders. In: KDD (2018)
31. Zeng, H., Zhou, H., Srivastava, A., Kannan, R., Prasanna, V.: Graphsaint: graph sampling based inductive learning method. In: ICLR (2020)
32. Zhang, X., Tang, S., Zhao, Y., Wang, G., Zheng, H., Zhao, B.Y.: Cold hard e-cash: friends and vendors in the venmo digital payments system. In: ICWSM (2017)
33. Zheng, C., Zhang, Q., Long, G., Zhang, C., Young, S.D., Wang, W.: Measuring time-sensitive and topic-specific influence in social networks with lstm and self-attention. IEEE Access 8, 82481–82492 (2020)
34. Zheng, C., et al.: Robust graph representation learning via neural sparsification. In: ICML (2020)
35. Zhou, L.K., Yang, Y., Ren, X., Wu, F., Zhuang, Y.: Dynamic network embedding by modeling triadic closure process. In: AAAI (2018)

# DyHGCN: A Dynamic Heterogeneous Graph Convolutional Network to Learn Users' Dynamic Preferences for Information Diffusion Prediction

Chunyuan Yuan[1,2], Jiacheng Li[1,2], Wei Zhou[1(✉)], Yijun Lu[3], Xiaodan Zhang[1], and Songlin Hu[1,2]

[1] Institute of Information Engineering, Chinese Academy of Sciences, Beijing, China
{yuanchunyuan,lijiacheng,zhouwei,zhangxiaodan,husonglin}@iie.ac.cn
[2] School of Cyber Security, University of Chinese Academy of Sciences, Beijing, China
[3] Alibaba Cloud Computing Co. Ltd., Hangzhou, China
yijun.lyj@alibaba-inc.com

**Abstract.** Information diffusion prediction is a fundamental task for understanding the information propagation process. It has wide applications in such as misinformation spreading prediction and malicious account detection. Previous works either concentrate on utilizing the context of a single diffusion sequence or using the social network among users for information diffusion prediction. However, the diffusion paths of different messages naturally constitute a dynamic diffusion graph. For one thing, previous works cannot jointly utilize both the social network and diffusion graph for prediction, which is insufficient to model the complexity of the diffusion process and results in unsatisfactory prediction performance. For another, they cannot learn users' dynamic preferences. Intuitively, users' preferences are changing as time goes on and users' personal preference determines whether the user will repost the information. Thus, it is beneficial to consider users' dynamic preferences in information diffusion prediction.

In this paper, we propose a novel dynamic heterogeneous graph convolutional network (DyHGCN) to jointly learn the structural characteristics of the social graph and dynamic diffusion graph. Then, we encode the temporal information into the heterogeneous graph to learn the users' dynamic preferences. Finally, we apply multi-head attention to capture the context-dependency of the current diffusion path to facilitate the information diffusion prediction task. Experimental results show that DyHGCN significantly outperforms the state-of-the-art models on three public datasets, which shows the effectiveness of the proposed model.

**Keywords:** Data mining · Information diffusion prediction · Dynamic diffusion graph · Graph convolutional network

© Springer Nature Switzerland AG 2021
F. Hutter et al. (Eds.): ECML PKDD 2020, LNAI 12459, pp. 347–363, 2021.
https://doi.org/10.1007/978-3-030-67664-3_21

# 1  Introduction

Online social media has become an indispensable part of our daily life, on which people can deliver or repost interesting news easily. The information diffusion prediction task aims at studying how information spread among users and predicting the future infected user. The modeling and prediction of the information diffusion process play an important role in many real-world applications, such as predicting social influence [17], analyzing how misinformation spreads [20,30] and detecting malicious accounts [14,31].

Previous studies either concentrate on the utilization of the diffusion sequence [2,22,23,26,28] or using the social network among users for diffusion prediction [6,25,27]. Some studies [8,22,23,26] proposed the diffusion path based models to learn user representation from the past diffusion records. For example, TopoL-STM [22] extended the standards LSTM model to learn the chain structure of the information diffusion sequence. CYAN-RNN [23] modeled the diffusion path as a tree structure and attention-based RNN to capture the cross-dependence based on the observed sequence. The diffusion path can reflect the information trends, so these models can achieve success in formulating the observed sequence.

Apart from utilizing the diffusion path, some studies apply the social network among users to facilitate information diffusion predication. An intuition behind it is that people have some common interests with their friends [29]. If their friends repost the information, they have a higher probability to repost it. Based on this assumption, many recent studies [1,24,27,29] exploit the structure of the social network to learn the social influence among users for improving prediction performance.

However, existing methods including state-of-the-art models [25,27] do not consider two important aspects: For one thing, they cannot jointly utilize both the social network and diffusion graph for prediction, which is insufficient to model the complexity of the diffusion process and results in unsatisfactory prediction performance. For another, they cannot learn users' dynamic preferences. Intuitively, users' preference is changing as time goes on and users' personal preference influences information diffusion. Thus, it is beneficial to consider the user's dynamic preference, which can be reflected by the dynamic diffusion structure at different points of diffusion time.

To take advantage of these aspects, we propose a novel dynamic heterogeneous graph convolutional network (DyHGCN) to utilize both the social network and dynamic diffusion graph for prediction. Firstly, we design a heterogeneous graph algorithm to learn the representation of the social network and diffusion relations. Then, we encode the temporal information into the heterogeneous representation to learn the users' dynamic preference. Finally, we capture the context-dependency of the current diffusion path to solve the information diffusion prediction problem.

The main contributions of this paper can be summarized as follows:

- We design a dynamic heterogeneous graph convolutional network (DyHGCN) to jointly model users' social graph and diffusion graph for learning complex diffusion processes.

– We encode the temporal information into the heterogeneous representation to learn the users' dynamic preferences. As far as we know, it is the first work to utilize users' dynamic preferences for information diffusion prediction.
– Experimental results suggest that DyHGCN outperforms the state-of-the-art models on three public datasets, which shows the effectiveness and efficiency of DyHGCN.

## 2 Related Work

Current information diffusion prediction methods can be categorized into two categories: diffusion path based methods and social network based methods.

### 2.1 Diffusion Path Based Methods

The diffusion path based methods infer the interpersonal influence based on given observed diffusion sequences. Early work assumed that there is a prior diffusion model in the information diffusion process, such as the independent cascade model [9] or linear threshold model [5]. Although these models [9,18] achieve success in formulating the implicit influence relations between users, the effectiveness of these methods relies on the hypothesis of prior information diffusion model, which is hard to specify or verify in practice [23].

With the development of neural network models, such as the recurrent neural network (RNN) and convolutional neural network (CNN), some studies [4,19,26] apply deep learning to automatically learn a representation of the underlying path from the past diffusion sequence for diffusion prediction, without requiring an explicit underlying diffusion model. For example, TopoLSTM [22] extended the standards LSTM model to learn the information diffusion path to generate a topology-aware node embedding. DeepDiffuse [8] employed embedding technique and attention model to utilize the infection timestamp information. The model can predict when and who is going to be infected in a social media based on previously observed cascade sequence. NDM [26] built a microscope cascade model based on self-attention and convolution neural networks to alleviate the long-term dependency problem.

Most diffusion path based methods treat the problem as a sequence prediction task, which aims at predicting diffusion users sequentially and explore how historical diffusion sequences affect future diffusion trends. However, social relations among users are one of the critical channels of information diffusion, which is not applied in these methods. Thus, it is hard to accurately identify and predict the direction of information flow without considering the structure of social networks.

### 2.2 Social Graph Based Methods

Apart from utilizing the diffusion path, some studies leverage the structure of the social network for diffusion prediction. An intuition behind it is that people have some common interests with their friends [29]. If their friends repost the news or microblogs, they have a higher probability to repost it too. Based on this assumption, many previous studies [1,13,24,27,29,32] have been exploring to improve prediction performance from the view of social relations. For example, [29] studied the interplay between users' social roles and their influence on information diffusion. They proposed a role-aware information diffusion model that integrated social role recognition and diffusion modeling into a unified framework. [25] explored both the sequential nature of an information diffusion process and structural characteristics of the user connection graph and employed a RNN-based framework to model the historical sequential diffusion. [27] proposed a multi-scale diffusion prediction model based on reinforcement learning. The model incorporated the macroscopic diffusion size information into the RNN-based microscopic diffusion model.

However, these graph based methods mainly focus on the current diffusion sequence and ignore the diffusion path of other messages in the meantime, which cannot capture global reposting relations. Thus, it is insufficient to model the complexity of the diffusion processes. Different from social graph based models, we jointly learn the global structure of the social graph and dynamic diffusion graph. Moreover, our model considers dynamic individual preference based on this heterogeneous graph.

## 3   Problem Formulation

Suppose that a collection of message $\mathcal{D}$ will be propagated among a set of users $\mathcal{U}$. In this paper, we regard a piece of message as a document. An explicit way to describe an information diffusion process can be viewed as a successive activation of nodes indicating when people share or repost a document. Most often, real cascades are recorded as single-chain structure sequences.

The diffusion process of document $d_m$ is recorded as a sequence of reposting behaviors $\mathcal{S}^m = \{s_1^m, s_2^m, \ldots, s_{N_c}^m\}$, where $N_c$ is the cascade number for message $d_m$. $N_c$ is the maximum length of diffusion sequence. The reposting behavior $s_k^m = \{(u_k^m, t_k^m) | u_k^m \in \mathcal{U}, t_k^m \in [0, +\infty)\}$ is a tuple, referring to that user $u_k^m$ reposted the message $d_m$ at a certain timestamp $t_k^m$. As shown in Fig. 1(a), for document $d_1$ is recorded as $\{(u_1, t_1^1), (u_2, t_2^1), (u_3, t_3^1), \ldots\}$ order by timestamp.

Given the observed diffusion traces, the target of the information diffusion prediction task is to predict the diffusion behavior at a future timestamp $t'$. As shown in Fig. 1(b), we have known that user $u_5$ delivers a piece of information $d_?$ and we should predict which users would be interested in it and will repost it in the future timestamp $t'$. The diffusion probability can be formulated as $P(s_{t'}^m | \mathcal{S}_t)$, where $t' > t$.

In this paper, we jointly encode the users' social graph and the diffusion graph to solve the diffusion prediction problem. The motivation is that both graphs provide different useful information for diffusion prediction and combining them will make predictions more accurate. As shown in Fig. 1(b), both $u_2$ and $u_6$ have higher probability to repost message from $u_5$ than $u_3$, because we can observe that $u_2$ and $u_6$ follow $u_5$ from social graph, whereas $u_3$ did not directly

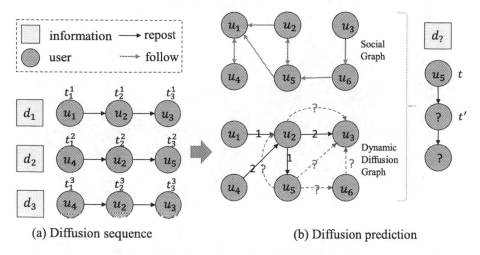

(a) Diffusion sequence                    (b) Diffusion prediction

**Fig. 1.** (a) An example of diffusion process of documents $d_1$, $d_2$, $d_3$ (marked as yellow squares). The edges denote that the user (marked as red circles) reposted a document at a certain timestamp. (b) Illustration of the information diffusion prediction task. The red dashed lines indicate possible reposting behavior and potential activation users. (Color figure online)

follow $u_5$. However, the message would be likely to be propagated to $u_3$ by $u_2$ or $u_6$, because $u_3$ follows $u_6$ and $u_3$ has twice reposting records from $u_2$. From this example, we can see that it is beneficial to combine both graphs for the diffusion prediction problem.

## 4   Framework

In this section, we will introduce the DyHGCN, a deep learning-based model with three stages, to elaborately learn the individual dynamic preference and the neighbors' influence. The overview architecture of the proposed model is shown in Fig. 2. Firstly, we construct a heterogeneous social and diffusion graph. Then, we design the dynamic heterogeneous graph to learn the node representations of the graph with users' dynamic preferences. Finally, we combine these representations to predict future infected users.

### 4.1   Heterogeneous Graph Construction

Intuitively, people would repost the message or microblog when they are interested in it. Users usually follow someone if they like his or her microblog. Therefore, social relations in the social graph would be useful to predict whether the user will repost the message. Furthermore, we can analyze the history of reposting behaviors at different diffusion period. In this way, we can capture the dynamic changes in users' preferences. Based on these motivations, we propose

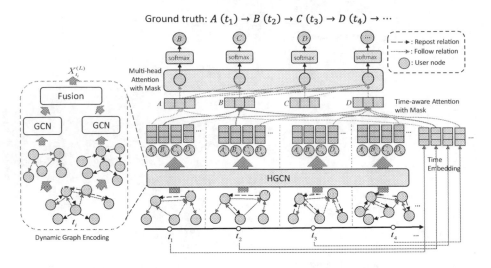

**Fig. 2.** The architecture of dynamic heterogeneous graph convolutional network.

to jointly model the social relations and dynamic repost relations to learn better user representations for information diffusion prediction.

In this paper, we utilize the social relations among users to construct a user social graph $\mathcal{G}^f$, which is a directed and unweighted graph. Then, we split the historical diffusion timeline into $n$ time intervals. At each time interval $t_i$, we use the repost relations among users to construct a diffusion graph $\mathcal{G}^r_{t_i}$, which is a directed and weighted graph.

### 4.2    Heterogeneous Graph Convolutional Network (HGCN)

As shown in the left part of Fig. 2, the heterogeneous graph has one type of node (user node) and two types of relations: follow relation and repost relation. At time interval $t_i, i \in [1, n]$, we use both relations to construct the adjacency matrix $\mathbf{A} = \{\mathbf{A}^F, \mathbf{A}^R_{t_i}\}$. $\mathbf{A}^F \in \mathbb{R}^{|U| \times |U|}$ is the adjacency matrix extracted from social relations, and $\mathbf{A}^R_{t_i} \in \mathbb{R}^{|U| \times |U|}$ is extracted from the repost relations. $|U|$ denotes the amount of users.

For each kind of relation, we apply a multi-layer graph convolutional network (GCN) [11] to learn the node representation from the graph. The layer-wise propagation rule can be defined as follows:

$$\begin{aligned} \mathbf{X}^{(l+1)}_F &= \sigma(\mathbf{A}^F \mathbf{X}^{(l)} \mathbf{W}^{(l)}_F), \\ \mathbf{X}^{(l+1)}_R &= \sigma(\mathbf{A}^R_{t_i}(\mathbf{X}^{(l)} + \mathbf{t}_i) \mathbf{W}^{(l)}_R), \end{aligned} \tag{1}$$

where $\mathbf{X}^{(0)} \in \mathbb{R}^{|U| \times d}$ is randomly initialized user embeddings by a normal distribution [3], and $\mathbf{W}^{(l)}_F, \mathbf{W}^{(l)}_R \in \mathbb{R}^{d \times d}$ are learnable parameters. $\mathbf{t}_i \in \mathbb{R}^d$ is the time interval embedding initialized by a random distribution. $d$ is the dimensionality of user embeddings. $\sigma(\cdot)$ is the $\mathbf{ReLU}(x) = \max(0, x)$ activation function. $l$ denotes the layers of GCN.

We can obtain the user representations $\mathbf{X}_F^{(l+1)} \in \mathbb{R}^{|U| \times d}$ from follow relations and $\mathbf{X}_R^{(l+1)} \in \mathbb{R}^{|U| \times d}$ from repost relations. To fuse both relations for generating better user representation, we apply a heuristic strategy [15] to make both $\mathbf{X}_F^{(l+1)}$ and $\mathbf{X}_R^{(l+1)}$ interact with each other:

$$
\begin{aligned}
\mathbf{X}_{FR}^{(l+1)} &= [\mathbf{X}_F^{(l+1)}; \mathbf{X}_R^{(l+1)}; \mathbf{X}_F^{(l+1)} \odot \mathbf{X}_R^{(l+1)}; \mathbf{X}_F^{(l+1)} - \mathbf{X}_R^{(l+1)}], \\
\mathbf{X}_{t_i}^{(l+1)} &= \mathbf{X}_{FR}^{(l+1)} \mathbf{W}_1,
\end{aligned}
\tag{2}
$$

where $\odot$ denotes the element-wise product and $\mathbf{W}_1 \in \mathbb{R}^{4d \times d}$ is a learnable parameter. $\mathbf{X}_{t_i}^{(l+1)}$ is the learnt user representation at time $t_i$.

## 4.3 Dynamic Graph Encoding

As discussed above, users' dynamic preference is important for diffusion prediction. In this section, we will describe how to learn user representations from the dynamic graphs at different time intervals.

---

**Algorithm 1:** The dynamic graph encoding algorithm.

**Input**: The heterogeneous graph $\mathcal{G}(\mathcal{V}, \mathcal{E})$;
Time intervals $T = \{t_1, t_2, \ldots, t_n\}$;
Adjacency matrix $\mathbf{A}^F, and [\mathbf{A}_{t_1}^R, \mathbf{A}_{t_2}^R, \ldots, \mathbf{A}_{t_n}^R]$.
**Output**: A list of user representations at all time intervals.

1 **for** *each $t_i \in T$* **do**
2     Construct adjacency matrix $A_{t_i}^R$ from diffusing sub-graph.
3     Construct adjacency matrix $A^F$ from social sub-graph.
4     **for** $l = 1, \ldots, L$ **do**
5         $\mathbf{X}_F^{(l+1)} = \sigma(\mathbf{A}^F X^{(l)} W_F^{(l)})$ ;
6         $\mathbf{X}_R^{(l+1)} = \sigma(\mathbf{A}_{t_i}^R (X^{(l)} + \mathbf{t}_i) W_R^{(l)})$ ;
7         Fuse $\mathbf{X}_F^{(l+1)}$ and $\mathbf{X}_R^{(l+1)}$ to form $\mathbf{X}_{t_i}^{(l+1)}$ by Equation (2).
8     **end**
9     Collect user representation $\mathbf{X}_{t_i}^{(L)}$ from HGCN.
10 **end**
11 **return** $\mathbf{X}_{t_1}^{(L)}, \mathbf{X}_{t_2}^{(L)}, \ldots, \mathbf{X}_{t_n}^{(L)}$

---

The overall process of the dynamic graph encoding algorithm is shown in Algorithm 1. Firstly, we split the historical diffusion timeline as $n$ intervals to construct $n$ dynamic heterogeneous graphs. Then, we apply the heterogeneous graph convolutional network to learn the user embeddings at each time interval. Finally, we collect all user representations and send them to the next stage.

## 4.4    Time-Aware Attention

After the above procedures, we obtain user representations from different heterogeneous graph snapshots of different time intervals. Then, we can generate the final user representation by fusing these user representations of different time intervals. In this subsection, we design two kinds of strategies to produce final user representations.

### Hard Selection Strategy

For every user in the diffusion trace, they all have a timestamp during repost. We can directly determine to which time interval a given timestamp belongs. Then we use the user embedding of that time interval as final user embedding. For example, given user id $u$, we look up user representation from all user representations $[\mathbf{X}_{t_1}^{(L)}, \mathbf{X}_{t_2}^{(L)}, \ldots, \mathbf{X}_{t_n}^{(L)}]$. We will obtain $n$ user representations $[\mathbf{u}_{t_1}, \mathbf{u}_{t_2}, \ldots, \mathbf{u}_{t_n}]$. Supposing user $u$ reposts the information at timestamp $t'$ and $t' \in [t_3, t_4)$, then we use $\mathbf{u}_{t_3}$ as final representations of user $u$.

### Soft Selection Strategy

The hard selection strategy only uses the user representation belonging to the repost time interval, which can not fully utilize the user representations produced by historical information. Thus, we design a time-aware attention module to fuse the historical user representations as final user representation.

Specifically, given user id $u$, we look up user representation from all user representations $[\mathbf{X}_{t_1}^{(L)}, \mathbf{X}_{t_2}^{(L)}, \ldots, \mathbf{X}_{t_n}^{(L)}]$, and obtain user representations $\mathbf{U}_t = [\mathbf{u}_{t_1}, \mathbf{u}_{t_2}, \ldots, \mathbf{u}_{t_n}] \in \mathbb{R}^{n \times d}$. Supposing user $u$ reposts the information at timestamp $t'$ and $t' \in [t_3, t_4)$, then we define the time-aware attention as follows:

$$t' = \mathbf{Lookup}(t_3),$$

$$\alpha = \mathbf{softmax}(\frac{\mathbf{U}_t^T \mathbf{t}'}{\sqrt{d}} + \mathbf{m}),$$  (3)

$$\tilde{\mathbf{u}} = \sum_{i=1}^{n} \alpha_i \mathbf{U}_{t_i}$$

where $\mathbf{m}_j = \begin{cases} 0 & t' \geq t_j, \\ -\infty & \text{otherwise.} \end{cases}$ is a mask matrix and $\mathbf{m} \in \mathbb{R}^n$. When $\mathbf{m}_j = -\infty$, the softmax function results in a zero attention weight, which can switch off the attention when $t' < t_j$ to avoid leaking labels of future timestamp. $\mathbf{Lookup}(\cdot)$ function is applied to transform the time interval id into time embedding. The time embeddings are initialized by a normal distribution [3]. $\tilde{\mathbf{u}}$ is the final representation of user $u$.

## 4.5    Information Diffusion Prediction

To capture the context-dependency information, we can apply the learned user representations to construct the current diffusion sequence $\mathbf{U} = [\tilde{\mathbf{u}}_A, \tilde{\mathbf{u}}_B, \tilde{\mathbf{u}}_C, \ldots]$

for future diffusion prediction. Instead of using a recurrent neural network (RNN) to model the current diffusion sequence, we apply masked multi-head self-attention module [21] to parallelly attend to each other for context encoding. Compared with RNN, a multi-head attention module is much faster and easier to learn the context information. It is worth noting that we also apply mask matrix as before to mask the future information to avoid leaking labels. The process can be formulated as follows:

$$
\begin{aligned}
\textbf{Attention}(\mathbf{Q}, \mathbf{K}, \mathbf{V}) &= \textbf{softmax}\left(\frac{\mathbf{Q}\mathbf{K}^T}{\sqrt{d_k}} + \mathbf{M}\right)\mathbf{V}, \\
\mathbf{h}_i &= \textbf{Attention}\left(\tilde{\mathbf{U}}\mathbf{W}_i^Q, \tilde{\mathbf{U}}\mathbf{W}_i^K, \tilde{\mathbf{U}}\mathbf{W}_i^V\right), \\
\mathbf{Z} &= [\mathbf{h}_1; \mathbf{h}_2; \ldots; \mathbf{h}_H]\mathbf{W}^O
\end{aligned}
\tag{4}
$$

where $W_i^Q, W_i^K, W_i^V \in \mathbb{R}^{d \times d_k}$ and $W^O \in \mathbb{R}^{H d_k \times d_Q}$; $d_k = d/H$; $H$ is the number of heads of attention module. The mask matrix $\mathbf{M}$, which is defined as:

$$
\mathbf{M}_{ij} = \begin{cases} 0 & i \leq j, \\ -\infty & \text{otherwise.} \end{cases}
\tag{5}
$$

is used to switch off attention weights of the future time step.

We obtain user representations $\mathbf{Z} \in \mathbb{R}^{L \times d}$ on $L$ diffusion time steps. Then, we use two layers fully-connected neural network to compute the diffusion probability:

$$
\hat{\mathbf{y}} = \mathbf{W}_3 \textbf{ReLU}(\mathbf{W}_2 \mathbf{Z}^T + \mathbf{b}_1) + \mathbf{b}_2,
\tag{6}
$$

where $\hat{\mathbf{y}} \in \mathbb{R}^{L \times |U|}$, and $\mathbf{W}_2 \in \mathbb{R}^{d \times d}, \mathbf{W}_3 \in \mathbb{R}^{|U| \times d}, \mathbf{b}_1, \mathbf{b}_2$ are the learnable parameters.

Finally, we apply the cross entropy loss as the objective function, which is formulated as:

$$
\mathcal{J}(\theta) = -\sum_{i=2}^{L}\sum_{j=1}^{|U|} \mathbf{y}_{ij} \log(\hat{\mathbf{y}}_{ij})
\tag{7}
$$

where $\mathbf{y}_{ij} = 1$ denotes that the diffusion behavior happened, otherwise $\mathbf{y}_{ij} = 0$. $\theta$ denotes all parameters needed to be learned in the model. The parameters are updated by Adam optimizer with mini-batch.

## 5    Experiments

### 5.1    Datasets

Following the previous studies [25,27], we conduct experiments on three public datasets to quantitatively evaluate the proposed model. The detailed statistics are presented in Table 1. #Links denotes the amount of follow relations of users in the social network. #Cascades denotes the amount of diffusion sequence in the dataset. Avg. Length indicates the average length of the information diffusion sequence.

**Table 1.** Statistics of the Twitter, Douban, and Memetracker datasets.

| Datasets | Twitter | Douban | Memetracker |
|---|---|---|---|
| # Users | 12,627 | 23,123 | 4,709 |
| # Links | 309,631 | 348,280 | - |
| # Cascades | 3,442 | 10,602 | 12,661 |
| Avg. Length | 32.60 | 27.14 | 16.24 |

**Twitter**[1] dataset [7] records the tweets containing URLs during October 2010. Each URL is interpreted as an information item spreading among users. The social relation of users is the follow relation on Twitter.

**Douban**[2] dataset [33] is collected from a social website where users can update their book reading statuses and follow the statuses of other users. Each book or movie is considered an information item and a user is infected if she reads or watches it. The social relation of users is the co-occurrence relation. If two users take part in the same discussion more than 20 times, they are considered a friend.

**Memetracker** dataset [12] collects millions of news stories and blog posts from online websites and tracks the most frequent quotes and phrases, i.e. memes, to analyze the migration of memes among people. Each meme is regarded as an information item and each URL of websites is treated as a user. Note that this dataset has no underlying social graph.

As in previous studies [25,27], we randomly sample 80% of cascades for training, 10% for validation and the rest 10% for test. The statistics of datasets are listed in Table 1.

## 5.2 Comparison Methods

The comparison baselines can be divided into diffusion path based models and social network based models. To evaluate the effectiveness of DyHGCN, we use five very recent models as baselines for a thorough comparison. The models are shown as follows:

### Based on Diffusion Path

- **TopoLSTM** [22]: models the information diffusion path as a dynamic directed acyclic graph and extends the standard LSTM model to learn a topology-aware user embedding for diffusion prediction.
- **DeepDiffuse** [8]: employs the embedding technique and attention model to utilize the infection timestamp information. The model can predict when and who is going to be infected in a social network based on previously observed cascade sequence.

---

[1] http://www.twitter.com.
[2] http://www.douban.com.

- **NDM** [26]: builds a microscopic cascade model based on the self-attention mechanism and convolution neural networks to alleviate the long-term dependency problem.

**Based on Social Network**

- **SNIDSA** [25]: is a sequential neural network with structure attention to model information diffusion. The recurrent neural network framework is employed to model the sequential information. The attention mechanism is incorporated to capture the structural dependency among users. A gating mechanism is developed to integrate sequential and structural information.
- **FOREST** [27]: is a multi-scale diffusion prediction model based on reinforcement learning. The model incorporates the macroscopic diffusion size information into the RNN-based microscopic diffusion model. It is the latest sequential model and achieves state-of-the-art performance.

**Our methods (DyHGCN-H, DyHGCN-S):** DyHGCN-H is the model with hard selection strategy, and DyHGCN-S is the model with soft selection strategy (time-aware attention).

## 5.3 Evaluation Metrics and Parameter Settings

Following the settings of previous studies [23, 25, 27], we consider the next infected user prediction as a retrieval task by ranking the uninfected users by their infection probabilities. We evaluate the performance of DyHGCN with state-of-the-art baselines in terms of Mean Average Precision (MAP) on top k (Map@k) and HITS scores on top k (Hits@k).

Our model is implemented by PyTorch [16]. The parameters are updated by Adam algorithm [10] and the parameters of Adam, $\beta_1$ and $\beta_2$ are 0.9 and 0.999 respectively. The learning rate is initialized as 1e-3. The batch size of the training set is set to 16. The dimensionality of user embedding and temporal interval embedding are set to $d = 64$. We use two layer of GCN to learn the graph structure. The kernel size is set to 128. The number of heads in multi-head attention $H$ is chosen from $\{2, 4, 6, 8, 10, 12, 14, 16, 18, 20\}$ and finally set to 14. We split the dynamic diffusion graph into $n$ time intervals, where $n = \{1, 2, 4, 6, 8, 10, 12, 14, 16, 18, 20\}$. Finally, we use $n = 8$ in the experiment. We select the best parameter configuration based on performance on the validation set and evaluate the configuration on the test set.

## 5.4 Experimental Results

We evaluate the effectiveness of DyHGCN on three public datasets for information diffusion prediction task. Table 2, 3 and 4 show the performance of all methods.

**Table 2.** Experimental results on Twitter dataset (%). All experimental results of baselines are cited from paper [27]. FOREST [27] is the state-of-the-art model until this submission. Improvements of DyHGCN are statistically significant with $p < 0.01$ on paired t-test.

| Models | Twitter | | | | | |
|---|---|---|---|---|---|---|
| | hits@10 | hits@50 | hits@100 | map@10 | map@50 | map@100 |
| DeepDiffuse | 4.57 | 8.80 | 13.39 | 3.62 | 3.79 | 3.85 |
| TopoLSTM | 6.51 | 15.48 | 23.68 | 4.31 | 4.67 | 4.79 |
| NDM | 21.52 | 32.23 | 38.31 | 14.30 | 14.80 | 14.89 |
| SNIDSA | 23.37 | 35.46 | 43.39 | 14.84 | 15.40 | 15.51 |
| FOREST | 26.18 | 40.95 | 50.39 | 17.21 | 17.88 | 18.02 |
| DyHGCN-H | 28.48 | 47.18 | 58.48 | 16.78 | 17.63 | 17.79 |
| DyHGCN-S | **28.98** | **47.89** | **58.85** | **17.46** | **18.30** | **18.45** |

**Table 3.** Experimental results on Douban dataset (%). All experimental results of baselines are cited from paper [27]. FOREST [27] is the state-of-the-art model until this submission. Improvements of DyHGCN are statistically significant with $p < 0.01$ on paired t-test.

| Models | Douban | | | | | |
|---|---|---|---|---|---|---|
| | hits@10 | hits@50 | hits@100 | map@10 | map@50 | map@100 |
| DeepDiffuse | 9.02 | 14.93 | 19.13 | 4.80 | 5.07 | 5.13 |
| TopoLSTM | 9.16 | 14.94 | 18.93 | 5.00 | 5.26 | 5.32 |
| NDM | 10.31 | 18.87 | 24.02 | 5.54 | 5.93 | 6.00 |
| SNIDSA | 11.81 | 21.91 | 28.37 | 6.36 | 6.81 | 6.91 |
| FOREST | 14.16 | 24.79 | 31.25 | 7.89 | 8.38 | 8.47 |
| DyHGCN-H | 15.69 | **28.95** | **36.45** | 8.42 | 9.03 | 9.13 |
| DyHGCN-S | **16.34** | 28.91 | 36.13 | **9.10** | **9.67** | **9.78** |

**Table 4.** Experimental results on Memetracker dataset (%). We exclude TopoLSTM and SNIDSA for Memetracker because of the absence of underlying social graph. Improvements of DyHGCN are statistically significant with $p < 0.01$ on paired t-test.

| Models | Memetracker | | | | | |
|---|---|---|---|---|---|---|
| | hits@10 | hits@50 | hits@100 | map@10 | map@50 | map@100 |
| DeepDiffuse | 13.93 | 26.50 | 34.77 | 8.14 | 8.69 | 8.80 |
| NDM | 25.44 | 42.19 | 51.14 | 13.57 | 14.33 | 14.46 |
| FOREST | 29.43 | 47.41 | 56.77 | 16.37 | 17.21 | 17.34 |
| DyHGCN-H | 29.63 | **48.78** | **58.78** | 16.33 | 17.21 | 17.36 |
| DyHGCN-S | **29.90** | 48.30 | 58.43 | **17.64** | **18.48** | **18.63** |

From the table, we can see that DyHGCN (DyHGCN-H and DyHGCN-S) consistently outperforms the state-of-the-art methods by an absolute improvement of more than 5% in terms of hits@100 and map@100 scores. Specifically, we have the following observations:

(1) Compared with TopoLSTM, DeepDiffuse, and NDM, DyHGCN-S achieves about 5% absolute improvement on hits@10, and over 10% improvement on hits@100. Moreover, the prediction precision also achieves about 2% absolute improvement. These baseline models mainly model the diffusion path as a sequence or graph structure, which ignores the social network information. However, the social network can reflect user preference. The experimental results show that it is important to consider the user social network for information diffusion prediction.

(2) Compared with SNIDSA and FOREST, DyHGCN-S achieves over 2% absolute improvement on hits@10, and over 5% improvement on hits@100 on Twitter and Douban datasets. Both SNIDSA and FOREST exploit user social relations to facilitate diffusion prediction. However, when predicting the diffusion path, they only model the history diffusion path as a sequential pattern, which is insufficient to model the complex diffusion behavior and users' dynamic preference. The improvement of DyHGCN shows that it is necessary to model the diffusion path as a graph rather than a sequence or tree structure.

(3) Compared with DyHGCN-H, DyHGCN-S also shows better performance on three datasets. DyHGCN-H only uses the current state of diffusion graph to learn user embedding, which cannot capture the user's dynamic preference well. DyHGCN-S utilizes the time-aware attention module to fuse the history and current diffusion graph for producing better user representations for diffusion prediction.

## 6    Further Study

### 6.1    Ablation Study

To figure out the relative importance of every module in DyHGCN, we perform a series of ablation studies over the different parts of the model. The experimental results are presented in Table 5. The ablation studies are conducted as following orders:

- **w/o time-aware attention:** Replace the time-aware attention module with a hard selection strategy.
- **w/o social graph:** Removing the social graph convolutional network.
- **w/o diffusion graph:** Removing the diffusion graph convolutional network.
- **w/o heterogeneous graph:** Removing heterogeneous graph encoding modules and randomly initializing the user representations.

Table 5 shows the overall performance on several variant methods of DyHGCN. Referring to the experimental results in the table, we can observe that:

**Table 5.** Ablation study on Twitter and Douban datasets (%).

| Models | hits@10 | hits@50 | hits@100 | hits@10 | its@50 | hits@100 |
|---|---|---|---|---|---|---|
| **DyHGCN-S** | 28.98 | 47.89 | 58.85 | 16.34 | 28.91 | 36.13 |
| w/o time-aware attention | 28.48 | 47.18 | 58.48 | 15.69 | 28.95 | 36.45 |
| w/o social graph | 27.76 | 45.31 | 56.60 | 14.28 | 25.72 | 33.63 |
| w/o diffusion graph | 28.27 | 46.53 | 57.22 | 14.62 | 26.14 | 34.28 |
| w/o heterogeneous graph | 27.63 | 42.31 | 51.13 | 13.58 | 23.16 | 31.15 |

(1) When replacing the time-aware attention with a hard selection strategy, the performance drops a little compared with DyHGCN-S. The experimental results show that time-aware attention can effectively fuse the user representations produced by historical information to generate better user representation.

(2) When we remove the social graph encoding modules, the performance degrades a lot compared with DyHGCN. A similar phenomenon can be seen when removing the diffusion graph. The results indicate that both the social and repost relations encoding modules in DyHGCN are essential for information diffusion prediction.

(3) When removing the heterogeneous graph, the performance further decays a lot compared with removing the social graph or diffusion graph. The phenomenon shows that both relations contain complementary information and combining them does help to improve the performance.

## 6.2    Parameter Analysis

In this section, we conduct some sensitivity analysis experiments of hyper-parameters on the Twitter dataset. We analyze how different choices of the hyper-parameter may affect performance.

**Number of Time Intervals $n$.** In Sect. 4.2, we split the diffusion graph into $n$ snapshots according to the diffusion timeline. The size of $n$ may affect the performance. When $n$ is larger, the diffusion graph is split many pieces and the model can learn more fine-grained changes in the dynamic graph. Referring to Fig. 3(a), we can observe that: (1) Learning the dynamic characteristics of diffusion graph is helpful for information diffusion prediction, because performance is increasing when $n$ increases before $n = 8$. (2) When $n$ is too large, the further improvement of performance is very limited.

**Number of Heads $H$.** From the Fig. 3(b), we can see that the performance is improved a little as the increasing number of heads of multi-head attention. The model can capture more abundant information with an increase of the number of heads. However, when using too many heads, the performance drops significantly due to overfitting. We can observe that $H = 14$ is the most suitable number of attention heads.

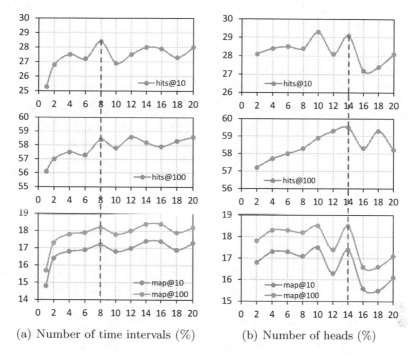

(a) Number of time intervals (%)          (b) Number of heads (%)

**Fig. 3.** Parameter analysis on the Twitter dataset.

# 7   Conclusion and Future Work

In this paper, we study the information diffusion prediction problem. To learn the dynamic preference of users for prediction, we propose a novel dynamic heterogeneous graph network to encode both the social and dynamic diffusion graph structure. We conduct experiments on three real-world datasets. The experimental results show that our model achieves significant improvements over state-of-the-art models, which shows the effectiveness and feasibility of the model for real-world applications.

For future work, we will study the text content of the diffused information, which is not applied in this work. If the users show preference about some particular topic or content, it is likely for users to repost them. So whether a user will repost the message is also determined by the content. Thus, it is worth further studying to help improve diffusion prediction performance.

**Acknowledgements.** We gratefully thank the anonymous reviewers for their insightful comments. This research is supported in part by the National Key Research and Development Program of China under Grant 2018YFC0806900.

# References

1. Bourigault, S., Lamprier, S., Gallinari, P.: Representation learning for information diffusion through social networks: an embedded cascade model. In: Proceedings of the Ninth ACM International Conference on Web Search and Data Mining, pp. 573–582 (2016)
2. Du, N., Dai, H., Trivedi, R., Upadhyay, U., Gomez-Rodriguez, M., Song, L.: Recurrent marked temporal point processes: embedding event history to vector. In: Proceedings of the 22nd ACM SIGKDD International Conference on Knowledge Discovery and Data Mining, pp. 1555–1564 (2016)
3. Glorot, X., Bengio, Y.: Understanding the difficulty of training deep feedforward neural networks. In: Proceedings of the Thirteenth International Conference on Artificial Intelligence and Statistics, pp. 249–256 (2010)
4. Gomez-Rodriguez, M., Leskovec, J., Krause, A.: Inferring networks of diffusion and influence. ACM Trans. Knowl. Discovery Data (TKDD) 5(4), 1–37 (2012)
5. Granovetter, M.: Threshold models of collective behavior. Am. J. Sociol. 83(6), 1420–1443 (1978)
6. Guille, A., Hacid, H.: A predictive model for the temporal dynamics of information diffusion in online social networks. In: Proceedings of the 21st International Conference on World Wide Web, pp. 1145–1152 (2012)
7. Hodas, N.O., Lerman, K.: The simple rules of social contagion. Scientific Reports 4, 4343 (2014)
8. Islam, M.R., Muthiah, S., Adhikari, B., Prakash, B.A., Ramakrishnan, N.: Deepdiffuse: Predicting the 'who' and 'when' in cascades. In: IEEE International Conference on Data Mining (ICDM), pp. 1055–1060. IEEE (2018)
9. Kempe, D., Kleinberg, J.M., Tardos, É.: Maximizing the spread of influence through a social network. In: Proceedings of the ninth ACM SIGKDD International Conference on Knowledge Discovery and Data Mining, pp. 137–146 (2003)
10. Kingma, D.P., Ba, J.: Adam: a method for stochastic optimization. arXiv preprint arXiv:1412.6980 (2014)
11. Kipf, T.N., Welling, M.: Semi-supervised classification with graph convolutional networks. In: ICLR 2017 (2017)
12. Leskovec, J., Backstrom, L., Kleinberg, J.: Meme-tracking and the dynamics of the news cycle. In: Proceedings of the 15th ACM SIGKDD International Conference on Knowledge Discovery and Data Mining. pp. 497–506 (2009)
13. Li, D., Zhang, S., Sun, X., Zhou, H., Li, S., Li, X.: Modeling information diffusion over social networks for temporal dynamic prediction. IEEE Trans. Knowl. Data Eng. 29(9), 1985–1997 (2017)
14. Liu, Z., Chen, C., Yang, X., Zhou, J., Li, X., Song, L.: Heterogeneous graph neural networks for malicious account detection. In: Proceedings of the 27th ACM International Conference on Information and Knowledge Management, pp. 2077–2085. ACM (2018)
15. Mou, L., et al.: Natural language inference by tree-based convolution and heuristic matching. In: ACL'16, pp. 130–136 (2016)
16. Paszke, A., et al.: Automatic differentiation in pytorch. In: NIPS-W (2017)
17. Qiu, J., Tang, J., Ma, H., Dong, Y., Wang, K., Tang, J.: Deepinf: social influence prediction with deep learning. In: Proceedings of the 24th ACM SIGKDD International Conference on Knowledge Discovery and Data Mining, pp. 2110–2119 (2018)

18. Saito, K., Kimura, M., Ohara, K., Motoda, H.: Learning continuous-time information diffusion model for social behavioral data analysis. In: Zhou, Z.-H., Washio, T. (eds.) ACML 2009. LNCS (LNAI), vol. 5828, pp. 322–337. Springer, Heidelberg (2009). https://doi.org/10.1007/978-3-642-05224-8_25

19. Saito, K., Ohara, K., Yamagishi, Y., Kimura, M., Motoda, H.: Learning diffusion probability based on node attributes in social networks. In: Kryszkiewicz, M., Rybinski, H., Skowron, A., Raś, Z.W. (eds.) ISMIS 2011. LNCS (LNAI), vol. 6804, pp. 153–162. Springer, Heidelberg (2011). https://doi.org/10.1007/978-3-642-21916-0_18

20. Tambuscio, M., Ruffo, G., Flammini, A., Menczer, F.: Fact-checking effect on viral hoaxes: a model of misinformation spread in social networks. In: Proceedings of the 24th international conference on World Wide Web, pp. 977–982 (2015)

21. Vaswani, A., et al.: Attention is all you need. In: Advances in Neural Information Processing Systems, pp. 5998–6008 (2017)

22. Wang, J., Zheng, V.W., Liu, Z., Chang, K.C.C.: Topological recurrent neural network for diffusion prediction. In: 2017 IEEE International Conference on Data Mining (ICDM), pp. 475–484. IEEE (2017)

23. Wang, Y., Shen, H., Liu, S., Gao, J., Cheng, X.: Cascade dynamics modeling with attention-based recurrent neural network. In: IJCAI. pp. 2985–2991 (2017)

24. Wang, Z., Chen, C., Li, W.: Attention network for information diffusion prediction. In: Companion Proceedings of the The Web Conference 2018, pp. 65–66 (2018)

25. Wang, Z., Chen, C., Li, W.: A sequential neural information diffusion model with structure attention. In: Proceedings of the 27th ACM International Conference on Information and Knowledge Management, pp. 1795–1798 (2018)

26. Yang, C., Sun, M., Liu, H., Han, S., Liu, Z., Luan, H.: Neural diffusion model for microscopic cascade prediction. arXiv preprint arXiv:1812.08933 (2018)

27. Yang, C., Tang, J., Sun, M., Cui, G., Liu, Z.: Multi-scale information diffusion prediction with reinforced recurrent networks. In: Proceedings of the 28th International Joint Conference on Artificial Intelligence, pp. 4033–4039. AAAI (2019)

28. Yang, J., Leskovec, J.: Modeling information diffusion in implicit networks. In: IEEE International Conference on Data Mining, pp. 599–608. IEEE (2010)

29. Yang, Y., et al.: Rain: social role-aware information diffusion. In: Proceedings of the Twenty-Ninth AAAI Conference on Artificial Intelligence, pp. 367–373 (2015)

30. Yuan, C., Ma, Q., Zhou, W., Han, J., Hu, S.: Jointly embedding the local and global relations of heterogeneous graph for rumor detection. In: IEEE International Conference on Data Mining (ICDM), pp. 796–805. IEEE (2019)

31. Yuan, C., Zhou, W., Ma, Q., Lv, S., Han, J., Hu, S.: Learning review representations from user and product level information for spam detection. In: The 19th IEEE International Conference on Data Mining, pp. 1444–1449. IEEE (2019)

32. Zhang, Y., Lyu, T., Zhang, Y.: Cosine: community-preserving social network embedding from information diffusion cascades. In: Thirty-Second AAAI Conference on Artificial Intelligence, pp. 2620–2627 (2018)

33. Zhong, E., Fan, W., Wang, J., Xiao, L., Li, Y.: Comsoc: adaptive transfer of user behaviors over composite social network. In: Proceedings of the 18th ACM SIGKDD International Conference on Knowledge Discovery and Data Mining, pp. 696–704 (2012)

# A Self-attention Network Based Node Embedding Model

Dai Quoc Nguyen$^{(\boxtimes)}$, Tu Dinh Nguyen, and Dinh Phung

Monash University, Melbourne, Australia
{dai.nguyen,dinh.phung}@monash.edu, nguyendinhtu@gmail.com

**Abstract.** Despite several signs of progress have been made recently, limited research has been conducted for an inductive setting where embeddings are required for newly unseen nodes – a setting encountered commonly in practical applications of deep learning for graph networks. This significantly affects the performances of downstream tasks such as node classification, link prediction or community extraction. To this end, we propose SANNE – a novel unsupervised embedding model – whose central idea is to employ a transformer self-attention network to iteratively aggregate vector representations of nodes in random walks. Our SANNE aims to produce plausible embeddings not only for present nodes, but also for newly unseen nodes. Experimental results show that the proposed SANNE obtains state-of-the-art results for the node classification task on well-known benchmark datasets.

**Keywords:** Node embeddings · Transformer · Self-attention network · Node classification

## 1 Introduction

Graph-structured data appears in plenty of fields in our real-world from social networks, citation networks, knowledge graphs and recommender systems to telecommunication networks, biological networks [3–5,11]. In graph-structured data, nodes represent individual entities, and edges represent relationships and interactions among those entities. For example, in citation networks, each document is treated as a node, and a citation link between two documents is treated as an edge.

Learning node embeddings is one of the most important and active research topics in representation learning for graph-structured data. There have been many models proposed to embed each node into a continuous vector as summarized in [29]. These vectors can be further used in downstream tasks such as node classification, i.e., using learned node embeddings to train a classifier to predict node labels. Existing models mainly focus on the *transductive* setting where a model is trained using the entire input graph, i.e., the model requires all nodes with a fixed graph structure during training and lacks the flexibility in inferring embeddings for unseen/new nodes, e.g., DeepWalk [22], LINE [24], Node2Vec [8],

F. Hutter et al. (Eds.): ECML PKDD 2020, LNAI 12459, pp. 364–377, 2021.
https://doi.org/10.1007/978-3-030-67664-3_22

SDNE [27] and GCN [15]. By contrast, a more important setup, but less mentioned, is the *inductive* setting wherein only a part of the input graph is used to train the model, and then the learned model is used to infer embeddings for new nodes [28]. Several attempts have additionally been made for the inductive settings such as EP-B [6], GraphSAGE [10] and GAT [26]. Working on the inductive setting is particularly more difficult than that on the transductive setting due to lacking the ability to generalize to the graph structure for new nodes.

One of the most convenient ways to learn node embeddings is to adopt the idea of a word embedding model by viewing each node as a word and each graph as a text collection of random walks to train a Word2Vec model [19], e.g., DeepWalk, LINE and Node2Vec. Although these Word2Vec-based approaches allow the current node to be directly connected with $k$-hops neighbors via random walks, they ignore feature information of nodes. Besides, recent research has raised attention in developing graph neural networks (GNNs) for the node classification task, e.g., GNN-based models such as GCN, GraphSAGE and GAT. These GNN-based models iteratively update vector representations of nodes over their $k$-hops neighbors using multiple layers stacked on top of each other. Thus, it is difficult for the GNN-based models to infer plausible embeddings for new nodes when their $k$-hops neighbors are also unseen during training.

The transformer self-attention network [25] has been shown to be very powerful in many NLP tasks such as machine translation and language modeling. Inspired by this attention technique, we present SANNE – an unsupervised learning model that adapts a transformer self-attention network to learn node embeddings. SANNE uses random walks (generated for every node) as inputs for a stack of attention layers. Each attention layer consists of a self-attention sub-layer followed by a feed-forward sub-layer, wherein the self-attention sub-layer is constructed using query, key and value projection matrices to compute pairwise similarities among nodes. Hence SANNE allows a current node at each time step to directly attend its $k$-hops neighbors in the input random walks. SANNE then samples a set of (1-hop) neighbors for each node in the random walk and uses output vector representations from the last attention layer to infer embeddings for these neighbors. As a consequence, our proposed SANNE produces the plausible node embeddings for both the transductive and inductive settings.

In short, our main contributions are as follows:

– Our SANNE induces a transformer self-attention network not only to work in the transductive setting advantageously, but also to infer the plausible embeddings for new nodes in the inductive setting effectively.
– The experimental results show that our unsupervised SANNE obtains better results than up-to-date unsupervised and supervised embedding models on three benchmark datasets CORA, CITESEER and PUBMED for the transductive and inductive settings. In particular, SANNE achieves relative error reductions of more than 14% over GCN and GAT in the inductive setting.

## 2   Related Work

DeepWalk [22] generates unbiased random walks starting from each node, considers each random walk as a sequence of nodes, and employs Word2Vec [19] to learn node embeddings. Node2Vec [8] extends DeepWalk by introducing a biased random walk strategy that explores diverse neighborhoods and balances between exploration and exploitation from a given node. LINE [24] closely follows Word2Vec, but introduces node importance, for which each node has a different weight to each of its neighbors, wherein weights can be pre-defined through algorithms such as PageRank [21]. DDRW [17] jointly trains a DeepWalk model with a Support Vector Classification [7] in a supervised manner.

SDNE [27], an autoencoder-based supervised model, is proposed to preserve both local and global graph structures. EP-B [6] is introduced to explore the embeddings of node attributes (such as words) with node neighborhoods to infer the embeddings of unseen nodes. Graph Convolutional Network (GCN) [15], a semi-supervised model, utilizes a variant of convolutional neural networks (CNNs) which makes use of layer-wise propagation to aggregate node features (such as profile information and text attributes) from the neighbors of a given node. GraphSAGE [10] extends GCN in using node features and neighborhood structures to generalize to unseen nodes. Another extension of GCN is Graph Attention Network (GAT) [26] that uses a similar idea with LINE [24] in assigning different weights to different neighbors of a given node, but learns these weights by exploring an attention mechanism technique [2]. These GNN-based approaches construct multiple layers stacked on top of each other to indirectly attend $k$-hops neighbors; thus, it is not straightforward for these approaches to infer the plausible embeddings for new nodes especially when their neighbors are also not present during training.

## 3   The Proposed SANNE

Let us define a graph as $\mathcal{G} = (\mathcal{V}, \mathcal{E})$, in which $\mathcal{V}$ is a set of nodes and $\mathcal{E}$ is a set of edges, i.e., $\mathcal{E} \subseteq \{(\mathsf{u}, \mathsf{v}) | \mathsf{u}, \mathsf{v} \in \mathcal{V}\}$. Each node $\mathsf{v} \in \mathcal{V}$ is associated with a feature vector $\mathbf{x}_\mathsf{v} \in \mathbb{R}^d$ representing node features. In this section, we detail the learning process of our proposed SANNE to learn a node embedding $\mathbf{o}_\mathsf{v}$ for each node $\mathsf{v} \in \mathcal{V}$. We then use the learned node embeddings to classify nodes into classes.

**SANNE Architecture.** Particularly, we follow DeepWalk [22] to uniformly sample random walks of length $N$ for every node in $\mathcal{V}$. For example, Fig. 1 shows a graph consisting of 6 nodes where we generate a random walk of length $N = 4$ for node 1, e.g., $\{1, 2, 3, 4\}$; and then this random walk is used as an input for the SANNE learning process.

Given a random walk $w$ of $N$ nodes $\{\mathsf{v}_{w,i}\}_{i=1}^N$, we obtain an input sequence of vector representations $\{\mathbf{u}_{\mathsf{v}_{w,i}}^0\}_{i=1}^N$: $\mathbf{u}_{\mathsf{v}_{w,i}}^0 = \mathbf{x}_{\mathsf{v}_{w,i}}$. We construct a stack of $K$ attention layers [25], in which each of them has the same structure consisting of a multi-head self-attention sub-layer followed by a feed-forward sub-layer together with additionally using a residual connection [12] and layer normalization [1]

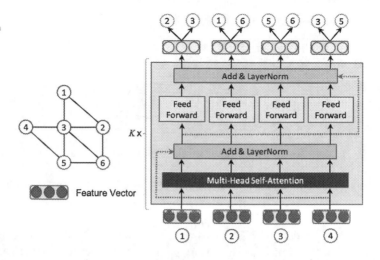

**Fig. 1.** Illustration of our SANNE learning process with $d = 3$, $N = 4$ and $M = 2$.

around each of these sub-layers. At the $k$-th attention layer, we take an input sequence $\{\mathbf{u}_{v_{w,i}}^{(k-1)}\}_{i=1}^{N}$ and produce an output sequence $\{\mathbf{u}_{v_{w,1}}^{(k)}\}_{i=1}^{N}$, $\mathbf{u}_{v_{w,1}}^{(k)} \in \mathbb{R}^d$ as:

$$\mathbf{u}_{v_{w,i}}^{(k)} = \text{WALK-TRANSFORMER}\left(\mathbf{u}_{v_{w,i}}^{(k-1)}\right)$$

In particular, $\quad \mathbf{u}_{v_{w,i}}^{(k)} = \text{LAYERNORM}\left(\mathbf{y}_{v_{w,i}}^{(k)} + \text{FF}\left(\mathbf{y}_{v_{w,i}}^{(k)}\right)\right)$

with $\quad \mathbf{y}_{v_{w,i}}^{(k)} = \text{LAYERNORM}\left(\mathbf{u}_{v_{w,i}}^{(k-1)} + \text{ATT}\left(\mathbf{u}_{v_{w,i}}^{(k-1)}\right)\right)$

where $\text{FF}(.)$ and $\text{ATT}(.)$ denote a two-layer feed-forward network and a multi-head self-attention network respectively:

$$\text{FF}\left(\mathbf{y}_{v_{w,i}}^{(k)}\right) = \mathbf{W}_2^{(k)}\text{ReLU}\left(\mathbf{W}_1^{(k)}\mathbf{y}_{v_{w,i}}^{(k)} + \mathbf{b}_1^{(k)}\right) + \mathbf{b}_2^{(k)}$$

where $\mathbf{W}_1^{(k)}$ and $\mathbf{W}_2^{(k)}$ are weight matrices, and $\mathbf{b}_1^{(k)}$ and $\mathbf{b}_2^{(k)}$ are bias parameters. And:

$$\text{ATT}\left(\mathbf{u}_{v_{w,i}}^{(k-1)}\right) = \mathbf{W}^{(k)}\left[\mathbf{h}_{v_{w,i}}^{(k),1}; \mathbf{h}_{v_{w,i}}^{(k),2}; ...; \mathbf{h}_{v_{w,i}}^{(k),H}\right]$$

where $\mathbf{W}^{(k)} \in \mathbb{R}^{d \times Hs}$ is a weight matrix, $H$ is the number of attention heads, and $[;]$ denotes a vector concatenation. Regarding the $h$-th attention head, $\mathbf{h}_{v_{w,i}}^{(k),h} \in \mathbb{R}^s$ is calculated by a weighted sum as:

$$\mathbf{h}_{v_{w,i}}^{(k),h} = \sum_{j=1}^{N} \alpha_{i,j,h}^{(k)}\left(\mathbf{W}_h^{(k),V}\mathbf{u}_{v_{w,j}}^{(k-1)}\right)$$

where $\mathbf{W}_h^{(k),V} \in \mathbb{R}^{s \times d}$ is a value projection matrix, and $\alpha_{i,j,h}^{(k)}$ is an attention weight. $\alpha_{i,j,h}^{(k)}$ is computed using the softmax function over scaled dot products between $i$-th and $j$-th nodes in the walk $w$ as:

$$\alpha_{i,j,h}^{(k)} = \mathsf{softmax}\left(\frac{\left(\mathbf{W}_h^{(k),Q}\mathbf{u}_{\mathsf{v}_{w,i}}^{(k-1)}\right)^{\mathsf{T}}\left(\mathbf{W}_h^{(k),K}\mathbf{u}_{\mathsf{v}_{w,j}}^{(k-1)}\right)}{\sqrt{k}}\right)$$

where $\mathbf{W}_h^{(k),Q}$ and $\mathbf{W}_h^{(k),K} \in \mathbb{R}^{s \times d}$ are query and key projection matrices respectively.

We randomly sample a fixed-size set of $M$ neighbors for each node in the random walk $w$. We then use the output vector representations $\mathbf{u}_{\mathsf{v}_{w,i}}^{(K)}$ from the $K$-th last layer to infer embeddings $\mathbf{o}$ for sampled neighbors of $\mathsf{v}_{w,i}$. Figure 1 illustrates our proposed SANNE where we set the length $N$ of random walks to 4, the dimension size $d$ of feature vectors to 3, and the number $M$ of sampling neighbors to 2. We also sample different sets of neighbors for the same input node at each training step.

---

**Algorithm 1:** The SANNE learning process.

---

1 **Input**: A network graph $\mathcal{G} = (\mathcal{V}, \mathcal{E})$.
2 **for** $\mathsf{v} \in \mathcal{V}$ **do**
3 $\quad \lfloor$ SAMPLE $T$ random walks of length $N$ rooted by $\mathsf{v}$.

4 **for** *each* random walk $w$ **do**
5 $\quad$ **for** $k = 1, 2, ..., K$ **do**
6 $\quad\quad$ $\forall \mathsf{v} \in w$
7 $\quad\quad$ $\mathbf{y}_{\mathsf{v}}^{(k)} \leftarrow$ LAYERNORM $\left(\mathbf{u}_{\mathsf{v}}^{(k-1)} + \text{ATT}\left(\mathbf{u}_{\mathsf{v}}^{(k-1)}\right)\right)$
8 $\quad\quad$ $\mathbf{u}_{\mathsf{v}}^{(k)} \leftarrow$ LAYERNORM $\left(\mathbf{y}_{\mathsf{v}}^{(k)} + \text{FF}\left(\mathbf{y}_{\mathsf{v}}^{(k)}\right)\right)$

9 $\quad$ **for** $\mathsf{v} \in w$ **do**
10 $\quad\quad$ SAMPLE a set $\mathsf{C}_{\mathsf{v}}$ of $M$ neighbors of node $\mathsf{v}$.
11 $\quad\quad$ $\mathbf{o}_{\mathsf{v}'} \leftarrow \mathbf{u}_{\mathsf{v}}^{(K)}, \forall \mathsf{v}' \in \mathsf{C}_{\mathsf{v}}$

---

**Training SANNE:** We learn our model's parameters including the weight matrices and node embeddings by minimizing the sampled softmax loss function [13] applied to the random walk $w$ as:

$$\mathcal{L}_{\mathsf{SANNE}}(w) = -\sum_{i=1}^{N}\sum_{\mathsf{v}' \in \mathsf{C}_{\mathsf{v}_{w,i}}} \log \frac{\exp(\mathbf{o}_{\mathsf{v}'}^{\mathsf{T}}\mathbf{u}_{\mathsf{v}_{w,i}}^{(K)})}{\sum_{\mathsf{u} \in \mathcal{V}'} \exp(\mathbf{o}_{\mathsf{u}}^{\mathsf{T}}\mathbf{u}_{\mathsf{v}_{w,i}}^{(K)})}$$

where $\mathsf{C}_{\mathsf{v}}$ is the fixed-size set of $M$ neighbors randomly sampled for node $\mathsf{v}$, $\mathcal{V}'$ is a subset sampled from $\mathcal{V}$, and $\mathbf{o}_{\mathsf{v}} \in \mathbb{R}^d$ is the node embedding of node $\mathsf{v}, \forall \mathsf{v} \in \mathcal{V}$. Node embeddings $\mathbf{o}_{\mathsf{v}}$ are learned implicitly as model parameters.

We briefly describe the learning process of our proposed SANNE model in Algorithm 1. Here, the learned node embeddings $\mathbf{o}_v$ are used as the final representations of nodes $v \in \mathcal{V}$. We explicitly aggregate node representations from both left-to-right and right-to-left sides in the walk for each node in predicting its neighbors. This allows SANNE to infer the plausible node embeddings even in the inductive setting.

---

**Algorithm 2:** The embedding inference for new nodes.

1 **Input**: A network graph $\mathcal{G} = (\mathcal{V}, \mathcal{E})$, a trained model $\text{SANNE}_{trained}$ for $\mathcal{G}$, a set $\mathcal{V}_{new}$ of new nodes.
2 **for** $v \in \mathcal{V}_{new}$ **do**
3     SAMPLE $Z$ random walks $\{w_i\}_{i=1}^{Z}$ of length $N$ rooted by $v$.
4     **for** $i \in \{1, 2, ..., Z\}$ **do**
5        $\mathbf{u}_{v,i}^{(K)} \leftarrow \text{SANNE}_{trained}(w_i)[0]$
6     $\mathbf{o}_v \leftarrow \text{AVERAGE}\left(\{\mathbf{u}_{v,i}^{(K)}\}_{i=1}^{Z}\right)$

---

**Inferring Embeddings for New Nodes in the Inductive Setting:** After training our SANNE on a given graph, we show in Algorithm 2 our method to infer an embedding for a new node $v$ adding to this given graph. We randomly sample $Z$ random walks of length $N$ starting from $v$. We use each of these walks as an input for our trained model and then collect the first vector representation (at the index 0 corresponding to node $v$) from the output sequence at the $K$-th last layer. Thus, we obtain $Z$ vectors and then average them into a final embedding for the new node $v$.

## 4 Experiments

Our SANNE is evaluated for the node classification task as follows: (i) We train our model to obtain node embeddings. (ii) We use these node embeddings to learn a logistic regression classifier to predict node labels. (iii) We evaluate the classification performance on benchmark datasets and then analyze the effects of hyper-parameters.

### 4.1 Datasets and Data Splits

**Datasets.** We use three well-known benchmark datasets CORA, CITESEER [23] and PUBMED [20] which are citation networks. For each dataset, each node represents a document, and each edge represents a citation link between two documents. Each node is assigned a class label representing the main topic of the document. Besides, each node is also associated with a feature vector of a bag-of-words. Table 1 reports the statistics of these three datasets.

**Table 1.** Statistics of the experimental datasets. |Vocab.| denotes the vocabulary size. Avg.W denotes the average number of words per node.

| Dataset | $|\mathcal{V}|$ | $|\mathcal{E}|$ | #Classes | |Vocab.| | Avg.W |
|---------|------|------|----------|---------|-------|
| CORA | 2,708 | 5,429 | 7 | 1,433 | 18 |
| CITESEER | 3,327 | 4,732 | 6 | 3,703 | 31 |
| PUBMED | 19,717 | 44,338 | 3 | 500 | 50 |

**Data Splits.** We follow the same settings used in [6] for a fair comparison. For each dataset, we uniformly sample 20 random nodes for each class as training data, 1000 different random nodes as a validation set, and 1000 different random nodes as a test set. We repeat 10 times to have 10 training sets, 10 validation sets, and 10 test sets respectively, and finally report the mean and standard deviation of the accuracy results over 10 data splits.

## 4.2    Training Protocol

**Feature Vectors Initialized by Doc2Vec.** For each dataset, each node represents a document associated with an existing feature vector of a bag-of-words. Thus, we train a PV-DBOW Doc2Vec model [16] to produce new 128-dimensional embeddings $\mathbf{x}_v$ which are considered as new feature vectors for nodes v. Using this initialization is convenient and efficient for our proposed SANNE compared to using the feature vectors of bag-of-words.

**Positional Embeddings.** We hypothesize that the relative positions among nodes in the random walks are useful to provide meaningful information about the graph structure. Hence we add to each position $i$ in the random walks a pre-defined positional embedding $\mathbf{t}_i \in \mathbb{R}^d, i \in \{1, 2, ..., N\}$ using the sinusoidal functions [25], so that we can use $\mathbf{u}^0_{v_{w,i}} = \mathbf{x}_{v_{w,i}} + \mathbf{t}_i$ where $t_{i,2j} = \sin(i/10000^{2j/d})$ and $t_{i,2j+1} = \cos(i/10000^{2j/d})$. From preliminary experiments, adding the positional embeddings produces better performances on CORA and PUBMED; thus, we keep to use the positional embeddings on these two datasets.

**Transductive Setting.** *This setting is used in most of the existing approaches where we use the entire input graph, i.e., all nodes are present during training.* We fix the dimension size $d$ of feature vectors and node embeddings to 128 ($d = 128$ with respect to the Doc2Vec-based new feature vectors), the batch size to 64, the number $M$ of sampling neighbors to 4 ($M = 4$) and the number of samples in the sampled loss function to 512 ($|\mathcal{V}'| = 512$). We also sample $T$ random walks of a fixed length $N = 8$ starting from each node, wherein $T$ is empirically varied in $\{16, 32, 64, 128\}$. We vary the hidden size of the feed-forward sub-layers in $\{1024, 2048\}$, the number $K$ of attention layers in $\{2, 4, 8\}$ and the number $H$ of attention heads in $\{4, 8, 16\}$. The dimension size $s$ of attention heads is set to satisfy that $Hs = d$.

**Inductive Setting.** We use the same inductive setting as used in [6,28]. *Specifically, for each of 10 data splits, we first remove all 1000 nodes in the test set from the original graph before the training phase, so that these nodes are becoming unseen/new in the testing/evaluating phase.* We then apply the standard training process on the resulting graph. From preliminary experiments, we set the number $T$ of random walks sampled for each node on CORA and PUBMED to 128 ($T = 128$), and on CITESEER to 16 ($T = 16$). Besides, we adopt the same value sets of other hyper-parameters for tuning as used in the transductive setting to train our SANNE in this inductive setting. After training, we infer the embedding for each unseen/new node v in the test set as described in Algorithm 2 with setting $Z = 8$.

**Training SANNE to Learn Node Embeddings.** For each of the 10 data splits, to learn our model parameters in the transductive and inductive settings, we use the Adam optimizer [14] to train our model and select the initial learning rate in $\{1e^{-5}, 5e^{-5}, 1e^{-4}\}$. We run up to 50 epochs and evaluate the model for every epoch to choose the best model on the validation set.

### 4.3   Evaluation Protocol

We also follow the same setup used in [6] for the node classification task. For each of the 10 data splits, we use the learned node embeddings as feature inputs to learn a L2-regularized logistic regression classifier [7] on the training set. We monitor the classification accuracy on the validation set for every training epoch, and take the model that produces the highest accuracy on the validation set to compute the accuracy on the test set. We finally report the mean and standard deviation of the accuracies across 10 test sets in the 10 data splits.

**Baseline Models:** We compare our unsupervised SANNE with previous unsupervised models including DeepWalk (DW), Doc2Vec and EP-B; and previous supervised models consisting of Planetoid, GCN and GAT. Moreover, as reported in [9], GraphSAGE obtained low accuracies on CORA, PUBMED and CITESEER, thus we do not include GraphSAGE as a strong baseline.

The results of DeepWalk (DW), DeepWalk+BoW (DW+BoW), Planetoid, GCN and EP-B are taken from [6].[1] Note that DeepWalk+BoW denotes a concatenation between node embeddings learned by DeepWalk and the bag-of-words feature vectors. Regarding the inductive setting for DeepWalk, [6] computed the embeddings for new nodes by averaging the embeddings of their neighbors. In addition, we provide our new results for Doc2Vec and GAT using our experimental setting.

---

[1] As compared to our experimental results for Doc2Vec and GAT, showing the statistically significant differences for DeepWalk, Planetoid, GCN and EP-B against our SANNE in Table 2 is justifiable.

## 4.4   Main Results

Table 2 reports the experimental results in the transductive and inductive settings where the best scores are in bold, while the second-best in underline. As discussed in [6], the experimental setup used for GCN and GAT [15, 26] is not fair enough to show the effectiveness of existing models when the models are evaluated only using the fixed training, validation and test sets split by [28], thus we do not rely on the GCN and GAT results reported in the original papers. Here, we do include the accuracy results of GCN and GAT using the same settings used in [6].

**Table 2.** Experimental results on the CORA, PUBMED and CITESEER test sets in the transductive and inductive settings across the 10 data splits. The best score is in bold, while the second-best score is in underline. "Unsup" denotes a group of unsupervised models. "Sup" denotes a group of supervised models using node labels from the training set during training. "Semi" denotes a group of semi-supervised models also using node labels from the training set together with node feature vectors from the entire dataset during training. * denotes the statistically significant differences against our SANNE at $p < 0.05$ (using the two-tailed *paired t-test*). Numeric subscripts denote the relative error reductions over the baselines. *Note that the inductive setting* [6,28] *is used to evaluate the models when we do not access nodes in the test set during training. This inductive setting was missed in the original GCN and GAT papers which relied on the semi-supervised training process* for CORA, PUBMED and CITESEER. Regarding the inductive setting on Cora and Citeseer, many neighbors of test nodes also belong to the test set, thus these neighbors are unseen during training and then become new nodes in the testing/evaluating phase.

| Transductive | | Cora | Pubmed | Citeseer |
|---|---|---|---|---|
| Unsup | DW [22] | $71.11 \pm 2.70^*_{33.6}$ | $73.49 \pm 3.00^*_{23.3}$ | $47.60 \pm 2.34^*_{43.0}$ |
| | DW+BoW | $76.15 \pm 2.06^*_{19.6}$ | $77.82 \pm 2.19^*_{8.3}$ | $61.87 \pm 2.30^*_{21.8}$ |
| | Doc2Vec [16] | $64.90 \pm 3.07^*_{45.4}$ | $76.12 \pm 1.62^*_{14.9}$ | $64.58 \pm 1.84^*_{15.8}$ |
| | EP-B [6] | $\underline{78.05 \pm 1.49}^*_{12.6}$ | $79.56 \pm 2.10_{0.5}$ | $\mathbf{71.01 \pm 1.35}_{-2.9}$ |
| | Our **SANNE** | $\mathbf{80.83 \pm 1.94}$ | $\mathbf{79.67 \pm 1.28}$ | $\underline{70.18 \pm 2.12}$ |
| Semi | GCN [15] | $79.59 \pm 2.02_{6.1}$ | $77.32 \pm 2.66^*_{10.4}$ | $69.21 \pm 1.25_{3.1}$ |
| | GAT [26] | $81.72 \pm 2.93_{-4.8}$ | $79.56 \pm 1.99_{0.5}$ | $70.80 \pm 0.92_{-2.1}$ |
| | Planetoid [28] | $71.90 \pm 5.33^*_{31.7}$ | $74.49 \pm 4.95^*_{20.3}$ | $58.58 \pm 6.35^*_{28.0}$ |
| Inductive | | Cora | Pubmed | Citeseer |
| Unsup | DW+BoW | $68.35 \pm 1.70^*_{26.5}$ | $74.87 \pm 1.23^*_{20.6}$ | $59.47 \pm 2.48^*_{23.1}$ |
| | EP-B [6] | $\underline{73.09 \pm 1.75}^*_{13.6}$ | $79.94 \pm 2.30_{0.5}$ | $\underline{68.61 \pm 1.69}_{0.7}$ |
| | Our **SANNE** | $\mathbf{76.75 \pm 2.45}$ | $\mathbf{80.04 \pm 1.67}$ | $\mathbf{68.82 \pm 3.21}$ |
| Sup | GCN [15] | $67.76 \pm 2.11^*_{27.9}$ | $73.47 \pm 2.48^*_{24.8}$ | $63.40 \pm 0.98^*_{14.8}$ |
| | GAT [26] | $69.37 \pm 3.81^*_{24.1}$ | $71.29 \pm 3.56^*_{30.5}$ | $59.55 \pm 4.21^*_{22.9}$ |
| | Planetoid [28] | $64.80 \pm 3.70^*_{33.9}$ | $75.73 \pm 4.21^*_{17.8}$ | $61.97 \pm 3.82^*_{18.0}$ |

Regarding the transductive setting, SANNE obtains the highest scores on CORA and PUBMED, and the second-highest score on CITESEER in the group of unsupervised models. In particular, SANNE works better than EP-B on CORA, while both models produce similar scores on PUBMED. Besides, SANNE produces high competitive results compared to the up-to-date semi-supervised models GCN and GAT. Especially, SANNE outperforms GCN with relative error reductions of 6.1% and 10.4% on CORA and PUBMED respectively. Furthermore, it is noteworthy that there is no statistically significant difference between SANNE and GAT at $p < 0.05$ (using the two-tailed paired t-test)on these datasets.

EP-B is more appropriate than other models for CITESEER in the transductive setting because (i) EP-B simultaneously learns word embeddings from the texts within nodes, which are then used to reconstruct the embeddings of nodes from their neighbors; (ii) CITESEER is quite sparse; thus word embeddings can be useful in learning the node embeddings. But we emphasize that using a significant test, there is *no* difference between EP-B and our proposed SANNE on CITESEER; hence the results are comparable.

More importantly, regarding the inductive setting, SANNE obtains the highest scores on three benchmark datasets, hence these show the effectiveness of SANNE in inferring the plausible embeddings for new nodes. Especially, SANNE outperforms both GCN and GAT in this setting, e.g., SANNE achieves absolute improvements of 8.9%, 6.6% and 5.4% over GCN, and 7.3%, 8.7% and 9.3% over GAT, on CORA, PUBMED and CITESEER, respectively (with relative error reductions of more than 14% over GCN and GAT).

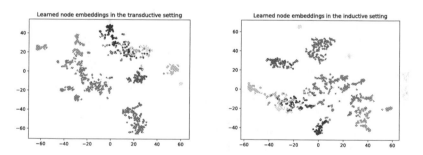

**Fig. 2.** A visualization of the learned node embeddings in the transductive and inductive settings on CORA.

Compared to the transductive setting, the inductive setting is particularly difficult due to requiring the ability to align newly observed nodes to the present nodes. As shown in Table 2, there is a significant decrease for GCN and GAT from the transductive setting to the inductive setting on all three datasets, while by contrast, our SANNE produces reasonable accuracies for both settings. To qualitatively demonstrate this advantage of SANNE, we use t-SNE [18] to visualize the learned node embeddings on one data split of the CORA dataset in Fig. 2. We see a similarity in the node embeddings (according to their labels) between two settings, verifying the plausibility of the node embeddings learned by our SANNE in the inductive setting.

## 4.5   Effects of Hyper-parameters

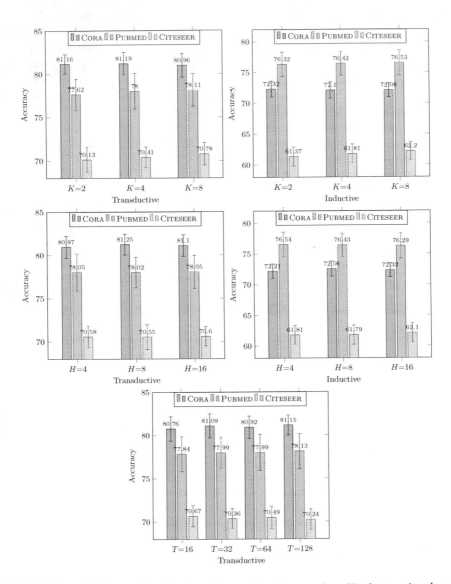

**Fig. 3.** Effects of the number $T$ of random walks, the number $K$ of attention layers and the number $H$ of attention heads on the validation sets. We fixed the same value for one hyper-parameter and tune other hyper-parameters for all 10 data splits of each dataset.

We investigate the effects of hyper-parameters on the CORA, PUBMED, and CITE-SEER validation sets of 10 data splits in Fig. 3, when we use the same value for one hyper-parameter and then tune other hyper-parameters for all 10 data splits of each dataset. Regarding the transductive setting, we see that the high accuracies can be generally obtained when using $T = 128$ on CORA and PUBMED, and $T = 16$ on CITESEER. This is probably because CITESEER are more sparse than CORA and PUBMED, especially the average number of neighbors per node on CORA and PUBMED are 2.0 and 2.2 respectively, while it is just 1.4 on CITE-SEER. This is also the reason why we set $T = 128$ on CORA and PUBMED, and $T = 16$ on CITESEER during training in the inductive setting. Besides, regarding the number $K$ of attention layers for both the transductive and inductive settings, using a small $K$ produces better results on CORA. At the same time, there is an accuracy increase on PUBMED and CITESEER along with increasing $K$. Regarding the number $H$ of attention heads, we achieve higher accuracies when using $H = 8$ on CORA in both the settings. Besides, there is not much difference in varying $H$ on PUBMED and CITESEER in the transductive setting. But in the inductive setting, using $H = 4$ gives high scores on PUBMED, while the high scores on CITESEER are obtained by setting $H = 16$.

## 4.6   Ablation Analysis

**Table 3.** Ablation results on the validation sets in the transductive setting. (i) Without using the feed-forward sub-layer: $\mathbf{u}_v^{(k)} = \text{LAYERNORM}\left(\mathbf{u}_v^{(k-1)} + \text{ATT}\left(\mathbf{u}_v^{(k-1)}\right)\right)$ (ii) Without using the multi-head self-attention sub-layer: $\mathbf{u}_v^{(k)} = \text{LAYERNORM}\left(\mathbf{u}_v^{(k-1)} + \text{FF}\left(\mathbf{u}_v^{(k-1)}\right)\right)$. * denotes the statistically significant differences at $p < 0.05$ (using the two-tailed paired t-test).

| Transductive | Cora | Pubmed | Citeseer |
|---|---|---|---|
| Our **SANNE** | 81.32 ± 1.20 | 78.28 ± 1.24 | 70.77 ± 1.18 |
| (i) w/o FF | 80.77 ± 1.34 | 77.90 ± 1.76 | 70.36 ± 1.32 |
| (ii) w/o ATT | 77.87 ± 1.09* | 74.52 ± 2.66* | 65.68 ± 1.31* |

We compute and report our ablation results on the validation sets in the transductive setting over two factors in Table 3. There is a decrease in the accuracy results when not using the feed-forward sub-layer, but we do not see a significant difference between with and without using this sub-layer (at $p < 0.05$ using the two-tailed paired t-test). More importantly, without the multi-head self-attention sub-layer, the results degrade by more than 3.2% on all three datasets, showing the merit of this self-attention sub-layer in learning the plausible node embeddings. Note that similar findings also occur in the inductive setting.

# 5    Conclusion

We introduce a novel unsupervised embedding model SANNE to leverage from the random walks to induce the transformer self-attention network to learn node embeddings. SANNE aims to infer plausible embeddings not only for present nodes but also for new nodes. Experimental results show that our SANNE obtains the state-of-the-art results on CORA, PUBMED, and CITESEER in both the transductive and inductive settings. Our code is available at: https://github.com/daiquocnguyen/Walk-Transformer.[2]

**Acknowledgements.** This research was partially supported by the ARC Discovery Projects DP150100031 and DP160103934.

# References

1. Ba, J.L., Kiros, J.R., Hinton, G.E.: Layer normalization. arXiv:1607.06450 (2016)
2. Bahdanau, D., Cho, K., Bengio, Y.: Neural machine translation by jointly learning to align and translate. arXiv preprint arXiv:1409.0473 (2015)
3. Battaglia, P.W., et al.: Relational inductive biases, deep learning, and graph networks. arXiv:1806.01261 (2018)
4. Cai, H., Zheng, V.W., Chang, K.: A comprehensive survey of graph embedding: problems, techniques and applications. IEEE Trans. Knowl. Data Eng. **30**(9), 1616–1637 (2018)
5. Chen, H., Perozzi, B., Al-Rfou, R., Skiena, S.: A tutorial on network embeddings. arXiv:1808.02590 (2018)
6. Duran, A.G., Niepert, M.: Learning graph representations with embedding propagation. In: Advances in Neural Information Processing Systems, pp. 5119–5130 (2017)
7. Fan, R.E., Chang, K.W., Hsieh, C.J., Wang, X.R., Lin, C.J.: LIBLINEAR: a library for large linear classification. J. Mach. Learn. Res. **9**, 1871–1874 (2008)
8. Grover, A., Leskovec, J.: Node2Vec: scalable feature learning for networks. In: The ACM SIGKDD Conference on Knowledge Discovery and Data Mining, pp. 855–864 (2016)
9. Guo, J., Xu, L., Chen, E.: SPINE: structural identity preserved inductive network embedding. arXiv:1802.03984 (2018)
10. Hamilton, W.L., Ying, R., Leskovec, J.: Inductive representation learning on large graphs. In: Advances in Neural Information Processing Systems, pp. 1024–1034 (2017)
11. Hamilton, W.L., Ying, R., Leskovec, J.: Representation learning on graphs: methods and applications. arXiv:1709.05584 (2017)
12. He, K., Zhang, X., Ren, S., Sun, J.: Deep residual learning for image recognition. In: Conference on Computer Vision and Pattern Recognition (CVPR), pp. 770–778 (2016)
13. Jean, S., Cho, K., Memisevic, R., Bengio, Y.: On using very large target vocabulary for neural machine translation. In: ACL, pp. 1–10 (2015)

---

[2] Our accuracy results are obtained using the implementation based on Tensorflow 1.6, and now this implementation is out-of-date. We have released the SANNE implementation based on Pytorch 1.5 for future works.

14. Kingma, D., Ba, J.: Adam: a method for stochastic optimization. arXiv preprint arXiv:1412.6980 (2015)
15. Kipf, T.N., Welling, M.: Semi-supervised classification with graph convolutional networks. In: International Conference on Learning Representations (ICLR) (2017)
16. Le, Q., Mikolov, T.: Distributed representations of sentences and documents. In: International Conference on Machine Learning, pp. 1188–1196 (2014)
17. Li, J., Zhu, J., Zhang, B.: Discriminative deep random walk for network classification. In: Proceedings of the 54th Annual Meeting of the Association for Computational Linguistics (Volume 1: Long Papers), pp. 1004–1013 (2016)
18. Van der Maaten, L., Hinton, G.: Visualizing data using t-SNE. J. Mach. Learn. Res. **9**, 2579–2605 (2008)
19. Mikolov, T., Sutskever, I., Chen, K., Corrado, G.S., Dean, J.: Distributed representations of words and phrases and their compositionality. In: Advances in Neural Information Processing Systems, pp. 3111–3119 (2013)
20. Namata, G.M., London, B., Getoor, L., Huang, B.: Query-driven active surveying for collective classification. In: Workshop on Mining and Learning with Graphs (2012)
21. Page, L., Brin, S., Motwani, R., Winograd, T.: The pagerank citation ranking: bringing order to the web. Tech. rep, Stanford InfoLab (1999)
22. Perozzi, B., Al-Rfou, R., Skiena, S.: DeepWalk: online learning of social representations. In: The ACM SIGKDD Conference on Knowledge Discovery and Data Mining, pp. 701–710 (2014)
23. Sen, P., Namata, G., Bilgic, M., Getoor, L., Galligher, B., Eliassi-Rad, T.: Collective classification in network data. AI Mag. **29**(3), 93 (2008)
24. Tang, J., Qu, M., Wang, M., Zhang, M., Yan, J., Mei, Q.: LINE: large-scale information network embedding. In: Proceedings of the 24th International Conference on World Wide Web, pp. 1067–1077 (2015)
25. Vaswani, A., et al.: Attention is all you need. In: Advances in Neural Information Processing Systems, pp. 5998–6008 (2017)
26. Veličković, P., Cucurull, G., Casanova, A., Romero, A., Liò, P., Bengio, Y.: Graph attention networks. arXiv preprint arXiv:1710.10903 (2018)
27. Wang, D., Cui, P., Zhu, W.: Structural deep network embedding. In: The ACM SIGKDD Conference on Knowledge Discovery and Data Mining, pp. 1225–1234 (2016)
28. Yang, Z., Cohen, W.W., Salakhutdinov, R.: Revisiting semi-supervised learning with graph embeddings. In: International Conference on Machine Learning, pp. 40–48 (2016)
29. Zhang, D., Yin, J., Zhu, X., Zhang, C.: Network representation learning: a survey. IEEE Trans. Big Data **6**, 3–28 (2020)

# Graph-Revised Convolutional Network

Donghan Yu$^{(\boxtimes)}$, Ruohong Zhang, Zhengbao Jiang, Yuexin Wu,
and Yiming Yang

Carnegie Mellon University, Pittsburgh, PA 15213, USA
{dyu2,ruohongz,zhengbaj,yuexinw,yiming}@cs.cmu.edu

**Abstract.** Graph Convolutional Networks (GCNs) have received increasing attention in the machine learning community for effectively leveraging both the content features of nodes and the linkage patterns across graphs in various applications. As real-world graphs are often incomplete and noisy, treating them as ground-truth information, which is a common practice in most GCNs, unavoidably leads to sub-optimal solutions. Existing efforts for addressing this problem either involve an over-parameterized model which is difficult to scale, or simply re-weight observed edges without dealing with the missing-edge issue. This paper proposes a novel framework called Graph-Revised Convolutional Network (GRCN), which avoids both extremes. Specifically, a GCN-based graph revision module is introduced for predicting missing edges and revising edge weights w.r.t. downstream tasks via joint optimization. A theoretical analysis reveals the connection between GRCN and previous work on multigraph belief propagation. Experiments on six benchmark datasets show that GRCN consistently outperforms strong baseline methods, especially when the original graphs are severely incomplete or the labeled instances for model training are highly sparse. (Our code is available at https://github.com/Maysir/GRCN).

**Keywords:** Graph convolutional network · Graph learning · Semi-supervised learning

## 1 Introduction

Graph Convolutional Networks (GCNs) have received increasing attention in recent years as they are highly effective in graph-based node feature induction and belief propagation, and widely applicable to many real-world problems, including computer vision [17,26], natural language processing [16,18], recommender systems [19,31], epidemiological forecasting [28], and more.

However, the power of GCNs has not been fully exploited as most of the models assume that the given graph perfectly depicts the ground-truth of the relationship between nodes. Such assumptions are bound to yield sub-optimal results as real-world graphs are usually highly noisy, incomplete (with many missing edges), and not necessarily ideal for different downstream tasks. Ignoring these issues is a fundamental weakness of many existing GCN methods.

© Springer Nature Switzerland AG 2021
F. Hutter et al. (Eds.): ECML PKDD 2020, LNAI 12459, pp. 378–393, 2021.
https://doi.org/10.1007/978-3-030-67664-3_23

Recent methods that attempt to modify the original graph can be split into two major streams: 1) Edge reweighting: GAT [25] and GLCN [14] use attention mechanism or feature similarity to reweight the existing edges of the given graph. Since the topological structure of the graph is not changed, the model is prone to be affected by noisy data when edges are sparse. 2) Full graph parameterization: LDS [9], on the other hand, allows every possible node pairs in a graph to be parameterized. Although this design is more flexible, the memory cost is intractable for large datasets, since the number of parameters increases quadratically with the number of nodes. Therefore, finding a balance between model expressiveness and memory consumption remains an open challenge.

To enable flexible edge editing while maintaining scalability, we develop a GCN-based graph revision module that performs edge addition and edge reweighting. In each iteration, we calculate an adjacency matrix via GCN-based node embeddings, and select the edges with high confidence to be added. Our method permits a gradient-based training of an end-to-end neural model that can predict unseen edges. Our theoretical analysis demonstrates the effectiveness of our model from the perspective of multigraph [2], which allows more than one edges from different sources between a pair of vertices. To the best of our knowledge, we are the first to reveal the connection between graph convolutional networks and multigraph propagation. Our contributions can be summarized as follows:

- We introduce a novel structure that simultaneously learns both graph revision and node classification through different GCN modules.
- Through theoretical analysis, we show our model's advantages in the view of multigraph propagation.
- Comprehensive experiments on six benchmark datasets from different domains show that our proposed model achieves the best or highly competitive results, especially under the scenarios of highly incomplete graphs or sparse training labels.

## 2   Background

We first introduce some basics of graph theory. An undirected graph $G$ can be represented as $(V, E)$ where $V$ denotes the set of vertices and $E$ denotes the set of edges. Let $N$ and $M$ be the number of vertices and edges, respectively. Each graph can also be represented by an adjacency matrix $A$ of size $N \times N$ where $A_{ij} = 1$ if there is an edge between $v_i$ and $v_j$, and $A_{ij} = 0$ otherwise. We use $A_i$ to denote the $i$-th row of the adjacency matrix. A graph with adjacency matrix $A$ is denoted as $G_A$. Usually each node $i$ has its own feature $x_i \in \mathbb{R}^F$ where $F$ is the feature dimension (for example, if nodes represent documents, the feature can be a bag-of-words vector). The node feature matrix of the whole graph is denoted as $X \in \mathbb{R}^{N \times F}$.

Graph convolutional networks generalize the convolution operation on images to graph structure data, performing layer-wise propagation of node features

[6,12,16,25]. Suppose we are given a graph with adjacency matrix $A$ and node features $H^{(0)} = X$. An $L$-layer Graph Convolution Network (GCN) [16] conducts the following inductive layer-wise propagation:

$$H^{(l+1)} = \sigma \left( \widetilde{D}^{-\frac{1}{2}} \widetilde{A} \widetilde{D}^{-\frac{1}{2}} H^{(l)} W^{(l)} \right), \tag{1}$$

where $l = 0, 1, \cdots, L-1$, $\widetilde{A} = A + I$ and $\widetilde{D}$ is a diagonal matrix with $D_{ii} = \sum_j \widetilde{A}_{ij}$. $\{W^{(0)}, \cdots, W^{(L-1)}\}$ are the model parameters and $\sigma(\cdot)$ is the activation function. The node embedding $H^{(L)}$ can be used for downsteam tasks. For node classification, GCN defines the final output as:

$$\widehat{Y} = \text{softmax} \left( H^{(L)} W^{(L)} \right). \tag{2}$$

where $\widehat{Y} \in \mathbb{R}^{N \times C}$ and $C$ denotes the number of classes. We note that in the GCN computation, $A$ is directly used as the underlining graph without any modification. Additionally, in each layer, GCN only updates node representations as a degree-normalized aggregation of neighbor nodes.

To allow for an adaptive aggregation paradigm, GLCN [14] learns to reweight the existing edges by node feature embeddings. The reweighted adjacancy matrix $\widetilde{A}$ is calculated by:

$$\widetilde{A}_{ij} = \frac{A_{ij} \exp \left( \text{ReLU} \left( a^T |x_i P - x_j P| \right) \right)}{\sum_{k=1}^n A_{ik} \exp \left( \text{ReLU} \left( a^T |x_i P - x_k P| \right) \right)}, \tag{3}$$

where $x_i$ denotes the feature vector of node $i$ and $a, P$ are model parameters. Another model GAT [25] reweights edges by a layer-wise self-attention across node-neighbor pairs to compute hidden representations. For each layer $l$, the reweighted edge is computed by:

$$\widetilde{A}_{ij}^{(l)} = \frac{A_{ij} \exp \left( a(W^{(l)} H_i^{(l)}, W^{(l)} H_j^{(l)}) \right)}{\sum_{k=1}^n A_{ik} \exp \left( a(W^{(l)} H_i^{(l)}, W^{(l)} H_k^{(l)}) \right)} \tag{4}$$

where $a(\cdot, \cdot)$ is a shared attention function to compute the attention coefficients. Compared with GLCN, GAT uses different layer-wise maskings to allow for more flexible representation. However, neither of the methods has the ability to add edges since the revised edge $\widetilde{A}_{ij}$ or $\widetilde{A}_{ij}^{(l)} \neq 0$ only if the original edge $A_{ij} \neq 0$.

In order to add new edges into the original graph, LDS [9] makes the entire adjacency matrix parameterizable. Then it jointly learns the graph structure $\theta$ and the GCN parameters $W$ by approximately solving a bilevel program as follows:

$$\min_{\theta \in \overline{\mathcal{H}}_N} \mathbb{E}_{A \sim \text{Ber}(\theta)} \left[ \zeta_{val} \left( W_\theta, A \right) \right], \tag{5}$$

$$\text{such that } W_\theta = \arg \min_W \mathbb{E}_{A \sim \text{Ber}(\theta)} [\zeta_{train}(W, A)],$$

where $A \sim \text{Ber}(\theta)$ means sampling adjacency matrix $A \in \mathbb{R}^{N \times N}$ from Bernoulli distribution under parameter $\theta \in \mathbb{R}^{N \times N}$. $\overline{\mathcal{H}}_N$ is the convex hull of the set of all

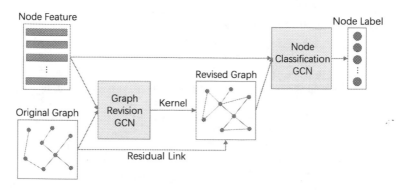

**Fig. 1.** Architecture of the proposed GRCN model for semi-supervised node classification. The node classification GCN is enhanced with a revised graph constructed by the graph revision GCN module.

adjecency matrices for $N$ nodes. $\zeta_{train}$ and $\zeta_{val}$ denote the node classification loss on training and validation data respectively. However, this method can hardly scale to large graphs since the parameter size of $\theta$ is $N^2$ where $N$ is the number of nodes. In the next section, we'll present our method which resolves the issues in previous work.

## 3   Proposed Method

### 3.1   Graph-Revised Convolutional Network

Our Graph-Revised Convolutional Network (GRCN) contains two modules: a *graph revision* module and a *node classification* module. The graph revision module adjusts the original graph by adding or reweighting edges, and the node classification module performs classification using the revised graph. Specifically, in our graph revision module, we choose to use a GCN to combine the node features and the original graph input, as GCNs are effective at fusing data from different sources [29]. We first learn the node embedding $Z \in \mathbb{R}^{N \times D}$ as follows:

$$Z = GCN_g(A, X) \tag{6}$$

where $GCN_g$ denotes the graph convolutional network for graph revision, $A$ is the original graph adjacency matrix and $X$ is node feature. Then we calculate a similarity graph $S$ based on node embedding using certain kernel function $k : \mathbb{R}^D \times \mathbb{R}^D \to \mathbb{R}$:

$$S_{ij} = k(z_i, z_j). \tag{7}$$

The revised adjacency matrix is formed by an elementwise summation of the original adjacency matrix and the calculated similarity matrix: $\widetilde{A} = A + S$. Compared with the graph revision in GAT and GLCN which use entrywise product, we instead adopt the entrywise addition operator "+" in order for new edges to be considered. In this process, the original graph $A$ is revised by the similarity graph $S$, which can insert new edges to $A$ and potentially reweight or delete existing edges in $A$. In practice, we apply a *sparsification* technique on dense matrix $S$ to reduce computational cost and memory usage, which will be introduced in the next section. Then the predicted labels are calculated by:

$$\widehat{Y} = GCN_c(\widetilde{A}, X) \tag{8}$$

where $GCN_c$ denotes the graph convolutional network for the downstream node classification task. Note that to prevent numerical instabilities, renormalization trick [16] is also applied on the revised adjacency matrix $\widetilde{A}$. Figure 1 provides an illustration of our model. Finally, we use cross-entropy loss as our objective function:

$$\zeta = -\sum_{i \in \mathcal{Y}_L} \sum_{j=1}^{C} Y_{ij} ln \widehat{Y}_{ij} \tag{9}$$

where $\mathcal{Y}_L$ is the set of node indices that have labels $Y$ and $C$ is the number of classes. It's worth emphasizing that our model does not need other loss functions to guide the graph revision process.

Overall, our model can be formulated as:

$$\begin{aligned} GRCN(A, X) &= GCN_c(\widetilde{A}, X), \\ \widetilde{A} &= A + K(GCN_g(A, X)), \end{aligned} \tag{10}$$

where $K(\cdot)$ is the kernel matrix computed from the node embeddings in Eq. (6). In our implementation, we use dot product as kernel function for simplicity, and we use a two-layer GCN [16] in both modules. Note that our framework is highly flexible for other kernel functions and graph convolutional networks, which we leave for future exploration.

## 3.2   Sparsification

Since the adjacency matrix $S$ of similarity graph is dense, directly applying it in the classification module is inefficient. Besides, we only want those edges with higher confidence to avoid introducing too much noise. Thus we conduct a $K$-nearest-neighbour (KNN) sparsification on the dense graph: for each node, we keep the edges with top-$K$ prediction scores. The adjacancy matrix of the KNN-sparse graph, denoted as $S^{(K)}$, is computed as:

$$S_{ij}^{(K)} = \begin{cases} S_{ij}, & S_{ij} \in topK(S_i), \\ 0, & S_{ij} \notin topK(S_i). \end{cases} \tag{11}$$

where $topK(S_i)$ is the set of top-$K$ values of vector $S_i$. Finally, in order to keep the symmetric property, the output sparse graph $\widehat{S}$ is calculated by:

$$\widehat{S}_{ij} = \begin{cases} \max(S_{ij}^{(K)}, S_{ji}^{(K)}), & S_{ij}^{(K)}, S_{ij}^{(K)} \geq 0 \\ \min(S_{ij}^{(K)}, S_{ji}^{(K)}), & S_{ij}^{(K)}, S_{ij}^{(K)} \leq 0 \end{cases} \tag{12}$$

Now since both original graph $A$ and similarity graph $\widehat{S}$ are sparse, efficient matrix multiplication can be applied on both GCNs as in the training time, gradients will only backpropagate through the top-$K$ values. By sparsification, the memory cost of similarity matrix is reduced from $O(N^2)$ to $O(NK)$.

## 3.3    Fast-GRCN

To further reduce the training time, we introduce a faster version of the propose model: Fast-GRCN. Note that GRCN needs to compute the dense adjacency matrix $S$ in every epoch, where the computational time complexity is $O(N^2)$. While in Fast-GRCN, the whole matrix $S$ is only calculated in the first epoch, and then the indices of the non-zero values of the KNN-sparse matrix $S^{(K)}$ are saved. For the remaining epochs, to obtain $S^{(K)}$, we only compute the values of the saved indices while directly setting zeros for other indices, which reduces the time complexity to $O(NK)$.

The intuition behind is that the top-$K$ important neighbours of each node may remain unchanged, while the weight of edges between them should still be adjusted during training.

## 3.4    Theoretical Analysis

In this section, we show the effectiveness of our model in the view of Multi-graph [2] propagation. The major observation is that for existing methods, the learned function from GCNs can be regarded as a linear combination of limited pre-defined kernels where the flexibility of kernels have a large influence on the final prediction accuracy.

We consider the simplified graph convolution neural network $GCN_s$ for the ease of analysis. That is, we remove feature transformation parameter $W$ and non-linear activation function $\sigma(\cdot)$ as:

$$GCN_s(A, X) = A^k X \tag{13}$$

where $k$ is the number of GCN layers. For simplicity we denote $A$ as the adjacency matrix with self-loop after normalization. The final output can be acquired by applying a linear or logistic regression function $f(\cdot)$ on the node embeddings above:

$$\widehat{Y} = f(GCN_s(A, X)) = f(A^k X) \tag{14}$$

where $\widehat{Y}$ denotes the predicted labels of nodes. Then the following theorem shows that under certain conditions, the optimal function $f^*$ can be expressed as a linear combination of kernel functions defined on training samples.

**Theorem 1 (Representer Theorem [22]).** *Consider a non-empty set $\mathcal{P}$ and a positive-definite real-valued kernel: $k : \mathcal{P} \times \mathcal{P} \rightarrow \mathbb{R}$ with a corresponding reproducing kernel Hilbert space $H_k$. If given: a. a set of training samples $\{(p_i, y_i) \in \mathcal{P} \times \mathbb{R} | i = 1, \cdots, n\}$, b. a strictly monotonically increasing real-valued* *function $g : [0, \infty) \to \mathbb{R}$, c. an error function $E : (\mathcal{P} \times \mathbb{R}^2)^n \to \mathbb{R} \cup \{\infty\}$, which together define the following regularized empirical risk functional on $H_k$:*

$$f \mapsto E\left((p_1, y_1, f(p_1)), \ldots, (p_n, y_n, f(p_n))\right) + g(\|f\|).$$

*Then, any minimizer of the empirical risk admits a representation of the form:*

$$f^*(\cdot) = \sum_{i=1}^{n} \alpha_i k(\cdot, p_i)$$

*where $a_i \in \mathbb{R} \ \forall i = 1, \cdots, n$.*

In our case, $p_i \in \mathbb{R}^D$ is the embedding of node $i$. As shown in the theorem, the final optimized output is the linear combination of certain kernels on node embeddings. We assume the kernel function to be dot product for simplicity, which means $k(p_i, p_j) = p_i^T p_j$. The corresponding kernel matrix can be written as:

$$K(GCN_s(A, X)) = A^k X X^T A^k = A^k B A^k \tag{15}$$

where $B = XX^T$ is the adjacency matrix of graph induced by node features. Now we have two graphs based on the same node set: original graph $G_A$ (associated with adjacency matrix $A$) and feature graph $G_B$ (associated with adjacency matrix $B$). They form a multigraph [2] where multiple edges is permitted between the same end nodes. Then the random-walk-like matrix $A^k B A^k$ can be regarded as one way to perform graph label/feature propagation on the multigraph. Its limitation is obvious: the propagation only happens once on the feature graph $G_B$, which lacks flexibility. However, for our method, we have:

$$\begin{aligned}
GRCN(A, X) &= (A + K(GCN_s(A, X)))^k X \\
&= (A + A^m X X^T A^m)^k X \\
&= (A + A^m B A^m)^k X, \\
K(GRCN(A, X)) &= (A + A^m B A^m)^k B \\
&\quad (A + A^m B A^m)^k,
\end{aligned} \tag{16}$$

where labels/features can propagate multiple times on the feature graph $G_B$. Thus our model is more flexible and more effective especially when the original graph $G_A$ is not reliable or cannot provide enough information for downstream tasks. In Eq. (16), $A + A^m B A^m$ can be regarded as a combination of different edges in the multigraph. To reveal the connection between GRCN and GLCN [14], we first consider the special case of our model that $m = 0$: $GRCN(A, X) = (A + B)^k X$. The operator "+" is analogous to the operator $OR$ which incorporates information from both graph $A$ and $B$. While GLCN [14] takes another combination denoted as $A \circ B$ using Hadamard (entrywise) product "$\circ$", which can be analogous to $AND$ operation.

**Table 1.** Data statistics

| Dataset | #Nodes | #Edges | #Feature | #Class |
|---|---|---|---|---|
| Cora | 2708 | 5429 | 1433 | 7 |
| CiteSeer | 3327 | 4732 | 3703 | 6 |
| PubMed | 19717 | 44338 | 500 | 3 |
| CoraFull | 19793 | 65311 | 8710 | 70 |
| Amazon computers | 13381 | 245778 | 767 | 10 |
| Coauthor CS | 18333 | 81894 | 6805 | 15 |

We can further extend our model to a layer-wise version for comparison to GAT [25]. More specifically, for the $l$-th layer, we denote the input as $X_l$. The output $X_{l+1}$ is then calculated by:

$$\begin{aligned} X_{l+1} &= (A + K(GCN_s(A, X_l)))X_l \\ &= (A + A^m X_l X_l^T A^m)X_l \qquad\qquad (17) \\ &= (A + A^m B_l A^m)X_l, \end{aligned}$$

where $B_l = X_l X_l^T$. Similar to the analysis mentioned before, if we consider the special case of GRCN that $m = 0$ and change the edge combination operator from entrywise sum "+" to entrywise product "∘", we have $X_{l+1} = (A \circ B_l)X_l$, which is the key idea behind GAT [25]. Due to the property of entrywise product, the combined edges of both GAT and GLCN are only the reweighted edges of $A$, which becomes ineffective when the original graph $G_A$ is highly sparse. Through the analysis above, we see that our model is more general in combining different edges by varying the value of $m$, and also has more robust combination operator "+" compared to previous methods.

## 4   Experiments

### 4.1   Datasets and Baselines

We use six benchmark datasets for semi-supervised node classification evaluation. Among them, Cora, CiteSeer [23] and PubMed [20] are three commonly used datasets. The data split is conducted by two ways. The first is the fixed split originating from [30]. In the second way, we conduct 10 random splits while keeping the same number of labels for training, validation and testing as previous work. This provides a more robust comparison of the model performance. To further test the scalability of our model, we utilize three other datasets: Cora-Full [4], Amazon-Computers and Coauthor CS [24]. The first is an extended version of Cora, while the second and the third are co-purchase and co-authorship graphs respectively. On these three datasets, we follow the previous work [24] and take 20 labels of each classes for training, 30 for validation, and the rest for

**Table 2.** Mean test classification accuracy and standard deviation in percent averaged for all models and all datasets. For each dataset, the highest accuracy score is marked in **bold**, and the second highest score is marked using ∗. N/A stands for the datasets that couldn't be processed by the full-batch version because of GPU RAM limitations.

| Models | Cora (fix. split) | CiteSeer (fix. split) | PubMed (fix. split) |
|---|---|---|---|
| GCN | $81.4 \pm 0.5$ | $70.9 \pm 0.5$ | $\mathbf{79.0 \pm 0.3}$ |
| SGC | $81.0 \pm 0.0$ | $71.9 \pm 0.1$ | $78.9 \pm 0.0^*$ |
| GAT | $83.2 \pm 0.7$ | $72.6 \pm 0.6$ | $78.8 \pm 0.3$ |
| LDS | $84.0 \pm 0.4^*$ | $\mathbf{74.8 \pm 0.5}$ | N/A |
| GLCN | $81.8 \pm 0.6$ | $70.8 \pm 0.5$ | $78.8 \pm 0.4$ |
| Fast-GRCN | $83.6 \pm 0.4$ | $72.9 \pm 0.6$ | $\mathbf{79.0 \pm 0.2}$ |
| GRCN | $\mathbf{84.2 \pm 0.4}$ | $73.6 \pm 0.5^*$ | $\mathbf{79.0 \pm 0.2}$ |
| Models | Cora (rand. splits) | CiteSeer (rand. splits) | PubMed (rand. splits) |
| GCN | $81.2 \pm 1.9$ | $69.8 \pm 1.9$ | $77.7 \pm 2.9^*$ |
| SGC | $81.0 \pm 1.7$ | $68.9 \pm 2.0$ | $75.8 \pm 3.0$ |
| GAT | $81.7 \pm 1.9$ | $68.8 \pm 1.8$ | $77.7 \pm 3.2^*$ |
| LDS | $81.6 \pm 1.0$ | $71.0 \pm 0.9$ | N/A |
| GLCN | $81.4 \pm 1.9$ | $69.8 \pm 1.8$ | $77.2 \pm 3.2$ |
| Fast-GRCN | $\mathbf{83.8 \pm 1.6}$ | $72.3 \pm 1.4^*$ | $77.6 \pm 3.2$ |
| GRCN | $83.7 \pm 1.7^*$ | $\mathbf{72.6 \pm 1.3}$ | $\mathbf{77.9 \pm 3.2}$ |
| Models | Cora-Full | Amazon computers | Coauthor CS |
| GCN | $\mathbf{60.3 \pm 0.7}$ | $81.9 \pm 1.7$ | $\mathbf{91.3 \pm 0.3}$ |
| SGC | $59.1 \pm 0.7$ | $81.8 \pm 2.3$ | $\mathbf{91.3 \pm 0.2}$ |
| GAT | $59.9 \pm 0.6$ | $81.8 \pm 2.0$ | $89.5 \pm 0.5$ |
| LDS | N/A | N/A | N/A |
| GLCN | $59.1 \pm 0.7$ | $80.4 \pm 1.9$ | $90.1 \pm 0.5$ |
| Fast-GRCN | $60.2 \pm 0.5^*$ | $83.5 \pm 1.6^*$ | $91.2 \pm 0.4^*$ |
| GRCN | $\mathbf{60.3 \pm 0.4}$ | $\mathbf{83.7 \pm 1.8}$ | $\mathbf{91.3 \pm 0.3}$ |

testing. We also delete the classes with less than 50 labels to make sure each class contains enough instances. The data statistics are shown in Table 1.

We compare the effectiveness of our GRCN model with several baselines, where the first two models are vanilla graph convolutional networks without any graph revision: GCN [16], SGC [27], GAT [25], LDS [9], and GLCN [14].

## 4.2   Implementation Details

Transductive setting is used for node classification on all the datasets. We train GRCN for 300 epochs using Adam [15] and select the model with highest validation accuracy for test. We set learning rate as $1 \times 10^{-3}$ for graph refinement module and $5 \times 10^{-3}$ for label prediction module. Weight decay and sparsification parameter $K$ are tuned by grid search on validation set, with the search space $[1 \times 10^{-4}, 5 \times 10^{-4}, 1 \times 10^{-3}, 5 \times 10^{-3}, 1 \times 10^{-2}, 5 \times 10^{-2}]$ and $[5, 10, 20, 30, 50, 100, 200]$ respectively. Our code is based on Pytorch [21] and one geometric deep learning extension library [8], which provides implementation for GCN, SGC and GAT. For LDS, the results were obtained using the publicly available code. Since an implementation for GLCN was not available, we report the results based on our own implementation of the original paper.

### 4.3   Main Results

Table 2 shows the mean accuracy and the corresponding standard deviation for all models across the 6 datasets averaged over 10 different runs[1]. We see that our proposed models achieve the best or highly competitive results for all the datasets. The effectiveness of our model over the other baselines demonstrates that taking the original graph as input for GCN is not optimal for graph propagation in semi-supervised classification. It's also worth noting that Fast-GRCN, with much lower computational time complexity, performs nearly as well as GRCN .

To further test the superiority of our model, we consider the edge-sparse scenario when a certain fraction of edges in the given graph is randomly removed. Given an edge retaining ratio, we randomly sample the retained edges 10 times and report the mean classification accuracy and standard deviation. For the Cora, CiteSeer and PubMed dataset, we conduct experiments under random data split. Figure 2 shows the results under different ratios of retained edges. There are several observations from this figure. First, GRCN and Fast-GRCN achieve notable improvements on almost all the datasets, especially when edge retaining ratio is low. For instance, when edge retaining ratio is 10%, GRCN outperforms the second best baseline model by $6.5\%, 2.5\%, 1.1\%, 11.0\%, 4.6\%, 2.3\%$ on each dataset. Second, the GAT and GLCN models which reweight the existing edges do not perform well, indicating that such a reweighting mechanism is not enough when the original graph is highly incomplete. Third, our method also outperforms the over-parameterized model LDS in Cora and CiteSeer because of our restrained edge editing procedure. Though LDS achieves better performances than other baseline methods in these two datasets, its inability to scale prevents us from testing it on four of the larger datasets.

### 4.4   Robustness on Training Labels

We also show that the gains achieved by our model are very robust to the reduction in the number of training labels for each class, denoted by $T$. We compare all the models on the Cora-Full, Amazon Computers and Coauthor CS

---

[1] Note that the performance difference of baseline models between fixed split and random split is also observed in previous work [24], where they show that different splits can lead to different ranking of models, and suggest multiple random splits as a better choice. Our later experiments are based on the random split setting.

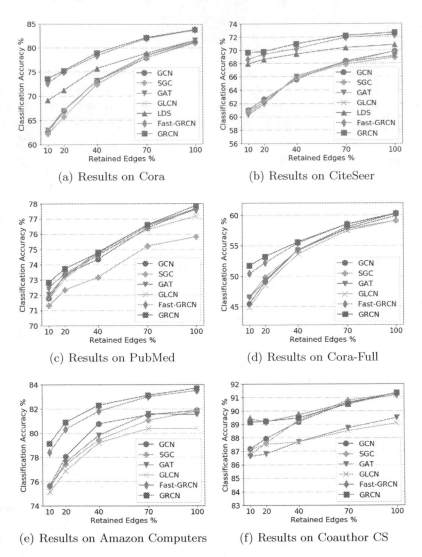

(a) Results on Cora

(b) Results on CiteSeer

(c) Results on PubMed

(d) Results on Cora-Full

(e) Results on Amazon Computers

(f) Results on Coauthor CS

**Fig. 2.** Mean test classification accuracy on all the datasets under different ratios of retained edges over 10 different runs.

datasets and fix the edge sampling ratio to 20%. We reduce $T$ from 15 to 5 and report the results in Table 3. While containing more parameters than vanilla GCN, our model still outperforms others. Moreover, it wins by a larger margin when $T$ is smaller. This demonstrates our model's capability to handle tasks with sparse training labels.

**Table 3.** Mean test classification accuracy and standard deviation on Cora-Full, Amazon Computers and Coauthor CS datasets under different number of training labels for each class. The edge retaining ratio is 20% for all the results. The highest accuracy score is marked in **bold**, and the second highest score is marked using *.

| Cora-Full | 5 labels | 10 labels | 15 labels |
|---|---|---|---|
| GCN | $31.3 \pm 1.5$ | $41.1 \pm 1.3$ | $46.0 \pm 1.1$ |
| SGC | $31.5 \pm 2.1$ | $42.0 \pm 1.5$ | $46.8 \pm 1.3$ |
| GAT | $32.5 \pm 2.1$ | $41.2 \pm 1.4$ | $45.5 \pm 1.2$ |
| GLCN | $30.9 \pm 1.9$ | $41.0 \pm 0.6$ | $45.0 \pm 0.9$ |
| Fast-GRCN | $41.2 \pm 1.2^*$ | $47.7 \pm 0.8^*$ | $50.6 \pm 1.0^*$ |
| GRCN | $\mathbf{42.3 \pm 0.8}$ | $\mathbf{48.2 \pm 0.7}$ | $\mathbf{51.8 \pm 0.6}$ |
| Amazon computers | 5 labels | 10 labels | 15 labels |
| GCN | $70.5 \pm 3.3$ | $74.6 \pm 2.3$ | $77.2 \pm 2.2$ |
| SGC | $67.2 \pm 5.0$ | $74.6 \pm 4.6$ | $77.1 \pm 1.6$ |
| GAT | $64.6 \pm 8.9$ | $72.5 \pm 4.5$ | $74.2 \pm 2.7$ |
| GLCN | $66.9 \pm 7.1$ | $73.8 \pm 3.6$ | $75.8 \pm 2.2$ |
| Fast-GRCN | $74.1 \pm 1.9^*$ | $78.6 \pm 2.1^*$ | $79.8 \pm 1.5^*$ |
| GRCN | $\mathbf{75.3 \pm 1.2}$ | $\mathbf{79.1 \pm 1.9}$ | $\mathbf{79.9 \pm 1.6}$ |
| Coauthor CS | 5 labels | 10 labels | 15 labels |
| GCN | $82.2 \pm 1.5$ | $86.1 \pm 0.5$ | $87.1 \pm 0.9$ |
| SGC | $81.5 \pm 1.6$ | $85.7 \pm 0.9$ | $86.7 \pm 0.9$ |
| GAT | $80.7 \pm 1.1$ | $84.8 \pm 1.0$ | $86.0 \pm 0.7$ |
| GLCN | $82.7 \pm 0.7$ | $85.7 \pm 0.6$ | $87.0 \pm 0.8$ |
| Fast-GRCN | $\mathbf{86.2 \pm 2.1}$ | $\mathbf{88.2 \pm 0.7}$ | $\mathbf{88.5 \pm 0.6}$ |
| GRCN | $86.1 \pm 0.7^*$ | $87.9 \pm 0.4^*$ | $88.2 \pm 0.5^*$ |

## 4.5   Hyperparameter Analysis

We investigate the influence of the hyperparameter $K$ in this section. Recall that after calculating the similarity graph in GRCN , we use $K$-nearest-neighbour specification to generate a sparse graph out of the dense graph. This is not only benificial to efficiency, but also important for effectiveness. Figure 3 shows the results of node classification accuracy vs. sampling ratio on Cora dataset, where we vary the edge sampling ratio from 10% to 100% and change $K$ from 5 to 200. From this figure, increasing the value of $K$ helps improve the classification accuracy at the initial stage. However, after reaching a peak, further increasing $K$ lowers the model performance. We conjecture that this is because a larger $K$ will introduce too much noise and thus lower the quality of the revised graph.

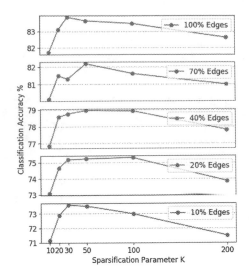

**Fig. 3.** Results of GRCN under different settings of sparsification parameter $K$ on Cora dataset, with different edge retaining ratios.

## 4.6 Ablation Study

To further examine the effectiveness of our GCN-based graph revision module, we conduct an ablation study by testing four different simplifications of the graph revision module:

- Truncated SVD Reconstruction [10]: $\widetilde{A} = \mathrm{SVD}_k(A)$
- Feature-Only (FO): $\widetilde{A} = K(X)$
- Feature plus Graph (FG): $\widetilde{A} = A + K(X)$
- Random Walk Feature plus Graph (RWFG): $\widetilde{A} = A + K(A^2 X)$

where SVD and FO only use the original graph structure or node features to construct the graph. They are followed by the FG method, which adds the original graph to the feature similarity graph used in FO. Our model is most closely related to the third method, RWFG, which constructs the feature graph with similarity of node features via graph propagation, but without feature learning.

We conduct the ablation experiment on Cora dataset with different edge retaining ratios and report the results in Fig. 4. The performance of SVD method indicates that simply smoothing the original graph is insufficient. The comparison between FO and FG shows that adding original graph as residual links is helpful for all edge retaing ratios, especially when there are more known edges in the graph. Examining the results of FG and RWFG, we can also observe a large improvement brought by graph propagation on features. Finally, the performance of our model and RWFG underscores the importance of feature learning, especially in the cases of low edge retraining ratio.

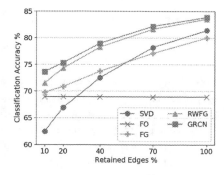

**Fig. 4.** Results of GRCN and its simplified versions on Cora dataset with different ratios of retained edges

# 5 Related Work

## 5.1 Graph Convolutional Network

Graph Convolution Networks (GCNs) were first introduced in the work by [6], with subsequent development and improvements from [13]. Overall, GCNs can be categorized into two categories: spectral convolution and spatial convolution. The spectral convolution operates on the spectral representation of graphs defined in the Fourier domain by the eigen-decomposition of graph Laplacian [7,16]. The spatial convolution operates directly on the graph to aggregate groups of spatially close neighbors [1,12]. GraphSage [12] computes node representations by sampling a fixed-size number of adjacent nodes. Besides those methods that are directly applied to an existing graph, GAT [25], GLCN [14] use attention mechanism or feature similarity to reweight the original graph for better GCN performance, while LDS [9] reconstructs the entire graph via a bilevel optimization. Although our work is related to these methods, we develop a different strategy for graph revision that maintains both efficiency and high flexibility.

## 5.2 Link Prediction

Link prediction aims at identifying missing links, or links that are likely to be formed in a given network. Previous line of work uses heuristic methods based on local neighborhood structure of nodes, including first-order heuristics by common neighbors and preferential attachment [3], second-order heuristics by Adamic-Adar and resource allocation [33], or high-order heuristics by PageRank [5]. To loose the strong assumptions of heuristic method, a number of neural network based methods [11,32] are proposed, which are capable to learn general structure features. The problem we study in this paper is related to link prediction since we try to revise the graph by adding or reweighting edges. However, instead of treating link prediction as an objective, our work focus on improving node classification by feeding the revised graph into GCNs.

# 6 Conclusion

This paper presents Graph-Revised Convolutional Network, a novel framework for incorporating graph revision into graph convolution networks. We show both theoretically and experimentally that the proposed way of graph revision can significantly enhance the prediction accuracy for downstream tasks. GRCN overcomes two main drawbacks in previous approaches to graph revision, which either employ over-parameterized models and consequently face scaling issues, or fail to consider missing edges. In our experiments with node classification tasks, the performance of GRCN stands out in particular when the input graphs are highly incomplete or if the labeled training instances are very sparse.

In the future, we plan to explore GRCN in a broader range of prediction tasks, such as knowledge base completion, epidemiological forecasting and aircraft anomaly detection based on sensor network data.

**Acknowledgments.** We thank the reviewers for their helpful comments. This work is supported in part by the National Science Foundation (NSF) under grant IIS-1546329.

# References

1. Atwood, J., Towsley, D.: Diffusion-convolutional neural networks. In: Advances in Neural Information Processing Systems, pp. 1993–2001 (2016)
2. Balakrishnan, V.: Graph theory (schaum's outline) (1997)
3. Barabási, A.L., Albert, R.: Emergence of scaling in random networks. Science **286**(5439), 509–512 (1999)
4. Bojchevski, A., Günnemann, S.: Deep Gaussian embedding of graphs: unsupervised inductive learning via ranking. In: International Conference on Learning Representations, pp. 1–13 (2018)
5. Brin, S., Page, L.: The anatomy of a large-scale hypertextual web search engine. Comput. Netw. ISDN Syst. **30**(1–7), 107–117 (1998)
6. Bruna, J., Zaremba, W., Szlam, A., LeCun, Y.: Spectral networks and locally connected networks on graphs. arXiv preprint arXiv:1312.6203 (2013)
7. Defferrard, M., Bresson, X., Vandergheynst, P.: Convolutional neural networks on graphs with fast localized spectral filtering. In: Advances in Neural Information Processing Systems, pp. 3844–3852 (2016)
8. Fey, M., Lenssen, J.E.: Fast graph representation learning with pytorch geometric. arXiv preprint arXiv:1903.02428 (2019)
9. Franceschi, L., Niepert, M., Pontil, M., He, X.: Learning discrete structures for graph neural networks. arXiv preprint arXiv:1903.11960 (2019)
10. Golub, G.H., Reinsch, C.: Singular value decomposition and least squares solutions. In: Linear Algebra, pp. 134–151. Springer (1971)
11. Grover, A., Leskovec, J.: Node2vec: scalable feature learning for networks. In: Proceedings of the 22nd ACM SIGKDD International Conference on Knowledge Discovery and Data Mining, pp. 855–864. ACM (2016)
12. Hamilton, W., Ying, Z., Leskovec, J.: Inductive representation learning on large graphs. In: Advances in Neural Information Processing Systems, pp. 1024–1034 (2017)
13. Henaff, M., Bruna, J., LeCun, Y.: Deep convolutional networks on graph-structured data. arXiv preprint arXiv:1506.05163 (2015)

14. Jiang, B., Zhang, Z., Lin, D., Tang, J., Luo, B.: Semi-supervised learning with graph learning-convolutional networks. In: Proceedings of the IEEE Conference on Computer Vision and Pattern Recognition, pp. 11313–11320 (2019)

15. Kingma, D.P., Ba, J.: Adam: a method for stochastic optimization. arXiv preprint arXiv:1412.6980 (2014)

16. Kipf, T.N., Welling, M.: Semi-supervised classification with graph convolutional networks. arXiv preprint arXiv:1609.02907 (2016)

17. Landrieu, L., Simonovsky, M.: Large-scale point cloud semantic segmentation with superpoint graphs. In: Proceedings of the IEEE Conference on Computer Vision and Pattern Recognition, pp. 4558–4567 (2018)

18. Marcheggiani, D., Titov, I.: Encoding sentences with graph convolutional networks for semantic role labeling. In: Proceedings of the 2017 Conference on Empirical Methods in Natural Language Processing, pp. 1506–1515 (2017)

19. Monti, F., Boscaini, D., Masci, J., Rodola, E., Svoboda, J., Bronstein, M.M.: Geometric deep learning on graphs and manifolds using mixture model cnns. In: Proceedings of the IEEE Conference on Computer Vision and Pattern Recognition, pp. 5115–5124 (2017)

20. Namata, G., London, B., Getoor, L., Huang, B., EDU, U.: Query-driven active surveying for collective classification. In: 10th International Workshop on Mining and Learning with Graphs, p. 8 (2012)

21. Paszke, A., et al.: Automatic differentiation in PyTorch. In: NIPS Autodiff Workshop (2017)

22. Schölkopf, B., Herbrich, R., Smola, A.J.: A generalized representer theorem. In: Helmbold, D., Williamson, B. (eds.) COLT 2001. LNCS (LNAI), vol. 2111, pp. 416–426. Springer, Heidelberg (2001). https://doi.org/10.1007/3-540-44581-1_27

23. Sen, P., Namata, G., Bilgic, M., Getoor, L., Galligher, B., Eliassi-Rad, T.: Collective classification in network data. AI Mag. **29**(3), 93–93 (2008)

24. Shchur, O., Mumme, M., Bojchevski, A., Günnemann, S.: Pitfalls of graph neural network evaluation. arXiv preprint arXiv:1811.05868 (2018)

25. Veličković, P., Cucurull, G., Casanova, A., Romero, A., Lio, P., Bengio, Y.: Graph attention networks. arXiv preprint arXiv:1710.10903 (2017)

26. Wang, Y., Sun, Y., Liu, Z., Sarma, S.E., Bronstein, M.M., Solomon, J.M.: Dynamic graph cnn for learning on point clouds. arXiv preprint arXiv:1801.07829 (2018)

27. Wu, F., Zhang, T., Souza Jr, A.H.d., Fifty, C., Yu, T., Weinberger, K.Q.: Simplifying graph convolutional networks. arXiv preprint arXiv:1902.07153 (2019)

28. Wu, Y., Yang, Y., Nishiura, H., Saitoh, M.: Deep learning for epidemiological predictions. In: The 41st International ACM SIGIR Conference on Research & Development in Information Retrieval, pp. 1085–1088. ACM (2018)

29. Wu, Z., Pan, S., Chen, F., Long, G., Zhang, C., Yu, P.S.: A comprehensive survey on graph neural networks. arXiv preprint arXiv:1901.00596 (2019)

30. Yang, Z., Cohen, W.W., Salakhutdinov, R.: Revisiting semi-supervised learning with graph embeddings. arXiv preprint arXiv:1603.08861 (2016)

31. Ying, R., He, R., Chen, K., Eksombatchai, P., Hamilton, W.L., Leskovec, J.: Graph convolutional neural networks for web-scale recommender systems. In: Proceedings of the 24th ACM SIGKDD International Conference on Knowledge Discovery & Data Mining, pp. 974–983. ACM (2018)

32. Zhang, M., Chen, Y.: Link prediction based on graph neural networks. In: Advances in Neural Information Processing Systems, pp. 5165–5175 (2018)

33. Zhou, T., Lü, L., Zhang, Y.C.: Predicting missing links via local information. Euro. Phys. J. B **71**(4), 623–630 (2009)

# Robust Training of Graph Convolutional Networks via Latent Perturbation

Hongwei Jin$^{(\boxtimes)}$ and Xinhua Zhang

University of Illinois at Chicago, Chicago, IL 60607, USA
{hjin25,zhangx}@uic.edu

**Abstract.** Despite the recent success of graph convolutional networks (GCNs) in modeling graph structured data, its vulnerability to adversarial attacks has been revealed and attacks on both node feature and graph structure have been designed. Direct extension of defense algorithms based on adversarial samples meets with immediate challenge because computing the adversarial network costs substantially. We propose addressing this issue by perturbing the latent representations in GCNs, which not only dispenses with generating adversarial networks, but also attains improved robustness and accuracy by respecting the latent manifold of the data. This new framework of latent adversarial training on graphs is applied to node classification, link prediction, and recommender systems. Our empirical experimental results confirm the superior robustness performance over strong baselines.

**Keywords:** Graph neural network · Adversarial training ·
Representation learning

## 1 Introduction

Neural networks have achieved great success on Euclidean data for image recognition, machine translation, and speech recognition, etc. However, modeling with non-Euclidean data—such as complex networks with geometric information and structural manifold—is more challenging in terms of data representation. Mapping non-Euclidean data to Euclidean, which is also referred to as embedding, is one of the most prevalent techniques. Recently, graph convolutional networks (GCNs) have received increased popularity in machine learning for structured data. The first phenomenal work of GCN was presented by Bruna *et al.* (2013), which developed a set of graph convolutional operations based on spectral graph theory. The conventional GCN was introduced by Kipf and Welling (2017) for the task of node classification, and they represent nodes by repeated multiplication of augmented normalized adjacency matrix and feature matrix, which can be interpreted as the first order approximation of localized spectral filters on graphs (Hammond *et al.* 2011; Defferrard *et al.* 2016). GCNs have been widely applied to a variety of machine learning tasks, including node classification (Kipf and Welling 2017), graph clustering (Duvenaud *et al.* 2015), link prediction (Kipf and Welling 2016; Schlichtkrull *et al.* 2018), recommender systems (Berg *et al.* 2018), etc.

© Springer Nature Switzerland AG 2021
F. Hutter et al. (Eds.): ECML PKDD 2020, LNAI 12459, pp. 394–411, 2021.
https://doi.org/10.1007/978-3-030-67664-3_24

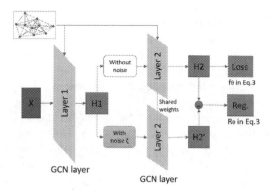

**Fig. 1.** LAT-GCN framework

Despite the advantages in efficiently and effectively learning representations and predictions, GCNs have been shown vulnerable to adversarial attacks. Although adversarial learning has achieved significant progress in recent years (Szegedy *et al.* 2014), the graph structure in GCNs constitutes an additional source of vulnerability. The conventional approaches based on adversarial samples (a.k.a. attacks) are typically motivated by adding imperceptible perturbation to images, followed by enforcing the invariance of prediction outputs (Kurakin *et al.* 2017). This corresponds to perturbing node features in GCNs and has received very recent study. Feng *et al.* (2019) introduced the graph adversarial training as a dynamic regularization scheme based on graph structure. Deng *et al.* (2019) proposed a sample-based batch virtual adversarial training to promote the smoothness of model.

However, the graph topology itself can be subject to attacks such as adding or deleting edges or nodes. Zügner *et al.* (2018) and Dai *et al.* (2018) constructed effective structural attacks at both training time (poisoning) and testing time (evasion). Finding the adversarial input graph is indeed a combinatorial optimization problem that is typically NP-hard. Dai *et al.* (2018) proposed a reinforcement learning based attack that learns a generalizable attack policy to misclassify a target in both graph classification and node classification. Zügner *et al.* (2018) introduced a surrogate model to approximate the perturbed graph structure and feature. Both of these methods considered attacks at the test stage, i.e., evasion attacks. By solving a bilevel problem from meta-learning, Zügner and Günnemann (2019) generated adversarial examples that are graph structures, and in contrast to the previous attacks, they do not need to specify the target. Their attack modifies the training data to undermine the performance, hence a poison attack. Xu *et al.* (2019) proposed two attack methods: PGD topology attack and min-max attack methods, which relaxed $\{0,1\}^n$ to $[0,1]^n$, leading to a continuous optimization. They also proposed a robust training model by solving a min-max problem.

Although the technique of adversarial sample can be directly applied to defend against structural attacks, an immediate obstacle arises from computational complexity. All the aforementioned structural attacks are much more computation intensive than conventional attacks in image classification. Therefore alternating between model optimization and adversarial sample generation can be prohibitively time consuming, making a new solution principle in demand.

The goal of this paper, therefore, is to develop a new adversarial training algorithm that defends against attacks on both node features *and* graph structure, while at the same time maintaining or even improving the generalization accuracy. Our intuition draws upon two prior works. Firstly, in the context of adversarial training on word inputs, Miyato *et al.* (2017) noted that words are discrete and are not amenable to infinitesimal perturbation. So they resorted to perturbing the word embeddings that are continuous.

A straightforward analogy of word embeddings in GCNs is the first layer output $\mathbf{H}^{(1)}$ after graph convolution, which blends the information of both node features and the graph. So we propose injecting adversarial perturbations to $\mathbf{H}^{(1)}$, as shown by $\zeta$ in Fig. 1. This leads to indirect perturbations to the graph, which implicitly enforces robustness to structural attacks. As shown in Sect. 3.1, this can be achieved via a regularization term on $\mathbf{H}^{(1)}$, completely circumventing the requirement of generating adversarial attacks on the graph structure. We will refer to the approach as latent adversarial training (LAT-GCN).

However, Miyato *et al.* (2017) also noted that "the perturbed embedding does not map to any word" and "we thus propose this approach exclusively as a means of regularizing a text classifier". To address the analogous concern in GCNs, we leverage the observation by Stutz *et al.* (2019) that adversarial examples can benefit both robustness and accuracy if they are on the manifold of low-dimensional embeddings. Using $\mathbf{H}^{(1)}$ as a proxy of the latent manifold, LAT-GCN manages to generate "on-manifold"perturbations, which, as our experiments show in Sect. 4, help to reduce the success rate of adversarial attacks for GCNs while preserving or improving the accuracy of the model.

## 2    Related Work

Vulnerability of deep neural networks has received intensive study in machine learning (Szegedy *et al.* 2014; Huang *et al.* 2017; Song *et al.* 2018; Jia and Liang 2017; He *et al.* 2017). Some certifications of robustness have been established, such as Sinha *et al.* (2018), Raghunathan *et al.* (2018), and Wong and Kolter (2018). Recently, investigations have been made on the important trade-off between adversarial robustness and generalization performance. Tsipras *et al.* (2019) showed that robustness may be at odds with accuracy, and a principled trade-off was studied by Zhang *et al.* (2019), which decomposed the prediction error for adversarial examples (robust error) into classification error and boundary error, and a differentiable upper bound was derived by leveraging classification calibration. A number of works also analyzed the robust error *in terms of* generalization error (e.g., Schmidt *et al.* 2018; Cullina *et al.* 2018; Yin *et al.* 2018) .

However all these analyses are on *continuous* input domains, typically with "imperceptible perturbations". In contrast, our focus is on adversarial robustness with respect to *discrete* objects such as network structures. Although it is hard to extend the aforementioned theoretical trade-off to the new setting, we demonstrate empirically that our novel adversarial training algorithm is able to improve both robustness and accuracy for graph data learning algorithms that are based on node embeddings, e.g., through GCNs.

## 2.1   Adversarial Attacks on Link Prediction

Since GCNs embed both the structure and feature information simultaneously, different adversarial attacks have been proposed recently. For the task of link prediction, Kipf and Welling (2016) introduced a GCN-based graph auto-encoder (GAE) which reconstructs the adjacency matrix from node embeddings produced by GCNs, and it outperformed spectral clustering (Tang and Liu 2011) and DeepWalk (Perozzi *et al.* 2014). Following GAE, Tran (2018) proposed an architecture to learn a joint representation of local graph structure and available node features for multi-task learning in both link prediction and semi-supervised node classification.

Unfortunately, GAEs are still short of robustness under adversarial link attacks. Chen *et al.* (2018b) proposed an iterative gradient attack based on the gradient information in the trained model, and GAEs were shown vulnerable to just a few link perturbations. In order to improve the robustness, Pan *et al.* (2019) designed an adversarial training method by virtually attacking the node features so that their latent representation matches a prior Gaussian distribution. Xu *et al.* (2019) proposed two attack methods: PGD topology attack and min-max attack methods, which relaxed $\{0,1\}^n$ to $[0,1]^n$, leading to a continuous optimization. They also proposed a robust training model by solving a min-max problem. All the above-mentioned approaches rely on adversarial examples to improve the robustness of the model for either node classification or link prediction. However, their methods need to generate the adversarial instance, either at the attack stage or at the training stage. In contract, LAT-GCN does not look for the exact adversarial examples, which is much less time consuming.

Naturally, the same idea can be applied to link prediction amounting to similar robustness and improvement in generalization accuracy, because GAE reconstructs the adjacency matrix based on *node embeddings*, which in turn employs GCNs as the encoder.

Parallel to our work, Wang *et al.* (2019) proposed DefNet to defend against adversarial attacks on GCNs in a conditional manner. It investigates vulnerabilities via graph encoder refining, and addresses discrete attacks via adversarial contractive learning. Both our method and theirs can reduce the attack success rate to about 60% on node classification tasks. However, our approach is much simpler and is more generally applicable to other tasks such as link prediction.

*Notation.* We denote the set of vertices for a graph $G$ as $\mathcal{V}$, and denote the feature matrix as $\mathbf{X} \in \mathbb{R}^{n \times d}$, where $n = |\mathcal{V}|$ and $d$ is the number of node features.

Let $\mathcal{E}$ be the set of existing edges, and $\mathbf{A} \in \mathbb{R}^{n \times n}$ be the adjacency matrix, whose entries are 0 or 1 for unweighted graphs. After adding a self loop to each node, we have $\hat{\mathbf{A}} = \mathbf{A} + \mathbf{I}$, and we construct a diagonal matrix with $\hat{\mathbf{D}}_{ii} = \sum_j \hat{\mathbf{A}}_{ij}$. The augmented normalized adjacency matrix is then defined as $\tilde{\mathbf{A}} = \hat{\mathbf{D}}^{-1/2} \hat{\mathbf{A}} \hat{\mathbf{D}}^{-1/2}$.

## 3    Latent Adversarial Training of GCNs

The original GCN model employs a forward propagation of representation as:

$$\mathbf{H}^{(l+1)} = \sigma(\tilde{\mathbf{A}} \mathbf{H}^{(l)} \mathbf{W}^{(l+1)}), \quad l \geq 0, \tag{1}$$

where the initial node representation is $\mathbf{H}^{(0)} = \mathbf{X}$. Without loss of generality, let us consider a two-layer GCN, which is commonly used in practice. Then the standard GCN tries to find the optimal weights $\boldsymbol{\theta} := (\mathbf{W}^{(1)}, \mathbf{W}^{(2)})$ that minimize some loss $f_{\boldsymbol{\theta}}$ (e.g., cross-entropy loss) over the latent representation $\mathbf{H}^{(2)}$. That is, $\min_{\boldsymbol{\theta}} f_{\boldsymbol{\theta}}(G, \mathbf{X})$.

In order to improve the robustness to the perturbation in input feature $\mathbf{X}$, Feng *et al.* (2019) and Deng *et al.* (2019) considered *generative adversarial training* over $\mathbf{X}$, which at a high level, optimizes $\min_{\boldsymbol{\theta}} \max_{\mathbf{r} \in C} f_{\boldsymbol{\theta}}(G, \mathbf{X} + \mathbf{r})$. Here $\mathbf{r}$ is the perturbation chosen from a constrained domain $C$.

Naturally, it is also desirable to defend against attacks on the graph topology of $G$. Such structural attacks have been studied in Dai *et al.* (2018); Zügner *et al.* (2018); Zügner and Günnemann (2019), but no corresponding defense algorithm has been proposed yet. Different from attacks on $\mathbf{X}$, here the attacks on $G$ are discrete (hence not imperceptible per se), and finding the most effective attacks can be NP-hard. This creates new obstacles to generative adversarial training, and therefore we resort to a regularization approach based on the *latent layer* $\mathbf{H}^{(1)}$.

Specifically, considering that the information in $G$ and $\mathbf{X}$ is summarized in the first layer output $\mathbf{H}^{(1)}$, we can adopt generative adversarial training directly on $\mathbf{H}^{(1)}$:

$$\min_{\boldsymbol{\theta}} \max_{\zeta \in D} f_{\boldsymbol{\theta}}(\mathbf{H}^{(1)} + \zeta), \tag{2}$$

where the symbol $f_{\boldsymbol{\theta}}$ is overloaded to denote the loss based on $\mathbf{H}^{(1)}$ with perturbation $\zeta$. The benefits are two folds. Firstly, $\mathbf{H}^{(1)}$ is continuously valued, making it meaningful to apply small perturbations to it, which indirectly represent the perturbations in $\mathbf{X}$ *and* $G$. Secondly, Stutz *et al.* (2019) argued that perturbation at latent layers (such as $\mathbf{H}^{(1)}$) are more likely to generate "on-manifold" samples that additionally benefits generalization, while perturbations in the raw input space (e.g., $\mathbf{X}$) is likely to deviate from the original data manifold and hence irrelevant to generalization.

Unfortunately, the perturbation $\zeta$ in (2) is chosen *jointly* over *all* nodes in the graph setting, which is different from the common adversarial setting where each individual example seeks its own perturbation independently. As a result, the

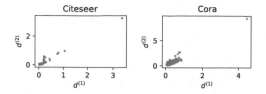

**Fig. 2.** Scatter plot of the $L_2$ displacement of hidden representation before and after Nettack, for both first and second layers. Each dot corresponds to a node. Citeseer: $\mathbf{d}^{(1)} : 0.048 \pm 0.084$, $\mathbf{d}^{(2)} : 0.038 \pm 0.083$, Cora: $\mathbf{d}^{(1)} : 0.216 \pm 0.122$, $\mathbf{d}^{(2)} : 0.486 \pm 0.265$. The single outlier (top right) corresponds to the targeted node.

computational cost is higher, which is further exacerbated by the nested min-max optimization. To alleviate this problem, we adopt the standard regularization variant of adversarial training, which aims to promote the smoothness of model predictions with respect to the perturbations:

$$\min_{\boldsymbol{\theta}} \mathcal{L}_{\boldsymbol{\theta}}(\tilde{\mathbf{A}}, \mathbf{X}) := f_{\boldsymbol{\theta}}(\mathbf{H}^{(1)}) + \gamma \mathcal{R}_{\boldsymbol{\theta}}(\mathbf{H}^{(1)}), \tag{3}$$

where $\gamma \geq 0$ is a trade-off parameter, and the regularizer $\mathcal{R}_{\boldsymbol{\theta}}$ is defined as the Frobenius distance between the original second layer output and that after perturbing $\mathbf{H}^{(1)}$:

$$\mathcal{R}_{\boldsymbol{\theta}}(\mathbf{H}^{(1)}) = \max_{\zeta \in D} \left\| \tilde{\mathbf{A}} \left( \mathbf{H}^{(1)} + \zeta \right) \mathbf{W}^{(2)} - \tilde{\mathbf{A}} \mathbf{H}^{(1)} \mathbf{W}^{(2)} \right\|_F^2 . \tag{4}$$

Here we constrain the perturbed noise to be imperceptible via $D := \{\zeta : \|\zeta_{i:}\|_2 \leq \epsilon, \forall i \in \{1, \ldots, n\}\}$, where $\zeta_{i:}$ stands for the $i$-th row of $\zeta$. This **row-wise** $L_{2,\infty}$ regularization on $\zeta$ is critical for good performance. As we observed in practice, changing it to the Frobenius norm of $\zeta$ leads to significantly inferior performance.

To gain a deeper insight into why this $L_{2,\infty}$ norm on $\zeta$ outperforms Frobenius norm, we empirically examined the change of hidden representations after applying Nettack (Zügner *et al.* 2018). With a target node randomly selected, we define

$$\mathbf{d}_i^{(l)} = \left\| \mathbf{H}_{i:}^{(l)} - \mathbf{H}_{i:}^{'(l)} \right\|_2 \quad \forall i \in \mathcal{V},$$

where $\mathbf{H}^{(l)}$ and $\mathbf{H}^{'(l)}$ are the hidden representations of the $l$-th layer upon the completion of vanilla GCN training, based on the graphs before and after applying Nettack, respectively. The changes in $L_2$ norm are demonstrated in Fig. 2 where each dot corresponds to a node, and the outlier in the top-right corner corresponds to the target node. Interestingly, most nodes undergo very little change (including neighbors of the target), while only the target node suffers a large change. Such an imbalanced distribution indicates that it is more reasonable to consider the largest norm of changes, rather than their sum or average (as in the Frobenius norm).

Although (3) is still a min-max problem, the inner maximization problem in $\mathcal{R}_\theta$ is now decoupled with the loss $f_\theta$, hence solvable with much ease. Indeed, both the objective and the constraint are quadratic, permitting efficient computation of gradient and projection. We also note in passing that turning the adversarial objective (2) into a regularized objective (3) is a commonly adopted technique, and their relationship has been studied by, e.g., Shafieezadeh-Abadeh *et al.* (2017).

### 3.1   Optimization

Since the objective (3) intrinsically couples all nodes, we apply ADAM to find the optimal $\theta$ using the entire dataset as the mini-batch (Kingma and Ba 2015). The major complexity stems from $\nabla_\theta \mathcal{R}_\theta(\mathbf{H}^{(1)})$, which can be readily computed using the Danskin's theorem (Bertsekas 1995). To this end, we first simplify (4) by

$$\mathcal{R}_\theta(\mathbf{H}^{(1)}) = \max_\zeta \left\| \tilde{\mathbf{A}}\zeta\mathbf{W}^{(2)} \right\|_F^2 \quad \text{s.t.} \quad \|\zeta_{i:}\| \leq \epsilon. \tag{5}$$

Once we find the optimal $\zeta^*$, the gradient in $\theta$ is simply $\nabla_\theta \left\| \tilde{\mathbf{A}}\zeta^*\mathbf{W}^{(2)} \right\|_F^2$. To find $\zeta^*$, note the gradient in $\zeta$ is

$$\nabla_\zeta \operatorname{tr}(\tilde{\mathbf{A}}\zeta\mathbf{W}\zeta^\top \tilde{\mathbf{A}}^\top) = \left(\mathbf{W}^\top \zeta + \mathbf{W}\zeta^\top\right) \tilde{\mathbf{A}}\tilde{\mathbf{A}}^\top, \tag{6}$$

where $\mathbf{W} = \mathbf{W}^{(2)}\mathbf{W}^{(2)\top}$. Although the constraint $\|\zeta_{i:}\| \leq \epsilon$ is convex and projection to it is trivial, the objective *maximizes* a convex function. So we simply use ADAM to approximately solve for $\zeta$. Empirical results in Table 1 show that the additional optimization does not incur much computation. ADV-GCN is the classical adversarial training where we sampled 20 nodes from the training set, which were successively taken as the target for Nettack Zügner *et al.* (2018) to find 20 adversarial graphs. And MIN-MAX GCN is the adversarial training algorithm in Xu *et al.* (2019) which is based on a node attack algorithm CE-PDG with $\epsilon = 5\%$. Since the noise $\zeta$ is applied to all the nodes, Eq. (5) becomes more expensive to solve as the number of node grows. The overall procedure is summarized in Algorithm 1.

## 4   Experiment 1: Node Classification

We tested the performance of LAT-GCN on a range of standard citation datasets for node classification, including CiteSeer, Cora, and PubMed (Sen *et al.* 2008). The competing baselines include the vanilla GCN, FastGCN (Chen *et al.* 2018a), SGCN (Chen *et al.* 2017), SGC (Wu *et al.* 2019) and GAT (Veličković *et al.* 2018). All hyperparameters in respective models followed from the original implementation, including step size, width of layers, etc. Since the optimal objective value

---

**Algorithm 1.** Latent adversarial training for GCN

---

**input A, X**

1: **while** not converged for (3) **do**
2:    **while** not converged for (5) **do**
3:       Apply ADAM to find $\zeta^*$ using the gradient in $\zeta$ computed from Eq (6).
4:    **end while**
5:    Take one step of ADAM in $\boldsymbol{\theta}$ with the gradient computed by $\nabla_{\boldsymbol{\theta}} f_{\boldsymbol{\theta}}(\mathbf{H}^{(1)}) + \gamma \nabla_{\boldsymbol{\theta}} \left\| \tilde{\mathbf{A}} \zeta^* \mathbf{W}^{(2)} \right\|_F^2$.
6: **end while**

---

**Table 1.** CPU wall-clock time in seconds (200 epochs on CPU only)

| | $|\mathcal{V}|$ | $|\mathcal{E}|$ | GCN | ADV-GCN | MIN-MAX GCN | LAT-GCN |
|---|---|---|---|---|---|---|
| Citeseer | 2110 | 3668 | 7.01 | 13.3 | 4012.2 | 25.49 |
| Cora | 2485 | 5069 | 6.26 | 15.7 | 1823.7 | 23.90 |
| PubMed | 19717 | 44324 | 31.18 | 53.4 | Out of memory | 133.64 |

in (5) is quadratic in $\epsilon$, only the value of $\gamma \epsilon^2$ matters for LAT-GCN. So we fixed $\gamma = 0.1$, and only tuned $\epsilon$.

Finally, all algorithms were applied in a transductive setting, where the graph was constructed by combining the nodes for training and testing. Accordingly, perturbation $\zeta$ was applied to both training and test nodes in (4) for training LAT-GCN.

*Comparison of Accuracy.* We first compared the test accuracy as shown in Table 2. Each test was based on randomly partitioning the nodes into training and testing for 20 times. Clearly, LAT-GCN delivers similar accuracy compared with all the competing algorithms.

In order to study the influence of $\epsilon$ on the performance of LAT-GCN, we plotted in Fig. 3 how the test accuracy changes with the value of $\epsilon$. We again used the Cora dataset for training. Interestingly, the accuracy tends to increase as $\epsilon$ grows from 0 to 0.17, and then drops for larger $\epsilon$. This is consistent with the observation in Stutz *et al.* (2019) where robustness is shown to be positively correlated with generalization, as long as the data points are not perturbed away from the latent manifold. In addition, we also measured the influence of attacking two layers instead of just the first one. Table 3 shows the test accuracy under the four combinations of settings. Perturbing both layers has similar performance as when only the first layer is perturbed. So for the benefit of computational cost, our model only perturbs the first layer.

We also varied the training set size in $\{10\%, 20\%, \ldots, 80\%\}$ of the entire dataset, and plotted how the hyperparameter $\epsilon$ affects the test accuracy. There is no perturbation when $\epsilon = 0$ or $\gamma = 0$.

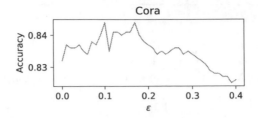

**Fig. 3.** Test accuracy of LAT-GCN on Cora as a function of $\epsilon$

**Table 2.** Test accuracy (%) for node prediction

|         | CiteSeer       | Cora           | PubMed         |
|---------|----------------|----------------|----------------|
| GCN     | 70.9 ± .5      | 81.4 ± .4      | **79.0 ± .4**  |
| FastGCN | 68.8 ± .6      | 79.8 ± .3      | 77.4 ± .3      |
| SGCN    | 70.8 ± .1      | 81.7 ± .5      | **79.0 ± .4**  |
| SGC     | 71.9 ± .1      | 81.0 ± .0      | 78.9 ± .0      |
| GAT     | **72.5 ± .7**  | **83.0 ± .7**  | **79.0 ± .3**  |
| LAT-GCN | 72.1 ± .4      | 82.3 ± .3      | 78.8 ± .7      |

**Table 3.** Test accuracy (%) with different settings of LAT-GCN. Subscript indicates the layer(s) perturbed in LAT-GCN.

|          | GCN  | LAT-GCN$_1$ | LAT-GCN$_2$ | LAT-GCN$_{1,2}$ |
|----------|------|-------------|-------------|------------------|
| Citeseer | 71.6 | 72.5        | 71.9        | 72.2             |
| Cora     | 83.4 | 84.4        | 84.0        | 84.2             |
| PubMed   | 82.8 | 85.6        | 84.2        | 84.4             |

Similar trends as above were observed in the results for Cora and CiteSeer, relegated to Figs. 4 and 5, respectively.

*Comparison of Robustness.* We next evaluated the robustness of LAT-GCN under Nettack, an evasion attack on graphs that was specifically designed for GCNs (Zügner *et al.* 2018). The goal is to perform small perturbations on graph $G = (\mathbf{A}, \mathbf{X})$, leading to a new graph $G' = (\mathbf{A}', \mathbf{X}')$, such that the classification performance drops as much as possible. Analogous to "imperceptible perturbations", Nettack imposes a budget on the number of allowed changes:

$$\sum_u \sum_i |\mathbf{X}_{ui} - \mathbf{X}'_{ui}| + \sum_{u<v} |\mathbf{A}_{uv} - \mathbf{A}'_{uv}| \le \Delta.$$

Since the logits in prediction are complicated by the nonlinear activation function $\sigma$, Zügner *et al.* (2018) first introduced a surrogate model that replaces $\sigma$ with the identity function, leading to a linearized GCN:

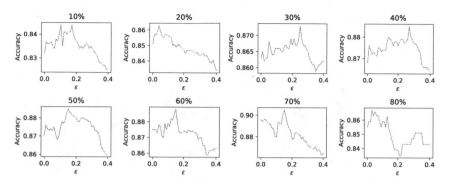

**Fig. 4.** Test accuracy of LAT-GCN on Cora as a function of $\epsilon$, over different sizes of training data.

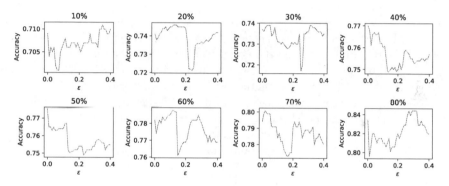

**Fig. 5.** Test accuracy of LAT-GCN on Citeseer as a function of $\epsilon$, over different sizes of training data.

$$\mathcal{L}_s(\mathbf{A}, \mathbf{X}; \mathbf{W}, v_o) = \max_{c \neq c_o} \left[ \hat{\mathbf{A}}^2 \mathbf{X} \mathbf{W} \right]_{v_o, c} - \left[ \hat{\mathbf{A}}^2 \mathbf{X} \mathbf{W} \right]_{v_o, c_o},$$

where $v_o$ is the target node of attack, and $c_o$ is the original label of $v_o$. Now the aim becomes to solve $\text{argmax}_{(\mathbf{A}', \mathbf{X}')} \mathcal{L}_s(\mathbf{A}', \mathbf{X}'; \mathbf{W}, v_o)$.

However, this problem is still intractable because the domain is discrete. To simplify, they defined scoring functions that evaluate the surrogate loss obtained after adding/deleting an edge $e$ or a feature $f$:

$$\mathcal{S}_{struct}(e; G, v_o) := \mathcal{L}_s(\mathbf{A}', \mathbf{X}; \mathbf{W}, v_o),$$
$$\mathcal{S}_{feature}(f; G, v_o) := \mathcal{L}_s(\mathbf{A}, \mathbf{X}'; \mathbf{W}, v_o).$$

Then Nettack proceeds iteratively, where each step greedily selects an edge $e$ or a feature $f$ which allow either $\mathcal{S}_{struct}$ or $\mathcal{S}_{feature}$ to be maximally increased. The algorithm terminates when the budget is depleted. Since our focus here is on structural attacks, we only considered attacking the structure $\mathbf{A}$, or attacking both structure $\mathbf{A}$ and feature $\mathbf{X}$. We will refer to our results as LAT-GCN-A and LAT-GCN-AX, respectively. $\epsilon$ was set to 0.1 for LAT-GCN.

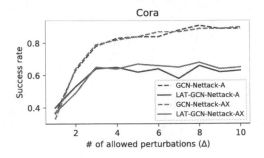

**Fig. 6.** Success rate on Cora with increasing value of $\Delta$

We first compared LAT-GCN to vanilla GCN when the perturbation budget $\Delta$ in Nettack was increased from 1 to 10. The test procedure is detailed in Algorithm 2, where the performance metric is the attack success rate. Since Nettack is targeted, we randomly sampled 100 nodes from the test set as the target.

From Fig. 6, it is clear that LAT-GCN enjoys significantly lower success rate than GCN on the Cora dataset, under both attack strategies.

Moreover, we compare the success rate with different robust training models. We compared our method (LAT-GCN) with the robust training method in Xu *et al.* (2019) (MIN-MAX GCN) which is based on a node attack algorithm CE-PDG with $\epsilon = 5\%$. We obtained robust models from both methods, and then

**Table 4.** Success rate (%)

|          | GCN | ADV-GCN | MIN-MAX GCN   | LAT-GCN |
| -------- | --- | ------- | ------------- | ------- |
| Cora     | 87  | 84      | 71            | 62      |
| Citeseer | 84  | 82      | 73            | 67      |
| PubMed   | 85  | 79      | Out of memory | 75      |

**Table 5.** AUC and AP scores on the test set for link prediction. GCN-1 and GCN-2 stand for GCN with one and two hidden layers, respectively. The second to the fourth columns under GCN-1 show the test performance when edges in $\mathcal{E}_{train}$ are added or removed randomly under a budget $\rho$ in the perturbation of $\mathbf{A}$, i.e., $\|\mathbf{A}' - \mathbf{A}\|_1 \leq \rho|\mathcal{V}|^2$.

|          |     | SC    | DW    | ARGA      | GCN-1        |               |               |              | GCN-2  | LAT-GCN   |
| -------- | --- | ----- | ----- | --------- | ------------ | ------------- | ------------- | ------------ | ------ | --------- |
|          |     |       |       |           | $\rho = 0$   | $\rho = 0.01$ | $\rho = 0.05$ | $\rho = 0.1$ |        |           |
| CiteSeer | AUC | 0.805 | 0.805 | 0.924     | 0.692        | 0.713         | 0.694         | 0.635        | 0.8618 | **0.945** |
|          | AP  | 0.850 | 0.836 | 0.930     | 0.702        | 0.723         | 0.670         | 0.636        | 0.867  | **0.952** |
| Cora     | AUC | 0.846 | 0.831 | 0.924     | 0.876        | 0.892         | 0.846         | 0.815        | 0.902  | **0.931** |
|          | AP  | 0.885 | 0.850 | 0.926     | 0.873        | 0.896         | 0.849         | 0.817        | 0.911  | **0.943** |
| PubMed   | AUC | 0.842 | 0.844 | **0.968** | 0.954        | 0.949         | 0.900         | 0.879        | 0.962  | 0.957     |
|          | AP  | 0.878 | 0.841 | **0.971** | 0.954        | 0.950         | 0.905         | 0.882        | 0.963  | 0.958     |

evaluated them by running Algorithm 2. In particular, we set the hidden unit to 32 and $\epsilon = 5\%$ when training the robust models. We reported the results in Table 4 where LAT-GCN clearly outperforms other methods in robustness.

## 5    Experiment 2: Link Prediction

Besides predicting the properties of each node, another important application of graph is to predict whether a pair of nodes should be connected by an edge. Applications include social network, recommender system, and knowledge graph.

Followed by the work from Kipf and Welling (2016), assuming we have a graph $G = (\mathcal{V}, \mathcal{E})$, our goal can be written as reconstructing the adjacency matrix $\mathbf{P}$ by using GCN's output $\mathbf{Z}$:

$$\mathbf{P} = \sigma\left(\mathbf{Z}\mathbf{Z}^{\top}\right), \quad (\sigma \text{ applied element-wise}) \tag{7}$$

$$\mathbf{Z} = \text{GCN}(\mathbf{A}, \mathbf{X}) = \tilde{\mathbf{A}}\sigma\left(\tilde{\mathbf{A}}\mathbf{X}\mathbf{W}^{(1)}\right)\mathbf{W}^{(2)}, \tag{8}$$

where $\sigma$ can be the sigmoid function to represent the probability of connection, or simply a sign function. Intuitively, since each row of $\mathbf{Z}$ corresponds to the embedding of a node, (7) uses their pairwise inner product to determine whether they are to be connected.

We partition the set of edges $\mathcal{E}$ into $\mathcal{E}_{train}$, $\mathcal{E}_{val}^{+}$, $\mathcal{E}_{val}^{-}$, $\mathcal{E}_{test}^{+}$, $\mathcal{E}_{test}^{-}$, where $+$ and $-$ represent the existent and non-existent edges. In order to ensure proper node embedding, we enforce that $\mathcal{E}_{train}$ covers all the nodes in the link prediction task.

Similar to node prediction, adversarial training can be achieved by incorporating a regularizer which penalizes the change of the prediction under the perturbation, i.e.,

$$f(\mathbf{A}, \mathbf{X}) = \mathcal{L}_{\theta}(\mathbf{P}, \mathbf{A}) + \lambda \max_{\zeta \in D}\left\|\sigma\left(\mathbf{Z}\mathbf{Z}^{\top}\right) - \sigma\left(\mathbf{Z}_{\zeta}\mathbf{Z}_{\zeta}^{\top}\right)\right\|_{F}^{2},$$

where $\mathbf{Z}_{\zeta} = \tilde{\mathbf{A}}\left(\sigma\left(\tilde{\mathbf{A}}\mathbf{X}\mathbf{W}^{(1)}\right) + \zeta\right)\mathbf{W}^{(2)}$ and $\|\zeta_{i:}\| \leq \epsilon$ for all $i \in |\mathcal{V}|$. It is natural to base the regularizer on the variation of $\mathbf{Z}\mathbf{Z}^{\top}$ because ultimately the

**Table 6.** Success rate for Nettack-link and random attack under a budget of $\Delta$ edges for structure attack

| Model | Attacker | CiteSeer | | | | Cora | | | | PubMed | | | |
|---|---|---|---|---|---|---|---|---|---|---|---|---|---|
| | $\Delta$ | 1 | 5 | 10 | 20 | 1 | 5 | 10 | 20 | 1 | 5 | 10 | 20 |
| GCN | Nettack-link | 0.12 | 0.33 | 0.54 | 0.86 | 0.09 | 0.27 | 0.45 | 0.83 | 0.10 | 0.24 | 0.41 | 0.73 |
| LAT-GCN-A | Nettack-link | 0.15 | 0.25 | 0.43 | 0.69 | 0.08 | 0.21 | 0.38 | 0.65 | 0.08 | 0.21 | 0.34 | 0.67 |
| GCN | Random | 0.05 | 0.13 | 0.17 | 0.23 | 0.01 | 0.08 | 0.17 | 0.31 | 0.00 | 0.03 | 0.11 | 0.18 |
| LAT-GCN-A | Random | 0.06 | 0.14 | 0.15 | 0.20 | 0.02 | 0.09 | 0.17 | 0.29 | 0.00 | 0.05 | 0.09 | 0.20 |

---

**Algorithm 2.** Evaluation of robustness

---

**input** $\mathbf{A}, \mathbf{X}, \Delta$ (budget of Nettack)
 1: $s_{\text{GCN}} = s_{\text{LAT-GCN}} = 0$
 2: $\mathcal{T} :=$ sample 100 nodes from test set to attack
 3: **for** $u \in \mathcal{T}$ (target of attack) & $g \in \{\text{GCN}, \text{LAT-GCN}\}$ **do**
 4:    $c_g := evaluate(u, g(\mathbf{A}, \mathbf{X}))$
 5:    **if** $c_g = c_{true}(u)$ **then**
 6:       $\mathbf{A}', \mathbf{X}' \leftarrow Nettack(\mathbf{A}, \mathbf{X}, u, \Delta)$
 7:       $s_g := s_g + \delta(c_g \neq c_{true}(u))$
 8:    **end if**
 9. **end for**
**output** success rates: $s_{\text{GCN}}/|\mathcal{T}|, s_{\text{LAT-GCN}}/|\mathcal{T}|$.

---

adjacency matrix $\mathbf{P}$ is based on $\mathbf{ZZ}^{\top}$. In practice, however, we noticed that the performance does not change noticeably if we directly regularize the variation of $\mathbf{ZZ}^{\top}$:

$$f(\mathbf{A}, \mathbf{X}) = \mathcal{L}_{\theta}(\mathbf{P}, \mathbf{A}) + \lambda \max_{\zeta \in D} \left\| \mathbf{ZZ}^{\top} - \mathbf{Z}_{\zeta} \mathbf{Z}_{\zeta}^{\top} \right\|_F^2. \tag{9}$$

Here $\mathbf{Z}_{\zeta}$ simplifies to $\tilde{\mathbf{A}}\mathbf{X}\mathbf{W}^{(1)} + \zeta$ when a one-layer GCN is used. Similar to (4), the inner maximization over $\zeta$ can be solved by ADAM as in Sect. 3.1. We will again refer to the adversarial objective (9) as LAT-GCN. Although GCNs used in link prediction are often referred to as graph auto-encoders (GAE), we will stick with the term "GCN" to simplify terminology. Indeed, many of the discussions and approaches based on GCNs in this paper are applicable to more general graph neural networks.

*Comparison of Accuracy.* We compared one-layer LAT-GCN against vanilla GCN and three more baselines. Spectral clustering (Tang and Liu 2011) is an effective approach to learn node embedding. DeepWalk (Perozzi *et al.* 2014) is also a popular approach to represent nodes into a continuous vector space. Adversarial regularized auto-encoder (Pan *et al.* 2019) generates adversarial examples from node embedding. Table 5 shows that with a slight perturbation on the latent layer in GCN, the LAT-GCN outperforms GCN in area under ROC curve (AUC) and average precision (AP) on CiteSeer and Cora datasets. Here the results are reported based on the optimal tuned hyperparameter, with $\epsilon = 0.1$ and $\gamma = 0.02$ for both CiteSeer and Cora. Moreover, Figure 7 shows how the varied value of $\epsilon$ impacts the performance of LAT-GCN ($\gamma$ fixed to 0.1). Clearly only small values of $\epsilon$ can be helpful.

Although LAT-GCN does not perform the best on PubMed, our further experiment suggests that robustness is probably not positively correlated with generalization performance on this dataset. In particular, analogous to standard data augmentation, we randomly added or removed $\rho |\mathcal{V}|^2$ number of edges in

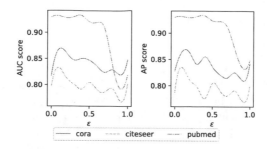

**Fig. 7.** AUC and AP curves of LAT-GCN with varied values of $\epsilon$

$\mathcal{E}_{train}$, where the budget $\rho \subset [0.01, 0.05, 0.1]$. So the resulting unnormalized adjacency matrix $\mathbf{A}'$ deviates from $\mathbf{A}$ in $L_1$ norm by at most $\rho |\mathcal{V}|^2$. We call it random attack (RA). Interestingly, as shown in the four columns underneath GCN-1 (one-layer GCN), increasing $\rho$ from 0 to 0.01 does improve the test AUC and AP for CiteSeer and Cora, but further increasing $\rho$ starts to be harmful. In contrast, the performance on PubMed decays monotonically as $\rho$ grows, suggesting that enforcing robustness to adversarial attacks on this dataset may be conflicting with generalization.

*Comparison of Robustness.* Since no specialized algorithm is available that adversarially attacks link prediction under GCNs, we designed a novel extension of Nettack, called Nettack-link, for this setting in addition to random attacks. Given a targeted link $(u, v)$, denote $\delta_{u,v} = 1$ if $(u, v)$ is in $\mathcal{E}$, and $-1$ otherwise. Nettack-link considers the change of $\mathbf{P}_{u,v}$ under perturbations of node features and graph structure, i.e.,

$$s_{struct}(e; \mathbf{A}, (u, v)) := \delta_{u,v} \cdot \left( \log \mathbf{P}_{u,v} - \log \mathbf{P}'_{u,v} \right),$$

for perturbing another edge $e$. Here we can restrict the candidate edge $e$ to one-hop or two-hop away from either $u$ or $v$. By greedily picking an edge $e$ whose addition/removal maximizes $s_{struct}(e; \mathbf{A}, (u, v))$, the algorithm can progressively find $\Delta$ number of edges to attack the target link prediction $(u, v)$. To facilitate an efficient computation which avoids computing $s_{struct}(e; \mathbf{A}, (u, v))$ from scratch every time, we followed the linearization recipe from Zügner *et al.* (2018), and identified an incremental procedure.

Table 6 shows the success rate of the two attacks, where both the set of existing and non-existing edges get a budget of $\Delta$ edges to remove or add, respectively. The test protocol followed Algorithm 2, except that we randomly picked 100 (existent or nonexistent) edges to attack in step 2. Clearly Nettack-list is much more effective than random attack. Both GCN and LAT-GCN suffer an increased rate of being successfully attacked as the budget grows, but LAT-GCN is more robust than GCN with a lower success rate in general.

# 6    Experiment 3: Recommendation

Finally we studied another effective application of GCNs: recommendation by matrix completion (GC-MC, Berg *et al.* 2018). Compared with the general structured graph above, here the graphs are bipartite with two groups of nodes such as users and items, and links only span between nodes of different groups. To model the $R$ levels of ratings (or other discrete recommendation levels), a set of adjacency matrices $\{\mathbf{A}_1, \cdots, \mathbf{A}_R\}$ are constructed for each rating type $r \in R$, where $\mathbf{A}_r$ is sized $N_u$-by-$N_v$ for $N_u$ users and $N_v$ items, respectively. The $(u, v)$-th entry of $\mathbf{A}_r$ is 1 if and only if the user $u$ gives a rating of $r$ to the item $v$, and 0 else. Normalization of $\mathbf{A}_r$ follows as above.

After GCN propagation is completed on each graph separately, the representation of an item or user is formed by aggregating its corresponding embeddings in the $R$ graphs, using concatenation or summation. The final rating for a (user, item) pair can then be predicted based on the inner product between the embeddings of the user and item, akin to link prediction.

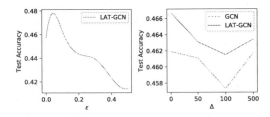

**Fig. 8.** Accuracy for MovieLen 100k with respect to $\epsilon$ in LAT-GCN (left), and the attack budget $\Delta$ (right)

Since there is no customized attacking algorithm for this setting, we resorted to random attack where a rating is randomly selected and flipped to another level; for simplicity, we did not consider just removing or adding ratings. The fact that random attack is not targeted allows us to present robustness and generalization performance in one plot. We followed the same setting as the GC-MC model, including the learning rate, dimensions of hidden layers, etc. Summation was applied to aggregate embeddings from five graphs, and we tested on the Movielens 100k dataset[1].

Figure 8 (left) shows that the test accuracy increases when LAT-GCN imposes a small perturbation on the GCN layer in training, but eventually the test accuracy decays. This is similar to Fig. 7. Furthermore, we varied the budget $\Delta$ for randomly flipping the ratings in the training data. As Fig. 8 (right) shows, the test accuracy of LAT-GCN (with $\epsilon$ fixed to 0.1) keeps higher than that of vanilla GCN for a range of $\Delta$.

---

[1] https://grouplens.org/datasets/movielens/.

# 7  Conclusion

In this work, we proposed a new regularization technique for GCN which not only improves generalization, but also defends against attacks in both node feature and graph structure. Superior empirical performance is achieved on node classification and link prediction problems. The method, which is based on perturbing latent representations, can be extended to adversarial training of other graph learning problems, such as recommender systems, and dynamic and multimodal graphs.

**Acknowledgements.** The Titan V used for this research was donated by the NVIDIA Corporation. We gratefully acknowledge the support of NVIDIA Corporation with the donation of the Titan V GPU used for this research.

# References

Berg, R.V.D., Kipf, T.N., Welling, W.: Graph convolutional matrix completion. In: KDD (2018)

Bertsekas, D.P.: Nonlinear Programming. Athena Scientific, Belmont, MA (1995)

Bruna, J., Zaremba, W., Szlam, A., LeCun, Y.: Spectral networks and locally connected networks on graphs. In: ICLR (2013)

Chen, J., Zhu, J., Song, L.: Stochastic training of graph convolutional networks with variance reduction. In: ICML (2017)

Chen, J., Ma, T., Xiao, C.: Fastgcn: fast learning with graph convolutional networks via importance sampling. In: ICLR (2018)

Chen, J., Shi, Z., Wu, Y., Xu, X., Zheng, H.: Link prediction adversarial attack. arXiv preprint arXiv:1810.01110(2018)

Cullina, D., Bhagoji, A.N., Mittal, P.: PAC-learning in the presence of adversaries. In: NeurIPS (2018)

Dai, H., et al.: Adversarial attack on graph structured data. In: ICML, pp. 1123–1132 (2018)

Defferrard, M., Bresson, X., Vandergheynst, P.: Convolutional neural networks on graphs with fast localized spectral filtering. In: NeurIPS, pp. 3844–3852 (2016)

Deng, Z., Dong, Y., Zhu, J.: Batch virtual adversarial training for graph convolutional networks. arXiv preprint arXiv:1902.09192 (2019)

Duvenaud, D.K., et al.: Convolutional networks on graphs for learning molecular fingerprints. In: NeurIPS, pp. 2224–2232 (2015)

Feng, F., He, X., Tang, J., Chua, T.-S.: Graph adversarial training: Dynamically regularizing based on graph structure. arXiv preprint arXiv:1902.08226 (2019)

Hammond, D.K., Vandergheynst, P., Gribonval, R.: Wavelets on graphs via spectral graph theory. Appl. Comput. Harmonic Anal. **30**(2), 129–150 (2011)

He, W., Wei, J., Chen, X., Carlini, N., Song, D.: Adversarial example defenses: Ensembles of weak defenses are not strong. arXiv preprint arXiv:1706.04701 (2017)

Huang, S., Papernot, N., Goodfellow, I., Duan, Y., Abbeel, P.: Adversarial attacks on neural network policies. arXiv preprint arXiv:1702.02284 (2017)

Jia, R., Liang, P.: Adversarial examples for evaluating reading comprehension systems. In: ICLR (2017)

Kingma, D.P., Ba, J. L.: Adam: a method for stochastic optimization. In: ICLR (2015)

Kipf, T.N., Welling, M.: Variational graph auto-encoders. arXiv preprint arXiv:1611.07308 (2016)

Kipf, T.N., Welling, M.: Semi-supervised classification with graph convolutional networks. In: ICLR (2017)

Kurakin, A., Goodfellow, I., Bengio, S.: Adversarial machine learning at scale. In: ICLR (2017)

Miyato, T., Dai, A.M., Goodfellow, I.: Adversarial training methods for semi-supervised text classification. In: ICLR (2017)

Pan, S., Hu, R., Fung, S.-F., Long, G., Jiang, J., Zhang, C.: Learning graph embedding with adversarial training methods. arXiv preprint arXiv:1901.01250 (2019)

Perozzi, B., Al-Rfou, R., Skiena, S.: Deepwalk: online learning of social representations. In: KDD, pp. 701–710. ACM (2014)

Raghunathan, A., Steinhardt, J., Liang, P.: Certified defenses against adversarial examples. In: ICLR (2018)

Schlichtkrull, M., Kipf, T.N., Bloem, P., van den Berg, R., Titov, I., Welling, M.: Modeling relational data with graph convolutional networks. In: Gangemi, A., et al. (eds.) ESWC 2018. LNCS, vol. 10843, pp. 593–607. Springer, Cham (2018). https://doi.org/10.1007/978-3-319-93417-4_38

Schmidt, L., Santurkar, S., Tsipras, D., Talwar, K., Madry, A.: Adversarially robust generalization requires more data. In: NeurIPS (2018)

Sen, P., Namata, G., Bilgic, M., Getoor, L., Galligher, B., Eliassi-Rad, T.: Collective classification in network data. AI Mag. 29(3), 93–93 (2008)

Shafieezadeh-Abadeh, S., Kuhn, D., Esfahani, P.M.: Regularization via mass transportation. arXiv preprint arXiv:1710.10016 (2017)

Sinha, A., Namkoong, H., Duchi, J.: Certifying some distributional robustness with principled adversarial training. In: ICLR (2018)

Song, Y., Kim, T., Nowozin, S., Ermon, S., Kushman, N.: Pixeldefend: leveraging generative models to understand and defend against adversarial examples. In: ICLR (2018)

Stutz, D., Hein, M., Schiele, B.: Disentangling adversarial robustness and generalization. In: CVPR (2019)

Szegedy, C., et al.: Intriguing properties of neural networks. In: ICLR (2014)

Tang, L., Liu, H.: Leveraging social media networks for classification. Data Min. Knowl. Discov. 23(3), 447–478 (2011)

Tran, P.V.: Learning to make predictions on graphs with autoencoders. In: 5th IEEE International Conference on Data Science and Advanced Analytics (2018)

Tsipras, D., Santurkar, S., Engstrom, L., Turner, A., Madry, A.: Robustness may be at odds with accuracy. In: ICLR (2019)

Veličković, P., Cucurull, G., Casanova, A., Romero, A., Lio, P., Bengio, Y.: Graph attention networks. In: ICLR (2018)

Wang, S., et al.: Adversarial defense framework for graph neural network. arXiv preprint arXiv:1905.03679 (2019)

Wong, E., Kolter, J.Z.: Provable defenses against adversarial examples via the convex outer adversarial polytope. In: ICML (2018)

Wu, F., Zhang, T., Souza Jr, A.H.D., Fifty, C., Yu, T., Weinberger, K.Q.: Simplifying graph convolutional networks. arXiv preprint arXiv:1902.07153 (2019)

Xu, K., et al.:. Topology attack and defense for graph neural networks: an optimization perspective. In: Proceedings of the 28th International Joint Conference on Artificial Intelligence, pp. 3961–3967. AAAI Press (2019)

Yin, D., Ramchandran, K., Bartlett, P.: Rademacher complexity for adversarially robust generalization. arXiv preprint arXiv:1810.11914 (2018)

Zhang, H., Yu, Y., Jiao, J., Xing, E.P., Ghaoui, L.E., Jordan, M.I.: Theoretically principled trade-off between robustness and accuracy. In: ICML (2019)

Zügner, D., Günnemann, S.: Certifiable robustness and robust training for graph convolutional networks. In: KDD, pp. 246–256. ACM (2019)

Zügner, D., Akbarnejad, A., Günnemann, S.: Adversarial attacks on neural networks for graph data. In: KDD, pp. 2847–2856. ACM (2018)

# Enhancing Robustness of Graph Convolutional Networks via Dropping Graph Connections

Lingwei Chen, Xiaoting Li, and Dinghao Wu[✉]

College of Information Sciences and Technology, The Pennsylvania State University,
University Park, PA 16802, USA
duw12@psu.edu

**Abstract.** Graph convolutional networks (GCNs) have emerged as one of the most popular neural networks for a variety of tasks over graphs. Despite their remarkable learning and inference ability, GCNs are still vulnerable to adversarial attacks that imperceptibly perturb graph structures and node features to degrade the performance of GCNs, which poses serious threats to the real-world applications. Inspired by the observations from recent studies suggesting that edge manipulations play a key role in graph adversarial attacks, in this paper, we take those attack behaviors into consideration and design a biased graph-sampling scheme to drop graph connections such that random, sparse and deformed subgraphs are constructed for training and inference. This method yields a significant regularization on graph learning, alleviates the sensitivity to edge manipulations, and thus enhances the robustness of GCNs. We evaluate the performance of our proposed method, while the experimental results validate its effectiveness against adversarial attacks.

**Keywords:** Graph convolutional networks · Adversarial attacks · Graph sampling · Robustness

## 1 Introduction

Graph structured data plays an important role in many real-world applications, which represents natural yet complex relationships between objects in different domains [27,33], such as social networks, biological networks, and citation networks. Inspired by the great success of applying convolutional neural networks (CNNs) to computer vision tasks [2,7,14,18], many research efforts have been devoted to a paradigm shift in graph learning that generalizes convolutions to the graph domain to address such a challenge [13,24,28]. While recent works in this direction are making progress on spatial and spectral approaches respectively, graph convolutional networks (GCNs) have been emerging as one of the most popular and significant graph neural networks [1,17,19,30,32]. These GCNs take the connectivity structure of the graphs as the filter to perform neighborhood information aggregation so as to extract high-level features from the nodes and

© Springer Nature Switzerland AG 2021
F. Hutter et al. (Eds.): ECML PKDD 2020, LNAI 12459, pp. 412–428, 2021.
https://doi.org/10.1007/978-3-030-67664-3_25

their neighborhoods [6], which have thus boosted the state-of-the-arts for a variety of tasks (e.g., node classification, clustering, and matching) over graphs.

Despite their remarkable graph representation learning and inference ability, GCNs are still faced with the same inherent learning-security challenge of lacking adversarial robustness existing in other deep neural networks (DNNs) [4,12,22]. Recent studies [9,26,27,34,35] have shown that GCNs remain vulnerable to adversarial attacks that carefully design imperceptible perturbations to graph structures and/or node features; these attacks may taint the node neighborhoods, and thus drastically degrade the node representations and the corresponding performance of GCNs. This poses serious threats to real-world applications, especially for those security-critical scenarios such as traffic, financial and healthcare systems [33]. However, compared to the regular DNNs targeting individual objects, GCNs with "message passing" framework have to cope with correlated objects; therefore, the question of how to improve the robustness of GCNs against adversarial attacks is more complicated and less studied.

Since GCNs deal with graph structured data where node representations are learned by propagating node features through neighborhoods, their vulnerabilities may naturally come from feature perturbations and edge manipulations. However, some recent works [9,26,34] suggested that node feature perturbations may not directly impact on the learning performance compromise, while edge manipulations are more effective to formulate successful graph adversarial attacks due to the facts that (1) graph connectivity plays a more crucial role in graph representation learning, and (2) small perturbations over high-dimensional feature space may not induce significant output variations for the aggregation layers. In other words, to enhance the robustness of GCNs against adversarial attacks, a feasible way is to alleviate the impacts of graph connections such that the adversarial edge manipulations will be penalized. When we revisit typical defensive methods over shallow or deep learning models where they are either using weight evenness to restrict weight values to a very narrow range [10], applying ensemble over random feature subspaces [3,8], or adopting adversarial training to bound the abilities of other adversarial feature perturbations [12], we can observe that these strategies yield the regularization of different degrees to enforce the learning models less sensitive to changes in features and thus improve their robustness. This means that a well-formulated regularization technique over target space can be leveraged to penalize the corresponding space perturbations. Generally, dropout [21] is devised into GCN layers to regularize feature learning, but it gives insufficient penalty to feature perturbations and invalid regularization for edge manipulations. As such, it calls for a more effective regularization method over graph connectivity to confer a significant reduction in GCNs' vulnerability to adversarial attacks.

To address this challenge, in this paper, we propose to exploit graph sampling to drop graph connections. More specifically, instead of constructing GCNs on the complete graph that yields a high variance, we drop out part of the input graph edges to form a set of random and sparse subgraphs and train the GCNs over them. Different from the current graph sampling techniques [5,6,11,20,31], we take the behaviors of graph adversarial attacks into consideration, and elaborate a biased graph-sampling scheme over well-defined edge sampling probabilities with respect to nodes' degrees and feature similarity. In this regard, the sampled subgraphs may provide more randomness, sparseness and deformation to alleviate the reliance on the graph connections, and hence penalize the adversarial edge manipulations. Moreover, in order to generate an additional regularization benefit, in the inference stage, we further employ an ensemble approach to aggregate the individual predictions over the subgraphs sampled in the same graph-sampling scheme that are rooted by the test nodes. We summarize our main contributions as follows:

- Graph-sampling mechanisms have been recently proposed to address GCN training over-fitting and deeper-layer computation issues, but never been discussed in the adversarial setting. We leverage this idea and design a biased graph sampling to regularize the graph connections, alleviate the sensitivity to the edge manipulations and hence enhance the robustness of GCNs.
- Instead of using the full graph to infer the test data, we further utilize bagging-like ensemble over sampled subgraphs as is done in parameter learning to approximate the final inference output and provide an additional regularization beyond that provided by model training.
- We conduct comprehensive experimental studies on a number of datasets, which demonstrate that our proposed method can effectively improve the GCN performance against different kinds of adversarial attacks.

The rest of the paper is organized as follows. Section 2 introduces the preliminaries. Section 3 presents our proposed method. Section 4 evaluates the effectiveness of our method. Section 5 discusses related work. Section 6 concludes.

## 2    Preliminaries

**Graph Convolutional Networks.** Let $\mathcal{G} = (\mathcal{V}, \mathcal{E}, X)$ be a graph, where $\mathcal{V}$ is the set of nodes $\{v_1, \cdots, v_n\}$ with $|\mathcal{V}| = n$, $\mathcal{E}$ is the set of edges, and $X = [x_1, x_2, \cdots, x_n]^T \in \mathbb{R}^{n \times c}$ is feature matrix; for each node $v \in \mathcal{V}$, its feature $x_v \in \mathbb{R}^c$ is a $c$-dimensional row vector. Given $\mathcal{V}$ and $\mathcal{E}$, the adjacency matrix $A$ can be accordingly formulated, where $A \in \mathbb{R}^{n \times n}$ and $A_{ij} = \{0, 1\}$, i.e., if $(v_i, v_j) \in \mathcal{E}$, then $A_{ij} = 1$; otherwise, $A_{ij} = 0$. Based on the adjacency matrix $A$, the diagonal degree matrix $D$ is defined as $D_{ii} = \sum_{j=1}^{n} A_{ij}$. In this work, the graph convolution network (GCN) proposed by Kipf and Welling [17] is exploited as a base model to facilitate the analysis and understanding of our further proposed approach; therefore, we briefly present its architecture here, where the graph convolutional layer is defined as:

$$H^{(l)} = \sigma \left( \tilde{A} H^{(l-1)} W^{(l)} \right) \tag{1}$$

where $H^{(l-1)}$ and $H^{(l)}$ are the input and output for layer $l$ $(l \geq 1)$, $W^{(l)} \in \mathbb{R}^{c_{l-1} \times c_l}$ is the learnable weight matrix for layer $l$, $\sigma$ is the non-linear activation function (e.g., ReLU), and $\tilde{A} = \hat{D}^{-\frac{1}{2}}\hat{A}\hat{D}^{-\frac{1}{2}}$. $\tilde{A}$ is a symmetrically normalized adjacency matrix where $\hat{A} = A + I$ is the adjacency matrix of the graph $\mathcal{G}$ with self connections added, and $\hat{D}$ is the diagonal degree matrix of $\hat{A}$. Given $H^{(0)} = X$, the GCN model with $l$ layers can be thus defined as:

$$Z = f(A, X) = \mathrm{softmax}\left(\tilde{A} \cdots \sigma\left(\tilde{A}XW^{(1)}\right) \cdots W^{(l)}\right). \qquad (2)$$

GCNs can be applied under inductive and transductive settings. In this paper, we focus on transductive classification where all node connections and features are accessible during training. To train a GCN model on such a task with labels $\mathcal{C} = \{1, 2, \cdots, c_l\}$, the softmax function normalizes the final output matrix $Z \in \mathbb{R}^{n \times c_l}$, where each row represents the probability of $c_l$ labels for a node. The cross-entropy loss $\mathcal{L} = \sum_{v \in \mathcal{V}_{tr}} \log Z_{v,c_v}$ can be accordingly evaluated between the output and the corresponding nodes' true labels, while the weights are updated using some gradient descent optimization algorithms (e.g., Adam [16]).

**Graph Adversarial Attacks.** The adversarial attacks over graph data and GCNs aim to perturb graph structures and/or node features in an imperceptible way to enforce GCNs to incorrectly classify certain nodes. Specifically, given a graph $\mathcal{G} = (A, X)$, it is often possible to find a similar graph $\mathcal{G}' = (A', X')$ such that for a set of nodes $v \in \mathcal{V}' \subset \mathcal{V}$, $y_v = \arg\max_{i \in \mathcal{C}} Z_{v,i}$, $y_v' = \arg\max_{i \in \mathcal{C}} Z_{v,i}'$, and $y_v \neq y_v'$, while $\mathcal{G}$ and $\mathcal{G}'$ are close according to specific measure metric. Such adversarial attacks may be implemented through directly manipulating the edges or features of the target nodes, or indirectly manipulating other nodes to influence the target nodes' classification results [9,34,35]. In this work, our goal is to build a robust GCN model that can improve its classification performance on the attacked $\mathcal{G}'$.

## 3   Proposed Method

In this section, we present the detailed method of how we drop graph connections to formulate subgraphs using biased graph sampling and how we leverage such subgraphs to enhance GCNs' robustness against adversarial attacks.

### 3.1   Regularization on Graph Connections

As observed in recent works [9,26,34], those effective graph adversarial attacks tend to manipulate edges rather than features, where the edge manipulations particularly fall into edge additions (e.g., connecting the target node to the nodes with very different features and labels). The reason behind this could be that GCNs with "message-passing" framework collectively aggregate feature information from the neighborhood of each node at each layer, where adding such an edge may more essentially impact on the target node's full feature dimensions than perturbing individual feature values [26]. As such, a potential defense

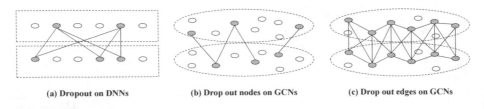

(a) Dropout on DNNs          (b) Drop out nodes on GCNs          (c) Drop out edges on GCNs

**Fig. 1.** Dropout examples: (a) a standard dropout on DNNs, (b) a potential way to drop out nodes on GCNs, and (c) another potential way to drop out edges on GCNs.

against adversarial attacks over GCNs should regularize the graph connections to prevent the model learning from over relying on some specific edges and thus cause the adversarial attacks crafted using edge manipulations over the vanilla GCNs less effective.

Dropout is a regularization technique widely used in different neural networks to prevent over-fitting through dropping out units in the hidden and visible layers, which is shown in Fig. 1(a). Srivastava et al. [21] explained that dropout may enforce each unit not to rely on a large set of units but to learn to work with a small random set of other units such that the complex co-adaptations would be reduced and each unit is more robust as well. Similarly, if we drop out the graph connections (e.g., dropping out the nodes in Fig. 1(b) or dropping out the edges in Fig. 1(c)) at each training time, feature propagation may not be able to rely on the full neighborhood while driving each edge to learn to work with a randomly chosen sample of edges and create useful features on its own. In this respect, those manipulated edges would become less effective and the aggregated features could be more robust against adversarial attacks. Although this dropout is shown promising to prevent over-fitting for DNN training, later work pointed out that it doesn't provide sufficient regularization to confer a significant reduction in model's vulnerability to adversarial examples [12]. One possible explanation for this vulnerability could be that each unit is dropped out with the uniform probability; over the long-term training, the hidden units within a layer may still manage to learn the mix-ability of all units to approximate the full feature information including the perturbations. According to the empirical analysis and observations on adversarial attacks presented in [26,34] that (1) nodes with higher degrees are more difficult to be manipulated than those with lower degrees and (2) adversarial attacks more likely connect the target node to the nodes that are dissimilar in terms of features, this implies that an edge should be dropped out in a different probability with respect to its associated nodes' degrees and feature similarity. That is to say, to better regularize the graph connections, the rationale is to reduce the possibility of those adversarial edges being selected for constructing each training graph batch, while those edges that connect nodes with extremely low similarity may never be selected; in this way, the node neighborhoods will be intuitively smoothed, which makes the aggregated features from nodes more robust against the adversarial attacks.

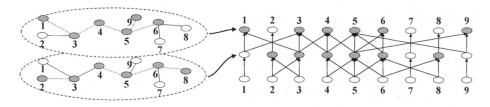

Fig. 2. Layer-wise graph sampling.

## 3.2   Training Through Dropping Graph Connections

A severe challenge for GCNs is that recursive feature information aggregation from neighborhoods across layers enforces exponential growth of computations in training batches, and thus makes GCNs difficult to be trained efficiently. To address such neighbor explosion, different state-of-the-art graph-sampling methods, either neighbor sampling [5,13], layer-wise sampling [6,15], or edge sampling [20,31], have been proposed to not only speed up the training process, but also reduce the model variance. In this respect, we leverage graph sampling idea to perform dropping graph connections. Clearly, when we sample a certain number of edges from the input graph, those excluded will be naturally dropped out. To make the defense more effective, we take the behaviors of adversarial attacks into consideration and design a biased graph-sampling scheme to construct the subgraphs at each training time with well-defined edge sampling probabilities.

**Biased Graph Sampling.** Given an input graph $\mathcal{G} = (A, X)$, our goal is to construct a set of subgraphs as training batches, each of which extracts a certain number of edges to formulate a new adjacency matrix. Formally, we introduce a Boolean matrix $Q \in \{0,1\}^{n \times n}$ to encode whether or not an edge in $\mathcal{G}$ is selected, i.e., if the edge connecting $v_i$ and $v_j$ is selected, then $Q_{ij} = 1$; otherwise, $Q_{ij} = 0$. We also define an edge sampling probability matrix $P \in \mathbb{R}^{n \times n}$ ($P_{ij} \in [0,1]$) to specify the probability of an edge being selected. If the graph $\mathcal{G}$ is indirect, then both matrices $Q$ and $P$ are symmetric. For each training batch, we utilize a pseudo random function $r(\cdot) \in (0,1)$ to activate the edge sampling probabilities and decide which edges will be sampled, where

$$Q_{ij} = \begin{cases} 1 & r(\cdot) \le P_{ij} \\ 0 & \text{otherwise} \end{cases} \tag{3}$$

If we denote the resulting adjacency matrix for training time $t$ as $A_t$, then the subgraph can be represented as

$$A_t = Q \circ A \tag{4}$$

where $\circ$ denotes the element-wise product. Following the processing steps in Sect. 2, we always perform the symmetrical normalization on $A_t$ to obtain $\tilde{A}_t$.

---

**Algorithm 1:** GCN training through dropping graph connections.

---

**Input**: Graph $\mathcal{G} = (A, X)$, epochs $T$, GCN layers $L$, training nodes $\mathcal{V}_{tr}$.
**Output**: GCN model with trained weights $W$.

Pre-calculate edge sampling probability matrix $P$;
**for** *each training epoch $t \leq T$* **do**

> Pseudorandomly sample $L$ subgraphs $A_t^{(l)}$ ($l \in [1, L]$) using Eq. (3) and (4);
>
> $Z = \text{softmax} \left( A_t^{\tilde{(L)}} \cdots \sigma \left( A_t^{\tilde{(1)}} X W^{(1)} \right) \cdots W^{(L)} \right)$;
>
> Back propagation from cross-entropy loss $\mathcal{L} = -\sum_{v \in \mathcal{V}_{tr}} \log Z_{v,c_v}$;
> Update weights $W$;

**end**

---

Considering that the layer-wise graph sampling provides more randomness and deformation of the input graph [20], which may yield more desirable regularization for the model training, here, we further sample subgraph $A_t^{(l)}$ for each layer $l$ at training time $t$. Afterwards, training proceeds by iterative weight updates through Eq. (2), where each iteration includes independently sampled subgraphs $A_t^{(l)}$ with respect to the number of GCN layers. Figure 2 and Algorithm 1 illustrate the training algorithm. This graph sampling scheme may inevitably introduce bias into training batches, while we consider such a bias as an advantage to penalize the adversarial edge manipulations.

**Edge Sampling Probability.** The biased graph sampling method is performed independently on each edge, while the only determinant for an edge being selected for subgraph construction is its sampling probability. Since each edge is associated with two nodes, edge sampling probability should consider the joint information from both nodes in terms of node degrees and features. Based on the variance reduction analysis in [33], the feature aggregation at each training batch can be computed as:

$$\zeta = \sum_l \sum_{i,j} \frac{\tilde{A}_{ij} X_j^{(l-1)} W^{(l-1)} + \tilde{A}_{ji} X_i^{(l-1)} W^{(l-1)}}{P_{ij}} \delta_{ij}^{(l)} \tag{5}$$

where $\delta_{ij}^{(l)} = 1$ if $(v_i, v_j) \in \mathcal{E}_t^{(l)}$; otherwise, $\delta_{ij}^{(l)} = 0$. As proved in [33], we can accordingly summarize a theorem that under independent edge sampling with a specified budget, the optimal edge probabilities $P_{ij}$ to minimize the sum of variance of feature aggregation $\zeta$ can be approximately given by $P_{ij} \propto \tilde{A}_{ij} + \tilde{A}_{ji}$. It suggests that the edge sampling probability for $(v_i, v_j)$ should be higher if two nodes $v_i$ and $v_j$ have fewer neighbors and are more likely to be influenced by each other. As aforementioned, in the adversarial setting, nodes with more neighbors are more difficult to be attacked than those with fewer neighbors and thus have higher classification accuracy in both the clean and attacked graphs [26]. We can interpret it in the way that (1) if either one of two connected nodes $v_i$ and $v_j$ has few neighbors, this edge is more likely to be manipulated by the attacks, and the edge sampling probability should hence be low, while (2) if both $v_i$ and $v_j$ have many neighbors, the edge tends to already exist in the original input graph. Therefore, in order to gain relatively low variance as well as high adversarial robustness, we define our edge sampling probability with respect to node degrees as

$$P_{ij} \propto \frac{1}{\tilde{A}_{ij} + \tilde{A}_{ji}} = \frac{\deg(v_i)\deg(v_j)}{\deg(v_i) + \deg(v_j)} = P_{ij}^{\deg} \tag{6}$$

We utilize a probability mass function for all the edges in the given graph:

$$P_{ij}^{\deg} = P_{ij}^{\deg} / \sum_{(v_a, v_b) \in \mathcal{E}} P_{ab}^{\deg}, \ (v_i, v_j) \in \mathcal{E} \tag{7}$$

As for node features, considering that the attackers more likely connect the target node to the nodes with very different features, we further calculate feature similarity score $S_{ij}$ between $v_i$ and $v_j$ to restrict the edge sampling probability. Based on different node feature spaces, the feature similarity measures may vary. For example, Jaccard similarity can be employed to measure binary feature space, while cosine similarity can be used for numeric feature space. Accordingly, the edge sampling probability can be updated as

$$P_{ij} \propto P_{ij}^{\deg} \cdot S_{ij} \tag{8}$$

which implies that the edges that connect nodes with more neighbors and larger feature similarity score have higher sampling probability; if $S_{ij} = 0$, then the corresponding edges would never be selected. Wu et al. [26] indicated that even though there may be a few legitimate edges with very low similarity score in the clean graph, it has little impact on the predictions of the target nodes to remove these edges. We take $P$ as our final edge sampling probability matrix (note that, $P_{ii} = 0$, $\forall v_i \in \mathcal{V}$ and $P_{ij} = 0$, $\forall (v_i, v_j) \notin \mathcal{E}$), which leads to better regularization and robustness since it enables the GCNs to explore the random and deformed graph connections and node features. As stated in Algorithm 1, the probability matrix $P$ can be pre-calculated, which would not significantly increase the computational complexity for the GCN training, except for some extra computations on adjacency matrix formulation and normalization.

### 3.3   Inference Through Ensemble

In the previous section, a biased graph sampling method is presented for constructing subgraphs, which essentially drops the graph connections, and penalizes the adversarial edge manipulations. Based on the trained GCNs with updated weights using such subgraph training batches, we can easily proceed with inference for test data. Generally, a full neighborhood architecture is fed to

---

**Algorithm 2:** GCN inference through ensemble.

---

**Input**: Graph $\mathcal{G} = (A, X)$, epochs $T$, trained weights $W$, testing nodes $\mathcal{V}_{ts}$.

**Output**: Labels for testing nodes.

Pre-calculate edge sampling probability matrix $P$;

Initialize the output matrix $O \in \mathbb{R}^{n \times c_l}$ and counter array $N \in \mathbb{R}^n$;

**for** *each testing epoch $t \leq T$* **do**

    Pseudorandomly sample a subgraph $A_t$ using Eq. (3) and (4);

$$Z = \text{softmax}\left(\tilde{A}_t \cdots \sigma\left(\tilde{A}_t X W^{(1)}\right) \cdots W^{(L)}\right);$$

    **for** $v \in \mathcal{V}_{ts}$ *and $v$ is sampled* **do**

        $O_v \leftarrow O_v + Z_v$;

        $N_v \leftarrow N_v + 1$;

    **end**

**end**

$O_v \leftarrow O_v / N_v, \forall v \in \mathcal{V}_{ts}$;

**return** *Labels for testing nodes $\mathcal{V}_{ts}$ using $O$.*

---

the model for testing, which is simple and straightforward [6]. However, considering that the adversarial attacks may inevitably manipulate the edges over test data, in this work, instead, we first sample subgraphs from the full graph for test data as is done in GCN training, and then utilize a bagging-like ensemble to average the predictions from all the individual outputs $Z$ over the subgraphs including the connections to the corresponding test nodes, and hence approximate the final inference results. With the use of ensemble, the inference may provide an additional regularization beyond that provided by model training. The inference algorithm is illustrated in Algorithm 2, which is similar to Algorithm 1, with three differences: (1) the inference only applies a sampled subgraph cross all the layers in a testing epoch for easier test node tracking; (2) the trained weights $W$ will not be updated anymore but directly used for forward aggregation and prediction; and (3) all the outputs $Z$ with sampled target test nodes are further averaged to return the inferred labels for the corresponding test nodes.

## 4     Experimental Results and Analysis

In this section, we evaluate the performance of our proposed robust GCN model on a number of datasets to defend against adversarial attacks. We perform experiments on the graph-based benchmark node classification tasks with the random splits [29] of each dataset, and compare our model with state-of-the-art baselines.

**Table 1.** Statistics of the datasets used in our experiments.

| Dataset | Nodes | Edges | Classes | Features |
|---|---|---|---|---|
| Cora | 2,708 | 5,429 | 7 | 1,433 |
| Citeseer | 3,327 | 4,732 | 6 | 3,703 |

## 4.1 Experimental Setup

**Datasets.** We test our model on two citation network benchmark datasets: *Cora* and *Citeseer* [17,29] - in both of these datasets, nodes represent documents and edges denote citation links; node features correspond to elements of a bag-of-words representation of a document *i.e.*, 0/1 values indicating the absence/presence of a certain word, while each node has a class label [25]. The dataset statistics are summarized in Table 1. For dataset random split, we adjust Cora and Citeseer respectively to align with our experimental data setting: 20 instances for each class are randomly sampled as training data while another 500 and 1000 instances are selected as validation and test data.

**Baselines and Adversarial Attacks.** We compare our proposed method (named DropCONN throughout the experiments) with some state-of-the-art baselines, including:

- GCN [17]: this is the original GCN model introduced by Kipf and Welling, which is described in Sect. 2.
- DropEdge [20]: this is a sampling-based GCN model, which drops out a certain rate of edges of the input graph by random and shares the same perturbed adjacency matrix across all the layers.
- GraphSAINT [31]: this is another sampling-based GCN model, which extracts a connected subgraph for each training batch with the edge sampling rate $P_{ij} \propto \tilde{A}_{ij} + \tilde{A}_{ji}$.
- RGCN [33]: this model defends against graph adversarial attacks by adopting Gaussian distributions as the hidden representations of nodes to absorb the effects of adversarial changes.

To assess the robustness of our proposed method against different attack methods, we choose three representative graph adversarial attacks as follows:

- Nettack [34]: this is a targeted attack method to enforce misclassification on the target nodes using edge and feature perturbations, which can handle both direct and influence attacks. We focus on direct poisoning attack here.
- Meta-gradient attack (Metattack) [35]: this is a non-targeted attack method to reduce the overall classification performance of GCNs using meta-gradients.
- Random attack [33]: it randomly selects some non-adjacent node pairs and add fake edges.

**Table 2.** Classification results (accuracy %) on clean datasets.

| Dataset | GCN | DropEdge | GraphSAINT | RGCN | DropCONN |
|---------|-----|----------|------------|------|----------|
| Cora | 81.5 | 82.8 | 81.1 | 82.8 | 80.5 |
| Citeseer | 70.3 | 72.3 | 71.0 | 71.2 | 69.8 |

**Parameter Setting.** Experiments are under the transductive, semi-supervised learning setting. All the GCNs are set as a two-layer structure with 16 hidden units, which are trained using Adam [16] with 0.01 initial learning rate, $5 \times 10^{-4}$ L2 regularization on the weights, and 0.5 dropout rate for the input and hidden layers. For DropEdge, the edge sampling rate is 0.5. For GraphSAINT, we use edge sampler. The parameter settings of Nettack, Metattack, and Random attack are directly taken from [34, 35] and [33]: (1) Nettack performs the perturbations ranging in $\{1, 2, 3, 4, 5\}$, (2) the edge perturbation rate for Metattack varies in $\{1\%, 5\%, 10\%, 15\%, 20\%\}$, and (3) the perturbation rate for random attack is set as $\{20\%, 40\%, 60\%, 80\%, 100\%\}$. Since the data feature space is binary, we use Jaccard similarity for feature measure in our method. Besides, we report the mean classification accuracy of 10 runs on the test nodes for all the experiments.

### 4.2  Robustness Evaluation Against Adversarial Attacks

**Performance Without Attack.** We first compare our proposed method (i.e., DropCONN) with the original GCN and other variant models on the clean datasets to assess the classification performance when the datasets are not attacked. The experimental results are illustrated in Table 2. From the results, we can see that DropCONN slightly underperforms the baseline models on both datesets. The reason behind this is that the graph sampling using our formulated edge sampling probability may inevitably enforce some very sparse subgraph generations and some edge information loss for those that connect nodes with low feature similarity score as well, while averaging the testing output may further smooth the classification accuracy. Compared to the original GCN, DropCONN degrades the accuracy from 81.5% to 80.5% on Cora and from 70.3% to 69.8% on Citeseer; the performance degradation introduced by our model is not significant. Considering that DropCONN is designed to be resilient against different graph adversarial attacks, which will be thoroughly evaluated later, this comparable performance on clean data shows that our proposed robust solution can be applied in a realistic setting with or without adversarial attacks since we cannot really tell if perturbations happen or not in real-world graphs.

**Robustness Against Adversarial Attacks.** In this section, we conduct experiments to evaluate the effectiveness of our defense strategy against three adversarial attacks (i.e., nettack, metattack, and random attack) to see (1) if the regularization on graph connections through biased graph sampling and ensemble inference contributes to robust graph learning and (2) if DropCONN can

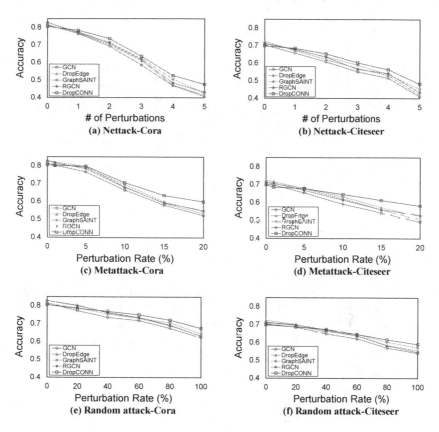

**Fig. 3.** Classification results (accuracy) under nettack, metattack, and random attack.

outperform other graph-sampling and defensive GCN models. More specifically, for non-targeted attacks (i.e., metattack and random attack), we perturb a certain rate of edges to reduce the overall classification of the model where this perturbation rate varies from 0% to 20% for metattack and from 0% to 100% for random attack. Note that, metattack is performed using meta-gradient approach with self-training [35]. For targeted attack, we set the target nodes as all nodes in the test data and perform nettack to generate different numbers of perturbations (ranging from 1 to 5) for the target nodes. After either non-targeted attacks or targeted attack, we train different GCN models on the modified graph and evaluate the classification accuracy on the test nodes.

We report the comparative evaluation results against three attacks in Fig. 3(a)–Fig. 3(f) respectively. From Fig. 3(a) and Fig. 3(b), we can observe that direct nettack attack on the target nodes is a strong attack method, where five adversarial perturbations over either graph structure or node features have already drastically degraded the testing accuracy by around 40% on Cora and 30% on Citeseer for most of GCN models; clearly, the original GCN suffers from the biggest drop from 81.5% to 40.8% on Cora and from 70.3% to 41.1%

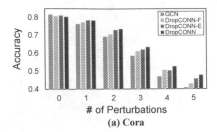 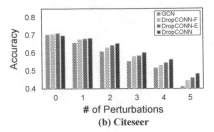

(a) Cora                                    (b) Citeseer

**Fig. 4.** Evaluations on importance of feature similarity and ensemble.

on Citeseer. Despite its apparent attack effect, different GCN variants still show some different defensive potentials against nettack, in which our proposed method DropCONN successfully outperforms other state-of-the-art baselines on both datasets, especially when more perturbations are injected to the graph. Compared to GCN, DropCONN improves the accuracy by a margin of 1.9% to 6.9% and 2.7% to 7.2% respectively, while compared to the best performances from DropEdge, GraphSAINT and RGCN, DropCONN still surpasses them by 0.9% to 4.4% and 0.5% to 2.9% on the corresponding datasets. This is because that DropCONN follows the behaviors and impact of adversarial attacks on node connections and features to design the biased graph sampling and yields a higher regularization on graph connections than other graph sampling methods like DropEdge and GraphSAINT to better penalize the adversarial edge manipulations and thus achieves higher performance against the adversarial attacks.

The same observations can be learned from Fig. 3(c)–(d) and Fig. 3(e)–(f). Though metattack shows less destructiveness than nettack, it still manages to enforce the GCN models to decay rapidly as the perturbation rate increases, which is stronger than random attack. The experimental results illustrate that DropCONN outperforms all the compared methods under metattack and random attack by a significant margin of performance improvements, which demonstrates its better robustness. Resting on these experimental results and observations, we conclude that DropCONN does not utilize any information of the attack methods and can hence be used in practice to improve the overall robustness of the GCN model against different adversarial attacks.

## 4.3 Ablation Study

In this section, we conduct ablation studies to further investigate how different components affect the robustness of our proposed method DropCONN. While designing the biased graph sampling scheme, we claim that both node degrees and feature similarity need to be considered for edge sampling probability formulation to adhere to adversarial perturbation behaviors and thus better penalize the edge manipulations. Based on the trained GCN model, we further introduce ensemble to average the outputs over the subgraphs rooted by test nodes to provide an additional regularization. To verify their contributions, we analyze the effect of the proposed method from two aspects: (1) DropCONN-F: since graph sampling methods generally consider the node degrees [6, 13, 31] but ignore the feature similarity, we remove the feature similarity measure from edge sampling probability; and (2) DropCONN-E: we omit ensemble and directly feed the full graph architecture for inference. We validate DropCONN-F and DropCONN-E on both datasets and their performances against nettack are reported in Fig. 4. As we can see, feature similarity plays a crucial role in graph sampling to limit negative effects from perturbed edges, while its absence enforces a performance drop by 2.7% of accuracy on Cora and 2.2% on Citeseer on average. DropCONN-E performs closely to DropCONN when the perturbation is small, but the gap between them widens as the perturbation number increases, which implies that ensemble yields an additional advantage for inference. These observations reaffirm the effectiveness of our design to enhance the robustness of GCNs against adversarial attacks.

## 5  Related Work

In recent years, graph neural networks (GNNs) on graph structured data have shown outstanding results in various applications [13, 24, 28]. Kipf and Welling [17] further introduced the GCNs via limiting the spectral filters to first-order neighbors for each node. Many GCN variants have been then proposed to leverage the graph structure to capture different properties [1, 19, 30, 32]. However, all of these GCN models lack adversarial robustness. To show the vulnerabilities of GCNs, different graph adversarial attack methods were proposed [9, 26, 27, 34, 35] to perturb graph structure and node features to enforce classification errors. Accordingly, a few attempts have been made to improve the robustness of GCNs against adversarial attacks [23, 26, 27, 33]. For example, Zhu et al. [33] proposed RGCN by adopting Gaussian distributions as the hidden representations of nodes to absorb the effects of adversarial changes; Wu et al. [26] removed the edges from graphs that connects the nodes with low feature similarity score; Tang et al. [23] utilized transfer learning over clean graphs to penalize the perturbations. Differently, in this paper, we leverage graph sampling idea [5, 6, 11, 20, 31] and design a biased graph sampling scheme to drop graph connections, which yields a strong regularization on the model and thus improves the robustness of GCNs.

## 6  Conclusion

In this paper, we propose to use graph sampling idea to formulate a secure learning solution that enhances the robustness of GCNs against adversarial attacks. By considering the properties of adversarial perturbations, we elaborate a biased graph sampling method over well-defined edge sampling probabilities with respect to nodes' degrees and feature similarity to drop graph connections and construct random, sparse and deformed subgraphs for training, which yields a significant penalty on edge manipulations. Based on the trained GCNs, we further use ensemble to average the outputs over sampled subgraphs rooted by test data to approximate the final inference results, which provides an additional regularization. We evaluate the performance of our proposed method on Cora and Citeseer datasets, while the experimental results validate its effectiveness against adversarial attacks.

**Acknowledgments.** We thank the reviewers for their valuable feedback. The work was supported in part by the National Science Foundation (NSF) under grant CNS-1652790 and a seed grant from Penn State Center for Security Research and Education (CSRE).

# References

1. Abu-El-Haija, S., et al.: Mixhop: higher-order graph convolutional architectures via sparsified neighborhood mixing. In: International Conference on Machine Learning, pp. 21–29 (2019)
2. Badrinarayanan, V., Kendall, A., Cipolla, R.: Segnet: a deep convolutional encoder-decoder architecture for image segmentation. IEEE Trans. Pattern Analy. Mach. Intell. **39**(12), 2481–2495 (2017)
3. Biggio, B., Fumera, G., Roli, F.: Multiple classifier systems for robust classifier design in adversarial environments. Int. J. Mach. Learn. Cybern. **1**(1–4), 27–41 (2010)
4. Carlini, N., Wagner, D.: Towards evaluating the robustness of neural networks. In: 2017 IEEE Symposium on Security and Privacy (sp), pp. 39–57. IEEE (2017)
5. Chen, J., Zhu, J., Song, L.: Stochastic training of graph convolutional networks with variance reduction. arXiv preprint arXiv:1710.10568 (2017)
6. Chen, J., Ma, T., Xiao, C.: Fastgcn: fast learning with graph convolutional networks via importance sampling. arXiv preprint arXiv:1801.10247 (2018)
7. Chen, L.C., Papandreou, G., Kokkinos, I., Murphy, K., Yuille, A.L.: Deeplab: semantic image segmentation with deep convolutional nets, atrous convolution, and fully connected crfs. IEEE Trans. Pattern Anal. Mach. Intell. **40**(4), 834–848 (2017)
8. Chen, L., Hou, S., Ye, Y.: Securedroid: enhancing security of machine learning-based detection against adversarial android malware attacks. In: Proceedings of the 33rd Annual Computer Security Applications Conference, pp. 362–372 (2017)
9. Dai, H., et al.: Adversarial attack on graph structured data. arXiv preprint arXiv:1806.02371 (2018)
10. Demontis, A., et al.: Yes, machine learning can be more secure! a case study on android Malware detection. IEEE Trans. Depend. Secure Comput. **16**(4), 711–724 (2017)
11. Gao, H., Wang, Z., Ji, S.: Large-scale learnable graph convolutional networks. In: Proceedings of the 24th ACM SIGKDD International Conference on Knowledge Discovery & Data Mining, pp. 1416–1424 (2018)
12. Goodfellow, I.J., Shlens, J., Szegedy, C.: Explaining and harnessing adversarial examples. arXiv preprint arXiv:1412.6572 (2014)
13. Hamilton, W., Ying, Z., Leskovec, J.: Inductive representation learning on large graphs. In: Advances in Neural Information Processing Systems, pp. 1024–1034 (2017)

14. Huang, G., Liu, Z., Van Der Maaten, L., Weinberger, K.Q.: Densely connected convolutional networks. In: Proceedings of the IEEE Conference on Computer Vision and Pattern Recognition, pp. 4700–4708 (2017)
15. Huang, W., Zhang, T., Rong, Y., Huang, J.: Adaptive sampling towards fast graph representation learning. In: Advances in Neural Information Processing Systems, pp. 4558–4567 (2018)
16. Kingma, D.P., Ba, J.L.: Adam: a method for stochastic optimization. In: International Conference on Learning Representations (ICLR) (2015)
17. Kipf, T.N., Welling, M.: Semi-supervised classification with graph convolutional networks. In: International Conference on Learning Representations (ICLR) (2017)
18. Lin, T.Y., Dollár, P., Girshick, R., He, K., Hariharan, B., Belongie, S.: Feature pyramid networks for object detection. In: Proceedings of the IEEE Conference on Computer Vision and Pattern Recognition, pp. 2117–2125 (2017)
19. Liu, S., Chen, L., Dong, H., Wang, Z., Wu, D., Huang, Z.: Higher order weighted graph convolutional networks. arXiv preprint arXiv:1911.04129 (2019)
20. Rong, Y., Huang, W., Xu, T., Huang, J.: Dropedge: towards deep graph convolutional networks on node classification. In: International Conference on Learning Representations (2020)
21. Srivastava, N., Hinton, G., Krizhevsky, A., Sutskever, I., Salakhutdinov, R.: Dropout: a simple way to prevent neural networks from overfitting. J. Mach. Learn. Res. **15**(1), 1929–1958 (2014)
22. Szegedy, C., et al.: Intriguing properties of neural networks. arXiv preprint arXiv:1312.6199 (2013)
23. Tang, X., Li, Y., Sun, Y., Yao, H., Mitra, P., Wang, S.: Transferring robustness for graph neural network against poisoning attacks. In: Proceedings of the 13th International Conference on Web Search and Data Mining, pp. 600–608 (2020)
24. Veličković, P., Cucurull, G., Casanova, A., Romero, A., Lio, P., Bengio, Y.: Graph attention networks. arXiv preprint arXiv:1710.10903 (2017)
25. Veličković, P., Cucurull, G., Casanova, A., Romero, A., Liò, P., Bengio, Y.: Graph attention networks. In: International Conference on Learning Representations (ICLR) (2018)
26. Wu, H., Wang, C., Tyshetskiy, Y., Docherty, A., Lu, K., Zhu, L.: Adversarial examples for graph data: deep insights into attack and defense. In: International Joint Conference on Artificial Intelligence, IJCAI, pp. 4816–4823 (2019)
27. Xu, K., et al.: Topology attack and defense for graph neural networks: An optimization perspective. arXiv preprint arXiv:1906.04214 (2019)
28. Xu, K., Hu, W., Leskovec, J., Jegelka, S.: How powerful are graph neural networks? ICLR (2019)
29. Yang, Z., Cohen, W., Salakhudinov, R.: Revisiting semi-supervised learning with graph embeddings. In: International Conference on Machine Learning, pp. 40–48 (2016)
30. Ying, R., He, R., Chen, K., Eksombatchai, P., Hamilton, W.L., Leskovec, J.: Graph convolutional neural networks for web-scale recommender systems. In: Proceedings of the 24th ACM SIGKDD International Conference on Knowledge Discovery & Data Mining, pp. 974–983 (2018)
31. Zeng, H., Zhou, H., Srivastava, A., Kannan, R., Prasanna, V.: Graphsaint: graph sampling based inductive learning method. In: International Conference on Learning Representations (2020)
32. Zhang, M., Cui, Z., Neumann, M., Chen, Y.: An end-to-end deep learning architecture for graph classification. In: Thirty-Second AAAI Conference on Artificial Intelligence (2018)

33. Zhu, D., Zhang, Z., Cui, P., Zhu, W.: Robust graph convolutional networks against adversarial attacks. In: Proceedings of the 25th ACM SIGKDD International Conference on Knowledge Discovery & Data Mining, pp. 1399–1407 (2019)
34. Zügner, D., Akbarnejad, A., Günnemann, S.: Adversarial attacks on neural networks for graph data. In: Proceedings of the 24th ACM SIGKDD International Conference on Knowledge Discovery & Data Mining, pp. 2847–2856 (2018)
35. Zügner, D., Günnemann, S.: Adversarial attacks on graph neural networks via meta learning. arXiv preprint arXiv:1902.08412 (2019)

# Gaussian Processes

# Automatic Tuning of Stochastic Gradient Descent with Bayesian Optimisation

Victor Picheny$^{(\boxtimes)}$, Vincent Dutordoir, Artem Artemev, and Nicolas Durrande

PROWLER.io, 72 Hills Road, Cambridge CB2 1LA, UK
{victor,vincent,artem,nicolas}@prowler.io

**Abstract.** Many machine learning models require a training procedure based on running stochastic gradient descent. A key element for the efficiency of those algorithms is the choice of the learning rate schedule. While finding good learning rates schedules using Bayesian optimisation has been tackled by several authors, adapting it dynamically in a data-driven way is an open question. This is of high practical importance to users that need to train a single, expensive model. To tackle this problem, we introduce an original probabilistic model for traces of optimisers, based on latent Gaussian processes and an auto-/regressive formulation, that flexibly adjusts to abrupt changes of behaviours induced by new learning rate values. As illustrated, this model is well-suited to tackle a set of problems: first, for the on-line adaptation of the learning rate for a cold-started run; then, for tuning the schedule for a set of similar tasks (in a classical BO setup), as well as warm-starting it for a new task.

**Keywords:** Learning rate · Gaussian process · Variational inference

## 1  Introduction

The great recent successes of machine learning generally rely on models with high complexity (e.g. deep models) and extensive datasets. Those models usually require running a training procedure over a (large) set of parameters, which often amounts to minimising a loss function with an iterative algorithm such as stochastic gradient descent (SGD) or the non-linear conjugate gradient method. In the deep learning community, a particular focus has been given to SGD algorithms such as Adagrad [9], RMSProp [39], and in particular Adam [1], which despite recent discussions and improvements [23,30] can be considered as the state-of-the-art and is widely used in practice [12]. Unfortunately, this training procedure is often extremely time-consuming due to the high model complexity and the amount of data at hand; hence, performing it with the best possible efficiency is paramount for many applications, in particular those for which training is done repeatedly, or continuously, for example in streaming models.

The performance of nearly all SGD variants depend critically on the choice of the *learning rate* level [3], which in short tunes by how much the algorithm should follow the (noisy) gradient signal. The default choice for most algorithms

© Springer Nature Switzerland AG 2021
F. Hutter et al. (Eds.): ECML PKDD 2020, LNAI 12459, pp. 431–446, 2021.
https://doi.org/10.1007/978-3-030-67664-3_26

is a constant learning rate, although recent experiments showed that a time-varying value can be extremely beneficial [2,33,34]. In any case, learning rate values require to be set, and tuning them by hand can be excessively burdensome. Bayesian optimisation (BO), on the other hand, is now a well-established tool for model selection and parameter tuning [4,35]. While learning rates are sometimes included in the parameters to be tuned [4], to our knowledge no work has been dedicated to speeding up SGD algorithms on-the-fly using BO: this is the purpose of the present work.

Let $\mathcal{L} : \Theta \times \Omega \to \mathbb{R}$ be the objective optimised by SGD; typically in deep learning, $\mathcal{L}$ can be a mean squared error or an evidence lower bound (ELBO)[1]. $\theta \in \Theta$ denotes a set of parameters, while $\omega \in \Omega$ defines the learning task at hand, that is, a particular model structure and a dataset. $\Omega$ is typically a singleton (when a single model is fitted to a single dataset), or a discrete set, for instance when the same model is used to fit different datasets. Given $\Omega$, an SGD algorithm produces a sequence of parameters $\theta_0, \ldots, \theta_T$, with $T \in \mathbb{N}^*$ a pre-defined number of iterations (i.e. optimisation steps).

In our setup we assume that the learning rate varies from one iteration to the next. One way to parametrise it, which avoids working directly on a high-dimensional vector of size $T$, is to use a piecewise constant form. Defining a sequence of length $d + 1 \ll T$: $T_0 = 0 < T_1 < \ldots < T_d = T$, the learning rate curve $\gamma(t)$ is defined by $d$ constants $x_0, \ldots, x_{d-1}$ such that $\gamma(t) = x_k$ for $T_k \le t < T_{k+1}$, $\forall t \in [0, T]$, $\forall 0 \le k < d$. Assuming lower and upper bounds for $x_k$, without loss of generality our tuning parameters $\mathbf{x} \in \mathbb{X}$ can be rescaled to $[0,1]^d$. For any task $\omega \in \Omega$, our objective is to seek the best possible learning rate, that is, the one that maximises $\mathcal{L}$ after $T$ steps:

$$\mathbf{x}_\omega^* = \arg\max_{\mathbf{x} \in \mathbb{X}} \mathcal{L}[\theta_T(\mathbf{x}, \omega), \omega], \tag{1}$$

with $\theta_T(\mathbf{x}, \omega)$ the parameters returned after $T$ iterations (now written as a function of the learning rate schedule $\mathbf{x}$ and task $\omega$).

Now, BO classically relies on Gaussian process (GP) surrogate models of the function to optimise. While predicting trace (that is, the value of $\mathcal{L}$ for all $t$) has been already addressed in related settings [8,19,28], adapting those models to make them dependent on a learning rate that changes over time is often not possible. Moreover, changing the learning rate often induces drastic changes on the trace behaviour, which makes typical parametric models unfit for the task. Our first contribution is the design of an auto-regressive model for the trace of a SGD algorithm, that flexibly adjusts to abrupt changes of behaviours. The model, which serves as a base for our sampling strategies, is presented in Sect. 2.

In the classical BO setting [32], first an initial set of experiments, typically with space-filling properties, $\{\mathbf{x}_1, \omega_1\}, \ldots, \{\mathbf{x}_{N_0}, \omega_{N_0}\}$ is run. The gathered data would then be used to build a predictive model for $\mathcal{L}[\theta_T(\mathbf{x}, \omega), \omega]$, which would allow us to estimate $\mathbf{x}_\omega^*$. In BO, such estimate is gradually improved by designing

---

[1] In the following, without loss of generality we use the convention that $\mathcal{L}$ should be maximised.

a sequence of additional experiments $\mathbf{x}_{N_0+1}, \ldots, \mathbf{x}_N$ that enhances the prediction capability of the model, following an exploration/exploitation trade-off.

However, often times the user is faced with training a single model, without much prior information on an appropriate choice of the learning rate schedule. In that case, the relevance of BO is somehow debatable, as it is uncertain that the outcome of $N$ SGD runs would outperform a single but much longer one with an empirically chosen learning rate. To overcome this, [10,22,38] used forecasting models to stop early unpromising runs and drastically reduce the computational cost of the BO loop. Yet, such approaches can only provide long-term recommendation at the price of repeated long runs, and are mostly relevant when other parameters are tuned simultaneously with the learning rate. We propose here an alternative strategy, unexplored in the BO literature, which is to adapt a single run *on the fly* rather than terminate some and start new ones from scratch. In addition, as modern hardware architecture favours parallel computing, we want to leverage the fact that training procedures can be run in parallel (say, one on each available GPU). Section 3 is dedicated to this problem.

In other practical situations, training is done repeatedly over similar tasks: for instance the same model fitted to different datasets or different variations of a model fitted to a dataset. In that case, one seeks an optimal mapping [or profile optima, 11] $\Phi : \Omega \rightarrow \mathbb{X}$ (from the space of tasks to the space of learning rates), such that $\forall \omega \in \Omega, \Phi(\omega) = \mathbf{x}_\omega^*$. Transferring information from one task to another allows designing strategies much more efficient than running independent BO loops for each task [37]. This framework is considered in Sect. 4.

# 2   A GP-Based NARX Model for Optimisation Traces

For simplicity, we focus for now on the case where $\Omega$ is a singleton (i.e. we consider a single dataset and model) and remove the dependence on $\omega$ from our notations. The generalisation to multiple tasks is deferred to Sect. 4.

## 2.1   Modelling of Optimisation Traces

Define first $y(\mathbf{x}, t) = \mathcal{L}[\theta(\mathbf{x}, t)]$, the trace of an optimisation run and denote by $Y_k = y(\mathbf{x}, T_k)$ the trace value at each changing point for the learning rate.

There exist many options to fit a parametric model to $y(\mathbf{x}, t)$: for instance, in [8], 11 models for traces are proposed, including exponential forms $(c - \exp(-at^\alpha + b))$ and power ones $(c - at^\alpha)$. While those forms make sense with a constant learning rate, they cannot fit properly a varying one, as changing the learning rate drastically modifies the trace dynamics. This is illustrated in Fig. 2, left, where the visible trace trend abruptly changes every time the learning rate changes (see also Fig. 4). Hence, instead of fitting a single parametric model, we propose to use a composite one, based on an auto-regressive formulation.

First, we model $Y_0$ as an i.i.d. Gaussian variable $\mathcal{N}(m_0, \sigma_0^2)$, to account for randomness in the starting point. Then, we propose to model the trace using a non-linear auto-regressive model with exogenous inputs [NARX(q), 21], which general expression is: $Y_{k+1} = \Gamma\left(\{Y_{k-l}\}_{l=0}^{q-1}, \{x_{k-l}\}_{l=0}^{q-1}\right) + \varepsilon_k$, where the new state is a non-linear function of the $q$ past states and exogenous inputs plus some independent noise. We propose a specific function for $\Gamma$ as follow:

$$Y_{k+1} = Y_k + \eta\left[f\left(\{Y_{k-l}\}_{l=0}^{q-1}, \{x_{k-l}\}_{l=0}^{q-1}\right), T_{k+1} - T_k\right] + \varepsilon_k,$$

where $f : \mathbb{R}^{2q} \to \mathbb{R}^p$ is a latent function that modulates the increments according to the current and past trace values $\{Y_{k-l}\}_{l=0}^{q-1}$ and learning rates $\{x_{k-l}\}_{l=0}^{q-1}$ and $\eta : \mathbb{R}^{p+1} \to \mathbb{R}$ (or $\mathbb{R}^+$ to ensure the monotonicity of the trace) is a link function that returns the trace increment for any given time $t$ according to its parameters. As the trace may be recorded for $t$ values outside $T_0, T_1, \ldots$, our model for the trace is:

$$y(\mathbf{x}, t) = Y_k + \eta\left[f\left(\{Y_{k-l}\}_{l=0}^{q-1}, \{x_{k-l}\}_{l=0}^{q-1}\right), t - T_k\right] + \varepsilon(t),$$

where $T_k \leq t < T_{k+1}$ and $\varepsilon(t)$ represents the noise in the case where the objective function is not evaluated exactly and/or for unaccounted-for deviations between observations and the model.

In the following, we use either piecewise linear or piecewise exponential forms for the traces. With respectively $f = f_1 : \mathbb{R}^{2q} \to \mathbb{R}$ defining the linear slope and $f = (f_1, f_2) : \mathbb{R}^{2q} \to \mathbb{R}^2$ with $f_1$ corresponding to a logit offset and $f_2$ to a logit rate, this gives:

$$\eta_{\lin}(f, t) = \phi(f_1)\, t, \tag{2}$$

$$\eta_{\exp}(f, t) = \phi(f_1)\, (1 - \exp[-\phi(f_2)\, t]), \tag{3}$$

where $f$ is a (multi-output) GP and $\phi(\cdot)$ is here the softplus function, $\phi(u) = \log(1+e^u)$, to ensure monotonicity. Note that any of the parametric models of e.g. [8] may be used here as $\eta$. In a sense, our model extend those for a time-varying learning rate.

Focusing on the linear case, and setting $q = 1$, we have: $y(\mathbf{x}, t) = Y_k + \phi[f_1(Y_k, x_k)](t - T_k)$. We directly see the dynamic implied by our model: increments are linear with respect to time, but the slope depends non-linearly on an exogenous input (the learning rate) and on the current state (intuitively, we expect flatter slopes if $Y_k$ is high, as the model is close to convergence). Considering the exponential case (Eq. (3)) and setting $q = 1$, we show GP surfaces in Fig. 1 for $\phi(f_1)$ (left) and $\phi(f_2)$ (right).

Importantly, the same $\eta$ is used over all the time intervals and only depends on $t - T_k$. Hence, in the NARX(1) case, our model implies that the same initial conditions and learning rates would lead to the same outcome, regardless of the time the algorithm has been run before. In that sense, the model can be considered as Markovian, as it is independent of the path followed by the optimiser before $T_k$. While our model is general for higher-order Markov chains (i.e. NARX(q)), we found empirically that the current formulation provides the right trade-off between accuracy, robustness and ease of inference, as the link function $\eta$ is only defined over $\mathbb{R}^2$.

Finally, we follow the *chained GP* framework [31] for $f$, and choose a GP prior with independent components.

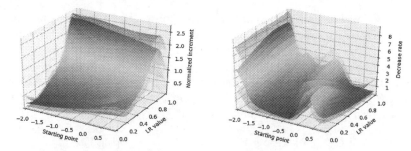

**Fig. 1.** Consider the NARX(1) model of Sect. 2 with the exponential link function of Eq. (3). On the left, we plot the mean $\pm$ the empirical std. dev. of the GP for the increment factor $\phi(f_1)$. On the right, we plot the GP surface of $\phi(f_2)$ which models the decay rate of the exponential. Given $q - 1$, both GPs are only a function of the most previous trace value 'starting point' and the current learning rate 'LR value'. These surfaces are learned as part of experiment Sect. 4.3. We notice, for example, that for low starting values of the objective, the response $\eta(f, t)$ behaves like a step-function.

## 2.2 Learning $f$ Using Variational Inference

We consider here a fixed link function, so the inference task boils down to estimating the parameters of the function $\eta$ and $Y_0$ (the starting value of the trace). Assume that we have run the optimiser $N \geq 1$ times for different learning rate schedules $\mathbf{x}^1, \ldots, \mathbf{x}^N$ and recorded the trace of each run $i$ at times $t_1^i, \ldots, t_{m_i}^i$ (in $[T_0, T_d]$). Let

$$y_j^i = y(\mathbf{x}^i, t_j^i) \quad (1 \leq i \leq N, 1 \leq j \leq m_i)$$

denote an observation of the trace at time $t_j^i$ for a learning rate schedule $\mathbf{x}^i$. We assume that each trace is evaluated at the beginning of each time interval (i.e. $\forall i, k, \exists j$ s.t. $t_j^i = T_k$), which gives explicit observations of $Y_k$, which we denote henceforth $Y_k^i$.

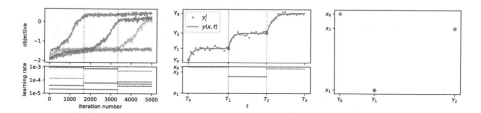

**Fig. 2.** Left: four traces (top) corresponding to the training of an SVGP model using Adam and corresponding learning rates. Middle: notations and model. Right: corresponding data point locations for the latent GP.

The parameters of $Y_0 \sim \mathcal{N}(m_0, \sigma_0^2)$ can be inferred by maximum likelihood from $\{Y_0^1, \ldots, Y_0^N\}$. Now, we focus on learning the latent GP functions

$f_1, f_2, \ldots f_p$. Without loss of generality, we assume in the following that our link function only requires a single GP parameter $f$ (i.e. p = 1). When a link function depends on more than one GP we simply apply the same procedure in parallel and model every function independently.

We follow a fully Bayesian approach to infer $f$ from the given dataset. We place a GP prior on the latent function $f$ and assume that $\varepsilon$ is i.i.d. Gaussian noise with zero mean and $\sigma^2$ variance, so that:

$$f \sim \mathcal{GP}(0, k(\cdot, \cdot)) \quad \text{and} \quad y_j^i \mid f, Y_k^i, x_k^i, t_j^i \sim \mathcal{N}(y_j^i \mid \eta(f(x_k^i, Y_k^i), t_j^i - T_k^i), \sigma^2),$$

with $k$ such that $T_k \leq t_j^i < T_{k+1}$.

For general link functions $\eta$ is exact inference of the latent function $f$ not possible due to the non-linear transformation. A classical solution is to follow the Sparse Variational GP (SVGP) framework [15,40], that relies on two components. First, it introduces a set of *inducing variables*, which main use is to specify the function value of the posterior GP at a specific set of $m$ pseudo inputs $Z = \{\mathbf{z}_i\}_{i=1}^m$, denoted as $\mathbf{u} = f(Z)$. The distribution of the inducing variables is specified by a fully parameterised Gaussian $q(\mathbf{u}) = \mathcal{N}(\mathbf{m}, S)$ with mean $\mathbf{m} \in \mathbb{R}^m$ and covariance $S \in \mathbb{R}^{m \times m}$, which are the variational parameters we want to learn. The prior GP on $f$ can then be conditioned on $\mathbf{u}$, which leads to a marginal posterior $q(f)$ with mean $\mu(\cdot)$ and the variance $\Sigma(\cdot, \cdot)$:

$$\mu(\cdot) = \mathbf{k}_Z^\top(\cdot) K_{ZZ}^{-1} \mathbf{m} \quad \text{and}$$
$$\Sigma(\cdot, \cdot) = k(\cdot, \cdot) + \mathbf{k}_Z^\top(\cdot) K_{ZZ}^{-1} (S - K_{ZZ}) K_{ZZ}^{-1} \mathbf{k}_Z(\cdot), \tag{4}$$

where $\mathbf{k}_Z(\cdot) := [k(\mathbf{z}_i, \cdot)]_{i=1}^m \in \mathbb{R}^m$ and $[K_{ZZ}]_{ij} = k(\mathbf{z}_i, \mathbf{z}_j)$. With this approximation in place we can set up our model's optimisation objective, which is a lower bound on the log marginal likelihood [ELBO, 16], equal to

$$\sum_{i=1}^N \sum_{j=1}^{m_i} \mathbb{E}_{q(f)} \left[ \log \mathcal{N}(y_j^i \mid \eta(f(x_k^i, Y_k^i), t_j^i - T_k^i), \sigma^2) \right] - \text{KL}\left[ q(\mathbf{u}) \| p(\mathbf{u}) \right], \tag{5}$$

where $k$ s.t. $T_k \leq t_j^i < T_{k+1}$, and KL is the Kullback-Leibler divergence between the approximate and the prior of $f$ [24]. It can be calculated analytically given the Gaussianity of both the prior and posterior on $\mathbf{u}$. The expectation can be estimated in an unbiased way using Monte-Carlo, by sampling $q(f)$ (Eq. (4)) and propagating the samples through the link-function. Optimising Eq. (5) with respect to the model parameters can be done by gradient descent thanks to automatic differentiation toolkits.

In our experiments we made use of the GPflow library [25]. More precisely, we used the provided multi-output framework for GPs [41], which is well-suited for implementing and optimising of these complex, composite GP models.

Note that as the approximation is sparse (i.e. it relies on a few inducing points), it can handle much larger datasets than classical GP models. This is a decisive advantage here as traces typically contain thousands of datapoints.

## 2.3   Generating Trace Predictions

Since the main objective is to maximise $\mathcal{L}$ after $T$ iterations, we would like to predict $Y_d = y(\mathbf{x}, T) = \eta(f(Y_{d-1}, x_{d-1}), T_d - T_{d-1})$, which requires access to $Y_{d-1}$. Recursively, we see that predicting $y(\mathbf{x}, T)$ is achieved by predicting the corresponding sequence $\{Y_0, \ldots, Y_d\}$. Importantly, the distribution of $\{Y_0, \ldots, Y_d\}$ is not available analytically because of the arbitrary link function. Hence, we must resort to sampling. Given the recursive structure (the value of $Y_i$ is necessary to draw from $Y_{i+1}$) we sample first $Y_0$, then $Y_1$ after conditioning $f$ on $f(Y_0, x_0)$, and recursively sampling $Y_{i+1}$ $(1 \leq i < d)$ after conditioning $f$ on $f(Y_0, x_0), \ldots, f(Y_i, x_i)$. Drawing multiple samples for a given $\mathbf{x}$ may be used to provide any statistic of $y(\mathbf{x}, t)$, such as mean, variance and quantiles.

# 3   Dynamic Tuning of the Learning Rate

We address now the question of dynamically tuning the learning rate for a single model and task, using the model previously defined. To do so, we depart from the standard BO approach, by modifying a small set of runs on the fly instead of starting repeatedly new ones.

## 3.1   Proposed Strategy

We consider the following framework. We assume that $Q$ SGD runs are performed in parallel with different learning rates. For simplicity, we assume synchronicity (all runs progress with the same speed). Each run is conveyed independently over time intervals $[T_k, T_{k+1}]$. At $t = T_{k+1}$, the $Q$ traces are fed to the model, which is then used to schedule $Q$ new learning rates for the next interval.

The learning rates are chosen as follows. At initialisation, as there is no model to help making decisions, $x_0^1, \ldots, x_0^Q$ are taken on a uniform grid between bounds. At any $T_k$ $(k > 0)$, the current trace observations are first integrated into the model. Then, each new learning rate $x_k^i$ is chosen to maximise the $\alpha_i$-quantile $q_{\alpha_i}$ (computed here empirically by sampling) of the trace at the end of the next interval:

$$x_k^i = \arg\max_{x \in [0,1]} q_{\alpha_i} \left[ \eta\left( f\left(y_i^j, x\right), T_k - T_{k-1} \right) \right]. \tag{6}$$

The $\alpha_i$'s are used here to balance exploitation and exploration: $\alpha_i$'s close to one may lead to very optimistic choices (learning rates for which the outcome is highly uncertain) while $\alpha_i$'s close to zero result in risk-averse choices (guaranteed immediate performance). Hence, to maximise our diversity of choices we set $\alpha_1 = \frac{1}{2Q}, \alpha_2 = \frac{3}{2Q}, \ldots, \alpha_Q = 1 - \frac{1}{2Q}$. In the case $Q = 1$, this forces to choose $\alpha = 0.5$, which is a risk-neutral strategy. Following an optimistic strategy (say, $\alpha = 0.75$) instead may enhance exploration and improve long-term performance.

In addition here, at each $T_i$ we greedily select the run with highest current trace and duplicate it $Q$ times while discarding the others. While this was found to accelerate significantly the performance, keeping each run may prove a valid alternative on problems that require more learning rate scheduling exploration.

The pseudo-code of the strategy is given in Algorithm 1. Note that a relevant choice of $d$ is problem-dependent: a large $d$ allows more changes of the learning rate value, but increases the computational overhead due to model fitting and solving Eq. (6). Besides, to facilitate inference the trace may not be sliced in too many parts (Fig. 2). In the experiment reported below, using $d = 20$ resulted with a negligible BO overhead.

---

Choose $Q$, $d$, $T_d$, set $k = 0$;

Take $x_0^1, \ldots, x_0^Q$ on a regular grid;

Run $Q$ SGDs for $T_1$ steps;

Gather $Q$ traces, read $Y_1^i$'s, and build the model;

**for** $k \leftarrow 1$ *to* $d-1$ **do**

> Get $i^* = \arg\max_{1 \le i \le Q} Y_k^i$;
>
> Duplicate $i^*$'s SGD run $Q$ times;
>
> $\forall 1 \le i \le Q$, find $x_k^i$ by solving Eq. (6);
>
> Pursue duplicated run for $T_{k+1} - T_k$ iterations and learning rates $x_k^1, \ldots, x_k^Q$, resp.;
>
> Gather $Q$ traces, collect $Y_k^i$'s, and update model;

**end**

Find $\arg\max_{1 \le i \le Q} Y_d^i$, return corresponding parameters;

**Algorithm 1:** Single task tuning

---

## 3.2    Experiment: Dynamic Tuning of Learnig Rate on CIFAR

We apply our approach to the training of a vanilla ResNet [14] neural network with 56 layers on the classification dataset CIFAR-10 [20] that contains 60,000 $32 \times 32$ colour images. We use an implementation of the ResNet model for the CIFAR dataset available in Keras [7]. We first split the dataset into 50,000 training and 10,000 testing images. The Adam optimiser is used for 100 epochs to maximise the log cross-entropy for future predictions.

Our BO setup is as follows: the 100 epochs are divided into $d = 20$ equal intervals and $Q = 5$ optimisations are ran in parallel. The objective is recorded every 50 iterations. The GP model for $f$ uses a Matérn-5/2 kernel, a linear link function and 100 inducing points. A practical issue we face is that increasing abruptly the learning rate sometimes causes aberrant behaviour (which can be seen by large peaks in the trace in Fig. 3). To avoid this problem, we use a GP classification model [15] to predict which runs are likely to fail based on $\{Y, x\}$ values. The optimisation of Eq. (6) is then restricted to the values of $x$ for which the probability of failure is lower than $\alpha_j$. We set the threshold for failure for a given trace inverse proportional to its quantile $\alpha_j$ as we want traces with larger $\alpha_j$ be more explorative. In addition, we limit the maximum change to one order of magnitude.

As baselines, we use five constant learning rates schedules uniformly spread in log space between $10^{-5}$ and $10^{-2}$, and 12 learning rates with exponential decay (three initial values $\gamma_0$ between $10^{-4}$ and $10^{-2}$ and four decay rates $\gamma$, 0.5, 0.63, 0.77 and 0.9), such that the learning rate in each epoch equals $\gamma_0 \times \gamma^{-\text{epoch}/10}$.

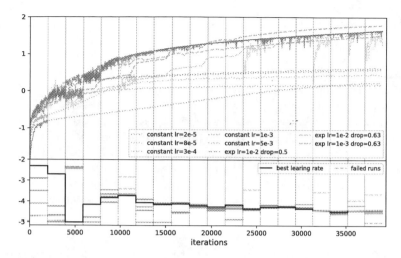

**Fig. 3.** Learning rates (bottom) and corresponding traces (top), following Algorithm 1. Dashed lines highlight 'failed' runs according to our classifier. The baselines (constant learning rates) are shown by dotted lines and the optimal learning rate schedule is given in black.

Figure 3 shows the dynamic of our approach. The initial interval shows the large performance differences when using different learning rates. Here, a very large learning rate is best at first, but almost immediately becomes sub-optimal. After a quarter of the optimisation budget, the optimal learning rates always takes values around $10^{-4}$, slowly decreasing over time. The algorithm behaviour captures this, by being very exploratory at first and much less towards the last intervals.

Comparing to constant learning rate schedules, our approach largely outperforms any of them. In the case where no parallel computation is performed, our approach would still outperform any constant learning rate, as those seem to have converged already to sub-optimal values after 40,000 iterations. Our approach also outperforms all exponential decay schedules but one. For this problem, a properly tuned exponential decay seems like a very efficient solution, and our dynamic tuning captures this solution. Arguably, five runs with different exponential decays might outperform our dynamic approach, but this would critically depend on the chosen bounds for the parameters and luck in the design of experiments. Standard BO (over the parameters) might be an alternative, but five observations would be too small to run it.

## 4   Multi-task Scheduling

### 4.1   Multi-task Learning

We now consider the case where $\Omega$ is discrete and relatively small (say, $|\Omega| = M \leq 10$), but can be increased when a new task needs to be solved, similarly to

[29]. The objective is then to find an optimal set of learning rates rather than a single one. However, as an efficient learning rate schedule for a task is often found to perform well for another, we assume that the values of the set share some resemblance.

Several sampling strategies have been proposed recently in this context [11,27,29,37]. However, all exploit the fact that posterior distributions are available in closed form. As our model is sampling-based, using those approaches would be either impractical, or overly expensive computationally. Hence, we propose a new principled strategy, adapted from the TRUVAR algorithm of [6], originally proposed for optimisation and level-set estimation.

We first extend our model to multiple tasks, by indexing the latent GP $f$ on $\omega$ on top of $Y$ and $x$. Then, following [37], we assume a product kernel for $f$:

$$k_f[(Y, x, \omega)(Y', x', \omega')] = k_Y(Y, Y')k_x(x, x')k_\Omega(\omega, \omega').$$

To facilitate inference, we assume further that the tasks can be embedded in a low-dimensional latent space, $\omega \to w \in \mathbb{R}^L$. This results in a set of $L \times M$ parameters to infer (the locations of the tasks in the latent space), independently of the number of runs.

## 4.2  Sequential Infill Strategy

In a nutshell, TRUVAR repeatedly applies two steps: 1) select a set of reference points $\mathcal{M} \in \mathbb{X}$ (e.g. for optimisation, potential maximisers), then 2) find the observation that greedily shrinks the sum of prediction variances at reference points.

We adapt here this strategy to uncover profile optima, that is:

$$\mathbb{X}^* = \{\mathbf{x}^{i*} = \arg\max_{\mathbf{x} \in \mathbb{X}} y(\mathbf{x}, T_d, \omega_i)\}_{i=1}^M.$$

The original algorithm selects as reference points $\mathcal{M}$ all the points for which an upper confidence bound (UCB) of the objective is higher than a threshold. As we work with continuous design spaces, we decided to simplify this step and consider for $\mathcal{M}$ the maximisers $\hat{\mathbb{X}}^* = \{\hat{\mathbf{x}}^{1*}, \ldots, \hat{\mathbf{x}}^{M*}\}$ of the UCB of the final trace value for each task, that is:

$$\hat{\mathbf{x}}^{i*} = \arg\max_{\mathbf{x} \in \mathbb{X}} q_\alpha[y(\mathbf{x}, T_d, \omega_i)], 1 \leq i \leq M, \tag{7}$$

with $\alpha \in (0.5, 1)$ so that the quantile defines a UCB for $y(\mathbf{x}, t)$. Note that to ensure theoretical guarantees, UCB strategies generally require quantile orders that increase with time [17]. However, a constant value usually works best in practice [5,36], so we focus on this case here.

Due to the lack of data, the performance at $\hat{\mathbb{X}}^*$ is uncertain, which can be quantified by the mean of variances at $\hat{\mathbb{X}}^*$:

$$J = \frac{1}{M} \sum_{i=1}^M \text{Var}\left[y\left(\hat{\mathbf{x}}^{i*}, T_d, \omega_i\right)\right].$$

Note that as $v$ is chosen using a UCB, it is likely to correspond to values for which the model has a high prediction variance. So, $J$ may increase monotonically with $\alpha$, which acts as a tuning parameter for the exploration/exploitation trade-off.

Now, we would like to find the run (learning rate and task) that reduces $J$ the most. Assume a potential candidate $(\mathbf{x}, \omega)$, that would provide, if evaluated, an additional set of observations $\mathbf{y}_{\mathbf{x}, \omega}$. Conditioning the model on this new data would reduce the prediction uncertainty at $\hat{\mathbb{X}}^*$ (by law of total variance), which we can measure with

$$\bar{J}(\mathbf{x}, \omega) = \frac{1}{M} \sum_{i=1}^{M} \mathrm{Var}\big[y\big(\hat{\mathbf{x}}^{i*}, T_d, \omega_i\big)|\mathbf{y}_{\mathbf{x}, \omega}\big] \leq J,$$

where $\mathrm{Var}(.|\mathbf{y}_{\mathbf{x}, \omega})$ denotes the variance conditionally on $\mathbf{y}_{\mathbf{x}, \omega}$. In the case of regular GP models, $\bar{J}$ is actually available in closed form independently of the values of $\mathbf{y}_{\mathbf{x}, \omega}$ [6]. This is not the case here, so we replace $\bar{J}(\mathbf{x}, \omega)$ by its expectation over the values of $\mathbf{y}_{\mathbf{x}, \omega}$, which leads to the following sampling strategy:

$$\{\mathbf{x}_{\mathrm{new}}, \omega_{\mathrm{new}}\} = \arg\min_{\{\mathbf{x}, \omega\}} \mathbb{E}_{\mathbf{y}_{\mathbf{x}, \omega}}\big(\bar{J}(\mathbf{x}, \omega)\big). \tag{8}$$

In practice, this criterion is not available in closed form, and must be computed using a double Monte-Carlo loop. However, conditioning on $\mathbf{y}_{\mathbf{x}, \omega}$ can be approximated simply, as follow. First, samples of $\{f(Y_0^{\mathrm{new}}, x_1^{\mathrm{new}}), \ldots, f(Y_{d-1}^{\mathrm{new}}, x_d^{\mathrm{new}})\}$ are obtained recursively, as in Sect. 2.3. Then, conditioning $f$ on each of those samples and computing the new conditional variance as in Sect. 2.3 allows to compute Eq. (8).

Once $\mathbf{x}_{\mathrm{new}}$ and $\omega_{\mathrm{new}}$ are obtained, the corresponding experiment is run and the model is updated, which in turn leads to a new set $\hat{\mathbb{X}}^*$, etc. Once the budget is exhausted, the final set $\hat{\mathbb{X}}^*$ may be chosen using a different $\alpha$ (either 0.5 for a risk-neutral solution or $\leq 0.5$ for a risk-averse one). The pseudo-code of the strategy is given in Algorithm 2.

---

Choose $N_0$, $N$, $d$, $T_d$, $L$, $\alpha$;
Select initial set of experiments $\{\mathbf{x}_0, \omega_0\}, \ldots, \{\mathbf{x}_{N_0}, \omega_{N_0}\}$;
Run $N_0$ SGDs for $T_d$ steps;
Gather $N_0$ traces, build trace model;
**for** $i \leftarrow N_0 + 1$ *to* $N$ **do**
   Find the optimistic set $\hat{\mathbb{X}}^*$ by solving Eq. (6);
   Find the experiment $\{\mathbf{x}_{\mathrm{new}}, \omega_{\mathrm{new}}\}$ that reduces the uncertainty related to $\hat{\mathbb{X}}^*$ by solving Eq. (8);
   Run SGD with $\{\mathbf{x}_{\mathrm{new}}, \omega_{\mathrm{new}}\}$;
   Gather new trace, update trace model;
**end**
Return the set of optimal learning rates $\hat{\mathbb{X}}^*$;

---

**Algorithm 2:** Multi-task tuning

Following [42], we use the so-called *reparametrisation trick* when generating samples in order to solve Eq. (8) with respect to $\mathbf{x}_{new}$ using gradient-based optimisation. Note that $\omega_{new}$ is found by exhaustive search.

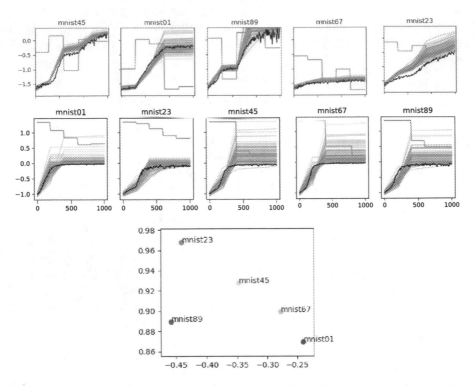

**Fig. 4.** Top: Actual traces (black), predictions (color) and learning rates (grey, in log-scale between $10^{-5}$ and $10^{-3}$). The learning rates are randomly chosen for the top row, and set to their optimal estimates on the bottom right. Bottom: Latent variable values for each dataset. (Color figure online)

### 4.3 Experiment: Multi-Task Setting with SVGP on MNIST

To illustrate our strategy, we consider the following setup. We use the MNIST digit dataset, split into five binary classification problems (0 against 1, 2 against 3 and so on until 8 against 9). Our goal is to fit a sparse GP classification model to each of these datasets, by maximising its ELBO using the Adam algorithm.

We choose here $T_d = 1,000$ Adam iterations, and $d = 5$ different learning rate values. The learning rates are bounded by $[10^{-5}, 10^{-3}]$. $N_0 = 5$ initial runs are performed with learning rates chosen by Latin Hypercube sampling (in the logarithmic space), and 15 runs are added sequentially according to our TRUVAR strategy. The GP model for $f$ uses a Matérn-5/2 kernel, an exponential link function, 50 inducing points and a dimension $L = 2$ for the latent space of tasks. To ease the resolution of Eq. (7), we follow a greedy approach by searching for one learning rate value at a time, starting from $x_0$, which is in line with the Markov assumption of the model.

Figure 4 shows the learning rate profiles and corresponding runs obtained after running the procedure, as well as some predictions for randomly chosen learning rates. We first observe the flexibility of our model, which is able to capture complex traces while providing relevant uncertainty estimates (top plots). Then, for all tasks, the learning rates found reach the upper bound at first and decreases when the trace reaches a plateau. The optimal way of decreasing the learning rate depends on the task. One can see that the predictions are uncertain, but only on one side (some samples largely overestimate the true trace but median ones are quite close to the truth).

The right panel of Fig. 4 shows the estimated proximity between tasks. Here, all the tasks have been found relatively similar, as they all load to close learning rate schedules. One may notice for instance that mnist01 is at an edge of the domain, which can be imputed to a different initial trace behaviour.

### 4.4   Extension: Warm-Starting SGD for a New Task

Now, assume that a new task $\omega_{new}$ is added to the current set $\Omega$. Unfortunately, our model cannot be used directly as the value of the corresponding latent variables $\omega_{new}$ is unknown. The first solution is to find a "universal" tuning: this can be obtained by maximising the prediction of $y(\mathbf{x}, T_d, \omega_{new})$ averaged over all possible values for $\omega_{new}$. This average can be calculated by Monte-Carlo assuming a probability measure for $\omega_{new}$, for instance the Lebesgue measure over the convex hull of $w_1, \ldots, w_M$.

Alternatively, one might want to spend some computing budget (say, $T_i$, $i > 0$) to learn $\omega_{new}$ and achieve then a better learning rate tuning. Assuming again a measure for $\omega_{new}$, an informative experiment would correspond to a learning rate for which the proportion of the variance of the predictor due to the uncertainty on $\omega_{new}$ is maximal. Averaging over all time steps, we define our sampling strategy (with again a criterion computable by Monte-Carlo) as:

$$\mathbf{x}_{:i}^* = \arg\max_{\mathbf{x}_{:i} \in [0,1]^i} \sum_{t=0}^{T_i} \frac{\mathbb{E}_w[\mathrm{Var}(Y(\mathbf{x}_{:i}, t, w)|w)]}{\mathrm{Var}(Y(\mathbf{x}_{:i}, t, w))},$$

with $\mathbf{x}_{:i} = [x_0, \ldots, x_i]$. Note that once $\omega_{new}$ is estimated, it is possible to apply Algorithm 1, exploiting the flexibility of dynamic tuning while leveraging information from previous runs. A more integrated approach would use directly Algorithm 1 while measuring and accounting for uncertainty in $\omega_{new}$; this is left for future work.

## 5   Concluding Comments

We proposed a probabilistic model for the traces of optimisers, which input parameters (the choice of learning rate values) correspond to particular periods of the optimisation. This allowed us to define a versatile framework to tackle a set of problems: tuning the optimiser for a set of similar tasks, warm-starting it for a new task or on-line adaptation of the learning rate for a cold-started run.

Convergence proof for the multitask strategy has not been considered here. We believe that the results of [6] may be adapted to our case: this is left for future work. Other possible extensions are to apply our framework to other optimisers: for instance, to control the population sizes of evolutionary strategy algorithms such as CMAES [13], for which adaptation mechanisms have been found promising [26]. Finally, additional efficiency could be achieved by leveraging the use of varying dataset size, in the spirit of [10, 18] for instance.

# References

1. Andrychowicz, M., et al.: Learning to learn by gradient descent by gradient descent. In: Advances in Neural Information Processing Systems, pp. 3981–3989 (2016)
2. Baydin, A.G., Cornish, R., Rubio, D.M., Schmidt, M., Wood, F.: Online learning rate adaptation with hypergradient descent. arXiv preprint arXiv:1703.04782 (2017)
3. Bengio, Y.: Practical recommendations for gradient-based training of deep architectures. In: Montavon, G., Orr, G.B., Müller, K.-R. (eds.) Neural Networks: Tricks of the Trade. LNCS, vol. 7700, pp. 437–478. Springer, Heidelberg (2012). https://doi.org/10.1007/978-3-642-35289-8_26
4. Bergstra, J.S., Bardenet, R., Bengio, Y., Kégl, B.: Algorithms for hyper-parameter optimization. In: Advances in Neural Information Processing Systems, pp. 2546–2554 (2011)
5. Bogunovic, I., Scarlett, J., Jegelka, S., Cevher, V.: Adversarially robust optimization with Gaussian processes. In: Advances in Neural Information Processing Systems, pp. 5760–5770 (2018)
6. Bogunovic, I., Scarlett, J., Krause, A., Cevher, V.: Truncated variance reduction: a unified approach to Bayesian optimization and level-set estimation. In: Advances in Neural Information Processing Systems, pp. 1507–1515 (2016)
7. Chollet, F.: Keras implementation of ResNet for CIFAR. https://keras.io/examples/cifar10_resnet/ (2009)
8. Domhan, T., Springenberg, J.T., Hutter, F.: Speeding up automatic hyperparameter optimization of deep neural networks by extrapolation of learning curves. In: Twenty-Fourth International Joint Conference on Artificial Intelligence (2015)
9. Duchi, J., Hazan, E., Singer, Y.: Adaptive subgradient methods for online learning and stochastic optimization. J. Mach. Learn. Res. **12**(7), 2121–2159 (2011)
10. Falkner, S., Klein, A., Hutter, F.: Bohb: Robust and efficient hyperparameter optimization at scale. arXiv preprint arXiv:1807.01774 (2018)
11. Ginsbourger, D., Baccou, J., Chevalier, C., Perales, F., Garland, N., Monerie, Y.: Bayesian adaptive reconstruction of profile optima and optimizers. SIAM/ASA J. Uncertainty Quantification **2**(1), 490–510 (2014)
12. Gugger, S., Howard, J.: Adamw and super-convergence is now the fastest way to train neural nets (2018). https://www.fast.ai/2018/07/02/adam-weight-decay/
13. Hansen, N., Ostermeier, A.: Completely derandomized self-adaptation in evolution strategies. Evol. Comput. **9**(2), 159–195 (2001)

14. He, K., Zhang, X., Ren, S., Sun, J.: Deep residual learning for image recognition. In: Proceedings of the IEEE Conference on Computer Vision and Pattern Recognition, pp. 770–778 (2016)
15. Hensman, J., Matthews, A.G.D.G., Ghahramani, Z.: Scalable variational Gaussian process classification. In: Proceedings of the Eighteenth International Conference on Artificial Intelligence and Statistics (2015)
16. Hoffman, M.D., Blei, D.M., Wang, C., Paisley, J.: Stochastic variational inference. Journal of Machine Learning Research (2013)
17. Kaufmann, E., Cappé, O., Garivier, A.: On Bayesian upper confidence bounds for bandit problems. In: Artificial Intelligence and Statistics, pp. 592–600 (2012)
18. Klein, A., Falkner, S., Bartels, S., Hennig, P., Hutter, F.: Fast Bayesian optimization of machine learning hyperparameters on large datasets. In: International Conference on Artificial Intelligence and Statistics (AISTATS 2017), pp. 528–536. PMLR (2017)
19. Klein, A., Falkner, S., Springenberg, J.T., Hutter, F.: Learning curve prediction with Bayesian neural networks. In: ICLR (2017)
20. Krizhevsky, A., Hinton, G.: Learning multiple layers of features from tiny images. Technical report, Citeseer (2009)
21. Leontaritis, I., Billings, S.A.: Input-output parametric models for non-linear systems part ii: stochastic non-linear systems. Int. J. Control **41**(2), 329–344 (1985)
22. Li, L., Jamieson, K., DeSalvo, G., Rostamizadeh, A., Talwalkar, A.: Hyperband: a novel bandit-based approach to hyperparameter optimization. J. Mach. Learn. Res. **18**(185), 1–52 (2018)
23. Loshchilov, I., Hutter, F.: Decoupled weight decay regularization. In: ICLR (2019)
24. Matthews, A.G.D.G., Hensman, J., Turner, R., Ghahramani, Z.: On sparse variational methods and the kullback-leibler divergence between stochastic Processes. J. Mach. Learn. Res. **51**, 231–239 (2016)
25. Matthews, A.G.D.G., et al.: Gpflow: a Gaussian process library using tensorflow. J. Mach. Learn. Res. **18**(1), 1299–1304 (2017)
26. Nishida, K., Akimoto, Y.: PSA-CMA-ES: CMA-ES with population size adaptation. In: Proceedings of the Genetic and Evolutionary Computation Conference, pp. 865–872 (2018)
27. Pearce, M., Branke, J.: Continuous multi-task Bayesian optimisation with correlation. Eur. J. Oper. Res. **270**(3), 1074–1085 (2018)
28. Picheny, V., Ginsbourger, D.: A nonstationary space-time Gaussian process model for partially converged simulations. SIAM/ASA J. Uncertainty Quantification **1**(1), 57–78 (2013)
29. Poloczek, M., Wang, J., Frazier, P.I.: Warm starting Bayesian optimization. In: Proceedings of the 2016 Winter Simulation Conference, pp. 770–781. IEEE Press (2016)
30. Reddi, S.J., Kale, S., Kumar, S.: On the convergence of ADAM and beyond. In: ICLR (2018)
31. Saul, A.D., Hensman, J., Vehtari, A., Lawrence, N.D., et al.: Chained Gaussian processes. In: AISTATS, pp. 1431–1440 (2016)
32. Shahriari, B., Swersky, K., Wang, Z., Adams, R.P., De Freitas, N.: Taking the human out of the loop: a review of Bayesian optimization. Proc. IEEE **104**(1), 148–175 (2016)
33. Smith, L.N.: Cyclical learning rates for training neural networks. In: 2017 IEEE Winter Conference on Applications of Computer Vision (WACV), pp. 464–472. IEEE (2017)

34. Smith, L.N., Topin, N.: Super-convergence: very fast training of neural networks using large learning rates. In: Artificial Intelligence and Machine Learning for Multi-Domain Operations Applications. vol. 11006, p. 1100612. International Society for Optics and Photonics (2019)
35. Snoek, J., Larochelle, H., Adams, R.P.: Practical Bayesian optimization of machine learning algorithms. In: Advances in Neural Information Processing Systems, pp. 2951–2959 (2012)
36. Srinivas, N., Krause, A., Kakade, S., Seeger, M.: Gaussian Process optimization in the bandit setting: no regret and experimental design. In: Proceedings of the 27th International Conference on International Conference on Machine Learning, pp. 1015–1022. Omnipress (2010)
37. Swersky, K., Snoek, J., Adams, R.P.: Multi-task Bayesian optimization. In: Advances in Neural Information Processing Systems, pp. 2004–2012 (2013)
38. Swersky, K., Snoek, J., Adams, R.P.: Freeze-thaw Bayesian optimization. arXiv preprint arXiv:1406.3896 (2014)
39. Tieleman, T., Hinton, G.: Lecture 6.5-rmsprop: divide the gradient by a running average of its recent magnitude. COURSERA: Neural Netw. Mach. Learn. **4**(2), 26–31 (2012)
40. Titsias, M.: Variational learning of inducing variables in sparse Gaussian processes. In: Artificial Intelligence and Statistics (2009)
41. van der Wilk, M., Dutordoir, V., John, S., Artemev, A., Adam, V., Hensman, J.: A framework for interdomain and multioutput Gaussian processes. arXiv:2003.01115 (2020). https://arxiv.org/abs/2003.01115
42. Wilson, J., Hutter, F., Deisenroth, M.: Maximizing acquisition functions for Bayesian optimization. In: Advances in Neural Information Processing Systems, pp. 9884–9895 (2018)

# MUMBO: MUlti-task Max-Value Bayesian Optimization

Henry B. Moss[1]([⊠]), David S. Leslie[2], and Paul Rayson[3]

[1] STOR-i Centre for Doctoral Training, Lancaster University,
Lancashire, England
h.moss@lancaster.ac.uk
[2] Department of Mathematics and Statistics, Lancaster University,
Lancashire, England
[3] School of Computing and Communications, Lancaster University,
Lancashire, England

**Abstract.** We propose MUMBO, the first high-performing yet computationally efficient acquisition function for multi-task Bayesian optimization. Here, the challenge is to perform efficient optimization by evaluating low-cost functions somehow related to our true target function. This is a broad class of problems including the popular task of multi-fidelity optimization. However, while information-theoretic acquisition functions are known to provide state-of-the-art Bayesian optimization, existing implementations for multi-task scenarios have prohibitive computational requirements. Previous acquisition functions have therefore been suitable only for problems with both low-dimensional parameter spaces and function query costs sufficiently large to overshadow very significant optimization overheads. In this work, we derive a novel multi-task version of entropy search, delivering robust performance with low computational overheads across classic optimization challenges and multi-task hyperparameter tuning. MUMBO is scalable and efficient, allowing multi-task Bayesian optimization to be deployed in problems with rich parameter and fidelity spaces.

**Keywords:** Bayesian optimization · Gaussian processes

## 1 Introduction

The need to efficiently optimize functions is ubiquitous across machine learning, operational research and computer science. Many such problems have special structures that can be exploited for efficient optimization, for example gradient-based methods on cheap-to-evaluate convex functions, and mathematical programming for heavily constrained problems. However, many optimization problems do not have such clear properties.

---

Supported by EPSRC and the STOR-i Centre for Doctoral Training.

F. Hutter et al. (Eds.): ECML PKDD 2020, LNAI 12459, pp. 447–462, 2021.
https://doi.org/10.1007/978-3-030-67664-3_27

**Bayesian optimization** (BO) is a general method to efficiently optimize 'black-box' functions for which we have weak prior knowledge, typically characterized by expensive and noisy function evaluations, a lack of gradient information, and high levels of non-convexity (see [1] for a comprehensive review). By sequentially deciding where to make each evaluation as the optimization progresses, BO is able to direct resources into promising areas and so efficiently explore the search space. In particular, a highly effective and intuitive search is achieved through **information-theoretic** BO, where we seek to sequentially reduce our uncertainty (measured in terms of differential entropy) in the location of the optima with each successive function evaluation [2,3].

For optimization problems where we can evaluate low-cost functions somehow related to our true objective function, **Multi-task** (MT) BO (as first introduced by [4]) provides additional efficiency gains. A popular subclass of MT BO problems is **multi-fidelity** (MF) BO, where the set of related functions can be meaningfully ordered by their similarity to the objective function. Unfortunately, performing BO over MT spaces has previously required complicated approximation schemes that scale poorly with dimension [4,5], limiting the applicability of information-theoretic arguments to problems with both low-dimensional parameter spaces and function query costs sufficiently large to overshadow very significant optimization overheads. Therefore, MT BO has so far been restricted to considering simple structures at a large computational cost. Despite this restriction, MT optimization has wide-spread use across physical experiments [6–8], environmental modeling [9], and operational research [10–12].

For expositional simplicity, this article focuses primarily on examples inspired by tuning the hyper-parameters of machine learning models. Such problems have large environmental impact [13], requiring multiple days of computation to collect even a single (often highly noisy) performance estimate. Consequently, these problems have been proven a popular and empirically successful application of BO [14]. MF applications for hyper-parameter tuning dynamically control the reliability (in terms of bias and noise) of each hyper-parameter evaluation [15–19] and can reduce the computational cost of tuning complicated models by orders of magnitude over standard BO. Orthogonal savings arise from considering hyper-parameter tuning in another MT framework; FASTCV [4] recasts tuning by $K$-fold cross-validation (CV) [20] into the task of simultaneously optimizing the $K$ different evaluations making a single $K$-fold CV estimate.

Information-theoretic arguments are particularly well suited to such MT problems as they provide a clear measure of the utility (the information gained) of making an evaluation on a particular subtask. This utility then can be balanced with computational cost, providing a single principled decision [4,5,17,21]. Despite MT BO being a large sub-field in its own right, there exist only a few alternatives to information-theoretic acquisition functions. Alternative search strategies include extensions of standard BO acquisition functions, including knowledge gradient (KG) [22,23], expected improvement (EI) [4,16,24], and upper-confidence bound (UCB) [18,19]. KG achieves efficient optimization but incurs a high computational overhead. The MT extensions of EI and UCB, although computationally cheap, lack a clear notion of utility and consequently rely on two-stage heuristics, where a hyper-parameter followed by a task are chosen as two separate decisions. Moreover, unlike our proposed work, the performance of MT variants of UCB and EI depends sensitively on problem-specific parameters which require careful tuning, often leading to poor performance in practical tasks. Information-theoretic arguments have produced the MF BO hyper-parameter tuner FABOLAS [17], out-competing approaches across richer fidelity spaces based on less-principled acquisitions [19]. This success motivates our work to provide scalable entropy reduction over MT structures.

We propose **MUMBO**, a novel, scalable and computationally light implementation of information-theoretic BO for general MT frameworks. Inspired by the work of [25], we seek reductions in our uncertainty in the value of the objective function at its optima (a single-dimensional quantity) rather than our uncertainty in the location of the optima itself (a random variable with the same dimension as our search space). MUMBO enjoys three major advantages over current information-theoretic MT approaches:

- MUMBO has a simple and scalable formulation requiring routine one-dimensional approximate integration, irrespective of the search space dimensions,
- MUMBO is designed for general MT and MF BO problems across both continuous and discrete fidelity spaces,
- MUMBO outperforms current information-theoretic MT BO with a significantly reduced computational cost.

Parallel work [26] presents essentially the same acquisition function but restricted to discrete multi-fidelity problems from the material sciences. Our article provides a different derivation and general presentation of the method which enables deployment with both discrete and continuous fidelity spaces in general MT BO (including MF). We also provide an implementation in a major BO toolbox and examples across synthetic and hyper-parameter tuning tasks.

## 2    Problem Statement and Background

We now formalize the goal of MT BO, introducing the notation used throughout this work. The goal of BO is to find the maximzer

$$\mathbf{x}^* = \underset{\mathbf{x} \in \mathcal{X}}{\operatorname{argmax}}\, g(\mathbf{x}) \tag{1}$$

of a function $g$ over a $d$-dimensional set of feasible choices $\mathcal{X} \subset \mathbb{R}^d$ spending as little computation on function evaluations as possible.

Standard BO seeks to solve (1) by sequentially collecting noisy observations of $g$. By fitting a Gaussian process (GP) [27], a non-parametric model providing regression over all functions of a given smoothness (to be controlled by a choice of kernel $k$), we are able to quantify our current belief about which areas of the search space maximize our objective function. An acquisition function $\alpha_n(\mathbf{x})$ :

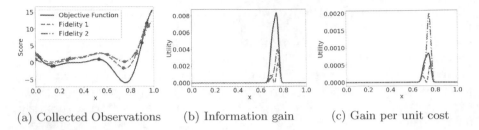

(a) Collected Observations     (b) Information gain     (c) Gain per unit cost

**Fig. 1.** Seeking the minimum of the 1D Forrester function (blue) with access to two low-fidelity approximations at $\frac{1}{2}$ (red) and $\frac{1}{5}$ (green) the cost of querying the true objective. Although we learn the most from directly querying the objective function, we can learn more per unit cost by querying the roughest fidelity. (Color figure online)

$\mathcal{X} \to \mathbb{R}$ uses this belief to predict the utility of making any given evaluation, producing large values at 'reasonable' locations. A standard acquisition function [2] is the expected amount of information provided by each evaluation about the location of the maximum. Therefore after making $n$ evaluations, BO will automatically next evaluate $\mathbf{x}_{n+1} = \operatorname{argmax}_{\mathbf{x} \in \mathcal{X}} \alpha_n(\mathbf{x})$.

### 2.1 Multi-task Bayesian Optimization

Suppose that instead of querying $g$ directly, we can alternatively query a (possibly infinite) collection of related functions indexed by $\mathbf{z} \in \mathcal{Z}$ (henceforth referred to as our fidelity space). We then collect the (noisy) observations $D_n = \{(\mathbf{x}_t, \mathbf{z}_t, y_t)\}$ for $y_t = f(\mathbf{x}_t, \mathbf{z}_t) + \epsilon_t$, where $f(\mathbf{x}, \mathbf{z})$ is the result of querying parameter $\mathbf{x}$ on fidelity $\mathbf{z}$, and $\epsilon_t$ is Gaussian noise. If these alternative functions $f$ are cheaper to evaluate and we can learn their relationship with $g$, then we have access to cheap sources of information that can be used to help find the maximizer of the true task of interest.

### 2.2 Multi-task Acquisition Functions

The key difference between standard BO and MT BO is that our acquisition function must be able to not only choose the next location, but also which fidelity to evaluate, balancing computational cost with how much we expect to learn about the maximum of $g$. Therefore, we require an extended acquisition function $\alpha_n : \mathcal{X} \times \mathcal{Z} \to \mathbb{R}$ and a cost function $c : \mathcal{X} \times \mathcal{Z} \to \mathbb{R}^+$, measuring the utility and cost of evaluating location $\mathbf{x}$ at fidelity $\mathbf{z}$ (as demonstrated in Fig. 1c). In Sect. 4, we consider problems both where this cost function is known *a priori* and where it is unknown but estimated using an extra GP [14]. In this work, we seek to make the evaluation that provides the largest information gain per unit cost, i.e. maximizing the ratio

$$(\mathbf{x}_{n+1}, \mathbf{z}_{n+1}) = \operatorname*{argmax}_{(\mathbf{x}, \mathbf{z}) \in \mathcal{X} \times \mathcal{Z}} \frac{\alpha_n(\mathbf{x}, \mathbf{z})}{c(\mathbf{x}, \mathbf{z})}. \tag{2}$$

## 2.3  Multi-task Models

To perform MT BO, our underlying Gaussian process model must be extended across the fidelity space. By defining a kernel over $\mathcal{X} \times \mathcal{Z}$, we can learn predictive distributions after $n$ observations with means $\mu^n(\mathbf{x}, \mathbf{z})$ and co-variances $\Sigma^n((\mathbf{x}, \mathbf{z}), (\mathbf{x}', \mathbf{z}'))$ from which $\alpha_n(\mathbf{x}, \mathbf{z})$ can be calculated. Although increasing the dimension of the kernel for $\mathcal{X}$ to incorporate $\mathcal{Z}$ provides a very flexible model, it is argued by [19] that overly flexible models can harm optimization speed by requiring too much learning, restricting the sharing of information across the fidelity space. Therefore, it is common to use more restrictive separable kernels that better model specific aspects of the given problem.

A common kernel for discrete fidelity spaces is the intrinsic coregionalization kernel of [28] (as used in Fig. 1). This kernel defines a co-variance between hyper-parameter and fidelity pairs of

$$k((\mathbf{x}, z)(\mathbf{x}', z')) = k_{\mathcal{X}}(\mathbf{x}, \mathbf{x}') \times B(z, z'), \tag{3}$$

for a base kernel $k_{\mathcal{X}}$ and a positive semi-definite $|\mathcal{Z}| \times |\mathcal{Z}|$ matrix $B$ (set by maximizing the model likelihood alongside the other kernel parameters). $B$ represents the correlation between different fidelities, allowing the sharing of information across the fidelity space. See Sect. 4 for additional standard MF kernels.

## 2.4  Information-Theoretic MT BO

Existing methods for information-theoretic MT BO seek to maximally reduce our uncertainty in the location of the maximizer $\mathbf{x}^* = \mathrm{argmax}_{\mathbf{x} \in \mathcal{X}} \, g(\mathbf{x})$. Following the work of [2], uncertainty in the value of $\mathbf{x}^*$ is measured as its differential entropy $H(\mathbf{x}^*) = -\mathbb{E}_{\mathbf{x} \sim p_{\mathbf{x}^*}} (\log p_{\mathbf{x}^*}(\mathbf{x}))$, where $p_{\mathbf{x}^*}$ is the probability density function of $\mathbf{x}^*$ according to our current GP model. For MT optimization, we require knowledge of the amount of information provided about the location of $\mathbf{x}^*$ from making an evaluation at $\mathbf{x}$ on fidelity $\mathbf{z}$, measured as the mutual information

$$I(y(\mathbf{x}, \mathbf{z}); \mathbf{x}^* | D_n) = H(\mathbf{x}^* | D_n) - \mathbb{E}_y [H(\mathbf{x}^* | y(\mathbf{x}, \mathbf{z}), D_n)]$$

between an evaluation $y(\mathbf{x}, \mathbf{z}) = f(\mathbf{x}, \mathbf{z}) + \epsilon$ and $\mathbf{x}^*$, where the expectation is over $p(y(\mathbf{x}, \mathbf{z}) | D_n)$ (see [29] for an introduction to information theory).

Successively evaluating the parameter-fidelity pair that provides the largest information gain per unit of evaluation cost provides the entropy search acquisition function used by [4] and [17], henceforth referred to as the MTBO acquisition function. Unfortunately, the calculation of MTBO relies on sampling-based approximations to the non-analytic distribution of $\mathbf{x}^* | D_n$. Such approximations scale poorly in both cost and performance with the dimensions of our search space (as demonstrated in Sect. 4). A modest computational saving can be made for standard BO problems by exploiting the symmetric property of mutual information, producing the predictive entropy search (PES) of [3]. However, PES still requires approximations of $\mathbf{x}^* | D_n$ and it is unclear how to extend this approach across MT frameworks.

## 3   MUMBO

In this work, we extend the computationally efficient information-theoretic acquisition function of [25] to MT BO. With their max-value entropy-search acquisition function (MES), they demonstrate that seeking to reduce our uncertainty in the value of $g^* = g(\mathbf{x}^*)$ provides an equally effective search strategy as directly minimizing the uncertainty in the location $\mathbf{x}^*$, but with significantly reduced computation. Similarly, MUMBO seeks to compute the information gain

$$\alpha_n^{MUMBO}(\mathbf{x}, \mathbf{z}) = H(y(\mathbf{x}, \mathbf{z}) \mid D_n) - \mathbb{E}_{g^*}[H(y(\mathbf{x}, \mathbf{z}) \mid g^*, D_n)], \qquad (4)$$

which can then be combined with the evaluation cost $c(\mathbf{x}, \mathbf{z})$ (following (2)). Here the expectation is over our current uncertainty in the value of $g^* \mid D_n$.

### 3.1   Calculation of MUMBO

Although extending MES to MT scenarios retains the intuitive formulation and the subsequent principled decision-making of the original MES, we require a novel non-trivial calculation method to maintain its computational efficiency for MT BO. We now propose a strategy for calculating the MUMBO acquisition function that requires the approximation of only single-dimensional integrals irrespective of the dimensions of our search space.

The calculation of our MUMBO acquisition function (4) for arbitrary $\mathbf{x}$ and $\mathbf{z}$ must be efficient as each iteration of BO requires a full maximization of (4) over $\mathbf{x}$ and $\mathbf{z}$ (i.e 2). For ease of notation we drop the dependence on $\mathbf{x}$ and $\mathbf{z}$, so that $g$ denotes the target function value at $\mathbf{x}$, $f$ denotes the evaluation of $\mathbf{x}$ at fidelity $\mathbf{z}$, and $y$ denotes the (noisy) observed value of $f(\mathbf{x}, \mathbf{z})$. Since BO fits a Gaussian process to the underlying functions, our assumptions about $g$ and $y$ imply that their joint predictive distribution is a bivariate Gaussian; with expectation, variance and correlation derived from our GP (as shown in Appendix A) and denoted by $(\mu_g, \mu_f)$, $(\sigma_g^2, \sigma_f^2 + \sigma^2)$ and $\rho$ respectively. These values summarize our current uncertainty in $g$ and $f$ and how useful making an evaluation $y$ will be for learning about $g$. Note that access to this simple two-dimensional predictive distribution is all that is needed to calculate MUMBO (4).

The first term of (4) is the differential entropy of a Gaussian distribution and so can be calculated analytically as $\frac{1}{2}\log(2\pi e(\sigma_f^2 + \sigma^2))$. The second term of (4) is an expectation over the maximum value of the true objective $g^*$, which can be approximated using a Monte Carlo approach; we use [25]'s method to approximately sample a set of $N$ samples $G = \{g_1, \ldots, g_N\}$ from $g^* \mid D_n$, using a mean-field approximation and extreme value theory.

It remains to calculate the quantity inside the expectation for a given value of $g^*$. The equivalent quantity in the original MES (without fidelity considerations) was analytically tractable, but we show that for MUMBO this term is intractable. In particular, we show that $y \mid g < g^*$ follows an extended-skew Gaussian (ESG) distribution [30,31] in Appendix A. Unfortunately, [32] have shown that there is no analytical form for the differential entropy of an ESG. Therefore, after manipulations presented also in Appendix A and reintroducing dependence on $\mathbf{x}$ and $\mathbf{z}$, we re-express (4) as

$$\alpha_n^{MUMBO}(\mathbf{x}, \mathbf{z}) = \frac{1}{N} \sum_{g^* \in G} \left[ \rho(\mathbf{x}, \mathbf{z})^2 \frac{\gamma_{g^*}(\mathbf{x}) \phi(\gamma_{g^*}(\mathbf{x}))}{2\Phi(\gamma_{g^*}(\mathbf{x}))} - \log(\Phi(\gamma_{g^*}(\mathbf{x}))) \right.$$

$$\left. + \mathbb{E}_{\theta \sim Z_{g^*}(\mathbf{x}, \mathbf{z})} \left[ \log \left( \Phi \left\{ \frac{\gamma_{g^*}(\mathbf{x}) - \rho(\mathbf{x}, \mathbf{z})\theta}{\sqrt{1 - \rho^2(\mathbf{x}, \mathbf{z})}} \right\} \right) \right] \right], \qquad (5)$$

where $\Phi$ and $\phi$ are the standard normal cumulative distribution and probability density functions, $\gamma_{g^*}(\mathbf{x}) = \frac{g^* - \mu_g(\mathbf{x})}{\sigma_g(\mathbf{x})}$ and $Z_{g^*}(\mathbf{x}, \mathbf{z})$ is an ESG (with probability density function provided in Appendix A).

Expression (5) is analytical except for the final term, which must be approximated for each of the $N$ samples of $g^*$ making up the Monte Carlo estimate. Crucially, this is just a single-dimensional integral of an analytic expression and, hence, can be quickly and accurately approximated using standard numerical integration techniques. We present MUMBO within a BO loop as Algorithm 1.

---

**Algorithm 1.** MUlti-fidelity and MUlti-task Max-value Bayesian Optimization: MUMBO

---

1: **function** MUMBO(budget $B$, $N$ samples of $g^*$)
2:    Initialize $n \leftarrow 0$, $b \leftarrow 0$
3:    Collect initial design $D_0$
4:    **while** $b < B$ **do**
5:        Begin new iteration $n \leftarrow n + 1$
6:        Fit GP to the collected observations $D_{n-1}$
7:        Simulate $N$ samples of $g^* | D_{n-1}$
8:        Prepare $\alpha_{n-1}^{MUMBO}(\mathbf{x}, \mathbf{z})$ as given by Eq. (5)
9:        Find  the  next  point  and  fidelity  to  query  $(\mathbf{x}_n, \mathbf{z}_n)$  $\leftarrow$
      $\text{argmax}_{(\mathbf{x}, \mathbf{z})} \frac{\alpha_{n-1}^{MUMBO}(\mathbf{x}, \mathbf{z})}{c(\mathbf{x}, \mathbf{z})}$
10:       Collect the new evaluation $y_n \leftarrow f(\mathbf{x}_n, \mathbf{z}_n) + \epsilon_n$, $\epsilon_n \sim N(0, \sigma^2)$
11:       Append new evaluation to observation set $D_n \leftarrow D_{n-1} \bigcup \{(\mathbf{x}_n, \mathbf{z}_n), y_n\}$
12:       Update spent budget $b \leftarrow b + c(\mathbf{x}_n, \mathbf{z}_n)$
13:    **return** Believed optimum across $\{\mathbf{x}_1, .., \mathbf{x}_n\}$

---

## 3.2   Interpretation of MUMBO

We provide intuition for (5) by relating MUMBO to an established BO acquisition function. In the formulation of MUMBO (5), we see that for a fixed parameter choice $\mathbf{x}$ (and ignoring evaluation costs) this acquisition is maximized by choosing the fidelity $z$ that provides the largest $|\rho(\mathbf{x}, \mathbf{z})|$, meaning that the stronger the correlation (either negatively or positively) the more we can learn about the true objective. In fact, if we find a fidelity $\mathbf{z}^*$ that provides evaluations that agree completely with $g$, then we would have $\rho(\mathbf{x}, \mathbf{z}^*) = 1$ and (5) would collapse to

$$\alpha_n(\mathbf{x}, \mathbf{z}^*) = \frac{1}{N} \sum_{g^* \in G} \left[ \frac{\gamma_{g^*}(\mathbf{x})\phi(\gamma_{g^*}(\mathbf{x}))}{2\Phi(\gamma_{g^*}(\mathbf{x}))} - \log(\Phi(\gamma_{g^*}(\mathbf{x}))) \right].$$

This is exactly the same expression presented by [25] in the original implementation of MES, appropriate for standard BO problems where we can only query the function we wish to optimize.

### 3.3  Computational Cost of MUMBO

The computational complexity of any BO routine is hard to measure exactly, due to the acquisition maximization (2) required before each function query. However, the main contributor to computational costs are the resources required for each calculation of the acquisition function with respect to problem dimension $d$ and the $N$ samples of $g^*$. Each prediction from our GP model costs $O(d)$, and single-dimensional numerical integration over a fixed grid is $O(1)$. Therefore, a single evaluation of MUMBO can be regarded as an $O(Nd)$ operation. Moreover, as MUMBO relies on the approximation of a single-dimensional integral, we do not require an increase in $N$ to maintain performance as the problem dimension $d$ increases (as demonstrated in Sect. 4) and so MUMBO scales linearly with problem dimension. In contrast, the MT BO acquisition used by [4] and [17] for information-theoretic MT BO relies on sampling-based approximations of $d$-dimensional distributions, therefore requiring exponentially increasing sample sizes to maintain performance as dimension increases, rendering them unsuitable for even moderately-sized BO problems. In addition, we note that these current approaches require expensive sub-routines and the calculation of derivative information, making their computational cost for even small $d$ much larger than that of MUMBO.

## 4    Experiments

We now demonstrate the performance of MUMBO across a range of MT scenarios, showing that MUMBO provides superior optimization to all existing approaches, with a significantly reduced computational overhead compared to current state-of-the-art. As is common in the optimization literature, we first consider synthetic benchmark functions in a discrete MF setting. Next, we extend the challenging continuous MF hyper-parameter tuning framework of FABO-LAS and use MUMBO to provide a novel information-theoretic implementation of the MT hyper-parameter tuning framework FASTCV, demonstrating that the performance of this simple MT model can be improved using our proposed fully-principled acquisition function. Finally, we use additional synthetic benchmarks to compare MUMBO against a wider range of existing MT BO acquisition functions.

Alongside the theoretical arguments of this paper, we also provide a software contribution to the BO community of a flexible implementation of MUMBO with support for Emukit [35]. We use a DIRECT optimizer [36] for the acquisition maximization at each BO step and calculate the single-dimensional integral in our acquisition (5) using Simpson's rule over appropriate ranges (from the known expressions of an ESG's mean and variance derived in Appendix A.1).

## 4.1    General Experimental Details

Overall, the purpose of our experiments is to demonstrate how MUMBO compares to other acquisition functions when plugged into a set of existing MT problems, focusing on providing a direct comparison with the existing state-of-the-art in information-theoretic MT BO used by [4] and [17] (which we name MTBO). Our main experiments also include the performance of popular low-cost MT acquisition functions MF-GP-UCB [18] and MT expected improvement [4]. In Sect. 4.5 we expand our comparison to include a wider range of existing BO routines, chosen to reflect popularity and code availability. We include the MF knowledge gradient (MISO-KG) [22][1], an acquisition function with significantly larger computational overheads than MUMBO (and MTBO), as-well as the low-cost acquisition functions of BOCA [19] and MF-SKO [10]. Due to a lack of provided code, and the complexity of their proposed implementations, we were unable to implement multi-fidelity extensions of PES [5,21] or the variant of knowledge-gradient for continuous fidelity spaces [23]. As both PES and knowledge gradient require approximations of quantities with dimensionality equal to the search space, their MT extensions will suffer the same scalability issue as MTBO (and MISO-KG). Finally, to demonstrate the benefit of considering MT frameworks, we also present the standard BO approaches of expected improvement (EI) and max-value entropy search (MES) which query only the true objective.

To test the robustness of the information-theoretic acquisitions we vary the number of Monte Carlo samples $N$ used for both MUMBO and MTBO (denoted as MUMBO-N and MTBO-N). We report both the time taken to choose the next location to query (referred to as the optimization overhead) and the performance of the believed objective function optimizer (the incumbent) as the optimization progresses. For our synthetic examples, we measure performance after $n$ evaluations as the simple regret $R_n = g(\mathbf{x}^*) - g(\hat{\mathbf{x}}_n)$, representing the sub-optimality of the current incumbent $\hat{\mathbf{x}}_n$. Experiments reporting wall-clock timings were performed on single core Intel Xeon 2.30 GHz processors. Detailed implementation details are provided in Appendix B.

## 4.2    Discrete Multi-fidelity BO

First, we consider the optimization of synthetic problems, using the intrinsic coregionalization kernel introduced earlier (3). Figure 2 demonstrates the superior performance and light computational overhead of MUMBO across these test

---

[1] As implemented by the original authors at *https://github.com/misokg/NIPS2017*.

functions when we have access to continuous or discrete collections of cheap low-fidelity functions at lower query costs. Although MTBO and MUMBO initially provide comparable levels of optimization, MUMBO quickly provides optimization with substantially higher precision than MTBO and MF-GP-UCB. We delve deeper into the low performance of MF-GP-UCB in Appendix B.1. In addition, MUMBO is able to provide high-precision optimization even when based on a single sample of $g^*$, whereas MTBO requires 50 samples for reasonable performance on the 2D optimization task, struggles on the 6D task even when based on 200 samples (requiring 20 times the overhead cost of MUMBO), and proved computationally infeasible to provide reasonable 8D optimization (and is therefore not included in Fig. 2d).

Note that MUMBO based on a single sample of $g^*$ is a more aggressive optimizer, as we only consider a single (highly-likely) max-value. Although less robust than MUMBO-10 on average across our examples, this aggressive behavior can allow faster optimization, but only for certain problems (Fig. 2(c)).

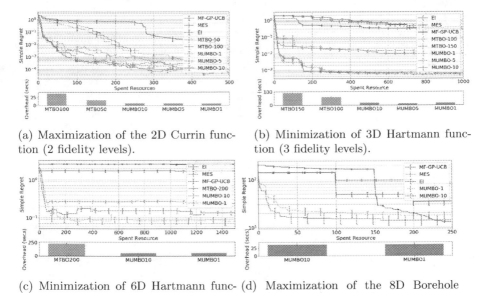

(a) Maximization of the 2D Currin function (2 fidelity levels).

(b) Minimization of 3D Hartmann function (3 fidelity levels).

(c) Minimization of 6D Hartmann function (4 fidelity levels).

(d) Maximization of the 8D Borehole function (2 fidelity levels).

**Fig. 2.** MUMBO provides high-precision optimization with low computational overheads for discrete MF optimization. We show the means and standard errors across 20 random initializations.

### 4.3  Continuous Multi-fidelity BO: FABOLAS

FABOLAS [17] is a MF framework for tuning the hyper-parameter of machine learning models whilst dynamically controlling the proportion of available data $z \in (0, 1]$ used for each hyper-parameter evaluation. By using the MTBO acquisition and imposing strong assumptions on the structure of the fidelity space,

FABOLAS is able to achieve highly efficient hyper-parameter tuning. The use of a 'degenerate' kernel [27] with basis function $\phi(z) = (z, (1 - z)^2)^T$ (i.e performing Bayesian linear regression over this basis) enforces monotonicity and strong smoothness across the fidelity space, acknowledging that when using more computational resources, we expect less biased and less noisy estimates of model performance. These assumptions induce a product kernel over the whole space $\mathcal{X} \times \mathcal{Z}$ of:

$$k((\mathbf{x}, z)(\mathbf{x}', z')) = k_\mathcal{X}(\mathbf{x}, \mathbf{x}')(\phi(z)^T \Sigma_1 \phi(z')),$$

where $\Sigma_1$ is a matrix in $\mathbb{R}^{2 \times 2}$ to be estimated alongside the parameters of $k_\mathcal{X}$. Similarly, evaluation costs are also modeled in log space, with a GP over the basis $\phi_c(z) = (1, z)^T$ providing polynomial computational complexity of arbitrary degree. We follow the original FABOLAS implementation exactly, using MCMC to marginalize kernel parameters over hyper-priors specifically chosen to speed up and stabilize the optimization.

In Fig. 3 we replace the MTBO acquisition used within FABOLAS with a MUMBO acquisition, demonstrating improved optimization on two examples from [17]. As the goal of MF hyper-parameter tuning is to find high-performing hyper-parameter configurations after using as few computational resources as possible, including both the fitting of models and calculating the next hyper-parameter and fidelity to query, we present incumbent test error (calculated offline after the full optimization) against wall-clock time. Note that the entire time span considered for our MNIST example is still less than required to try just four hyper-parameter evaluations on the whole data and so we cannot include standard BO approaches in these figures. MUMBO's significantly reduced computational overhead allows twice as many hyper-parameter evaluations as MTBO for the same wall clock time, even though MUMBO consistently queries larger proportions of the data (on average 30% rather than 20% by MTBO). Moreover, unlike MTBO, with an overhead that increases as the optimization progresses, MUMBO remains computationally light-weight throughout and has substantially less variability in the performance of the chosen hyper-parameter configuration. While we do not compare FABOLAS against other hyper-parameter tuning methods, we have demonstrated that, for this well-respected tuner and complicated MF BO problem, that MUMBO provides an improvement in efficiency and a substantial reduction in computational cost.

### 4.4  Multi-task BO: FASTCV

We now consider the MT framework of FASTCV [4]. Here, we seek the simultaneous optimization of the $K$ performance estimates making up $K$-fold CV. Therefore, our objective function $g$ is the average score across a categorical fidelity space $\mathcal{Z} = \{1, .., K\}$. Each hyper-parameter is evaluated on a single fold, with the corresponding evaluations on the remaining folds inferred using the learned between-fold relationship. Therefore, we can evaluate $K$ times as many distinct hyper-parameter choices as when tuning with full $K$-fold CV whilst retaining the precise performance estimates required for reliable tuning [37,38].

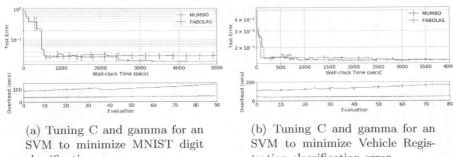

(a) Tuning C and gamma for an SVM to minimize MNIST digit classification error.

(b) Tuning C and gamma for an SVM to minimize Vehicle Registration classification error.

**Fig. 3.** MUMBO provides MF hyper-parameter tuning with a much lower overhead than FABOLAS. We show the means and standard errors based on 5 runs.

Unlike our other examples, this is not a MF BO problem as our fidelities have the same query cost (at $1/K^{th}$ the cost of the true objective). Recall that all we require to use MUMBO is the predictive joint (bi-variate Gaussian) distribution between an objective function $g(\mathbf{x})$ and fidelity evaluations $f(\mathbf{x}, z)$ for each choice of $\mathbf{x}$. For FASTCV, $g$ corresponds with the average score across folds and so (following our earlier notation) our underlying GP provides;

$$\mu_g(\mathbf{x}) = \frac{1}{K} \sum_{z \in \mathcal{Z}} \mu^n(\mathbf{x}, z), \quad \sigma_g(\mathbf{x})^2 = \frac{1}{K^2} \sum_{z \in \mathcal{Z}} \sum_{z' \in \mathcal{Z}} \Sigma^n((\mathbf{x}, z)(\mathbf{x}, z')),$$

where $\mu^n(\mathbf{x}, z)$ is the predictive mean performance of $\mathbf{x}$ on fold $z$ and $\Sigma^n((\mathbf{x}, z), (\mathbf{x}, z'))$ is the predictive co-variance between the evaluations of $\mathbf{x}$ on folds $z$ and $z'$ after $n$ hyper-parameter queries. Similarly, we have the correlation between evaluations of $\mathbf{x}$ on fold $z$ with the average score $g$ as

$$\rho(\mathbf{x}, z) = \frac{\frac{1}{K} \sum_{z' \in \mathcal{Z}} \Sigma^n((\mathbf{x}, z)(\mathbf{x}, z'))}{\sqrt{\sigma_g^2(\mathbf{x}) \Sigma^n((\mathbf{x}, z)(\mathbf{x}, z))}},$$

providing all the quantities required to use MUMBO.

In the original implementation of FASTCV, successive hyper-parameter evaluations are chosen using a two-step heuristic based on expected improvement. Firstly they choose the next hyper-parameter $\mathbf{x}$ by maximizing the expected improvement of the predicted average performance and secondly choosing the fold that has the largest fold-specific expected improvement at this chosen hyper-parameter. We instead propose using MUMBO to provide a principled information-theoretic extension to FASTCV. Figure 4 demonstrates that MUMBO provides an efficiency gain over FASTCV, while finding high-performing hyper-parameters substantially faster than standard BO tuning by $K$-fold CV (where we require $K$ model evaluations for each unique hyper-parameter query).

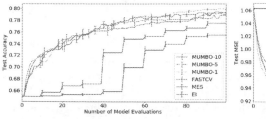

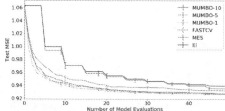

(a) Tuning two SVM hyper-parameters to maximize sentiment classification accuracy for IMDB movie reviews by 10-fold CV.

(b) Tuning four hyper-parameters for probabilistic matrix factorization to minimize mean reconstruction error for movie recommendations using 5-fold CV.

**Fig. 4.** MUMBO provides faster hyper-parameter tuning than the MT framework of FASTCV. We show the mean and standard errors across 40 runs. To measure total computational cost we count each evaluation by $K$-fold CV as $k$ model fits. Experimental details are included in Appendix B.3.

### 4.5 Wider Comparison with Existing Methods

Finally, we make additional comparisons with existing MT acquisition functions in Figs. 5 and 6. Knowledge-gradient search strategies are designed to provide particularly efficient optimization for noisy functions, however this high performance comes with significant computational overheads. Although providing reasonable early performance on a synthetic noisy MF optimization task (Fig. 5), we see that MUMBO is able to provide higher-precision optimization and that, even for this simple 2-d search space, MISO-KG's optimization overheads are magnitudes larger than MUMBO (and MTBO). Figure 6 shows that MUMBO substantially outperforms existing approaches on a continuous MF benchmark. MF-SKO, MF-UCB and BOCA's search strategies are guided by estimating $g^*$ (rather than $\mathbf{x}^*$) and so have comparable computational cost to MUMBO, however, only MUMBO is able to provide high-precision optimization with this low-computational overhead.

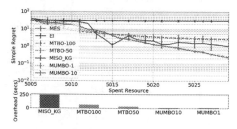

**Fig. 5.** The 2D noisy Rosenbrock function (2 fidelities).

**Fig. 6.** The 2-d Currin function (1-d continuous fidelity space)

# 5   Conclusions

We have derived a novel computationally light information-theoretic approach for general discrete and continuous multi-task Bayesian optimization, along with an open and accessible code base that will enable users to deploy these methods and improve replicability.

MUMBO reduces uncertainty in the optimal value of the objective function with each subsequent query, and provides principled decision-making across general multi-task structures at a cost which scales only linearly with the dimension of the search space. Consequently, MUMBO substantially outperforms current acquisitions across a range of optimization and hyper-parameter tuning tasks.

# References

1. Shahriari, B., Swersky, K., Wang, Z., Adams, R.P., De Freitas, N.: Taking the human out of the loop: a review of Bayesian optimization. Proc. IEEE **104**(1), 148–175 (2016)
2. Hennig, P., Schuler, C.J.: Entropy search for information-efficient global optimization. Journal of Machine Learning Research (2012)
3. Hernández-Lobato, J.M., Hoffman, M.W., Ghahramani, Z.: Predictive entropy search for efficient global optimization of black-box functions. In: Neural Information Processing Systems (2014)
4. Swersky, K., Snoek, J., Adams, R.P.: Multi-task Bayesian optimization. In: Neural Information Processing Systems (2013)
5. Zhang, Y., Hoang, T.N., Low, B.K.H., Kankanhalli, M.: Information-based multi-fidelity Bayesian optimization. In: Neural Information Processing Systems: Workshop on Bayesian Optimization (2017)
6. Nguyen, N.V., et al.: Multidisciplinary unmanned combat air vehicle system design using multi-fidelity model. Aerosp. Sci. Technol. **26**(1), 200–210 (2013)
7. Zheng, L., Hedrick, T.L., Mittal, R.: A multi-fidelity modelling approach for evaluation and optimization of wing stroke aerodynamics in flapping flight. J. Fluid Mech. **721**, 118 (2013)
8. Pilania, G., Gubernatis, J.E., Lookman, T.: Multi-fidelity machine learning models for accurate bandgap predictions of solids. Comput. Mater. Sci. **129**, 156–163 (2017)
9. Prieß, M., Koziel, S., Slawig, T.: Surrogate-based optimization of climate model parameters using response correction. J. Comput. Sci. **2**(4), 335–344 (2011)
10. Huang, D., Allen, T.T., Notz, W.I., Zeng, N.: Global optimization of stochastic black-box systems via sequential kriging meta-models. Journal of Global Optimization (2006)
11. Xu, J., Zhang, S., Huang, E., Chen, C.H., Lee, L.H., Celik, N.: Mo2tos: multi-fidelity optimization with ordinal transformation and optimal sampling. Asia Pac. J. Oper. Res. **33**(03), 1650017 (2016)
12. Yong, H.K., Wang, L., Toal, D.J.J., Keane, A.J., Stanley, F.: Multi-fidelity Kriging-assisted structural optimization of whole engine models employing medial meshes. Struct. Multidisciplinary Optim. **60**(3), 1209–1226 (2019). https://doi.org/10.1007/s00158-019-02242-6

13. Strubell, E., Ganesh, A., McCallum, A.: Energy and policy considerations for deep learning in NLP. In: Proceedings of the 57th Annual Meeting of the Association for Computational Linguistics (2019)
14. Snoek, J., Larochelle, H., Adams, R.P.: Practical Bayesian optimization of machine learning algorithms. In: Neural Information Processing Systems (2012)
15. Kennedy, M.C., O'Hagan, A.: Predicting the output from a complex computer code when fast approximations are available. Biometrika 87(1), 1–13 (2000)
16. Lam, R., Allaire, D.L., Willcox, K.E.: Multifidelity optimization using statistical surrogate modeling for non-hierarchical information sources. In: Structures, Structural Dynamics, and Materials Conference (2015)
17. Klein, A., Falkner, S., Bartels, S., Hennig, P., Hutter, F.: Fast Bayesian optimization of machine learning hyperparameters on large datasets. In: International Conference on Artificial Intelligence and Statistics (2017)
18. Kandasamy, K., Dasarathy, G., Oliva, J.B., Schneider, J., Póczos, B.: Gaussian process bandit optimisation with multi-fidelity evaluations. In: Neural Information Processing Systems (2016)
19. Kandasamy, K., Dasarathy, G., Schneider, J., Póczos, B.: Multi-fidelity Bayesian optimisation with continuous approximations. In: International Conference in Machine Learning (2017)
20. Kohavi, R.: A study of cross-validation and bootstrap for accuracy estimation and model selection. In: International Joint Conference on Artificial Intelligence (1995)
21. McLeod, M., Osborne, M.A., Roberts, S.J.: Practical bayesian optimization for variable cost objectives. arXiv preprint arXiv:1703.04335 (2017)
22. Poloczek, M., Wang, J., Frazier, P.: Multi-information source optimization. In: Neural Information Processing Systems (2017)
23. Wu, J., Toscano-Palmerin, S., Frazier, P.I., Wilson, A.G.: Practical multi-fidelity bayesian optimization for hyperparameter tuning. arXiv preprint arXiv:1903.04703 (2019)
24. Picheny, V., Ginsbourger, D., Richet, Y., Caplin, G.: Quantile-based optimization of noisy computer experiments with tunable precision. Technometrics 55(1), 2–13 (2013)
25. Wang, Z., Jegelka, S.: Max-value entropy search for efficient Bayesian optimization. In: International Conference on Machine Learning (2017)
26. Takeno, S., et al.: Multi-fidelity bayesian optimization with max-value entropy search. arXiv preprint arXiv:1901.08275 (2019)
27. Rasmussen, C.E.: Gaussian processes in machine learning. In: Bousquet, O., von Luxburg, U., Rätsch, G. (eds.) ML-2003. LNCS (LNAI), vol. 3176, pp. 63–71. Springer, Heidelberg (2004). https://doi.org/10.1007/978-3-540-28650-9_4
28. Álvarez, M.A., Lawrence, N.D.: Computationally efficient convolved multiple output Gaussian processes. J. Mach. Learn. Res. 12, 1459–1500 (2011)
29. Cover, T.M., Thomas, J.A.: Elements of Information Theory. John Wiley & Sons, New Jersey (2012)
30. Azzalini, A.: A class of distributions which includes the normal ones. Scandinavian Journal of Statistics (1985)
31. Arnold, B.C., Beaver, R.J., Groeneveld, R.A., Meeker, W.Q.: The nontruncated marginal of a truncated bivariate normal distribution. Psychometrika 58, 471–488 (1993)
32. Arellano-Valle, R.B., Contrera-Reyes, J.E., Genton, M.G.: Shannon entropy and mutual information for multivariate skew-elliptical distributions. Scand. J. Stat. 40(1), 42–62 (2013)

33. The GPyOpt authors: GPyOpt: A Bayesian optimization framework in Python (2016). http://github.com/SheffieldML/GPyOpt
34. Klein, A., Falkner, S., Mansur, N., Hutter, F.: RoBo: a flexible and robust Bayesian optimization framework in Python. In: Neural Information Processing Systems: Bayesian Optimization Workshop (2017)
35. Paleyes, A., Pullin, M., Mahsereci, M., Lawrence, N., González, J.: Emulation of physical processes with emukit. In: Second Workshop on Machine Learning and the Physical Sciences, NeurIPS (2019)
36. Jones, D.R., Perttunen, C.D., Stuckman, B.E.: Lipschitzian optimization without the lipschitz constant. J. Optim. Theory Appl. **79**, 157–181 (1993)
37. Moss, H., Leslie, D., Rayson, P.: Using J-K-fold cross validation to reduce variance when tuning NLP models. In: International Conference on Computational Linguistics (2018)
38. Moss, H., Moore, A., Leslie, D., Rayson, P.: FIESTA: fast identification of state-of-the-art models using adaptive bandit algorithms. In: Proceedings of the 57th Annual Meeting of the Association for Computational Linguistics (2019)
39. Xiong, S., Qian, P.Z., Wu, C.J.: Sequential design and analysis of high-accuracy and low-accuracy computer codes. Technometrics **55**(1), 37–46 (2013)
40. Forrester, A., Sobester, A., Keane, A.: Engineering Design via Surrogate Modelling: A Practical Guide. John Wiley & Sons, New Jersey (2008)
41. Eggensperger, K., et al.: Towards an empirical foundation for assessing Bayesian optimization of hyperparameters. In: Neural Information Processing Systems: Workshop on Bayesian Optimization (2013)
42. Deng, L.: The MNIST database of handwritten digit images for machine learning research. IEEE Sign. Process. Mag. **29**(6), 141–142 (2012)
43. Siebert, J.: Vehicle recognition using rule based methods. Turing Institute, Glasgow (1987)
44. Maas, A.L., Daly, R.E., Pham, P.T., Huang, D., Ng, A.Y., Potts, C.: Learning word vectors for sentiment analysis. In: Association for Computational Linguistics (2011)
45. Mnih, A., Salakhutdinov, R.R.: Probabilistic matrix factorization. In: Neural Information Processing Systems (2008)
46. Hoffman, M., Bach, F.R., Blei, D.M.: Online learning for latent Dirichlet allocation. In: Neural Information Processing Systems (2010)
47. Bengio, Y., Grandvalet, Y.: No unbiased estimator of the variance of k-fold cross-validation. J. Mach. Learn. Res. **5**(9), 1089–1105 (2004)

# Interactive Multi-objective Reinforcement Learning in Multi-armed Bandits with Gaussian Process Utility Models

Diederik M. Roijers[1,2(✉)], Luisa M. Zintgraf[3], Pieter Libin[2],
Mathieu Reymond[2], Eugenio Bargiacchi[2], and Ann Nowé[2]

[1] HU University of Applied Sciences Utrecht, Utrecht, The Netherlands
diederik.yamamoto-roijers@hu.nl
[2] Vrije Universiteit Brussel, Brussels, Belgium
{pieter.libin,mathieu.reymond,eugenio.bargiacchi,ann.nowe}@vub.ac.be
[3] University of Oxford, Oxford, UK
luisa.zintgraf@cs.ox.ac.uk

**Abstract.** In interactive multi-objective reinforcement learning (MORL), an agent has to simultaneously learn about the environment and the preferences of the user, in order to quickly zoom in on those decisions that are likely to be preferred by the user. In this paper we study interactive MORL in the context of multi-objective multi-armed bandits. Contrary to earlier approaches to interactive MORL that force the utility of the user to be expressed as a weighted sum of the values for each objective, we do not make such stringent a priori assumptions. Specifically, we not only allow non-linear preferences, but also obviate the need to specify the exact model class in the utility function must fall. To achieve this, we propose a new approach called Gaussian-process Utility Thompson Sampling (GUTS). GUTS employs parameterless Bayesian learning to allow any type of utility function, exploits monotonicity information, and limits the number of queries posed to the user by ensuring that questions are statistically significant. We show empirically that GUTS can learn non-linear preferences, and that the regret and number of queries posed to the user are highly sub-linear in the number of arm pulls. (A preliminary version of this work was presented at the ALA workshop in 2018 [20]).

**Keywords:** Multiple objectives · Multi-armed bandits · Thompson sampling · Reinforcement learning

## 1 Introduction

Real-world decision problems often require learning about the effects of various decisions by interacting with an environment. When these effects can be measured in terms of a single scalar objective, such problems can be modelled as a *multi-armed bandit (MAB)* [2]. However, many real-world decision problems have

© Springer Nature Switzerland AG 2021
F. Hutter et al. (Eds.): ECML PKDD 2020, LNAI 12459, pp. 463–478, 2021.
https://doi.org/10.1007/978-3-030-67664-3_28

multiple possibly conflicting objectives [18]. may want to minimise response time, while also minimising fuel cost and the stress levels of the drivers. For example, an agent learning the best strategy to control epidemics may want to minimise number of infections and minimise the burden on society [13]. When the user is unable to accurately and a priori specify preferences with respect to such objectives for all hypothetically possible trade-offs, the user needs to be informed about the values of actually available trade-offs between objectives in order to make a well-informed decision. For such problems, MABs no longer suffice, and need to be extended to *multi-objective multi-armed bandits (MOMABs)* [3,7], in which the effects of alternative decisions are measured as a vector containing a value for each objective.

Most research on MOMABs [3,8,28] focusses on producing *coverage sets* [18], i.e., sets of all possibly optimal alternatives given limited *a priori* information about the possible utility functions of the user, as the solution to a MOMAB. A minimal assumption is that utility functions are monotonically increasing in all objectives. This leads to the popular *Pareto front* concept as the coverage set.

Typically, such research assumes an offline learning setting. I.e., there is a learning phase in which there is no interaction with the user, after which the coverage set is presented to the user in order to select one final alternative. The rewards attained during learning are assumed to be unimportant; only the expected rewards of the final alternative is assumed to be relevant to the utility of the user. However, this is a limiting assumption. For example, the ambulances discussed above may well be deployed while still learning about the expected value of the deployment strategies in the different objectives. In such cases, we need to use our interactions with the environment efficiently, as we care about the rewards accrued during learning. When the learning agent can interact with the user and *elicit* preferences from the user during learning [19], the agent can focus its learning on strategies that the user finds most appealing quickly. This is thus an *online* and *interactive* learning scenario.

Roijers et al. [19] recently proposed the *interactive Thompson sampling (ITS)* algorithm for online interactive reinforcement learning in MOMABs, and show that the number of interactions with the environment in an multi-objective interactive online setting can be reduced to approximately the same amount of interactions as in single-objective state-of-the-art algorithms such as Thompson sampling. However, they make highly restrictive assumptions about the utility functions of users. Specifically, they assume that user utility is a weighted sum of the values in each objective.

In real-life multi-objective RL, it is essential to be able to deal with users that have non-linear preferences. Therefore, we propose the *Gaussian-process Utility Thompson Sampling (GUTS)* algorithm. GUTS can deal with non-linear utility functions, enabling interactive reinforcement learning in MOMABs for the full range of multi-objective RL settings. To do so, we make the following key improvements over ITS. First, rather than using a pre-specified model-space for utility functions, we employ parameterless Bayesian learning, i.e., *Gaussian processes* (GPs) [5,6,15], to be able to learn arbitrary utility functions. GPs are data-efficient, and can thus quickly learn the relevant part of the utility function. Secondly, because GPs cannot enforce monotonicity with respect to each objective (i.e., the minimal assumption in multi-objective RL), as GPs cannot incorporate monotonicity constraints, GUTS enforces this monotonicity separately from the GP model of the utility by filtering provably dominated actions. Finally, we propose *statistical significance testing* to only ask the user about the preferences between statistically significantly different arms, to prevent irrelevant questions in the earlier iterations, and show empirically that this does not increase regret.

## 2    Background

In this section we define our problem setting, i.e., multi-objective multi-armed bandits (MOMABs), and how we model user utility. First however, we define the scalar version of a MOMAB, i.e, a MAB.

**Definition 1.** *A scalar multi-armed bandit (MAB) [25] is a tuple ⟨$\mathcal{A}, \mathcal{P}$⟩ where*

- *$\mathcal{A}$ is a set of actions or arms, and*
- *$\mathcal{P}$ is a set of probability distributions $P_a(r) : \mathbb{R} \to [0,1]$ over scalar rewards $r$ associated with each arm $a \in \mathcal{A}$.*

*We refer to the the mean reward of an arm as $\mu_a = \mathbb{E}_{P_a}[r] = \int_{-\infty}^{\infty} rP_a(r)dr$, and to the optimal reward as the mean reward of the best arm $\mu^* = \max_a \mu_a$.*

The goal of an agent interacting with a MAB is to minimise the expected regret.

**Definition 2.** *The expected cumulative regret of pulling a sequence of arms for each timestep between $t = 1$ and a time horizon $T$ (following the definition of Agrawal et al. [1], is*

$$\mathbb{E}\left[\sum_{t=1}^{T} \mu^* - \mu_{a(t)}\right],$$

*where $a(t)$ is the arm pulled at time $t$.*

In scalar MABs, agents aim to find the arm $a$ that maximises the expected scalar reward. In contrast, in multi-objective problems there are $n$ objectives that are all desirable. Hence, the stochastic rewards $\mathbf{r}(t)$ and the expected rewards for each alternative $\boldsymbol{\mu}_a$ are vector-valued.

**Definition 3.** *A multi-objective multi-armed bandit (MOMAB) [3, 28] is a tuple ⟨$\mathcal{A}, \mathcal{P}$⟩ where*

- *$\mathcal{A}$ is a finite set of actions or arms, and*
- *$\mathcal{P}$ is a set of probability distributions, $P_a(\mathbf{r}) : \mathbb{R}^d \to [0,1]$ over vector-valued rewards $\mathbf{r}$ of length d, associated with each arm $a \in \mathcal{A}$.*

Rather than having a single optimal alternative as in a MAB, MOMABs can have multiple arms whose value vectors are optimal for different preferences that users may have. Following the utility-based approach to MORL [18], we assume such preferences can be expressed using a utility function, $u(\boldsymbol{\mu})$ that returns the *scalarised* value of $\boldsymbol{\mu}$. Using this utility function, the online regret for the interactive multi-objective reinforcement learning setting is as follows.

**Definition 4.** *The expected cumulative* user regret *of pulling a sequence of arms for each timestep between $t = 1$ and a time horizon $T$ in a MOMAB is*

$$\mathbb{E}\left[\sum_{t=1}^{T} u(\boldsymbol{\mu}^*) - u(\boldsymbol{\mu}_{a(t)})\right],$$

*where $a(t)$ is the arm pulled at time $t$, $\boldsymbol{\mu}^* = \arg\max_a u(\boldsymbol{\mu}_a)$, and $n_a(T)$ is the number of times arm $a$ is pulled until timestep $T$.*

For the aforementioned ITS algorithm, Roijers et al. [19] require that $u(\boldsymbol{\mu})$ is of the form $\mathbf{w} \cdot \boldsymbol{\mu}$, where $\mathbf{w}$ is a vector of positive weights per objective that sum to one. However, in many cases, users cannot be assumed to have such simple linear preferences. Instead, users often exhibit some degree of non-linearity. To accommodate any user, we want to only make the minimal assumption, i.e., that $u(\boldsymbol{\mu})$ is monotonically increasing in all objectives. This assumption implies that more is better in each objective, which follows from the definition of an objective.

## 2.1    Modelling User Preference Using Gaussian Processes

Because the user's utility $u(\boldsymbol{\mu})$ can be of any form, we use a general-purpose *Gaussian process* (GP) [15] due to its flexibility for a wide range of tasks. GPs capture uncertainty by assigning to each point a normal distribution – the mean reflecting the expected value, and the variance reflecting the uncertainty at that point. GPs are specified by a mean function $m$ and a kernel $k$,

$$u(\boldsymbol{\mu}) \sim GP(m(\boldsymbol{\mu}), k(\boldsymbol{\mu}, \boldsymbol{\mu}')) , \tag{1}$$

where the kernel defines how data points influence each other. Since we do not have any prior information on the shape of the utility function of the user, we use the common zero-mean prior, $m(\boldsymbol{\mu}) = 0$. We assume the user's utility function to be smooth (i.e., it does not make sudden large jumps if the reward vector $\boldsymbol{\mu}$ changes only slightly), and therefore use the squared exponential kernel [15], $k(x, x') = \exp\left(-(2\phi)^{-1}||x - x'||^2\right)$, which is appropriate for modelling smooth functions, with $\phi$ controlling the distance over which neighbouring points influence each other. For large $\phi$, the expected values of the data points are going to be highly correlated, while for small $\phi$ the GP can have larger fluctuations.

Given observation data $C$ about the user's preferences, we update our prior belief $P(u)$ about the utility function $u(\boldsymbol{\mu})$ using Bayes' rule $P(u|C) \propto P(u)P(C|u)$, where $P(C|u)$ is the likelihood of the data given $u(\boldsymbol{\mu})$. The posterior $P(u|C)$ is the updated belief about the user's utility, and the current approximation of the utility function.

In this paper, the data for the GP, $C$, is given in terms of *pairwise comparisons* between vectors,

$$C = \{\boldsymbol{\mu}_m \succ \boldsymbol{\mu}'_m\}_{m=1}^M , \tag{2}$$

where $\succ$ denotes a noisy comparison $u(\boldsymbol{\mu}_m)+\varepsilon > u(\boldsymbol{\mu}'_m)+\varepsilon'$, where $\varepsilon \sim \mathcal{N}(0, \sigma^2)$ accounts for noise in the user's utility (i.e., the user is allowed to make small errors). We opt for binary comparison because it is highly difficult—and thus error-prone—for humans to provide numerical utility evaluations of single items; it is known that such evaluations can differ based on a number of external factors that are not relevant to the data, making the valuations time-dependent, while pairwise comparisons (i.e., asking the question "Which of these two vectors do you prefer?") are more stable [9, 21, 22, 24].

To update the GP using this data, we need to define a likelihood function. Chu et al. [6] introduce a probit likelihood for pairwise comparison data with Gaussian noise, which can be used to update the GP. Given this likelihood, the posterior becomes analytically intractable and has to be approximate (e.g., using Laplace approximation as in [5]).

The main advantages of GPs for our setting are that they do not assume a model-class of functions for $u(\boldsymbol{\mu})$, and that they are sample-efficient.

Note that strictly enforcing monotonicity constraints on Gaussian processes is not possible. Some approximate solutions exist [11, 16, 26], however, none of these directly apply to the probit likelihood [6] function we use. In this paper, we instead use the prior knowledge that Pareto-dominated arms cannot be optimal and enforce the monotonicity constraint by using a pruning strategy (introduced in more detail in the next section).

# 3   Gaussian-Process Utility Thompson Sampling

In this section, we propose our main contribution: *Gaussian-process Utility Thompson Sampling (GUTS)* (Algorithm 1). GUTS interacts with the environment by pulling arms, and storing the resulting reward-vectors together with the associated arm numbers in a dataset $D$. GUTS also interacts with the user by posing comparison queries to the user, leading to a dataset $C$ (as in Eq. 2).

GUTS selects the arms to pull by sampling the posterior mean reward distributions as in Thompson Sampling (TS) [25] for single-objective MABs. In contrast to single-objective MABs, the means sampled from the posteriors in MOMABs are vector-valued (line 5). As there is typically not one arm that has the optimal reward for all objectives, minimising the regret (Definition 4) is only possible by also estimating the utility function, $u(\boldsymbol{\mu})$. To model this utility function, GUTS employs a Gaussian Process (GP) (Eq. 1), which is a posterior distribution over utility functions, given data. Because the GP is a posterior over utility functions, GUTS can sample $u(\boldsymbol{\mu})$ from this posterior. We assume that the user's utility function does not depend on the values of the arms, and can thus be learnt and sampled independently (lines 7–8). When such a sampled $u(\boldsymbol{\mu})$ is applied to the reward vectors, the action that maximises the utility according to $u(\boldsymbol{\mu})$ can be found (lines 9–10) and executed (line 18). Aside from

---

**Algorithm 1:** GP Utility Thompson Sampling

---

**Input**: Parameter priors on reward distributions, and GP prior
parameters $m(\boldsymbol{\mu})$ and $k(\boldsymbol{\mu}, \boldsymbol{\mu}')$, `countdown`, $\alpha$

1  $C \leftarrow \emptyset$ ;                                    // previous comparisons
2  $D \leftarrow \emptyset$ ;                                    // observed reward data
3  $cd \leftarrow$ `countdown`
4  **for** $t = 1, ..., T$ **do**
5  $\quad$ $\theta_D, \leftarrow$ draw set of samples from $P(\theta | D)$
6  $\quad$ $\theta'_D \leftarrow$ PPrune$(\theta_D)$
7  $\quad$ $u_1 \leftarrow$ sample utility function from GP posterior given $C$
8  $\quad$ $u_2 \leftarrow$ sample utility function from GP posterior given $C$
9  $\quad$ $a_1(t) \leftarrow \arg\max_a \; u_1(\boldsymbol{\mu}_{\theta'_D, a})$
10 $\quad$ $a_2(t) \leftarrow \arg\max_a \; u_2(\boldsymbol{\mu}_{\theta'_D, a})$
11 $\quad$ $significant \leftarrow$ HotellingT2$(D[a_1(t)], D[a_2(t)]) \leq \alpha$
12 $\quad$ **if** $a_1(t) \neq a_2(t) \;\wedge\; cd \leq 0 \;\wedge\; significant$ **then**
13 $\quad\quad$ Perform user comparison for $\boldsymbol{\mu}_{\theta'_D, a_1(t)}$ and $\boldsymbol{\mu}_{\theta'_D, a_2(t)}$ and add result
$\quad\quad$ (in the format, $\boldsymbol{\mu}_x \succ \boldsymbol{\mu}_y$) to C
14 $\quad\quad$ $cd \leftarrow$ `countdown`
15 $\quad$ **else**
16 $\quad\quad$ $cd$--
17 $\quad$ **end**
18 $\quad$ $\mathbf{r}(t) \leftarrow$ play $a_1(t)$ and observe reward
19 $\quad$ append $(\mathbf{r}(t), a_1(t))$ to $D$
20 **end**

---

needing to sample a utility function from the GP to find the utility-maximising arm, learning the optimal arm to pull given the GP is thus highly similar to TS. The main difficulty is in learning to estimate $u(\boldsymbol{\mu})$ accurately and efficiently.

To learn $u(\boldsymbol{\mu}_a)$, we need data about the preferences of the user. As mentioned in Sect. 2 we gather data about the preferences of users in terms of binary comparisons between two value-vectors (Eq. 2). In addition to a dataset about the value-vectors associated with the arms, $D$ (line 2), GUTS thus keeps a dataset of outcomes of pairwise preference queries, $C$ (line 1). To fill $C$ with meaningful and relevant data, GUTS needs to pose queries to a (human) decision maker. Because a comparison consists of two arms, GUTS draws one sample set of mean reward vectors from the posteriors of the mean reward vectors given $D$, $\theta_D$ (line 5), and subsequently samples two sample utility functions from the posterior over utility functions given $C$ (lines 7–8). When these two sets of samples, paired with their respective sampled utility function, disagree on what the optimal arm should be, a preference comparison between the sampled mean vectors of these arms are a candidate query to the user. Please note that this sampling is different from ITS [19], where the mean reward vectors are sampled twice as well; we argue that this is not necessary as the utility function is independent from the reward vectors, and we are only interested in possible differences between sampled utilities for a single given set of mean reward vectors.

GUTS uses a GP to estimate $u(\boldsymbol{\mu})$ from the data in $C$. GPs are highly general, data-efficient, and as non-parametric function approximators can fit any function. However, in multi-objective decision-making we have one more piece of information that we can exploit. We know that $u(\boldsymbol{\mu})$ is monotonically increasing in all objectives; i.e., increasing the value of a mean reward vector $\boldsymbol{\mu}$ in one objective, without diminishing it in the other objectives can only have a positive effect on utility. This corresponds to the minimal assumption about utility in multi-objective decision making [18], and leads the notion of *Pareto domination*. An arm $a$ is Pareto-dominated, $\boldsymbol{\mu}_{a'} \succ_P \boldsymbol{\mu}_a$, when there is another arm $a'$ with an equal or higher mean in all objectives, and a higher mean in at least one objective:

$$\boldsymbol{\mu}_{a'} \succ_P \boldsymbol{\mu}_a \stackrel{\text{def}}{=} (\forall i)\ \mu_{a',i} \geq \mu_{a,i}\ \wedge\ (\exists j)\ \mu_{a',j} > \mu_{a,j}.$$

Because of monotonicity, Pareto-dominated arms cannot be optimal [18]. GUTS enforces this condition as follows: Pareto-dominated arms are excluded from the sets of sampled mean reward vectors, by using a pruning operator PPrune[1] resulting in a smaller sets of arms $\theta'_D$ (line 6). This ensures that the user will not be queried w.r.t. her preferences for dominating-dominated pairs of vectors from $\theta'_D$, as we already know the answer without posing the question to the user.

Using $C$ we sample utility functions and maximise over the available actions in $\theta'_D$ (line 7–10). We note that because we estimate $u$ using GPs, GUTS can directly sample values for the values in $\theta'_D$ without fully specifying the entire function $u(\boldsymbol{\mu})$ for all possible $\boldsymbol{\mu}$ in $\mathbb{R}^n$.

Because interactions with a decision maker are an expensive resource, we need to select preference queries between value vectors carefully. Therefore, a) these queries should be relevant with respect to finding the optimal arm (preferences between arms that are both suboptimal are less relevant), b) the arms in the queries should be statistically significantly different from each other, and c) queries should be progressively infrequent, as we assume that the user is decreasingly motivated to provide more information to the system as time progresses. To make sure that only relevant questions are asked (a), GUTS only poses queries to the user when the two sampled and utility functions disagree on the arm that maximises the utility for the user. As more information comes in, we expect the two sampled utility functions to disagree on which arm is optimal less (c). Additionally, we also want to assure that queries only concern arms for which the difference between values is statistically significant (b). For this, when we are dealing with multi-variate Gaussian-distributed rewards, we can use Hotellings-$T^2$ test [10], which is the multi-variate version of the Students-$T$ test. The input for this test is all the data with respect to both arms we want to compare, i.e., $a_1(t)$ and $a_2(t)$, in the data gathered so far ($D$). The threshold p-value with respect to when we say the difference is not insignificant, $\alpha$, can be set to a low value. In the experiments we used $\alpha = 0.01$.

---

[1] See e.g. [17] for a reference implementation of PPrune.

Finally, one might argue that a human decision maker would not always be able to answer all the queries the algorithm might require. For example, the algorithm might simply generate too many questions for the user to answer while it continues to learn about the environment, and when the user has answered a question, the subsequent questions might already be deprecated. To simulate this effect we introduce a simple cool-down of the number of arm-pulls that will be taken before a user is again able to answer another query (line 14 and 16). We hypothesise that while more queries are desirable, GUTS will still be able to achieve sub-linear regret. In the most positive scenario, a higher cool-down could even lead to the algorithm being able to ask more informative queries because it learned more about the environment before it asks the queries.

## 4   Experiments

To test the performance of GUTS in terms of user regret (Definition 4) and the number of queries posed to the user, and to test the effects of different parameter settings, we perform experiments on two types of problems. In Sect. 4.1, we use a 5-arm MOMAB that requires learning non-linear preferences to show that GUTS is indeed able to handle such non-linear preferences. In Sect. 4.2 we use a 5-arm real-world inspired benchmark with Poisson-distributed rewards to show that GUTS can handle different reward distributions as well as utility functions. In Sect. 4.3, random MOMAB instances are used to perform more extensive experiments. The code for all experiments can be found at https://github.com/Svalorzen/morl_guts.

We compare GUTS to single-objective Thompson sampling provided with the ground truth utility functions of the user. We refer to this baseline as cheat-TS. While this is an unfair comparison (putting GUTS at a disadvantage), as our setting does not actually allow algorithms to know the ground truth utility function, it does provide insight into how much utility is lost due to having to estimate the utility function via pairwise comparison queries. Furthermore, in Sect. 4.1 we compare GUTS to ITS [19].

We note that cheat-TS does not have the same regret guarantees as Thompson Sampling for e.g., single-objective Gaussian-distributed rewards. This is because the ground-truth utility function is non-linear. So even though the reward vectors may be distributed as multi-variate Gaussian, and therefore the posteriors over mean reward vectors are also multi-variate Gaussians, the posterior distribution over utilities that result from these means is not Gaussian. As the theoretical regret bounds for Thompson sampling rely on the underlying reward distributions of the arms, cheat-TS does not inherit these regret bounds.

In Sects. 4.1 and 4.3 the MOMABs have multi-variate Gaussian reward distributions, as common in many real-world environments. This stands in contrast to earlier work, i.e., [19], that uses independent Bernoulli-distributions for each objective. For multi-variate Gaussian distributions, given the rewards $r_1, \ldots r_n$, the posterior distribution is also a multi-variate Gaussian [4]:

$$\mu \sim \mathcal{N}(\frac{\mu_0 \kappa_0 + n\bar{r}}{n + \kappa_0}(\Lambda(n + \kappa_0))^{-1}),$$

where $\Lambda$ is Wishart-distributed:

$$\Lambda \sim \mathcal{W}i(n_0 + \frac{n}{2}, \Lambda_0 + \frac{1}{2}[\bar{\Sigma} + \frac{\kappa_0}{\kappa_0 + n}(\bar{r} - \mu_0)(\bar{r} - \mu_0)^T])$$

In Sect. 4.2, we use a real-world inspired MOMAB that has independent Poisson distributions for each objective, to test whether GUTS performs well for differently distributed reward vectors.

For the GPs we use the common zero-mean prior, $m(\mu) = 0$, and squared exponential kernel [15]. When significance testing is used, we use $\alpha = 0.01$ as the significance threshold.

## 4.1   A 5-Arm MOMAB

In this subsection, we employ a 5-arm MOMAB (depicted in Fig. 1), with the following ground truth mean vectors: $(0, 0.8)$, $(0.1, 0.9)$, $(0.4, 0.4)$, $(0.8, 0.0)$, and $(0.9, 0.1)$. Note that $(0, 0.8)$ and $(0.8, 0.0)$ are Pareto-dominated. We employ a (monotonically increasing) utility function:

Fig. 1. A 5-arm MOMAB

$$u_5(\mu) = 6.25 \ \max(\mu_0, 0) \ \max(\mu_1, 0),$$

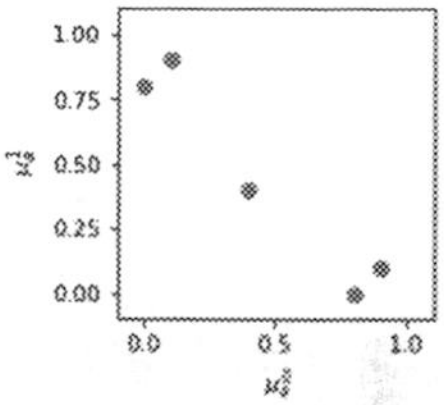

leading to the arm with mean vector $(0.4, 0.4)$ being optimal with utility 1. Each reward distribution is a multi-variate Gaussian with correlations 0 and in-objective variance 0.005.

To find the optimal arm, a MOMAB-learning algorithm must be able to handle the non-linearity of $u_5(\mu)$. In order to test whether GUTS can indeed handle this, we run GUTS on the MOMAB of Fig. 1, and compare it to ITS [19], and Thompson Sampling provided with the ground truth utility function in Fig. 2 (top). The results confirm that GUTS can handle non-linear preferences and has a highly sub-linear regret curve. In contrast, ITS is limited to the space of linear utility functions and does not manage to attain sub-linear regret.

As previously stated, a user may not have time (or is not willing) to answer all questions posed by GUTS, as the user will need time to answer the queries, and may be preoccupied with other tasks as well. Therefore, we introduced the count-down parameter in the GUTS algorithm to simulate the user answering a smaller portion of the questions. We compare different settings for the count-down parameter (Fig. 2, top) we observe that, as expected, no count-down (i.e., $cd = 0$), performs best in terms of regret, the regret closely approximates the regret of Thompson sampling supplied with the ground truth (Fig. 2, top left). The number of queries to the user can be reduced by increasing count-down (Fig. 2, top right). For example, at count-down 3, the user answers to a query once in at most 4 arm-pulls. We observe that while this comes at the cost of some regret, the learning curves remain highly sub-linear, and GUTS is still able to learn the right preferences, albeit more slowly.

We further test the difference in sampling strategy between ITS and GUTS. Specifically, we test how GUTS performance is affected if it does not sample

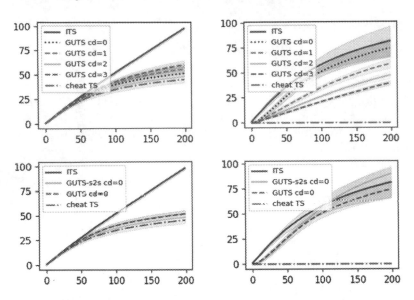

**Fig. 2.** 5-arm MOMAB results for ITS, ground-truth (cheat) TS, and GUTS with different count-down settings and significance testing enabled (top), and with count-down 0 comparing the GUTS sampling strategy versus that of ITS, i.e., sampling two sets of means (s2s) (bottom): regret (left) and the corresponding number of queries posed (right), as a function of arm-pulls. The error regions indicate $\pm 1\sigma$.

the mean rewards once, but twice, as in ITS. The results in Fig. 2 bottom-left, indicate that the regret is not affected by this difference. This is as expected, as the shape of the utility function (the user) and the values of the mean rewards (the environment) are independent. However, Fig. 2 bottom-right shows that the number of queries is significantly improved by using our sampling strategy. We therefore conclude that our sampling strategy is superior to that of ITS.

We tested the effect of (disabling) significance testing. We found no significant effect on regret in this simple problem, and a small effect on the number of queries in favour of the doing significance testing. We did observe a slight increase in the number of queries when significance testing was disabled. We show the effect on more complex MOMABs in the next subsection.

In summary, we conclude that GUTS can effectively approximate the utility function for the purpose of selecting the optimal arm in this problem. Furthermore, we conclude that sampling the means of the rewards for each arm once (the GUTS sampling strategy) is better than sampling them twice (as in ITS).

## 4.2    Organic MOMAB

To test GUTS on different reward distributions and utility functions, we adapt a real-world inspired benchmark environment [12]. This environment is motivated by a research question that stems from organic agriculture, i.e., to investigate strategies that maximize the prevalence of certain insect species on farmland to

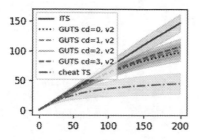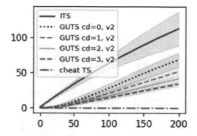

**Fig. 3.** Regret (left) and number of queries posed (right), as a function of arm-pulls, averaged over 30 trials of the MOMAB of Fig. 1, with count-downs $0, 1, 2, 3$.

predate on a pest insect species [23]. Maximising the occurrence of predatory insect species will reduce pest species and will benefit crop yield, without using artificial pesticides. This prevalence distribution follows a Poisson distribution [23]. We consider the case where there are two species of insects that need to be controlled, and therefore we aim to maximize the prevalence of two species that predate on the insect species that is to be controlled. As such, the utility function is defined as:

$$u_{\min}(\boldsymbol{\mu}) = \min(\mu_0, \mu_1)$$

The benchmark environment consists of a 5-arm MOMAB, with the same mean vectors as our first experiment (Sect. 4.1). Each reward distribution consists of independent Poisson distributions per objective. For a Poisson distribution, the conjugate Jeffreys prior is a gamma distribution [14], $\mathcal{G}\text{amma}(\alpha_0 = 0.5, \beta_0 = 0)$. Given rewards $\mathbf{r} = \langle r_1, ..., r_n \rangle$, this leads to posterior

$$\mu \sim \mathcal{G}\text{amma}(\alpha_0 + \sum_{i=1}^{n} r_i, \beta_0 + n).$$

As $\beta_0 = 0$, this posterior needs to be initialized one time for it be proper.

For this experiment, not only does GUTS need to cope with the nonlinearity of $u_{\min}(\boldsymbol{\mu})$, it also needs to learn on a more complicated problem due to each reward following a Poison distribution (where the variance equals the mean). Despite that, the results on Fig. 3 show that GUTS has a sub-linear regret curve, regardless of the cooldown used. Moreover, the number of queries required is lower than ITS, even without any cooldown.

We thus conclude that GUTS can cope with different reward distributions, as we experimented with both Gaussian and Poisson distributions.

### 4.3   Random MOMABs

Random MOMABs are generated using the number of objectives $d$, and the number of arms, $|\mathcal{A}|$, as parameters, following the procedure of [19]: $|\mathcal{A}|$ samples $\boldsymbol{\mu}'_a$ are drawn from a $d$-dimensional Gaussian distribution $\mathcal{N}(\boldsymbol{\mu}'_a | \boldsymbol{\mu}_{rand}, \Sigma_{rand})$,

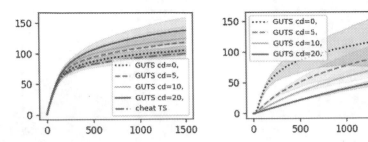

**Fig. 4.** Regret (left) and number of queries posed to the user (right), as a function of arm-pulls, averaged over 30 random, 2-objective, 20-arm MOMAB instances with random utility functions or order 3, with noise $\varepsilon = 0.1$ on the comparisons of the user (about 2% of the optimal utility), for GUTS with different count-downs.

where $\boldsymbol{\mu}_{rand} = 1$ (vector of ones), and $\Sigma_{rand}$ is a diagonal matrix with $\sigma^2_{rand} = (\frac{1}{2})^2$ for each element on the diagonal; this set is normalised such that all $\boldsymbol{\mu}_a$ fall into the $d$-dimensional unit hypercube, $[0,1]^d$. Each arm has an associated multi-variate Gaussian reward distribution, with the aforementioned true mean vectors, $\boldsymbol{\mu}_a$, and randomly drawn covariance matrices with covariances ranging from perfectly negatively correlated to perfectly positively correlated, with an average variance magnitude of $\sigma^2_{P,a} = 0.05$ unless otherwise specified.

To test whether GUTS can learn to identify the optimal arm, under different settings of the count-down parameter (i.e., when a user cannot answer all the questions otherwise generated by the algorithm), we run GUTS on random 20-arm MOMABs with randomly generated polynomial utility functions of order 3. In Fig. 4, we observe that the regret curves of GUTS (on the left) are all sub-linear, and only a small amount worse than Thompson Sampling provided with the ground-truth utility function (cheat-TS); at 1500 arm-pulls, GUTS with count-down 0 has only 5.8% more cumulative regret than cheat-TS. Furthermore, we observe that the number of questions for count-down behaves as we expect: first queries are not yet significant; then the queries become significant leading to a higher number of queries; but the number of queries quickly goes down as the approximation of $u(\boldsymbol{\mu})$ becomes better. We thus conclude that GUTS can adequately approximate the utility function in order to identify the optimal arm.

For count-down 5, GUTS has 18.6% more regret than cheat-TS at 500 arm pulls, for count-down 10, 25.3% and, for count-down 20, 38.8%. However, count-down 5 requires only 78.9% of the queries with respect to count-down 0, and count-downs 10 and 20 only 64.1% resp. 45.4%. I.e., the increase in regret is less steep than the reduction in the number of queries. We hypothesise that this is because GUTS continues learning about the arms before the next question is asked, yielding more information per query. We thus conclude that GUTS remains an effective algorithm when the user cannot answer a query every arm-pull.

To test whether GUTS can learn effectively in random MOMABs with an increasing number of objectives, and the effect of (disabling) significance

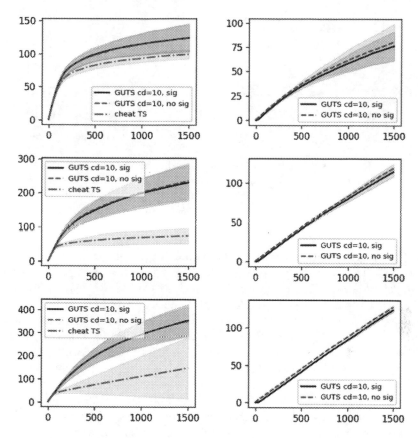

**Fig. 5.** Regret (left) and number of queries posed (right) as a function of arm-pulls, averaged over 30 random, 20-arm MOMAB instances with 2 (top), 4 (middle) and 6 (bottom) objectives, with random utility functions or order 3, with noise $\varepsilon = 0.1$ on the comparisons of the user ($\sim 2\%$ of the optimal utility), for GUTS with count-down 10, and with or without significance testing (s1/s0).

testing, we run GUTS with cooldown 10 on 2-, 4-, and 6-objective 20-arm random MOMABs with randomly generated polynomial utility functions of order 3 (Fig. 5). As can be seen, the curves remain sub-linear, but less so for more objectives, and the difference in regret with cheat-TS increases. This is because GUTS suffers from the curse of dimensionality, as it has to estimate higher-dimension utility functions, while cheat-TS does not (it is provided with the ground-truth utility function, so it only has to optimise a scalar utility). This was to be expected; cheat-TS has an unfair advantage, and is only included as a theoretical reference. We observe that for 6 objectives, there is a lot of variance in the performance of cheat-TS. This is due to several runs where the utility function is steeper (has a higher gradient) at the mean of a suboptimal arm than that of the optimal arm, so that smaller sampling differences in the mean reward vector get amplified by the utility function. This is a clear example of the

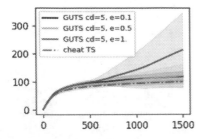 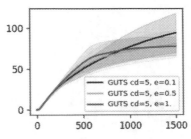

**Fig. 6.** (left) Regret as a function of arm-pulls (right) number of queries posed to the user as a function of arm-pulls, averaged over 30 random, 2-objective, 20-arm MOMABs with random utility functions or order 3, with noise $\varepsilon = 0.1$, 0.5 and 1.0 (resp. 2%, 10%, 20% of the optimal utility) on the comparisons, for GUTS with count-down 5.

theoretical regret bounds for TS for Gaussian reward distributions not applying to cheat-TS. In other words, the MOMAB with non-linear utility function over multi-variate Gaussian reward distributions can be a considerably harder problem than a single-objective MAB with Gaussian reward distributions.

Finally, to check how GUTS handles different levels of noise in the comparisons of the user, we run GUTS on 20, 2-objective, 20-arm, random MOMAB instances with randomly generated polynomial utility functions of order 3 (Fig. 6). These randomly generated utility functions have utilities ranging from about 1 for the worst arm to 5 for the best arm. We use 2%, 10%, 20% of this utility as noise, i.e., $\varepsilon = 0.1$, 0.5 and 1.0. For $\varepsilon = 0.1$ and 0.5, we observe very similar regret curves, and no significant difference in the number of queries. For $\varepsilon = 1.0$ we observe the regret and number of query curves suddenly going up; this is because in many of the runs GUTS suddenly samples a non-optimal arm more often after a series of erroneous comparisons. In one of these runs we observe that GUTS does converge back to the optimal arm after a while before the run ends. We thus conclude that GUTS is robust against reasonable levels of noise, but becomes unstable when the noise in the comparisons is too large.

## 5    Related Work

Several papers exist that study *offline* multi-objective reinforcement learning in MOMABs [3,8,28]. These papers focus on producing a coverage set in a separate learning phase, without user interaction. After this learning phase, the interaction with the user needs to be done in a separate selection phase. In contrast, we study an online interactive setting, and aim to minimise online regret by estimating the user's utility function during learning.

*Relative bandits* [27,30] are a related setting to online interactive learning for MOMABs. Similar to interactive online MORL for MOMABs, relative bandits assume a hidden utility function which can be queried with pairwise comparisons. Contrary our setting, these rewards (i.e., reward vectors) cannot be observed, and the comparisons are made regarding *single* arm pulls, rather than the aggregate information over all pulls of a given arm as in GUTS (and ITS).

# 6    Discussion

In this paper we proposed Gaussian-process Utility Thompson Sampling (GUTS) for interactive online multi-objective reinforcement learning in MOMABs. Contrary to earlier methods, GUTS can handle non-linear utility functions, which is an essential feature for MORL algorithms. We have shown empirically that GUTS can effectively approximate non-linear utility functions for the purpose of selecting the optimal arm, leading to only little extra regret if a query can be posed to the user at every arm pull w.r.t. Thompson Sampling when provided a priori with the ground truth utility function. We show the effects of limiting the number of user queries. In this case—even if the number of queries is strongly reduced—the regret remains sub-linear, albeit higher than when queries can be posed at every time-step. Therefore, we conclude that GUTS is robust against less questions being answered than generated. Furthermore, our experiments indicate that GUTS is robust against reasonable levels of noise in the user comparisons.

We note that all conclusions in this paper are based on empirical evaluations rather than theoretical bounds. This is due to the usage of Gaussian Processes (GPs) which makes a theoretical analysis hard. However, we believe it essential to use GPs, because GPs can fit any utility function without assuming a model-space. A limitation of our regret metric is that it assumes the utility is derived from the expected values ($\mu$) at each timestep. We intend to provide new regret metrics in future work. We also aim to test GUTS on real-world problems, and integrate the algorithm with an effective query-answering interface for decision makers. Furthermore, we aim to investigate the potential of asking more complex queries than just pairwise comparisons, such as ranking and clustering [29].

**Acknowledgment.** This research received funding from the Flemish Government (AI Research Program). Pieter Libin and Eugenio Bargiacchi were supported by a PhD grant of the FWO (Fonds Wetenschappelijk Onderzoek – Vlaanderen).

# References

1. Agrawal, S., Goyal, N.: Analysis of Thompson sampling for the multi-armed bandit problem. In: COLT, pp. 39–1 (2012)
2. Auer, P., Cesa-Bianchi, N., Fischer, P.: Finite-time analysis of the multiarmed bandit problem. Mach. Learn. **47**(2–3), 235–256 (2002)
3. Auer, P., Chiang, C.K., Ortner, R., Drugan, M.M.: Pareto front identification from stochastic bandit feedback. In: AISTATS, pp. 939–947 (2016)
4. Bishop, C.M.: Pattern Recognition and Machine Learning (2006)
5. Brochu, E., Cora, V.M., De Freitas, N.: A tutorial on Bayesian optimization of expensive cost functions, with application to active user modeling and hierarchical reinforcement learning. arXiv:1012.2599 (2010)

6. Chu, W., Ghahramani, Z.: Preference learning with Gaussian processes. In: ICML, pp. 137–144 (2005)
7. Drugan, M.M., Nowé, A.: Designing multi-objective multi-armed bandits algorithms: a study. In: IJCNN, pp. 1–8. IEEE (2013)
8. Drugan, M.M.: PAC models in stochastic multi-objective multi-armed bandits. In: GEC, pp. 409–416 (2017)
9. Forgas, J.P.: Mood and judgment: the affect infusion model (aim). Psychol. Bull. 117(1), 39 (1995)
10. Hotelling, H.: The generalization of Student's ratio. In: Annals of Mathematical Statistics ii, pp. 360–378 (1931)
11. Lampinen, J.: Gaussian processes with monotonicity constraint for big data (2014)
12. Libin, P., Verstraeten, T., Roijers, D.M., Wang, W., Theys, K., Nowé, A.: Bayesian anytime m-top exploration. In: ICTAI, pp. 1422–1428 (2019)
13. Libin, P.J., et al.: Bayesian best-arm identification for selecting influenza mitigation strategies. In: ECML-PKDD, pp. 456–471 (2018)
14. Lunn, D., Jackson, C., Best, N., Thomas, A., Spiegelhalter, D.: The BUGS Book: A Practical Introduction to Bayesian Analysis. CRC Press, Boca Raton (2012)
15. Rasmussen, C.E.: Gaussian processes for machine learning (2006)
16. Riihimäki, J., Vehtari, A.: Gaussian processes with monotonicity information. In: Proceedings of the Thirteenth International Conference on Artificial Intelligence and Statistics, pp. 645–652 (2010)
17. Roijers, D.M.: Multi-Objective Decision-Theoretic Planning. Ph.D. thesis, University of Amsterdam (2016)
18. Roijers, D.M., Vamplew, P., Whiteson, S., Dazeley, R.: A survey of multi-objective sequential decision-making. JAIR 48, 67–113 (2013)
19. Roijers, D.M., Zintgraf, L.M., Nowé, A.: Interactive Thompson sampling for multi-objective multi-armed bandits. In: Algorithmic Decision Theory, pp. 18–34 (2017)
20. Roijers, D.M., Zintgraf, L.M., Libin, P., Nowé, A.: Interactive multi-objective reinforcement learning in multi-armed bandits for any utility function. In: ALA workshop at FAIM (2018)
21. Siegel, S.: Nonparametric statistics for the behavioral sciences (1956)
22. Sirakaya, E., Petrick, J., Choi, H.S.: The role of mood on tourism product evaluations. Ann. Tourism Res. 31(3), 517–539 (2004)
23. Soulsby, R.L., Thomas, J.A.: Insect population curves: modelling and application to butterfly transect data. Methods Ecol. Evol. 3(5), 832–841 (2012)
24. Tesauro, G.: Connectionist learning of expert preferences by comparison training. NeurIPS 1, 99–106 (1988)
25. Thompson, W.R.: On the likelihood that one unknown probability exceeds another in view of the evidence of two samples. Biometrika 25(3/4), 285–294 (1933)
26. Ustyuzhaninov, I., Kazlauskaite, I., Ek, C.H., Campbell, N.D.: Monotonic Gaussian process flow. arXiv preprint arXiv:1905.12930 (2019)
27. Wu, H., Liu, X.: Double thompson sampling for dueling bandits. In: NeurIPS, pp. 649–657 (2016)
28. Yahyaa, S.Q., Drugan, M.M., Manderick, B.: Thompson sampling in the adaptive linear scalarized multi objective multi armed bandit. In: ICAART, pp. 55–65 (2015)
29. Zintgraf, L.M., Roijers, D.M., Linders, S., Jonker, C.M., Nowé, A.: Ordered preference elicitation strategies for supporting multi-objective decision making. In: AAMAS, pp. 1477–1485 (2018)
30. Zoghi, M., Whiteson, S., Munos, R., De Rijke, M.: Relative upper confidence bound for the k-armed dueling bandit problem. In: ICML, pp. 10–18 (2014)

# Deep Gaussian Processes Using Expectation Propagation and Monte Carlo Methods

Gonzalo Hernández-Muñoz, Carlos Villacampa-Calvo,
and Daniel Hernández-Lobato[✉]

Computer Science Department, Universidad Autónoma de Madrid, Francisco Tomás
Y Valiente 11, 28049 Madrid, Spain
{gonzalo.hernandez,carlos.villacampa,daniel.hernandez}@uam.es

**Abstract.** Deep Gaussian processes (DGPs) are the natural extension
of Gaussian processes (GPs) to a multi-layer architecture. DGPs are powerful probabilistic models that have shown better results than standard
GPs in terms of generalization performance and prediction uncertainty
estimation. Nevertheless, exact inference in DGPs is intractable, making these models hard to train. For this task, current approaches in the
literature use approximate inference techniques such as variational inference or approximate expectation propagation. In this work, we present
a new method for inference in DGPs using an approximate inference
technique based on Monte Carlo methods and the expectation propagation algorithm. Our experiments show that our method scales well to
large datasets and that its performance is comparable or better than
other state of the art methods. Furthermore, our training method leads
to interesting properties in the predictive distribution of the DGP. More
precisely, it is able to capture output noise that is dependent on the
input and it can generate multimodal predictive distributions. These two
properties, which are not shared by other state-of-the-art approximate
inference methods for DGPs, are analyzed in detail in our experiments.

**Keywords:** Deep Gaussian processes · Expectation propagation ·
Monte Carlo methods

## 1 Introduction

Gaussian processes (GPs) are non-parametric models with nice features [18].
They have few parameters to fix (learned via maximum marginal-likelihood) and
are often less prone to over-fitting; it is possible to encode high level properties of
the problem in the covariance function; and critically, GPs provide uncertainty
associated to the predictions made. These usability and flexibility features make
them appealing and powerful models in several application areas [15,17,23].

**Electronic supplementary material** The online version of this chapter (https://
doi.org/10.1007/978-3-030-67664-3_29) contains supplementary material, which is
available to authorized users.

F. Hutter et al. (Eds.): ECML PKDD 2020, LNAI 12459, pp. 479–494, 2021.
https://doi.org/10.1007/978-3-030-67664-3_29

GPs are, however, limited by the kernel or covariance function. More precisely, the global smoothness implied by popular kernels such as the squared exponential, can reduce the performance in problems that, *e.g.*, present discontinuities in the data. Some authors proposed to model non-smooth data by carefully combining kernels to create complex covariance functions [8,26] or by using a deep neural network to encode the covariance function [3,27]. These approaches are limited to simple base kernels and/or run a high risk of over-fitting.

Deep Gaussian process (DGP) can overcome the limitations of standard GPs while maintaining the properties that make them attractive [2,4,19]. DGPs are a hierarchical composition of GPs, which makes them more flexible models. Thus, they reduce the assumptions made about the underlying process that generated the data. They can also repair the loss of flexibility produced by sparse approximations in standard GPs, simply by increasing the number of layers in the model [2]. Finally, DGPs can output better calibrated uncertainty estimates than standard GPs, leading to more accurate predictive distributions.

A difficulty of DGPs is, however, that inference becomes more challenging. Nevertheless, there are several approximate techniques for this task [2,4,5,19]. Most of them rely on variational inference (VI) and only [2] considers expectation propagation as an alternative. Here, we show that the approach described in [2] makes strong assumptions that are not appropriate to obtain an accurate approximation. In particular, in [2] the output distribution at each hidden layer of the DGP (given by a GP predictive distribution in which the inputs are random) is approximated using a Gaussian distribution. This distribution can, in practice, be very different from a Gaussian, and may include features such as multiple modes, skewness or heavy tails. To overcome this limitation and increase the flexibility of the approximate method we consider Monte Carlo samples to approximate the distribution described. This is an efficient and practical alternative to the inaccurate Gaussian approximation of [2].

Extensive experiments show the beneficial properties of the proposed approach, which not only improves over the method of [2], but is also able to capture complex properties of the data such as output noise that is input dependent or target values that have been generated from a multi modal distribution (*e.g.*, several processes explain the observed data). Critically, we show that these properties are not shared by inference methods for DGPs based on VI, which is currently the most commonly used method to train DGPs.

## 2   Gaussian Process Regression

Assume a training set of $N$ pairs of $d$-dimensional inputs $\{\mathbf{x}_i\}_{i=1}^N$ and their corresponding output values $\{y_i\}_{i=1}^N$, that include some random additive Gaussian noise, *i.e.*, $y_i = f(\mathbf{x}_i) + \epsilon_i$. The GP regression setting follows a Bayesian approach for inferring $f(\cdot)$. For this, we set a factorizing Gaussian likelihood function $p(\mathbf{y}|\mathbf{f})$ and the GP prior $p(\mathbf{f})$, where $\mathbf{f} = (f(\mathbf{x}_1), \dots, f(\mathbf{x}_N))$ is the vector of function values [18]. Importantly, the Gaussian prior depends on a covariance function $k : \mathbb{R}^d \times \mathbb{R}^d \to \mathbb{R}$ that makes strong assumptions about $f$.

We are interested in computing the distribution of the output values $\mathbf{f}_\star$ at an input test location $\mathbf{x}_\star$. Although these computations are tractable, they scale cubically with the number of training points. Therefore, in practice, approximate methods must be used. For this, a common approach is to introduce a set of $M$ ($D$-dimensional) pseudo-inputs, also called inducing points $\mathbf{Z} = \{\mathbf{z}_m\}_{m=1}^M$, with their corresponding outputs $\mathbf{u} = f(\mathbf{Z})$ [22,24]. Given $\mathbf{u}$ the distribution of $\mathbf{f}$, i.e., $p(\mathbf{f}|\mathbf{u})$ is a conditional multi-variate Gaussian distribution [18]. For simplicity, we have omitted the dependence on $\mathbf{Z}$. For inference, the posterior of $\mathbf{f}$ and $\mathbf{u}$, i.e., $p(\mathbf{f}, \mathbf{u}|\mathbf{y}) \propto p(\mathbf{y}|\mathbf{f})p(\mathbf{f}|\mathbf{u})p(\mathbf{u})$, is approximated, where $p(\mathbf{u})$ is the GP prior for $\mathbf{u}$. Popular choices for the approximate posterior distribution constrain this distribution to be $q(\mathbf{f}, \mathbf{u}) = p(\mathbf{f}|\mathbf{u})q(\mathbf{u})$, where $q(\mathbf{u})$ is a tunable multi-variate Gaussian with mean $\boldsymbol{\mu}$ and variance $\boldsymbol{\Sigma}$ while $p(\mathbf{f}|\mathbf{u})$ is fixed.

Two popular techniques to find the parameters of $q(\mathbf{u})$ are expectation propagation (EP) and variational inference (VI) [2,19]. VI minimizes the Kullback-Leibler (KL) divergence between $q$ and the exact posterior. In contrast, EP minimizes the KL divergence between the exact posterior and $q$ in approximate way. Both EP and VI estimate of the marginal likelihood, and all hyper-parameters (inducing points locations, length-scales, etc.) can be found by simply maximizing that quantity. An approximate predictive distribution for $f^\star$ at $\mathbf{x}_\star$ is:

$$p(f^\star|\mathbf{x}^\star, \mathbf{y}) \approx \int p(f^\star|\mathbf{u})q(\mathbf{u})d\mathbf{u} = \mathcal{N}(f^\star|m_\star, v_\star), \tag{1}$$

where $m_\star = \mathbf{k}_{\mathbf{x}^\star,\mathbf{u}}^{\mathrm{T}}\mathbf{K}_{\mathbf{u},\mathbf{u}}^{-1}\boldsymbol{\mu}$, and $v_\star = k(\mathbf{x}^\star, \mathbf{x}^\star) + \mathbf{k}_{\mathbf{x}^\star,\mathbf{u}}^{\mathrm{T}}\mathbf{K}_{\mathbf{u},\mathbf{u}}^{-1}(\boldsymbol{\Sigma} - \mathbf{K}_{\mathbf{u},\mathbf{u}})$ $\mathbf{K}_{\mathbf{u},\mathbf{u}}^{-1}\mathbf{k}_{\mathbf{u},\mathbf{x}^\star}$. Here, $\mathbf{k}_{\mathbf{x}^\star,\mathbf{u}}^{\mathrm{T}}$ are the covariances between the test point and the inducing points and $\mathbf{K}_{\mathbf{u},\mathbf{u}}$ are the covariances among the inducing points $\mathbf{Z}$. This vector and matrix can be obtained in terms of the GP covariance function $k(\cdot,\cdot)$ (see [18] for further details). Therefore, (1) has cost $\mathcal{O}(NM^2)$, better than $\mathcal{O}(N^3)$ if $M \ll N$.

## 3    Deep Gaussian Processes

Deep Gaussian Processes (DGPs) [5] are defined as a composition of GPs. These models have multiple layers of hidden variables connected in a feed-forward architecture [13]. In a DGP the nodes represent GPs in which the output of each layer is used as the input to the next one. See Fig. 1 (left) for an illustrative example. For efficiency, a sparse GP approximation, such as the one described in Sect. 2, is used for each GP in the network. The output for a layer $l$ is represented as a $N \times D_l$ matrix $\mathbf{H}^l$. For simplicity we will ignore dimensions in the notation and write $\mathbf{H}^l$ as $\mathbf{h}^l$. The same applies for the inducing outputs $\mathbf{U}^l$. We write $\mathbf{u}^l$. The joint distribution of $\mathbf{y}$ and $\{\mathbf{u}^l, \mathbf{h}^l\}_{l=1}^L$, is:

$$p(\mathbf{y}, \{\mathbf{u}, \mathbf{h}\}_{l=1}^L) = p(\mathbf{y}|\mathbf{h}^L) \prod_{l=1}^L p(\mathbf{h}^l|\mathbf{h}^{l-1}, \mathbf{u}^l)p(\mathbf{u}^l), \tag{2}$$

where $\mathbf{h}^0 = \mathbf{X}$; $p(\mathbf{h}^l|\mathbf{h}^{l-1}, \mathbf{u}^l)$ is a conditional Gaussian distribution, as the one described in Sect. 2; and each $p(\mathbf{u}^l)$ is the GP prior for the inducing outputs of

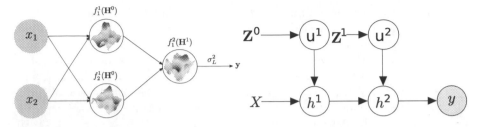

**Fig. 1.** (left) DGP example for a model with two layers L = 2 and two nodes in the hidden layer. The 2-dimensional training points are fed to the first layer (hidden). The output at each layer is a non-linear GP function contaminated with Gaussian noise. (right) Graphical model for the Deep GP with L – 2. The output of the hidden layers not only depends on the output of the previous layer, but also on the inducing values.

layer $l$. In general, we assume independence among GPs and all these distributions factorize across the GPs of each layer. We have omitted the dependence on noise level term between layers and the parameters of the kernel function. Figure 1 (right) shows the graphical model for a DGP with two layers ($L = 2$).

In (2) one can identify two factors [19]. Namely, the likelihood $p(\mathbf{y}|\mathbf{h}^L) = \prod_{i=1}^N p(y_i|\mathbf{h}_i^L)$ and the DGP prior $\prod_{l=1}^L p(\mathbf{h}^l|\mathbf{h}^{l-1},\mathbf{u}^l)p(\mathbf{u}^l)$. Exact inference in this model is intractable and approximate techniques have to be used in practice. In particular, computing the output distribution in each layer $p(\mathbf{h}^l|\mathbf{u}^l,\mathbf{h}^{l-1})$ requires the marginalization of the input variables $\mathbf{h}^{l-1}$, which are random (except at the first layer). This is intractable as a consequence of the non-linear GP covariance function. Ideally, we would like to approximate the posterior distribution of all the latent variables of the model. That is, $p(\{\mathbf{u}^l,\mathbf{h}^l\}_{l=1}^L|\mathbf{y})$. The approximate distribution can then be used for making predictions on new data points, as in Sect. 2. Another quantity of interest is the marginal likelihood. That is, the denominator in Bayes theorem, $p(\mathbf{y})$, where the latent variables have been marginalized. This quantity can be used to find good values of the model hyper-parameters $\boldsymbol{\alpha} = \{\mathbf{Z}^{l-1},\theta^l,\sigma_l^2\}_{l=1}^L$, including the inducing points locations $\mathbf{Z}^{l-1}$, the parameters of the covariance function of each GP, $\theta^l$, and the noise variance at the output of each layer, $\sigma_l^2$. To simplify the notation, we consider that the layer dimensionality is 1, but the same derivations apply in general.

## 4     Expectation Propagation and Monte Carlo Methods

We describe how to approximate $p(\{\mathbf{u}^l,\mathbf{h}^l\}_{l=1}^L|\mathbf{y})$ in the model of the previous section. For this, we consider a similar approximation to the one suggested in [19]. This approximation factorizes within layers, but not across layers. However, for simplicity this is omitted in the notation to improve clarity. Specifically,

$$q(\{\mathbf{u}^l,\mathbf{h}^l\}_{l=1}^L) = \prod_{l=1}^L p(\mathbf{h}^l|\mathbf{h}^{l-1},\mathbf{u}^l)q(\mathbf{u}^l)\,, \tag{3}$$

where each $p(\mathbf{h}^l|\mathbf{h}^{l-1},\mathbf{u}^l)$ is fixed and equal to the corresponding factors in the exact posterior (2) and $q(\mathbf{u}^l)$ is a product of multi-variate Gaussian distributions (one Gaussian per each dimension of layer $l$). Again, assuming that the layer dimensionality is 1 for simplicity (otherwise the distributions simply factorize across the layer's dimension):

$$q(\mathbf{u}^l) = \mathcal{N}(\mathbf{u}^l | \boldsymbol{\mu}^l, \boldsymbol{\Sigma}^l) \, , \qquad (4)$$

where $\boldsymbol{\mu}^l$ and $\boldsymbol{\Sigma}$ are tunable parameters. To adjust the parameters of $q$, we propose a new method. Namely, DGP-AEPMCM, which is based on the work of [2], and approximates the inducing points posterior using a variant of the expectation propagation (EP) combined with an stochastic approximation [16].

EP approximates each likelihood factor $p(y_i | \mathbf{h}_i^L)$ using a multi-variate Gaussian factor on $\mathbf{u}^l$, for $l = 1, \ldots, L$. This factor factorizes across layers and layers dimensions. Because storing each of these individual factors is inefficient in terms of memory, in [2] it is proposed to use a variant of EP called stochastic EP (SEP), which only stores the product of all the approximate factors. This is equivalent to having $N$ factors that are equal, replicated $N$ times, one per each likelihood factor. We denote these factors as $g(\mathbf{u}^l)$.

The posterior approximation is then obtained by replacing the exact likelihood factors by the approximate ones, i.e.,

$$q(\{\mathbf{h}^l, \mathbf{u}^l\}_{l=1}^L) \propto \prod_{l=1}^L p(\mathbf{h}^l | \mathbf{u}^l, \mathbf{h}^{l-1}) p(\mathbf{u}^l) g(\mathbf{u}^l)^N \, . \qquad (5)$$

Because $p(\mathbf{u}^l)$ and $g(\mathbf{u}^l)^N$ are Gaussian factors, their product is also Gaussian, and hence the posterior approximation of $\{\mathbf{u}^l\}_{l=1}^L$ is Gaussian. The particular parameters of the global factors $g(\mathbf{u}^l)$ are found by optimizing the EP energy function instead of doing the standard EP updates [2]. The reason is that this allows to use standard optimization techniques. In general, optimizing the full EP energy function, as an alternative to doing the EP updates, requires a double-loop algorithm [12]. With the tied factors constraint this is no longer the case. The SEP approximation to the marginal likelihood and the EP energy is:

$$\ln p(\mathbf{y} | \boldsymbol{\alpha}, \mathbf{X}) \approx \mathcal{F}(\boldsymbol{\alpha}, \{\theta^l\}_{l=1}^L) = \sum_{i=1}^N \log \mathcal{Z}_i +$$
$$\sum_{l=1}^L \left[ (1 - N) \Phi(\theta_q^l) - \Phi(\theta_{\text{prior}}^l) + N \Phi(\theta^{\setminus l}) \right] \, , \qquad (6)$$

with

$$\mathcal{Z}_i = \int p(y_i | \mathbf{h}_i^L) \prod_{l=1}^L p(\mathbf{h}_i^l | \mathbf{h}_i^{l-1}, \mathbf{u}^l) q^{\setminus l}(\mathbf{u}^l) d\mathbf{u}^l \, d\mathbf{h}_i^l \, , \qquad (7)$$

where the integrals in (7) involve all the latent variables $\{\mathbf{u}^l, \mathbf{h}_i^l\}_{l=1}^L$; $\theta_q^l, \theta^{\setminus l}, \theta_{\text{prior}}^l$ are the natural parameters of the corresponding distribution: either the approximate posterior $q(\mathbf{u}^l)$, the cavity $q^{\setminus l}(\mathbf{u}^l)$ (this is the approximate posterior where the influence of a single training point has been eliminated) or the prior $p(\mathbf{u}^l)$, respectively; and $\Phi(\cdot)$ is the log-normalizer of the corresponding Gaussian distribution. Note that (6) includes a sum across the data. This objective is hence suitable for stochastic optimization using mini-batches (to scale to large datasets).

The optimization of (6) does not yield the same results as optimizing the exact EP energy. Thus, this technique is an Approximate Expectation Propagation (AEP) method. The tied factor $g(\mathbf{u}^l)$ is Gaussian. Therefore, the natural

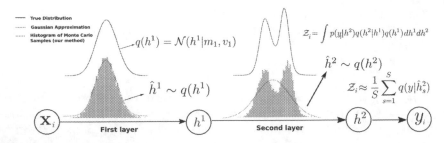

**Fig. 2.** Comparison of two different approaches to calculate $\mathcal{Z}_i$ for a DGP model with 2 layers. The DGP-AEP method proposed in [2] approximates the output distribution at each layer with a Gaussian distribution (in red). The histogram of the samples from our method used to calculate $\mathcal{Z}_i$ are shown in green. The true predictive distribution at each layer is shown in blue. The Gaussian approximation fails to capture the bimodality of the predictive distribution of the second layer. The Monte Carlo approximation (based on samples) that we propose can capture the bimodal distribution. The dependence of $h^l$ on $i$ has been omitted to improve the readability. Best seen in color. (Color figure online)

parameters of $q$ and $q^\setminus$ can be obtained from the natural parameters $\theta$ of $g(\mathbf{u}^l)$ [21]:

$$\theta_q^l = \theta_{\text{prior}}^l + N\theta^l, \qquad \theta^{\setminus l} = \theta_{\text{prior}}^l + (N-1)\theta^l. \qquad (8)$$

Unfortunately, the integral required to calculate $\tilde{\mathcal{Z}}_i$, although tractable with respect to each $\mathbf{u}^l$, cannot be calculated in closed form w.r.t to each $\mathbf{h}^l$ when $L > 1$. The work in [2] proposes an iterative process to approximate its value. More precisely, the predictive distribution of $\mathbf{h}^l$ for $l \geq 2$ at each layer is approximated using a Gaussian distribution with the same moments as the actual distribution. These moments can be computed when the input variables to a GP are Gaussian distributed for some kernels such as the squared exponential. Nevertheless, the Gaussian approximation can be a poor choice, as illustrated by Fig. 2. In this case, the predictive distribution at the second layer is bimodal (potentially due to the non-linearity of the GP and the Gaussian random inputs coming from the output of the previous layer). A Gaussian approximation cannot capture this.

To calculate $\mathcal{Z}_i$, we first marginalize the inducing points values, $\mathbf{u}^l$, in (7). The distribution of these variables is Gaussian and the same for all training points. Namely, the cavity distribution. $q^{\setminus l}(\mathbf{u}^l) = \mathcal{N}(\mathbf{u}^l|\boldsymbol{\mu}^{\setminus l}, \boldsymbol{\Sigma}^{\setminus l})$. This marginalization can then be calculated exactly:

$$\mathcal{Z}_i = \int p(y_i|\mathbf{h}_i^L)\prod_{l=1}^{L} q(\mathbf{h}_i^l|\mathbf{h}_i^{l-1})d\mathbf{h}^l = \int \mathcal{N}(y_i|\mathbf{h}_i^L, \sigma_L^2)\prod_{l=1}^{L} \mathcal{N}(\mathbf{h}_i^l|m_l, v_l)d\mathbf{h}^l \quad (9)$$

with $q(\mathbf{h}_i^l|\mathbf{h}_i^{l-1}) = \mathcal{N}(\mathbf{h}_i^l|m_l, v_l + \sigma_l^2)$ where $m_l = \mathbf{k}_{h^l,\mathbf{u}^l}\mathbf{K}_{\mathbf{u}^l,\mathbf{u}^l}^{-1}\boldsymbol{\mu}^{\setminus l}$, and $v_l = \mathbf{k}_{\mathbf{h}^l,\mathbf{h}^l} + \mathbf{k}_{h^l,\mathbf{u}^l}\mathbf{K}_{\mathbf{u}^l,\mathbf{u}^l}^{-1}(\boldsymbol{\Sigma}^{\setminus l} - \mathbf{K}_{\mathbf{u}^l,\mathbf{u}^l})\mathbf{K}_{\mathbf{u}^l,\mathbf{u}^l}^{-1}\mathbf{k}_{\mathbf{u}^l,\mathbf{h}^l}$, and we have omitted the dependence of $\mathbf{h}^l$ on $i$ for simplicity. We have assumed that the layer dimensionality is

1. For higher dimensions, the distributions factorize across dimensions. Here the covariance matrices have as sub-indices the corresponding outputs. *E.g.*, $\mathbf{k}_{\mathbf{h}^l,\mathbf{u}^l}$ is the covariance vector for the output of the layer and the inducing points values, which takes as arguments the input to the layer $\mathbf{h}_i^{l-1}$ and the inducing points locations respectively $\mathbf{Z}^{l-1}$. Nevertheless, (9) is still intractable when $L > 1$.

We propose a Monte Carlo method to estimate $\mathcal{Z}_i$. For a layer $l$, given a deterministic input $\hat{\boldsymbol{h}}_i^{l-1}$, the distribution of $\hat{\boldsymbol{h}}_i^l$ can be calculated analytically. It is a Gaussian distribution with parameters given by $m_l$ and $v_l$, as defined above. We know that $\hat{\boldsymbol{h}}_i^0 = \mathbf{x}_i$. Therefore, we can generate a sample from $q(\mathbf{h}_i^l|\hat{\mathbf{h}}_i^{l-1})$ by sampling from a standard Gaussian and then using the reparametrization trick [19]. More precisely, we can sample $\hat{h}_i^l$ from $q(\mathbf{h}_i^l|\hat{\mathbf{h}}_i^{l-1})$ for $l = 1,\ldots,L$ given that $\hat{h}_i^{l-1}$ is deterministic (either other samples or $\mathbf{x}_i$) by setting: $\epsilon \sim \mathcal{N}(0,1)$, $\hat{h}_i^l = m_l + \epsilon\sqrt{v_l}$, $\hat{\mathbf{h}}_i^0 = \mathbf{x}_i$. Thus, $\mathcal{Z}_i$, approximated with $S$ Monte Carlo samples, is given by a Gaussian mixture across the samples of the last layer:

$$\mathcal{Z}_i \approx \tilde{\mathcal{Z}}_i = S^{-1}\sum_{s=1}^S q(y_i|\hat{h}_{i,s}^L) = S^{-1}\sum_{s=1}^S \mathcal{N}(y_i|\hat{h}_{i,s}^L,\sigma_L^2),\qquad(10)$$

where $\hat{h}_{i,s}^L$ denotes the $s$-th sample taken from $q(\hat{h}_i^L|\hat{\mathbf{h}}_i^{L-1})$. Critically, as opposed to the procedure proposed in [2] this method can capture the complex distributions between the DGP layers as shown in Fig. 2. This is expected to be translated into a better estimation of $\mathcal{Z}_i$ and to provide significantly better results. In practice, we use stochastic optimization to maximize a noisy estimate of the objective in (6) obtained via the Monte Carlo approximation described [14]. Of course, $\log \tilde{\mathcal{Z}}_i$ is going to be biased, as a consequence of the non-linearity of the logarithm (*i.e.*, $\mathbb{E}[\log \tilde{\mathcal{Z}}_i] \neq \log \mathcal{Z}_i$). However, the bias can be reduced just by increasing the number of samples $S$ used in the Monte Carlo approximation.

Assume as many units in each layer as data dimensions $D$. The cost of our method is $\mathcal{O}(SN_bM^2DL)$, with $S$ the number of samples, $N_b$ the mini-batch size, $M$ the number of inducing points and $L$ the number of layers. This is the cost of the method described in [19]. However, our method is slightly slower since it requires a few more matrix operations. The cost of the AEP method in [2] is $\mathcal{O}(N_bM^2DL)$, which seems better but is in practice worse as it involves more expensive operations. See Sect. 6.3 for a comparison in terms of CPU time.

## 5   Related Work

Other works have considered a general sampling algorithm to evaluate the log-likelihood of the model parameters in the context of DGPs [25]. The difference with respect to our method is that in [25] the sampling algorithm does not set any distribution over the inducing outputs, which are fixed, and regarded as model parameters. Training is done via MAP (maximum a posterior) estimation, instead of following a more principled Bayesian approach. The experimental evaluation in [25] is not very solid (only synthetic datasets are considered), which casts doubts on the practical utility of the method. Furthermore, no comparison with other techniques for training DGPs is carried out, and no improvement is observed when the number of layers grows.

Another approach for training DGPs is considered in [4]. In that work, the covariance function $k(\mathbf{x}, \mathbf{x}')$ of a GP is approximated using random feature vectors $\phi(\cdot)$. The covariance function is simply approximated as an inner product in an extended feature space. Namely, $k(\mathbf{x}, \mathbf{x}') \approx \phi(\mathbf{x})^T \phi(\mathbf{x}')$. The result is that the DGP model can be seen as a Bayesian deep neural network (BNN). Calculating the posterior distribution of the weights in a BNN is very challenging and requires approximate techniques. A factorized Gaussian approximation (without dependencies among the random weights) is used as the approximate posterior distribution of the weights. The independence assumption is un-realistic and is expected to give sub-optimal results.

In [5], it is suggested to use variational inference to approximate the posterior distribution of the latent variables of the model. Nevertheless, to keep the computations tractable, a fully factorized variational posterior is considered in which independence is assumed between the input and output variables of each hidden layer in the DGP. This assumption is of course not fulfilled in practice and this technique has been observed to give in general sub-optimal results [19].

The approach proposed in [2] uses SEP to approximate the posterior distribution of the inducing outputs using a Gaussian distribution. This technique also faces the difficulty of computing the value of $\mathcal{Z}_i$, as in our proposed method. The approximation suggested is based on an iterative procedure in which the output distribution at each layer is approximated by a Gaussian distribution. As shown by Fig. 2 this can ignore important properties of the actual predictive distribution, such as multi-modality, skewness or heavy tails. By contrast, a Monte Carlo approximation of $\mathcal{Z}_i$, as considered in the method we propose, is expected to give better results. Furthermore, an important drawback is that computing the moments of the predictive distribution (under Gaussian random inputs) is only feasible for some covariance functions, and although tractable, the computations are very expensive, as shown in our experiments.

In [19] the approximate distribution $q$ is also given by (3), but variational inference (VI), instead of AEP, is used to tune its parameters. For this, the following variational lower bound on the log-marginal-likelihood is maximized w.r.t the parameters of $q$ and any other model hyper-parameters:

$$\log p(\mathbf{y}) \geq \mathcal{L}(q, \boldsymbol{\alpha}) = \sum_{i=1}^{N} \mathbb{E}_{q(\{\mathbf{h}^l\}_{l=1}^{L})} \left[ \log p(y_i | \mathbf{h}_i^L) \right] - \sum_{l=1}^{L} \mathrm{KL}(q^l | p^l), \quad (11)$$

where $\mathrm{KL}(q^l | p^l)$ is the Kullback-Leibler divergence between $q(\mathbf{u}^l)$ and $p(\mathbf{u}^l)$. In the expectation, the inducing values $\mathbf{u}^l$, for $l = 1, \ldots, L$, have been marginalized out. The terms $\mathbb{E}_{q(\{\mathbf{h}^l\}_{l=1}^{L})} \left[ \log p(y_i | \mathbf{h}_i^L) \right]$ in $\mathcal{L}(q, \boldsymbol{\alpha})$ are intractable, however, they can be approximated by Monte Carlo as in our method. In this case, the Monte Carlo estimate of $\mathbb{E}_{q(\{\mathbf{h}^l\}_{l=1}^{L})} \left[ \log p(y_i | \mathbf{h}_i^L) \right]$ and its gradients will be unbiased. The objective also allows for mini-batch training. This method is similar to our proposed approach. However, we have found that (11) sometimes leads to worse predictive distributions than the objective we consider in (6) for AEP. More precisely, $\mathbb{E}_{q(\{\mathbf{h}^l\}_{l=1}^{L})} \left[ \log p(y_i | \mathbf{h}_i^L) \right]$ only considers the squared error in the case of regression problems since $p(y_i | \mathbf{h}_i^L)$ is a Gaussian factor. By contrast, the $\log \mathcal{Z}_i$

terms in (6) are closer to the data log-likelihood since the cavity distributions $q^\backslash$ is expected to be similar to the approximate posterior $q$.

In [10] it has been proposed to use Markov Chain Monte Carlo techniques for approximate inference in DGPs. Instead of using a parametric approximation $q$, stochastic gradient Hamilton Monte Carlo is employed to approximately draw samples from the target posterior. This approach has shown promising results, but it has to face the difficulty of adjusting the model hyper-parameters, including the length-scales and inducing points locations, since Monte Carlo methods lack an objective function to adjust these parameters. A moving window Markov Chain expectation maximization algorithm is used for that purpose.

In [9] it is proposed an improvement over [19] based on using a different number of inducing points in the individual GPs of the DGP for the mean and for the variance. This leads to faster training times. Such an improvement is orthogonal to the method we propose here and could be used to further reduce its computational cost.

An alternative to VI called importance-weighted VI has been used to train a DGP in [20]. That method does not assume any additive noise in the DGP hidden layers, unlike the models we consider here, but receives the noise as an input in the form of latent covariates. Importance-weighted VI improves the results of VI and the latent covariates allow to capture complex patterns in the predictive distribution. It requires, however, the inclusion of extra data-dependent latent input attributes in the model and the computation of an approximate posterior distribution for each of them, with different parameters for each data point in the training set. The method we propose here is able to capture complex patterns in the predictive distribution without using latent covariates.

A non-Gaussian approximate distribution $q$ has been considered in the context of DGPs in [28]. There, VI is used for approximate inference and an implicit distribution is specified for $q$. Namely, $q$ is a flexible distribution, easy to sample from, but that lacks a closed-form density. This makes difficult the evaluation of the KL divergence in the objective in (11). The output of a classifier that discriminates between samples from $q(\mathbf{u}^l)$ and the prior $p(\mathbf{u}^l)$ is used as a proxy. This method, although more flexible, has the extra cost of training the required classifiers. This improvement is again orthogonal to the method we propose here.

## 6   Experiments

We have implemented the proposed method, DGP-AEPMCM, in Tensorflow[1] [1]. We evaluate the performance of the proposed approach in three main experiments and compare results with: ① DGPs using variational inference, from [19], abbreviated as DGP-VI; ② DGPs using Approximate Expectation propagation, from [2], abbreviated as DGP-AEP; ③ the proposed method in this work, DGPs using Approximate Expectation propagation and Monte Carlo Methods, abbreviated as DGP-AEPMCM; ④ as a baseline, we have also included a standard sparse GP (single-layer DGPs), as described in [24].

---

[1] Available at https://github.com/Gonzalo933/dgp-aepmcm.

## 6.1 Prediction Performance on UCI Datasets

We compare each method on 8 regression datasets and 7 binary classification datasets for the UCI repository [7]. For binary classification we use a probit likelihood [18]. Table 1 shows a summary of the datasets considered. Following [19], we use the same architecture for all the models: a mini-batch size of 100; we use Adam [14] for optimization, with the default parameters, except the learning rate (set to 0.01). The number of epochs is 2,000 in all the datasets except for Naval, Power and Protein, where use use 500. All the models use the RBF kernel with ARD and 100 inducing points for each GP. We use 20 random splits of the data into training and test sets, with 90% and 10% of the instances, respectively. The number of Monte Carlo samples, $S$, is 10 in DGP-VI and 20 in DGP-AEPMCM. A smaller value for $S$ is used in DGP-VI because the stochastic objective of this method (11) is unbiased and the performance does not improve with $S$. See Sect. 6.3. There we show that DGP-AEPMCM with $S = 20$ gives better results (in terms of the test log-likelihood) as function of training time than DGP-AEPMCM with $S = 10$, which means that $S = 20$ is preferred. Larger values for $S$ do not seem to improve results. In all methods inducing point locations are shared within layers. We consider 2, 3, 4 and 5 layers for all the methods. The number of hidden units is the minimum between the problem dimension and 30, as in [19]. We also assume a linear mean function for each GP (except for the last layer) as in [19].

**Table 1.** Summary of the UCI datasets used in our experiments.

| Regression | | | Binary Classification | | |
|---|---|---|---|---|---|
| Dataset | # Inst. | # Dims | Dataset | # Inst. | # Dims |
| Boston | 506 | 13 | Australian | 690 | 14 |
| Concrete | 1,030 | 8 | Breast | 683 | 10 |
| Energy | 768 | 8 | Crabs | 200 | 7 |
| Kin8 | 8,191 | 8 | Ionosphere | 351 | 34 |
| Naval | 11,934 | 16 | Pima | 768 | 8 |
| Power | 9,568 | 4 | Sonar | 208 | 60 |
| Protein | 45,730 | 9 | Banknote | 1,372 | 4 |
| Wine | 1,599 | 11 | | | |

**Regression:** Figure 3 shows the average test log-likelihood for each method on each dataset (RMSE results are found in the supplementary material). Results for DGP-AEP on the protein dataset are not reported due to the excessive computation time of this method which cannot be trained for the same number of epochs as the other methods (see the following sections for an analysis of performance with respect to the computational time). We observe that DGP-AEPMCM gives similar and often better results than the other methods.

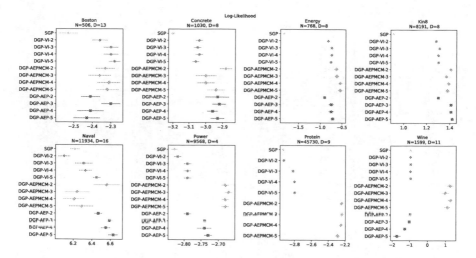

**Fig. 3.** Average test log-likelihood of each method on the UCI datasets for regression. The proposed method, DGP-AEPMCM, is shown in brown with a diamond marker for the mean. The closer to the right size the better the result. Best seen in color. DGP-AEPMCM gives similar or better results than the other methods. (Color figure online)

**Binary Classification:** Figure 4 shows the average test log-likelihood for each method on each dataset (prediction errors are found in the supplementary material). In this case, however, it is not that clear the advantage of using DGPs over a single sparse GP. In spite of this, again, we observe that DGP-AEPMCM gives similar and sometimes better results than the other DGP methods.

**Multi-class Classification:** The supplementary material has extra results involving multi-class problems. The results obtained in these extra experiments also show that sometimes DGP-AEPMCM gives similar or better results than the other methods. However, the differences among the methods are less clear.

## 6.2    Properties of the Predictive Distribution

Following [6], we have generated synthetic data to study the properties of the predictive distribution for the DGP models using VI and AEP combined with Monte Carlo methods. The data has only one dimension to improve visualization. For the first experiment we draw uniformly $x$ in the interval $[-2, 2]$ and generate $y$ values from two different functions with probability 0.5:

$$y = 10\sin x + \epsilon, \qquad\qquad y = 10\cos x + \epsilon, \qquad (12)$$

where $\epsilon \sim \mathcal{N}(0, 1)$. The $y$ values generated for this problem follow a bimodal distribution for each $x$. See Fig. 5. For the second experiment, $x$ is sampled uniformly in the interval $[-4, 4]$ and $y$ is obtained as:

$$y = 7\sin x + 3 \left| \cos(x/2) \right| \epsilon, \qquad (13)$$

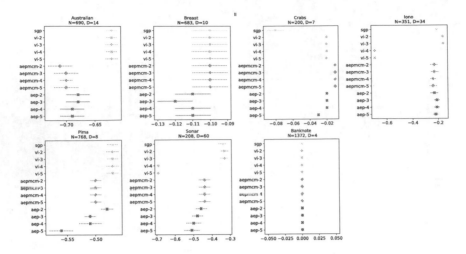

**Fig. 4.** Average test log-likelihood of each method on the UCI datasets for binary classification. The proposed method, DGP-AEPMCM, is shown in brown with a diamond marker for the mean. The closer to the right size the better the result. Best seen in color. DGP-AEPMCM gives similar or better results than the other methods. (Color figure online)

where $\epsilon \sim \mathcal{N}(0,1)$. The $y$ values generated from this function are heteroscedastic, *i.e.*, the noise variance depends on $x$. See Fig. 5. We train on these datasets DGP-VI and DGP-AEPMCM using 3 layers and 3 units in each hidden layer. The mini-batch size, learning rate and number of inducing points is set to 50, 0.01 and 50, respectively. We also removed the mean linear function of each GP and train each method for 500 epochs using Adam [14]. We do not compare results here with the single GP and the DGP-AEP method since the resulting predictive distribution is always Gaussian for those methods.

Figure 5 shows a comparison of samples taken from the predictive distribution of DGP-VI and DGP-AEPMCM (our method). Even though they perform similarly in terms of squared error, DGP-AEPMCM captures the bimodal predictive distribution of the first problem and the heteroscedastic noise of the second. This robustness when estimating the distribution from which the data is generated can explain the results from Sect. 6.1, where our model seems to give better results in terms of the test log likelihood. Table 2 shows average root mean squared error (RMSE) and test log-likelihood results on an independent test set.

DGP-AEPMCM seems to obtain better test log-likelihood results than DGP-VI. This can be explained by the objective function optimized in each method. DGP-VI optimizes the ELBO, (11), which includes terms like $\mathbb{E}_q\left[\log p(y_i|\mathbf{h}_i^L)\right]$. In regression problems, a Gaussian likelihood is used. Therefore, the previous expression results in something that resembles the squared error. On the other hand, DGP-AEPMCM optimizes the AEP energy (6) that has terms of the form $\log \mathbb{E}_{q\setminus}\left[p(y_i|\mathbf{h}_i^L)\right]$, *i.e.*, each $\log \mathcal{Z}_i$. This last expression is closer to the

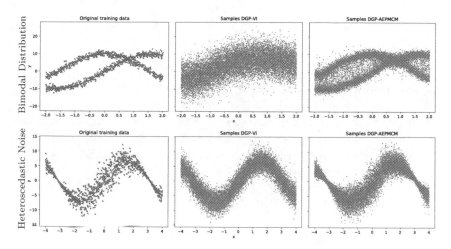

**Fig. 5.** (left) Training data for each problem. In red the target function. (middle) Samples from the predictive distribution of DGP-AEPMCM for each $x$. (right) Samples from DGP-VI for each $x$. Both predictive distributions give similar results in terms of squared error, but DGP-VI fails to capture the bimodal nature of the first problem and the heteroscedastic noise of the second. Best seen in color. (Color figure online)

**Table 2.** RMSE and test log-likelihood on the synthetic datasets.

| Problem | Method | RMSE | Log-likelihood |
|---|---|---|---|
| Bimodal | DGP-VI | 5.293 | −3.084 |
| Distribution | DGP-AEPMCM | **5.288** | **−2.255** |
| Heteroscedastic | DGP-VI | **1.905** | −2.069 |
| Noise | DGP-AEPMCM | 2.016 | **−1.773** |

log-likelihood of the data since $q^{\backslash} \approx q$. Therefore, DGP-AEPMCM is expected to give better predictive distributions in terms of the log-likelihood.

## 6.3   Experiments on Bigger Datasets

To show that DGP-AEPMCM is suitable for big data problems, we have compared training times of the three state of the art methods for DGPs. For this experiment we have used the Airline dataset (N = 2,082,007; D = 8) [11]. The supplementary material has results for another large dataset. A test set of 10% of the data instances is used for validation. The rest of the instances are used for training. We have trained all models with a mini-batch size of 100, using 100 inducing points and all models have a 3 layer architecture with 8 units in each layer. For the models that propagate samples (VI and AEPMCM) we have compared two different setups. Namely, propagating 10 and 20 Monte Carlo samples for training. We trained each method with Adam with a learning rate of 0.01. We measure training times and performance results after 100 gradient steps.

These experiments are carried out on the same CPU for all models: "Intel(R) Xeon(R) CPU E5-2630 v3 @ 2.40 GHz". We do not compare results here with the standard GP since Sect. 6.1 shows that it tends to give sub-optimal results.

Figure 6 shows the average RMSE and test log-likelihood of each method as a function of the training time. We observe that DGP-AEPMCM gives the best results as a function of the training time. Furthermore, using 20 samples seems to slightly improve the test log-likelihood and to give better results in terms of the computational time. DGP-VI does not improve when $S$ is increased from 10 to 20. The reason is that the objective of this method is unbiased. DGP-AEP is the slowest method as a consequence of the high cost of computing the moments of the predictive distribution when the inputs to each GP are Gaussian random variables. Table 3 shows the final RMSE and test log-likelihood of each method and the average seconds that takes to process each mini-batch. DGP-VI takes less time to compute a gradient step. The reason is that our code requires a few more matrix operations and, although not used in the experiments, it allows for the inducing points to be different for each unit within the same hidden layer, unlike the code of [19], which assumes shared locations.

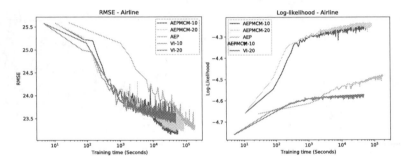

**Fig. 6.** (left) Average RMSE results for each method as a function of the training time (in seconds, log scale) measured after every 100 gradient steps for the Airline dataset. (right) Average test log-likelihood results for each method. Best seen in color. (Color figure online)

**Table 3.** Average final results for the airline dataset.

| Method | # Seconds per mini-batch | RMSE | Log-Likelihood |
|---|---|---|---|
| DGP-AEPMCM-10 | 0.0221 | **23.22** | −4.25 |
| DGP-AEPMCM-20 | 0.0347 | 23.32 | **−4.24** |
| DGP-VI-10 | 0.0202 | 23.55 | −4.58 |
| DGP-VI-20 | 0.0388 | 23.47 | −4.57 |
| DGP-AEP | 0.2914 | 23.32 | −4.48 |

# 7   Conclusions

We have proposed a new method for training deep Gaussian processes (DGPs) that is based on approximate expectation propagation and Monte Carlo methods, DGP-AEPMCM. DGP-AEPMCM, reduces the strong approximations made by previous methods for training DGPs [2]. Namely, using a Gaussian distribution to approximate the output distribution of a GP with random Gaussian input variables. Instead of this, we simply propagate Monte Carlo samples through the DGP network. This allows to better capture properties of the output distribution, like skewness or multiple modes.

Extensive regression and classification experiments show that DGP-AEPMCM improves the results of state-of-the-art inference techniques for DGPs in terms of the test log-likelihood. More precisely, DGP-AEPMCM is able to output complicated predictive distributions that may have multiple modes or input-dependent noise. Alternative methods based on VI do not capture these properties and perform worse. Finally, we have shown that our method is suitable for training on large datasets by making use of sparse GPs and stochastic gradient descent (*i.e.*, mini-batch training), leading to competitive training times.

**Acknowledgment.** We acknowledge using the facilities of Centro de Computación Científica at UAM and support from the Spanish Plan Nacional I+D+i (grants TIN2016-76406-P, TEC2016-81900-REDT and PID2019-106827GB-I00) and from Comunidad de Madrid (grants PEJD-2016-TIC-238 and PEJD-2017-PRE-TIC-3776).

# References

1. Abadi, M., et al.: TensorFlow: Large-scale machine learning on heterogeneous systems (2015)
2. Bui, T.D., Hernández-Lobato, J.M., Hernández-Lobato, D., Li, Y., Turner, R.E.: Deep Gaussian processes for regression using approximate expectation propagation. In: International Conference on International Conference on Machine Learning, pp. 1472–1481 (2016)
3. Calandra, R., Peters, J., Rasmussen, C.E., Deisenroth, M.P.: Manifold Gaussian processes for regression. In: International Joint Conference on Neural Networks, pp. 3338–3345 (2016)
4. Cutajar, K., Bonilla, E.V., Michiardi, P., Filippone, M.: Random feature expansions for deep Gaussian processes. In: International Conference on Machine Learning, pp. 884–893 (2017)
5. Damianou, A., Lawrence, N.: Deep Gaussian processes. In: International Conference on Artificial Intelligence and Statistics, pp. 207–215 (2013)
6. Depeweg, S., Hernández-Lobato, J.M., Doshi-Velez, F., Udluft, S.: Learning and policy search in stochastic dynamical systems with Bayesian neural networks. CoRR abs/1605.07127 (2016)
7. Dua, D., Graff, C.: UCI repository (2017). http://archive.ics.uci.edu/ml
8. Duvenaud, D., Lloyd, J., Grosse, R., Tenenbaum, J., Zoubin, G.: Structure discovery in nonparametric regression through compositional kernel search. In: Proceedings of the 30th International Conference on Machine Learning. Proceedings of Machine Learning Research, vol. 28, pp. 1166–1174. PMLR (17–19 June 2013)

9. Havasi, M., Hernández-Lobato, J.M., Murillo-Fuentes, J.J.: Deep Gaussian processes with decoupled inducing inputs. In: arXiv:1801.02939 [stat.ML] (2018)
10. Havasi, M., Hernández-Lobato, J.M., Murillo-Fuentes, J.J.: Inference in deep Gaussian processes using stochastic gradient Hamiltonian Monte Carlo. In: Advances in Neural Information Processing Systems, pp. 7517–7527 (2018)
11. Hensman, J., Fusi, N., Lawrence, N.D.: Gaussian processes for big data. In: Uncertainty in Artificial Intellegence, pp. 282–290 (2013)
12. Heskes, T., Zoeter, O.: Expectation propagation for approximate inference in dynamic Bayesian networks. In: Uncertainty in Artificial Intelligence, pp. 216–223 (2002)
13. Hinton, G.E., Salakhutdinov, R.R.: Reducing the dimensionality of data with neural networks. Science **313**, 504–507 (2006)
14. Kingma, D.P., Ba, J.: Adam: A method for stochastic optimization. In: International Conference on Learning Representations, pp. 1–15 (2015)
15. Ko, J., Fox, D.: GP-BayesFilters: Bayesian filtering using Gaussian process prediction and observation models. Auton. Robots **27**, 75–90 (2009)
16. Li, Y., Hernandez-Lobato, J.M., Turner, R.E.: Stochastic expectation propagation. In: Neural Information Processing Systems, pp. 2323–2331 (2015)
17. Pope, C.A., Gosling, J.P., Barber, S., Johnson, J., Yamaguchi, T., Feingold, G., Blackwell, P.: Modelling spatial heterogeneity and discontinuities using voronoi tessellations. ArXiv e-prints (2018)
18. Rasmussen, C.E., Williams, C.K.I.: Gaussian Processes for Machine Learning. The MIT Press, Cambridge (2006)
19. Salimbeni, H., Deisenroth, M.: Doubly stochastic variational inference for deep Gaussian processes. In: Advances in Neural Information Processing Systems, pp. 4588–4599 (2017)
20. Salimbeni, H., Dutordoir, V., Hensman, J., Deisenroth, M.: Deep Gaussian processes with importance-weighted variational inference. In: International Conference on Machine Learning, pp. 5589–5598 (2019)
21. Seeger, M.: Expectation propagation for exponential families. Technical report (2005)
22. Snelson, E., Ghahramani,: Sparse Gaussian processes using pseudo-inputs. In: Advances in Neural Information Processing Systems, pp. 1257–1264 (2006)
23. Snoek, J., Larochelle, H., Adams, R.P.: Practical Bayesian optimization of machine learning algorithms. In: Advances in Neural Information Processing Systems, pp. 2951–2959 (2012)
24. Titsias, M.: Variational learning of inducing variables in sparse Gaussian processes. In: International Conference on Artificial Intelligence and Statistics, pp. 567–574 (2009)
25. Vafa, K.: Training and inference for deep Gaussian processes. Technical report (last year project), Harvard College (2016)
26. Wilson, A.G., Adams, R.P.: Gaussian process kernels for pattern discovery and extrapolation. In: International Conference on International Conference on Machine Learning, pp. 1067–1075 (2013)
27. Wilson, A.G., Hu, Z., Salakhutdinov, R., Xing, E.P.: Deep kernel learning. In: International Conference on Artificial Intelligence and Statistics, pp. 370–378 (2016)
28. Yu, H., Chen, Y., Low, B.K.H., Jaillet, P., Dai, Z.: Implicit posterior variational inference for deep Gaussian processes. Adv. Neural Inf. Process. Syst. **32**, 14502–14513 (2019)

# Computer Vision and Image Processing

# Companion Guided Soft Margin for Face Recognition

Yingcheng Su[1], Yichao Wu[1(✉)], Zhenmao Li[1], Qiushan Guo[1], Ken Chen[1], Junjie Yan[1], Ding Liang[1], and Xiaolin Hu[2]

[1] Sensetime Group Limited, Hong Kong, China
{wuyichao,liuxuebo,qinhaoyu,yanjunjie,liangding}@sensetime.com,
kenchen1024@gmail.com
[2] Tsinghua University, Beijing, China
xlhu@mail.tsinghua.edu.cn

**Abstract.** Face recognition has achieved remarkable improvements with the help of the angular margin based softmax losses. However, the margin is usually manually set and kept constant during the training process, which neglects both the optimization difficulty and the informative similarity structures among different instances. Although some works have been proposed to tackle this issue, they adopt similar methods by simply changing the margin for different classes, leading to limited performance improvements. In this paper, we propose a novel sample-wise adaptive margin loss function from the perspective of the hypersphere manifold structure, which we call companion guided soft margin (CGSM). CGSM introduces the information of distribution in the feature space, and conducts teacher-student optimization within each mini-batch. Samples of better convergence are considered as teachers, while students are optimized with extra soft penalties, so that the intra-class distances of inferior samples can be further compacted. Moreover, CGSM does not require sophisticated mining techniques, which makes it easy to implement. Extensive experiments and analysis on MegaFace, LFW, CALFW, IJB-B and IJB-C demonstrate that our approach outperforms state-of-the-art methods using the same network architecture and training dataset.

**Keywords:** Face recognition · Companion guided soft margin · Sample-wise adaptive margin

## 1 Introduction

We have witnessed the tremendous achievements of deep convolutional neural networks (CNNs) in recent years. Face recognition (FR) also benefits from the representation power of CNNs, and has made great breakthroughs [1,2,6,11,12, 19]. Open-set FR is essentially a typical scenario of metric learning [5], where

---

Y. Su and Y. Wu—Contributed equally to this work.

© Springer Nature Switzerland AG 2021
F. Hutter et al. (Eds.): ECML PKDD 2020, LNAI 12459, pp. 497–514, 2021.
https://doi.org/10.1007/978-3-030-67664-3_30

features of different classes are expected to have large discriminative margins
rather than to be only separable.

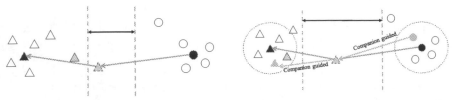

(a) Traditional angular margin based softmax losses optimization mode.

(b) The proposed companion guided optimization mode.

**Fig. 1.** Geometrical interpretation of the AMS loss and the proposed CGSM. Different
shapes stand for different classes. The black solid shapes indicate the class centers.
The green shapes indicates the selected companion samples. The yellow shapes stand
for samples to be supervised. Samples with superior convergence are inside the dashed
circles. The proposed CGSM enforces the intra-class compactness and the inter-class
discrepancy by introducing companions. (Color figure online)

The training loss plays an important role in learning robust face features. One
can train CNNs followed by softmax loss to get the representing features [19].
However, the traditional softmax loss is insufficient to gain the discriminating
power for FR. Therefore, the angular margin based softmax (AMS) losses [2,3,
12,13,21,23,29] introduce extra margins to further reduce intra-class distance
and enlarge inter-class distance, which outperform previous Euclidean distance
based losses [17,18,25,26] suffering unstable or inefficient training [24]. Existing
AMS losses embed the face image into a hypersphere manifold, and incorporate
the angular margin in different kinds of manners (usually multiplicative [12] or
additive [2,23]). Whereas, the margin is usually set manually and kept fixed along
the training process, which is obviously inappropriate for practical large-scale FR
training. It is well-known that the convergence rates are different for a variety
of samples, while constant margin neglects both the optimization difficulty and
the informative similarity structures among different instances, leading to a sub-
optimum solution for FR.

Some pioneer works [11,27] are proposed to adaptively adjust the hyperparameter of AMS during the training process. [11] proposed learnable margins for
different classes to overcome the data unbalance problem. As far as we know,
they propose a class-wise method that simply changes the margin for different
classes, leading to limited performance improvements. In this study, to make
better use of the hypersphere manifold structure, we attempt to incorporate a
new kind of sample-wise adaptive margin into AMS, which we call *companion
guided soft margin* (CGSM).

Our idea is motivated by the observation of the distribution in the FR feature
space. Common AMS losses are supervised by the angular difference between

the current sample with homogeneous class-center and inhomogeneous class-centers, as shown in Fig. 1a. However, in real application scenarios, the similarity score is obtained by the comparison with another sample instead of the FC layers' weights, which means we should integrate the knowledge of manifold structures among different instances and identities into the loss function for better modeling. Therefore, we supply sample-wise supervision to the traditional AMS optimization objective, which we call *companion guided*. CGSM conducts teacher-student optimization within each mini-batch, as shown in Fig. 1b. Since the convergence rate varies between different samples, samples of superior convergence are considered as teachers, then students are optimized with extra penalties, so that the intra-class variances can be further reduced. As the training proceeds, each student has different teachers in each step. Thus the margins can be adjusted dynamically. It can be easily inferred that samples of inferior optimization have larger soft margins.

The objective of CGSM is to build a connection between easy samples and hard samples, and to force the hard samples to gain similar convergence with the easy samples. Compared with previous adaptive AMS works, CGSM utilizes the structure information of the feature space rather than local knowledge to make the feature more robust. Extensive experiments on MegaFace [10], LFW [9], CALFW [30], IJB-B and IJB-C [14] show that our approach outperforms state-of-the-art (SOTA) methods with the same network architecture and training dataset.

To sum up, our major contribution can be summarized as follows:

- We propose CGSM for FR from the novel perspective of the manifold structure, which generates an extra sample-wise soft margin to each sample according to their degree of convergence.
- The proposed CGSM introduces a novel teacher-student supervision pattern within each mini-batch. During training, no complex sample mining techniques are required.
- We conduct a series of experiments on several face benchmarks. Experimental results show that our method outperforms SOTA methods.

## 2   Related Works

**Euclidean Distance Based Loss.** In practical applications, FR is an open-set problem[1] [5]. Although some early works adopt softmax with cross-entropy loss [19,20] for FR, the features are not discriminative enough, especially in large-scale datasets. To reduce intra-variance and enlarge inter-variance, some works including contrastive loss [18,26], triplet loss [17] and center loss [25] attempt to embed face features into Euclidean space. Triplet loss [17] proposes to learn embedding features by selecting triplets to separate positive pairs from negative ones. Center loss [25] introduces an additional intra-class distance penalty to

---

[1] For open-set protocol, the testing identities are usually disjoint from the training set [12].

enhance the discriminative power of the learned features. However, these methods usually require rigorous selection of effective training samples, otherwise, the training would become unstable and inefficient [24].

**Angular Margin Based Softmax Loss.** Based on the previous works [16,22], many works [2,3,12,13,21,23] have introduced margin penalty to improve the feature discrimination, which are referred to as angular margin based softmax (AMS) loss. AMS losses incorporate the angular margin in different ways and help to achieve great improvements, which are considered as a milestone of FR. SphereFace [12] introduces the angular margin to face recognition task. However, the margin term is a multiplicative angular, leading to unstable training. CosFace [23] and AM-Softmax [21] add a cosine margin term to $L_2$ normalized features and weight vectors, and further improve the intra-class compactness and inter-class discrepancy. ArcFace [2] directly adds the angular margin to the target angle and obtains a constant linear decision boundary in angular space. Other techniques are proposed to further improve the FR performance. He et al. [8] study the terminal optimization point of the current face losses and decouple the softmax loss into independent intra-class and inter-class objective. Zhang et al. [28] reveal the sub-optimal gradient lengths of cosine softmax losses and design a hyperparameter-free optimization gradient. Duan et al. [4] propose a new supervision objective called UniformFace to learn deep equidistributed representations for FR.

On the other hand, some works point out that the performance of original angular margin losses heavily relys on hyperparameter, leading to insufficient training. Zhang et al. [27] propose hyperparameter-free loss by leveraging an adaptive scale parameter to strengthen the training process automatically. AdaptiveFace [11] focuses on solving the unbalanced face training dataset, and proposes adaptive margin softmax for different classes and adaptive data sampling. Although these works claim to be effective during the training process, they conduct adaptive margin by simply changing the magnitude of margins for different classes. In this work, we supply extra structure information of the feature space to solve the adaptive margin problem on sample level.

## 3  Preliminaries

Existing angular margin based softmax (AMS) losses [2,12,23] add a constant margin term with different forms including addictive and multiplicative type. We generalize these AMS losses as follows:

$$\mathcal{L}_1 = -\frac{1}{N}\sum_{i=1}^{N}\log\frac{e^{s(\cos(m_1\theta_i+m_2)-m_3)}}{e^{s(\cos(m_1\theta_i+m_2)-m_3)}+\sum_{j\neq i}e^{s\cos\theta_j}}, \quad (1)$$

$$\theta_i = \arccos(\hat{\boldsymbol{w}}_{y_i}^T\hat{\boldsymbol{x}}_i), \quad \theta_j = \arccos(\hat{\boldsymbol{w}}_{y_j}^T\hat{\boldsymbol{x}}_i), \quad (2)$$

$$\hat{\boldsymbol{x}}_i = \frac{\boldsymbol{x}_i}{\|\boldsymbol{x}_i\|}, \hat{\boldsymbol{w}}_{y_i} = \frac{\boldsymbol{w}_{y_i}}{\|\boldsymbol{w}_{y_i}\|}, \hat{\boldsymbol{w}}_{y_j} = \frac{\boldsymbol{w}_{y_j}}{\|\boldsymbol{w}_{y_j}\|}, \quad (3)$$

where $\boldsymbol{x}_i \in \mathbb{R}^d$ denotes the d-dimension feature of the $i$-th sample from class $y_i$, $\boldsymbol{w}_{y_i} \in \mathbb{R}^d$, $\boldsymbol{w}_{y_j} \in \mathbb{R}^d$ denote the parameters for class $y_i, y_j$ in the last fully-connection layer of a model, $N$ is the batch size, $s$ is the scale factor, $\theta_i, \theta_j$ are the angles between $\boldsymbol{x}_i$ and $\boldsymbol{w}_{y_i}$, $\boldsymbol{w}_{y_j}$ respectively. $m_1, m_2, m_3$ are margin terms corresponding to SphereFace, ArcFace and CosFace respectively. These extra margin penalties enforce the CNN model to learn intra-class features more compact and inter-class features more discriminative.

For clear illustrations in the next section, we reformulate Eq. (1) as follows:

$$\mathcal{L}_2 = -\frac{1}{N} \sum_{i=1}^{N} \log P_i, \tag{4}$$

$$P_i = \frac{e^{f_1(\theta_i)}}{e^{f_1(\theta_i)} + \sum_{j\neq i} e^{f_2(\theta_j)}}, \tag{5}$$

$$P_j = \frac{e^{f_2(\theta_j)}}{e^{f_1(\theta_i)} + \sum_{j'\neq i} e^{f_2(\theta_{j'})}}, \tag{6}$$

where $f_1(\theta) = s(\cos(m_1\theta + m_2) - m_3)$, $f_2(\theta) = s\cos\theta$, $P_i$ is the probability that $x_i$ belongs to class $y_i$, while $P_j$ is the probability that $x_i$ belongs to other class $y_j \neq y_i$.

## 4   Proposed Approach

### 4.1   Limitations of AMS Losses

Although AMS losses help to achieve great improvements for FR, there still exist some intrinsic problems, which we will introduce in the following.

**Gradient Analysis for AMS Losses.** Here, we analyze the behavior of gradients while optimizing with Arcface. It should be mentioned that we can acquire similar conclusions using other AMS losses. We set $f_1(\theta) = s\cos(\theta+m)$, $f_2(\theta) = s\cos(\theta)$, thus, the gradient of the $\mathcal{L}_2$ respect to $\boldsymbol{x}_i, \boldsymbol{w}_{y_i}, \boldsymbol{w}_{y_j}$ are:

$$\frac{\partial \mathcal{L}_2(\boldsymbol{x}_i)}{\partial \boldsymbol{x}_i} = \frac{\partial \mathcal{L}_2(\boldsymbol{x}_i)}{\partial \theta_i}\frac{\partial \theta_i}{\partial \boldsymbol{x}_i} + \sum_{j\neq i} \frac{\partial \mathcal{L}_2(\boldsymbol{x}_i)}{\partial \theta_j}\frac{\partial \theta_j}{\partial \boldsymbol{x}_i}$$

$$= s(1 - P_i)\frac{\sin(\theta_i + m)}{\sin\theta_i\|\boldsymbol{x}_i\|}(\hat{\boldsymbol{x}}_i\cos\theta_i - \hat{\boldsymbol{w}}_{y_i})$$

$$+ \sum_{j\neq i} s(-P_j)\frac{1}{\|\boldsymbol{x}_i\|}(\hat{\boldsymbol{x}}_i\cos\theta_j - \hat{\boldsymbol{w}}_{y_j}), \tag{7}$$

$$\frac{\partial \mathcal{L}_2(\boldsymbol{w}_{y_i})}{\partial \boldsymbol{w}_{y_i}} = \frac{\partial \mathcal{L}_2(\boldsymbol{w}_{y_i})}{\partial \theta_i}\frac{\partial \theta_i}{\partial \boldsymbol{w}_{y_i}}$$

$$= s(1 - P_i)\frac{\sin(\theta_i + m)}{\sin\theta_i\|\boldsymbol{w}_{y_i}\|}(\hat{\boldsymbol{w}}_{y_i}\cos\theta_i - \hat{\boldsymbol{x}}_i), \tag{8}$$

$$\frac{\partial \mathcal{L}_2(\boldsymbol{w}_{y_j})}{\partial \boldsymbol{w}_{y_j}} = \frac{\partial \mathcal{L}_2(\boldsymbol{w}_{y_j})}{\partial \theta_j} \frac{\partial \theta_j}{\partial \boldsymbol{w}_{y_j}}$$

$$= s(-P_j) \frac{1}{\|\boldsymbol{w}_{y_j}\|} (\hat{\boldsymbol{w}}_{y_j} \cos \theta_j - \hat{\boldsymbol{x}}_i), \tag{9}$$

where $\hat{\boldsymbol{x}}_i, \hat{\boldsymbol{w}}_{y_i}, \hat{\boldsymbol{w}}_{y_j}$ are the unit vectors of $\boldsymbol{x}_i, \boldsymbol{w}_{y_i}$ and $\boldsymbol{w}_{y_j}$ respectively. From the above computations, two limitations of updating mechanism for popular AMS losses can be observed: **1)** There is no gradient from samples of other classes for updating $\boldsymbol{x}_i$, which neglects informative similarity structures between different instances; **2)** The gradient to update $\boldsymbol{w}_{y_i}$ is only from $\boldsymbol{x}_i$, resulting to no interactive between $\boldsymbol{w}_{y_i}$ and $\boldsymbol{w}_{y_j}$, which does not take relative information of different identities into consideration.

**Objective for FR.** We rethink the optimization objective of FR. In real application scenarios, face recognition compares samples utilizing the cosine similarity of features extracted by CNN models. Reasonably, the intrinsic target of FR is to maximize the objective defined as follows:

$$\mathcal{O}_{FR} = \sum_{(y_i = y_j)} \cos \theta_{ij} - \sum_{(y_i \neq y_j)} \cos \theta_{ij}, \tag{10}$$

where $\theta_{ij}$ is the angle of $i$-th and $j$-th samples from class $y_i$ and $y_j$. This objective is based on sample pairs, which increases similarities of sample pairs belong to the same classes while decreases similarities of sample pairs belong to different classes. However, for AMS losses, models are supervised by the angular difference between the current sample with homogeneous class-center and inhomogeneous class-centers, which means that there is no explicit optimization for sample pairs in angular margin based loss in Eq. (1), which is not fully equivalent to $\mathcal{O}_{FR}$ and may cause sub-optimal results.

## 4.2   CGSM: Companion Guided Soft Margin

To alleviate the limitations of AMS losses, we add an extra dynamic margin term, which is decided by sample pairs of different instances and center features from different identities. In this way, we add interactions for different samples and different identities that contain relative information of similarity structure in the feature space. Furthermore, the extra margin is adaptive to the convergence degree of different samples in different training stages.

The proposed CGSM method is illustrated in Fig. 2. In detail, within current mini-batches, let $\boldsymbol{x}_i$ be the $i$-th sample, we define a set of samples $S_T$ as companions, which have larger cosine similarities with their centers[2] compared with $\boldsymbol{x}_i$. For each sample in $S_T$, we have $\cos(\boldsymbol{x}_t, \boldsymbol{w}_{y_t}) > \cos(\boldsymbol{x}_i, \boldsymbol{w}_{y_i})$, where $\boldsymbol{w}_{y_t}, \boldsymbol{w}_{y_i}$ are centers of $\boldsymbol{x}_t$ and $\boldsymbol{x}_i$ respectively. Since the companions converge better than the

---

[2] For simplicity, in this paper, we use the weight of FC layers as the center.

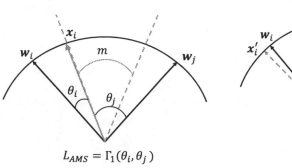

$$L_{AMS} = \Gamma_1(\theta_i, \theta_j)$$

(a) Current AMS losses are supervised by the discrepancy between $\theta_i$ and $\theta_j$.

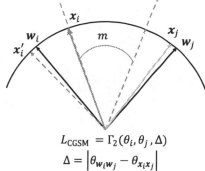

$$L_{CGSM} = \Gamma_2(\theta_i, \theta_j, \Delta)$$
$$\Delta = \left| \theta_{w_i w_j} - \theta_{x_i x_j} \right|$$

(b) CGSM builds a connection of different samples by forcing the hard sample $x_i$ to learn from the easy sample $x_j$.

**Fig. 2.** Illustration the difference between current AMS losses and the proposed CGSM.

specified sample, we call them teachers. Our desire is to construct a "teacher-student" pattern within each mini-batch and make teacher samples with more compact intra-class distances supervise student samples (samples with larger intra-class distances), in order to make the student samples obtain the topological information of teacher samples with dynamic margins and better optimizations. Therefore, we design $\delta_{x_i, x_t}$ in Eq. (12) that calculates difference values of angles between student-teacher pair samples and angles between corresponding centers. We select top-$K$ of teacher samples which are much more superior optimized than $x_i$. Then we calculate the mean of the absolute values of $\delta_{x_i, x_t}$ according to Eq. (11), where $T$ is the number of samples in $S_T$ and $0 \leq \rho \leq 1$ is the selected proportion.

$$\Delta_{m_i} = \frac{1}{K} \sum_{t=1}^{K} |\delta_{x_i, x_t}|, K = \rho * T \qquad (11)$$

$$\delta_{x_i, x_t} = \theta(x_i, x_t) - \theta(w_{y_i}, w_{y_t}) \qquad (12)$$

$$\text{s.t. } \cos(x_t, w_{y_t}) > \cos(x_i, w_{y_i}) \qquad (13)$$

Finally, we add $\Delta_{m_i}$ into the AMS and get the CGSM loss as shown in Eq. (14).

$$\mathcal{L} = -\frac{1}{N} \sum_{i=1}^{N} \log \frac{e^{s \cos(\theta_i + m + \lambda \Delta_{m_i})}}{e^{s \cos(\theta_i + m + \lambda \Delta_{m_i})} + \sum_{j \neq i} e^{s \cos \theta_j}}, \qquad (14)$$

where $m$ is the original margin term, $\lambda$ is a constant to control the strength of the extra margin. Inspired by [2], we set $m = 0.5$ as a basic margin. From the perspective of the angle between the sample and its class center, our method can be regarded as adding a dynamic margin according to the current sample's convergence and punishing the insufficiently trained samples, which is realized by supervision from samples optimized more superiorly, and makes the training process more effective.

## 4.3   Analysis of CGSM

**Sample-Wise Adaptive Margin.** To compare with Arcface [2], we expand the cosine part in Eq. (15).

$$\cos(\theta_i + m + \lambda\Delta_{m_i}) = \cos(\theta_i + m)\cdot\cos(\lambda\Delta_{m_i}) - \sin(\theta_i + m)\cdot\sin(\lambda\Delta_{m_i}). \tag{15}$$

Adding an extra dynamic delta margin leads to two extra effective elements: one is the multiplication factor $\cos(\lambda\Delta_{m_i})$, and the other is a subtraction factor like CosFace [23]. For teacher samples, $\Delta_{m_i}$ is close to zero: $\Delta_{m_i} \to 0$ and CGSM acts like Arcface. For student samples, $\Delta_{m_i} > 0$ leads to an extra dynamic penalty for samples trained much more inferior. During the training procedure, the teacher samples would be optimized rapidly and then provide supervision for student samples with larger dynamic margins, which lead to more sufficient training of student samples.

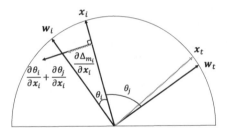

**Fig. 3.** The optimization direction w.r.t $x_i$. The CGSM inherits the advantage of traditional AMS and promotes its superiority.

**Gradient Analysis.** In addition to adding an extra dynamic margin term, $\Delta_{m_i}$ also imports interaction between samples of different classes. We replace $f_1(\theta)$ in Eq. (5) with $s\cos(\theta + m + \lambda\Delta_{m_i})$ and set $K = 1, \delta_{x_i x_t} > 0$ for concise expressions. The corresponding gradients of CGSM loss respect to $x_i, w_{y_i}, w_{y_j}$ are:

$$\frac{\partial\mathcal{L}(x_i)}{\partial x_i} = \frac{\partial\mathcal{L}(x_i)}{\partial\theta_i}\frac{\partial\theta_i}{\partial x_i} + \sum_{j\neq i}\frac{\partial\mathcal{L}(x_i)}{\partial\theta_j}\frac{\partial\theta_j}{\partial x_i} + \frac{\partial\mathcal{L}(x_i)}{\partial\Delta m_i}\frac{\partial\Delta m_i}{\partial x_i}$$

$$= s(1 - P_i)\frac{\sin(\theta_i + m + \lambda\Delta_{m_i})}{\sin\theta_i\|x_i\|}(\hat{x}_i\cos\theta_i - \hat{w}_{y_i})$$

$$+ \sum_{j\neq i}s(-P_j)\frac{1}{\|x_i\|}(\hat{x}_i\cos\theta_j - \hat{w}_{y_j})$$

$$\underbrace{+ \, s(1-P_i)\frac{\lambda \sin(\theta_i + m + \lambda\Delta_{m_i})}{\sin\theta_{x_ix_t}\|x_i\|}(\hat{x}_i\cos\theta_{x_ix_t} - \hat{x}_t)}_{\nabla CGSM\_x_i} \qquad (16)$$

$$\frac{\partial\mathcal{L}(w_{y_i})}{\partial w_{y_i}} = \frac{\partial\mathcal{L}(w_{y_i})}{\partial\theta_i}\frac{\partial\theta_i}{\partial w_{y_i}} + \frac{\partial\mathcal{L}(w_{y_i})}{\partial\Delta m_i}\frac{\partial\Delta m_i}{\partial w_{y_i}}$$

$$= s(1-P_i)\frac{\sin(\theta_i + m + \lambda\Delta_{m_i})}{\sin\theta_i\|w_{y_i}\|}(\hat{w}_{y_i}\cos\theta_i - \hat{x}_i)$$

$$\underbrace{+ \, s(1-P_i)\frac{-\lambda\sin(\theta_i + m + \lambda\Delta_{m_i})}{\sin\theta_{w_{y_i}w_{y_t}}\|w_{y_i}\|}(\hat{w}_{y_i}\cos\theta_{w_{y_i}w_{y_t}} - \hat{w}_{y_t})}_{\nabla CGSM\_w_{y_i}} \qquad (17)$$

$$\frac{\partial\mathcal{L}(w_{y_j})}{\partial w_{y_j}} = \frac{\partial\mathcal{L}(w_{y_i})}{\partial\theta_j}\frac{\partial\theta_j}{\partial w_{y_j}}$$

$$= s(-P_j)\cdot\frac{1}{\|w_{y_j}\|}(\hat{w}_{y_j}\cos\theta_j - \hat{x}_i), \qquad (18)$$

where $\theta_{x_ix_t}, \theta_{w_{y_i}w_{y_t}}$ are angles between corresponding vectors. As illustrated in Eq. (16) and Eq. (17), compared with Eq. (7) and (8), the gradients of the CGSM loss inherit the property of ArcFace to optimize intra-class distances and inter-class distances. Besides, CGSM adds extra items $\nabla CGSM\_x_i$ and $\nabla CGSM\_w_{y_i}$.

Based on the above analysis of the gradients, we can answer the question: *Why does CGSM work?* According to $\nabla CGSM\_x_i$ for $x_i$, CGSM adds gradients from teacher samples $x_t$ with a direction orthogonal to $x_i$ which increase interactions between different instances. For $w_{y_i}$ according to $\nabla CGSM\_w_{y_i}$, CGSM adds gradients from teacher samples' center features $w_{y_t}$ and increases the interaction between center features. Figure 3 illustrates the optimizing direction of CGSM. All the components of the gradient w.r.t $x_i$ are vertical to $x_i$ itself to get the fastest direction for updating $x_i$.

In practical, there are two different circumstances where CGSM plays different roles for the optimization. One is that the student samples and teacher samples are from the same class. In this circumstances, $w_{y_i} = w_{y_t}$, the extra propulsive force from teacher samples push $x_i$ to be close to $x_t$, which compacts the features of same classes by decreasing $\theta_{x_ix_t}$ directly. The other is that the student samples and teacher samples are from the different classes. In this situation, we have $w_{y_i} \neq w_{y_t}$, the extra propulsive force from teacher samples separate $w_{y_i}$ from $w_{y_t}$ by increasing $\theta_{w_{y_i}w_{y_t}}$. Overall, the CGSM adds an extra margin for every inferior optimized sample according to the superior optimized samples adaptively, which makes the training smoother and raises the performance effectively.

**Time Complexity.** The increased computation of CGSM comes from the selection of teacher samples and the calculation of the additional soft margin. We ignore the complexity of normalization operation and arccosine computation for simplicity. Let $N$ be the batch size, $C$ be the number of class, $d$ be the dimension of the feature. The computational complexity of ArcFace is $O(N \cdot C \cdot d)$. For each sample, the additional computation caused by the CGSM consists of several parts. Firstly, it takes $O(N)$ to filter teacher samples. Secondly, it takes $O(N \log_2 K)$ to select the topK teacher samples. Lastly, the computation of the similarity of the current samples and teacher samples and the similarity of the current class center and teacher samples' class center is $O(2 \cdot K \cdot d)$. Thus, the additional computation of the CGSM in each minibatch is $N \cdot (N + N \log_2 K + 2 \cdot K \cdot d)$. In this paper, We may set $N = 128, C = 85,000, K = 64, d = 256$, the increased proportion is $\frac{N + N \cdot \log_2 K + 2K \cdot d}{C \cdot d} = 0.16\%$, which is almost negligible.

**Companion Selection.** One of the key points of CSGM is how to select companions for the specified instance. In our method, we choose those with a fast convergence rate within mini-batch as companions. We can make use of other strategies as alternatives. One simple method is to choose from the batch randomly. Although the margin is dynamic as well, this method will introduce noise to the training as the sample may learn from the companions of inferior convergence. The other method is to choose $K$ nearest neighbors (KNN) as the companions. We will compare different companion selection strategies in our experiments.

## 5 Experiments

### 5.1 Experimental Settings

**Training Data.** We separately use CASIA [26] and MS1MV2 [6] as the training datasets in our experiments. CASIA is a small face training benchmark. It consists of about 0.5 million images of 10572 identities. MS1MV2 is a refined version of the MS-Celeb-1M [6]. It consists 85 K identities, and 5.8 million face images, which is one of the most commonly used face training datasets. We use the same alignment as [2], and all the training sample is cropped and resize to $112 \times 112$. We employ random flip with a ratio of 0.5, and no other data augmentation tricks are introduced.

**Evaluation.** In the evaluation step, we first align and crop the image by using five facial landmarks just the same as the training dataset. We extract both the original image and the flipped image and sum them up in an element-wise manner. The score is obtained by the cosine similarity. The results of face verification and identification are conducted by thresholding and ranking the scores.

**Other Implementation Details.** We implement the CGSM based on ArcFace [2], where the constant margin $m$ is set to 0.5 in all our experiments. We use ResNet50 (R50) and ResNet100 (R100) [7] as the backbone, and the BN-Dropout-FC-BN structure is employed to get the final 256-D embedding features.

We set the momentum to 0.9, and the weight decay is 1e-5. We train models on CASIA by 30k iterations. The learning rate is initialized to 0.1 and decayed by a factor of 10 at 20k, 25k iterations. For the training on MS1MV2, we set the initial learning to 0.1, and the decaying steps are 90k, 140k 170k iterations. The training is finished at 180k iterations.

## 5.2  Exploratory Experiments

**Effect of $\rho$ and $\lambda$.** We employ CASIA [26] for training, while for face identification and verification, MegaFace [10] and AgeDB [15] are respectively used to explore the effect of $\lambda$ and the selected proportion $\rho$ of teacher samples in each mini-batch. In order to explore the maximal superiority of the proposed CGSM, we vary $\rho$ from 0.1 to 1 and $\lambda$ from 0.3 to 1.5 Figure 4 shows the MegaFace identification results under different parameter settings. The baseline is the situation of $\rho = 0, \lambda = 0$, which is also the same as ArcFace [2]. We can see that the proposed CGSM is superior to the baseline of large margins under various settings, except that $\lambda$ is greater than 1.5. For each $\rho$, we observe that the performance gradually grows as $\lambda$ increasing at the beginning. There exist a peak point that as $\lambda$ keeps increasing, the performance goes down. We speculate that it is because the penalty is too harsh for training. Despite that, the performance of CGSM maintains robust in wide parameters range.

**Effect of the Constant Margin $m$.** In order to figure out the effect of the constant margin $m$, we employ ArcFace to train the CNN models with margins of 0.5, 0.55. Besides, we select the best $\lambda$ of each $\rho$ and test the corresponding models on AgeDB [15] as shown in Table 1. It is obvious that the margin term has an impact on the final performance. However, the impact can be negative if the margin is too large and too hard to train. In Table 1, the promotion by simply increasing the margin is tiny. The proposed CGSM can maximal gain 2.18% on MegaFace and 0.76% on AgeDB while training on the small CASIA dataset. Inspired by this, we set $\lambda = 0.8$ and $\rho = 0.5$ in the rest of our experiments.

**Different Companion Selection Strategies.** We employ another two companion selection strategies i.e. random selection and KNN selection. We apply different training configures with various $\rho$ and $\lambda$ and report the best result of each strategy. In Table 2, the Top-K well-optimized companion selection obtains better results than the others, which demonstrates our perspective in the previous discussion.

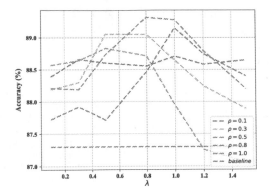

**Fig. 4.** Results of MegaFace [10] (FAR@$10^{-6}$) under different parameters setting. For the baseline model, we set $\rho = \lambda = 0$. (ResNet50 [7] trained on CASIA [26].)

**Table 1.** Verification and identification results of different $\rho$ and corresponding best $\lambda$. (ResNet50 trained on CASIA.)

| Models | AgeDB [15] (%) | MF1 Rank1 [10] (%) |
|---|---|---|
| ArcFace, $m = 0.5$ | 93.36 | 87.30 |
| ArcFace, $m = 0.55$ | 93.33 | 87.14 |
| $\rho = 0.1, \lambda = 1.0$ | 93.82 | 89.21 |
| $\rho = 0.3, \lambda = 0.8$ | 93.55 | 89.41 |
| $\rho = 0.5, \lambda = 0.8$ | 93.97 | **89.48** |
| $\rho = 0.8, \lambda = 0.5$ | **94.03** | 89.00 |
| $\rho = 1.0, \lambda = 1.0$ | 93.83 | 88.99 |

**Average of $\theta_{y_i}$ and $\Delta_{m_i}$ w.r.t. Iterations.** In order to better understand the superiority of the proposed CGSM for face recognition, we visualize the average angle between training samples and their class-centers, i.e. $\theta_{y_i}$ as shown in Fig. 5a, and the average of the additional margin, i.e. $\Delta_{m_i}$ as shown in Fig. 5b. Even though ArcFace provides faster convergence in compacting intra-class at the beginning of training, the proposed CGSM catches up at the end of the first stage and surpasses ArcFace at the last two stages resulting in a more compact intra-class distance. The reason is that the teacher samples are not well-trained at the beginning. Thus they can not provide useful guided for the student samples.

**Table 2.** Result of different companion selection strategies on MegaFace. (ResNet50 trained on CASIA.)

| Strategy | Random | KNN | Top-K well-optimized |
|---|---|---|---|
| Accuracy (%) | 88.93 | 89.05 | **89.48** |

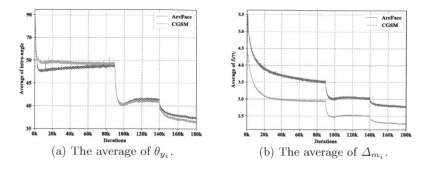

(a) The average of $\theta_{y_i}$.    (b) The average of $\Delta_{m_i}$.

**Fig. 5.** The visualization of $\theta_{y_i}$ and $\Delta_{m_i}$ with respect to training iterations.

When the model is training to a certain degree, the convergence rate becomes faster with the help of companion guided.

Figure 5b shows the change of $\Delta_{m_i}$ during training. The proposed CGSM obtains a more convergent $\Delta_{m_i}$ comparing to ArcFace. $\Delta_{m_i}$ is calculated from Eq. (11) and (12), which is also a reflection of the distribution of the embedding features across different identities. Smaller $\Delta_{m_i}$ indicates a more united manifold structure of different instances. In other words, the proposed CGSM enforces different samples to have similar topological distribution around their class-center in the hyperspace, which makes the embedding features much more robust and discriminative than the traditional AMS methods.

**Discussion About the Topology Generalization.** In CGSM, we find that the easy samples' topology generalize enough to guide the hard examples. It is widely accepted the optimization process would be easy if the distribution of data is smooth. Take the pose as an example, if the distribution of the pose is smooth, like $0°$, $10°$, $20°$, $30°$, $40°$, $50°$, $60°$. The easiest samples are normally faces of $0°$ while the hardest samples could be faces of $60°$. With CGSM, faces of $20°$ would be optimized guided by the faces of $0°$, and faces of $30°$ would be optimized guided by the faces of both $20°$ and $0°$. The hardest samples like $60°$ would be guided by all easy samples. This conclusion can be demonstrated in our experiments. Figure 5b visualizes the convergence gap between the teacher samples and student samples during training. It can be seen that CGSM greatly reduces the gap compared with ArcFace.

## 5.3   Benchmark Comparisons

In this section, we compare the proposed CGSM loss with other losses on several public face recognition testing benchmarks. For fair comparison, we implement the $L_2$-softmax [16], CosFace [23] and ArcFace [23] using the same hyperparameters in the original papers. The CNN models are trained on MS1MV2 [6].

**Table 3.** Face verification (%) on LFW and CALFW.

| Method | LFW | CALFW |
|---|---|---|
| SphereFace [12] | 99.42 | – |
| CosFace [23] | 99.73 | – |
| AcrFace [2] | 99.82 | 95.45 |
| R50, $L_2$-softmax | 99.35 | 94.88 |
| R50, CosFace | 99.36 | 95.53 |
| R50, ArcFace | 99.36 | 95.55 |
| R50, CGSM | **99.48** | **96.03** |
| R100, ArcFace | 99.72 | 95.91 |
| R100, CGSM | **99.83** | **96.20** |

**Evaluation on LFW and CALFW.** LFW [9] is one of most common face verification testing dataset. It consists of 13,233 images of 5,749 identities.There are 3,000 negative pairs and 3,000 positive pairs equally divided into 10 subsets. CALFW [30] is more difficult testing dataset comparing to LFW, which consists of 3,000 positive face pairs with age gaps and 3,000 negative face pairs with gender, race variance. The testing protocol is just the same as LFW. Table 3 shows the results with the same training configuration of different methods. As the LFW is an easy dataset and has been quite well optimized, the proposed CGSM still outperforms other methods. The results on CALFW are clearer to show the performance gaps between CGSM and the other approaches, demonstrating that the proposed CGSM has advantage over the others in handling age gaps, gender and race et al. problems.

**Evaluation on MegaFace.** The MegaFace dataset [10] consists of 1 million distractors of 690 K unique identities and probe set 100 K images of 530 unique identities. It is a widely used face recognition testing benchmark for both verification and identification. We follow the cleaned and testing protocol in [2] and report our results on the identification challenge. In Table 4, Our method beats other SOTA methods, including the strong baseline CosFace [23] and ArcFace [2]. Specifically, we reproduce the accuracy of CosFace and Arc-Face which already surpassed some improved methods of AMS, such as Ada-tiveFace [11][3], P2SGrad [28]. With such a high baseline, our method achieves 98.10% on ResNet50 and 98.57% on ResNet100 and builds an upper boundary of MegaFace identification challenge. Figure 6 shows the recognition curves on MegaFace dataset by ResNet50 models trained with MS1MV2 dataset. Clearly, the CGSM is superior than the other two AMS methods at all distractor scales.

---

[3] In fact, we try to reimplement this method but get non-convergence result. The open source project claimed by [11] is blank, and many researchers also report the same problem on github.

**Table 4.** Face identification results on MegaFace. We report the rank1 face identification accuracy with 1 million distractors.

| Method | Protocol | MF1 Rank1 |
|---|---|---|
| R50, AdaptiveFace [11] | Large | 95.02 |
| R50, P2SGrad [28] | Large | 97.25 |
| R50, $L_2$-softmax | Large | 90.84 |
| R50, CosFace | Large | 97.75 |
| R50, ArcFace | Large | 97.87 |
| R50, CGSM | Large | **98.10** |
| FaceNet [17] | Large | 70.49 |
| UniformFace [4] | Large | 79.98 |
| CosFace [23] | Large | 82.72 |
| Inception-ResNet, AdaCos [27] | Large | 97.41 |
| R100, ArcFace [2] | Large | 98.35 |
| R100, ArcFace | Large | 98.12 |
| R100, CGSM | Large | **98.57** |

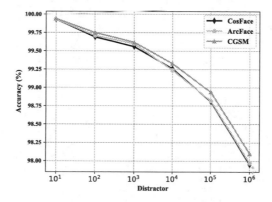

**Fig. 6.** Comparison with other AMS methods on MegaFace identification challenge. (ResNet50 models trained on MS1MV2.)

**Evaluation on IJB-B and IJB-C.** The IJB-B dataset [14] consists of 1,845 identities with both images and video frames. There are 12,115 templates with 10,270 matching pairs and 8M non-matching pairs for face verification. The IJB-C dataset [14] is an extension of IJB-B, which contains 3,531 identities with 23,124 templates. In total, there are 19,557 matching pairs and 15,638,932 non-matching pairs. We follow the same feature fusion protocol with [2]. We compare the true accept rate (TPR) at false accept rate of $10^{-4}$ against other methods as shown in Table 5. Again, CGSM outperforms other methods. CGSM trained with ResNet50 further boosts the TAR (FAR@$10^{-4}$) to 93.74% and 95.26% on IJB-B

**Table 5.** Face verification TAR on IJB-B and IJB-C at FAR $= 10^{-4}$.

| Method | IJB-B | IJB-C |
|---|---|---|
| D-softmax [8] | – | 88.17 |
| VGG2, R50, ArcFace [2] | 89.8 | 92.1 |
| R50, $L_2$-softmax | 88.98 | 91.64 |
| R50, CosFace | 92.89 | 94.58 |
| R50, ArcFace | 93.22 | 94.67 |
| R50, CGSM | **93.74** | **95.26** |
| Inception-ResNet, AdaCos [27] | – | 92.4 |
| R100, ArcFace [2] | 94.2 | 95.6 |
| R100, ArcFace | 93.62 | 95.35 |
| R100, CGSM | **94.08** | **95.80** |

and IJB-C. When CGSM is trained with the deeper network ResNet100, we can get 95.8% TAR (FAR@$10^{-4}$) on IJB-C, which is a new record. Figure 7 draws the ROC curves of different methods with ResNet50 trained on the MS1MV2 dataset, which demonstrates the superiority of the proposed CGSM.

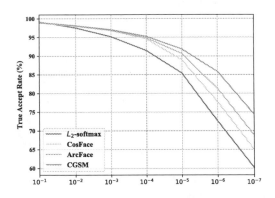

**Fig. 7.** ROC curves of different methods on IJB-C.

## 6    Conclusions

In this paper, we propose a companion guided softmax margin (CGSM) for face recognition, which builds a connection between the easy samples and hard samples within each mini-batch. The proposed CGSM introduces a sample-wise dynamic margin to each samples according to their degree of convergence, which can make the training more sufficient, and improve the disctiminative capability of the embedding features. Experimental results demonstrate the effectiveness of the proposed CGSM on several public benchmarks.

**Acknowledgement.** This work was supported in part by Beijing Postdoctoral Research Foundation, the National Natural Science Foundation of China under Grant Nos. U19B2034 and 61836014.

# References

1. Chen, K., Wu, Y., Qin, H., Liang, D., Liu, X., Yan, J.: R3 adversarial network for cross model face recognition. In: Proceedings of the IEEE Conference on Computer Vision and Pattern Recognition, pp. 9868–9876 (2019)
2. Deng, J., Guo, J., Xue, N., Zafeiriou, S.: Arcface: additive angular margin loss for deep face recognition. In: Proceedings of the IEEE Conference on Computer Vision and Pattern Recognition, pp. 4690–4699 (2019)
3. Deng, J., Zhou, Y., Zafeiriou, S.: Marginal loss for deep face recognition. In: Proceedings of the IEEE Conference on Computer Vision and Pattern Recognition Workshops, pp. 60–68 (2017)
4. Duan, Y., Lu, J., Zhou, J.: Uniformface: learning deep equidistributed representation for face recognition. In: Proceedings of the IEEE Conference on Computer Vision and Pattern Recognition, pp. 3415–3424 (2019)
5. Gunther, M., Cruz, S., Rudd, E.M., Boult, T.E.: Toward open-set face recognition. In: Proceedings of the IEEE Conference on Computer Vision and Pattern Recognition Workshops, pp. 71–80 (2017)
6. Guo, Y., Zhang, L., Hu, Y., He, X., Gao, J.: MS-Celeb-1M: a dataset and benchmark for large-scale face recognition. In: Leibe, B., Matas, J., Sebe, N., Welling, M. (eds.) ECCV 2016. LNCS, vol. 9907, pp. 87–102. Springer, Cham (2016). https://doi.org/10.1007/978-3-319-46487-9_6
7. He, K., Zhang, X., Ren, S., Sun, J.: Deep residual learning for image recognition. In: Proceedings of the IEEE Conference on Computer Vision and Pattern Recognition, pp. 770–778 (2016)
8. He, L., Wang, Z., Li, Y., Wang, S.: Softmax dissection: towards understanding intra-and inter-clas objective for embedding learning. arXiv preprint arXiv:1908.01281 (2019)
9. Huang, G.B., Mattar, M., Berg, T., Learned-Miller, E.: Labeled faces in the wild: a database for studying face recognition in unconstrained environments (2008)
10. Kemelmacher-Shlizerman, I., Seitz, S.M., Miller, D., Brossard, E.: The megaface benchmark: 1 million faces for recognition at scale. In: Proceedings of the IEEE Conference on Computer Vision and Pattern Recognition, pp. 4873–4882 (2016)
11. Liu, H., Zhu, X., Lei, Z., Li, S.Z.: Adaptiveface: adaptive margin and sampling for face recognition. In: Proceedings of the IEEE Conference on Computer Vision and Pattern Recognition, pp. 11947–11956 (2019)
12. Liu, W., Wen, Y., Yu, Z., Li, M., Raj, B., Song, L.: Sphereface: deep hypersphere embedding for face recognition. In: Proceedings of the IEEE Conference on Computer Vision and Pattern Recognition, pp. 212–220 (2017)
13. Liu, W., Wen, Y., Yu, Z., Yang, M.: Large-margin softmax loss for convolutional neural networks. In: ICML, vol. 2, p. 7 (2016)
14. Maze, B., et al.: Iarpa janus benchmark-c: Face dataset and protocol. In: 2018 International Conference on Biometrics (ICB), pp. 158–165. IEEE (2018)
15. Moschoglou, S., Papaioannou, A., Sagonas, C., Deng, J., Kotsia, I., Zafeiriou, S.: Agedb: the first manually collected, in-the-wild age database. In: Proceedings of the IEEE Conference on Computer Vision and Pattern Recognition Workshops, pp. 51–59 (2017)

16. Ranjan, R., Castillo, C.D., Chellappa, R.: L2-constrained softmax loss for discriminative face verification. arXiv preprint arXiv:1703.09507 (2017)
17. Schroff, F., Kalenichenko, D., Philbin, J.: Facenet: a unified embedding for face recognition and clustering. In: Proceedings of the IEEE Conference on Computer Vision and Pattern Recognition, pp. 815–823 (2015)
18. Sun, Y., Chen, Y., Wang, X., Tang, X.: Deep learning face representation by joint identification-verification. In: Advances in Neural Information Processing Systems, pp. 1988–1996 (2014)
19. Sun, Y., Wang, X., Tang, X.: Deep learning face representation from predicting 10,000 classes. In: Proceedings of the IEEE Conference on Computer Vision and Pattern Recognition, pp. 1891–1898 (2014)
20  Taigman, Y., Yang, M., Ranzato, M., Wolf, L.: Deepface: closing the gap to human-level performance in face verification. In: Proceedings of the IEEE Conference on Computer Vision and Pattern Recognition, pp. 1701–1708 (2014)
21. Wang, F., Cheng, J., Liu, W., Liu, H.: Additive margin softmax for face verification. IEEE Signal Process. Lett. **25**(7), 926–930 (2018)
22. Wang, F., Xiang, X., Cheng, J., Yuille, A.L.: Normface: l2 hypersphere embedding for face verification. In: Proceedings of the 25th ACM International Conference on Multimedia, pp. 1041–1049. ACM (2017)
23. Wang, H., et al.: Cosface: large margin cosine loss for deep face recognition. In: Proceedings of the IEEE Conference on Computer Vision and Pattern Recognition, pp. 5265–5274 (2018)
24. Wang, M., Deng, W.: Deep face recognition: a survey. arXiv preprint arXiv:1804.06655 (2018)
25. Wen, Y., Zhang, K., Li, Z., Qiao, Yu.: A discriminative feature learning approach for deep face recognition. In: Leibe, B., Matas, J., Sebe, N., Welling, M. (eds.) ECCV 2016. LNCS, vol. 9911, pp. 499–515. Springer, Cham (2016). https://doi.org/10.1007/978-3-319-46478-7_31
26. Yi, D., Lei, Z., Liao, S., Li, S.Z.: Learning face representation from scratch. arXiv preprint arXiv:1411.7923 (2014)
27. Zhang, X., Zhao, R., Qiao, Y., Wang, X., Li, H.: Adacos: adaptively scaling cosine logits for effectively learning deep face representations. In: Proceedings of the IEEE Conference on Computer Vision and Pattern Recognition, pp. 10823–10832 (2019)
28. Zhang, X., et al.: P2sgrad: refined gradients for optimizing deep face models. In: Proceedings of the IEEE Conference on Computer Vision and Pattern Recognition, pp. 9906–9914 (2019)
29. Zhao, K., Xu, J., Cheng, M.M.: Regularface: deep face recognition via exclusive regularization. In: Proceedings of the IEEE Conference on Computer Vision and Pattern Recognition, pp. 1136–1144 (2019)
30. Zheng, T., Deng, W., Hu, J.: Cross-age LFW: a database for studying cross-age face recognition in unconstrained environments. CoRR abs/1708.08197 (2017). http://arxiv.org/abs/1708.08197

# Soft Labels Transfer with Discriminative Representations Learning for Unsupervised Domain Adaptation

Manliang Cao[1(✉)], Xiangdong Zhou[1], and Lan Lin[2]

[1] School of Computer Science, Fudan University, Shanghai, China
{mlcao17,xdzhou}@fudan.edu.cn
[2] Department of Electronic Science and Technology, Tongji University, Shanghai, China
linlan@tongji.edu.cn

**Abstract.** Domain adaptation aims to address the challenge of transferring the knowledge obtained from the source domain with rich label information to the target domain with less or even no label information. Recent methods start to tackle this problem by incorporating the hard-pseudo labels for the target samples to better reduce the cross-domain distribution shifts. However, these approaches are vulnerable to the error accumulation and hence unable to preserve cross-domain category consistency. Because the accuracy of pseudo labels cannot be guaranteed explicitly. To address this issue, we propose a Soft Labels transfer with Discriminative Representations learning (SLDR) framework to jointly optimize the class-wise adaptation with soft target labels and learn the discriminative domain-invariant features in a unified model. Specifically, to benefit each other in an effective manner, we simultaneously explore soft target labels by label propagation for better condition adaptation and preserve the important properties of inter-class separability and intra-class compactness for reducing more domain shifts. Extensive experiments are conducted on several unsupervised domain adaptation datasets, and the results show that SLDR outperforms the state-of-the-art methods.

**Keywords:** Unsupervised domain adaptation · Distribution alignment · Discriminative feature · Soft Labels transfer

## 1 Introduction

Machine learning models can be trained very well with sufficient labeled data, which have shown great success in many practical applications. However, manually labeling datasets is exceedingly expensive and time-consuming. Domain adaptation (DA) [3,10,19,36], can leverage the off-the-shelf data from a different but related source domain. This will boost the task in the new target domain and reduce the cost of data labeling as well. Depending on the availability of

© Springer Nature Switzerland AG 2021
F. Hutter et al. (Eds.): ECML PKDD 2020, LNAI 12459, pp. 515–530, 2021.
https://doi.org/10.1007/978-3-030-67664-3_31

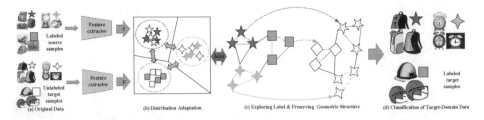

**Fig. 1.** Overview of SLDR. The intuition of SLDR is to jointly optimize the target labels and learn the discriminative features in a unified model. (a) We jointly apply two projections to map the original data into a domain-invariant space. (b) With soft target labels gradually learning, class-wise adaptation and discriminative features learning are able to realize (green arrows separate different classes, color circles cluster the same classes). (c) With discriminative shared features iteratively learning, more domain shifts will be mitigated and then an effective graph based on cross-domain samples can be constructed so that more robust source label information will propagate to target data. (d) Finally, (b) and (c) will benefit each other to enhance target classification results (white unlabeled signs turn to color labeled ones). (Color figure online)

labeled data in the target domain, domain adaptation can be divided into semi-supervised and unsupervised domain adaptation (UDA).

The most commonly used domain adaptation methods include instance-based adaptation [4,30] and feature transformation [2,22]. The instance-based methods directly adjust the weights of source instances to match the target domain. However, it requires the strict assumption that the conditional distributions between domains are identical. In contrast, feature transformation methods are more flexible, which try to incorporate some regularizers, *i.e.*, maximum mean discrepancy (MMD) [13] and adversarial structure [12], into the models to minimize the domain shifts. Despite their efficacy, these methods only focus on exploring the global domain distributions by matching marginal distributions, while ignoring the importance of the discriminative properties between classes.

Alternatively, a growing number of recent UDA methods [11,24,26,34] start to adopt the iterative pseudo-label guided scheme to further explore the conditional distributions across domains. Due to the lack of labeled target samples for UDA, adaptation of conditional distributions becomes a none trivial problem. To tackle this issue, existing methods implement conditional distributions by pseudo labeling, which we term hard-pseudo labeling, since only one pseudo label is assigned to each data sample. However, because of domain discrepancy, the accuracy of hard-pseudo labels assigned by source classifiers cannot be guaranteed explicitly, which can deteriorate the following feature transformation optimization process. Hence, how to properly transfer the source-domain label information to the target domain data becomes a challenging task.

In this paper, we propose an effective Soft Labels transfer with Discriminative Representations learning (SLDR) framework as shown in Fig. 1, where we simultaneously explore the structural information of both domains to optimize the target labels and keep the discriminative properties among different classes.

Specifically, Our method aims at seeking a domain-invariant feature space in matching cross-domain marginal and conditional feature distributions. Instead of using the hard-pseudo target labels, we assign the target samples with soft ones, namely label probability distribution. We deal with the original problem of UDA by solving a label-propagation based optimization task. Such strategy introduces additional robustness to alleviate wrong information caused by those false pseudo-target labels and thus allows us to better associate cross-domain feature distributions. Moreover, SLDR encourages the learned features to be discriminative, which explicitly minimizes the distance between every two projected samples with the same label, and maximizes the distance between every pair of projected samples in different classes.

The Discriminative Feature Learning (DFL) and Soft-target Label Learning (SLL) can complement each other. DFL will facilitate the prediction which boosts the robustness of SLL by providing reliable soft labels, and the cross-domain features learned by SLL can effectively promote DFL to introduce more target label information. As training goes on, an increasing number of reliable soft-target labels will be introduced to learn the model. Such progressive learning can encourage SLDR to capture more accurate statistics of data distributions. We summarize our contributions as follows:

- We propose graph-based label propagation to assign soft labels for target samples. This will explore more intrinsic structure information across domain samples and get better label information inferred from the source domain.
- SLDR is to learn both domain-invariant and class discriminative feature representations for the source and target data, which could effectively minimize the considerable divergence between two domains and fully exploit the discriminative information of both domains to boost the final classification performance.
- The experimental results show that compared with the state-of-the-art methods, our method achieves the competitive performance on the Office31 benchmark and the challenging Office-Home benchmark.

## 2 Related Work

A common practice for unsupervised domain adaptation is to reduce the domain shifts between the source and target domain distributions [1,28,29,32] so as to obtain domain-invariant features. In this section, we briefly review on domain adaptation methods, which are closely related to our work.

Early instance-based methods [7,17] address the domain adaptation by using instance re-weighting strategy to select the source samples, which share similar distribution with the target domain, such as covariate shift and sample selection bias. Later, feature space transformation methods [20,25] start to take some strategies to match the domain distributions. For example, the Transfer Component Analysis (TCA) [25] tries to learn the domain-invariant features by using the MMD metric to match marginal distribution. Moreover, the proposed model joint distribution adaptation (JDA) [20] further reduces the domain shifts across two domains by jointly adapting both marginal and conditional distributions. [21,33] explore the benefits from jointly optimizing marginal distribution alignment and sample re-weighting. These studies ignore the discriminative properties for domain classes.

Recently, researchers start to consider the importance of discriminative properties between different classes for domain adaptation. In [5], the authors apply the center loss [31] to guarantee the class relationships in the source domain with better intra-class compactness and inter-class separability. [34] proposes Joint Geometrical and Statistical Alignment (JGSA) model, which simultaneously matches the domain distributions and considers the discriminative information of the source classes. However, these methods only concentrate on exploring the discriminative information of source data, none of them have ever investigated this property for the target data. We will show the effectiveness of reinforcing the discriminative properties for both domains.

Moreover, several works [6,15,18,23,24] utilize the hard-pseudo labels to compensate the lack of categorical information in the target domain. Since no label information in target domain for matching conditional distributions, these methods iteratively select pseudo-labeled target samples based on the source-domain classifier from the previous training epoch and then update the model by using the enlarged training set. Nevertheless, the direct use of such hard-pseudo labels might not be preferable due to possible domain mismatch. We resort to label propagation (LP) [35] strategy to solve this issue. [9] performs LP on a large image dataset with CNN descriptors for few shot learning. [16] applies LP to assign labels for unlabeled samples, and then choose the samples with reliable labels to retrain the model iteratively. However, these methods only apply LP to a single domain of semi-supervised learning scenario, we will explore it to cross-domain problems. [8] applies LP strategy for domain adaptation while it ignores the importance of discriminative feature properties between domains.

By jointly learning soft target labels and discriminative information of both domain data in a unified framework, improved recognition performance can be expected. Our method, applied to learned transferable features extracted from fine-tuned models and deep domain adaptation approaches like JAN [22], MADA [26], achieves better performance to these more complex methods and is expected to be incorporated directly into these end-to-end models.

## 3   The Proposed Method

### 3.1   Problem Settings and Notations

A domain $\mathcal{D}$ is composed of a feature space $\chi$ and a marginal probability distribution $P(x)$, $i.e.$, $x \in \chi$. For a specific domain, a task $\mathcal{T}$ consists of a $C$-cardinality label set $\mathcal{Y}$ and a classifier $f(x)$, $i.e.$, $\mathcal{T} = \{\mathcal{Y}, f(x)\}$, where $y \in \mathcal{Y}$, and $f(x) = \mathcal{Q}(y|x)$ is the conditional probability distribution.

In unsupervised domain adaptation, given a labeled source domain $\mathcal{D}_s = \{x_i^s, y_i^s\}_{i=1}^{n_s} = \{X_s, Y_s\}$ with $n_s$ labeled samples, where $x_i^s \in \mathbb{R}^{d_s}$ is the feature vector. Define an unlabeled target domain $\mathcal{D}_t = \{x_j^t\}_{j=1}^{n_t} = X_t$ with $n_t$ unlabeled target samples, where $x_i^t \in \mathbb{R}^{d_t}$. Note that $\mathcal{Y}_s = \mathcal{Y}_t$, $\chi_S \neq \chi_T$, $\mathcal{P}(\chi_s) \neq \mathcal{P}(\chi_t)$, $\mathcal{Q}(\mathcal{Y}_s|\chi_s) \neq \mathcal{Q}(\mathcal{Y}_t|\chi_t)$. The goal of SLDR is to project the source and the target domain into a shared feature space by projection matrices A and B with the following properties: 1) matching feature distributions; 2) making the domain classes more discriminative; 3) preserving the data manifolds and learning the soft target labels.

## 3.2    Graph-Based Label Propagation

To explore the conditional distributions, recent works [20,34] try to assign hard-pseudo labels for the target samples predicted by a sample Nearest Neighbor (NN) classifier. Although the distribution shifts can be reduced to some extent, NN still contains several limitations because NN is usually based on $L1$ or $L2$ distance. Thus, NN could fail to measure the similarity between domain samples which may be projected into a manifold space with complex geometric structure. Moreover, even in the latent feature space, the domain discrepancy still exists. To preserve the cross-domain data manifold structure and alleviate the bias caused by those underlying false-pseudo labels. We propose a graph-based label propagation optimization strategy for refining the target labels with soft ones. This term is defined as:

$$
\begin{aligned}
\mathcal{D}(Z) &= \frac{\nu}{2} \sum_{i,j=1}^{n_s+n_t} W_{ij} \left\| \frac{1}{\sqrt{D_{ii}}} z_i - \frac{1}{\sqrt{D_{jj}}} z_j \right\|^2 + (1-\nu) \sum_{j=1}^{C} \sum_{i=1}^{n_s+n_t} \| Z_{ij} - Y_{ij} \|^2 \\
&= \nu Z^T (I - D^{-\frac{1}{2}} W D^{-\frac{1}{2}}) Z + (1-\nu) \| Z - Y \|_F^2 ,
\end{aligned}
\tag{1}
$$

where $Y \in \mathbb{R}^{(n_s+n_t) \times C}$ is the initial label matrix, the first $n_s$ rows of $Y$ corresponding to labeled examples are one-hot encoded labels and the rest rows for the target data are zero. $Z \in \mathbb{R}^{(n_s+n_t) \times C}$ is the learning soft label matrix, $z_i \in \mathbb{R}^c$ is the $i$-th row of matrix $Z$, in which every element $z_{i,c}(z_{i,c} \geqslant 0 \text{ and } \sum_{c=1}^{C} z_{i,c} = 1)$ means the probability for the $i$-th data point belonging to the $c$-th category. $\|\bullet\|_F$ represents the Frobenius norm and $\nu \in [0,1)$ is a parameter . $W \in \mathbb{R}^{(n_s+n_t) \times (n_s+n_t)}$ is the symmetric nonnegative affinity matrix which is constructed by a k-nearest neighbors (k-NN) graph over source and target transformed samples (i.e., $A^T X_s$, $B^T X_t$). This matrix can preserve the geometric structure information among domain samples. $D$ denotes a diagonal matrix with $D_{ii} = \sum_{j=1}^{n_s+n_t} W_{ij}$. Note that in Eq. (1), the first term encourages smoothness such that nearby examples get the same predictions, while the second term attempts to maintain predictions for the labeled examples. In $W$, each entry $W_{ij}$ is formulated as:

$$
W_{ij} = \begin{cases} exp(\frac{-(\|h_i - h_j\|^2)}{2\sigma^2}), & if\ i \neq j \wedge h_i \in NN_k(h_j) \\ 0, & otherwise \end{cases}
\tag{2}
$$

where $h_i$ or $h_j$ is the transformed sample, $NN_k$ denotes the set of $k$ nearest neighbors of $h_j$, $\sigma$ is the length scale parameter and to be set as 1 in this paper.

### 3.3   Matching Domain Distributions

As aforementioned, the primary goal is to learn a domain-invariant feature space by matching domain distributions. We employ the nonparametric distance measurement MMD to adapt the distributions, which computes the distance between expectations of source and target data in the projected feature space. To fit different dimensionality between source and target feature, we apply two coupled projections ($A \in \mathbb{R}^{d_s \times d}$ for source domain, and $B \in \mathbb{R}^{d_t \times d}$ for target domain) to project both domain data into the shared feature subspace,

$$E_M(A, B) - \left\| \frac{1}{n_s} \sum_{i=1}^{n_s} A^T x_i^s - \frac{1}{n_t} \sum_{j=1}^{n_t} B^T x_j^t \right\|^2. \tag{3}$$

The MMD strategy in Eq. (3) can reduce the difference of the marginal distributions, but it fails to guarantee that the divergency of conditional distributions is minimized. We propose to predict soft labels for unlabeled target data. This will develop a probabilistic class-wise adaptation formula to effectively guide the intrinsic knowledge transfer. Formally,

$$E_C(A, B, Z_t) = \sum_{c=1}^{C} \left\| \frac{1}{n_s^c} \sum_{x_i \in \mathcal{D}_s^c} A^T x_i - \frac{1}{n_t^c} \sum_{x_j \in \mathcal{D}_t} z_{j,c}^t B^T x_j \right\|^2, \tag{4}$$

where $\mathcal{D}_s^c = \{x_i : x_i \in \mathcal{D}_s \wedge y(x_i) = c\}$ denotes the set of source domain samples belonging to the $c$-th class , $y(x_i)$ represents the true label of $x_i$, and $n_s^c = |\mathcal{D}_s^c|$. $z_{j,c}^t$ is the probability for the $j$-th unlabeled target sample belonging to the $c$-th class. Thus, the $c$-th target sample size $n_t^c$ can be approximately computed by $n_t^c = \sum_{j=1}^{n_t} z_{j,c}^t$. $Z_t \in \mathbb{R}^{n_t \times C}$ is the target soft label matrix constructed by the target $n_t$ rows of the learning soft label matrix $Z$. We denote $Z_t$ as $Z_t = [\alpha^1, ..., \alpha^c, ..., \alpha^C]$ , $\alpha^c = [z_{1,c}^t; ...; z_{n_t,c}^t] \in \mathbb{R}^{n_t}$.

Then, we get the distribution shifts measurement term,

$$E_{MC}(A, B, Z_t) = Tr \left( [A^T \; B^T] \begin{bmatrix} M_{ss} & M_{st} \\ M_{ts} & M_{tt} \end{bmatrix} \begin{bmatrix} A \\ B \end{bmatrix} \right), \tag{5}$$

where

$$M_{ss} = X_s \left( H_{sm} + \sum_{c=1}^{C} H_{sc} \right) X_s^T, H_{sm} = \frac{1}{n_s^2} 1_s 1_s^T,$$

$$(H_{sc})_{ij} = \begin{cases} \frac{1}{(n_s^c)^2}, & x_i, x_j \in \mathcal{D}_s^c \\ 0, & otherwise \end{cases}$$

$$M_{st} = X_s \left( H_{stm} + \sum_{c=1}^{C} H_{stc} \right) X_t^T, \tag{6}$$

$$H_{stm} = -\frac{1}{n_s n_t} 1_s 1_t^T, H_{stc} = -\frac{\beta^c \alpha^{cT}}{n_s^c \sum_{j=1}^{n_t} z_{j,c}^t},$$

$$M_{tt} = X_t \left( H_{tm} + \sum_{c=1}^{C} H_{tc} \right) X_t^T, H_{tm} = \frac{1}{n_t^2} 1_t 1_t^T,$$

$$H_{tc} = \frac{\alpha^c \alpha^{cT}}{\sum_{i=1}^{n_t} z_{i,c}^t \sum_{j=1}^{n_t} z_{j,c}^t}, M_{ts} = X_t \left( H_{tsm} + \sum_{c=1}^{C} H_{tsc} \right) X_s^T, \tag{7}$$

$$H_{tsm} = -\frac{1}{n_s n_t} 1_t 1_s^T, H_{tsc} = -\frac{\alpha^c \beta^{cT}}{n_s^c \sum_{j=1}^{n_t} z_{j,c}^t}.$$

Note that, $\beta^c \in \mathbb{R}^{n_s}, (\beta^c)_i = \begin{cases} 1, & if \ x_i^s \in \mathcal{D}_s^c \\ 0, & otherwise \end{cases}$. $1_t \in \mathbb{R}^{n_t}$ and $1_s \in \mathbb{R}^{n_s}$ are the column vectors with all ones.

### 3.4   Learning Discriminative Features

The goal of minimizing the marginal and conditional distributions is to obtain domain-invariant features. However, these features may not be discriminative enough, which will degrade the final classification results. In this paper, we propose two constraints of inter-class separability and intra-class compactness to keep the domain-invariant features discriminative enough, and finally to improve the classification performance.

**Inter-class Separability.** This constraint is expected to preserve the inter-class disparity as much as possible by maximizing the distance of different class instance in the source and target domains, respectively. The goal is to make the clusters of diverse classes to be discriminative enough. To be specific, for the source data, we can formulate the distance between samples in different categories as:

$$E_{dif}^s(A) = \sum_{c=1}^{C} \sum_{\substack{k=1, \\ k \neq c}}^{C} \left\| \frac{1}{n_s^c} \sum_{x_i \in \mathcal{D}_s^c} A^T x_i - \frac{1}{n_s^k} \sum_{x_j \in \mathcal{D}_s^k} A^T x_j \right\|^2$$

$$= \sum_{c=1}^{C} \sum_{\substack{k=1, \\ k \neq c}}^{C} Tr \left( A^T X_s H_{diff}^{ss} X_s^T A \right), \tag{8}$$

where $H_{dif}^{ss}$ is defined as:

$$(H_{dif}^{ss})_{ij} = \begin{cases} \frac{1}{n_s^c n_s^c}, & x_i, x_j \in \mathcal{D}_s^c \\ \frac{1}{n_s^k n_s^k}, & x_i, x_j \in \mathcal{D}_s^k \\ -\frac{1}{n_s^c n_s^k}, & \begin{cases} x_i \in \mathcal{D}_s^c, x_j \in \mathcal{D}_s^k \\ x_i \in \mathcal{D}_s^k, x_j \in \mathcal{D}_s^c \end{cases} \\ 0, & otherwise \end{cases} \tag{9}$$

Similar to Eq. (8), we calculate the distance term for the target samples with the learning soft labels,

$$
E_{dif}^t(B) = \sum_{\substack{c,k=1,\\k\neq c}}^{C} \left\| \frac{1}{n_t^c} \sum_{x_i\in\mathcal{D}_t} z_{i,c}^t B^T x_i - \frac{1}{n_t^k} \sum_{x_j\in\mathcal{D}_t} z_{j,k}^t B^T x_j \right\|^2
$$

$$
= \sum_{c=1}^{C} \sum_{\substack{k=1,\\k\neq c}}^{C} Tr(B^T X_t H_{dif}^{tt} X_t^T B), \tag{10}
$$

where $H_{dif}^{tt}$ is defined as:

$$
H_{dif}^{tt} = \frac{\alpha^c \alpha^{cT}}{\sum_{i=1}^{n_t} z_{i,c}^t \sum_{j=1}^{n_t} z_{j,c}^t} - \frac{\alpha^c \alpha^{kT}}{\sum_{i=1}^{n_t} z_{i,c}^t \sum_{j=1}^{n_t} z_{j,k}^t} + \frac{\alpha^k \alpha^{kT}}{\sum_{i=1}^{n_t} z_{i,k}^t \sum_{j=1}^{n_t} z_{j,k}^t} - \frac{\alpha^k \alpha^{cT}}{\sum_{i=1}^{n_t} z_{i,k}^t \sum_{j=1}^{n_t} z_{j,c}^t} \tag{11}
$$

For clarity, we can rewrite the inter-class distance loss term of both domains in a whole formulation as:

$$
E_{dif} = E_{dif}^s + E_{dif}^t = Tr\left(A^T H_d^s A + B^T H_d^t B\right), \tag{12}
$$

where $H_d^s = \sum_{c=1}^{C} \sum_{\substack{k=1,\\k\neq c}}^{C} X_s H_{dif}^{ss} X_s^T$ and $H_d^t = \sum_{c=1}^{C} \sum_{\substack{k=1,\\k\neq c}}^{C} X_t H_{dif}^{tt} X_t^T$.

**Intra-class Compactness.** We further minimize the intra- class variation to make the clusters more discriminative. For the source domain, we use the following formula:

$$
E_{same}^s(A) = \sum_{c=1}^{C} \sum_{x_i,x_j\in\mathcal{D}_s^c} \left\| A^T x_i - A^T x_j \right\|^2
$$

$$
= \sum_{c=1}^{C} Tr\left(A^T X_s H_{same}^{ss} X_s^T A\right), \tag{13}
$$

where

$$
(H_{same}^{ss})_{ij} = \begin{cases} |\mathcal{D}_s^c|, & if\ i=j \\ -1, & if\ i\neq j \wedge x_i,x_j \in \mathcal{D}_s^c \\ 0 & otherwise \end{cases} \tag{14}
$$

A similar term for the target domain can be obtained:

$$
E_{same}^t(B) = \sum_{c=1}^{C} Tr\left(B^T X_t H_{same}^{tt} X_t^T B\right), \tag{15}
$$

$E_{same}^s$ and $E_{same}^t$ can also be intergraded as a whole term,

$$
E_{same} = E_{same}^s + E_{same}^t = Tr\left(A^T H_{sa}^s A + B^T H_{sa}^t B\right), \tag{16}
$$

where $H_{sa}^s = \sum_{c=1}^{C} X_s H_{same}^{ss} X_s^T$ and $H_{sa}^t = \sum_{c=1}^{C} X_t H_{same}^{tt} X_t^T$.

## 3.5   Learning Algorithm

**Overall Objective Function.** We formulate the SLDR approach by incorporating the above four Eqs. ((1), (5), (12) and (16)) into one object function as follows:

$$
\begin{aligned}
\min_{A,B} Tr & \left( A^T(M_{ss} + \mu(H_{sa}^s - H_d^s) + I_{d_s})A + A^T M_{st}B \right) \\
& + Tr \left( B^T(M_{tt} + \mu(H_{sa}^t - H_d^t) + I_{d_t})B + B^T M_{ts}A \right) \\
& + \nu Z^T(I - D^{-\frac{1}{2}}WD^{-\frac{1}{2}})Z + (1-\nu)\|Z - Y\|_F^2 \\
& s.t. \quad A^T X_s H_s X_s^T A = I_d, B^T X_t H_t X_t^T B = I_d,
\end{aligned}
\tag{17}
$$

where $\mu$ and $\nu$ are regularization parameters. $I$ is identity matrix, and $H_s$ and $H_t$ are the centering matrixes, which are defined as $H_s = I_{n_s} - \frac{1}{n_s}1_{n_s}1_{n_s}^T$ and $H_t = I_{n_t} - \frac{1}{n_t}1_{n_t}1_{n_t}^T$, respectively.

---

**Algorithm 1:** Soft Labels transfer with Discriminative Representations learning (SLDR) framework.

---

**Require:** Data and source labels: $X_s$, $X_t$, $Y_s$; Subspace $K$, iteration $T$; Regularization parameters: $\mu$, $\nu$.

**Ensure:** Adpatation matrices: $A$ and $B$; The final target labels $Y_t^T$.

1: **while** $\sim isempty(X_s, Y_s)$ and $t < T$ **do**
2:   **Step 1:** Construct $M_{ss}$, $M_{st}$, $M_{tt}$, $M_{ts}$, $H_{sa}^s$, $H_d^t$,
3:   $H_{sa}^t$, and $H_d^s$ according to corresponding equation;
4:   **Step 2:** Solve the generalized eigendecomposition
5:   problem as in Eq. (17) and obtain domain adaptation
6:   matrices $A$ and $B$, then embed data resort to the
7:   transformations: $E_s = A^T X_s$, $E_t = B^T X_t$;
8:   **Step 3:** Soft target labels learning;
9:   **if** $\sim isempty(Z_s, Y_s)$ **then**;
10:     (i) construct the label matrix $Y$;
11:     (ii) initialize the graph $G$, construct the affinity
12:   matrix $W$ and diagonal matrix $D$;
13:     (iii) obtain $Z^T$ in solving Eq.(20);
14:   **else**;
15:     **break**;
16:   **Step 4:** update soft target labels by:
17:   $Z_t^T = Z^T[:, (n_s + 1) : (n_s + n_t)]$;
18:   **Step 5:** Return to Step1; $t = t + 1$;
19: Obtain the final target labels $Y_t^T = argmax(Z_t^T)$.

---

**Optimization Solution.** It is easy to check directly optimizing this Eq. (17) is nontrivial. To address this optimization problem, we rewrite $[A^T \ B^T]$ as $P^T$.

Then, we divided the function into two sub-problems. The first one is:

$$\min_{P^TOP=I_{2d}} Tr\left(P^TQP\right),\tag{18}$$

where $Q = \begin{bmatrix} M_{ss} + \mu(H_{sa}^s - H_d^s) + I_{d_s} & M_{st} \\ M_{ts} & M_{tt} - \mu(H_{sa}^t - H_d^t) + I_{d_t} \end{bmatrix}$ and $O = \begin{bmatrix} X_sH_sX_s^T & 0 \\ 0 & X_tH_tX_t^T \end{bmatrix}$. The other sub-one is:

$$\min_{P^TOP=I_{2d}} \nu Z^T(I - D^{-\frac{1}{2}}WD^{-\frac{1}{2}})Z + (1-\nu)\|Z-Y\|_F^2.\tag{19}$$

These two sub-problems implement an EM-like optimization step to iteratively update the variables. For the E-step, we fix $P^T$ and update the soft label matrix $Z$ by:

$$Z = (D - \nu W)^{-1}Y,\tag{20}$$

where $Y$ is the initial label matrix. For the M-step, we update the subspace projection $P$ with the learned $Z$ by:

$$QP = O\Phi,\tag{21}$$

where $\Phi = diag(\phi_1, ..., \phi_{2d})$ is the Lagrange function, $(\phi_1, ..., \phi_{2d})$ are the $2d$ leading eigenvalues and $O = [O_1, ..., O_{2d}]$ contains the corresponding eigenvectors, which can be solved through eigenvalue decomposition. The detailed optimization process of the proposed method is shown in Algorithm 1.

## 4    Experiments

### 4.1    Datasets and Experimental Settings

We use two popular DA datasets: 1) **Office31** [27] contains 4652 images and 31 categories collected from 3 domains: Amazon(A), Webcam(W) and DSLR(D). The low resolution images in Webcam are captured with a web camera. The medium resolution images in Amazon are downloaded from amazon.com. DSLR consists of high resolution images collected by a SLR camera. 2) **Office-Home**[1] is a more challenge dataset for domain adaptation, which consists of 15,500 images in total from 65 categories of common objects in office and home settings. These images come from 4 significantly different domains: Artistic images (Ar), Clip Art (Cl), Product images (Pr) and Real-World images (Rw).

We compare our approach with several recent works. They are shallow UDA methods: JDA [20] and JGSA [34] utilize the hard pseudo-target labels for matching marginal and conditional distributions; GAKT [8] uses a single label propagation strategy to perform the class-wise adaptation. ResNet-50 [14] is adopted as the basic method. We further compare with several deep UDA approaches: JAN [22] matches the joint distributions of multiple domain-specific layers with

---

[1]  https://hemanthdv.github.io/officehome-dataset/.

joint kernel MMD loss, MADA [26] resorts to hard pseudo-target labels to capture multimode structures to enable fine-grained alignment, JDDA [5] introduces the discriminative properties in the source domain, PFAN [6] progressively refines the hard pseudo target labels for class-wise adaptation, TAT [19] generates transferable examples to fill in the domain gap, and then adversarially trains the deep classifiers to make consistent predictions over the these examples, DSR [2] employs a variational auto-encoder to reconstruct the semantic latent variables and domain latent variables for domain adaptation.

We follow the same protocols as previous studies [22, 26] in our experiments. We further exploit the features after the 5-th pooling layer fine-tuned on the source domain, making it fair to be compared with deep UDA methods via ResNet-50. In all our experiments, we adopt $k$-nearest neighbor graph ($k = 5$) with heat-kernel weight. We set the subspace dimension $K = 200$, and the iteration number is set as T = 10. The regularization parameters are set as $\mu = 1.0$, $\nu = 0.9$ for Office31, and $\mu = 0.5$, $\nu = 0.9$ for Office-Home, respectively.

### 4.2    Results and Discussion

**Results on Office31.** As shown in Table 1, SLDR outperforms all deep methods in terms of the average accuracy and consistently performs better than TAT and DSR in 4 out of 6 tasks. It is noteworthy that SLDR shows competitive results to the deep methods for some challenging adaptation tasks, such as D → A and W → A, where a small source domain is adapted to a large target domain. The encouraging results on these hard transfer tasks demonstrate the importance of optimizing the class-wise adaptation with soft target labels and suggest that our method can learn more transferable features.

We wonder whether deep adaptation approaches can be promoted with our method. Here we investigate this issue by exploring two classical deep adaptation methods, *i.e.*, JAN and MADA. Table 1 shows, with the help of SLDR, the average accuracy of these two models have been significantly improved, from 84.3% to 89.1% and 85.2% to 89.5%. MADA+SLDR even beats DSR.

**Results on Office-Home.** As shown in Table 2. Obviously, Our SLDR model outperforms the shallow comparison methods on all the transfer tasks . Compared with those state-of-the-art deep methods, *i.e.*, TAT and DSR, our model achieves quite competitive results and performs best in 11 out of 12 tasks. Moreover, in contrast to Office-31, the general accuracy results are much lower. The reason is that the Office-Home dataset contains more categories, which makes the error accumulation with much higher possibility.

All of the above results reveal several observations. (1) Taking class information of the target samples into account is beneficial to the adaptation. It can be seen that MADA and our method achieve better performance than those class-agnostic approaches, *i.e.*, JDDA. (2) JGSA, PFAN and MADA assign a single or hard label for every target sample, which may deteriorate the following training when the target samples are assigned with wrong labels. On the contrary, SLDR applies the soft target labels, which further adopts the fine-tuned deep features

**Table 1.** Accuracy (%) on Office-31 for unsupervised domain adaptation.

| Method | A → W | D → W | W → D | A → D | D → A | W → A | Avg |
|---|---|---|---|---|---|---|---|
| ResNet-50 [14] | 68.4 | 93.2 | 97.3 | 68.9 | 62.5 | 60.7 | 75.1 |
| JAN [22] | 85.4 | 97.4 | 99.7 | 84.7 | 68.6 | 70.0 | 84.3 |
| MADA [26] | 90.0 | 97.4 | 99.6 | 87.8 | 70.3 | 66.4 | 85.2 |
| JDDA [5] | 82.6 | 95.2 | 99.7 | 79.8 | 57.4 | 66.7 | 80.2 |
| PFAN [6] | 83.0 | 99.1 | 99.9 | 76.3 | 63.3 | 60.8 | 80.4 |
| TAT [19] | 92.5 | 99.3 | **100.0** | 93.2 | 73.1 | 72.1 | 88.4 |
| DSR [2] | 93.1 | 98.7 | 99.8 | 92.4 | 73.5 | 73.9 | 88.6 |
| SLDR | 93.3 | 99.3 | 99.2 | 92.4 | 73.7 | 74.1 | 88.7 |
| JAN+SLDR | 93.5 | 99.5 | **100.0** | 93.4 | 73.9 | 74.5 | 89.1 |
| MADA+SLDR | **94.2** | **99.8** | **100.0** | **93.8** | **74.1** | **74.8** | **89.5** |

**Table 2.** Accuracy (%) for cross-domain experiments on Office-Home.

| Method | Ar ↓ Cl | Ar ↓ Pr | Ar ↓ Rw | Cl ↓ Ar | Cl ↓ Pr | Cl ↓ Rw | Pr ↓ Ar | Pr ↓ Cl | Pr ↓ Rw | Rw ↓ Ar | Rw ↓ Cl | Rw ↓ Pr | Avg |
|---|---|---|---|---|---|---|---|---|---|---|---|---|---|
| JGSA [34] | 28.8 | 37.6 | 48.9 | 31.7 | 46.3 | 46.8 | 28.7 | 35.9 | 54.5 | 40.6 | 40.8 | 59.2 | 41.7 |
| GAKT [8] | 34.5 | 43.6 | 55.3 | 36.1 | 52.7 | 53.2 | 31.6 | 40.6 | 61.4 | 45.6 | 44.6 | 64.9 | 47.0 |
| ResNet-50 | 34.9 | 50.0 | 58.0 | 37.4 | 41.9 | 46.2 | 38.5 | 31.2 | 60.4 | 53.9 | 41.2 | 59.9 | 46.1 |
| JAN [22] | 45.9 | 61.2 | 68.9 | 50.4 | 59.7 | 61.0 | 45.8 | 43.4 | 70.3 | 63.9 | 52.4 | 76.8 | 58.3 |
| TAT [19] | 51.6 | 69.5 | 75.4 | **59.4** | 69.5 | 68.6 | 59.5 | 50.5 | 76.8 | 70.9 | 56.6 | 81.6 | 65.8 |
| DSR [2] | 53.4 | 71.6 | 77.4 | 57.1 | 66.8 | 69.3 | 56.7 | 49.2 | 75.7 | 68.0 | 54.0 | 79.5 | 64.9 |
| SLDR | **54.2** | **73.9** | **77.8** | 58.9 | **70.1** | **71.9** | **59.8** | **51.9** | **78.4** | **72.4** | **56.9** | **83.9** | **67.5** |

and explores the intrinsic structure to benefit our final performance. (3) SLDR achieves competitive results to some recent best performance shallow method GAKT and deep models like TAT and DSR. The reason is that only SLDR adopts the strategy of conducting discriminative feature learning and soft labels to benefit each other for effective knowledge transfer.

## 4.3 Empirical Analysis

**Ablation Study.** Table 3 examines two key components of SLDR, *i.e.*, label propagation (LP) and discriminative properties (DP). We conduct ablation study on several tasks with different components ablation. The method "SLDR-LP" means the model learns the discriminative domain-invariant features with hard-pseudo target labels. "SLDR-DP" learns the features with soft-target labels but without discriminative properties. Interestingly, "SLDR-DP" achieves better than "SLDR-LP", which verifies target samples assigned with soft labels can alleviate some wrong information caused by false pseudo-target labels. SLDR gets the best result, which confirms all the components are designed reasonably.

**Feature Visualization.** We apply t-SNE to visualize the features on task A → D (randomly selected 10 classes) learned by Source Only model (ResNet), MADA and SLDR respectively. In Fig. 2(a)–2(c), we plot the feature distribution from

**Table 3.** The effect of alternative optimization (LP) and (DP).

| Method | A → W | D → W | Ar → Pr | Ar → Rw |
|---|---|---|---|---|
| SLDR-LP | 91.5 | 98.0 | 73.1 | 75.9 |
| SLDR-DP | 92.4 | 98.8 | 73.5 | 76.8 |
| SLDR | 93.3 | 99.3 | 73.9 | 77.8 |

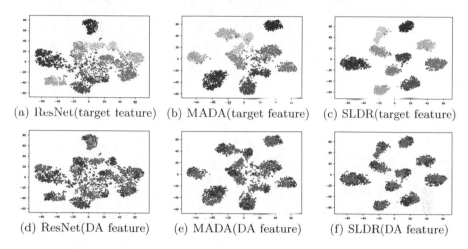

(a) ResNet(target feature)  (b) MADA(target feature)  (c) SLDR(target feature)

(d) ResNet(DA feature)  (e) MADA(DA feature)  (f) SLDR(DA feature)

**Fig. 2.** The t-SNE visualization of features extracted by different models for A → D task. (a)–(c) represent the distribution from category perspective and each color denotes a category. (d)–(f) represent the distribution from domain alignment (DA) perspective, where red and blue points represent samples of source and target domains, respectively. (Best Viewed in Color). (Color figure online)

category perspective and each color denotes a category. Figure 2(d)–2(f) show the domain alignment information, where red and blue points represent samples of source and target domain, respectively. The visualization results reveal several intuitive observations. (1) Compared with Fig. 2(b), the distributions of Fig. 2(a) have more scattered points distributed in the inter-class gap, which verifies features learned by the hard target label guided model MADA are discriminated much better than that learned by the no domain adaptation metric model (ReNet). Besides, Fig. 2(c) shows the representations learned by our method that features in the same class are much more compact and features with different classes are well separated. This makes the features more discriminative. (2) From domain alignment perspective, in Fig. 2(d) and 2(e), the source and target domain distributions are made indistinguishable, but different categories are not aligned very well across domains. However, with features learned by SLDR, the categories are aligned much better.

**Misclassified Samples Analysis.** To deeply explore the advantages of SLDR, Fig. 3 shows randomly selected misclassified samples of MADA for the task D →

**Fig. 3.** Misclassified samples analysis of MADA and SLDR for task D→A of Office-31 with respect to classes "Mobile Phone" and "Ring Binder". Red and black are the misclassified and correct samples. (Color figure online)

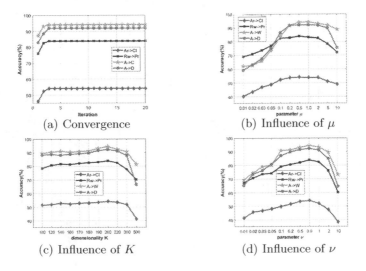

**Fig. 4.** (a)~(d): Parameter sensitivity.

A with respect to the classes "Mobile Phone" and "Ring Binder". For MADA, it will misclassify the target samples that are much similar to other classes in source domain, which means without considering the discriminative structures and alleviating the wrong information caused by those false pseudo-target labels, JAN will mix up source and target samples. By contrast, SLDR will distinguish those similar cross-domain samples.

**Parameter Sensitivity.** As shown in Fig. 4(a), for the convergence performance, SLDR can reach a steady performance in only a few $T < 5$ iterations. We further run SLDR with different values of $\mu$. Figure 4(b) shows the accuracy increases first and then decreases as the parameter increases. This confirms the validity of jointly adapting distributions and exploring soft class information. As can be observed in Fig. 4(c)–(d), SLDR is robust with regard to different values of $K$. For high-dimensional features via ResNet-50, $K \in [100, 200]$ is an optimal choice. We can also find the suitable value of parameter $\nu$ is 0.9.

## 4.4 Conclusion

In this paper, we propose a simple, yet effective method for visual domain adaptation. The key idea is to jointly optimize the target labels and learn the discriminative domain-invariant features in a unified model. Extensive experiments on several cross-domain datasets benefit from the interest of applying label propagation to explore the soft label information of target samples for class-wise adaptation and keeping the discriminative properties of the learned features.

**Acknowledgement.** This work was supported by the National Key Research and Development Program of China, No.2018YFB1402600.

# References

1. Bermúdez Chacón, R., Salzmann, M., Fua, P.: Domain-adaptive multibranch networks. In: ICLR. No. CONF (2020)
2. Cai, R., Li, Z., Wei, P., Qiao, J., Zhang, K.: Learning disentangled semantic representation for domain adaptation. In: IJCAI (2019)
3. Cao, M., Zhou, X., Xu, Y., Pang, Y., Yao, B.: Adversarial domain adaptation with semantic consistency for cross-domain image classification. In: CIKM (2019)
4. Cao, Z., Ma, L., Long, M., Wang, J.: Partial adversarial domain adaptation. In: ECCV (2018)
5. Chen, C., Chen, Z., Jiang, B., Jin, X.: Joint domain alignment and discriminative feature learning for unsupervised deep domain adaptation. In: AAAI (2019)
6. Chen, C., et al.: Progressive feature alignment for unsupervised domain adaptation. In: CVPR (2019)
7. Chen, M., Weinberger, K.Q., Blitzer, J.: Co-training for domain adaptation. In: NIPS, pp. 2456–2464 (2011)
8. Ding, Z., Li, S., Shao, M., Fu, Y.: Graph adaptive knowledge transfer for unsupervised domain adaptation. In: ECCV, pp. 37–52 (2018)
9. Douze, M., Szlam, A., Hariharan, B., Jégou, H.: Low-shot learning with large-scale diffusion. In: CVPR (2018)
10. Gautheron, L., Redko, I., Lartizien, C.: Feature selection for unsupervised domain adaptation using optimal transport. In: ECML-PKDD, pp. 759–776 (2018)
11. Ghifary, M., Balduzzi, D., Kleijn, W.B., Zhang, M.: Scatter component analysis: a unified framework for domain adaptation and domain generalization. IEEE Trans. Pattern Anal. Mach. Intell. **39**(7), 1414–1430 (2016)
12. Goodfellow, I., et al.: Generative adversarial nets. In: NIPS (2014)
13. Gretton, A., Borgwardt, K., Rasch, M., Schölkopf, B., Smola, A.J.: A kernel method for the two-sample-problem. In: NIPS, pp. 513–520 (2007)
14. He, K., Zhang, X., Ren, S., Sun, J.: Deep residual learning for image recognition. In: CVPR, pp. 770–778 (2016)
15. Huang, J., Zhou, Z.: Transfer metric learning for unsupervised domain adaptation. IET Image Proc. **13**(5), 804–810 (2019)
16. Iscen, A., Tolias, G., Avrithis, Y., Chum, O.: Label propagation for deep semi-supervised learning. In: CVPR (2019)
17. Jiang, J., Zhai, C.: Instance weighting for domain adaptation in NLP. In: ACL, pp. 264–271 (2007)

18. Li, S., Song, S., Huang, G., Ding, Z., Wu, C.: Domain invariant and class discriminative feature learning for visual domain adaptation. IEEE Trans. Image Process. **27**(9), 4260–4273 (2018)
19. Liu, H., Long, M., Wang, J., Jordan, M.: Transferable adversarial training: a general approach to adapting deep classifiers. In: ICML, pp. 4013–4022 (2019)
20. Long, M., Wang, J., Ding, G., Sun, J., Yu, P.S.: Transfer feature learning with joint distribution adaptation. In: ICCV (2013)
21. Long, M., Wang, J., Ding, G., Sun, J., Yu, P.S.: Transfer joint matching for unsupervised domain adaptation. In: CVPR, pp. 1410–1417 (2014)
22. Long, M., Zhu, H., Wang, J., Jordan, M.I.: Deep transfer learning with joint adaptation networks. In: ICML, pp. 2208–2217 (2017)
23. Lu, H., Shen, C., Cao, Z., Xiao, Y., van den Hengel, A.: An embarrassingly simple approach to visual domain adaptation. IEEE Trans. Image Process. **27**(7), 3403–3417 (2018)
24. Luo, Y., Zheng, L., Guan, T., Yu, J., Yang, Y.: Taking a closer look at domain shift: category-level adversaries for semantics consistent domain adaptation. In: CVPR, pp. 2507–2516 (2019)
25. Pan, S.J., Tsang, I.W., Kwok, J.T., Yang, Q.: Domain adaptation via transfer component analysis. IEEE Trans. Neural Networks **22**(2), 199–210 (2011)
26. Pei, Z., Cao, Z., Long, M., Wang, J.: Multi-adversarial domain adaptation. In: AAAI (2018)
27. Saenko, K., Kulis, B., Fritz, M., Darrell, T.: Adapting visual category models to new domains. In: ECCV (2010)
28. Soh, J.W., Cho, S., Cho, N.I.: Meta-transfer learning for zero-shot super-resolution. In: CVPR (2020)
29. Sohn, K., Shang, W., Yu, X., Chandraker, M.: Unsupervised domain adaptation for distance metric learning. In: ICLR (2019)
30. Wang, R., Utiyama, M., Liu, L., Chen, K., Sumita, E.: Instance weighting for neural machine translation domain adaptation. In: EMNLP (2017)
31. Wen, Y., Zhang, K., Li, Z., Qiao, Y.: A discriminative feature learning approach for deep face recognition. In: ECCV (2016)
32. Xie, S., Zheng, Z., Chen, L., Chen, C.: Learning semantic representations for unsupervised domain adaptation. In: ICML, pp. 5419–5428 (2018)
33. Zhang, J., Ding, Z., Li, W., Ogunbona, P.: Importance weighted adversarial nets for partial domain adaptation. In: CVPR, pp. 8156–8164 (2018)
34. Zhang, J., Li, W., Ogunbona, P.: Joint geometrical and statistical alignment for visual domain adaptation. In: CVPR, pp. 1859–1867 (2017)
35. Zhou, D., Bousquet, O., Lal, T.N., Weston, J., Schölkopf, B.: Learning with local and global consistency. In: NIPS (2004)
36. Zhuo, J., Wang, S., Cui, S., Huang, Q.: Unsupervised open domain recognition by semantic discrepancy minimization. In: CVPR, pp. 750–759 (2019)

# Information-Bottleneck Approach
# to Salient Region Discovery

Andrey Zhmoginov$^{(\boxtimes)}$, Ian Fischer, and Mark Sandler

Google Inc., Mountain View, CA, USA
azhmogin@google.com

**Abstract.** We propose a new method for learning image attention masks in a semi-supervised setting based on the Information Bottleneck principle. Provided with a set of labeled images, the mask generation model is minimizing mutual information between the input and the masked image while maximizing the mutual information between the same masked image and the image label. In contrast with other approaches, our attention model produces a Boolean rather than a continuous mask, entirely concealing the information in masked-out pixels. Using a set of synthetic datasets based on MNIST and CIFAR10 and the SVHN datasets, we demonstrate that our method can successfully attend to features known to define the image class.

## 1 Introduction

Information processing in deep neural networks is carried out in multiple stages and the data-processing inequality implies that the information content of the input signal decays as it undergoes consecutive transformations. Even though this applies to both information that is relevant and irrelevant for the task at hand, in a well-trained model, most of the useful information in the signal will be preserved up to the network output. However, standard objectives, such as the cross-entropy loss, do not constrain the irrelevant information that is retained in the output.

The Information Bottleneck (IB) framework [28,29] constrains the information content retained at the output by trading off between prediction and compression: $IB \equiv \min \beta \mathbb{I}(X; Z) - \mathbb{I}(Y; Z)$, where $X$ is the input, $Y$ is the target output, and $Z$ is the learned representation. This framework has been applied to numerous deep learning tasks including a search of compressed input representations [2,9,17], image segmentation [3], data clustering [24,26], generalized dropout [1], Generative Adversarial Networks [19] and others.

In this paper, we use the IB approach to generate saliency maps for image classification models, directing model attention away from distracting features and towards features that define the image label. The method is based on the

**Electronic supplementary material** The online version of this chapter (https://doi.org/10.1007/978-3-030-67664-3_32) contains supplementary material, which is available to authorized users.

© Springer Nature Switzerland AG 2021
F. Hutter et al. (Eds.): ECML PKDD 2020, LNAI 12459, pp. 531–546, 2021.
https://doi.org/10.1007/978-3-030-67664-3_32

observation that the information content of the image region that we want to "attend to" should ideally be minimized while still being descriptive of the image class.

The proposed technique can be thought of as a form of semi-supervised attention learning. The entire model consisting of the per-pixel mask generator and the classifier operating on the masked regions can also be viewed as a step towards "explainable models", which not only make predictions, but also assign importance to particular input components. This technique could potentially be useful for datasets that can be costly to annotate by experts, like for example in medical imaging applications where acquiring image labels can be easier than getting reliable annotation regions.

The paper is structured as follows. Section 2 describes prior work and relates our approach to other existing methods. In Sect. 3 we outline theoretical foundations of our method and the experimental results are summarized in Sect. 4. Section 5 discusses an alternative IB-based approach and finally, Sect. 6 summarizes our conclusions.

## 2  Prior Work

Semi-supervised image segmentation is a task of learning to identify object boundaries without access to the boundary groundtruth information. Object detection and reconstruction of object shape in the context of this task is frequently achieved based on the knowledge of image labels alone [10, 13, 15, 30, 32]. A similar task of identifying salient regions in input samples is frequently addressed in the studies of neural network interpretation [5, 7, 8, 20, 22, 23, 31]. A successfully trained model would effectively "know" which parts of the input image carry information defining the image class and which parts are irrelevant. Most of these methods use (in one or another way) a signal supplied by the classification model with a partially occluded input. By changing the attention mask and probing classifier performance it is possible to identify "salient" regions, as well as those regions that are not predictive of the object present in the image.

Most semi-supervised semantic segmentation approaches including those mentioned above tend to rely on hand-designed optimization objectives and supplementary techniques that are carefully tuned to work well in specialized domains. In contrast, more general frameworks, like those based on information theory, could provide a more elegant and universal alternative. In recent years, the Information Bottleneck method has been applied to generating instance-based input attention maps. Most notably, an information-theoretic generalization of dropout called Information Dropout [1] based on element-wise tensor masking was shown to successfully generate representations insensitive to nuisance factors present in the model input. Another novel approach called Info-Mask recently proposed in [27] independently of our work, applies IB-inspired approach to generating continuous attention masks for the image classification task. The authors demonstrated superior performance of InfoMask on the Chest Disease localization task compared to multiple other existing methods.

A similar IB approach has also been recently applied to the attribution problem [21]. Given a *pre-trained* classification model, this method produces an attribution heat map that emphasizes image regions responsible for assigning this image a particular label.

In this work, we propose an alternative approach using variational bounds on the full and conditional Information Bottleneck optimization objectives. In contrast to the approaches described above, we target the information content of the masked image and we do not multiply image pixels by a floating-point continuous mask, but instead use Boolean masks, thus completely preventing masked-out pixels from propagating any information to the model output. An alternative approach to blocking information coming from partially-masked pixels is based on adding pixel noise into the masked image [21], but we do not address this approach here.

## 3   Model

Consider a conventional image classification task trained on samples drawn from a joint distribution $p(I, C)$ with the random variable $I$ corresponding to images and $C$ being image classes. Let us tackle a complimentary task of learning a self-attention model that given an image $i$ produces such a Boolean pixel mask[1] $m_\zeta(i)$ that the masked image $i \odot m_\zeta(i)$ satisfies two following properties: (a) it captures as little information from the original image as possible, but (b) it contains enough information about the contents of the image for the model to predict the image class.

Using the language of information theory these two conditions can be satisfied by writing a single optimization objective:

$$\min_\zeta Q_\beta \equiv \min_\zeta \left[ \beta \mathbb{I}(I \odot M; I) - \mathbb{I}(I \odot M; C) \right], \qquad (1)$$

where $\beta$ is a constant and $M$ is a random mask variable governed by some learnable conditional distribution $p_\zeta(m|i)$. Being written in the form of Eq. (1), our task can be seen as a reformulation of the Information Bottleneck principle [28]. Alternative optimization objectives based on, for example, Deterministic Information Bottleneck [25] could also be of interest, but fall outside of the scope of this paper. Another optimization objective based on the Conditional Entropy Bottleneck [6] is discussed in Supplementary Materials.

Notice that Eq. (1) has one significant issue: it allows the masking model to deduce the class from the image and encode this class in the mask itself.[2] Consider a binary classification task. If the image belongs to the first class, the generated mask can be empty. On the other hand, for images belonging to the second class, the mask can be chosen to be just a single or a few pixels taken from the "low entropy" part of the image thus both minimizing $\mathbb{I}(I \odot M; I)$ and maximizing $\mathbb{I}(I \odot M; C)$.

---

[1] see Eq. 5 for one definition of image masking.

[2] Assuming it is sufficiently complex and has a receptive field covering the entire image.

This unwanted behavior can be avoided in practice by choosing mask models with a finite receptive field that is comparable to the size of the feature distinguishing one class from another. A more general approach, however, has to rely on a more nuanced definition of the attention mask and its desired properties. This section and Sect. 5 discuss several such definitions that are not equivalent, but each have their own justification and potentially distinct use cases.

Attention mask defined via Eq. (1) essentially requires that there is no label uncertainty once the masked image **and** mask model parameters $\zeta$ are specified. It is precisely the knowledge of $\zeta$ that allows one to deduce additional information about the image and its class even when not enough detail is presented in the masked pixels themselves. This issue can be alleviated by noticing that one of the desired properties of the mask $m_\zeta(i)$ is that uncovering more image pixels than we see in $i \odot m_\zeta(i)$ should not change our prediction of the image label. In other words, we require that $\mathbb{I}(I \odot M; C) \leq \mathbb{I}(I \odot M'; C)$ for Boolean masks $M'$ "larger" than $M$ in a sense that $m'_{x,y} = 0$ implies that $m_{x,y} = 0$. We can define $M'$ by, for example, specifying $p(m'_{x,y}|m_{x,y}, x, y)$ and restricting it via $p(m'_{x,y} = 0|m_{x,y} = 1) = 0$. Defined like this, our optimization objective (1) can be rewritten as:

$$\min_\zeta Q_\beta \equiv \min_\zeta \left[ \beta \mathbb{I}(I \odot M; I) - \mathbb{I}(I \odot M'; C) \right]. \qquad (2)$$

Notice that adding a stochastic perturbation to the mask, we make it more difficult for the model $m_\zeta(i)$ to communicate information through $M$ while still being guaranteed to see image pixels in $m_\zeta(i) \odot i$. Preliminary exploration of the effect that *mask randomization* technique has on attention regions is presented in Sect. 4.3. In the remainder of this section, we complete the description of this model.

## 3.1   Variational Upper Bound

Expanding the expressions for the mutual information in Eq. (2), we obtain:

$$Q_\beta = \beta \mathbb{H}(I \odot M) - \beta \mathbb{H}(I \odot M|I) - \mathbb{H}(C) + \mathbb{H}(C|I \odot M'). \qquad (3)$$

Since $\mathbb{H}(I)$ and $\mathbb{H}(C)$ are immutable properties of the dataset, they will frequently be omitted from the final optimization objectives. Other entropies of the form $\mathbb{H}(A)$ permit variational upper bounds of the form $-\mathbb{E}_a \log p_\phi(a)$ with $p_\phi(a)$ taken from an arbitrary family of distribution functions, and similarly for conditional entropies $\mathbb{H}(A|B)$. This allows us to formulate our variational optimization objective as [2]:

$$\mathcal{L} = \min_{\zeta,\theta,\psi} \Big\{ \mathbb{E}_{p(i,c)p_\zeta(m|i)} \left[ -\beta \log g_\theta(i \odot m) - \log h_\psi(c|i \odot m') \right] \qquad (4)$$

$$-\beta \mathbb{H}(I \odot M|I) \Big\},$$

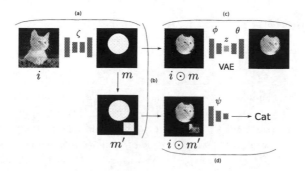

**Fig. 1.** Model diagram: (a) the image $i$ is used to produce masking probability $\rho_\zeta(i)$ and the mask $m$ is then sampled from Bernoulli($\rho_\zeta(i)$), (b) the mask $m$ is randomly augmented (grown) to produce $m'$, (c) the original masked image $i \odot m$ is autoencoded via $(g_\theta, q_\phi)$, (d) the masked image $i \odot m'$ is used as an input to a classification model $f_\psi : i \odot m' \mapsto c$. Parameters $\zeta$, $\phi$, $\theta$ and $\psi$ are optimized simultaneously.

where $g_\theta$ and $h_\psi$ are variational approximations of $p(i \odot m)$ and $p(c|i \odot m')$ correspondingly. As discussed in Sect. 3.3, the first term in this loss $\mathcal{L}$ becomes a modified variational autoencoder of the masked image (see Eq. (7) and Eq. (8)), the second term becomes a cross-entropy classification loss and $\mathbb{H}(I \odot M|I)$ can be computed explicitly for our choice of the mask model (see Eq. (6) and Eq. (9)).

## 3.2 Mask and Masked Image

Let $\rho_\zeta : X \to \mathbb{R}^{n \times n}$ be the "masking probability" model parameterized[3] by $\zeta$. Each $\rho_{x,y}(i)$ for $1 \le x, y \le n$ is assumed to satisfy $0 \le \rho_{x,y}(i) \le 1$. We introduce a discrete mask $m = \text{Bernoulli}(\rho)$ sampled according to $\rho$ independently for each pixel. The masked image $i \odot m$ can then be defined as follows:

$$(i \odot m)_{x,y} \equiv \begin{cases} (i_{x,y}, 1) & \text{if } m_{x,y} = 1, \\ (0,0) & \text{if } m_{x,y} = 0. \end{cases} \tag{5}$$

Given this definition, the entropy $\mathbb{H}(I \odot M|I)$ can be expressed as:

$$\mathbb{H}(I \odot M|I) = - \sum_{x,y=1}^{n} \left[ \rho_{x,y} \log \rho_{x,y} + (1 - \rho_{x,y}) \log(1 - \rho_{x,y}) \right]. \tag{6}$$

It is worth noticing here that the mask $\rho_\zeta(i)$ can be interpreted as an adaptive "continuous" downsampling of the image. Low values of $\rho$ cause most, but not all image pixels to be removed; the remaining pixels and the mere fact that the mask chose to partially remove them can still provide enough information to the image classification model.

---

[3] We will frequently be omitting $\zeta$ for brevity.

### 3.3 Loss Function

Having the expression for the last term in Eq. (4), we will now provide specific models for the first two.

Let us start with $-\log h_\psi(c|i\odot m')$. Consider a family of deep neural network models $f_\psi$ mapping masked images $i\odot m'$ to $\mathbb{R}^{|c|}$. We can define $h_\psi(i\odot m')$ to be softmax $(f_\psi(i\odot m'))$ allowing us to rewrite $-\log h_\psi(c|i\odot m')$ as a cross-entropy loss with respect to softmax $(f_\psi(i\odot m'))$. Recalling that mask $m$ is sampled from Bernoulli$(\rho_\zeta)$, we cannot simply back-propagate gradients all the way down to the parameters of the model $\rho_\zeta(i)$. We alleviate this problem by using the Gumbel-softmax reparametrization approach [11,16],[4] thus approximating $m(i)$ with a differentiable function.

Now let us consider the first term in Eq. (4). Since the space of masked images $i\odot m$ is generally very high-dimensional, we adapt the variational autoencoder approach [12], considering a space of marginal distribution functions $g_\theta(i\odot m) = g_\theta(i\odot m|z)p(z)$ with $p(z)$ being a tractable prior distribution for the latent variable space $Z$. Following [12], $-\log g_\theta(i\odot m)$ can be upper bounded by:

$$-\mathbb{E}_{z\sim q_\phi(z|i\odot m)}\left[\log g_\theta(i\odot m|z)\right] + D_{\mathrm{KL}}\left[q_\phi(z|i\odot m)\|p(z)\right], \qquad (7)$$

where $q_\phi$ is a variational approximation of $g_\theta(z|i\odot m)$. The encoder $q_\phi$ in our model receives both the input pixels $i_{x,y}$ (or 0 if $m_{x,y}=0$) and the mask $m_{x,y}$ as its inputs and produces a conventional embedding $z\in\mathbb{R}^d$. The decoder $g_\theta$, in turn, maps $z$ back to $\hat\rho$ and $\hat i$. In our model, we define $g_\theta(i\odot m|z)$ as a probability for a masked image to be sampled from a Bernoulli process with a probability $\hat\rho$ and the image to be sampled from a Gaussian random variable with the mean $\hat i$ and a constant covariance matrix. This allows us to rewrite $-\log g_\theta(i\odot m|z)$ as:

$$-\log g_\theta(i\odot m|z) = \sum_{x,y=1}^n\Big\{-(1-m_{x,y})\log(1-\hat\rho_{x,y}) \qquad (8)$$
$$-m_{x,y}\left[\log\hat\rho_{x,y} - \ell_2(i_{x,y},\hat i_{x,y})\right]\Big\}+C,$$

where $\ell_2(i,\hat i)=(i-\hat i)^2/2\sigma^2$ and $\sigma$, $C$ are constants. Given this choice, $\beta$ becomes an overall multiplier of the VAE objective in the full loss and $\sigma$ defines a weight of the image pixel reconstruction relative to the mask reconstruction. The entire model is illustrated in Fig. 1.

It is worth noticing that adopting the Gumbel-softmax trick we find that a discrete approximation of Eq. (6) reading

$$-\sum_{x,y=1}^n\left[m_{x,y}\log\rho_{x,y} + (1-m_{x,y})\log(1-\rho_{x,y})\right] \qquad (9)$$

---

[4] It is worth mentioning that the Gumbel temperature should be chosen with care; very small values lead to high-variance estimators, while low temperature would introduce bias.

leads to better convergence in our experiments. We hypothesize that better empirical performance of models using Eq. (9) rather than Eq. (6) can potentially be explained by the fact that the Gumbel-softmax reparametrization introduces bias and therefore, expression in Eq. (6) will not cancel on average with the corresponding term (8) in VAE even for the perfect mask reconstruction, i.e., $\hat{\rho} = \rho$.

## 4   Experimental Results

All our experiments were conducted for the original optimization objective (2) by optimizing the loss function derived in Sect. 3 using Eq. (9) instead of Eq. (6). We observed that the behaviour of the model was very sensitive to the constant $\beta$. If $\beta$ was too small, the mask $\rho_\zeta$ would monotonically approach $\rho_\zeta = 1$. Conversely, for sufficiently large $\beta$, $\rho_\zeta$ would vanish. We used two different techniques to improve behaviour of our model: (i) stop masking model gradients in variational autoencoders once $-\log g_\theta$ falls below a certain threshold, or (ii) change $\beta$ adaptively in such a way that $-\log g_\theta$ stays within a pre-defined range. Both of these approaches were able to guarantee in practice that the variational autoencoder loss reached a certain predefined value. For additional details of our model, see Supplementary Materials.

In all experiments discussed in this section, the groundtruth "features" that define image class are known in advance allowing us to interpret experimental results with ease. In a more general case, the quality of the model prediction can be judged based on the following three criteria: (a) accuracy of the trained classifier operating on masked images $I \odot M$ should be sufficiently close to the accuracy of a separate classifier trained on original images $I$; (b) VAE loss should fall into a predefined range; (c) the accuracy of the classifier prediction on $I \odot M'$ should be sufficiently close to the prediction on $I \odot M$ for any fixed $I$ and all sampled realizations of $M'$.

In the following subsections, we first discuss our results on synthetic datasets with "anomalies" and "distractors". These experiments were conducted without mask randomization, but we verified that experiments with mask randomization produced nearly identical results. We then discuss our experiments on a synthetic dataset designed to explore the effect that mask randomization has on produced masks. Finally, we show results on a realistic SVHN dataset with apriori known localized features defining the image class (number of digits in the image). For this dataset, mask randomization appears to play an important role.

### 4.1   Experiments with "Anomalies"

For the first series of experiments, we used images from CIFAR10 [14] and MNIST datasets augmented by adding randomly-placed rectangular "anomalies" (thus designed to be low-entropy). The anomaly was added with a probability of 1/2 and the classification task was to distinguish original images from the altered ones.

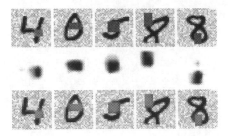

**Fig. 2.** Results for the MNIST dataset with rectangular patches: augmented images (top row); masks (middle row; white represents opaque regions, black transparent); mask on top of the augmented image (bottom row; see in color). (Color figure online)

For these datasets, our models learned to produce opaque masks for most images without anomalies. For images with anomalies, generated masks were opaque everywhere except for the regions around rectangles added into the image (see Fig. 2 and Fig. 3). As a result, the image classifiers reached almost perfect accuracy in both of these examples: approximately 98% test and train accuracy for MNIST and approximately 99% test and train accuracy for CIFAR10 dataset.

In both models, $\ell_1$ norm of the mask was a strong predictor of whether the "anomaly" was in the image (see Figs. 9 and 10 in Supplementary material). However, interestingly, the separation was much more visible for CIFAR10, while the masks predicted for the MNIST dataset were much better aligned with the actual anomalies. The latter fact can also be seen to be reflected in the mask averages inside and outside of the actual added rectangles (see Figs. 9 and 10 in Supplementary material).

We hypothesize that these properties of the trained models can be attributed to receptive fields of the masking models used in both examples. For the MNIST dataset, the masking model has a receptive field of about 40% of the image size, while for the CIFAR10 dataset, the receptive field covered nearly the entire input image.

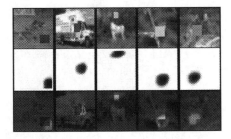

**Fig. 3.** Results for the CIFAR10 dataset with rectangular patches: augmented images (top row); masks (middle); mask on top of the augmented image (bottom).

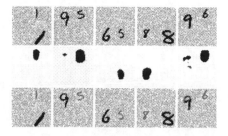

**Fig. 4.** Results for the double-digit MNIST-based dataset: original images (top row); learned masks (middle); mask on top of the original image (bottom; see in color). Images on the right demonstrate one of the failures of the model. (Color figure online)

## 4.2 Experiments with "Distractors"

In another set of experiments, we used two synthetic datasets based on MNIST, in which we combined: (a) two digits and (b) four digits in a single $56 \times 56$ image. In both datasets, one of the digits was always smaller and it defined the class of the entire image. The larger digits are thus "distractors".

For the vast majority of masks generated by the trained model, everything outside of the region around the small digit was masked-out (see Figs. 4, 5). In some rare cases, however, generated masks were also letting some pixels of the larger digits to pass through. In most of our experiments, the classifier training and test accuracy reached 95% and 90% for the two- and four-digits datasets correspondingly. However, there were some runs, in which the test accuracy could be lower than the training accuracy by 10% or 20%. We believe this is due to the greater capacity to overfit to the training data of the combined masking and classifier models.

## 4.3 Mask Randomization Experiments

We identified a simple MNIST-based synthetic example, in which it can be clearly seen that without mask randomization, generated masks can encode class information without using virtually any pixels from the actual digits. In our example,

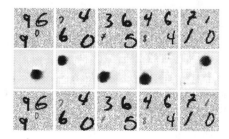

**Fig. 5.** Same as Fig. 4, but for the four-digit MNIST-based dataset.

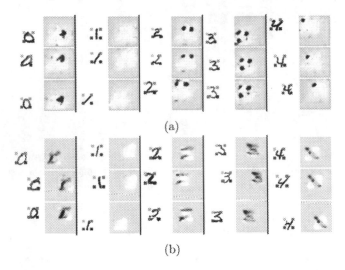

(a)

(b)

**Fig. 6.** Masked images and corresponding masks for models without (a) and with (b) *mask randomization*. Red shows pixels visible to the classifier and black pixels are masked-out by the learned attention region. Masks use white color to show pixels that are most likely to be masked-out and black for pixels that are most likely to be visible to the classifier. (a) Results for a model trained without mask randomization. Notice that the model rarely chooses actual digit pixels and is frequently seen to encode digit class into the "anchors"; (b) Results for a model trained with mask randomization. Now the attention is directed towards digit pixels. (Color figure online)

we use 5 MNIST digits (0 through 4) and add 4 solid rectangles (*"anchors"*) into the image thus allowing the mask to use them for encoding image label. Model trained without any mask randomization, i.e., $M' = M$ can be seen to produce attention regions selecting anchors, but frequently avoiding actual digit pixels altogether (see Fig. 6a). Trained classifier has almost perfect accuracy ($\sim$99%) on original masked images. However, once we start evaluating the same classifier on images with randomized masks (adding random transparent rectangular patches), the accuracy drops down to $\sim$33% for some of the digits. After the classifier is fine-tuned on images with randomized masks, the lowest accuracy for a digit goes up to 70.3% (for digit 2, which ends up being most frequently confused for 3).

We then conduct experiments with the same dataset and enable mask randomization during training (by selecting $M'$ to be equal to $M$ with a randomly placed transparent rectangle). New trained models now mainly concentrate on the digit pixels and seem to select discriminative parts of the image (see Fig. 6b). Evaluating the accuracy of this classifier with mask randomization, we observe that the average accuracy now stays above 93% for all digits.

## 4.4   SVHN Experiments

We chose the original SVHN dataset [18] for our experiments with realistic images. The task given to a classifier was to predict the number of digits in the street/house number shown in the image. With this task, the generated mask was expected to concentrate on areas of the image containing numbers.

We started our experiments without mask randomization, i.e., $m' = m$. We picked $\sigma = (1/8)^{1/2}$ and the target VAE loss objective was chosen in such a way that the mask was neither transparent, nor almost completely opaque. For intermediate values of the VAE objective, most of the observed solutions produced noticeable peaks of transparency around the digits. Results obtained for one of the models trained with sufficiently low VAE target are shown in Fig. 7.

For lower VAE loss targets, we frequently observed masks that used interleaved transparent and opaque lines (either vertical or horizontal) as means of minimizing VAE loss while still allowing the classifier to achieve high accuracy in predicting the number of digits in the image.

For even lower VAE thresholds, generated masks were no longer transparent around the digits, but instead were mostly opaque in these areas. This behavior can be understood by noticing that the digits containing many complex sharp edges may carry more information[5] than relatively featureless surrounding areas. In essence, the "negative space" outside of the number bounding box may be smaller-entropy, but its shape may still be enough to determine the number of digits in the image. In this case, the mask itself became a feature strongly correlated with the image label. Plotting histograms for $\ell_1$ mask norm, we observed that masks generated for 1-digit images were almost entirely transparent while the masks generated for 4-digit images were mostly opaque (see Fig. 12 in Supplementary Materials). We verified that using the $\ell_1$ mask norm alone, we could reach a 59% accuracy on the image classification task, just 3% lower than the actual trained classifier receiving the masked image.

If all digits had the same aspect ratio, attending to the "negative space" of the number could actually be a reasonable solution satisfying all conditions outlined in Sect. 3. In a more general case, observed masks that simply encode image class information do not seem to satisfy the condition $\mathbb{I}(I \odot M; C) \leq \mathbb{I}(I \odot M'; C)$. Implementing mask randomization by adding randomly-placed transparent rectangles to $m$, we verified that newly trained masking models were now nearly always concentrating on digits rather than the "negative space".

## 5   Alternative Approach Based on Conditional Mutual Information

In previous sections, we showed that the Information Bottleneck optimization objective (1) allows for the class information to be encoded in the mask itself. Previously, we used mask randomization to address this issue. However, there are alternative approaches to refining the model. For example, given a masked image

---

[5] also poorly approximated by VAEs, which tend to favor smooth reconstructions.

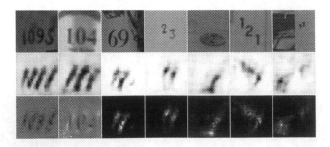

**Fig. 7.** Attention results for SVHN dataset: original image (top row); mask (middle row) and masked image (bottom row). First two columns show results for typical test images obtained using the same data augmentation procedure as the training images (masked images use Boolean mask for these images and are similar to masked images actually seen by the classifier during training); remaining columns show results on out-of-distribution samples obtained by cropping out $128 \times 128$ regions from high-resolution source images.

$i_* \odot m_\zeta(i_*)$ we can introduce a distribution $p_*(i, c; i_*)$ obtained by restricting the original input distribution $p(i, c)$ to a manifold $\{i | i \odot m_\zeta(i_*) = i_* \odot m_\zeta(i_*)\}$ of images that match $i_*$ within the mask. We can then request that the mask generator $m_\zeta$ is such that for every $i_*$, $p_*(i, c; i_*)$ contains images from a single class only, i.e., that $\mathbb{H}(C | I_*) = 0$ for $p_*(i, c; i_*)$.

Another approach to disallowing the generated mask to encode class information is based on modifying the Information Bottleneck objective by replacing $\mathbb{I}(I \odot M; C)$ with $\mathbb{I}(I \odot M; C | M)$ thus leading to the optimization objective:

$$\arg \min_\zeta \left[ \beta \mathbb{I}(I \odot M; I | M) - \mathbb{I}(I \odot M; C | M) \right], \qquad (10)$$

where we also chose to minimize $\mathbb{I}(I \odot M; I | M)$ instead of $\mathbb{I}(I \odot M; I)$ for consistency. Conditioning on the mask implies that for any realization of the mask, masked pixels should contain the entirety of the information about the image class. If, for example, the image class could be inferred just from the mask, the conditional mutual information $\mathbb{I}(I \odot M; C | M)$ would vanish.

In order to optimize this objective, we have to modify Eq. (10) by introducing a function $c'(i)$ that is chosen to approximate the groundtruth $C$:

$$\arg \min_\zeta \left[ \beta \mathbb{I}(I \odot M; I | M) - \mathbb{I}(I \odot M; C' | M) \right]. \qquad (11)$$

The exact form of $c'(i)$ will prove to be not important and in practice we frequently chose actual labels for our experiments assuming that the perfect groundtruth model $I \to C$ exists.

Optimization objective (11) can be transformed as follows. First, using the definition of the conditional mutual information, we see:

$$\mathbb{I}(I \odot M; I | M) = \mathbb{H}(I \odot M | M) - \mathbb{H}(I \odot M | I, M) = \mathbb{H}(I \odot M | M).$$

Next, noticing that $\mathbb{H}(M, I) = \mathbb{H}(M, I, C')$, we obtain:

$$\mathbb{H}(M|I) + \mathbb{H}(I) = \mathbb{H}(I|M, C') + \mathbb{H}(M, C')$$

and therefore

$$\mathbb{I}(C'; I \odot M|M) = \mathbb{H}(C'|M) - \mathbb{H}(C'|I \odot M, M)$$
$$= \mathbb{H}(C', M) - \mathbb{H}(M) - \mathbb{H}(C'|I \odot M)$$

can be rewritten as:

$$\mathbb{H}(M|I) + \mathbb{H}(I) - \mathbb{H}(I|M, C') - \mathbb{H}(M) - \mathbb{H}(C'|I \odot M).$$

Combining all terms together we then see that original optimization objective (11) is equivalent to:

$$\arg\min_{\zeta} \left[ \beta\mathbb{H}(I \odot M|M) + \mathbb{H}(M) - \mathbb{H}(M|I) \right. \tag{12}$$

$$\left. + \mathbb{H}(I|M, C') + \mathbb{H}(C'|I \odot M) \right].$$

Following an earlier discussion, we can explicitly calculate $\mathbb{H}(M|I)$ and use variational upper bounds for all four entropies and conditional entropies. The complete model will therefore include: (a) VAE on the masked portion of the image conditioned on the image mask; (b) VAE for the mask itself and (c) VAE auto-encoding the image and conditioned on the mask and the class approximation $C'$. Notice that it is the conditional entropy $\mathbb{H}(I|M, C')$ that is responsible for disentangling $M$ and $C'$. Indeed, trying to minimize the entropy of images conditioned on the mask and the image class, we effectively reward the mask for containing information from $I$ that is not encoded in $C'$.

In our first preliminary experiments, we trained model (12) on the MNIST-based synthetic datasets. For the dataset with "anchors", the model was able to generate masks that were: (a) covering the digits, (b) allowing the classification model to achieve 94% accuracy and (c) nearly independent of the image label (see Fig. 8), which is exactly what objective (12) was designed to achieve. Similarly, for the dataset with distractors, the generated masks were almost indistinguishable from those shown in Fig. 5.

For the dataset with anomalies, the experiments based on Eq. (12) failed to identify a proper mask and instead produced a mask transparent at the boundary and almost entirely opaque at the image center. Average masking probability $\langle \rho \rangle$ in the center encoded information about the image class so that the classifier could (with accuracy close to 100%) predict the presence of anomaly by just averaging values of visible pixels. This failure is not surprising if you notice that a mask transparent near an anomaly (see for example Fig. 2), but opaque for an image without one, does not optimize objective (12). Indeed, given such a mask, one would be able to predict image class by just looking at the mask itself. The optimal mask would have to have shape and location independent of the

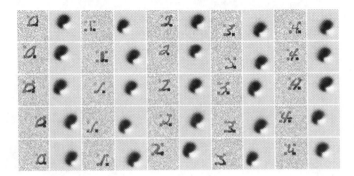

**Fig. 8.** Same as Fig. 6, but showing masked images and corresponding masks for the model based on Eq. (12) (in this dataset, we also added background noise to introduce a "price" of autoencoding background). Notice that the generated mask is almost independent of the image class.

image class and concentrate on anomaly if it is present in the image. Instead of finding this complex solution, our model identified a simpler one by producing a mask that is *almost* independent of the image class, but still conveys enough information about the presence of anomaly in the image.

Overall, while being conceptually sound, objective Eq. (12) is much more complex than Eq. (3) making it potentially less effective in practice. The disentanglement of $M$ and $C'$ critically relies on the upper bound for $\mathbb{H}(I|M, C')$ to be sufficiently tight and we suspect that it may be difficult to achieve this in practice for complex datasets containing realistic images. More complex density estimation models could, however, alleviate this problem.

## 6   Conclusions

In this work, we propose a novel universal semi-supervised attention learning approach based on the Information Bottleneck method. Supplied with a set of labeled images, the model is trained to generate discrete attention masks that occlude irrelevant portions of the image, but leave enough information for the classifier to correctly predict the image class. Using synthetic and real datasets based on MNIST, CIFAR10, and SVHN, we demonstrate that this technique can be used to identify image regions carrying information that defines the image class. In some special cases when the feature itself is high-entropy (for example, digits in SVHN images), but its shape is sufficient to determine the image class (number of digits in our SVHN example), we show that the generated mask may occlude the feature and use its "negative space" instead. Additionally, we identify a potential failure of this approach, in which the generated mask acts not as an attention map, but rather as an encoding of the image class itself. We then propose two techniques based on finite receptive fields and mask randomization that mitigate this problem. We believe this technique is a promising method to train explainable models in a semi-supervised manner.

# References

1. Achille, A., Soatto, S.: Information dropout: learning optimal representations through noisy computation. IEEE Trans. Pattern Anal. Mach. Intell. **40**(12), 2897–2905 (2018). https://doi.org/10.1109/TPAMI.2017.2784440
2. Alemi, A.A., Fischer, I., Dillon, J.V., Murphy, K.: Deep variational information bottleneck. In: 5th International Conference on Learning Representations, ICLR 2017, 24–26 April 2017, Toulon, France, Conference Track Proceedings. OpenReview.net (2017)
3. Bardera, A., Rigau, J., Boada, I., Feixas, M., Sbert, M.: Image segmentation using information bottleneck method. IEEE Trans. Image Process. **18**(7), 1601–1612 (2009). https://doi.org/10.1109/TIP.2009.2017823
4. Bengio, S., Wallach, H.M., Larochelle, H., Grauman, K., Cesa-Bianchi, N., Garnett, R. (eds.): Advances in Neural Information Processing Systems 31: Annual Conference on Neural Information Processing Systems 2018, NeurIPS 2018, 3–8 December 2018. Montréal, Canada (2018)
5. Dabkowski, P., Gal, Y.: Real time image saliency for black box classifiers. In: Advances in Neural Information Processing Systems, pp. 6967–6976 (2017)
6. Fischer, I.: The Conditional Entropy Bottleneck (2018). https://openreview.net/forum?id=rkVOXhAqY7
7. Fong, R., Patrick, M., Vedaldi, A.: Understanding deep networks via extremal perturbations and smooth masks. In: Proceedings of the IEEE International Conference on Computer Vision, pp. 2950–2958 (2019)
8. Fong, R.C., Vedaldi, A.: Interpretable explanations of black boxes by meaningful perturbation. In: Proceedings of the IEEE International Conference on Computer Vision, pp. 3429–3437 (2017)
9. Hjelm, R.D., Fedorov, A., Lavoie-Marchildon, S., Grewal, K., Trischler, A., Bengio, Y.: Learning deep representations by mutual information estimation and maximization. CoRR abs/1808.06670 (2018)
10. Hou, Q., Jiang, P., Wei, Y., Cheng, M.: Self-erasing network for integral object attention. In: Bengio et al. [4], pp. 547–557 (2018)
11. Jang, E., Gu, S., Poole, B.: Categorical reparameterization with Gumbel-Softmax. In: 6th International Conference on Learning Representations, ICLR 2017 (2016)
12. Kingma, D.P., Welling, M.: Auto-encoding variational bayes. In: Bengio, Y., LeCun, Y. (eds.) 2nd International Conference on Learning Representations, ICLR 2014, 14–16 April 2014, Banff, AB, Canada, Conference Track Proceedings (2014)
13. Kolesnikov, A., Lampert, C.H.: Seed, expand and constrain: three principles for weakly-supervised image segmentation. In: Leibe, B., Matas, J., Sebe, N., Welling, M. (eds.) ECCV 2016. LNCS, vol. 9908, pp. 695–711. Springer, Cham (2016). https://doi.org/10.1007/978-3-319-46493-0_42
14. Krizhevsky, A.: Learning multiple layers of features from tiny images. Technical report Citeseer (2009)
15. Li, K., Wu, Z., Peng, K., Ernst, J., Fu, Y.: Tell me where to look: guided attention inference network. In: 2018 IEEE Conference on Computer Vision and Pattern Recognition, CVPR 2018, 18–22 June 2018, Salt Lake City, UT, USA, pp. 9215–9223 (2018). https://doi.org/10.1109/CVPR.2018.00960
16. Maddison, C.J., Mnih, A., Teh, Y.W.: The concrete distribution: a continuous relaxation of discrete random variables. In: 6th International Conference on Learning Representations, ICLR 2017 (2016)

17. Moyer, D., Gao, S., Brekelmans, R., Galstyan, A., Steeg, G.V.: Invariant representations without adversarial training. In: Bengio et al. [4], pp. 9102–9111 (2018)
18. Netzer, Y., Wang, T., Coates, A., Bissacco, A., Wu, B., Ng, A.Y.: Reading digits in natural images with unsupervised feature learning. In: NIPS Workshop on Deep Learning and Unsupervised Feature Learning (2011)
19. Peng, X.B., Kanazawa, A., Toyer, S., Abbeel, P., Levine, S.: Variational discriminator bottleneck: improving imitation learning, inverse RL, and GANs by constraining information flow. CoRR abs/1810.00821 (2018)
20. Petsiuk, V., Das, A., Saenko, K.: RISE: randomized input sampling for explanation of black-box models. arXiv preprint arXiv:1806.07421 (2018)
21. Schulz, K., Sixt, L., Tombari, F., Landgraf, T.: Restricting the flow: information bottlenecks for attribution. CoRR abs/2001.00396 (2020). http://arxiv.org/abs/2001.00396
22. Selvaraju, R.R., Cogswell, M., Das, A., Vedantam, R., Parikh, D., Batra, D.: Grad-CAM: visual explanations from deep networks via gradient-based localization. In: Proceedings of the IEEE International Conference on Computer Vision, pp. 618–626 (2017)
23. Smilkov, D., Thorat, N., Kim, B., Viégas, F., Wattenberg, M.: SmoothGrad: removing noise by adding noise. arXiv preprint arXiv:1706.03825 (2017)
24. Still, S., Bialek, W., Bottou, L.: Geometric clustering using the information bottleneck method. In: Thrun, S., Saul, L.K., Schölkopf, B. (eds.) Advances in Neural Information Processing Systems 16 Neural Information Processing Systems, NIPS 2003, 8–13 December 2003, Vancouver and Whistler, British Columbia, Canada, pp. 1165–1172. MIT Press (2003)
25. Strouse, D., Schwab, D.J.: The deterministic information bottleneck. Neural Comput. **29**(6), 1611–1630 (2017). https://doi.org/10.1162/NECO_a_00961
26. Strouse, D., Schwab, D.J.: The information bottleneck and geometric clustering. Neural Comput. **31**(3) (2019). https://doi.org/10.1162/neco_a_01136
27. Taghanaki, S.A., et al.: InfoMask: masked variational latent representation to localize chest disease. CoRR abs/1903.11741 (2019)
28. Tishby, N., Pereira, F.C.N., Bialek, W.: The information bottleneck method. CoRR physics/0004057 (2000)
29. Tishby, N., Zaslavsky, N.: Deep learning and the information bottleneck principle. In: 2015 IEEE Information Theory Workshop, ITW 2015, 26 April – 1 May 2015, Jerusalem, Israel, pp. 1–5. IEEE (2015). https://doi.org/10.1109/ITW.2015.7133169
30. Wei, Y., Feng, J., Liang, X., Cheng, M., Zhao, Y., Yan, S.: Object region mining with adversarial erasing: a simple classification to semantic segmentation approach. In: 2017 IEEE Conference on Computer Vision and Pattern Recognition, CVPR 2017, 21–26 July 2017, Honolulu, HI, USA, pp. 6488–6496. IEEE Computer Society (2017). https://doi.org/10.1109/CVPR.2017.687
31. Zhang, J., Bargal, S.A., Lin, Z., Brandt, J., Shen, X., Sclaroff, S.: Top-down neural attention by excitation backprop. Int. J. Comput. Vis. **126**(10), 1084–1102 (2018)
32. Zhang, X., Wei, Y., Feng, J., Yang, Y., Huang, T.S.: Adversarial complementary learning for weakly supervised object localization. In: 2018 IEEE Conference on Computer Vision and Pattern Recognition, CVPR 2018, 18–22 June 2018, Salt Lake City, UT, USA, pp. 1325–1334 (2018). https://doi.org/10.1109/CVPR.2018.00144

# FAWA: Fast Adversarial Watermark Attack on Optical Character Recognition (OCR) Systems

Lu Chen[1], Jiao Sun[2], and Wei Xu[1(✉)]

[1] Institute for Interdisciplinary Information Sciences, Tsinghua University,
Beijing 100084, China
`lchen17@mails.tsinghua.edu.cn, weixu@mail.tsinghua.edu.cn`
[2] Department of Computer Science, University of Southern California,
Los Angeles, CA 90007, USA
`jiaosun@usc.edu`

**Abstract.** Deep neural networks (DNNs) significantly improved the accuracy of optical character recognition (OCR) and inspired many important applications. Unfortunately, OCRs also inherit the vulnerability of DNNs under adversarial examples. Different from colorful vanilla images, text images usually have clear backgrounds. Adversarial examples generated by most existing adversarial attacks are unnatural and pollute the background severely. To address this issue, we propose the *Fast Adversarial Watermark Attack (FAWA)* against sequence-based OCR models in the white-box manner. By disguising the perturbations as watermarks, we can make the resulting adversarial images appear natural to human eyes and achieve a perfect attack success rate. FAWA works with either gradient-based or optimization-based perturbation generation. In both letter-level and word-level attacks, our experiments show that in addition to natural appearance, FAWA achieves a 100% attack success rate with 60% less perturbations and 78% fewer iterations on average. In addition, we further extend FAWA to support full-color watermarks, other languages, and even the OCR accuracy-enhancing mechanism.

**Keywords:** Watermark · OCR model · Targeted White-Box Attack

## 1 Introduction

Optical Character Recognition (OCR) has been an important component in text processing applications, such as license-plate recognition, street sign recognition and financial data analysis. Deep neural networks (DNNs) significantly improve OCR's performance. Unfortunately, OCR inherits all counter-intuitive security problems of DNNs. OCR models are also vulnerable to *adversarial examples*, which are crafted by making human-imperceptible perturbations on original images to mislead the models. The wide application of OCR provides incentives

© Springer Nature Switzerland AG 2021
F. Hutter et al. (Eds.): ECML PKDD 2020, LNAI 12459, pp. 547–563, 2021.
https://doi.org/10.1007/978-3-030-67664-3_33

for adversaries to attack OCR models, thus damaging downstream applications, resulting in fake ID information, incorrect metrics readings and financial data.

Prior works have shown that we can change the prediction of DNNs in image classification tasks only by applying carefully-designed perturbations [8,19,20,22], or adding a small patch [3,13] to original images. But these methods are not directly applicable to OCR attacks. 1) Most text images are on white backgrounds. Perturbations added by existing methods appear too evident for human eyes to hide, and pollute clean backgrounds. 2) Instead of classifying characters individually, modern OCR models are segmentation-free, inputting entire variable-sized image and outputting a sequence of labels. It is called the *sequential labeling* task [9]. Since modern OCR models use the CNN [15] and the LSTM [12] as feature extractors, the internal feature representation also relies on contexts (i.e., nearby characters). Therefore, besides perturbing the target character, we also need to perturb its context, resulting in more perturbations. 3) Since text images are usually dense, there is no open space to add a patch.

In this paper, we propose a new attack method, *Fast Adversarial Watermark Attack* (FAWA), against modern OCR models. Watermarks are images or patterns commonly used in documents as backgrounds to identify things, such as marking a document proprietary, urgent, or merely for decoration. Human readers are so used to the watermark that they naturally ignore it. We generate natural watermark-style adversarial perturbations in text images. Such images appear natural to human eyes but misguide OCR models. Watermark perturbations are similar to patch-based attacks [3,13] where perturbations are confined to a region. Different from patches occupying a separate region, watermarks overlay on texts but not hinder the text readability. Laplace attack [10] and HAAM [11] try to generate seemingly smooth perturbations, which work well on colored photos. However, they do not solve the problem of background pollution.

FAWA is a *white-box* and *targeted* attack. We assume adversaries have perfect knowledge of the DNN architecture and parameters (white-box model) and generate specific recognition results (targeted). There are three steps in the perturbation generation. 1) We find a good position over the target character to add an initial watermark. 2) We generate perturbations inside the watermark. 3) Optionally, we convert the gray watermark into a colored one. To generate perturbations, we leverage either gradient-based or optimization-based methods. We evaluate FAWA on a state-of-the-art open-source OCR model, Calamari-OCR [26] for English texts with five fonts. FAWA generates adversarial images with 60% less perturbations and 78% fewer iterations than existing methods, while maintaining a 100% attack success rate. Our adversarial images also look quite similar to natural watermarked images. We evaluate the effects of colored watermarks and other languages under real-world application settings. Last, we propose a positive application of the FAWA, i.e., using the perturbations to enhance the accuracy of OCR models. The contributions of this paper include:

1. We propose an attack method, FAWA, of hiding adversarial perturbations in watermarks from human eyes. We implemented FAWA as the efficient adversarial attacks against the DNN-based OCR in sequential labeling tasks;

2. Extensive experiments show that FAWA performs targeted attacks perfectly, and generate natural watermarked images with imperceptible perturbations;
3. We demonstrate several applications of FAWA, such as colored watermarks as an attack mechanism, using FAWA as an accuracy-enhancing mechanism.

# 2  Background and Related Work

**Optical Character Recognition (OCR).** We can roughly categorize existing OCR models into character-based models and end-to-end models. *Character-based recognition models* segment the image into character images first, before passing them into the recognition engine. Obviously, the OCR performance heavily relies on the character segmentation. *End-to end recognition models* apply an unsegmented recognition engine, which recognizes the entire sequence of characters in a variable-sized image. [1,7] adopt sequential models such as Markov models and [2,24] utilize DNNs as feature extractors for the sequential recognition. [9] introduces a segmentation-free approach, connectionist temporal classification (CTC), which provides a sort of loss function of enabling us to recognize variable-length sequences without explicit segmentation while training DNN models. Thus, many state-of-the-art OCR models use CTC as the loss function.

**Attacking DNN-Based Computer Vision Models.** Attacking DNN-based models is a popular topic in both computer vision and security fields. Existing attack methods use the following two ways to generate perturbations. 1) Making perturbations small enough for evading human perception. For example, many projects aim at generating minor $L_p$-norm perturbations for the purpose of making them visually imperceptible. FGSM [8], L-BFGS [22], DeepFool [18], C&W $L_2/L_\infty$ [4], PGD [17] and EAD [5], all perform modifications at the pixel level by a small amount bounded by $\epsilon$. 2) Making perturbations in a small region of an image. For example, JSMA [20], C&W $L_0$ [4], Adversarial Patch [3] and LaVAN [13], all perturb a small region of pixels in an image, but their perturbations are not limited by $\epsilon$ at the pixel level. Though FAWA is similar to patches, in the text images, watermarks overlay on the text instead of covering the text.

**Generating Minimal Adversarial Perturbations.** FAWA generates adversarial perturbations using either gradient-based or optimization-based methods.

Gradient-based methods add perturbations generated from gradient against input pixels. *FGSM* [8] is a $L_\infty$-norm one-step attack. It is efficient but only provides a coarse approximation of the optimal perturbations. *BIM* [16] takes multiple smaller steps and the resulting image is clipped by the same bound $\epsilon$. Thus BIM produces superior results to FGSM. *MI-FGSM* [6] extends BIM with momentum. MI-FGSM can not only stabilize update directions but also escape from poor local maxima during iterations, and thus generates more transferable adversarial examples. Considering the efficiency, we adopt MI-FGSM in FAWA.

Optimization-based methods directly solve the box-constrained optimization problem to minimize the $L_p$-norm distance between the original image and the adversarial image, while yielding the targeted classification. *Box-constraint L-BFGS* [22] seeks the adversarial examples with L-BFGS. While L-BFGS constructs subtle perturbations, it is far less efficient in the generation of adversarial examples. Instead of applying cross-entropy as loss function, *C&W $L_2$ attack* [4] uses a new loss function and then solves it with the gradient descent. *OCR attack* [21] generates adversarial examples using CTC loss function for sequential labeling tasks. In FAWA, we use the same optimization setting as OCR attack.

**Perturbations with Other Optimization Goals.** Besides minimizing the perturbation level, many works make efforts to produce smooth and natural perturbations. Laplace attack [10] smooths perturbations relying on Laplacian smoothing. HAAM [11] creates edge-free perturbations using harmonic functions for disguising natural shadows or lighting. However, the smoothing and disguising are for photos but not for text images. Instead of manipulating pixel values directly, [27] produces perceptually realistic perturbations with spatial transformation. Though avoiding background pollution, it cannot guarantee the readability of text when the attack needs large deformation. [23] also utilizes the watermark idea but performs attacks only by scaling and rotating plain watermarks. Without adding pixel-level perturbations, it does not offer a high attack success rate.

# 3    Fast Adversarial Watermark Attack

In this section, we introduce *fast adversarial watermark attacks* (FAWA). FAWA consists of three steps. 1) We automatically determine a good position to add the initial watermark in the text image so that we can confine the perturbation generation to that region. 2) We generate the watermark-style perturbations with either the *gradient-based method* or the *optimization-based method*. 3) Optionally, we convert gray watermarks into full-color ones to improve the text readability.

## 3.1    Preliminaries

**Problem Definition of Adversarial Image Generation.** Given a text image $x = [x_1, x_2, ..., x_n]^T$ for any $x_i \in [0, 1]$, where $n$ is the number of pixels in the image, our goal is to generate an adversarial example $x'$ with minimum $L_p$-norm perturbations $\|x' - x\|_p$ against the white-box model $f$ with intent to trick model $f$ into outputting the targeted result $t$, $f(x') = t$. Formally, the problem of adversarial image generation is $\min_{x'} \|x' - x\|_p$ s.t. $f(x') = t$ and $x' \in [0, 1]^n$. Also, we define the *CTC loss function* with respect to the image $x$ as $\ell_{\text{CTC}}(x, t)$ and the target labels $t = [t_1, t_2, ..., t_N]$ for $t_i \in \mathcal{T}$, where $\mathcal{T}$ is the character set.

**Saliency Map.** The saliency map [20] is a versatile tool that not only provides us valuable information to cause the targeted misclassification in the threat model but also assists us in intuitively explaining some attack behaviors as a visualization tool. The saliency map indicates the output sensitivity, relevant to

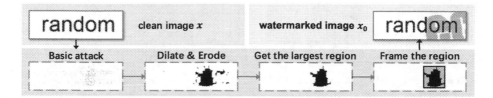

**Fig. 1.** Find the position of watermarks.

the adversarial targets, to the input features. In the saliency map, a larger value indicates a higher likelihood of fooling the model $f$ to output the target $t$ instead of the ground-truth $f(x)$. We can construct a saliency map of an image using the forward derivative with respect to each input component in the text image.

$L_p$-**norms.** $L_p$-norms are commonly-used metrics to measure the perceptual similarity between the clean image $x$ and the adversarial image $x'$, denoted by $\|x - x'\|_p = (\sum_{i=i}^n |x - x_i'|^p)^{\frac{1}{p}}$, $p = 2, \infty$. In gradient-based methods, we apply $L_p$-norms to prune the saliency map for generating $L_p$-norm perturbations. In optimization-based methods, $L_p$-norms are usually as an optimization term in the objective function. Particularly, $L_2$-*norm* is greatly useful to enhance the visual quality. $L_\infty$-*norm* is to measure the maximum variation of perturbations.

### 3.2 Finding the Position of Watermarks

To automatically determine a good position of watermarks, as shown in Fig. 1, we perform the following steps. 1) We produce adversarial perturbations of the basic attack (i.e., Grad-Basic or Opt-Basic in Sect. 3.3) in the text image. 2) We binarize such adversarial image, and get its perturbed regions $r = [r_1, r_2, ..., r_n]^T$, where $r_i = 1$ if $x_i \neq x_i'$ and 0 otherwise. 3) In order to find relatively complete perturbed regions $r'$, we apply a combination of erosion $\oplus$ and dilation $\ominus$ operations twice in the perturbed regions $r$, $r' = ((r \oplus \mathcal{K}) \ominus \mathcal{K})^2$, where we set the kernel $\mathcal{K} = \mathbf{1}_{3 \times 3}$. 4) After sorting the perturbed regions in $r'$ by their area, we obtain the largest perturbed region in $r'$. We can find that our target texts locate in the same position as the largest found region. 5) Last, in the text image, we place a watermark big enough to cover the found position, and obtain an initial watermarked image $x_0$ and a binary watermark mask $\Omega_w$ with the same shape of $x$, where $\Omega_{w,i} = 1$ if the position $i$ is inside the watermark and 0 otherwise.

### 3.3 Generating Adversarial Watermarks

After finding the position of watermarks, we need to generate the perturbations within the watermarks to mislead the OCR models to output targeted texts. Integrated with two popular methods, we use the following methods to attack.

**Gradient-Based Watermark Attack (Grad-WM).** Considering the high efficiency of MI-FGSM [6], we apply it as our basic gradient-based method

---

**Algorithm 1.** Gradient-based Watermark Attack

---

**Input:** A text image $\boldsymbol{x}$, OCR model $f$ with CTC loss $\ell_{\mathrm{CTC}}$, targeted text $\boldsymbol{t}$, $\epsilon$-bounded perturbation, attack step size $\alpha$, # of maximum iterations $I$, decay factor $\mu$.

**Output:** An adversarial example $\boldsymbol{x}'$ with $\|\boldsymbol{x}' - \boldsymbol{x}\|_p \leq \epsilon$ or attack failure $\perp$.

1: Initialization: $\boldsymbol{g}_0 = \boldsymbol{0}$; $\boldsymbol{x}'_0 = \boldsymbol{x}_0$
2: **for all** each iteration $i = 0$ to $I - 1$ **do**
3:     Input $\boldsymbol{x}'_i$ to $f$ and obtain the saliency map $\nabla_x \ell_{\mathrm{CTC}}(\boldsymbol{x}'_i, \boldsymbol{t})$
4:     Update $\boldsymbol{g}_{i+1}$ by accumulating $\boldsymbol{g}_i$ in the saliency map normalized by $\mathrm{L}_1$-norm as

$$\boldsymbol{g}_{i+1} = \mu \cdot \boldsymbol{g}_i + \frac{\nabla_x \ell_{\mathrm{CTC}}(\boldsymbol{x}'_i, \boldsymbol{t})}{\|\nabla_x \ell_{\mathrm{CTC}}(\boldsymbol{x}'_i, \boldsymbol{t})\|_1} \qquad (1)$$

5:     Update $\boldsymbol{x}'_{i+1}$ by applying watermark-bounded $\mathrm{L}_p$-norm perturbations as

$$\boldsymbol{x}'_{i+1} = \boldsymbol{x}'_i + \mathrm{clip}_\epsilon(\alpha \cdot (\Omega_w \odot \frac{\boldsymbol{g}_{i+1}}{\|\boldsymbol{g}_{i+1}\|_p})) \qquad (2)$$

6:     **if** $f(\boldsymbol{x}'_{i+1}) == \boldsymbol{t}$ **then**
7:         **return** $\boldsymbol{x}'_{i+1}$
8:     **end if**
9: **end for**
10: **return** attack failure $\perp$

---

(*Grad-Basic*). Each iterative update of the Grad-Basic is to 1) get the saliency map normalized by $\mathrm{L}_1$-norm with the cross-entropy loss, 2) adjust the update direction in the saliency map and update the momentum, and 3) update a new $\mathrm{L}_p$-norm adversarial image $\boldsymbol{x}'_{i+1}$ bounded by $\epsilon$ using the updated momentum.

There are two main differences between Grad-WM and Grad-Basic. 1) Different from off-the-shelf MI-FGSM, where it applies the cross-entropy as loss function in the image classification tasks, in our Grad-WM, we use the CTC loss function to compute the saliency map. CTC loss function fits better because it is widely used in OCR models to handle the sequential labeling tasks. 2) In Grad-WM, to hide perturbations in the watermarks, we confine the perturbations within the boundary of the watermark $\Omega_w$, rather than spreading perturbations over the entire image in the result of the background pollution like Grad-Basic.

Algorithm 1 summarizes the Grad-WM generation procedure. Notably, input images are the initial watermarked images $\boldsymbol{x}_0$ created in Sect. 3.2. Then the saliency map is produced with the CTC loss function relevant to the adversarial targets $\boldsymbol{t}$. Through element-wisely multiplying the watermark mask $\Omega_w$ with the updated $\mathrm{L}_p$-norm $\boldsymbol{g}_{i+1}$, we get rid of the perturbations outside the watermark to maintain the clean background. We gain a significant visual improvement in the Grad-WM than the Grad-Basic. Besides, for further improving the attack efficiency, we determine whether to stop the attacks in every few iterations.

**Optimization-Based Watermark Attack (Opt-WM).** We employ OCR attack [21] as our basic optimization-based method (Opt-Basic). Opt-Basic solves the following optimization problem: $\min_{x'}\ c \cdot \ell_{CTC}(x', t) + \|x' - x\|_2^2$ s.t. $x' \in [0,1]^n$, where $c$ is a hyper-parameter. To improve the visual quality of adversarial images, Opt-Basic adopts $L_2$-norm for penalizing perturbations. To eliminate the box-constraint $[0,1]$ of $x'$, it reformulates the problem using the change of variables [4] as $\min_\omega\ c \cdot \ell_{CTC}(\frac{\tanh(\omega)+1}{2}, t) + \|\frac{\tanh(\omega)+1}{2} - x\|_2^2$, which optimizes a new variable $\omega$ instead of the box-constrained $x'$. The fact $-1 \leq \tanh(\cdot) \leq 1$ implies that $\frac{\tanh(\cdot)+1}{2}$ satisfies the box-constraint $[0,1]$ automatically. Intuitively, we treat the variable $\omega$ as the adversarial example $x'$ without the box-constraint.

Similar to Grad-WM, we first get the initial watermarked input image $x_0$ and the watermark mask $\Omega_w$. To confine perturbations in watermarks more conveniently, we separate the perturbation term from $w$, and rewrite the original $w$ to $w + x_0$ that represents the adversarial image of combining the perturbation term $w$ and the initial watermarked image term $x_0$. The objective function changes to $\min_\omega\ c \cdot \ell_{CTC}(\frac{\tanh(\omega+x_0)+1}{2}, t) + \|\frac{\tanh(\omega+x_0)+1}{2} - x_0\|_2^2$. To constrain the manipulation region in the watermark fashion, we perform the element-wisely multiplication between the perturbation variable $\omega$ and the watermark mask $\Omega_w$. Formally, we reformulate the optimization problem as

$$\min_{\omega}\ c \cdot \ell_{CTC}\Big(\frac{\tanh(\Omega_w \odot \omega + x) + 1}{2}, t\Big) + \Big\|\frac{\tanh(\Omega_w \odot \omega + x) + 1}{2} - x\Big\|_2^2. \quad (3)$$

We adopt the Adam optimizer [14] for both Opt-Basic and Opt-WM to seek the watermark-style adversarial images. We utilize the binary search to adapt the tradeoff hyper-parameter $c$ between the loss function and the $L_2$-norm distance.

## 3.4   Improving Efficiency

To improve attack efficiency, we employ batch attack and early stopping by increasing the number of parallel attacks and reducing redundant attack iterations.

**Batch Attack.** Because OCR models only support batch image processing with the same size, when we attack a large number of variable-size text images, it will be time-consuming to attack them one by one. To improve attack parallelism (i.e., attack multiple images simultaneously), we 1) resize the variable-size images into the same height, 2) pad them into the maximum width among these images, 3) put the same-size images into a matrix. After the preprocessing, we could perform batch adversarial attacks in a single matrix to improve attack efficiency.

**Early Stopping.** For avoiding redundant attack iterations, we insert the early-stop mechanism in the attack iterations. We evaluate attack success rate (ASR) every few iterations. We stop attack iteration once ASR achieves 100%, indicating that all adversarial images are misidentified as the targeted texts by the threat model. To a certain extent, early stopping reduces the attack efficiency as it takes time to check attack status during iterations. This is a basic tradeoff between the cycle of ASR evaluation and the maximum attack iteration setting.

### 3.5   Improving Readability

Sometimes the gray watermarks still hinder the readability of the text as it reduces the contrast. To achieve better readability of the text, we employ the following two effective strategies: adding text masks and full-color watermark.

**Adding Text Masks.** When generating the initial watermarked image $x_0$ in Sect. 3.2, to prevent the text from being obscured by the watermarks, we add in watermarks outside the text. To achieve this, we binarize the text images $x$ by a threshold $\tau$ to get text masks $\Omega_t$ with the same size of $x$, where $\Omega_{t,i} = 1$ if $x_i > \tau$ and 0 otherwise. Then we get the initial watermarked images $x_0 = x \odot (1 - \Omega_w \odot \overline{\Omega_t}) + \beta \cdot \Omega_w \odot \overline{\Omega_t}$, where $\beta$ is the grayscale value of initial watermarks.

**Full-Color Watermark.** Modern OCR systems always preprocess colored images into gray ones before recognition. Compared to gray watermarks, colored watermarks have better visual quality on the black texts. To convert gray watermarks into RGB watermarks, we only manipulate the pixel $x_i$ inside the watermark, where $\Omega_{w,i} = 1$. Given a grayscale value $Gray$ in the gray watermark, we preset $R$ value and $B$ value to certain values. Next we make the color conversion with the transform equation: $Gray = R * 0.299 + G * 0.587 + B * 0.114$, to calculate the left $G$ value. Last we get RGB values from a grayscale value.

## 4   Experiments

### 4.1   Experiment Setup

**OCR Model.** We choose to attack Calamari-OCR[1] [26]. The OCR model has two convolutional layers, two pooling layers, and the following LSTM layer. We use the off-the-shelf English model as our threat model, trained with CTC loss.

**Generate Text Images.** After processing the corpus in the IMDB dataset, we generate a dataset with 97 paragraph, 1479 sentence and 1092 word images using the Text Recognition Data Generator[2]. We use five fonts to generate these images: Courier, Georgia, Helvetica, Times and Arial. We set the font size to 32 pixels. We verify the Calamari-OCR can achieve 100% accuracy on these images.

**Generating Letter-Level Attack Targets.** We choose a target word that is 1) a valid word, and 2) with edit distance 1 from the original. We evaluate replacement, insertion and deletion. More similar the replacement target is to the original, the easier the attack is. For example, replacing letter **t** with letter

---

[1] https://github.com/Calamari-OCR/calamari.
[2] https://github.com/Belval/TextRecognitionDataGenerator.

**Table 1.** Adversarial examples of letter-level and word-level attacks. Colored watermark examples are converted from Grad-WM examples.

| | original $x$ | WM$_0$ $x_0'$ | Grad-Basic | Grad-WM | Opt-Basic | Opt-WM | Color-WM | target |
|---|---|---|---|---|---|---|---|---|
| letter | year | year | year | year | year | year | year | hear |
| | here | here | here | here | here | here | here | there |
| | short | short | short | short | short | short | short | sort |
| word | tennis | tennis | tennis | tennis | tennis | tennis | tennis | amazon |
| | entry | entry | entry | entry | entry | entry | entry | waken |
| | hoy | hoy | hoy | hoy | hoy | hoy | hoy | jow |

f is easier than replacing t with letter j. We use the *logit* value from the output of the last hidden layer, as the similarity metric between a pair of letters. Given the logit value, we assign replacement attacks into easy, random and hard case.

**Generating Word-Level Attack Targets.** In this task, we replace the entire word in word, sentence and paragraph images. Different from letter-level attacks, we randomly choose a word in the English dictionary which has the same length as the original. However, we don't constrain the edit distance between them.

**Attack Implementation and Settings.** We implemented Grad-WM and Opt-WM attacks[3]. We normalize all input images to $[0, 1]$ and set $\beta$, the grayscale value of initial watermarks, to 0.682. Given color transform equation, fixing $R = 255$ and $B = 0$, RGB's upper bound $(255, 255, 0)$ is equal to $Gray = 0.882$. After adding $\epsilon$-bounded perturbations ($\epsilon = 0.2$), watermarks are still less than 0.882, the upper bound for valid conversion. We use the watermark style: the word "ecml" of Impact font with 78-pixel font size and 15-degree rotation.

For the gradient-based methods, we use the implementation of MI-FGSM in CleverHans Python library[4]. We set maximum iterations $I = 2000$, batch size = 100 and $\epsilon = 0.2$. For L$_2$-norm, $\alpha = 0.05$; for L$_\infty$-norm, $\alpha = 0.01$. The momentum decay factor $\mu$ is 1.0. For the optimization-based methods, we use the Adam optimizer [14] for 1000 steps with mini-batch size 100 and learning rate 0.01. We choose $c = 10$ as the tradeoff between the targeted loss and perturbation level.

**Evaluation Metrics.** We evaluate attack capability from the following aspects.

**1) Perturbation Level.** We quantify the perturbation level with three metrics: $MSE = \frac{1}{n}\|x - x'\|_2^2$, evaluates the difference between two images. $PSNR = 10\log\left(\frac{D^2}{MSE}\right)$, where $D$ is the range of pixel intensities. $SSIM$ [25] captures structural information and measures the similarity between two images. For these metrics, we take the average of all images in the following results. Smaller MSE, larger PSNR and SSIM closer to 1 indicate less perturbations.

---

[3] https://github.com/strongman1995/Fast-Adversarial-Watermark-Attack-on-OCR.
[4] https://github.com/tensorflow/cleverhan.

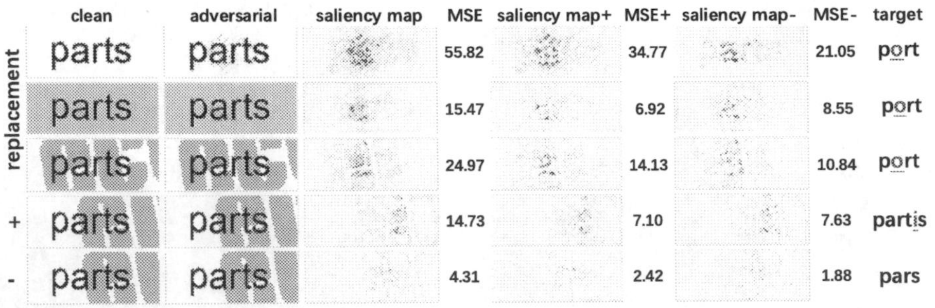

| | clean | adversarial | saliency map | MSE | saliency map+ | MSE+ | saliency map- | MSE- | target |
|---|---|---|---|---|---|---|---|---|---|
| replacement | parts | parts | | 55.82 | | 34.77 | | 21.05 | port |
| | parts | parts | | 15.47 | | 6.92 | | 8.55 | port |
| | parts | parts | | 24.97 | | 14.13 | | 10.84 | port |
| + | parts | parts | | 14.73 | | 7.10 | | 7.63 | partis |
| - | parts | parts | | 4.31 | | 2.42 | | 1.88 | pars |

**Fig. 2.** Saliency map visualization. First three lines are replacement. They have clean, gray and watermark backgrounds, respectively. Last two lines are insertion($+$) and deletion($-$). We fetch the positive part of saliency map to generate saliency map+ and the negative part to generate saliency map$-$. MSE, MSE+, MSE$-$ represent the perturbation level in the saliency map, saliency map+ and saliency map$-$, respectively.

2) **Success Rate.** Targeted attack success rate, $\text{ASR} = \frac{\#(f(x')=t)}{\#(x)}$, is the proportion of adversarial images of fooling OCR models to output targeted text.

3) **Attack Efficiency.** $I_{avg}$ is average iterations of images to reach 100% ASR.

### 4.2   Letter-Level Attack Performance

Table 1 illustrates generated adversarial examples. We find that the perturbations of basic attacks are quite obvious, especially when we want to replace the entire word, while the watermarks help hide the perturbations from human eyes.

To intuitively analyze the effects of watermarks on our attacks, in Fig. 2, we first illustrate the saliency map that highlights the influence of each pixel on the target output $t$. In replacement cases, we observe that the white background needs more perturbations, both positive ($+$) and negative ($-$), while the gray background requires 72% less perturbations due to reduced contrast. Watermarks approximate the gray background and look more natural. Specifically, the attack of the white background also adds significant a-shaped negative perturbations to weaken the letter a, while the other two require negative perturbations less than 50% due to lower contrast. In addition, the attacks add perturbations to neighboring letters, as the sequence-based OCR models also consider them.

Table 2 and Fig. 3 show the quantitative analysis of attack performance and use different targets in letter-level attacks. In Fig. 3, a sharper slope indicates a higher efficiency, that reaches a higher ASR in a fixed number of iterations. ASRs of all attacks are 100%. We observe that: 1) MSEs of watermark attacks are only 40% of basic attacks' on average, confirming to the intuitive analysis above. 2) Watermark attacks only require 78% fewer $I_{avg}$ on average, and have around 3 to 8 times sharper slope than basic attacks, showing the significant improvement of attack efficiency. 3) Not surprisingly, hard cases require both

**Table 2.** Letter-level attacks using Grad-Basic, Grad-WM, Opt-Basic, Opt-WM attacks. Last line is the target output. We denote each font with their first letter.

| | | Gradient-based | | | | | Optimization-based | | | | |
|---|---|---|---|---|---|---|---|---|---|---|---|
| | | replacement | | | insertion | deletion | replacement | | | insertion | deletion |
| | | easy | random | hard | | | easy | random | hard | | |
| | | MSE I_avg | MSE I_avg | MSE I_avg | MSE I_avg | MSE I_avg | MSE I_avg | MSE I_avg | MSE I_avg | MSE I_avg | MSE I_avg |
| Basic | C | 10.5 59 | 14.0 74 | 17.0 70 | 11.6 50 | 3.2 21 | 25.4 266 | 30.3 313 | 36.7 321 | 25.4 309 | 13.6 43 |
| | G | 27.4 43 | 32.8 99 | 37.3 104 | 22.1 83 | 17.3 55 | 52.0 292 | 59.4 318 | 67.5 328 | 41.6 337 | 45.0 169 |
| | H | 27.0 51 | 33.6 113 | 38.6 113 | 23.0 70 | 16.7 43 | 52.1 301 | 60.2 328 | 68.2 340 | 47.1 321 | 45.0 178 |
| | T | 26.4 62 | 31.5 85 | 35.8 109 | 20.3 98 | 17.2 68 | 49.9 294 | 56.1 324 | 61.6 345 | 41.7 314 | 44.3 172 |
| | A | 29.8 51 | 36.7 73 | 42.5 66 | 24.3 88 | 19.2 59 | 56.3 304 | 65.3 327 | 73.8 341 | 48.3 324 | 51.0 176 |
| | | parts | parts | parts | parts | parts | parts | parts | parts | parts | parts |
| WM | C | 2.8 30 | 3.6 18 | 4.3 27 | 3.6 21 | 0.7 8 | 16.7 116 | 20.1 96 | 20.4 95 | 31.1 29 | 3.2 13 |
| | G | 7.8 15 | 8.9 33 | 9.8 30 | 5.1 39 | 3.5 21 | 31.6 30 | 35.1 32 | 38.3 37 | 21.7 12 | 16.2 9 |
| | H | 8.4 9 | 10.0 52 | 11.2 52 | 6.3 23 | 3.7 19 | 33.3 31 | 37.0 42 | 38.8 53 | 25.1 13 | 16.5 9 |
| | T | 7.3 15 | 8.3 20 | 9.3 34 | 4.5 7 | 3.4 21 | 30.3 22 | 33.9 26 | 35.9 36 | 19.2 11 | 15.4 8 |
| | A | 9.4 13 | 11.1 14 | 12.7 25 | 6.2 33 | 4.4 20 | 37.2 30 | 40.4 45 | 43.6 50 | 25.4 16 | 19.4 10 |
| | | parts | parts | parts | parts | parts | parts | parts | parts | parts | parts |
| output | | pants | pacts | pasts | partis | pars | pants | pacts | pasts | partis | pars |

more perturbations (higher MSE) and more iterations (larger $I_{avg}$) with lower efficiency (flatter slope), due to lack of similarity between two letters. 4) In addition, deletions are easier than insertions and replacements. This is because OCR models are sensitive to perturbations, and classify fuzzy letters into blank tokens that will be ignored by CTC in the output. Intuitively, in the saliency map of Fig. 2, perturbations of deletion are much slighter than other cases. Thus few perturbations can achieve deletion. 5) Courier font is easier to attack because it is thinner than other fonts. Thus it requires less perturbations. 6) Comparing gradient-based and optimization-based methods, the perturbation level is higher in gradient-based methods, no matter if they have watermarks or not. Because perturbations of optimization-based methods are not constrained by $\epsilon$.

### 4.3 Word-Level Attack Performance

We can still achieve 100% ASR in the word-level attacks, but both MSE and $I_{avg}$ of word-level attacks are significantly higher than those of letter-level attacks.

We perform word-level attacks in word, sentence and paragraph images. Table 3 and Fig. 3(c) summarize results in word-level attacks. Due to limited space, we only show results of gradient-based methods. The optimization-based methods have similar insights. In word-level attacks, we have similar observations as letter-level attacks. 1) Watermark attacks spend 50% less $I_{avg}$ with higher efficiency (sharper slope) generating adversarial images than basic attacks. 2) We observe 56% lower MSE with watermarks averagely. 3) It is easier to attack Courier font than other fonts for less MSE and $I_{avg}$. 4) Due to a larger number of pixels in sentence and paragraph images, their MSEs are much lower than MSE of word images, but the absolute number of affected pixels stays the same.

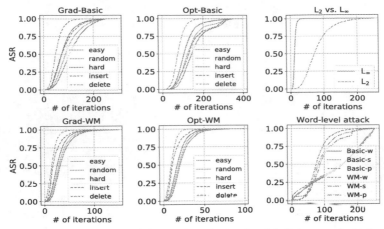

(a) Grad-based attacks. (b) Opt-based attacks. (c) Other comparisions

**Fig. 3.** Attack efficiency in word images with Arial font.

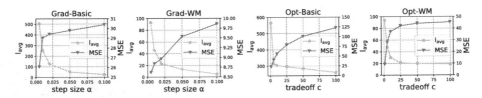

**Fig. 4.** In easy case of Arial word images, Grad-Basic and Grad-WM with different step size $\alpha = 0.005, 0.010, 0.020, 0.050, 0.100$. In easy case of Arial word images, Opt-Basic and Opt-WM with different tradeoff $c = 1, 5, 10, 25, 50, 100$.

### 4.4 Effects of Hyper-parameter Settings

To better understand these attacks, we evaluate the effects of hyper-parameter settings, such as tradeoff $c$, $L_p$-norms, font-weight, position of watermarks, etc.

**Effects of Plain Watermarks.** First, we evaluate the effects of plain watermarks, i.e., initial watermarks without perturbations. The accuracy of Courier, Georgia, Helvetica, Times, Arial drop to 0.066, 0.768, 0.531, 0.753 and 0.715, respectively, showing that it is quite trivial to launch *untargeted attacks* on OCR. Also, it further confirms that thinner font, Courier, is highly sensitive to plain watermarks. In samples of incorrect recognition, we count the percentage of output text shorter than the ground-truth text, Courier (0.65), Georgia (0.12), Helvetica (0.08), Times (0.09), Arial (0.14). It's easy to induce that thinner font is more sensitive to perturbations resulting in losing letters like the deletion case.

**Tradeoff Between Efficiency and Visual Quality.** In both gradient-based and optimization-based methods, we need to set tradeoff parameters to balance between efficiency ($I_{avg}$) and visual quality (MSE). In Fig. 4, we plot the change

**Table 3.** Word-level attacks in word, sentence and paragraph images.

| | | word image | | | | sentence image | | | | paragraph image | | | |
|---|---|---|---|---|---|---|---|---|---|---|---|---|---|
| | | MSE | PSNR | SSIM | $I_{avg}$ | MSE | PSNR | SSIM | $I_{avg}$ | MSE | PSNR | SSIM | $I_{avg}$ |
| Grad-Basic | Courier | 50.6 | 31.1 | 0.944 | 235 | 5.8 | 40.5 | 0.993 | 153 | 7.0 | 39.7 | 0.993 | 113 |
| | Georgia | 118.6 | 27.4 | 0.900 | 326 | 14.6 | 36.5 | 0.988 | 239 | 21.3 | 34.9 | 0.986 | 203 |
| | Helvetica | 124.0 | 27.2 | 0.894 | 254 | 14.2 | 36.6 | 0.988 | 238 | 22.5 | 34.6 | 0.984 | 233 |
| | Times | 114.4 | 27.5 | 0.904 | 291 | 13.2 | 36.9 | 0.989 | 201 | 17.4 | 35.7 | 0.989 | 164 |
| | Arial | 133.9 | 26.9 | 0.888 | 222 | 15.8 | 36.1 | 0.987 | 273 | 23.2 | 34.5 | 0.984 | 242 |
| | example | parts | | | | This one did exactly that. | | | | | | | |
| Grad-WM | Courier | 12.9 | 37.0 | 0.993 | 51 | 3.4 | 42.8 | 0.999 | 60 | 8.7 | 38.7 | 0.998 | 57 |
| | Georgia | 35.4 | 32.6 | 0.985 | 131 | 5.2 | 41.0 | 0.999 | 90 | 9.2 | 38.5 | 0.998 | 81 |
| | Helvetica | 40.0 | 32.1 | 0.984 | 124 | 5.4 | 40.8 | 0.999 | 91 | 11.8 | 37.4 | 0.998 | 96 |
| | Times | 34.8 | 32.7 | 0.985 | 160 | 4.5 | 41.6 | 0.999 | 79 | 6.7 | 39.9 | 0.990 | 74 |
| | Arial | 44.6 | 31.6 | 0.982 | 138 | 6.1 | 40.3 | 0.999 | 98 | 12.0 | 37.3 | 0.998 | 98 |
| | example | parts | | | | This one did exactly that. | | | | | | | |
| target output | | taupe | | | | Tale one did exactly that. | | | | | | | |

**Table 4.** Comparision of $L_2$-norm and $L_\infty$-norm in Grad-Basic. Examples' target output are "ports". Fonts are denoted by their initial letters. Bold fonts marked as 'b'.

| | $L_2$ | | | | | | $L_2$ | | | | | | $L_\infty$ | | | | |
|---|---|---|---|---|---|---|---|---|---|---|---|---|---|---|---|---|---|
| | MSE | PSNR | SSIM | $I_{avg}$ | example | | MSE | PSNR | SSIM | $I_{avg}$ | example | | MSE | PSNR | SSIM | $I_{avg}$ | example |
| Cb | 23.6 | 34.4 | 0.977 | 69 | parts | C | 10.5 | 37.9 | 0.988 | 59 | parts | | 72.6 | 29.5 | 0.771 | 11 | parts |
| Gb | 25.9 | 34.0 | 0.975 | 59 | parts | G | 27.4 | 33.7 | 0.974 | 43 | parts | | 163.6 | 26 | 0.645 | 11 | parts |
| Hb | 32.8 | 33.0 | 0.970 | 75 | parts | H | 27.0 | 33.8 | 0.975 | 51 | parts | | 159.4 | 26.1 | 0.658 | 10 | parts |
| Tb | 26.3 | 33.9 | 0.975 | 66 | parts | T | 26.4 | 33.9 | 0.975 | 62 | parts | | 156.6 | 26.2 | 0.653 | 11 | parts |
| Ab | 34.2 | 32.8 | 0.968 | 69 | parts | A | 29.8 | 33.4 | 0.972 | 51 | parts | | 169.3 | 25.8 | 0.656 | 11 | parts |

of $I_{avg}$ and MSE along with distinct tradeoff settings. In gradient-based methods, we can reduce perturbation level with smaller step size $\alpha$, at the cost of attack efficiency (raised $I_{avg}$). In optimization-based methods, the main quality-efficiency tradeoff parameter $c$ balances the targeted loss and the $L_p$-norm distance. We find that a smaller $c$ makes the optimized objective function pay more attention to reduce the perturbation level with larger $I_{avg}$. Thus we can use the binary search for $c$ to find adversarial examples with a lower perturbation level under the premise of maintaining perfect ASR. In addition, in various hyperparameter settings, the performance of watermark attacks is better than the basic attacks.

**Setting $L_p$-norms.** $L_\infty$-norm for measuring the maximum variation of perturbations has the same perturbed value at each perturbed pixel, and thus significantly narrows down operation space. Even more, observing $L_\infty$-norm examples in Table 4, $L_\infty$-norm causes severe background noise. $L_2$-norm yields perturbed values varying in perturbed pixels, and thus offers better flexibility to perform stronger adversarial attacks. Although $L_2$-norm is lower efficient than $L_\infty$-norm

**Table 5.** Non-overlapping Grad-WM in easy case of word images. Add a shift about 10 pixels right to the target letter.

| | MSE | $I_{avg}$ | ASR | example | output |
|---|---|---|---|---|---|
| Courier | 48.5 | 186 | 0.83 | love | hove |
| Georgia | 156.6 | 546 | 0.36 | move | mo_ve |
| Helvetica | 146.8 | 474 | 0.45 | parts | pa_nts |
| Times | 145.9 | 515 | 0.43 | hoy | boy |
| Arial | 153.5 | 463 | 0.44 | broad | bre_ad |

**Table 6.** Protection mechanism. Acc and Acc* are the prediction accuracy with and without protection, respectively.

| | MSE | $I_{avg}$ | Acc | Acc* | example |
|---|---|---|---|---|---|
| Courier | 0.6 | 5 | 1.0 | 0.066 | part |
| Georgia | 0.2 | 1 | 1.0 | 0.768 | part |
| Helvetica | 0.2 | 1 | 1.0 | 0.531 | part |
| Times | 0.2 | 1 | 1.0 | 0.753 | part |
| Arial | 0.2 | 1 | 1.0 | 0.715 | part |

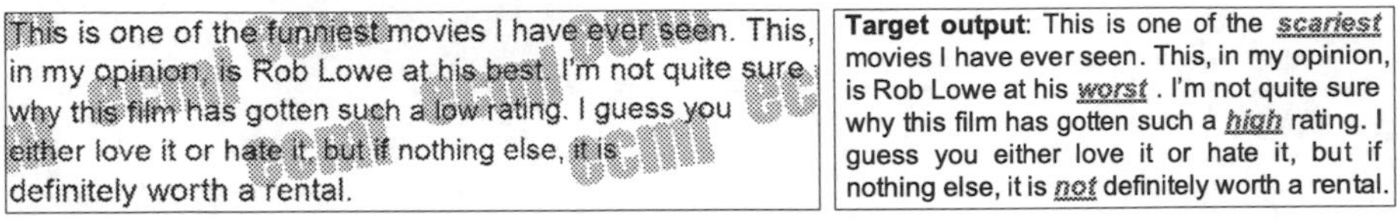

This is one of the funniest movies I have ever seen. This, in my opinion, is Rob Lowe at his best. I'm not quite sure why this film has gotten such a low rating. I guess you either love it or hate it, but if nothing else, it is definitely worth a rental.

**Target output**: This is one of the *scariest* movies I have ever seen. This, in my opinion, is Rob Lowe at his *worst* . I'm not quite sure why this film has gotten such a *high* rating. I guess you either love it or hate it, but if nothing else, it is *not* definitely worth a rental.

**Fig. 5.** Full-color watermarks on a paragraph image. MSE/PSNR/SSIM: 9.36/38.42/0.998. (Color figure online)

in Fig. 3(c), $L_2$-norm's image quality metrics, MSE, PSNR and SSIM, in Table 4, all are better than those of $L_\infty$-norm. Intuitively, $L_2$-norm examples also have better visual quality. Therefore, we choose the $L_2$-norm in our experiments.

**Bold Fonts Settings.** We also investigate the bold version of the five fonts. Table 4 shows most bold fonts require slightly more MSE and $I_{avg}$ to attack successfully, except for Courier requiring doubling MSE. This is not surprising, as bold fonts contain more useful pixels per letter. So they need more perturbations.

**Attacking Sequence-Based OCRs.** As OCR recognizes entire sequences rather than individual characters, we demonstrate that we can replace a letter even the watermark does not overlap with the letter, indicating we will not add perturbations over it. In this experiment, we shift the watermark about 10 pixels right to the target letter when generating initial watermarked images. Table 5 summarizes results that the attacks require 18 times MSE and 34 times $I_{avg}$, and ASRs drop to less than 50% except for the simplest Courier case. It confirms the fact that in sequential-based OCRs, the influence around the letter

9月17日止当周，美国原油每月进口量增加31.3万桶。小方一点下场，左侧配置资产。

**Target Output :** (Chinese) 9月12日上当周，美国原油每日进口量增加 31.3万桶。孙方二点不下场，右侧配置资产。

**Original:** *From September 17 to this week*, U.S. crude oil imports increased by *313,000* barrels per *month*. *Xiao* Fang *will* buy it at *1*'o clock, and she will do *left* transactions.
**Target:** *Last week, in September 12*, U.S. crude oil imports increased by *3,130,000* barrels per *day*. *Sun* Fang *won't* buy it at *2*'o clock, and she will do *right* transactions.

**Fig. 6.** A Chinese paragraph example. MSE/PSNR/SSIM: 735.34/19.46/0.697.

is not as important as that overlaying the letter. Also, it reveals the necessity of finding the position of watermarks accurately to perform strong adversarial attacks.

### 4.5 Extensions and Applications of Watermark Attacks

**Full-Color Watermarks.** Sometimes adding gray watermarks still hurts human readability. We use the fact that modern OCR first transforms colored images into gray ones before recognition. Colored watermarks significantly improve overall readability when mixed with black texts. Figure 5 shows a colored-watermark example of altering a positive movie review into a negative one by replacing and inserting words in it. Note that not all watermarks are malicious (e.g., the watermark on "I have ever seen" of Fig. 5 does not include adversarial perturbations). We evenly distribute watermarks over the paragraph image to make it look more similar to a naturally watermarked paragraph image in the real-world scenario.

**Attacking Chinese Characters.** In addition to English, we show that the method is applicable to other languages. Figure 6 shows a Chinese example where we almost altered all important information, and its perturbation level is much larger than that in Fig. 5, because of the complex structure of Chinese characters.

**Using FAWA to Enhance the OCR Readability of Watermarked Contents.** Sometimes we want people to notice vital watermarks (e.g., urgent, confidential), but we do not want them to affect OCR's accuracy. In such case, we generate accuracy-enhancing watermarks by setting the ground-truth as the target. In Table 6, with few $I_{avg}$ and MSEs, Acc (accuracy with protection mechanism) increases to 1.0 compared with the low Acc* (accuracy of initial watermarked images). It works because protective perturbations strengthen the features of the ground-truth and boost the confidence of the original, making it more "similar" to the target (i.e., the ground-truth). As the target is the same as the original, both MSE and $I_{avg}$ stay low. So we can use FAWA as a protection mechanism to produce both human-friendly images and OCR-friendly images.

## 5  Conclusion

DNN-based OCR systems are vulnerable to adversarial examples. On the text images, we hide perturbations in watermarks, making adversarial examples look natural to human eyes. Specifically, we develop FAWA that automatically generate the adversarial watermarks targeting sequence-based OCR models. In extensive experiments, while maintaining perfect attack success rate, FAWA exhibits the outstanding attack capability. For example, FAWA boosts the attack speed up to 8 times, and reduces the perturbation level lower than 40% on average. In word, sentence and paragraph contexts, FAWA works well with letter-level and word-level targets. We further extend our natural watermarked adversarial examples in many scenarios, such as full-color watermarks to increase the text readability, applicability for other languages, protection mechanism for enhancing OCR's accuracy. We believe human-eye-friendly adversarial samples are applicable in many other scenarios, and we plan to explore them as future work.

**Acknowledgments.** This work is supported in part by the National Natural Science Foundation of China (NSFC) Grant 61532001 and the Zhongguancun Haihua Institute for Frontier Information Technology.

# References

1. Bengio, Y., LeCun, Y., Nohl, C., Burges, C.: LeRec: a NN/HMM hybrid for on-line handwriting recognition. Neural Comput. **7**(6), 1289–1303 (1995)
2. Breuel, T.M., Ul-Hasan, A., Al-Azawi, M.A., Shafait, F.: High-performance OCR for printed English and fraktur using LSTM networks. In: 2013 12th International Conference on Document Analysis and Recognition, pp. 683–687. IEEE (2013)
3. Brown, T.B., Mané, D., Roy, A., Abadi, M., Gilmer, J.: Adversarial patch. arXiv preprint arXiv:1712.09665 (2017)
4. Carlini, N., Wagner, D.: Towards evaluating the robustness of neural networks. In: 2017 IEEE Symposium on Security and Privacy (SP), pp. 39–57. IEEE (2017)
5. Chen, P.Y., Sharma, Y., Zhang, H., Yi, J., Hsieh, C.J.: EAD: elastic-net attacks to deep neural networks via adversarial examples. In: Thirty-Second AAAI Conference on Artificial Intelligence (2018)
6. Dong, Y., et al.: Boosting adversarial attacks with momentum. In: Proceedings of the IEEE Conference on Computer Vision and Pattern Recognition, pp. 9185–9193 (2018)
7. Espana-Boquera, S., Castro-Bleda, M.J., Gorbe-Moya, J., Zamora-Martinez, F.: Improving offline handwritten text recognition with hybrid HMM/ANN models. IEEE Trans. Pattern Anal. Mach. Intell. **33**(4), 767–779 (2011)
8. Goodfellow, I.J., Shlens, J., Szegedy, C.: Explaining and harnessing adversarial examples. arXiv preprint arXiv:1412.6572 (2014)
9. Graves, A., Fernández, S., Gomez, F., Schmidhuber, J.: Connectionist temporal classification: labelling unsegmented sequence data with recurrent neural networks. In: Proceedings of the 23rd International Conference on Machine Learning, pp. 369–376. ACM (2006)
10. Zhang, H., Avrithis, Y., Furon, T., Amsaleg, L.: Smooth adversarial examples. arXiv preprint arXiv:1903.11862 (2019)
11. Heng, W., Zhou, S., Jiang, T.: Harmonic adversarial attack method. arXiv preprint arXiv:1807.10590 (2018)
12. Hochreiter, S., Schmidhuber, J.: Long short-term memory. Neural Comput. **9**(8), 1735–1780 (1997)
13. Karmon, D., Zoran, D., Goldberg, Y.: LaVAN: Localized and visible adversarial noise. arXiv preprint arXiv:1801.02608 (2018)
14. Kingma, D.P., Ba, J.: Adam: A method for stochastic optimization. arXiv preprint arXiv:1412.6980 (2014)
15. Krizhevsky, A., Sutskever, I., Hinton, G.E.: ImageNet classification with deep convolutional neural networks. In: Advances in Neural Information Processing Systems, pp. 1097–1105 (2012)

16. Kurakin, A., Goodfellow, I., Bengio, S.: Adversarial machine learning at scale. arXiv preprint arXiv:1611.01236 (2016)
17. Madry, A., Makelov, A., Schmidt, L., Tsipras, D., Vladu, A.: Towards deep learning models resistant to adversarial attacks. arXiv preprint arXiv:1706.06083 (2017)
18. Moosavi-Dezfooli, S.M., Fawzi, A., Frossard, P.: DeepFool: a simple and accurate method to fool deep neural networks. In: Proceedings of the IEEE Conference on Computer Vision and Pattern Recognition, pp. 2574–2582 (2016)
19. Nguyen, A., Yosinski, J., Clune, J.: Deep neural networks are easily fooled: high confidence predictions for unrecognizable images. In: Proceedings of the IEEE Conference on Computer Vision and Pattern Recognition, pp. 427–436 (2015)
20. Papernot, N., McDaniel, P., Jha, S., Fredrikson, M., Celik, Z.B., Swami, A.: The limitations of deep learning in adversarial settings. In: 2016 IEEE European Symposium on Security and Privacy (EuroS&P), pp. 372–387. IEEE (2016)
21. Song, C., Shmatikov, V.: Fooling OCR systems with adversarial text images (2018)
22. Szegedy, C., et al.: Intriguing properties of neural networks. arXiv preprint arXiv:1312.6199 (2013)
23. Wang, G., Chen, X., Xu, C.: Adversarial watermarking to attack deep neural networks. In: ICASSP 2019–2019 IEEE International Conference on Acoustics, Speech and Signal Processing (ICASSP), pp. 1962–1966. IEEE (2019)
24. Wang, T., Wu, D.J., Coates, A., Ng, A.Y.: End-to-end text recognition with convolutional neural networks. In: Proceedings of the 21st International Conference on Pattern Recognition (ICPR2012), pp. 3304–3308. IEEE (2012)
25. Wang, Z., Bovik, A.C., Sheikh, H.R., Simoncelli, E.P.: Image quality assessment: from error visibility to structural similarity. IEEE Trans. Image Process. **13**(4), 600–612 (2004)
26. Wick, C., Reul, C., Puppe, F.: Calamari-a high-performance tensorflow-based deep learning package for optical character recognition. arXiv preprint arXiv:1807.02004 (2018)
27. Xiao, C., Zhu, J.Y., Li, B., He, W., Liu, M., Song, D.X.: Spatially transformed adversarial examples. arXiv abs/1801.02612 (2018)

# Natural Language Processing

# Less Is More: Rejecting Unreliable Reviews for Product Question Answering

Shiwei Zhang[1,3], Xiuzhen Zhang[1]([✉]), Jey Han Lau[2], Jeffrey Chan[1], and Cecile Paris[3]

[1] RMIT University, Melbourne, Australia
{shiwei.zhang,xiuzhen.zhang,jeffrey.chan}@rmit.edu.au
[2] The University of Melbourne, Melbourne, Australia
jeyhan.lau@gmail.com
[3] CSIRO Data61, Sydney, Australia
cecile.paris@data61.csiro.au

**Abstract.** Promptly and accurately answering questions on products is important for e-commerce applications. Manually answering product questions (e.g. on community question answering platforms) results in slow response and does not scale. Recent studies show that product reviews are a good source for real-time, automatic product question answering (PQA). In the literature, PQA is formulated as a retrieval problem with the goal to search for the most relevant reviews to answer a given product question. In this paper, we focus on the issue of *answerability* and *answer reliability* for PQA using reviews. Our investigation is based on the intuition that many questions may not be answerable with a finite set of reviews. When a question is not answerable, a system should return nil answers rather than providing a list of irrelevant reviews, which can have significant negative impact on user experience. Moreover, for answerable questions, only the most relevant reviews that answer the question should be included in the result. We propose a conformal prediction based framework to improve the reliability of PQA systems, where we reject unreliable answers so that the returned results are more concise and accurate at answering the product question, including returning nil answers for unanswerable questions. Experiments on a widely used Amazon dataset show encouraging results of our proposed framework. More broadly, our results demonstrate a novel and effective application of conformal methods to a retrieval task.

**Keywords:** Product question answering · Unanswerable questions · Conformal prediction

© Springer Nature Switzerland AG 2021
F. Hutter et al. (Eds.): ECML PKDD 2020, LNAI 12459, pp. 567–583, 2021.
https://doi.org/10.1007/978-3-030-67664-3_34

# 1   Introduction

On e-commerce websites such as Amazon[1] and TaoBao[2], customers often ask product-specific questions prior to a purchase. With the number of product-related questions (queries) growing, efforts to answer these queries manually in real time is increasingly infeasible. Recent studies found that product reviews are a good source to extract helpful information as answers [1–3].

**Table 1.** Example of an answerable (Q1) and an unanswerable question (Q2). Green denotes high probability/confidence scores, and red otherwise.

| Q1: What is the chain for on the side? Top 3 Ranked Reviews | Prob | Conf | Accept |
|---|---|---|---|
| - This was driving me crazy but i see that another reviewer explained that grill has wire clip on chain to be used as extended match holder for igniting the gas if the spark mechanism fails to work or is worn out as sometimes happens with any gas grill. | 0.99 | 0.82 | ✔ |
| - PS Could not figure out the purpose of that little chain with the clip attached to the outside of the grill - even after reading entire manual. | 0.95 | 0.54 | ✘ |
| - It is to replace an old portable that I have been using for about 10 years.' | 0.91 | 0.40 | ✘ |

| Q2: Does this Dell Inspiron 14R i14RMT-7475s come with dell's warranty? Top 3 Ranked Reviews | Prob | Conf | Accept |
|---|---|---|---|
| - I don't really recommend the PC for people who wants install heavy games programs. | 0.74 | 0.48 | ✘ |
| - The computer is nice, fast, light, ok. | 0.12 | 0.01 | ✘ |
| - I bought the computer for my daughter. | 0.05 | 0.00 | ✘ |

To illustrate this, in Table 1 Q1 poses a question about the purpose of the chain on the side of a grill, and the first review addresses the question. The core idea of state-of-the-art PQA models is to take advantage of existing product reviews and find relevant reviews that answer questions automatically. In general, most PQA models implement a relevance function to rank existing reviews based on how they relate to a question. Some directly present a fixed number (typically 10) of the top ranked reviews as answers for the question [1–3], while others generate natural-language answers based on the relevant reviews  [4,5].

---

[1] https://www.amazon.com/.
[2] https://world.taobao.com/.

However, not every question can be answered by reviews: the existing set of reviews may not contain any relevant answers for the question, or a question may be poorly phrased and difficult to interpret and therefore requires additional clarification. Q2 in Table 1 is an example of an unanswerable question. The user wants to know whether the notebook comes with Dell's warranty, but none of the reviews discuss anything about warranty. In such a case, a system should abstain from returning any reviews and forward the question to the product seller. In the PQA literature, the issue of answerability is largely unexplored, and as such evaluation focuses on simple ranking performance without penalising systems that return irrelevant reviews.

That said, the question answerability issue has begun to draw some attention in machine comprehension (MC). Traditionally, MC models assume the correct answer span always exists in the context passage for a given question. As such, these systems will give an incorrect (and often embarrassing) answer when the question is not answerable. This motivates the development of better comprehension systems that can distinguish between answerable and unanswerable questions. Since SQuAD—a popular MC dataset—released its second version [6] which contains approximately 50,000 unanswerable questions, various MC models have been proposed to detect question-answerability in addition to predicting answer span [17–20]. The MC models are trained to first detect whether a question is answerable or unanswerable and then find answers to answerable questions. [23] proposed a risk controlling framework to increase the reliability of MC models, where risks are quantified based on the extent of incorrect predictions, for both answerable and unanswerable questions. Different from MC which always has one answer, PQA is a ranking problem where there can be a number of relevant reviews/answers. As such, PQA is more challenging, and risk models designed for MC cannot be trivially adapted for PQA.

In this paper, we focus on the problem of answer reliability for PQA. Answer reliability is generalisation of answerability. Answerability is a binary classification problem where answerable questions have only one answer (e.g. MC). In our problem setting, a product question can have a variable number of reliable answers (reviews), and questions with nil reliable answers are the unanswerable questions. The challenge is therefore on how we can estimate the reliability of answers. As our paper shows, the naive approach of thresholding based on the predicted probability from PQA models is not effective for estimating the reliability of candidate answers.

We tackle answer reliability by introducing a novel application of conformal predictors as a rejection model. The rejection model [7] is a technique proposed to reduce the misclassification rate for risk-sensitive classification applications. The risk-sensitive prediction framework consists of two models: a classifier that outputs class probabilities given an example, and a rejection model that measures the confidence of its prediction and rejects unconfident prediction. In our case, given a product question, the PQA model makes a prediction on the relevance for each review, and the rejection model judges the reliability for each prediction and returns only reviews that are relevant *and* reliable as answers. As an example, although the positive class probabilities given by the PQA model to the top 3 reviews are very high in Q1 (Table 1), the rejection model would reject the last two because their confidence scores are low. Similarly for Q2, even though the first review has high relevance, the rejection model will reject all reviews based on the confidence scores, and return an empty list to indicate that this is an unanswerable question.

For the PQA models, we explore both classical machine learning models [1] and BERT-based neural models [3]. For the rejection model, we use an Inductive Mondrian Conformal Predictor (IMCP) [8–11]. IMCP is a popular non-parametric conformal predictor used in a range of domains, from drug discovery [12] to chemical compound activity prediction [11]. The challenge of applying a rejection model to a PQA model is how we define the associated risks in PQA. Conventionally, a rejection mode is adopted to reduce the misclassification rate, but for PQA we need to reduce the error of including irrelevant results. We propose to use IMCP to transform review relevance scores (or probabilities) into confidence scores and tune a threshold to minimize risk (defined by an alternative ranking metric that penalises inclusion of irrelevant results), and reject reviews whose confidence scores are lower than the threshold.

To evaluate the effectiveness of our framework in producing relevant and reliable answers, we conduct a crowdsourcing study using Mechanical Turk[3] to acquire the relevance of reviews for 200 product-specific questions on Amazon [1]. We found encouraging results, demonstrating the applicability of rejection models for PQA to handle unanswerable questions. To facilitate replication and future research, we release source code and data used in our experiments.[4]

## 2    Related Work

In this section, we survey three related topics to our work: product question answering, question answerability and conformal predictors.

### 2.1    Product Question Answering

Existing studies on product-related question answering using reviews can be broadly divided into extractive approaches [1,3] and generative approaches [4,5]. For extractive approaches, relevant reviews or review snippets are extracted from reviews to answer questions, while for the generative approaches natural answers are further generated based on the review snippets. In both approaches, the critical step is to first identify relevant reviews that can answer a given question.

The key challenge in PQA is the lack of ground truth, i.e. there is limited data with annotated relevance scores between questions and reviews. Even with crowdsourcing, the annotation work is prohibitively expensive, as a product question may have a large number of reviews; and more so if we were to annotate at sentence level (i.e. annotating whether a sentence in a review is

---

[3] https://www.mturk.com/.
[4] https://github.com/zswvivi/ecml_pqa.

relevant to a query), which is typically the level of granularity that PQA studies work with [1,3]. For that reason, most methods adopt a distant supervision approach that uses existing question and answer pairs from an external source (e.g. the community question answer platform) as supervision. The first of such study is the Mixture of Opinions for Question Answering (MOQA) model proposed by [1], which is inspired by the mixture-of-experts classifier [14]. Using answer prediction as its objective, MOQA decomposes the task into learning two relationships: (1) relevance between a question and a review; and (2) relevance between a review and an answer, sidestepping the need for ground truth relevance between a question and a review. [3] extends MOQA by parameterising the relevance scoring functions using BERT-based models [15] and found improved performance.

Another recent work [2] learns deep representations of words between existing questions and answers. To expand the keywords of a query, the query words are first mapped to their continuous representations and similar words in the latent space are included in the expanded keywords. To find the most relevant reviews, the authors use a standard keyword-based retrieval method with the expanded query and found promising results.

## 2.2  Unanswerable Questions

There are few studies that tackle unanswerable questions in PQA. One exception is [16], where they develop a new PQA dataset with labelled unanswerable questions. That said, the author frame PQA as a classification problem, where the goal is to find an answer span in top review snippets (retrieved by a search engine), and as such the task is more closely related to machine comprehension.

In the space of MC, question answerability drew some attention when SQuAD 2.0 [6] was released, which includes over 50,000 unanswerable questions created by crowdworkers. Several deep learning MC systems have since been proposed to tackle these unanswerable questions. [17] proposed a read-then-verify system with two auxiliary losses, where the system detects whether a question is answerable and then checks the validity of extracted answers. [18] proposed a multi-task learning model that consists of three components: answer span prediction, question answerability detection and answer verification.

More generally in question answering (QA), Web QA is an open-domain problem that leverages Web resources to answer questions, e.g. TriviaQA [21] and SearchQA [22]. [23] introduced a risk control framework to manage the uncertainty of deep learning models in Web QA. The authors argue that there are two forms of risks, by returning: (1) wrong answers for answerable questions; and (2) any answers for unanswerable questions. The overall idea of their work is similar to ours, although their approach uses a probing method that involves intermediate layer outputs from a neural model, which is not applicable to non-neural models such as MOQA.

## 2.3   Conformal Predictors

To measure the reliability of prediction for an unseen example, conformal predictors (CP) compares how well the unseen example *conforms* to previously seen examples. Given an error probability $\epsilon$, CP is guaranteed to produce a prediction region with probability $1 - \epsilon$ of containing the true label $y$, thereby offering a means to control the error rate. CP has been applied to many different areas, from drug discovery [11,13,24] to image and text classification [25]. [13] proposed a neural framework using Monte Carlo Dropout [26] and CP to compute reliable errors in prediction to guide the selection of active molecules in retrospective virtual screen experiments. In [25], the authors replace the softmax layer with CP to predict labels based on a weighted sum of training instances for image and text classification.

## 3   Methodology

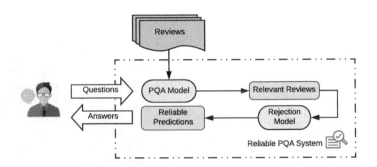

**Fig. 1.** Proposed PQA + a rejection model framework.

Our proposed framework consists of two components: a PQA model that predicts the relevance of reviews given a product query, and a rejection model that rejects unconfident reviews and produce only confident reviews as answers. An illustration of our framework is presented in Fig. 1.

More specifically, the PQA component (a binary classifier) models the function $\hat{y} = F(r, q)$ where $r$ is a review, $q$ is a product question, and $\hat{y}$ is the probability of the positive class, i.e. the review is relevant to the question.

The rejection model takes the output $\hat{y}$ from the PQA model and transforms it into a confidence score $y^*$. Given the confidence score, we can then set a significance level $\epsilon \in [0, 1]$ to reject/accept predictions. E.g. if the confidence score of the positive class (review is relevant to the question) is 0.6 and $\epsilon$ is 0.5, then we would accept the positive class prediction and return the review as relevant. On the other hand, if $\epsilon$ is 0.7, we would reject the review.

## 3.1   PQA Models

We explore 3 state-of-the-art PQA models for our task:

**MOQA** [1] is inspired by the mixture-of-experts classifier [14]. In a mixture of experts classifier, a prediction is made by a weighted combination of a number of weak classifiers. Using answer prediction ($P(a|q)$) as its objective, MOQA decomposes the problem into: $P(a|q) = \sum_r P(a|r)P(r|q)$, where $a, q, r$ = answer, query, review respectively. Each review can be interpreted as an "expert", where it makes a prediction on the answer ($P(a|r)$) and its prediction is weighted by its confidence ($P(r|q)$). The advantage of doing this decomposition is that the learning objective is now $P(a|q)$, which allows the model to make use of the abundance of existing questions and answers on e-commerce platforms. In practice, MOQA is optimised with a maximum margin objective: given a query, the goal is to score a real answer higher than a randomly selected non-answer. To model the two relevance functions ($P(a|r)$ and $P(r|q)$), MOQA uses off-the-shelf pairwise similarity function such as BM25+ and ROUGE-L and a learned bilinear scoring function that uses bag-of-words representation as input features. After the model is trained, the function of interest is $P(r|q)$, which can be used to score a review for a given query. We use the open source implementation and its default optimal configuration.[5]

**FLTR** [3] is a BERT classifier for answer prediction. Using existing question and answer pairs on e-commerce platforms, [3] fine-tunes a pre-trained BERT to classify answers given a question. After fine-tuning, FLTR is used to classify *reviews* for a given question. FLTR can be seen as a form of zero-shot domain transfer, where it is trained in the (answer, question) domain but at test time it is applied to the (review, question) domain. We use the open-source implementation and its default optimal configuration.[6]

**BERTQA** [3] is an extension of MOQA, which uses the same mixture of experts framework, but parameterises the relevance functions ($P(a|r)$ and $P(r|q)$) with neural networks: BERT. BERTQA addresses the vocabulary/language mismatch between different domains (e.g. answer vs. review, or review vs. query) using contextualised representations and fine-grained comparison between words via attention. [3] demonstrates that this neural parameterisation substantially improves review discovery compared to MOQA. The downside of BERTQA is its computational cost: while MOQA can be used to compute the relevance for every review for a given query (which can number from hundreds to thousands), this is impractical with BERTQA. To ameliorate this, [3] propose using FLTR to pre-filter reviews to reduce the set of reviews to be ranked by BERTQA.[7]

---

[5] https://cseweb.ucsd.edu/~jmcauley.

[6] https://github.com/zswvivi/icdm_pqa.

[7] The original implementation uses a softmax activation function to compute $P(r|q)$ (and so the probability of all reviews sum up to one); we make a minor modification to the softmax function and use a sigmoid function instead (and so each review produces a valid probability distribution over the positive and negative classes).

**Table 2.** An example of predicted labels with different choices of significance level. $p$-value for the positive (1) and negative (0) label is 0.65 and 0.45 respectively.

| $\epsilon$ | Predicted labels | Outcome |
|------|------------------|----------|
| 0.05 | $\{0, 1\}$ | Rejected |
| 0.45 | $\{1\}$ | Accepted |
| 0.75 | $\{\}$ | Rejected |

### 3.2    Rejection Model

In a classification task, a model could make a wrong prediction for a difficult instance, particularly when the positive class probability is around 0.5 in a binary task. For medical applications such as tumor diagnostic, misclassification can have serious consequences. In these circumstances, rejection techniques are used to reduce misclassification rate by rejecting unconfident or unreliable predictions [7].

To apply rejection techniques to PQA, we need to first understand the associated risks in PQA. There are forms of risks in PQA: (1) including irrelevant reviews as part of its returned results for an answerable question; and (2) returning any reviews for an unanswerable question. This is similar to the risks proposed for Web QA [23], although a crucial difference is that PQA is a ranking problem (output is a list of reviews). The implication is that when deciding the $\epsilon$ threshold to guarantee a "error rate", we are using a ranking metric as opposed to a classification metric. As standard ranking metric such as normalised discounted cumulative gain (NDCG) is unable to account for unanswerable questions, we explore alternative metrics (detailed in Sect. 4.2).

We propose to use conformal predictors (CP) [9,10] as the rejection model. Intuitively, for a test example, CP computes a nonconformity score that measures how well the new example *conforms* to the previously seen examples as a way to estimate the reliability of the prediction. CP is typically applied to the output of other machine learning models, and can be used in both classification and regression tasks. There are two forms of CP, namely Inductive CP (ICP) and Transductive CP (TCP). The difference between them is their training scheme: in TCP, the machine learning model is updated for each new examples (and so requires re-training) to compute the nonconformity score, while in ICP the model is trained once using a subset of the training data, and the other subset— the *calibration set*—is set aside to be used to compute nonconformity scores for a new example. As such, the associated computation cost for the transductive variant is much higher due to the re-training. We use the Inductive Mondrian Conformal Predictor (IMCP) for our rejection model, which is a modified ICP that is better at handling imbalanced data. When calculating a confidence score, IMCP additionally conditions it on the class label. In PQA, there are typically a lot more irrelevant reviews than relevant reviews for given a query, and so IMCP is more appropriate for our task.

Given a bag of calibration examples $\{(x_1, y_1), ...(x_n, y_n)\}$ and a binary classifier $\hat{y} = f(x)$, where $n$ is number of calibration examples, $x$ the input, $y$ the true label and $\hat{y}$ the predicted probability for the positive label, we compute a nonconformity score, $\alpha(x_{n+1})$, for a new example $x_{n+1}$ using its inverse probability:

$$\alpha(x_{n+1}) = -f(x_{n+1})$$

As IMCP conditions the nonconformity score on the label, there are 2 nonconformity scores for $x_{n+1}$, one for the positive label, and one for the negative label:

$$\alpha(x_{n+1}, 1) = -f(x_{n+1})$$
$$\alpha(x_{n+1}, 0) = -(1 - f(x_{n+1}))$$

We then compute the confidence score ($p$-value) for $x_{n+1}$ conditioned on label $k$ as follows:

$$p(x_{n+1}, k) = \frac{\sum_{i=0}^{n} I(\alpha(x_i, k) \geq \alpha(x_{n+1}, k))}{n + 1} \tag{1}$$

where $I$ is the indicator function.

Intuitively, we can interpret the $p$-value as a measure of how confident/reliable the prediction is for the new example by comparing its predicted label probability to that of the calibration examples.

Given the $p$-values (for both positive and negative labels), the rejection model accepts a prediction for all labels $k$ where $p(x_{n+1}, k) > \epsilon$, where $\epsilon \in [0, 1]$ is the significance level. We present an output of the rejection model in Table 2 with varying $\epsilon$, for an example whose $p$-values for the positive and negative labels are 0.65 and 0.45 respectively. Depending on $\epsilon$, the number of predicted labels ranges from zero to 2. For PQA, as it wouldn't make sense to have both positive and negative labels for an example (indicating a review is both relevant and irrelevant for a query), we reject such an example and consider it an unreliable prediction.

## 4 Experiments

### 4.1 Data

We use the Amazon dataset developed by [1] for our experiments. The dataset contains QA pairs and reviews for each product. We train MOQA, BERTQA and FLTR on this data, noting that MOQA and BERTQA leverages both QA pairs and reviews, while FLTR uses only the QA pairs. After the models are trained, we can use the $P(r|q)$ relevance function to score reviews given a product question.

To assess the quality of the reviews returned for a question by the PQA models, we ask crowdworkers on Mechanical Turk to judge how well a review answers a question. We randomly select 200 questions from four categories ("Tools and Home Improvement", "Patio Lawn and Garden", "Baby" and "Electronics"), and pool together the top 10 reviews returned by the 3 PQA models (MOQA, BERTQA

**Table 3.** Answerable question statistics.

| Relevance threshold | 2.00 | 2.25 | 2.50 | 2.75 | 3.00 |
|---|---|---|---|---|---|
| #Relevant reviews | 640 | 351 | 175 | 71 | 71 |
| #Answerable questions | 170 | 134 | 89 | 44 | 44 |
| %Answerable questions | 85% | 67% | 45% | 22% | 22% |

and FLTR),[8] resulting in approximately 20 to 30 reviews per question (total number of reviews = 4,691).[9]

In the survey, workers are presented with a pair of question and review, and they are asked to judge how well the review answers the question on an ordinal scale: 0 (completely irrelevant), 1 (related but does not answer the question), 2 (somewhat answers the question) and 3 (directly answers the question). Each question/review pair is annotated by 3 workers, and the final relevance score for each review is computed by taking the mean of 3 scores.

Given the annotated data with relevance scores, we can set a relevance threshold to define a cut-off when a review answers a question, allowing us to control how precise we want the system to be (i.e. a higher threshold implies a more precise system). We present some statistics in Table 3 with different relevance thresholds. For example, when the threshold is set to 2.00, it means a review with a relevance score less than 2.00 is now considered irrelevant (and so its score will be set to 0.00), and a question where all reviews have a relevance score less than 2.00 is now unanswerable. The varying relevance thresholds will produce a different distribution of relevant/irrelevant reviews and answerable/unanswerable questions; at the highest threshold (3.00), only a small proportion of the reviews are relevant (but they are all high-quality answers), and most questions are unanswerable.

### 4.2 Evaluation Metric

As PQA is a retrieval task, it is typically evaluated using ranking metrics such as normalised discounted cumulative gain (NDCG) [28]. Note, however, that NDCG is not designed to handle unanswerable queries (i.e. queries with no relevant documents), and as such isn't directly applicable to our task. We explore a variant, NDCG', that is designed to work with unanswerable queries [27]. The idea of NDCG' is to "quit while ahead": the returned document list should be truncated earlier rather than later, as documents further in the list are more likely to be irrelevant. To illustrate this, we present an answerable question in Table 4. Assuming it has three relevant documents (1 represents a relevant

---

[8] Following the original papers, a "review" is technically a "review sentence" rather than the full review.

[9] To control for quality, we insert a control question with a known answer (from the QA pair) in every 3 questions. Workers who consistently give low scores to these control questions are filtered out.

**Table 4.** NDCG' examples.

| Question type | Systems | Doc list | NDCG' |
|---|---|---|---|
| Answerable | System A | 111 | 1.000 |
| | System B | 11100 | 0.971 |
| | System C | 11 | 0.922 |
| Unanswerable | System A | ∅ | 1.000 |
| | System B | 00 | 0.500 |
| | System C | 000 | 0.431 |

document and 0 an irrelevant document), System A receives a perfect NDCG' score while System B is penalised for including 2 irrelevant documents. System C has the lowest NDCG' score as it misses one relevant document. The second example presents an unanswerable question. The ideal result is the empty list (∅) returned by System A, which receives a perfect score. Comparing System B to C, NDCG' penalises C for including one more irrelevant document.

NDCG' appends a terminal document $(t)$ to the end of the document list returned by a ranking system. For example, "111" → "111$t$", and "11100" → "11100$t$". The corresponding gain value for the terminal document $t$ is $r_t$, calculated as follows:

$$r_t = \begin{cases} 1 & \text{if } R = 0 \\ \sum_{i=1}^{d} r_i / R & \text{if } R > 0 \end{cases}$$

where $R$ is the total number of ground truth relevant documents, and $r_i$ is the relevance of document $i$ in the list. As an example, for the document list "11" produced by System C, $r_t = \frac{1}{3} + \frac{1}{3} = \frac{2}{3}$.

Given $r_t$, we compute NDCG' for a ranked list of $d$ items as follows:

$$\text{NDCG'}_d = \frac{\text{DCG}_{d+1} \langle r_1, r_2, ..., r_d, r_t \rangle}{\text{IDCG}_{d+1}}$$

With NDCG, for an unanswerable question like the second example, both System B ("00") and System C ("000") will produce a score of zero, and so it fails to indicate that B is technically better. NDCG' solves this problem by introducing the terminal document score $r_t$.

In practice, the relevance of our reviews is not binary (unlike the toy examples). That is, the relevance of each review is a mean relevance score from 3 annotators, and ranges from 0–3. Note, however, that given a particular relevance threshold (e.g. 2.0), we mark all reviews under the threshold as irrelevant by setting their relevance score to 0.0.

In our experiments, we compute NDCG' up to a list of 10 reviews $(d = 10)$, and separately for answerable $(\mathcal{N}_A)$ and unanswerable questions $(\mathcal{N}_U)$. To aggregate NDCG' over two question types $(\mathcal{N}_{A+U})$, we compute the geometric mean:

$$\mathcal{N}_{A+U} = \sqrt{\mathcal{N}_A \times \mathcal{N}_U}$$

We use geometric mean here because we want an evaluation metric that favours a balanced performance between answerable and unanswerable questions [29]. In preliminary experiments, we found that micro-average measures will result in selecting a system that always returns no results when a high relevance threshold is selected (e.g. $\geq 2.50$ in Table 3) since a large number of questions are unanswerable. This is undesirable in a real application where choosing a high relevance threshold means we want a very precise system, and not one that never gives any answers.

## 4.3   Experimented Methods

We compare the following methods in our experiments:

**Vanilla PQA Model**: a baseline where we use the top-10 reviews returned by a PQA model (MOQA, FLTR or BERTQA) without any filtering/rejection.

**PQA Model + THRS**: a second baseline where we tune a threshold based on the review score returned by a PQA model (MOQA, FLTR or BERTQA) to truncate the document list. We use leave-one-out cross-validation for tuning. That is, we split the 200 annotated questions into 199 validation examples and 1 test example, and find an optimal threshold for the 199 validation examples based on $\mathcal{N}_{A+U}$. Given the optimal threshold, we then compute the final $\mathcal{N}_{A+U}$ on the 1 test example. We repeat this 200 times to get the average $\mathcal{N}_{A+U}$ performance.

**PQA Model + IMCP**: our proposed framework that combines PQA and IMCP as the rejection model. Given a review score by a PQA model, we first convert the score into probabilities,[10] and then compute the $p$-values for both positive and negative labels (Eq. (1)). We then tune the significance level $\epsilon$ to truncate the document list, using leave-one-out cross-validation as before. As IMCP requires calibration data to compute the $p$-value, the process is a little more involving. We first split the 200 questions into 199 validation examples and 1 test examples as before, and within the 199 validation examples, we do another leave-out-out: we split them into 198 calibration examples and 1 validation example, and compute the $p$-value for the validation example based on the 198 calibration examples. We then tune $\epsilon$ and find the optimal threshold that gives the best $\mathcal{N}_{A+U}$ performance on the single validation performance, and repeat this 199 times to find the best overall $\epsilon$. Given this $\epsilon$, we then compute the $\mathcal{N}_{A+U}$ performance on the 1 test example. This whole process is then repeated 200 times to compute the average $\mathcal{N}_{A+U}$ test performance.

## 4.4   Results

We present the full results in   Table 5, reporting NDCG' performances over 4 relevance thresholds: 2.00, 2.25, 2.50, and 2.75.

---

[10] This step is only needed for MOQA, as BERTQA and FLTR produce probabilities in the first place. For MOQA, we convert the review score into a probability applying a sigmoid function to the log score.

**Table 5.** Model performance; boldface indicates optimal $\mathcal{N}_{A+U}$ performance for a PQA model.

| Relevance | Model | $\mathcal{N}_{A+U}$ | $\mathcal{N}_A$ | $\mathcal{N}_U$ |
|---|---|---|---|---|
| ≥2.00 | MOQA | 0.294 | 0.309 | 0.279 |
| | MOQA+THRS | **0.319** | 0.212 | 0.480 |
| | MOQA+IMCP | 0.318 | 0.212 | 0.477 |
| | FLTR | 0.372 | 0.495 | 0.279 |
| | FLTR+THRS | **0.516** | 0.400 | 0.666 |
| | FLTR+IMCP | 0.514 | 0.392 | 0.675 |
| | BERTQA | 0.360 | 0.464 | 0.279 |
| | BERTQA+THRS | 0.436 | 0.356 | 0.534 |
| | BERTQA+IMCP | **0.447** | 0.345 | 0.580 |
| ≥2.25 | MOQA | 0.264 | 0.249 | 0.279 |
| | MOQA+THRS | **0.296** | 0.179 | 0.489 |
| | MOQA+IMCP | 0.295 | 0.163 | 0.535 |
| | FLTR | 0.361 | 0.468 | 0.279 |
| | FLTR+THRS | 0.452 | 0.335 | 0.608 |
| | FLTR+IMCP | **0.482** | 0.329 | 0.705 |
| | BERTQA | 0.344 | 0.423 | 0.279 |
| | BERTQA+THRS | 0.373 | 0.293 | 0.477 |
| | BERTQA+IMCP | **0.405** | 0.310 | 0.530 |
| ≥2.50 | MOQA | 0.243 | 0.211 | 0.279 |
| | MOQA+THRS | **0.274** | 0.165 | 0.453 |
| | MOQA+IMCP | 0.265 | 0.155 | 0.452 |
| | FLTR | 0.359 | 0.462 | 0.279 |
| | FLTR+THRS | 0.439 | 0.326 | 0.592 |
| | FLTR+IMCP | **0.470** | 0.316 | 0.699 |
| | BERTQA | 0.340 | 0.414 | 0.279 |
| | BERTQA+THRS | **0.404** | 0.308 | 0.530 |
| | BERTQA+IMCP | 0.387 | 0.294 | 0.510 |
| ≥2.75 | MOQA | **0.235** | 0.199 | 0.279 |
| | MOQA+THRS | 0.229 | 0.129 | 0.407 |
| | MOQA+IMCP | 0.213 | 0.107 | 0.423 |
| | FLTR | 0.333 | 0.397 | 0.279 |
| | FLTR+THRS | 0.409 | 0.272 | 0.615 |
| | FLTR+IMCP | **0.416** | 0.299 | 0.577 |
| | BERTQA | 0.330 | 0.390 | 0.279 |
| | BERTQA+THRS | 0.349 | 0.279 | 0.435 |
| | BERTQA+IMCP | **0.388** | 0.296 | 0.509 |

**Table 6.** Reviews produced by FLTR, FLTR+THRS and FLTR+IMCP for an answerable (Q1) and unanswerable (Q2) question.

| | | |
|---|---|---|
| Q1: How long the battery lasts on X1 carbon touch? | | |
| | Ground Truth | [3, 3] |
| | FLTR | [0, 0, 0, 0, 3, 0, 0, 0, 0, 0] |
| | FLTR+THRS | [0, 0, 0, 0, 3, 0, 0, 0] |
| | FLTR+IMCP | [0, 0, 0, 0, 3] |
| Q2: What type of memory SD card should I purchase to go with this? | | |
| | Ground Truth | [ ] |
| | FLTR | [0, 0, 0, 0, 0, 0, 0, 0, 0, 0] |
| | FLTR+THRS | [0, 0] |
| | FLTR+IMCP | [ ] |

We'll first focus on the combined performances ($\mathcal{N}_{A+U}$). In general, all models (MOQA, FLTR and BERTQA) see an improvement compared to their vanilla model when we tune a threshold (+THRS or +IMCP) to truncate the returned review list, implying it's helpful to find a cut-off to discover a more concise set of reviews. Comparing between the simple thresholding method (+THRS) vs. conformal method (+IMCP), we also see very encouraging results: for both FLTR and BERTQA, +IMCP is consistently better than +THRS for most relevance thresholds, suggesting that +IMCP is a better rejection model. For MOQA, however, +THRS is marginally better than +IMCP. We hypothesize this may be due to MOQA producing an arbitrary (non-probabilistic) score for review, and as such is less suitable for conformal predictors. Comparing between the 3 PQA models, FLTR consistently produces the best performance: across most relevance thresholds, FLTR+IMCP maintains an NDCG' performance close to 0.5.

Looking at the $\mathcal{N}_U$ results, we notice all vanilla models produce the same performance (0.279). This is because there are no relevant reviews for these unanswerable questions, and so the top-10 returned reviews by any models are always irrelevant. When we introduce +THRS or +IMCP to truncate the reviews, we see a very substantial improvement ($\mathcal{N}_U$ more than doubled in most cases) for all models over different relevance thresholds. Generally, we also find that +IMCP outperforms +THRS, demonstrating that the conformal predictors are particularly effective for the unanswerable questions.

That said, when we look at the $\mathcal{N}_A$ performance, they are consistently worse when we introduce rejection (+THRS or +IMCP). This is unsurprising, as ultimately it is a trade-off between precision and recall: when we introduce a rejection model to truncate the review list, we may produce a more concise/shorter list (as we see for the unanswerable questions), but we could also inadvertently exclude some potentially relevant reviews. As such, the best system is one that can maintain a good balance between pruning unreliable reviews and avoiding discarding potentially relevant reviews.

Next, we present two real output of how these methods perform in Table 6. The first question (Q1) is an answerable question, and the ground truth contains two relevant reviews (numbers in the list are review relevance scores). FLTR returns 10 reviews, of which one is relevant. FLTR+THRS rejects the last two reviews, and FLTR+IMCP rejects three more reviews, producing a concise list of 5 reviews (no relevant reviews were discarded in this case). One may notice both FLTR+THRS and FLTR+IMCP reject predictions but do not modify the original ranking of the returned reviews by FLTR. +THRS tunes a threshold based on original relevance score, and in +IMCP the conversion of class probability to confidence score ($p$-value) is a monotonic transformation, and as such the original order is preserved in both methods. This also means that if the vanilla model misses a relevant review, the review will not be recovered by the rejection model, as we see here.

The second question (Q2) is an unanswerable question (ground truth is an empty list). FLTR always returns 10 reviews, and so there are 10 irrelevant reviews. FLTR+THRS discards most of the reviews, but there are still two irrelevant reviews. FLTR+IMCP returns an empty list, indicating existing reviews do not have useful information to answer Q2, detecting correctly that Q2 is an unanswerable question.

## 5  Conclusion

PQA is often formulated as a retrieval problem with the goal to find the most relevant reviews to answer a given product question. In this paper, we propose incorporating conformal predictors as a rejection model to a PQA model to reject unreliable reviews. We test 3 state-of-the-art PQA models, MOQA, FLTR and BERTQA, and found that incorporating conformal predictors as the rejection model helps filter unreliable reviews better than a baseline approach. More generally, our paper demonstrates a novel and effective application of conformal predictors to a retrieval task.

**Acknowledgement.** Shiwei Zhang is supported by the RMIT University and CSIRO Data61 Scholarships.

## References

1. McAuley, J., Yang, A.: Addressing complex and subjective product-related queries with customer reviews. In: WWW (2016)
2. Zhao, J., Guan, Z., Sun, H.: Riker: mining rich keyword representations for interpretable product question answering. In: SIGKDD (2019)
3. Zhang, S., Lau, J.H., Zhang, X., Chan, J., Paris, C.: Discovering Relevant Reviews for Answering Product-related Queries. In: ICDM (2019)
4. Gao, S., Ren, Z., et al.: Product-aware answer generation in e-commerce question-answering. In: WSDM (2019)
5. Chen, S., Li, C., et al.: Driven answer generation for product-related questions in e-commerce. In: WSDM (2019)

6. Rajpurkar, P., Jia, R., Liang, P.: Know what you don't know: unanswerable questions for SQuAD. In: ACL (2018)
7. Herbei, R., Wegkamp, M.H.: Classification with reject option. The Canadian Journal of Statistics/La Revue Canadienne de Statistique (2006)
8. Gammerman, A.: Conformal Predictors for Reliable Pattern Recognition. In: Computer Data Analysis and Modeling: Stochastics and Data Science (2019)
9. Vovk, V., Gammerman, A., Shafer, G.: Algorithmic Learning in a Random World. Springer, New York (2005)
10. Shafer, G., Vovk, V.: A tutorial on conformal prediction. J. Mach. Learn. Res. **9**, 371–421 (2008)
11. Toccaceli, P., Gammerman, A.: Combination of inductive mondrian conformal predictors. Mach. Learn. **108**(3), 489–510 (2018). https://doi.org/10.1007/s10994-018-5754-9
12. Carlsson, L., Bendtsen, C., Ahlberg, E.: Comparing performance of different inductive and transductive conformal predictors relevant to drug discovery. In: Conformal and Probabilistic Prediction and Applications (2017)
13. Cortes-Ciriano, I., Bender, A.: Reliable prediction errors for deep neural networks using test-time dropout. J. Chem. Inf. Model. **59**(7), 3330–3339 (2019)
14. Jacobs, R.A., Jordan, M.I., Nowlan, S.J., Hinton, G.E.: Adaptive mixtures of local experts. Neural Comput. **3**(1), 79–87 (1991)
15. Devlin, J., Chang, M.W., et al.: BERT: pre-training of deep bidirectional transformers for language understanding. In: NAACL (2019)
16. Gupta, M., Kulkarni, N., Chanda, R., et al.: AmazonQA: a review-based question answering task. In: IJCAI (2019)
17. Hu, M., Wei, F., Peng, Y., et al.: Read+ verify: machine reading comprehension with unanswerable questions. In: AAAI (2019)
18. Sun, F., Li, L., et al.: U-net: machine reading comprehension with unanswerable questions (2018)
19. Godin, F., Kumar, A., Mittal, A.: Learning when not to answer: a ternary reward structure for reinforcement learning based question answering. In: NAACL-HLT (2019)
20. Huang, K., Tang, Y., Huang, J., He, X., Zhou, B.: Relation module for non-answerable predictions on reading comprehension. In: CoNLL (2019)
21. Joshi, M., Choi, E., Weld, D.S., Zettlemoyer, L.: TriviaQA: a large scale distantly supervised challenge dataset for reading comprehension. In: ACL (2017)
22. Dunn, M., Sagun, L., Higgins, M., Guney, V.U., Cirik, V., Cho, K.: Searchqa: a new qa dataset augmented with context from a search engine (2017)
23. Su, L., Guo, J., Fan, Y., Lan, Y., Cheng, X.: Controlling risk of web question answering. In: SIGIR (2019)
24. Sun, J., Carlsson, L., Ahlberg, E., et al.: Applying mondrian cross-conformal prediction to estimate prediction confidence on large imbalanced bioactivity data sets. J. Chem. Inf. Model. **57**(7), 1591–1598 (2017)
25. Card, D., Zhang, M., Smith, N.A.: Deep weighted averaging classifiers. In: Proceedings of the Conference on Fairness, Accountability, and Transparency (2019)
26. Gal, Y., Ghahramani, Z.: Dropout as a bayesian approximation: Representing model uncertainty in deep learning. In: ICML (2016)
27. Liu, F., Moffat, A., Baldwin, T., Zhang, X.: Quit while ahead: Evaluating truncated rankings. In: SIGIR (2016)

28. Järvelin, K., Kekäläinen, J.: Cumulated gain-based evaluation of IR techniques. ACM Trans. Inf. Syst. (TOIS) **20**(4), 422–446 (2002)
29. Kubat, M., Holte, R.C., Matwin, S.: Machine learning for the detection of oil spills in satellite radar images. Mach. Learn. **30**, 195–215 (1998). https://doi.org/10.1023/A:1007452223027

# AMQAN: Adaptive Multi-Attention Question-Answer Networks for Answer Selection

Haitian Yang[1,2], Weiqing Huang[1,2(✉)], Xuan Zhao[3], Yan Wang[1(✉)],
Yuyan Chen[4], Bin Lv[1], Rui Mao[1], and Ning Li[1]

[1] Institute of Information Engineering, Chinese Academy of Sciences, Beijing, China
{yanghaitian,huangweiqing,wangyan,lvbin,maorui,lining01}@iie.ac.cn
[2] School of Cyber Security, University of Chinese Academy of Sciences,
Beijing, China
[3] York University, Ontario, Canada
xuanzhao1003@gmail.com
[4] Sun Yat-sen University, Guangzhou, Guangdong, China
chenyy387@mail2.sysu.edu.cn

**Abstract.** Community Question Answering (CQA) provides platforms for users with various background to obtain information and share knowledge. In the recent years, with the rapid development of such online platforms, an enormous amount of archive data has accumulated which makes it more and more difficult for users to identify desirable answers. Therefore, answer selection becomes a very important subtask in Community Question Answering. A posted question often consists of two parts: a question subject with summarization of users' intention, and a question body clarifying the subject with more details. Most of the existing answer selection techniques often roughly concatenate these two parts, so that they cause excessive noises besides useful information to questions, inevitably reducing the performance of answer selection approaches. In this paper, we propose AMQAN, an adaptive multi-attention question-answer network with embeddings at different levels, which makes comprehensive use of semantic information in questions and answers, and alleviates the noise issue at the same time. To evaluate our proposed approach, we implement experiments on two datasets, SemEval 2015 and SemEval 2017. Experiment results show that AMQAN outperforms all existing models on two standard CQA datasets.

**Keywords:** Answer selection · Adaptive multi-attention · Community Question Answering

## 1 Introduction

In recent years, Community Question Answering (CQA), such as Quora and Stack Overflow, has become more and more popular. Users can ask questions of their own concerns, search for desired information, and expect answers from

© Springer Nature Switzerland AG 2021
F. Hutter et al. (Eds.): ECML PKDD 2020, LNAI 12459, pp. 584–599, 2021.
https://doi.org/10.1007/978-3-030-67664-3_35

expert users. However, with the rapid increase of CQA users, a massive amount of questions and answers accumulate in the community leading to a proliferation of low-quality answers, which not only make CQA difficult to operate normally, but also take users a lot of time to find the most relevant answer from many candidates seriously affecting their experience. Therefore, answer selection in CQA which aims to select the most possibly relevant answer in community becomes very important.

Generally speaking, there are two application scenarios for an answer selection task. The first scenario is to use CQA information retrieval technology to identify whether a given query is semantically equivalent to an existing question in the repository. If these two are semantically equivalent after retrieval, the corresponding answers of the existing question are referred to users as the relevant candidates. The second scenario is treating the existing questions and answers in the archive as "question-answer pairs" to determine whether these question-answer pairs are best match, namely, converting the answer selection task to a classification task. In this research, we focus on improving the performance of answer selection in the second application scenario.

A typical CQA example is shown in Table 1. In this example, Answer 1 is a good answer because it gives "check it to the traffic dept", which is more suitable for the question, while Answer 2, although related to the question, does not provide useful information, considered as a bad answer.

**Table 1.** An example question and its related answers in CQA

| Question subject | Checking the history of the car |
|---|---|
| Question body | How can one check the history of the car like maintenance, accident or service history. In every advertisement of the car, people used to write "Accident Free", but in most cases, car have at least one or two accident, which is not easily detectable through Car Inspection Company. Share your opinion in this regard |
| Answer 1 | Depends on the owner of the car.. if she/he reported the accidents i believe u can check it to the traffic dept.. but some owners are not doing that especially if its only a small accident.. try ur luck and go to the traffic dept. |
| Answer 2 | How about those who claim a low mileage by tampering with the car fuse box? In my sense if you're not able to detect traces of an accident then it is probably not worth mentioning... For best results buy a new car :) |

Compared with questions in other QAs, the given example in Table 1 shows unique characteristics of questions in CQA that is they usually consists of two parts, question subjects and question bodies. A question subject is a brief summary of the question with key words in it while a question body is a detailed description of the question, including critical information and some extended information. In addition, for both questions and answers, there are a vast amount of redundant information, which affect the performance of answer selection solutions.

In this paper, we propose a novel model, which comprehensively uses question subjects, question bodies, and corresponding answers for the answer selection task. Specifically, the information of answers with their noise filtered out are used as external resources. In order to obtain hierarchical text features, we concatenate word embeddings, character embeddings, and syntactic features as the input of the representation layer. Moreover, we use three heterogeneous attention mechanisms, including self-attention, cross attention, and adaptive co-attention to integrate answer information and effectively obtain text relevance. In addition, we also develop a gated fusion module adaptively integrating answer-based features as well as adopt a filtration gate module as a filter to reduce the noise introduced by answers. Meanwhile, the interaction layer enhances the local semantic information between questions and their corresponding answers. Finally, predictions are made based on the similarity features extracted from question-answer pairs.

The main contributions of our work can be summarized as follows:

(1) We consider the noise issues generated from adding answer information, and study how to integrate answer information into neural network to perform the answer selection task.
(2) We propose a novel method that treats question subjects and question bodies separately, integrating answer information into neural attention model so as to reduce the noise from answers and improves performance for answer selection task.
(3) Our proposed model outperforms all state-of-the-art answer selection models on various datasets.

The remaining of this paper is organized as follows: Sect. 2 gives an overview of the existing techniques related to our work; In Section 3, we introduce our proposed answer selection model, AMQAN, describe the details of the structure; in Sect. 4, we discuss the implementation of AMQAN, explain various parameters and the training process, and compare our model with other baselines; Sect. 5 summarizes the conclusion and suggests the further research potentials in the future.

## 2    Related Work

Answer selection in community is a challenging and complicated problem. Some work on this task [1–4] have shown the effectiveness mainly in high-quality questions. In the early years, answer selection task highly relied on feature engineering, linguistics tools and other external resources. Nakov et al. [5] studied a wide range of feature types, including similarity features, content features, and meta-features, which are automatically extracted from the SemEval CQA model.

Filice et al. [6] also designed various heuristic features and thread-based features, and yielded better solutions. Although these techniques have achieved good performance, the highly dependence on feature engineering results in indispensability of domain knowledge and an enormous amount of manpower.

Moreover, to successfully accomplish an answer selection task, semantic and syntactic features are also necessary. The most relevant candidate answer can be obtained by loose syntactic alteration through studying syntactic matching between questions and answers. Wang et al. [7] designed a generative model based on the soft alignment of quasi-synchronous grammar by matching dependency trees of question answer pairs. Heilman et al. [8] used a tree kernel as a heuristic algorithm to search for the smallest edit sequence between parse trees and features extracted from these sequences are fed into logistic regression classifier to determine whether an answer is relevant to the given question.

Some researchers focus on studying different translation model, such as word-based translation model and phrase-based translation model. Murdock et al. [9] proposed a simple translation model for sentence retrieval in factoid question answering. Zhou et al. [10] proposed a phrase-based translation model aiming to find semantically equivalent questions in Q&A archives with new queries.

Although research described above have solved a majority of answer selection tasks, there are still a great quantity of issues in suspense mainly because these questions are short in length and various in vocabulary. And in other cases, answer selection approaches solely based on questions themselves, but the questions cannot provide enough useful information, making it difficult to identify the similarity between different questions. However, since answers usually explain questions in detail, they can be used as a very useful supplementary resource. But simply combining answers with their corresponding questions may cause redundancies, which might also reduce the performance of answer selection solutions.

## 3    Proposed Model

### 3.1    Task Description

In this research, the answer selection task about community question answering can be described as a tuple of four elements $(S, B, A, y)$. $S = [s^1, s^2, \cdots, s^m]$ represents a question subject whose length is $m$. $B = [b^1, b^2, \cdots, b^g]$ represents the corresponding question body whose length is $g$. $A = [a^1, a^2, \cdots, a^n]$ represents the corresponding answer whose length is $n$. And $y \in Y$ represents the relevance degree, namely, $Y = \{Good, PotentiallyUseful, Bad\}$ to determine whether a candidate can answer a question properly or not. More detailed, Good represents that the answer can provide a proper solution for the question, while PotentiallyUseful indicates that the answer might provide a useful solution to users and bad means the answer is not relevant to the question. Generally, our AMQAN model on the answer selection task in CQA can be summarized as assigning a label to each answer based on the conditional probability $Pr(y \mid S, B, A)$ with the given set $\{S, B, A\}$.

## 3.2   Overview of Proposed Model

The structure of AMQAN can be represented as six layers distributed layer by layer from bottom to top, including Embedding layer, Encoder layer, Self-Attention layer, Cross Attention layer, Adaptive Co-Attention layer, and Interaction and Prediction Layer. The pipeline of the proposed framework is demonstrated in Fig. 1.

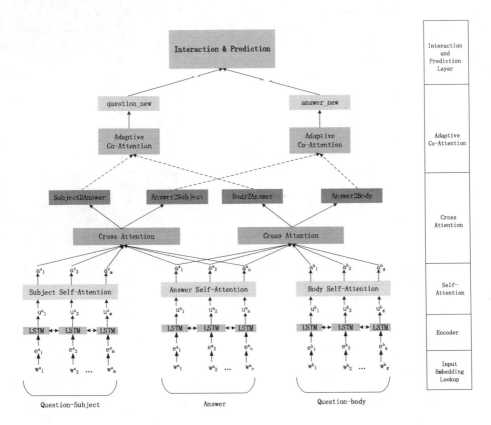

**Fig. 1.** Overview of our proposed AMQAN model

## 3.3   Word-Level Embedding

Word embeddings are comprised of three different modules: GloVe word representation trained on the Yahoo! Answers corpus proposed by Pennington et al. [11], character embeddings proposed by Kim et al. [12], and syntactic features based on one-hot encoding proposed by Chen et al. [12]. The relationship of words can be captured more accurately and precisely with the concatenation of three embeddings which are trained on the domain-specific corpus for the reason that texts in CQA are different mainly in grammar and spelling from others. It has also proved character embeddings are effective for OOV (out-of-vocabulary), especially in CQA tasks and syntactic features based on one-hot encoding can provide more grammar information to make a better query representation.

We define $\{w_t^{subject}\}_{t=1}^m$, $\{w_t^{body}\}_{t=1}^g$ and $\{w_t^{answer}\}_{t=1}^n$ as word sets of all candidate question subjects, all candidate question bodies and all candidate answers, respectively. Here, $m$, $g$ and $n$ are the length of each question subject, each question body and each answer, respectively. Through this layer, each candidate question-subject, question-body and answer are converted into vectors $\{e_t^{subject}\}_{t=1}^m$, $\{e_t^{body}\}_{t=1}^g$ and $\{e_t^{answer}\}_{t=1}^n$, respectively.

### 3.4   Encoder

We use bidirectional LSTM (Bi-LSTM) encoders to convert question-subjects, question-bodies, and answers into coded form based on temporal dependency. A H-dimensional contextual representation for each word is obtained by concatenating outputs of two layers whose directions are opposite. For a question-subject, the input of Bi-LSTM is the embedding of a question-subject, denoted as $\{e_t^{subject}\}_{t=1}^m$, and the outputs are returns from the Bi-LSTM, denoted as $U^{subject} = \{u_t^{subject}\}_{t=1}^m \in R^{m \times H}$. Therefore, the encoded-question-subject is calculated as follows:

$$u_t^{subject} = BiLSTM_{subject}(u_{t-1}^{subject}, e_t^{subject}) \tag{1}$$

Similarly, we get the encoded-question-body and the encoded-answer, respectively.

$$u_t^{body} = BiLSTM_{body}(u_{t-1}^{body}, e_t^{body}) \tag{2}$$

$$u_t^{answer} = BiLSTM_{answer}(u_{t-1}^{answer}, e_t^{answer}) \tag{3}$$

### 3.5   Self-attention

In this layer, we use a self-attention mechanism proposed in [14] to convert question-subjects, question-bodies and answers of different lengths into fix-length vectors. Since the significance of a certain word varies in a question (or an answer) and between questions (or answers) the self-attention technique will assign different weights to the certain word according to where it occurs. Let $A^{subject}$, $A^{body}$ and $A^{answer}$ be self-attention question-subject representation set, self-attention question-body representation set and self-attention answer representation set, respectively. Given a specific feature of a question-subject representation set $U^{subject} = \{u_t^{subject}\}_{t=1}^m$ as the input, a self-attention question-subject representation set $A^{subject} = \{a_t^{subject}\}_{t=1}^m$ is generated by this layer. The details are illustrated in the following Equations:

$$c_t^{subject} = w_{subject}^T \cdot (tanh(W_{subject} \cdot u_t^{subject})) \tag{4}$$

$$\alpha_t^{subject} = \frac{\exp(c_t^{subject})}{\sum_{j=1}^m \exp(c_j^{subject})} \tag{5}$$

Similarly, we get self-attention question-body representation set and self-attention answer representation set as follows.

$$\alpha_t^{body} = \frac{\exp(c_t^{body})}{\sum_{j=1}^{g} \exp(c_j^{body})} \tag{6}$$

$$\alpha_t^{answer} = \frac{\exp(c_t^{answer})}{\sum_{j=1}^{n} \exp(c_j^{answer})} \tag{7}$$

## 3.6   Cross Attention

The Cross Attention layer, which is proposed in [15, 16], aims at fusing words information in question-subjects with words information in answers. The cross attention mechanism used in this research has been proven to be a vital composition of the best model for reading comprehension task. Specifically, this Cross Attention layer is to compute the relevance between each word in question-subjects and each word in answers by bidirectional attention mechanism, including *Subject2Answer*, the attention mechanism from question-subjects to answers, and *Answer2Subject*,the attention mechanism from answers to question-subjects. By computing the similarity between question-subjects and answers, we get a matrix denoted as $SAS \in R^{m \times n}$. Then two similarity matrices, $\bar{S}_{Subject2Answer} \in R^{m \times n}$ and $\bar{S}_{Answer2Subject} \in R^{m \times n}$, are generated after normalization over each row and column by softmax function.

Let $s_{x,y} \in R$ be an element in similarity matrix $SAS \in R^{m \times n}$, where rows represent question-subjects and columns represent answers. Given $A^{subject}$ and $A^{answer}$ as inputs, we get two cross-attention matrices $A_{Subject2Answer} \in R^{m \times H}$ and $A_{Answer2Subject} \in R^{m \times H}$ as final outputs. The computation process can be described as the following equations:

$$s_{x,y} = w'^T_{subject} \cdot [a_x^{subject}; a_y^{answer}; a_x^{subject} \odot a_y^{answer}] \tag{8}$$

$$\bar{S}_{Subject2Answer} = softmax_{row}(SAS) \tag{9}$$

$$\bar{S}_{Answer2Subject} = softmax_{col}(SAS) \tag{10}$$

$$A_{Subject2Answer} = \bar{S}_{Subject2Answer} \cdot A^{answer} \tag{11}$$

$$A_{Answer2Subject} = \bar{S}_{Subject2Answer} \cdot \bar{S}_{Answer2Subject}^T \cdot A^{subject} \tag{12}$$

where $w'^T_{subject}$ is a parameter.

We can similarly obtain $A_{Body2Answer}$ and $A_{Answer2Body}$ as aboves.

## 3.7   Adaptive Co-attention

Inspired by previous works [18], we use adaptive co-attention, including question-guided co-attention and answer-guided co-attention to capture interaction between questions and their corresponding answers. Then we propose to use a gated fusion module to adaptively fuse features. In addition, a filtration gate is adopted to filter out useless information of question bodies so as to reduce the noise in question-answer fused features. Through cross attention layer, $A_{Subject2Answer_i}$ represents the $i$-th cross-attention word in a question subject while $A_{Body2Answer}$ is a cross-attention question body matrix. A single-layer neural network with a softmax function are used to generate attention distribution of question subjects over their question bodies.

$$z_i = tanh(W_{AB2A} \cdot A_{Body2Answer} \oplus (W_{AS2A_i} \cdot A_{Subject2Answer_i} + b_{AS2A_i})) \quad (13)$$

$$\alpha_i = softmax(W_{\alpha_i} \cdot z_i + b_{\alpha_i}) \quad (14)$$

Here, $W_{AB2A}, W_{AS2A_i}, W_{\alpha_i}, b_{AS2A_i}$ and $b_{\alpha_i}$ are parameters.

We use $\oplus$ to represent the concatenation of a question-body feature matrix and a question-subject feature vector. The concatenation between a matrix and a vector is achieved by connecting each column of the matrix to the vector.

$\alpha_i$ is the attention distribution, namely, the attention weight of each word in a question body. The $i$-th word in a question body related to the question subject can be computed by the following operation.

$$A'_{Body2Answer_i} = A_{Body2Answer} \cdot \alpha_i \quad (15)$$

Next, we use the new word vector of a question body $A'_{Body2Answer_i}$ to obtain the attention matrix of the question body.

$$\gamma_i = tanh(W_{AS2A} \cdot A_{Subject2Answer} \oplus (W_{AB2A_i} \cdot A'_{Body2Answer_i} + b_{AB2A_i})) \quad (16)$$

$$\beta_i = softmax(W_{\beta_i} \cdot \gamma_i + b_{\beta_i}) \quad (17)$$

Then, we obtain a new representation of the question subject related to the question body.

$$A'_{Subject2Answer_i} = A_{Subject2Answer} \cdot \beta_i \quad (18)$$

Afterwards, we propose to use a gated fusion module to fuse the new question subject with the new question body. Details are demonstrated as follows:

$$A''_{Body2Answer_i} = tanh(W_{B2A_i} \cdot A'_{Body2Answer_i} + b_{B2A_i}) \quad (19)$$

$$A''_{Subject2Answer_i} = tanh(W_{S2A_i} \cdot A'_{Subject2Answer_i} + b_{S2A_i}) \quad (20)$$

$$g_i = \sigma(W_{g_i} \cdot (A''_{Subject2Answer_i} \oplus A''_{Body2Answer_i}) \quad (21)$$

$$v_i = g_i \cdot A''_{Body2Answer_i} + (1 - g_i) \cdot A''_{Subject2Answer_i} \quad (22)$$

where $W_{B2A_i}$, $W_{S2A_i}$, $b_{B2A_i}$ and $b_{S2A_i}$ are parameters, $\sigma$ is the logistic sigmoid activation, $g_i$ is the gate fusion model applied to the new question-subject vector $A''_{Subject2Answer_i}$ and the new question-body vector $A''_{Body2Answer_i}$, and $v_i$ is the fusion features incorporating question-subject information and question-answer information.

Since the fusion features $v_i$ originates from two kinds of information which may cause extra noise, we use a filtration gate to combine fusion features with original features. The filtration gate is a scalar in the range of $[0, 1]$. When the fusion feature is helpful to improve the model performance, the filtration gate is 1; otherwise, the value of the filtration gate is set to 0. The filtration gate $s_i$ and the question-information-enhanced features of new questions $question\_new_i$ are defined as follows:

$$s_i = \sigma(W_{AS2A_i} \cdot A_{Subject2Answer_i} \oplus W_{AB2A_i} \cdot A_{Body2Answer_i}$$
$$\oplus (W_{v_i,s_i,b_i} \cdot v_i + b_{v_i,s_i,b_i})) \tag{23}$$
$$u_i = s_i \cdot tanh(W_{v_i} \cdot v_i + b_{v_i}) \tag{24}$$
$$question\_new_i = w_{question\_new_i} \cdot (A_{Subject2Answer_i} \oplus A_{Body2Answer_i} \oplus u_i) \tag{25}$$

where $W_{v_i,s_i,b_i}$, $W_{v_i}$, $b_{v_i,s_i,b_i}$ and $b_{v_i}$ are parameters, and $u_i$ is the filtered fusion features.

We can similarly derive the answer-information-enhanced features of new answers $answer\_new_i$ as aboves.

### 3.8   Interaction and Prediction Layer

Inspired by previous works [19, 20], we combine the new representation of question subjects with the new representation of question bodies to further enhance local semantic information. To be specific:

$$Q_i^m = [u_i^{subject}; u_i^{body}; question\_new_i; u_i^{subject} + u_i^{body} - question\_new_i;$$
$$u_i^{subject} \odot u_i^{body} \odot question\_new_i] \tag{26}$$
$$A_j^n = [u_j^{answer}; answer\_new_j; u_j^{answer} - answer\_new_j;$$
$$u_j^{answer} \odot answer\_new_j] \tag{27}$$

where $[\cdot; \cdot; \cdot; \cdot;]$ is the concatenation operation of vectors, and $\odot$ indicates element-wise product.

Next, we train our model with bidirectional GRU (Bi-GRU) to acquire context information between questions and their corresponding answers. The detailed acquisition methods are shown as follows.

$$Q_i^v = BiGRU(Q_i^m, Q_{i-1}^v, Q_{i+1}^v) \tag{28}$$
$$A_j^v = BiGRU(A_j^n, A_{j-1}^v, A_{j+1}^v) \tag{29}$$

After that, we use max pooling and mean pooling of $Q^v = (Q_1^v, Q_2^v, \cdots, Q_m^v)$ and $A^v = (A_1^v, A_2^v, \cdots, A_n^v)$ to acquire fixed-length vectors. Specifically, we compute max pooling vectors $r_Q^{max}$ and $r_A^{max}$ as well as mean pooling vectors $r_Q^{mean}$ and $r_A^{mean}$. Then, we concatenate these vectors to get the global representation $r$. The details are indicated as follows.

$$r_Q^{mean} = \Sigma_{i=1}^m \frac{Q_i^v}{m} \tag{30}$$
$$r_Q^{max} = max_{i=1}^m Q_i^v \tag{31}$$

$$r_A^{mean} = \Sigma_{i=1}^n \frac{A_j^v}{n} \qquad (32)$$

$$r_A^{max} = max_{i=1}^n A_j^v \qquad (33)$$

$$r = [r_Q^{mean}; r_Q^{max}; r_A^{mean}; r_A^{max}] \qquad (34)$$

Finally, we pass the global representation $r$ to the prediction layer which is consisted of a multi-layer perceptron (MLP) classifier to determine whether the semantic meaning of the give question-answer pair is equivalent or not.

$$\nu = tanh(W_r \cdot r + b_r) \qquad (35)$$

$$\hat{y} = softmax(W_\nu \cdot \nu + b_\nu) \qquad (36)$$

Here, $W_r$, $b_r$, $W_\nu$, and $b_\nu$ are trainable parameters. The entire model is trained end-to-end.

## 4    Experimental Setup

### 4.1    DataSet

The two corpora using to train and evaluate our model are CQA datasets SemEval2015 and SemEval2017, both containing two parts, questions and their corresponding answers. Each question comprises a brief title and informative descriptions. Detailed statistics of two corpora are shown in Table 2 and Table 3.

**Table 2.** Statistical information of SemEval2015 corpus

|  | Train | Dev | Test |
|---|---|---|---|
| Number of questions | 2376 | 266 | 300 |
| Number of answers | 15013 | 1447 | 1793 |
| Average length of subject | 6.36 | 6.08 | 6.24 |
| Average length of body | 39.26 | 39.47 | 39.53 |
| Average length of answer | 35.82 | 33.90 | 37.33 |

### 4.2    Training and Hyper Parameters

For the text preprocessing procedure, we exert NLTK toolkit over each question and its corresponding answers, including capitalization conversion, stemming, removal of stop words, etc. After preprocessing, we train two datasets with GloVe proposed by Pennington [11] to obtain 300-dimensional initialized word vectors and set the number of out-of-vocabulary(OOV) words to zero. Adam Optimizer is chosen for optimization with the first momentum coefficient $\beta_1$ 0.9 and the second momentum coefficient $\beta_2$ 0.999. In addition, initial learning rate is set to $[1 \times 10^{-4}, 4 \times 10^{-6}, 1 \times 10^{-5}]$, L2 regularization value is set to $[1 \times 10^{-5}, 4 \times 10^{-7}, 1 \times 10^{-6}]$, and batch-size is set to [64, 128, 256]. We select parameters which perform best on validation sets and evaluate model performance on test sets.

**Table 3.** Statistical information of SemEval2017 corpus

|  | Train | Dev | Test |
|---|---|---|---|
| Number of questions | 5124 | 327 | 293 |
| Number of answers | 38638 | 3270 | 2930 |
| Average length of subject | 6.38 | 6.16 | 5.76 |
| Average length of body | 43.01 | 47.98 | 54.06 |
| Average length of answer | 37.67 | 37.30 | 39.50 |

## 4.3  Results and Analysis

**Results and Analysis on SemEval 2015 Dataset.** We adopt two evaluation metrics, F1 and Acc (accuracy) to compare performance of AMQAN with other seven answer selection models on SemEval 2015 dataset as shown in Table 4 and Table 5.

As shown in Table 5, not all deep learning models can outperform conventional machine learning models. For example, machine learning model JAIST increase by 1.73% in F1 and 0.7% in Acc than deep learning model BGMN. Then, Graph-cut and FCCRF outperform BGMN, JAIST, and HITSZ-ICRC on F1 and Acc, which prove that feature-engineering-based model with effective features sometimes have better performance in the answer selection task.

Most importantly, the experiment results show that our model AMQAN has the best performance on this dataset, outperforming the current state-of-the-art model (7) by 0.48% in F1 and 1.28% in Acc ($p < 0.05$ on student t-test), which can be mainly attributed to multi-attention mechanisms. Specifically, self-attention focuses on modeling temporal interaction in long sentences, finding the most relevant words in answers to their corresponding questions. With cross attention between question subjects and answers, question bodies and answers, AMQAN can successfully identify words that represent the correlation between questions and answers. In adaptive co-attention module, question-driven attention and answer-driven attention are interacted to precisely capture the semantic meaning between questions and answers.

**Results and Analysis on SemEval 2017 Dataset.** We also adopt three evaluation metrics, F1, Acc (accuracy) and MAP (Mean Average Precision) to compare performance of AMQAN with other eight answer selection models, including two embeddings on SemEval 2017 dataset as shown in Table 6 and Table 7.

From Table 7, we can also find deep learning models not always achieve unsurpassable results. For example, model (1), (2) outperform two deep learning models (4) and (5) in MAP. Next, model (5) and (6) which treat question subjects and question bodies separately outperform model (4) which takes question subjects and question bodies as an entirety because mechanical connection of question-subjects and question-bodies introduces too much noise, resulting in

**Table 4.** Descriptions of answer selection models on SemEval 2015 dataset

| Model | Reference and description |
|---|---|
| (1) JAIST [21] | Use SVM to incorporate various kinds of features, including topic model based features and word vector features |
| (2) HITSZ-ICRC [22] | Propose ensemble learning and hierarchical classification to classify answers |
| (3) Graph-cut [23] | Model the relationship between answers in the same question thread, based on the idea that similar answers should have similar labels |
| (4) FCCRF [4] | Apply local learned classifiers and fully connected CRF to make global inference so as to predict the label of each individual node more precisely |
| (5) BGMN [24] | Use the memory mechanism to iteratively aggregate more relevant information in order to identify the relationship between questions and answers |
| (6) CNN-LSTM-CRF [25] | Propose multilingual hierarchical attention networks for learning document structures |
| (7) QCN [1] | A Question Condensing Network focusing on the similarity and disparities between question-subjects and question-bodies for answer selection in Community Question Answering, which is the state-of-the-art model for Answer Selection |
| (8) AMQAN (ours) | An adaptive multi-attention question-answer network, outperforming all current answer selection models |

**Table 5.** Comparisons on the SemEval 2015 dataset

| Model | F1 | Acc |
|---|---|---|
| (1) JAIST | 0.7896 | 0.7910 |
| (2) HITSZ-ICRC | 0.7652 | 0.7611 |
| (3) Graph-cut | 0.8055 | 0.7980 |
| (4) FCCRF | 0.8150 | 0.8050 |
| (5) BGMN | 0.7723 | 0.7840 |
| (6) CNN-LSTM-CRF | 0.8222 | 0.8224 |
| (7) QCN | 0.8391 | 0.8224 |
| **(8) AMQAN (ours)** | **0.8439** | **0.8352** |

the performance degradation. Then the comparison between model (7) and (8) indicates that using task-specific embeddings and character embeddings both contribute to model performance, especially help attenuate OOV [29] problems

**Table 6.** Descriptions of answer selection models on SemEval 2015 dataset

| Model | Reference and description |
|---|---|
| (1) KeLP [26] | Use syntactic tree kernels with relational links between questions and answers, and apply standard text similarity measures linearly combined with the tree kernel |
| (2) Beihang-MSRA [27] | Use gradient boosted regression trees to combine traditional linguistic features and neural network-based matching features |
| (3) ECUN [28] | Use traditional features and a convolutional neural network to represent question-answer pairs |
| (4) LSTM | A LSTM-based model applied on question subjects and question bodies respectively, and concatenate these two results to obtain final representation of questions |
| (5) LSTM-subject-body | Use the memory mechanism to iteratively aggregate more relevant information in order to identify the relationship between questions and answers |
| (6) QCN [1] | A Question Condensing Network focusing on the similarity and disparities between question-subjects and question-bodies for answer selection in CQA, which is the state-of-the-art model for Answer Selection |
| (7) w/o task-specific word embeddings | Word embeddings are initialized with the 300-dimensional GloVe trained on Wikipedia 2014 and Gigaword 5 |
| (8) w/o character embeddings | Word embeddings are initialized with the 600-dimensional GloVe trained on the domain-specific unannotated corpus |
| (9) AMQAN (ours) | An adaptive multi-attention question-answer network, outperforming all current answer selection models |

**Table 7.** Comparisons on the SemEval 2017 dataset

| Model | MAP | F1 | Acc |
|---|---|---|---|
| (1) KeLP | 0.8843 | 0.6987 | 0.7389 |
| (2) Beihang-MSRA | 0.8824 | 0.6840 | 0.5198 |
| (3) ECNU | 0.8672 | 0.7767 | 0.7843 |
| (4) LSTM | 0.8632 | 0.7441 | 0.7569 |
| (5) LSTM-subject-body | 0.8711 | 0.7450 | 0.7728 |
| (6) QCN | 0.8851 | 0.7811 | 0.8071 |
| (7) w/o task-specific word embeddings | 0.8896 | 0.7832 | 0.8085 |
| (8) w/o character embeddings | 0.8756 | 0.7768 | 0.7978 |
| **(9) AMQAN (ours)** | **0.8925** | **0.7868** | **0.8132** |

since CQA text is non-standard with abbreviations, typos, emoticons, and grammatical mistakes, etc.

The most important is our proposed model AMQAN increase by 0.74% in MAP, 0.57% in F1, and 0.61% in Acc ($p < 0.05$ on student t-test) compared with the best model (6), for the reason that AMQAN studies the relationship of question-bodies, question-subjects, and answers, fully using self-attention, cross-attention and adaptive co-attention, so that it greatly enhances the performance of the answer selection task.

## 5    Conclusion

In this study, we propose AMQAN, an adaptive multi-attention question answer networks for answer selection, which take the relationship between questions and answers as important information for answer selection. In order to effectively answer both factoid and non-factoid questions with various length, our model AMQAN applies deep attention mechanism at word, sentence, and document level, utilizing characteristics of linguistic knowledge to explore the complex relationship between different components. Further, we use multi-attention mechanism like self-attention, cross attention and adaptive co-attention to comprehensively capture more interactive information between questions and answers. In general, AMQAN outperforms all baseline models, achieving significantly better performance than all current state-of-the-art answer selection methods. In the future, our research group will mainly focus on improving the computing speed of AMQAN to further level up the performance of our solution.

## References

1. Wu, W., Sun, X., Wang, H., et al.: Question condensing networks for answer selection in community question answering. In: Meeting of the Association for Computational Linguistics, pp. 1746–1755 (2018)
2. Yao, X., Van Durme, B., Callisonburch, C., et al.: Answer extraction as sequence tagging with tree edit distance. In: North American Chapter of the Association for Computational Linguistics, pp. 858–867 (2013)
3. Tran, Q.H., Tran, D.V., Vu, T., Le Nguyen, M., Pham, S.B.: JAIST: Combining multiple features for Answer Selection in Community Question Answering. In: North American Chapter of the Association for Computational Linguistics, pp. 215–219 (2015)
4. Joty, S., Marquez, L., Nakov, P.: Joint learning with global inference for comment classification in community question answering. In: North American Chapter of the Association for Computational Linguistics, pp. 703–713 (2016)
5. Nakov, P., Marquez, L., Moschitti, A., et al. SemEval-2016 task 3: community question answering. In: North American Chapter of the Association for Computational Linguistics, pp. 525–545 (2016)
6. Filice, S., Croce, D., Moschitti, A., et al.: KeLP at SemEval-2016 task 3: Learning semantic relations between questions and answers. In: North American Chapter of the Association for Computational Linguistics, pp. 1116–1123 (2016)

7. Wang, M., Smith, N.A., Mitamura, T.: What is the jeopardy model? a quasisynchronous grammar for QA. In: EMNLP-CoNLL (2007)

8. Heilman, M., Smith, N.A.: Tree edit models for recognizing textual etailments, paraphrases, and answers to questions. In: Proceedings of NAACL (2010)

9. Murdock, V., Croft, W.B.: Simple translation models for sentence retrieval in factoid question answering. In: Proceedings of the SIGIR-2004 Workshop on Information Retrieval For Question Answering (IR4QA), pp. 31–35 (2004)

10. Zhou, G., Cai, L., Zhao, J., et al.: Phrase-based translation model for question retrieval in community question answer archives. In: Meeting of the Association for Computational Linguistics, pp. 653–662 (2011)

11. Pennington, J., Socher, R., Manning, C. D., et al.: Glove: global vectors for word representation. In: Empirical Methods in Natural Language Processing, pp. 1532–1543 (2014)

12. Kim, Y., Jernite, Y., Sontag, D., et al.: Character-aware neural language models. In: National Conference on Artificial Intelligence pp. 2741–2749 (2016)

13. Chen, Q., Zhu, X., Ling, Z., et al.: Neural natural language inference models enhanced with external knowledge. In: Meeting of the Association for Computational Linguistics, pp. 2406–2417 (2018)

14. Lin, Z.: A structured self-attentive sentence embedding. In: International Conference on Learning Representations (ICLR) (2017)

15. Weissenborn, D., Wiese, G., Seiffe, L., et al.: Making neural QA as simple as possible but not simpler. In: Conference on Computational Natural Language Learning, pp. 271–280 (2017)

16. Yu, A.W.: QANet: combining local convolution with global self-attention for reading comprehension. In: International Conference on Learning Representations (ICLR) (2018)

17. Seo, M., Kembhavi, A., Farhadi, A., Hajishirzi, H.: Bidirectional attention flow for machine comprehension. In: International Conference on Learning Representations (ICLR) (2017)

18. Lu, J., Xiong, C., Parikh, D., et al.: Knowing when to look: adaptive attention via a visual sentinel for image captioning. In: Computer Vision and Pattern Recognition, pp. 3242–3250 (2017)

19. Chen, Q., Zhu, X., Ling, Z., et al.: Enhanced LSTM for natural language inference. In: Meeting of the Association for Computational Linguistics, pp. 1657–1668 (2017)

20. Mou, L., Men, R., Li, G., et al.: Natural language inference by tree-based convolution and heuristic matching. In: Meeting of the Association for Computational Linguistics, pp. 130–136 (2016)

21. Tran, Q.H., Tran, V.D., Vu,T.T., et al.: JAIST: Combining multiple features for answer selection in community question answering. In: North American Chapter of the Association for Computational Linguistics, pp. 215–219 (2015)

22. Hou, Y., Tan, C., Wang, X., et al. HITSZ-ICRC: exploiting classification approach for answer selection in community question answering. In: North American Chapter of the Association for Computational Linguistics, pp. 196–202 (2015)

23. Joty, S.R., Barroncedeno, A., Martino, G.D., et al.: Global thread-level inference for comment classification in community question answering. In: Empirical Methods in Natural Language Processing, pp. 573–578 (2015)

24. Wei, W., Houfeng, W., Sujian, L.: Bidirectional gated memory networks for answer selection. In: Chinese Computational Linguistics and Natural Language Processing Based on Naturally Annotated Big Data, LNAI 10565, Springer, Cham pp. 251–262 (2017)

25. Xiang, Y., Zhou, X., Chen, Q., et al.: Incorporating label dependency for answer quality tagging in community question answering via CNN-LSTM-CRF. In: International conference on computational linguistics, pp. 1231–1241 (2016)
26. Filice, S., Martino, G.D., Moschitti, A., et al.: KeLP at SemEval-2017 task 3: learning pairwise patterns in community question answering. In: Meeting of the Association for Computational Linguistics, pp. 326–333 (2017)
27. Feng, W., Wu, Y., Wu, W., et al.: Beihang-MSRA at semEval-2017 task 3: a ranking system with neural matching features for community question answering. In: Meeting of the Association for Computational Linguistics, pp. 280–286 (2017)
28. Wu, G., Sheng, Y., Lan, M., et al.: ECNU at semEval-2017 task 3: using traditional and deep learning methods to address community question answering task. In: Meeting of the Association for Computational Linguistics, pp. 365–369 (2017)
29. Zhang, X., Li, S., Sha, L., et al.: Attentive interactive neural networks for answer selection in community question answering. In: National Conference on Artificial Intelligence, pp. 3525–3531 (2017)

# Inductive Document Representation Learning for Short Text Clustering

Junyang Chen[1], Zhiguo Gong[1(✉)], Wei Wang[1(✉)], Xiao Dong[2], Wei Wang[3], Weiwen Liu[4], Cong Wang[3], and Xian Chen[5]

[1] University of Macau, Macao, China
{yb77403,fstzgg,wwang}@umac.mo
[2] The University of Queensland, Brisbane, Australia
dx.icandoit@gmail.com
[3] Dalian University of Technology, Dalian, China
{WWLoveTransfer,congwang}@mail.dlut.edu.cn
[4] The Chinese University of Hong Kong, Hong Kong, People's Republic of China
wwliu@cse.cuhk.edu.hk
[5] The University of Hong Kong, Hong Kong, Hong Kong
chenxian@hku.hk

**Abstract.** Short text clustering (STC) is an important task that can discover topics or groups in the fast-growing social networks, e.g., Tweets and Google News. Different from the long texts, STC is more challenging since the word co-occurrence patterns presented in short texts usually make the traditional methods (e.g., TF-IDF) suffer from a sparsity problem of inevitably generating sparse representations. Moreover, these learned representations may lead to the inferior performance of clustering which essentially relies on calculating the distances between the presentations. For alleviating this problem, recent studies are mostly committed to developing representation learning approaches to learn compact low-dimensional embeddings, while most of them, including probabilistic graph models and word embedding models, require all documents in the corpus to be present during the training process. Thus, these methods inherently perform transductive learning which naturally cannot handle well the representations of unseen documents where few words have been learned before. Recently, Graph Neural Networks (GNNs) has drawn a lot of attention in various applications. Inspired by the mechanism of vertex information propagation guided by the graph structure in GNNs, we propose an inductive document representation learning model, called IDRL, that can map the short text structures into a graph network and recursively aggregate the neighbor information of the words in the unseen documents. Then, we can reconstruct the representations of the previously unseen short texts with the limited numbers of word embeddings learned before. Experimental results show that our proposed method can learn more discriminative representations in terms of inductive classification tasks and achieve better clustering performance than state-of-the-art models on four real-world datasets.

© Springer Nature Switzerland AG 2021
F. Hutter et al. (Eds.): ECML PKDD 2020, LNAI 12459, pp. 600–616, 2021.
https://doi.org/10.1007/978-3-030-67664-3_36

# 1   Introduction

Short text clustering (STC) has drawn a lot of attention for the explosive volume of short texts in the real-world applications, e.g., Tweets and Google News. STC aims to group semantically similar documents without supervision or manually assigned labels, which has been proven to be beneficial in various applications including topic discovery [23] and document summarization [4]. Compared with long documents, short texts containing few words make the traditional document representations sparse, like Term Frequency–Inverse Document Frequency (TF-IDF) or bag-of-words. Moreover, these sparse representations may lead to the inferior performance of clustering which essentially relies on calculating the distances between the representations.

Up to now, many approaches of short text representation learning have been proposed, which mainly focus on learning compact low-dimensional embeddings for alleviating the sparse representation problem. For example, probability graph models, such as GSDMM [32] and BTM [31], aim to learn the document-topic and word-topic representations (or called distributions) by exploiting the latent topic structures of short texts. More recently, embedding learning approaches, including Word2vec [22] and SIF [1], are devoted to maximizing the probabilities of word co-occurrences within a sliding window to learn the word representations. However, no matter the probability graph models which are mostly designed for specific clustering tasks, or the embedding learning models which employ classical clustering methods like K-means for document clustering, they are generally transductive models that require that all documents in a corpus are present during the training process. Though these models can perform well in mapping those visited data (i.e., training data) into a low-dimensional space, they could not be generalized to unseen documents, thereby being unable to support the tasks that involve the unseen documents for clustering.

In essence, the major limitation of the existing STC techniques is that they could not handle well the sparseness of words in the unseen documents. Compared with long texts containing rich contexts, it is more challenging to distinguish the clusters of short documents with few words occurring in the training set. Nevertheless, the previous models either only concentrate on learning representation from the local co-occurrences of words within the sliding window, e.g., Skip-Gram [21], or simply rely on the statistic models, e.g., bag-of-words introduced in [2]. As such, they ignore the word information propagation guided by the text structure which may benefit the clustering performance by generating the unseen document representations. Besides, these mentioned models often suffer from computational inefficiency because no parameters are shared between words in the encoder (each word is presented as a unique vector during the training).

In the past few years, Graph Neural Networks (GNNs), such as GCN [15], GraphSAGE [11] and GAT [27], have shown ground-breaking performance on many tasks of graph mining, while relatively little work has been applied to the document clustering problem. One of the advantages in GNNs is that the vertices of graphs can share the learning parameters in a neural network, thus the neighbor information of each vertex can be easily aggregated. However, the main difference between graphs and texts is that they have distinct data structures, specifically there are no explicit edges among words in documents. Moreover, how to incorporate GNNs into document representation learning for short text clustering is still challenging.

To address the aforementioned problems, in this paper, we propose an inductive document representation learning model (IDRL) that can map the short text structures into a graph network and aggregate the neighbor information of words in the unseen documents for representations. Specifically, starting from the original raw texts, we naturally construct a graph based on the word co-occurrence within a sliding window, i.e., building edges between words. Then, we can import the edges into a neural network for further operations. Here we follow the aggregation approaches in GraphSAGE [11] to generate representations of words by aggregating their neighbor information. In this way, the correlation between the local connectivity learned from word co-occurrence, and the global structure obtained from multi-layer aggregations can be uniformly preserved in our model, which also alleviates the sparsity problem in the short text representation learning. Finally, for the output layer, we adopt the objective function of negative sampling [21] which encourages a target word to be close to its neighbors while being far from its negative samples in the embedding space. To evaluate whether our proposed method can learn more discriminative representations than state-of-the-art models, we conduct experiments of inductive clustering and classification tasks on real-world datasets. The main contributions of this paper can be summarized as follows:

- We propose an inductive document representation learning model, referred as IDRL, that can efficiently generate embeddings for the unseen documents and benefit short text clustering.
- Specifically, we firstly map the raw texts into a graph based on the co-occurrences of words within a sliding window. Then, we construct a graph neural network that can generate word representations by aggregating their neighbor information. Finally, we perform unsupervised representation learning with the negative sampling method.
- We empirically evaluate the representations learned by IDRL on several real-world datasets. Experimental results show that our proposed model can learn more discriminative representations on inductive classification tasks and achieve better clustering performance than state-of-the-art models.

The code and datasets are available at https://github.com/junyachen/IDRL. The rest of this paper is organized as follows. In Sect. 2, we introduce the core idea of our proposed model and present the IDRL algorithm. We discuss experimental results in Sect. 3 and present the related work in Sect. 4. Section 5 makes a conclusion and provides our future work.

## 2    Proposed IDRL Model

In this section, we will first introduce the problem formulation and notations. Then, we will present an overview of the proposed IDRL model as shown in Fig. 1, followed by detail descriptions of each training step.

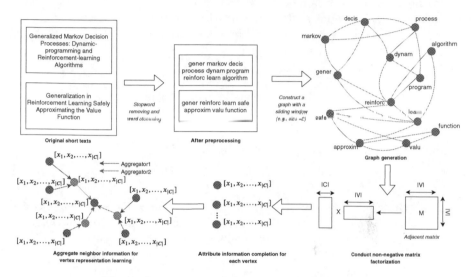

**Fig. 1.** An overview of the proposed IDRL model.

### 2.1    Problem Formulation and Notations

Since we aim to learn the representations of documents that consist of words, we denote a corpus as $C = (D, V)$, where $D$ denotes a set of documents and $V$ represents the vocabulary of the documents. For each word $v \in V$, we want to learn a low-dimensional embedding $\mathbf{v} \in \mathbb{R}^d$ which preserves the word correlations, where $d \ll |V|$ represents the embedded dimension of the representation space.

Moreover, to generate representations of the unseen documents, we need to aggregate the neighbor information of the words that are unlearned before. Thus, we map the short texts into a graph network for easily mining the neighbor structures of words. Without loss of generality, we denote the network as $G = (V, E)$, where $V$ is the set of vertices (**note that we use the same notation $V$ to represent the vertices and words because they are referred to the same object**), and $E \subseteq V \times V$ denotes the set of edges generated based on the word co-occurrence within a sliding window.

### 2.2    Step One: Graph Generation

As shown in the upper part of Fig. 1, before constructing a graph, we firstly conduct preprocessing on the input short texts. Here we adopt the documents of paper titles as the examples and use the standard NLTK[1] tool for stopword removing and word stemming. Then we present all words as vertices and build edges between them within a sliding window. For example, for the target word 'decis', we can obtain the edges between 'decis' and 'gener, markov, process, dynam' accordingly when the window size is two. It is worth mentioning that there is no need to use a large window size to capture the word relations with long distances, because our IDRL can learn the information propagation guided by the graph structure (the details will be introduced in Sect. 2.4).

## 2.3   Step Two: Vertex Attribute Completion

In the next step, to preserve the global information of vertices in the graph, we perform non-negative matrix factorization (NMF) [17] to obtain the vertex-community distributions as the vertex attribute features. Let $\mathbf{M} \in \mathbb{R}^{|V| \times |V|}$ be the symmetric adjacency matrix of a network, NMF can be performed by solving:

$$\min_{\mathbf{X} \geq 0} ||\mathbf{M} - \mathbf{X} \cdot \mathbf{X}^T||_F^2 + \alpha ||\mathbf{X}||_F^2, \tag{1}$$

where $\mathbf{X} \in \mathbb{R}^{|V| \times |C|}$ indicates the vertex-community distribution which encodes the global understanding of the network state, $|V|$ denotes the number of vertices, $|C|$ represents the latent community number, $|| \cdot ||_F$ is Frobenius norm of the matrix and $\alpha$ is a harmonic factor to balance two components. In general, the reason why not simply using $\mathbf{X}$ as the final vertex representations is that we want to preserve the correlation of vertex-community from global network structures and vertex-vertex from local connectivity in the embedding space (more details will be introduced in the next section), which may alleviate the sparse problem in the short text representation learning. Experimental results in Sect. 3.4 and Sect. 3.5 also validate that the proposed jointly learning method can outperform NMF on the subsequent tasks. Please note that the framework of IDRL is general enough to adopt other feature extraction methods to replace the NMF approach for the vertex attribute feature learning. We employ NMF for its popularity and fast training (its time complexity is linear to the size of vertices).

## 2.4   Step Three: Aggregation of Neighbor Information for Representation Learning

To perform inductive learning, the designed model needs to allow embeddings to be efficiently generated for the unseen vertices (i.e., words, which represent the same objects in this paper). GCN [15] firstly introduces an effective variant of convolutional neural networks that can operate directly on graphs. For inductive learning on large-scale networks, an improved version of GCN [11] is derived. We follow their work and adopt a localized spectral convolutional aggregator for representation learning, which is defined as follows:

$$\mathbf{v}^l \leftarrow \sigma(\mathbf{W}^l \cdot \text{MEAN}(\{\mathbf{v}^{l-1}\} \cup \{\mathbf{v}_p^{l-1}, \forall v_p \in \mathcal{N}(v)\})), \tag{2}$$

---

[1] https://www.nltk.org/.

where MEAN denotes the element-wise mean of the vectors, $\mathbf{W}^l$ is the weight matrices of layer $l$, $\mathbf{v}^l$ is the embedding vector in layer $l$, $\sigma$ is the sigmoid function, and $\mathcal{N}(v)$ denotes the neighbors of vertex $v$. Note that we let $\mathbf{v}^0 \leftarrow \mathbf{x}_v$, where $\mathbf{x}$ denotes the vertex-community distributions learned from Eq. (1). In summary, from Eq. (2), we can see that the MEAN aggregator is able to aggregate the neighbor representations of vertices including the unseen ones for inductive learning. Then, to learn vertex representations in an unsupervised setting, we employ negative sampling (NS) [21] to the output representations, which encourages a target vertex to be close to its neighbors and meanwhile being far from its negative samples in the embedding space. The objective function is defined as follows:

$$\mathcal{J}(v_i) = -\log(\sigma(\mathbf{v}_p^T \cdot \mathbf{v}_i)) - \sum_{j=1}^{K} \mathbb{E}_{v_j \sim P_{NS}(v)} \log(\sigma(-\mathbf{v}_j^T \cdot \mathbf{v}_i)), \qquad (3)$$

where $v_i$ is a target vertex, $v_p$ is its context vertex, $\sigma$ is the sigmoid function, i.e., $\sigma(x) = 1/(1+\exp(-x))$, $P_{NS}(v)$ denotes the negative sampling distribution (the details will be given in the followings), $v_j$ is a negative sample drawn from $P_{NS}(v)$, $K$ represents the number of negative samples used for the estimation, and $\mathbf{v_p}$, $\mathbf{v_i}$, $\mathbf{v_j}$ are the representations aggregated from the features involving their local neighbors. This objective aims to encourage nearby vertices to have similar embeddings while being distinct to their negative vertices. Besides, the distribution of negative sampling $P_{NS}(v)$ is formulated as follows:

$$P_{NS}(v) = \frac{f_v^\beta}{\sum_{v \in V} f_v^\beta}. \qquad (4)$$

where $f_v$ is the degree of vertex $v$ in the graph, $\beta$ is an empirical degree power that is usually set to $3/4$ [21, 26]. In general, NS is the most widely used method to optimize the objective function of unsupervised representation learning for its low computational complexity [10, 21, 25, 26].

### 2.5   IDRL Algorithm

Till now, we have introduced the training process of IDRL shown in Fig. 1. To go into the details, we present Algorithm 1 for demonstration. Specifically, we firstly conduct preprocessing on the original short texts (Line 2). Secondly, we construct a graph to represent the text data (Line 3). Next, we adopt NMF to obtain the vertex attribute features (Line 4). Then, we perform forward propagation to obtain the aggregate embedding (Line 8–14). Last, we employ NS [21] as the objective function and use stochastic gradient descent [14] for optimization (Line 15). We repeat the last two steps until achieving convergence.

## 3   Experiments

In experiments, we evaluate the performance of document representations on real-world datasets with the tasks of inductive clustering and classification. Moreover, we also investigate the parameter sensitivity of IDRL on these tasks.

---

**Algorithm 1:** Training Process of IDRL

---

**Input**: Short text corpus $C = (D, V)$, dimension $d$, window size $t$, number of community $|C|$, and number of neural network layer $|L|$

**Result**: Vertex embedding $\mathbf{v}, \forall v \in V$

1 **begin**
2    Conduct preprocessing on the short texts including stopword
3     removing and word stemming;
4    Construct a graph based on the word co-occurrence in a sliding
5     window with size $t$;
6    Perform NMF to obtain the community-vertex distributions $\{\mathbf{x}_v, \forall v \in V\}$ according to Eq. (1);
7    Initialize weight matrices $\mathbf{W}^l, \forall l \in \{1, ..., |L|\}$ randomly;
8    $\mathbf{v}^0 \leftarrow \mathbf{x}_v, \forall v \in V$;
9    **while** *not converge* **do**
10      **for** $l \in [1, |L|]$ **do**
11        **for** $v \in V$ **do**
12          $\mathbf{v}^l \leftarrow \sigma(\mathbf{W}^l \cdot \text{MEAN}(\{\mathbf{v}^{l-1}\} \cup \{\mathbf{v}_p^{l-1}, \forall v_p \in \mathcal{N}(v)\}))$
13          according to Eq. (2);
14        **end**
15        $v^l \leftarrow v^l / ||v^l||_2, \forall v \in V$ ;
16      **end**
17      $\mathbf{v} \leftarrow \mathbf{v}^{|L|}, \forall v \in V$;
18      Employ NS [21] as the objective function (Eq.(3)) and use
19       stochastic gradient descent [14] for optimization;
20    **end**
21 **end**

---

### 3.1 Datasets

We conduct experiments on four widely used short text datasets with the statistics shown in Table 1.

**TweetSet**[2] consists of tweets from 2011 and 2012 microblog tracks published by Text REtrieval Conference. It contains 2,472 tweets with 89 topics.

**TSet**, **SSet** and **TSSet** are the three variants of the **Google News**[3] dataset, which represent titles, snippets, and the combination of them, respectively. In general, they contain 11,109 news including 152 topics.

### 3.2 Baseline Models

We employ several state-of-the-art approaches that are designed for short texts as the baselines, including embedding learning models and probability graph

---

[2] https://trec.nist.gov/data/microblog.html.
[3] https://news.google.com/news/.

**Table 1.** Statistics of datasets

| Dataset | Documents | Topics | Vocabulary | Avg. len |
|---------|-----------|--------|------------|----------|
| TweetSet | 2,472 | 89 | 5,098 | 8.56 |
| TSet | 11,109 | 152 | 8,111 | 6.23 |
| SSet | 11,109 | 152 | 18,478 | 22.20 |
| TSSet | 11,109 | 152 | 19,672 | 28.43 |

models. There are many other short text representation learning methods but we do not consider them here, because either their performance is inferior to these baselines as shown in corresponding papers or they are transductive models that are inappropriate for inductive document representation learning. The descriptions of the baselines are as follows.

**SIF** [1] proposes a relatively simple yet effective model for sentence embeddings. It achieves significantly better performance than baselines including some sophisticated supervised methods such as RNN [20] and LSTM [9] models.

**Word2vec** [22] is one of the most popular models which firstly incorporates a two-layer neural network to learn word embedding. The designed network structure is shallow and computationally effective on large text corpora.

**NMF** [17] exploits non-negative matrix factorization to obtain vertex-community distributions in a global understanding of networks. We include NMF as one of the baseline models because we apply it to gain the vertex attribute features for our model.

**GSDMM** [31] is one of the state-of-the-art probabilistic graph models designed for short text clustering. It adopts the mixture of Dirichlet and multinomial distributions to learn the document and word representations.

**BTM** [31] is a classical probabilistic model that uses an aggregated pattern (called biterms) for alleviating the sparse word co-occurrence problem at the document-level in the short texts.

### 3.3 Parameter Settings and Evaluation Metrics

For the models (i.e., NMF, GSDMM, BTM, and the proposed IDRL) requiring a predefined topic/community number, we provide the ground-truth setting shown in the datasets. Since SIF needs pre-trained embedding to initialize parameters, we adopt the GloVe word embedding [24] which is trained on several large social-network corpora. For Word2vec and IDRL using negative sampling for optimization, we follow [22] and uniformly set the number of negative samples as 5. For the other parameters of models, we follow the preferred settings in their corresponding papers. Besides, since IDRL exploits an aggregate neural network, we follow [11] by setting the number of network layers $l = 2$, the hidden dimension $d = 128$, and the neighborhood sample sizes of layers $S_1 = 25$ and $S_2 = 10$, respectively. In addition, we use Adam optimizer [14] with the initial learning rate 1e-3 to perform stochastic gradient descent.

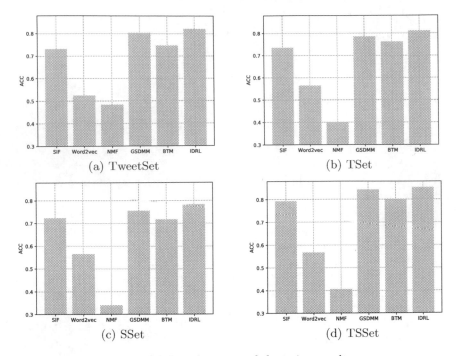

**Fig. 2.** ACC performance of clustering results.

To evaluate document clustering performance, we report two widely used metrics, the clustering accuracy (ACC) [13] and the normalized mutual information (NMI) [30]. When the clustering results perfectly match the ground-truth topics, both the values of ACC and NMI will be one. While the clustering results are randomly generated, the values will be close to zero. Moreover, to measure inductive classification performance, we adopt Liblinear package [7] with default settings to build the classifier and employ Micro-F1 and Macro-F1 metrics [8].

### 3.4   Evaluation on Inductive Clustering

In this section, we conduct inductive document clustering to verify the performance of our proposed IDRL. Specifically, we randomly select partial short texts (90%) as the training data for representation learning. After that, combined with the learned word embeddings, we can obtain each document embedding by recursively aggregating the neighbor information of its words. Then we run K-means on the rest documents (10%) for clustering where few words have been visited before. To ensure reliability, we take 10 runs and report the averages of experimental results. Figures 2 and 3 show the clustering performance of various models on four short text datasets. More concretely, we can obtain the following observations:

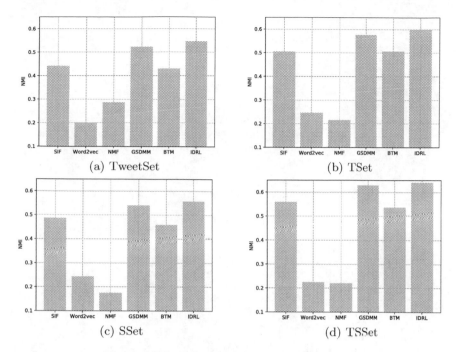

**Fig. 3.** NMI performance of clustering results.

(1) From Fig. 2, we can observe that IDRL can outperform the other base-line models in terms of ACC, which verifies the effectiveness of our pro-posed method. In general, the order of ACC performance except IDRL is $GSDMM > BTM > SIF > Word2vec > NMF$. One possible reason is that both GSDMM and BTM are specially designed for clustering tasks (performing transductive clustering in their papers) while SIF, Word2vec, and NMF are more general methods for representation learning.

(2) Besides, from Fig. 3, we can see that IDRL can also achieve consistent improvements over state-of-the-art models, which is supplementary to con-firm the superiority of IDRL for handling unseen document clustering. More-over, we note that IDRL performs better than NMF, which demonstrates that the advantage of IDRL comes beyond the prior knowledge from NMF (we apply it to obtain the vertex-community distributions as the vertex attribute features before inductive representation learning and the details can be referred to Sect. 2.3).

**Table 2.** Document classification result (%) of TweetSet

| Method | 30% | | 50% | | 70% | | 90% | |
|---|---|---|---|---|---|---|---|---|
| | Micro-F1 | Macro-F1 | Micro-F1 | Macro-F1 | Micro-F1 | Macro-F1 | Micro-F1 | Macro-F1 |
| GSDMM | 31.23 | 14.45 | 30.45 | 13.92 | 36.47 | 18.36 | 40.34 | 20.10 |
| BTM | 32.99 | 15.79 | 32.08 | 16.03 | 38.91 | 22.33 | 44.80 | 28.23 |
| NMF | 36.76 | 20.79 | 40.58 | 24.89 | 41.77 | 26.81 | 44.16 | 28.20 |
| Word2vec | 52.87 | 47.28 | 54.77 | 49.19 | 56.62 | 50.48 | 60.72 | 54.08 |
| SIF | 61.99 | 59.67 | 62.82 | 60.15 | 63.74 | 61.47 | 65.82 | 64.18 |
| **IDRL** | **68.44** | **64.50** | **71.28** | **67.91** | **72.32** | **69.58** | **72.82** | **70.26** |

**Table 3.** Document classification result (%) of TSet

| Method | 30% | | 50% | | 70% | | 90% | |
|---|---|---|---|---|---|---|---|---|
| | Micro-F1 | Macro-F1 | Micro-F1 | Macro-F1 | Micro-F1 | Macro-F1 | Micro-F1 | Macro-F1 |
| GSDMM | 36.98 | 18.53 | 40.30 | 24.30 | 42.66 | 27.05 | 45.51 | 29.64 |
| BTM | 42.09 | 25.34 | 44.86 | 29.72 | 46.79 | 32.09 | 48.91 | 34.98 |
| NMF | 37.85 | 23.97 | 39.91 | 27.97 | 40.55 | 29.11 | 41.16 | 29.48 |
| Word2vec | 63.73 | 59.67 | 65.46 | 61.54 | 65.06 | 61.13 | 65.28 | 61.03 |
| SIF | 59.91 | 56.11 | 60.47 | 56.95 | 60.37 | 56.93 | 60.24 | 56.89 |
| **IDRL** | **67.36** | **63.43** | **69.56** | **66.32** | **70.67** | **67.85** | **70.94** | **67.48** |

**Table 4.** Document classification result (%) of SSet

| Method | 30% | | 50% | | 70% | | 90% | |
|---|---|---|---|---|---|---|---|---|
| | Micro-F1 | Macro-F1 | Micro-F1 | Macro-F1 | Micro-F1 | Macro-F1 | Micro-F1 | Macro-F1 |
| GSDMM | 31.73 | 12.19 | 31.59 | 12.08 | 31.64 | 12.15 | 32.92 | 13.01 |
| BTM | 32.52 | 12.79 | 32.92 | 12.65 | 33.00 | 12.54 | 34.16 | 13.49 |
| NMF | 34.00 | 16.06 | 34.35 | 16.74 | 35.08 | 17.89 | 36.25 | 19.16 |
| Word2vec | 54.98 | 47.99 | 55.69 | 48.56 | 56.16 | 49.26 | 57.31 | 50.43 |
| SIF | 59.88 | 55.59 | 60.19 | 55.97 | 59.93 | 55.76 | 60.18 | 56.04 |
| **IDRL** | **59.81** | **56.17** | **61.93** | **57.47** | **62.71** | **58.36** | **64.36** | **59.91** |

### 3.5 Evaluation on Inductive Classification

To verify whether our proposed model can learn more discriminative representations, we perform inductive classification tasks on four real-world datasets. More concretely, we randomly select 30%, 50%, 70%, and 90% documents as the training set and the remained ones as the testing set. We report the classification results with Micro-F1 and Macro-F1 metrics in Tables 2, 3, 4, 5, where the highest scores are highlighted in boldface. From these tables, we have the following observations:

(1) Our proposed model (IDRL) significantly outperforms the other models on all datasets with different training ratios, which demonstrates the effectiveness of IDRL for aggregating the neighbor information of the words unlearned before to generate representations for the unseen documents.
(2) Specifically, the overall performance of the baselines follows a sequence as: $SIF, Word2vec > NMF > BTM > GSDMM$. Though BTM and

**Table 5.** Document classification result (%) of TSSet

| Method | 30% | | 50% | | 70% | | 90% | |
|---|---|---|---|---|---|---|---|---|
| | Micro-F1 | Macro-F1 | Micro-F1 | Macro-F1 | Micro-F1 | Macro-F1 | Micro-F1 | Macro-F1 |
| GSDMM | 32.99 | 12.42 | 32.95 | 12.45 | 32.87 | 12.40 | 33.91 | 12.60 |
| BTM | 34.32 | 12.97 | 33.70 | 12.87 | 34.41 | 12.78 | 35.24 | 13.16 |
| NMF | 36.76 | 14.33 | 38.73 | 17.28 | 40.34 | 20.32 | 41.04 | 21.79 |
| Word2vec | 59.54 | 53.58 | 60.25 | 53.92 | 60.64 | 54.22 | 61.97 | 55.98 |
| SIF | 50.51 | 45.52 | 51.06 | 46.17 | 51.61 | 46.48 | 51.27 | 46.04 |
| **IDRL** | **60.60** | **52.38** | **62.91** | **55.90** | **64.83** | **59.30** | **65.70** | **60.90** |

GSDMM perform well on the clustering tasks as shown before, the representations learned by them are less discriminative than the ones learned by SIF, Word2vec, and NMF. In general, IDRL can achieve 11.5%, 8.1%, 3.7%, and 4.9% performance gains over the second-place models on average of training ratios in TWeetSet, TSet, SSet, and TSSet, respectively. We can summarize that our proposed IDRL can obtain more benefits on the texts with shorter length (the statistics of datasets is shown in Table 1), which validates the capability of IDRL on short text representation learning.

## 3.6 Parameter Sensitivity

In this part, we analyze the sensitivity of IDRL to the parameters when conducting clustering and classification tasks.

Firstly, we try to investigate how the embedded representation dimension affects the performance of IDRL by varying its numbers in {100, 128, 160, 190}. We report the experimental results on TweetSet as shown in Fig. 4. Specifically, from Fig. 5a, we can see that the NMI performance increases slowly while ACC performance keeps stable with the growth of dimension. In general, the clustering performance of IDRL is robust to the dimension setting. Besides, Fig. 5b and 5c show that the curves of classification performance, including Micro-F1 and Macro-F1, tends to rise with slight fluctuations as the dimension number goes. Note that for easy presentation, we only present the performance of TweetSet with training ratios 30%, 50%, 70%, and 90%. In summary, our IDRL achieves relatively stable performance on different dimensions.

Secondly, as mentioned before, since we run K-means on the testing documents for clustering, we want to estimate how the settings of cluster numbers affect the clustering performance of IDRL. We vary the numbers in {70, 90, 110, 130, 150, 170, 190} and report the NMI results on four datasets as shown in Fig. 5. Note that we omit the ACC performance for similar trends. From Fig. 5, we can see that IDRL can achieve the best performance when the cluster number is set to 90 on TweetSet. Besides, on the other datasets, IDRL firstly grows with the cluster size. Then it achieves diminishing performances and becomes stable when the number is around 150. For larger values, IDRL begins to drop gradually. These detected cluster numbers are matched the ground truth of the total topics in the datasets, which indicates that IDRL can dynamically detect the cluster numbers in the short texts.

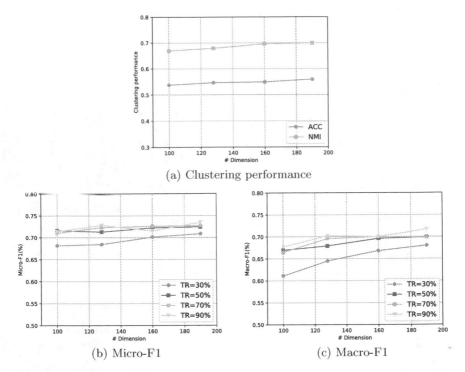

(a) Clustering performance

(b) Micro-F1                                    (c) Macro-F1

**Fig. 4.** Influence of dimension setting on clustering performance, Micro-F1, and Macro-F1, respectively. Note that we use 'TR' to represent training ratios.

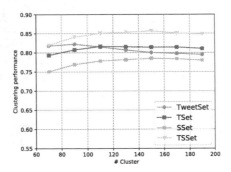

**Fig. 5.** Clustering performance with respect to the cluster number on four datasets.

## 4   Related Work

In short text clustering, compared with long texts containing rich contexts, one of the main challenges is the sparsity problem that a short document only covers a few words. To address this problem, a lot of work has been proposed.

Some of the previous work resort to external knowledge for enriching short text representations. For example, Hu et al. [12] exploit Wikipedia to extend short texts. Similarly, Wei et al. [29] incorporate ontologies to enrich texts. Therefore, these methods are on supervised learning and their performance may be affected by the quality of external sources. Instead, our proposed method is performed in an unsupervised setting for short text representation learning.

In the past few years, graph probabilistic models (e.g., LDA [2]) obtain a lot of attention for their excellent capability of interpretability. Specially designed for short text clustering, GSDMM [32] employs Dirichlet distributions and follows the assumption that all words in a document are related to the same topic. BTM [31] proposes an explicit way (biterms) to model the word co-occurrence pattern in the short texts. After that, variant of probabilistic models (e.g., GSDPMM [33] and DP-BMM [5]) are proposed for different applications of short texts. However, all these mentioned models are restricted by the sparse word co-occurrence pattern which may not be applicable to inductive short text clustering where few words have been visited before in the training process. Moreover, they are statistic models that lack consideration of word order information which could generate benefits in representation learning.

More recently, with the rise of neural network techniques, Word2vec [22] firstly incorporates a two-layer neural network to learn word embedding. Then the variants including sentence-based embedding (e.g., SIF [1] and STV [16]) and paragraph embedding (e.g., PV [18] and PV2 [6]) methods are presented and proven to be effective on a variety of tasks. Though these embedding approaches are context-aware and perform well in mapping training data into a low-dimensional space, they are transductive models that could not support the clustering tasks involving the unseen documents. Besides, most of the models are not specially designed for short texts, thus, it would be challenging for them to distinguish the clusters of short documents with few words occurring in the training set. Moreover, these previous models ignore the word information propagation guided by the text structure and they are computational inefficiency since no parameters are shared between words during the training.

Note that there are many other deep clustering methods [3,19,28,34] that have shown promising performance in graph mining. One important difference between these methods and our work is that we have distinct data structures, specifically, there are no explicit edges among words in documents. In addition, these methods are also transductive models that are still challenging for inductive short text clustering. In general, the main limitation of the existing methods for short text clustering is that they could not handle well the sparseness of words in the unseen documents.

## 5    Conclusion and Future Work

In this paper, we propose an inductive document representation learning model, called IDRL, for short text clustering. IDRL can efficiently generate embeddings for the unseen documents where few words have been visited before in the training process. Specifically, we firstly map the short texts into a graph network based on word co-occurrences within a sliding window. Then to preserve the global information of vertices (each word is represented as a vertex) in the graph, we conduct non-negative matrix factorization to obtain the vertex-community distributions as the vertex attribute features. Lastly, to perform inductive representation learning, we construct a graph neural network that can aggregate the neighbor information of the words in the unseen documents. In general, our proposed model is practical and the experimental results on the real-word datasets validate the effectiveness of IDRL in terms of inductive clustering and classification tasks. Besides, one of our possible future work is exploring the possibility of making inductive document clustering and conducting online representation learning simultaneously without completely retraining the whole dataset.

**Acknowledgement.** MOST (2019YFB1600704), FDCT (SKL-IOTSC-2018-2020, FDCT /0045/2019/A1, FDCT/0007/2018/A1), GSTIC (EF005/FST-GZG/2019/ GSTIC), University of Macau (MYRG2017-00212-FST, MYRG2018-00129-FST).

# References

1. Arora, S., Liang, Y., Ma, T.: A simple but tough-to-beat baseline for sentence embeddings. In: 5th International Conference on Learning Representations, ICLR 2017 (2017)
2. Blei, D.M., Ng, A.Y., Jordan, M.I.: Latent Dirichlet allocation. J. Mach. Learn. Res. **3**, 993–1022 (2003)
3. Bo, D., Wang, X., Shi, C., Zhu, M., Lu, E., Cui, P.: Structural deep clustering network. arXiv preprint arXiv:2002.01633 (2020)
4. Chen, J., Gong, Z., Liu, W.: A nonparametric model for online topic discovery with word embeddings. Inf. Sci. **504**, 32–47 (2019)
5. Chen, J., Gong, Z., Liu, W.: A dirichlet process biterm-based mixture model for short text stream clustering. Appl. Intell. **50**, 1–11 (2020)
6. Dai, A.M., Olah, C., Le, Q.V.: Document embedding with paragraph vectors. arXiv preprint arXiv:1507.07998 (2015)
7. Fan, R.E., Chang, K.W., Hsieh, C.J., Wang, X.R., Lin, C.J.: Liblinear: a library for large linear classification. J. Mach. Learn. Res. **9**, 1871–1874 (2008)
8. Gao, H., Pei, J., Huang, H.: Progan: network embedding via proximity generative adversarial network. In: Proceedings of the 25th ACM SIGKDD International Conference on Knowledge Discovery & Data Mining, pp. 1308–1316 (2019)
9. Gers, F.A., Schraudolph, N.N., Schmidhuber, J.: Learning precise timing with LSTM recurrent networks. J. Mach. Learn. Res. **3**, 115–143 (2002)
10. Grover, A., Leskovec, J.: node2vec: scalable feature learning for networks. In: Proceedings of the 22nd ACM SIGKDD International Conference on Knowledge Discovery and Data Mining, pp. 855–864. ACM (2016)
11. Hamilton, W., Ying, Z., Leskovec, J.: Inductive representation learning on large graphs. In: Advances in Neural Information Processing Systems, pp. 1024–1034 (2017)
12. Hu, X., Zhang, X., Lu, C., Park, E.K., Zhou, X.: Exploiting wikipedia as external knowledge for document clustering. In: Proceedings of the 15th ACM SIGKDD International Conference on Knowledge Discovery and Data Mining, pp. 389–396 (2009)

13. Huang, P., Huang, Y., Wang, W., Wang, L.: Deep embedding network for clustering. In: 2014 22nd International Conference on Pattern Recognition, pp. 1532–1537. IEEE (2014)
14. Kingma, D.P., Ba, J.: Adam: a method for stochastic optimization. arXiv preprint arXiv:1412.6980 (2014)
15. Kipf, T.N., Welling, M.: Semi-supervised classification with graph convolutional networks. arXiv preprint arXiv:1609.02907 (2016)
16. Kiros, R., et al.: Skip-thought vectors. In: Advances in Neural Information Processing Systems, pp. 3294–3302 (2015)
17. Kuang, D., Ding, C., Park, H.: Symmetric nonnegative matrix factorization for graph clustering. In: Proceedings of the 2012 SIAM International Conference on Data Mining, pp. 106–117. SIAM (2012)
18. Le, Q., Mikolov, T.: Distributed representations of sentences and documents. In: International Conference on Machine Learning, pp. 1188–1196 (2014)
19. Li, X., Zhang, H., Zhang, R.: Embedding graph auto-encoder with joint clustering via adjacency sharing. arXiv preprint arXiv:2002.08643 (2020)
20. Lipton, Z.C., Berkowitz, J., Elkan, C.: A critical review of recurrent neural networks for sequence learning. arXiv preprint arXiv:1506.00019 (2015)
21. Mikolov, T., Chen, K., Corrado, G., Dean, J.: Efficient estimation of word representations in vector space. arXiv preprint arXiv:1301.3781 (2013)
22. Mikolov, T., Sutskever, I., Chen, K., Corrado, G.S., Dean, J.: Distributed representations of words and phrases and their compositionality. In: Advances in Neural Information Processing Systems, pp. 3111–3119 (2013)
23. Nguyen, H.L., Woon, Y.K., Ng, W.K.: A survey on data stream clustering and classification. Knowl. Inf. Syst. 45(3), 535–569 (2015)
24. Pennington, J., Socher, R., Manning, C.D.: Glove: global vectors for word representation. In: Proceedings of the 2014 Conference on Empirical Methods in Natural Language Processing (EMNLP), pp. 1532–1543 (2014)
25. Ribeiro, L.F., Saverese, P.H., Figueiredo, D.R.: struc2vec: Learning node representations from structural identity. In: Proceedings of the 23rd ACM SIGKDD International Conference on Knowledge Discovery and Data Mining, pp. 385–394. ACM (2017)
26. Tang, J., Qu, M., Wang, M., Zhang, M., Yan, J., Mei, Q.: Line: large-scale information network embedding. In: Proceedings of the 24th International Conference on World Wide Web, pp. 1067–1077. International World Wide Web Conferences Steering Committee (2015)
27. Veličković, P., Cucurull, G., Casanova, A., Romero, A., Lio, P., Bengio, Y.: Graph attention networks. arXiv preprint arXiv:1710.10903 (2017)
28. Wang, C., Pan, S., Hu, R., Long, G., Jiang, J., Zhang, C.: Attributed graph clustering: A deep attentional embedding approach. arXiv preprint arXiv:1906.06532 (2019)
29. Wei, T., Lu, Y., Chang, H., Zhou, Q., Bao, X.: A semantic approach for text clustering using wordnet and lexical chains. Expert Syst. Appl. 42(4), 2264–2275 (2015)
30. Xu, J., Xu, B., Wang, P., Zheng, S., Tian, G., Zhao, J.: Self-taught convolutional neural networks for short text clustering. Neural Networks 88, 22–31 (2017)
31. Yan, X., Guo, J., Lan, Y., Cheng, X.: A biterm topic model for short texts. In: Proceedings of the 22nd International Conference on World Wide Web, pp. 1445–1456 (2013)

32. Yin, J., Wang, J.: A Dirichlet multinomial mixture model-based approach for short text clustering. In: Proceedings of the 20th ACM SIGKDD International Conference on Knowledge Discovery and Data Mining, pp. 233–242. ACM (2014)
33. Yin, J., Wang, J.: A model-based approach for text clustering with outlier detection. In: 2016 IEEE 32nd International Conference on Data Engineering (ICDE), pp. 625–636. IEEE (2016)
34. Zhang, X., Liu, H., Li, Q., Wu, X.M.: Attributed graph clustering via adaptive graph convolution. arXiv preprint arXiv:1906.01210 (2019)

# FUSE: Multi-faceted Set Expansion by Coherent Clustering of Skip-Grams

Wanzheng Zhu[1]([✉]), Hongyu Gong[1], Jiaming Shen[1], Chao Zhang[2],
Jingbo Shang[3], Suma Bhat[1], and Jiawei Han[1]

[1] University of Illinois at Urbana-Champaign, Champaign, IL, USA
{wz6,hgong6,js2,spbhat2,hanj}@illinois.edu
[2] Georgia Institute of Technology, Atlanta, GA, USA
chaozhang@gatech.edu
[3] University of California San Diego, San Diego, CA, USA
jshang@ucsd.edu

**Abstract.** Set expansion aims to expand a small set of seed entities into a complete set of relevant entities. Most existing approaches assume the input seed set is unambiguous and completely ignore the multi-faceted semantics of seed entities. As a result, given the seed set {"Canon", "Sony", "Nikon"}, previous models return *one mixed set* of entities that are either *Camera Brands* or *Japanese Companies*. In this paper, we study the task of **multi-faceted set expansion**, which aims to capture all semantic facets in the seed set and return *multiple sets* of entities, one for each semantic facet. We propose an unsupervised framework, FUSE, which consists of three major components: (1) *facet discovery module*: identifies all semantic facets of each seed entity by extracting and clustering its skip-grams, and (2) *facet fusion module*: discovers shared semantic facets of the entire seed set by an optimization formulation, and (3) *entity expansion module*: expands each semantic facet by utilizing a masked language model with pre-trained BERT models. Extensive experiments demonstrate that FUSE can accurately identify multiple semantic facets of the seed set and generate quality entities for each facet.

**Keywords:** Set expansion · Multi-facetedness · Word sense disambiguation

## 1 Introduction

The task of *set expansion* is to expand a small set of seed entities into a more complete set of relevant entities. For example, to explore all *Universities in the U.S.*, one can feed a seed set (*e.g.*, {"Stanford", "UCB", "Harvard"}) to a set expansion system and then expect outputs such as "Princeton", "MIT" and "UW". Those expanded entities can benefit numerous entity-aware applications, including query suggestion [1], taxonomy construction [30], recommendation [36], and information extraction [12,22,34,35]. Besides, the set expansion algorithm itself becomes a basic building block of many NLP-based systems [15,25].

© Springer Nature Switzerland AG 2021
F. Hutter et al. (Eds.): ECML PKDD 2020, LNAI 12459, pp. 617–632, 2021.
https://doi.org/10.1007/978-3-030-67664-3_37

Previous studies on set expansion focus on returning *one single set* of most relevant entities. Methods have been developed to incrementally and iteratively add the entities of high confidence scores into the set. A variety of features are extracted, including word co-occurrence statistics [17], unary patterns [24], or coordinational patterns [23], from different data sources such as query log [29], web table [31], and raw text corpus [15,24]. However, all these methods assume the given seed set is unambiguous and completely ignore the multi-faceted semantics of seed entities. As a result, given a seed set {"apollo", "artemis", "poseidon"} which has two semantic facets – *Major Gods in Greek Mythology* and *NASA Missions*, previous methods can only generate one mixed set of entities from these two facets, which inevitably hampers their applicabilities.

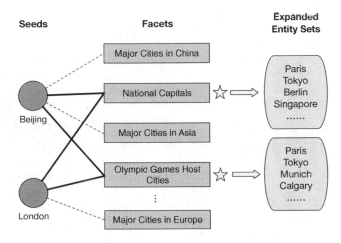

**Fig. 1.** An illustrative example of a multi-faceted seed set {"Beijing", "London"}. Facets (*e.g. Major Cities in China*) that do not appear in both seed entities should be eliminated in the set expansion process. As a result, we expect to output two separate entity sets: one for semantic facet *National Capitals* and the other one for semantic facet *Olympic Games Host Cities*.

In this paper, we approach the set expansion task from a new angle. Our study focuses on **multi-faceted set expansion** which aims to identify semantic facets shared by all seed entities and return *multiple* expanded sets, one for each semantic facet. The key challenge lies in the discovery of shared semantic facets from a seed set. However, the only initial attempt towards multi-facetedness, EgoSet [21], not only requires user-created ontologies as external knowledge, but also has no guarantee that their generated semantic facets are relevant to all seed entities. As an illustrative example in Fig. 1, EgoSet generates more than five facets, but only two of them are relevant to both seeds.

To handle the key challenge of multi-faceted set expansion, we propose a novel framework, FUSE, as illustrated in Fig. 2. First, we discover all possible facets of each seed by extracting and clustering its skip-grams. Second, we leverage an

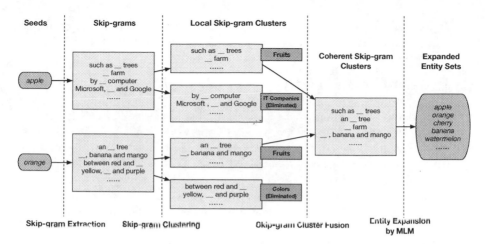

**Fig. 2.** Overview of FUSE. The key novelty is to discover coherent skip-gram clusters, whereas previous methods skip this stage and directly combine all skip-grams into the same pool for entity expansion.

optimization formulation to discover the shared semantic facets across all seeds as *coherent semantic facets*. This helps eliminate those facets relevant only to a partial set of seeds. Third, based on the coherent skip-gram clusters, we utilize a masked language model (MLM) with pre-trained BERT models to generate quality entities for each semantic facet.

It is considerably complicated to evaluate such multi-faceted set expansion task, mainly because we have no prior knowledge about the number of facets in a seed set. Therefore, we are likely to observe different number of facets between the generated result and the ground truth (*e.g.*, the ground truth may have 3 facets, while the generated result has 4 facets). Previously proposed metric Mean-MAP (MMAP) in [21] only measures how many entities and facets in the ground truth are covered by the generated result. However, it fails to measure how noisy those generated facets are and thus it biases toward methods that output as many facets as possible. To overcome the intrinsic limitation of MMAP, we propose a more comprehensive evaluation metric, Best-Matching Average Precision (BMAP), that can not only capture the purity of generated facets but also their coverage of ground truth facets.

Our contributions are highlighted as follows.

- We identify the key challenge of *multi-faceted set expansion* and develop FUSE to address it.[1]
- We propose to determine semantic facets by clustering skip-gram contexts, and utilize an optimization formulation to discover coherent semantic facets.
- We propose a novel evaluation metric for multi-faceted set expansion problem, which is shown to be a more comprehensive measure.

---

[1] The code is available at https://github.com/WanzhengZhu/FUSE.

– Extensive experiments demonstrate that our proposed framework outperforms state-of-the-art set expansion algorithms significantly.

## 2    Problem Formulation

A facet refers to one semantic aspect or sense of seed words. For example, *Fruit*
and *Technology Companies* are two facets of the word "apple". Previous works
study mostly single-faceted set expansion and ignore the seeds' multi-facetedness
nature. In this work, we explore a better coverage of all coherent semantic facets
of a seed set and study *corpus-based multi-faceted set expansion*.

More formally, given a seed set query $q = \{s_1, s_2, ..., s_m\}$ where $s_i$ is a seed
and a raw text corpus $D$, our set expansion system is to find all lists of entities
$\mathbb{E} = \{E^{(i)}, E^{(j)}, E^{(k)}, ...\}$, where $E^{(i)} = \{x_1^{(i)}, ..., x_n^{(i)}\}$ is relevant to the $i$-th
facet $f_i$ of query $q$, and $x_l^{(i)}$ denotes an expanded entity.

## 3    Model

Our proposed FUSE framework consists of three main steps: 1) extracting and
clustering skip-grams for each seed (c.f. Sect. 3.1); 2) discovering coherent semantic facets of a seed set (c.f. Sect. 3.2); and 3) expanding entities for each semantic
facet (c.f. Sect. 3.3). An overview of our approach is shown in Fig. 2 and the
algorithm is shown in Algorithm 1.

### 3.1    Skip-Gram Features Extraction and Clustering

We preprocess the raw corpus and extract skip-gram features of seed words as
[21,24] do. Here skip-gram features are a sequence of words surrounding the
seed word. Based on the distributional hypothesis [15], the semantics of a word
is reflected by its neighboring skip-grams. We can derive different facets of a seed
word by separating its skip-grams into different semantic clusters.

Embedding is commonly used in NLP applications to represent rich semantic
information of words and phrases. We obtain the embedding for each skip-gram
by simply averaging the embedding of its component words. The derivation of
skip-gram embedding is another interesting research question, but it is not our
focus in this work.

Now we cluster these skip-gram embeddings to discover different semantic
facets of a seed word. Most clustering algorithms require the number of clusters as input, which deviates from our problem setting. Also, we note that the
embedding usually lies in a high-dimension space (typically of dimension 100–
300), which leads to the poor and unstable performance of most existing non-
parametric clustering algorithms [27] (*e.g.*, MeanShift [4]).

To solve the two main issues of instability and hard coded cluster numbers as
mentioned above, we propose to tackle the high-dimensional embedding clustering problem by the affinity propagation algorithm [6]. Specifically, we construct

---

**Algorithm 1:** FUSE: Multi-faceted Set Expansion

---

    **Input**: Corpus $D$; a user query $q$.
    **Output**: a list of expanded entity lists $\mathbb{E}$.
1    ⊡ *Skip-gram Clustering*;
2    seedClusterDict = {};
3    **for** *seed in q* **do**
4         sgs ← extractSkipgrams(*seed, D*);
5         sgClusters ← clustering(sgs);
6         seedClusterDict[*seed*] ← sgClusters;
7    ⊡ *Clusters Fusion*;
8    refSeed ← q.pop();
9    refC ← seedClusterDict[refSeed];
10   **while** *q is not empty* **do**
11        curSeed ← q.pop();
12        curC ← seedClusterDict[curSeed];
13        coherentC ← fuseClusters(refC, curC);
14        refC ← coherentC;
15   ⊡ *Entity Expansion*;
16   $\mathbb{E}$ ← entityExpansion(refC);
17   return $\mathbb{E}$;

---

a complete weighted graph where each node represents a skip-gram, and the edge weight between each pair of nodes indicates the cosine similarity of their corresponding skip-gram embeddings. After the weighted graph of skip-grams is constructed, the affinity propagation algorithm [6] is applied to find the best skip-gram clusters.[2]

Empirical results demonstrate that the affinity propagation based skip-gram clustering is able to identify a reasonable number of semantic facets (c.f. Sect. 4.4). We think one possible reason is that affinity propagation takes a similarity graph as its input, while most other non-parametric clustering algorithms (*e.g.*, MeanShift [4]) take skip-gram embeddings as its input. In such a clustering task, we are only interested in semantic similarities between skip-grams and do not care about the complete information of the skip-grams (*e.g.*, semantic and syntactic information). Therefore, though skip-gram embeddings contain more information than a similarity graph, the information serves more as "noise" and less as useful information. Moreover, the robustness of affinity propagation is immune to the dimension of the embeddings, while others can be very sensitive to it. In our experiment, we find MeanShift is highly unstable if the dimension is greater than 30. Hence affinity propagation, which takes similarities between pairs of data points as input, serves for our needs well.

---

[2] We set the preference to be −60.

## 3.2   Discovering Coherent Semantic Facets of a Seed Set

After obtaining multiple skip-gram clusters for each seed, we then need to determine the coherent semantic facets among all seeds and generate the coherent skip-gram clusters. Take two seed words "apple" with facets *fruit* and *company*, and "orange" with facets *fruit* and *color* as an example, their coherent semantic facet is *fruit*.

The key is to determine whether a facet of seed word $A$ matches any facet of word $B$. Suppose that $A$ has $r$ skip-gram clusters $S_A = \{S_A^{(1)}, \ldots, S_A^{(r)}\}$, where cluster $S_A^{(i)}$ contains a set of skip-grams relevant to the $i$-th facet of $A$. Similarly, $B$ has $t$ skip-gram clusters $S_B = \{S_B^{(1)}, \ldots, S_B^{(t)}\}$. If $A$ and $B$ share $k$ facets, and they have $k$ pairs of matching clusters $\{(S_A^{(i_1)}, S_B^{(j_1)}), \ldots, (S_A^{(i_k)}, S_B^{(j_k)})\}$ accordingly. Therefore, these $k$ facets are jointly represented by these clusters:

$$S_{A,B} = \{S_A^{(i_1)} \bigcup S_B^{(j_1)}, \ldots, S_A^{(i_k)} \bigcup S_B^{(j_k)}\}.$$

We first measure the pairwise correlation of their skip-gram clusters (c.f. Sect. 3.2), and then make a matching decision on a pair of clusters (c.f. Sect. 3.2).

**Calculating Correlation Between Two Skip-Gram Clusters.** Suppose that facet $A_1$ (one facet of word $A$) corresponds to a skip-gram cluster $\mathbf{X} = [\mathbf{x}_1; \ldots; \mathbf{x}_m]$ with $m$ skip-gram vectors, where $\mathbf{x}_i \in \mathbb{R}^d$. Similarly, facet $B_1$ (one facet of word $B$) corresponds to a cluster $\mathbf{Y} = [\mathbf{y}_1; \ldots; \mathbf{y}_n]$ with $n$ skip-gram vectors, where $\mathbf{y}_j \in \mathbb{R}^d$. Two clusters $\mathbf{X}$ and $\mathbf{Y}$ are from different seed words, and we want to measure their correlation in order to decide whether they correspond to the same semantic facet.

To measure their correlation, we find the semantic sense which $\mathbf{X}$ and $\mathbf{Y}$ have in common. Inspired by the idea of compositional semantics [10,26], we set the sense vector to the linear combination of skip-gram vectors.

Suppose that the sense vector $\mathbf{u}$ from cluster $\mathbf{X}$ and the sense vector $\mathbf{v}$ from $\mathbf{Y}$ are the sense shared by the two clusters. Therefore, the common sense vectors should be highly correlated, *i.e.*, we want to find $\mathbf{u}$ and $\mathbf{v}$ so that their correlation is maximized. We formulate the following optimization problem (1).

$$\max_{a,b} \frac{\mathbf{u}^T \mathbf{v}}{\|\mathbf{u}\| \cdot \|\mathbf{v}\|}$$
$$\text{s.t.} \quad \mathbf{u} = \mathbf{X}\mathbf{a},$$
$$\mathbf{v} = \mathbf{Y}\mathbf{b}, \tag{1}$$

where $\mathbf{a} \in \mathbb{R}^m$ and $\mathbf{b} \in \mathbb{R}^n$ are coefficient vectors.

Solving the problem (1) by CCA [8], we can find their common sense vectors $\mathbf{u}^*$ and $\mathbf{v}^*$. The semantic correlation $corr(\mathbf{X}, \mathbf{Y})$ between cluster $\mathbf{X}$ and $\mathbf{Y}$ is defined as the correlation between these two sense vectors:

$$corr(\mathbf{X}, \mathbf{Y}) = \mathbf{u}^{*T} \mathbf{v}^* \tag{2}$$

**Matching Facets of All Seeds.** After quantifying correlation for two skip-gram clusters, we cast it as a binary decision whether the cluster $\mathbf{X}$ of facet $A_1$ matches semantically with any facet of word $B$.

We note that it is not a good way to decide the matching clusters by setting a hard correlation threshold, since the numerical correlation range is word-specific. It is easy to see that if a facet of seed $A$ (*e.g.*, $A_1$) is of the same semantic class with a facet of seed $B$ (*e.g.*, $B_2$), then $corr(A_1, B_2)$ is higher than the correlation between $A_1$ and any other facets of seed $B$. Otherwise, the correlation of $A_1$ and all facets of seed $B$ should be equally small.

Based on the intuition above, we define a relevance score below:

$$rele(A_1, B) = D_{KL}(\mathbf{Corr}(A_1, B), U) \tag{3}$$

where $U$ is uniform distribution, $D_{KL}$ is KL-divergence [13], and $\mathbf{Corr}(A_1, B) = softmax((corr(A_1, B_1), ..., corr(A_1, B_m)))$,

We then make the matching decision based on the relevance score $rele(A_1, B)$. The threshold of the relevance score is set to 0.25 empirically. Once the matching decision is satisfied, we find the best matching facet $B^*$ in word $B$ and generate one coherent skip-gram cluster $A_1 \bigcup B^*$. Finally, we do facet matching for all facets of word $A$ and obtain the resulting skip-gram clusters as coherent skip-gram clusters.

**Remarks:** If there are more than two seed words, we first discover coherent skip-gram clusters of two seeds and then use their coherent skip-gram clusters to match with the third seed and so on.

### 3.3 Entity Expansion

The coherent skip-gram clusters of different facets are used to expand the seed set. Traditionally, researchers expands entities from a group of skip-gram clusters based on graph-based approaches [32], entity matching approaches [24], distributional hypothesis [21] and iterative approaches [9,33]. As a distinction, we tackle the problem by a Masked Language Model (MLM), which leverages the discriminative power of the BERT model and contextual representations [5].

For each skip-gram denoted as $sg$, we compute the MLM probability $h_{c,sg}$ for each word candidate $c$ in the vocabulary by a pre-trained BERT model.[3] Therefore, given a set of skip-grams, the weight $w_c$ of a word candidate $c$ is calculated as: $w_c = \sum_{sg'} h_{c,sg'}$. The final entity expansion process simply ranks all word candidates by their weights.

## 4   Experiments

Our model targets the corpus-based entity set expansion problem, and thus we evaluate its performance on a local corpus.

---

[3] We use the 'bert-base-uncased' pre-trained model from https://huggingface.co/transformers/model_doc/bert.html#bertformaskedlm.

**Table 1.** Evaluation using MMAP ("recall"), PMAP ("precision") and BMAP ("F1 score").

| | | MMAP@$l$ | | | PMAP@$l$ | | | BMAP@$l$ | | |
|---|---|---|---|---|---|---|---|---|---|---|
| | | $l = 5$ | $l = 10$ | $l = 20$ | $l = 5$ | $l = 10$ | $l = 20$ | $l = 5$ | $l = 10$ | $l = 20$ |
| Single-faceted | word2vec | 0.323 | 0.283 | 0.252 | 0.552 | 0.499 | 0.448 | 0.390 | 0.352 | 0.316 |
| | SEISA | 0.345 | 0.301 | 0.268 | 0.550 | 0.503 | 0.455 | 0.408 | 0.368 | 0.331 |
| | SetExpan | 0.373 | 0.337 | 0.304 | 0.605 | 0.563 | **0.512** | 0.448 | 0.413 | 0.374 |
| Multi-faceted | Sensegram | 0.312 | 0.301 | 0.275 | 0.479 | 0.443 | 0.398 | 0.359 | 0.343 | 0.314 |
| | EgoSet | 0.446 | 0.390 | 0.325 | 0.306 | 0.261 | 0.206 | 0.335 | 0.292 | 0.236 |
| | FUSE | **0.477** | **0.414** | **0.366** | **0.643** | **0.573** | 0.507 | **0.531** | **0.469** | **0.414** |
| Ablations | FUSE-k ($k = 2$) | 0.420 | 0.364 | 0.326 | 0.607 | 0.540 | 0.494 | 0.478 | 0.422 | 0.383 |
| | FUSE-k ($k = 3$) | 0.454 | 0.406 | 0.360 | 0.624 | 0.562 | 0.505 | 0.504 | 0.455 | 0.407 |

**Dataset:** We evaluate our approach, FUSE, on the dataset in [21]. The dataset contains 56 million articles (1.2 billion words) retrieved from English Wikipedia 2014 Dump and 150 human-labeled multi-faceted queries.

## 4.1 Evaluation Metric

It is considerably complicated to properly evaluate multi-faceted set expansion task due to different number of facets between the generated result and the ground truth. Previous work [21] adopted the following Mean Mean Average Precision (MMAP) measure:

$$MMAP@l = \frac{1}{M_q} \sum_{m=1}^{M_q} AP_l(B_{qi^*}, G_{qm}),$$

where $M_q$ is the number of facets for query $q$ in the ground truth; $G_{qm}$ is the ground truth set of $m$-th facet for $q$; $B_{qi^*}$ is the output facet that best matches $G_{qm}$, and $AP_l(c, r)$ represents the average precision of top $l$ entities in a ranked list $c$ given an unordered ground truth set $r$. This metric measures the coverage of ground truth sets by the generated sets.

However, it does not penalize additional noisy facets in generated sets and thus it is biased towards the model that generates more facets. For example, a model generating 30 facets with 3 relevant facets achieves higher MMAP than another model generating 3 facets with 2 relevant facets. One can "cheat" the performance by generating as many facets as possible.

To overcome the intrinsic limitation of MMAP, we, inspired by [3,7], propose a new metric, Best-Matching Average Precision (BMAP) to capture both the purity of generated facets and their coverage of ground truth facets. Our metric is defined as follows:

$$BMAP@l = HMean(MMAP@l, PMAP@l),$$

$$PMAP@l = \frac{1}{F_q} \sum_{f=1}^{F_q} AP_l(B_{qf}; G_{qi^*}),$$

where $F_q$ is the number of facets in generated output; $B_{qf}$ is the $f$-th output ranked list for query $q$; $G_{qi^*}$ is the ground truth facet that best matches $B_{qf}$. Here $HMean(a,b) = \frac{2ab}{a+b}$ is the harmonic mean of $a$ and $b$.

Our proposed BMAP metric not only evaluates how well generated facets match the ground truth by $MMAP@l$ but also penalizes low-quality facets by $PMAP@l$. Intuitively, $MMAP@l$ measures "recall" to capture how many ground truth results has been discovered, while $PMAP@l$ measures "precision" to capture the fraction of good facets in the generated output. Accordingly, $BMAP@l$ measures "F1 score" to leverage "precision" and "recall". Results are reported by averaging all queries for each dataset.

## 4.2   Baselines

The following approaches are evaluated:

- **word2vec**[4] [16]: We use the "skip-gram" model in word2vec to learn the embedding vector for each entity, and then return $k$ nearest neighbors of the seed words.
- **SEISA** [9]: An entity set expansion algorithm based on iterative similarity aggregation. It uses the occurrence of entities in web list and query log as entity features. In our experiments, we replace the web list and query log with skip-gram features.
- **SetExpan**[5] [24]: A corpus-based set expansion that selects quality context features for entity-entity similarity calculation and expand the entity sets using rank ensemble.
- **EgoSet** [21]: The only existing work for multi-faceted set expansion. It expands word entities from skip-gram features, and then clusters the expanded entities into multiple sets by the Louvain community detection algorithm.
- **Sensegram**[6] [18]: We learn different embeddings for each word's different senses and return $k$ nearest neighbors for each embedding.
- **FUSE-k:** A variant of FUSE which replaces Affinity Propagation with $k$-means clustering algorithm for skip-gram clustering.

## 4.3   Results

We compare the performance of FUSE against all baselines using MMAP ("recall"), PMAP ("precision") and BMAP ("F1 score"). As shown in Table 1, FUSE achieves the highest scores in most cases and outperforms all other baselines in BMAP and MMAP.

It is worth mentioning that EgoSet achieves decent results in MMAP. However, it generates too many noisy facets, which deteriorate PMAP and the overall performance BMAP. We will further discuss this phenomenon in Sect. 4.4.

---

[4] https://code.google.com/p/word2vec.
[5] https://github.com/mickeystroller/SetExpan.
[6] https://github.com/uhh-lt/sensegram.

It is also interesting to note that single-faceted baselines (*i.e.*, SetExpan, SEISA) have much stronger PMAP performance than multi-faceted baselines. This is because by generating a single cluster of the most confident expansion results, they usually match with one ground truth cluster very well and thus achieve high PMAP ("precision") value.

In the ablation analysis, it is worth noting that FUSE, even without predetermined number of clusters, performs better than FUSE-k. We experiment the number of clusters $k$ to be 2 and 3, which are the mode and the mean of the number of clusters of the ground truth respectively. We think the poor performance is because forcing skip-grams into a fixed number of clusters will induce clustering noise. Furthermore, the noise will propagate and be enlarged in the skip-gram cluster fusion step and the entity expansion step, and therefore leads to bad performance.

### 4.4   Number of Facets Identified

We explore the number of facets identified by different multi-faceted set expansion methods. Specifically, we adopt $l_1$ and $l_2$ distances.

$$l_1 \text{ distance} = \sum_{q \in Q} |\text{GT}_q - \text{Gen}_q|$$

$$l_2 \text{ distance} = \sqrt{\sum_{q \in Q} (\text{GT}_q - \text{Gen}_q)^2}$$

Here $Q$ is all queries, $\text{GT}_q$ and $\text{Gen}_q$ are the number of facets that ground truth has and the number of facets that the corresponding model identifies for query $q$, respectively.

The distance measurement are summarized in Table 2. FUSE generates closer number of facets to the ground truth, compared to EgoSet, demonstrating about 65% reduction of the $l_1$ distances and 45% reduction of the $l_2$ distances.

**Table 2.** Distance between the number of facets identified and the number of facets the ground truth has.

|         | $l_1$ distance | $l_2$ distance |
|---------|----------------|----------------|
| EgoSet  | 783            | 78.02          |
| FUSE    | **277**        | **43.05**      |

To further explore identified facets between FUSE and EgoSet, we present one case study of query {"berkeley"} in Table 3. FUSE generates two facets and each facet has its distinctive semantic meaning (one for *Top universities in the US* and one for *Cities in California, US*), while EgoSet generates too many

**Table 3.** Case study on comparison between FUSE and EgoSet.

| Query | FUSE Identified Facets and Their Associated Top Skip-grams | EgoSet Identified Facets and Their Associated Top Skip-grams |
|---|---|---|
| {berkeley} | Facet 1 (*Top universities in the US*): columbia_university, harvard, new_york, yale_university, princeton, harvard_university, ucla, stanford_university, columbia, boston_college, nyu, cornell_university, indiana_university, johns_hopkins_university, georgetown, georgetown_university, ... **Top Skip-grams:** ('at __ , yale'), ('graduate school at __ ,'), ('former __ professor'), (', __ , harvard university'), (', __ campus .') | Facet 1: los_angeles, san_diego, san_francisco, santa_barbara, santa_cruz, riverside, sacramento, ... **Top Skip-grams:** ('in __ , ca.'), ('to __ , california .') |
| | | Facet 2: stanford, harvard_university, yale_university, columbia, cornell_university, mit, ucsd, ... **Top Skip-grams:** ('the __ campus.'), ('at __ 's school of') |
| | | Facet 3: chicago, san_leandro, huntington_beach, los_angeles_county, oxnard, van_nuys, ... **Top Skip-grams:** ('in __ , los'), ('in __ , los angeles') |
| | Facet 2 (*Cities in California, US*): los_angeles, san_diego, santa_barbara, san_francisco, irvine, santa_cruz, riverside, palo_alto, long_beach, oakland, pasadena, san_jose, fresno, ... **Top Skip-grams:** ('of california , __ .'), ('area of __ , california'), ('in __ , california is'), ('california , __ and the'), ('founded in __ , california') | Facet 4: culver_city, santa_ana, costa_mesa, alameda, redondo_beach, san_rafael, fullerton, redlands, ... **Top Skip-grams:** ('city of __ , california'), ('in __ , california where') |
| | | Facet 5: california_state_university, marine_corps_recruit_depot, howard_university, california_polytechnic_state_university, uc_santa_barbara, scripps, ... **Top Skip-grams:** ('at __ , san'), ('of california , __ ,') |
| | | Facet 6: ojai, yuba_city, whittier_college, pasadena_city_college, mather_air_force_base, moffett_field, beale_afb, march_field, beale_air_force_base, fort_ord, ... **Top Skip-grams:** ('at __ , california.'), ('at __ , california in') |

scattered facets, with less distinctiveness between facets. One of the reasons is that EgoSet performs clustering on expanded entities while FUSE performs clustering on skip-grams. Skip-grams, consisting of multiple words, are usually of more clear semantics compared to entities themselves (*e.g.*, the entity "columbia" can be either *Universities*, *Cities* or *Rivers*, while the skip-gram "graduate school at __" is more clear to be *Universities*). Therefore, clustering on skip-grams is an easier process and results in better performance.

### 4.5 Case Studies: Multi-faceted Setting

Table 4 shows intermediate results of FUSE by listing top skip-grams of each semantic facet. It is worth noting that even the ground truth may not present a full coverage of semantic facets of given seeds. For example, as shown in Case 4, the ground truth only includes semantic facet *Animals*. Our system also finds another meaningful semantic facet *Tributaries*.[7] The query { "Chongqing" } shown in Case 5 is another example, where the ground truth again fails to capture the semantic facet of *War-related Major Cities*.[8]

---

[7] Beaver River, Elk River and Bear River are tributaries of Pennsylvania, Mississippi River and the Great Salt Lake, respectively.

[8] Chongqing was the second capital of Chinese nationalist party during the war.

**Table 4.** Case studies on top skip-grams for each semantic facet. The concept name of each facet is in bold.

| ID | Query | FUSE Identified Facets: Associated Top Skip-grams | Ground Truth Facets: Example Entities |
|---|---|---|---|
| 1 | {hydrogen, uranium} | **Chemical Elements**: (helium , __ ,), (of __ and nitrogen.)<br>**Energy Sources**: (for __ energy), (in __ energy), (the __ fuel cell) | **Chemical Elements**: Helium, Carbon, Nitrogen, Oxygen<br>**Energy Sources**: Solar, Coal, Oil, Natural Gas |
| 2 | {apollo, artemis, poseidon} | **Major gods**: (the god __ and), (zeus , __ ,), (, athena , __ ,)<br>**NASA missions**: (nasa 's __ ,) | **Major Greek gods**: Aphrodite, Ares, Athena, Zeus<br>**NASA missions**: Juno, Voyager, InSight, NuSTAR |
| 3 | {Beijing} | **Chinese Major Cities**: (in __ , china), (, __ , shanghai)<br>**Capitals/International Major Cities**: (paris and __ .), ( __ capital)<br>**International Metropolitan/Art Cities**: (theater in __ .), (of music in __ .), (with the __ symphony orchestra)<br>**Olympic Games Host Cities**: (olympic games in __ .), (at the __ olympic) | **Chinese Major Cities**: Beijing, Shanghai, Wuhan, Harbin<br>**Province-level divisions of China**: Beijing, Jiangsu, Zhejiang<br>**Capital cities in the world**: Paris, Tokyo, Berlin<br>**Olympic Games Host Cities**: Paris, Tokyo, Munich, Calcary |
| 4 | {beaver, elk, bear} | **Animals**: (tailed deer , __ ,), (wolf , __ and)<br>**Tributaries**: (in __ river,), (along the __ river.) | **Animals**: alligator, bear, deer, pig |
| 5 | {Chongqing} | **Chinese Major Cities**: (of __ , china.), (based in __ , china.)<br>**War related Major Cities**: ( __ broadcasting), (congress of __ .), (party in __ led by) | **Chinese Major Cities**: Beijing, Shanghai, Wuhan, Harbin<br>**Province-level divisions of China**: Beijing, Jiangsu, Zhejiang, Guangdong |
| 6 | {fox} | **Animals**: (, __ , wolf ,), (, __ , badger), (deer , __ , and)<br>**Popular TV networks**: (in the __ sitcom), (on the __ tv), (the __ drama series)<br>**English Surnames**: (officer of __ .), (by david __ .) | **Animals**: alligator, bear, deer, pig<br>**Popular TV networks**: NBC, CBS, CNN<br>**Common English surnames**: Smith, Jones, Williams<br>**Snake species**: Blind snake, Sea snake, Python |

## 5    Single-Faceted Set Expansion

In previous sections, we have demonstrated that FUSE has favorable performance against state-of-the-art systems in expanding multiple semantic facets of a seed set. Yet, it's more common in real-life cases that a seed set has one single semantic facet (especially when more seeds are provided). In this subsection, we inspect the performance of FUSE on single-faceted cases.

In the singled-faceted set expansion task, there is exactly one semantic facet from a seed set. However, if one or more words in the seed set are ambiguous, such ambiguity will introduce entities related to noisy facets, and thus hurt the quality of the expanded set. For example, the seed set {"apple", "amazon"} has only one semantic facet corresponding to *Technology Companies*, however, the seed "apple" is an ambiguous word and has a noisy facet, *i.e.*, *Fruits*. Most existing systems [2,9,15,17,21,24,29,31–33] first extract all contextual features (*e.g.*, skip-grams) from the entire seed set and then rank the keyword candidates based on the contextual features. To the best of our knowledge, none of them denoises contextual features from the noisy facet (*i.e.*, *Fruits*). Therefore, they are likely to generate entities related to the facet of *Fruits* (*e.g.*, "pear", "banana"), despite the fact that the seed set {"apple", "amazon"} has only one semantic facet.

In contrast, FUSE is robust to such lexical ambiguity, since we discover the shared coherent semantic facet across all seeds and expand entities by relevant contextual features only. From this example, one can clearly see that even for single-faceted set expansion, it is also critical to resolve the lexical ambiguity and identify the common facet among seeds.

To gain deeper insights in the single-faceted setting, we present a case study on the seed set { "apple", "amazon" } in Table 5. We highlight those noisy entities resulting from seed ambiguity (*i.e.*, the *Fruits* sense) in bold, bright red. As expected, EgoSet has 2 facets that contain entites from the *Fruits* sense and SetExpan suffers from the noisy facet issue too. As it has shown, FUSE performs favorably against previous approaches in single-faceted set expansion too, in that it is robust to semantic ambiguity by extracting the coherent semantic facet shared by all seed words.

**Table 5.** Case studies on single-faceted set expansion: { "apple", "amazon" }.

| Approach | Expanded Entities of query ("apple", "amazon") |
|---|---|
| Ground Truth | *Technology Companies*: samsung electronics, foxconn, hp, lbm, amazon, microsoft, sony, panasonic, dell, intel, toshiba, ... |
| FUSE | google, microsoft, sony, ibm, intel, general_electric, motorola, ebay, hewlett_packard, coca_cola, syntel, idw_publishing, corus_entertainment, black_isle, independent_record_label, grunewald, facebook, ... |
| SetExpan | united_states, **pear**, blackberry, american, **strawberry**, palm, **blueberry**, green, **cherry**, google, company, **banana**, gnu/linux, ... |
| EgoSet | *Facet 1*: microsoft, google, ibm, compaq, oracle, avaya, ebay, motorola, netflix, aol, virtualization, novell, general_mills, digital, ... <br> *Facet 2*: amazon_mp3, napster, adobe_photoshop, imovie, windows_media_player, itunes, iphone, mozilla_application_suite, ... <br> *Facet 3*: cp_/_m, microsoft_windows, apple_inc, as/400, vms, openstep, oem, vax, qnx, android, solaris, unix, smartphone, ... <br> *Facet 4*: amiga, autodesk, gp, microcontroller, laptop, smithfield_foods, liberty_global, medtronic, digital_research, dr_pepper, ... <br> *Facet 5*: fasta, ntfs, wav, gemstone, nautilu <br> *Facet 6*: **avocado, fruit**, palm, blackberry, **tapioca**, manure, cochineal, musk, chocolate, horse_meat, symantec, **lime_juice**, ... <br> *Facet 7*: asia, milk, north_america, **blueberry**, **pomegranate**, **grapefruit**, flash, mobile, betty_crocker, xcode, cream_soda, pxe, ... <br> *Facet 8*: fish, french_fry, consumer_electronic, debian, burger_king, poultry, coconut_oil, kool_aid, lobster, starch, byte, beef, ... |

# 6   Related Work

Early work on entity set expansion, including *Google Sets* [29], *SEAL* [32], and *Lyretail* [2] submits a query consisting of seed entities to a general-domain search engine (*e.g.*, Google) and then mines the returned, top-ranked web pages. These approaches depend on the external search engine and require costly online data extraction.

Later studies shift to the *corpus-based* set expansion setting, where sets are expanded within a given domain-specific corpus. For example, [17] compute the semantic similarity between two entities based on their local contexts and treat the nearest neighbors around the seed entities as the expanded set. [9] further extend this idea by proposing an iterative similarity aggregation function to calculate entity-entity similarity using query logs and web lists besides free text. More recently, [24,25] propose to compute semantic similarity using only selected high-quality context features, and [14,15] develop SetExpander system to leverage multi-context term embedding for entity set expansion. All the above attempts, however, assume the input seed entities belong to one unique, clear semantic class, and thus largely suffer from the multi-faceted nature of these seeds – they could represent multiple semantic meanings.

To resolve the ambiguity of seeds, [31] propose to utilize the target semantic facet name and then retrieve its most relevant web tables. However, it requires the exact name of the target semantic facet and outputs the one semantic facet of entities only. This does not accomplish multi-faceted set expansion in nature.

The only attempt towards multi-faceted set expansion is EgoSet [21], to the best of our knowledge. EgoSet first extracts quality skip-gram features to construct an entity-level ego-network, and then perform the Louvain community detection algorithm on the ego-network to extract entities for different facets. Finally, it combines them with external knowledge (*i.e.*, user-generated ontologies) to generate final output. Although FUSE may appear to be similar to EgoSet at first glance, we highlight key differences and significance below:

- **Key challenges of multi-faceted set expansion:** We identify the key challenge of multi-faceted set expansion to be discovery of shared semantic facets from a seed set. While in EgoSet, noisy facets that are relevant only to partial seeds are also generated. As a result, it fails to solve multi-seed single-faceted cases (*e.g.*, {"apple", "amazon"}) and multi-seed multi-faceted cases (*e.g.*, Fig. 1).
- **External knowledge:** EgoSet requires user-created ontology (obtained from Wikipedia) as external knowledge. While these semi-structured web tables and ontologies are helpful for disambiguation, they are not always available for domain-specific corpus. FUSE relies on free text only and thus, can be applied in a more general setting.
- **Clustering over skip-grams:** EgoSet adopts clustering (community detection) over expanded entities, while FUSE adopts clustering over skip-grams. Clustering over entities usually leads to mediocre results in non-parametric settings, since any expanded entity can be ambiguous. However, skip-grams, consisting of multiple words, are usually of more clear semantics and much easier to be clustered compared to entities themselves (demonstrated in Sect. 4.4). In additional, EgoSet adopts hard clustering on entities, which ignores the nature that the same entity may fall into different facets (*e.g.*, "Paris" should appear in both sets of *National Capitals* and *Olympic Games Host Cities* in Fig. 1), while the design of FUSE naturally allows the same entities to appear in multiple facets.

In a more general way, our work is also related to word sense disambiguation [11,18–20,28]. The major difference is that our work aims to find the coherent semantic facets of all seed words and achieve entity expansion from the coherent skip-gram clusters.

## 7   Conclusion

We identify the key challenge of the problem – *multi-faceted set expansion* and have proposed a novel and effective approach, FUSE, to address it. By extracting and clustering skip-grams for each seed, identifying coherent semantic facets of all seeds, and expanding entity sets for each semantic facet, FUSE is capable of identifying coherent semantic facets, generating quality entity set for each facet, and therefore outperforms previous state-of-the-art approaches significantly.

The proposed framework FUSE is general in that it achieves quality set expansion in both multi-faceted and single-faceted settings. In particular, it, for the first time, solves the case where different seeds have different multi-facetedness. For future work, we plan to explore other skip-gram clustering approaches and coherent semantic facet discovery algorithms.

**Acknowledgments.** Research was sponsored in part by US DARPA KAIROS Program No. FA8750-19-2-1004 and SocialSim Program No. W911NF-17-C-0099, National Science Foundation IIS 16-18481, IIS 17-04532, and IIS-17-41317, and DTRA HDTRA11810026.

# References

1. Cao, H., et al.: Context-aware query suggestion by mining click-through and session data. In: KDD (2008)
2. Chen, Z., Cafarella, M., Jagadish, H.: Long-tail vocabulary dictionary extraction from the web. In: WSDM (2016)
3. Chinchor, N.: Muc-4 evaluation metrics. In: Proceedings of the 4th Conference on Message Understanding. Association for Computational Linguistics (1992)
4. Comaniciu, D., Meer, P.: Mean shift: a robust approach toward feature space analysis. IEEE Trans. Pattern Anal. Mach. Intell. **24**(5), 603–619 (2002)
5. Devlin, J., Chang, M.W., Lee, K., Toutanova, K.: Bert: pre-training of deep bidirectional transformers for language understanding. In: NAACL (2019)
6. Frey, B.J., Dueck, D.: Clustering by passing messages between data points. Science **315**(5814), 972–976 (2007)
7. Goldberg, M.K., Hayvanovych, M., Magdon-Ismail, M.: Measuring similarity between sets of overlapping clusters. In: 2010 IEEE Second International Conference on Social Computing (SoialCom), pp. 303–308. IEEE (2010)
8. Hardoon, D.R., Szedmak, S., Shawe-Taylor, J.: Canonical correlation analysis: an overview with application to learning methods. Neural Comput. **16**(12), 2639–2664 (2004)
9. He, Y., Xin, D.: SEISA: set expansion by iterative similarity aggregation. In: WWW (2011)
10. Hermann, K.M., Blunsom, P.: Multilingual models for compositional distributed semantics. arXiv preprint arXiv:1404.4641 (2014)
11. Iacobacci, I., Pilehvar, M.T., Navigli, R.: Embeddings for word sense disambiguation: an evaluation study. In: ACL (2016)
12. Jain, A., Pennacchiotti, M.: Open entity extraction from web search query logs. In: COLING (2010)
13. Kullback, S., Leibler, R.A.: On information and sufficiency. Ann. Math. Stat. **22**(1), 79–86 (1951)
14. Mamou, J., et al.: SetExpander: end-to-end term set expansion based on multi-context term embeddings. In: COLING (2018)
15. Mamou, J., et al.: Term set expansion based NLP architect by Intel AI Lab. In: EMNLP (2018)
16. Mikolov, T., Sutskever, I., Chen, K., Corrado, G.S., Dean, J.: Distributed representations of words and phrases and their compositionality. In: NIPS (2013)
17. Pantel, P., Crestan, E., Borkovsky, A., Popescu, A.M., Vyas, V.: Web-scale distributional similarity and entity set expansion. In: EMNLP (2009)

18. Pelevina, M., Arefyev, N., Biemann, C., Panchenko, A.: Making sense of word embeddings. In: Rep4NLP@ACL (2016)
19. Raganato, A., Bovi, C.D., Navigli, R.: Neural sequence learning models for word sense disambiguation. In: EMNLP (2017)
20. Raganato, A., Camacho-Collados, J., Navigli, R.: Word sense disambiguation: a unified evaluation framework and empirical comparison. In: EACL (2017)
21. Rong, X., Chen, Z., Mei, Q., Adar, E.: EgoSet: exploiting word ego-networks and user-generated ontology for multifaceted set expansion. In: WSDM (2016)
22. Sarker, A., Gonzalez, G.: Portable automatic text classification for adverse drug reaction detection via multi-corpus training. J. Biomed. Inform. **53**, 196–207 (2015)
23. Sarmento, L., Jijkoun, V., de Rijke, M., Oliveira, E.C.: "More like these": growing entity classes from seeds. In: CIKM (2007)
24. Shen, J., Wu, Z., Lei, D., Shang, J., Ren, X., Han, J.: SetExpan: corpus-based set expansion via context feature selection and rank ensemble. In: Ceci, M., Hollmén, J., Todorovski, L., Vens, C., Džeroski, S. (eds.) ECML PKDD 2017. LNCS (LNAI), vol. 10534, pp. 288–304. Springer, Cham (2017). https://doi.org/10.1007/978-3-319-71249-9_18
25. Shen, J., et al.: HiExpan: task-guided taxonomy construction by hierarchical tree expansion. In: Proceedings of the 24th ACM SIGKDD International Conference on Knowledge Discovery and Data Mining, pp. 2180–2189 (2018)
26. Socher, R., Karpathy, A., Le, Q.V., Manning, C.D., Ng, A.Y.: Grounded compositional semantics for finding and describing images with sentences. Trans. Assoc. Comput. Linguist. **2**(1), 207–218 (2014)
27. Steinbach, M., Ertöz, L., Kumar, V.: The challenges of clustering high dimensional data. In: Wille L.T. (eds.) New Directions in Statistical Physics. Springer, Berlin, Heidelberg (2004). https://doi.org/10.1007/978-3-662-08968-2_16
28. Taghipour, K., Ng, H.T.: Semi-supervised word sense disambiguation using word embeddings in general and specific domains. In: NAACL-HLT (2015)
29. Tong, S., Dean, J.: System and methods for automatically creating lists. US Patent 7,350,187 (2008)
30. Velardi, P., Faralli, S., Navigli, R.: Ontolearn reLoaded: a graph-based algorithm for taxonomy induction. Comput. Linguist. **39**(3), 665–707 (2013)
31. Wang, C., Chakrabarti, K., He, Y., Ganjam, K., Chen, Z., Bernstein, P.A.: Concept expansion using web tables. In: WWW (2015)
32. Wang, R.C., Cohen, W.W.: Language-independent set expansion of named entities using the web. In: ICDM (2007)
33. Wang, R.C., Cohen, W.W.: Iterative set expansion of named entities using the web. In: ICDM (2008)
34. Weikum, G., Theobald, M.: From information to knowledge: harvesting entities and relationships from web sources. In: PODS (2010)
35. Zhao, J., Liu, K., Zhou, G., Qi, Z., Liu, Y., Han, X.: Knowledge extraction from wikis/bbs/blogs/news web sites (2014)
36. Zhu, W., Zhang, C., Yao, S., Gao, X., Han, J.: A spherical hidden Markov model for semantics-rich human mobility modeling. In: AAAI (2018)

# Hierarchical Interaction Networks with Rethinking Mechanism for Document-Level Sentiment Analysis

Lingwei Wei[1,3], Dou Hu[2]($\boxtimes$), Wei Zhou[1]($\boxtimes$), Xuehai Tang[1], Xiaodan Zhang[1], Xin Wang[1], Jizhong Han[1], and Songlin Hu[1,3]

[1] Institute of Information Engineering, Chinese Academy of Sciences, Beijing, China
zhouwei@iie.ac.cn
[2] National Computer System Engineering Research Institute of China, Beijing, China
hudou18@mails.ucas.edu.cn
[3] School of Cyber Security, University of Chinese Academy of Sciences,
Beijing, China

**Abstract.** Document-level Sentiment Analysis (DSA) is more challenging due to vague semantic links and complicate sentiment information. Recent works have been devoted to leveraging text summarization and have achieved promising results. However, these summarization based methods did not take full advantage of the summary including ignoring the inherent interactions between the summary and document. As a result, they limited the representation to express major points in the document, which is highly indicative of the key sentiment. In this paper, we study how to effectively generate a discriminative representation with explicit subject patterns and sentiment contexts for DSA. A Hierarchical Interaction Networks (HIN) is proposed to explore bidirectional interactions between the summary and document at multiple granularities and learn subject-oriented document representations for sentiment classification. Furthermore, we design a Sentiment-based Rethinking mechanism (SR) by refining the HIN with sentiment label information to learn a more sentiment-aware document representation. We extensively evaluate our proposed models on three public datasets. The experimental results consistently demonstrate the effectiveness of our proposed models and show that HIN-SR outperforms various state-of-the-art methods.

**Keywords:** Document-level sentiment analysis · Rethinking mechanism · Document representation

## 1 Introduction

Document-level Sentiment Analysis (DSA), a subtask of Sentiment Analysis (SA), aims to understand user attitudes and identify sentiment polarity expressed at document-level. This task has grown to be one of the most active

---

L. Wei and D. Hu—Equal Contribution.

© Springer Nature Switzerland AG 2021
F. Hutter et al. (Eds.): ECML PKDD 2020, LNAI 12459, pp. 633–649, 2021.
https://doi.org/10.1007/978-3-030-67664-3_38

research areas in natural language processing and plays an important role in many real-world applications such as intent identification [1], recommender systems [2] and misinformation detection [3, 4].

Generally, DSA can be regarded as a text classification problem and thus traditional text classification approaches [5–8] can be adopted naturally. Different from other subtasks in SA, DSA is more challenging due to the large size of words, vague semantic links between sentences and abundance of sentiments. Hence, the research question that how to learn an expressive document representation to understand long documents for sentiment classification has been given a growing interest to researchers.

Inspired by the document structure, one of the earliest and most influential models HAN was proposed by [9] to encode the entire document, which suffered from attending explicit but irrelevant sentimental words. Subsequent works mainly dedicated to introducing latent topics [10] or global context [11] to tackle this limitation. However, most of the user-generated documents are very comprehensive and contain a wealth of sentiments, which makes it difficult to directly learn from the whole document without the understanding of the major points, especially in long documents. The above works attempted to explore the major points of the document by learning a global embedding from the document. Intuitively, the user-generated summary contains more accurate information about the major points of the document. These summaries describe the long document in a more specific way, which are highly indicative of the key sentiment and subject, and can facilitate further to identify important text present and sentiments.

To reduce the processing of the substantial text in the document and be well aware of the major idea of it, summary-based methods [12, 13] introducing the user-generated summary has been developed for DSA and achieved promising results, which brings brilliant processing for understanding complex documents. They refined the document with an abstractive summary to predict the sentiment effectively. Recently, some effective works [14, 15] concerned both the text summarization task and DSA, and jointly modeled them to boost from each other. Nevertheless, most of the joint models did not fully utilize user-generated summaries and ignored the interaction between summary and document, because they did not encode summaries explicitly during test time.

For example, the document of a product review in Amazon SNAP Review Dataset is "... They just sent a new camera and it showed up without any warning or communication about the bad one. Minimal *Customer Service*... The 1st camera was promising and worked so well for about two weeks..." and its length is 1134. The corresponding summary is "Quality is a reflection of *Customer Service*". The document contains complex sentiment expressed, such as *promising* or *bad* corresponding to positive or negative, respectively. The subject *Customer Service* can help better predict the key sentiment of the document, such as *bad* and *minimal*. Meanwhile the document can supplement ambiguously semantic features in the summary, such as the details of *Customer Service*. They are complementary. Therefore, the auxiliary of the summary is significant for subject mining and semantic understanding in DSA.

To tackle the aforementioned problems, we investigate how to effectively focus on more accurate subject information for DSA. In this paper, we present an end-to-end model, named Hierarchical Interaction Network (HIN), to encode the bidirectional interactions between summary and document. The method works by utilizing multiple granularities interactions between summary and document, accordingly to learn a subject-oriented document representation. Specifically, the interactions at character-level and the contextual semantic features can be captured via BERT [7]. Afterward, the segment-level interactions are encoded by gated interaction network and the context of the document is taken into account simultaneously. Finally, the document-level interactions are embedded via the attention mechanism to learn a more expressive document representation with the consideration of subject information for predicting sentiments.

Furthermore, because of the complex sentiment in the document, we attempt to learn the affective representation and alleviate the distraction from other sentiments. We introduce the sentiment label information into the model in a feedback way. Most existing structures learn document representation via only feedforward networks and have no chance to modify the invalid features of the document. Some effective works in image classification [16] and named entity recognition [17] added feedback structure to re-weight of feature embeddings and obtained gratifying results. Motivated by their works, we propose a Sentiment-based Rethinking mechanism (SR) and feedback the sentiment polarity label information. This rethinking mechanism can refine the weights of document embeddings to learn a more discerning low-level representation with the guidance from the high-level sentiment features, and relieve negative effects of noisy data simultaneously.

We evaluate our proposed models on three public datasets from news to reviews. Through experiments versus a suite of state-of-the-art baselines, we demonstrate the effectiveness of interactions and the rethinking mechanism, and the model HIN-SR can significantly outperform than baseline systems.

The main contributions of this paper are summarized as follows.

- We present HIN to incorporate multi-granularities interactions between summary and document-related candidates and learn subject-oriented document representations for document-level sentiment analysis. To the best of our knowledge, we are the first neural approach to DSA that models bidirectional interactions to learn the document representation.
- We propose SR to adaptively refine the HIN with high-level features to learn a more discriminative document representation and can relieve the negative impacts of noisy samples simultaneously.
- We evaluate our model on three public datasets. Experimental results demonstrate the significant improvement of our proposed models than state-of-the-art methods. The source code and dataset are available[1].

The remainder of this paper is structured as follows. We review the related works in Sect. 2. Then we explain the details of our contributions in Sect. 3.

---

[1] https://github.com/zerohd4869/HIN-SR.

Section 4 describes the experiments and the results. Further analysis and discussion are shown in Sect. 5. In the end, the conclusions are drawn in Sect. 6.

## 2 Related Works

### 2.1 Document-Level Sentiment Analysis

In the literature, traditional methods regarded document-level sentiment analysis (DSA) as a text classification problem. Therefore, typical text classification approaches have been naturally applied to solve the DSA task, such as Bi-LSTM [6], TextCNN [5] and BERT [7].

To learn a better document representation, some effective methods have been proposed. Tang et al. [18] encoded the intrinsic relations between sentences in the semantic meaning of a document. Xu et al. [19] proposed a cached LSTM model to capture the overall semantic information in a long text. Yang et al. [9] firstly proposed a hierarchical attention network (HAN) for document classification, which can aggregate differentially informative components of a document. Remy et al. [11] extended HAN to take content into account. Song et al. [20] integrated user characteristics, product and review information to the neural networks. Although most of the methods attempted to explore the major idea of the document consisting of comprehensive information via a learnable embedding, obtained the major points from user-generated summary directly would be more accurate than learning from the entire document.

Some effects based on text summarization have been devoted to predicting the sentiment label. The text summarization task naturally compresses texts for an abstractive summarization from the long document. Rush et al. [21] first proposed an abstractive based summarization model, adopting an attentive CNN encoder for distillation. Bhargava and Sharma [22] leveraged different techniques of machine learning to perform sentiment analysis of different languages. Recently, there are some studies concerning both text summarization task and document-level sentiment classification. Ma et al. [14] proposed a hierarchical end-to-end model HSSC for joint learning of text summarization and sentiment analysis, exploiting the sentiment classification label as the further summarization of text summarization output. Based on their work, Wang et al. [15] extended a variant hierarchical framework with self-attention to capture emotional information at text-level as well as summary-level. Although these methods boosted the performance by jointly encoding the summary and document, they neglected the bidirectional interactions between the user-generated summary and document.

In this paper, we attempt to capture interactions between the summary and document from multiple granularities to learn a subject-oriented document representation for DSA.

### 2.2 Rethinking Mechanism

Previous works attempting to use a rethinking mechanism have been demonstrated promising results in image classification [16] and named entity recognition [17]. Li et al. [16] applied the output posterior possibilities of a CNN to refine its intermediate feature maps. Gui et al. [17] used the high-level semantic feedback layer to refine the weights of embedded words and tackled the word conflicts problem in the lexicon.

Inspired by their works, we extend these concepts into sentiment analysis and design a rethinking mechanism to introduce the sentiment label information to learn an expressive document representation with more sentiment features and alleviate the negative impacts of the data noise.

# 3    Method

In this section, we describe a unified architecture for document-level sentiment analysis. Firstly, we define some notations in Sect. 3.1. Then, we introduce the architecture of the hierarchical interactions networks (HIN) model in Sect. 3.2. Finally, we show details of the sentiment-based rethinking (SR) mechanism in Sect. 3.3. An illustration of the overall architecture in this paper is shown in Fig. 1.

## 3.1    Problem Definition

Given a dataset that consists of $D$ data samples, the $i$-th data sample $(x^i, y^i, l^i)$ contains a user-generated document $x^i$, the corresponding summary $y^i$, and the corresponding sentiment label $l^i$. The model is applied to learn the mapping from the source text $x^i$ and $y^i$ to $l^i$. Before assigning the document into the model, we attempt to reconstruct the document to purify the complex document leveraging the user-generated summary. This preprocess can enable the model to pay more attention to shorter text relevant to the major idea. We compress the document $x^i$ into several segment candidates $x^i = \{x^{i,1}, x^{i,2}, ..., x^{i,T}\}$ with high text similarity given the query (i.e., summary $y^i$), where $T$ is the number of candidates. Specifically, each candidate $x^{i,j}$ and the summary $y^i$ are sequences of characters: $x^{i,j} = \{x_1^{i,j}, x_2^{i,j}, ..., x_{L_i}^{i,j}\}$, $y^i = \{y_1^i, y_2^i, ..., y_{M_i}^i\}$, where $L_i$ and $M_i$ denote the number of characters in the sequences $x^{i,j}$ and $y^i$, respectively. The label $l^i \in \{1, 2, ..., K\}$ denotes the sentiment polarities of the original document $x^i$, from the lowest rating 1 to the highest rating $K$.

## 3.2    Hierarchical Interactions Networks

As shown in Fig. 1, the proposed model HIN is composed of several components including *character-, segment-,* and *document-level interaction encoding* and *decoding*.

**Character-Level Interaction Encoding.** In this part, we employ the BERT [7] to model character-level interaction between the summary $y^i$ and segment candidates $x^i$ generated by the document. BERT is a masked language model

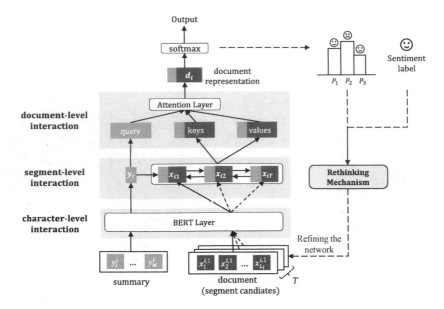

**Fig. 1.** An illustration of the proposed HIN-SR. It consists of hierarchical interactions network and a sentiment-aware rethinking mechanism. In the figure, we show that the document is distilled into 3 segment candidates with the guidance of the summary.

that apply multi-layer bidirectional transformer and large-scale unsupervised corpora to obtain contextual semantic representations, which has achieved the promising performance in many tasks. The summary $y^i$ and segment candidates $x^i$ are as an input-pair sequence (*summary, candidate*) to feed into BERT.

Formally, we concatenate two character-based sequences of the summary $y^i = \{y_1^i, y_2^i, ..., y_{M_i}^i\}$ and $j$-th candidate $x^{i,j} = \{x_1^{i,j}, x_2^{i,j}, ..., x_{L_i}^{i,j}\}$ into a sentence-pair text sequence $\{[CLS], y^i, [SEP], x^{i,j}, [SEP]\}$. $[CLS]$ and $[SEP]$ represent the begining and ending of the sequence, respectively. Then we pad each text sequence in a mini-batch to $N$ tokens, where $N$ is the maximum length in the batch. After finetuning the training parameters, the outputs of BERT are denoted as the summary vector $\boldsymbol{y}_i^c$ and the candidates' vectors $\boldsymbol{x}_{ij}, j \in [1, T]$, where $T$ is the number of the candidate.

**Segment-Level Interaction Encoding.** Then, we design the gated interaction network by applying a bidirectional gated recurrent unit(Bi-GRU) to embed the segment-level interactions between the summary and candidate. The vector including both of summary and candidate features propagates through the hidden units of GRU. For each document $d_i$, we concatenate the summary vector $\boldsymbol{y}_i^c$ and $j$th candidate vector $\boldsymbol{x}_{ij}$ as input to compute the update hidden state by

$$\overrightarrow{\boldsymbol{h}_{ij}} = \overrightarrow{\mathrm{GRU}}([\boldsymbol{y}_i^c; \boldsymbol{x}_{ij}]), j \in [1, T], \tag{1}$$

$$\overleftarrow{\boldsymbol{h}_{ij}} = \overleftarrow{\mathrm{GRU}}([\boldsymbol{y}_i^c; \boldsymbol{x}_{ij}]), j \in [T, 1]. \tag{2}$$

We concatenate $\overrightarrow{\boldsymbol{h}_{ij}}$ and $\overleftarrow{\boldsymbol{h}_{ij}}$ to get an annotation of $j$th candidate, $i.e.$, $\boldsymbol{h}_{ij} = [\overrightarrow{\boldsymbol{h}_{ij}}; \overleftarrow{\boldsymbol{h}_{ij}}]$, which summaries contextual segment-level interactive features among summary and each candidate globally. Simultaneously, the module aggregates contextual semantic information of segments and captures structural features of the document. The embedding of summary $\boldsymbol{y}_i^s$ at segment-level is defined by an average-pooling operation of the output of BERT:

$$\boldsymbol{y}_i^s = \frac{1}{T} \sum_{j=1}^{T} tanh(\boldsymbol{W}_s \boldsymbol{h}_{ij} + \boldsymbol{b}_s). \tag{3}$$

**Document-Level Interaction Encoding.** Not all segments contribute equally to the sentiment of the document. Hence, we extract the embedding of summary from text encoder and adopt an attention mechanism to formulate the document-level interaction by extracting subject-oriented segments that are essential to the sentiment:

$$\boldsymbol{u}_{ij} = tanh(\boldsymbol{W}_d \boldsymbol{h}_{ij} + \boldsymbol{b}_d), \tag{4}$$

$$\alpha_{ij} = \frac{exp(\boldsymbol{u}_{ij} \boldsymbol{y}_i^s)}{\sum_{j=1}^{T} exp(\boldsymbol{u}_{ij} \boldsymbol{y}_i^s)}, \tag{5}$$

$$\boldsymbol{d}_i = \sum_{j=1}^{T} \alpha_{ij} \boldsymbol{h}_{ij}. \tag{6}$$

Specifically, we first feed the candidate annotation $\boldsymbol{h}_{ij}$ through a dense layer to get $\boldsymbol{u}_{ij}$ as a hidden representation of $\boldsymbol{h}_{ij}$. We evaluate the contribution of each candidate in the document as the similarity with the vector of summary $\boldsymbol{y}_i$ and formulate a normalized importance weight $\alpha_{ij}$ through a *softmax* function. And then, the representations of informative segments are aggregated to form a document vector $\boldsymbol{d}_i$.

**Decoding and Training.** Given a document $x$ and its corresponding summary $y$, we obtain final document vector $\boldsymbol{d}$ with subject and context sentiment information. We can use a *softmax* function to project it into sentiment distribution. The probability distribution of the sentiment label can be computed as:

$$P(l|x, y) = softmax(\boldsymbol{W}_c \boldsymbol{d} + \boldsymbol{b}_c). \tag{7}$$

The logistics layer makes the final prediction with the top probability of the sentiment label.

We employ the cross-entropy error to train the model by minimizing the loss. The loss function can be defined as:

$$L = -\sum_{l=1}^{K} \hat{p}(l, x, y) log P(l|x, y), \tag{8}$$

where $P(l|x, y)$ is the probability distribution of labels, and $\hat{p}(l, x, y)$ is the gold one-hot distribution of sample $(x, y)$.

### 3.3    Sentiment-Based Rethinking Mechanism

Although HIN can effectively learn an expressive document representation, it still has insufficiency to capture sentiment features for sentiment analysis. Intuitively, when two samples of different sentiment labels have similar posterior probabilities, it is not easy to classify them.

In this part, we introduce the gold sentiment label information as the high-level features to guide the previous layers based on the current posterior probabilities of these confusable categories. As a result, the bottom layers can be strengthened or weakened to capture more discriminative features specifically for those categories difficult to distinguish. When an input sample passes through IIIN, instead of immediately making a classification based on the predicted posterior probability of the sample belonging to a specific sentiment, a rethinking mechanism is deployed to propagate the predicted posterior probability to the bottom layers to update the network.

Formally, we denote the state vector as $s$ to represent the current document representation information and input it to the rethinking mechanism. The state of $i$-th document is computed by:

$$s_i = d_i, \tag{9}$$

where $d_i$ is the document representation by Eq. (6). A feedback layer is a fully connected layer, that is:

$$\hat{P}(l|x,y) = softmax(W_r s_i + b_r), \tag{10}$$

where $W_r$ and $b_r$ are the parameters.

Then we introduce the sentiment label information as high-level features to feedback a reward of the current state. In the $t$-th episode, the reward of the sample is defined:

$$r^{(t)} = \lambda r^{(t-1)} + (1 - \lambda)log\hat{P}(\hat{l}|x,y), \tag{11}$$

where $r^{(t-1)}$ represents the reward of last episode and the reward initialization of training samples is $r^0 = 1$, $\lambda \in [0,1]$ is a hyperparameter to weight the importance of the sentiment label of the current $t$-th episode against the previous $t-1$ step, and $\hat{l}$ is the gold label of the document. We utilize the reward to reweight the lower-level features in the HIN to enable it to selectively emphasize some discriminative sentiment features, and suppress the feature causing confusion in the classification.

Specifically, we rethink cross-entropy error between gold sentiment distribution and the sentiment distribution of HIN with cumulative rewards. In $t$-th episode, the loss function is:

$$L' = -\sum_{l=1}^{K} r^{(t)}\hat{p}(l,x,y)log\hat{P}(l|x,y), \tag{12}$$

where $\hat{p}(l,x,y)$ is the gold one-hot distribution of sample $(x,y)$.

# 4    Experiments

In this section, we first describe the experimental setup. We then report the results of experiments and demonstrate the effectiveness of the proposed modules.

## 4.1    Experimental Setup

**Datasets.** We conduct experiments on three public datasets, one from the online news and another two from the online reviews. The details are shown in Table 1.

- **Toys & Games** and **Sports & Outdoors** are parts of the Stanford Network Analysis Project(SNAP)[2], provided by [23]. These datasets consist of reviews from Amazon spanning May 1996-July 2014. Raw data includes review content, product, user information, ratings, and summaries. The rating of product is from 1 to 5. Following existing works [14,15], we select the first 1,000 samples for validation, the following 1,000 samples for testing, and the rest for training.
- **Online News Datasets (News)** is from *Emotional Analysis of Internet News* task[3]. The dataset is collected from websites including news websites, WeChat, blogs, Baidu Tieba, etc. Each document consists of a title, main-body content and sentiment polarity. Here, the title is used as the summary. We randomly select 80% of data for training, 10% for validation and the remaining 10% for testing [9].

**Table 1.** Statistics of three datasets. # denotes the average length.

| Dataset | Total size | Classes | # Summary | # Document |
|---|---|---|---|---|
| Toys & Games | 167,597 | 5 | 4.4 | 99.9 |
| Sports & Outdoors | 296,337 | 5 | 4.2 | 87.2 |
| News | 14,696 | 3 | 23.8 | 1216.1 |

**Implementation Details.** In the implementation, we first preprocess to compress the document into 3 segment candidates empirically, i.e., $T = 3$. We use the bert-based model (uncased, 12-layer, 768-hidden, 12-heads, 110M parameters) for reviews dataset, and the bert-base-chinese model (12-layer, 768-hidden, 12-heads, 110M parameters) for news dataset [7]. We train the model using Adam optimizer with the learning rate of 5e-6 for 2 epochs. The dropout probability of Bi-GRU is 0.1. We set the hyperparameter $\lambda$ to $\{0.6, 0.8\}$ and maximum length $N$ to 256.

---

[2] http://snap.stanford.edu/data/web-Amazon.html.
[3] https://www.datafountain.cn/competitions/350.

## 4.2  Compared Methods

We compare HIN and HIN-SR with several baselines, including traditional methods and summary-based methods for DSA. The details are as follows.

- **TextCNN** [5]. The method used CNN to learn document representations.
- **Bi-LSTM** [6]. This was the baseline that directly took the whole document as a single sequence using a bidirectional LSTM network for DSA.
- **BERT** [7]. This was a pre-trained model with deep bidirectional transformer.
- **BERT(head+tail)** [8]. The model finetuned BERT with different truncated methods.
- **HAN** [9]. The method classified documents via hierarchical attention networks.
- **CAHAN** [11]. The method extended the HAN by making sentence encoder context-aware.
- **HSSC** [14]. The method applied a hierarchical end-to-end model for joint learning of text summarization and sentiment classification.
- **SAHSSC** [15]. The method jointly established text summarization and sentiment classification via self-attention.

## 4.3  Experimental Results

Experimental results are illustrated in Table 2, divided into three blocks: traditional methods, summary-based methods, and our proposed models. Our proposed models perform the best among all baselines on three datasets, which reveals the effectiveness and advancement for DSA.

The first block shows the comparative results with traditional methods. HIN and HIN-SR outperform traditional methods, and the benefit mainly comes from the full utilization of the user-generated summary, boosting a more discriminative document representation. As expected, HAN and CAHAN using Bi-LSTM with hierarchical attention achieve better results than traditional classification methods TextCNN and Bi-LSTM, which confirms the effectiveness of hierarchically modeling documents. It is observed that BERT and BERT(head+tail) achieve mediocre results, which reveals that they can capture contextual semantic information effectively.

Then, we compare our models with summary-based methods including HSSC and SAHSSC. The results in the second block of Table 2 indicate that HIN-SR outperforms among these summary-based baselines on both Toys & Games and Sports & Outdoors datasets. It is not surprising since HIN takes advantage of the user-generated summary via modeling the interactions between the summary and document at multiple granularities levels and explores more discriminative features.

Moreover, HIN-SR outperforms HIN among three datasets. The result demonstrates the positive impacts of the sentiment-based rethinking mechanism, which can further leverage the sentiment-aware document representation to refine the weights of document features and tackle the negative impacts on data noise.

**Table 2.** Experimental results on three datasets. Evaluation metric is accuracy. - means not available. The best results are in bold. * represents that the result is reported in the corresponding reference.

| Models | Toys & Games | Sports & Outdoors | News |
|---|---|---|---|
| TextCNN [5] | 70.5 | 72.0 | 77.5 |
| Bi-LSTM [6] | 70.7 | 71.9 | 76.5 |
| BERT [7] | 75.5 | 74.2 | 85.7 |
| BERT(head+tail) [8] | 75.9 | 74.0 | 86.1 |
| HAN [9] | 69.1 | 72.3 | 78.6 |
| CAHAN [11] | 70.8 | 73.0 | 79.9 |
| HSSC [14] | 71.9* | 73.2* | - |
| SAHSSC [15] | 72.5* | 73.6* | - |
| **HIN** | **77.5** | **76.7** | **89.0** |
| **HIN-SR** | **78.1** | **77.2** | **89.3** |

## 4.4   Ablation Study of HIN

For further study, we conduct a series of ablation experiments to examine the relative contributions of each component in HIN on accuracy (Acc) and macro-F1 (F1). To this end, HIN is compared with the following variants:

- **w/o Doc.** indicates the HIN model without modeling the document-level interactions via removing attention mechanism.
- **w/o Doc. and Seg.** indicates the HIN model removing Bi-GRU and without modeling both segment-level and document-level interactions.
- **w/o Interactions** indicates the model performs the sentiment classification by directly concatenating the summary and document embeddings.
- **w/o Summary** indicates the model performs the sentiment classification by BERT without the summary and corresponding interactions.

The results are shown in Table 3. We observe that compared with these partial models, the full model yields significant improvements on F1. Noted that HIN without both document- and segment-level interactions achieve the best accuracy in Toys & Games dataset, a reasonable explanation is due to the imbalanced distribution of categories. Therefore, consideration the imbalanced distribution, F1 is a comprehensive method for evaluation.

HIN without document-level interactions obtained a worse result than the complete model. The fact can be attributed that the module can aggregate multiple segments with rich subject information with the consideration of the summary representation via the attention mechanism. Moreover, HIN without both document- and segment-level interactions reports the relatively poor performance as Bi-GRU was able to capture more contextual information to generate the better summary and segment representations via the propagation in the hidden states. Without modeling any interactions, the performance of HIN drops

**Table 3.** An ablation study of the proposed model. Evaluation metrics are accuracy (Acc) and F1. The best results are in bold.

| Ablation Settings | Toys & Games | | Sports & Outdoors | | News | |
|---|---|---|---|---|---|---|
| | Acc | F1 | Acc | F1 | Acc | F1 |
| **HIN** | 77.5 | **54.8** | **76.7** | **62.7** | **89.0** | **81.3** |
| - w/o Doc. | 76.9 | 53.0 | 76.2 | 62.3 | 88.9 | 81.1 |
| - w/o Doc. and Seg. | **77.8** | 40.0 | 75.9 | 62.2 | 87.9 | 80.2 |
| - w/o Interactions | 76.0 | 50.6 | 75.3 | 61.7 | 87.8 | 79.6 |
| - w/o Summary | 75.5 | 49.3 | 74.2 | 61.0 | 85.7 | 78.1 |

off significantly, which demonstrates the effectiveness of interactions between the summary and document and the encoding can indeed capture more effective context semantic features with the utilization of user-generated summary. Besides, for HIN without any interactions, modeling with the summary outperforms modeling without the summary, which reveals that the summary can effectively serve the identification of sentiment polarity.

## 5  Discussion

In this section, we give some analyses about training episodes and document length on predicting sentiment labels, and discuss the effect of the summary with visualization.

### 5.1  Analysis of Training Episodes

The number of episodes is an important parameter in the sentiment-based rethinking mechanism (SR). We investigate its effects on three datasets in Fig. 2.

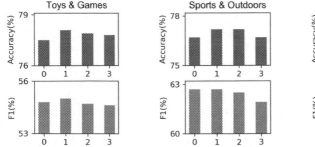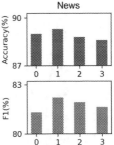

**Fig. 2.** Results against training episodes. The X-axis is the the number of training episodes and the Y-axis represents accuracy or F1.

From the results in Fig. 2, we observe that as the number of training episodes increases, the performance grows significantly and then decline slightly on both accuracy and F1. And the best performance is when the number of training episodes is set to 1.

Superior performance when training episode is set to 1 than 0 demonstrates the effectiveness of gold sentiment label information. Rethinking these high-level sentiment patterns guides the model to capture more discriminative features specifically for different classes. However, drop performance when the number of training episodes increases reveals that, an overload of sentiment label information would limit the original feature expression of HIN and result in slightly descending performance. Besides, three datasets show different meliorations of performance when training episodes is set to 1 than 0, Especially, the model achieves improvements of 0.9% on in News dataset where more vague semantic links in Chinese bring much noise samples. This reveals that the SR can release the negative impacts of noisy samples simultaneously.

## 5.2   Analysis of Document Length

We further investigate the performance of HIN-SR, HIN, and several baselines when analyzing sentiment polarity with different document lengths. The results of different datasets are shown in Fig. 3.

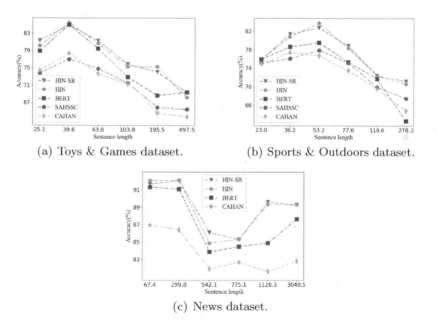

(a) Toys & Games dataset.          (b) Sports & Outdoors dataset.

(c) News dataset.

**Fig. 3.** Results against document length. We divide the testing dataset into six parts according to quantiles of document length. The X-axis is the average length of each part and the Y-axis indicates accuracy.

As document length increases, the performance of all models clearly decreases after a certain range. The result reveals that document-level sentiment analysis is more challenging to the longer document usually containing more vague semantic links and complicated sentiment information. In particular, the curve of accuracy presents a partial upward trend over an extremely long length in the News datasets. And we speculate that more semantic information can partially alleviate the above phenomenon.

From the figures, it is seen that HIN achieves a considerable improvement than other baselines on three datasets over almost any range of document length. Compared with other baselines system, HIN takes the multiple-granularity interactions between the summary and document into considerations. After modeling the interactions, HIN can adequately exploit the semantic and subject information of the summary and thus learn subject-oriented document representation for document-level sentiment analysis. In addition, in Fig. 3(a) and (c), the results of BERT are slightly higher than HIN at a few of length ranges. Through the analysis of the abnormal samples, we conclude that this is probably due to the uneven distribution of samples.

Moreover, HIN-SR slightly outperforms than HIN over most length ranges. This indicates that the rethinking mechanism introducing sentiment label information can adapt and refine the HIN with high-level sentiment features to boost the performance. In addition, HIN-SR is not sensitive to document length and is adaptive in both short and long documents. It illustrates the robustness of the model becomes better with the help of reweighting document representations. We contribute it to this rethinking mechanism can indeed alleviate the data noise via feedbacking the sentiment label information.

## 5.3   Case Study

We visualize multi-granularity interactions encoding modules in Fig. 4. The review describes the disappointing experience of using a simulator, containing complex sentiment polarity expressed such as *awesome* and *nice*. According to the summary, the major point of the review can conclude as *a simulator*.

From the blue marks, we can observe that after encoding character- and segment-level interactions between the summary and document, our model performs by consciously selecting texts carrying the major point (i.e., *simulator*). Note that the contexts of these subject-related tokens contain pivotal sentimental information, such as *awesome*. While the sentiment expressed *nice* is neglected because it is irrelevant to the subject *simulator*. Subsequently, document-level interaction features are captured to further aggregate the document representation for sentiment classification.

In conclusion, HIN successfully tackles the challenge of vague semantic links and complicate sentiment information and makes correct classification. Evidenced by the visualization, subject information is affirmative for capturing accurate sentiment information. Our proposed model can explicitly explore multi-granularities interactions between the summary and document to learn a subject-oriented document representation.

| Summary | Rating | Reviews |
|---------|--------|---------|
| simulator? | 1 | 1   "a simulater replicates what the real thing is ever used a rc controller? well throttle is on the left pitch yaw, elevator.. on right.. right? well to tally reversed on this .. |
| | | 2   so I'm trying build ..instincts on the sticks .. well learn onthis when you got 200 or more on a model, crash your cash its backwards!!! i have flown real airplanes have 83 hrs solo. |
| | | 3   this would kill you in the real world.. nowsims are awesome this will waste your model... because... the inputs on this.. mirror revers from spektrum.. futuba.. nice game though...lol forget this #$% ok get a real one.. ok?" |

**Fig. 4.** The review is from Toys & Games dataset with a negative sentiment label. Each block represents one segment candidate of the document. Blue color denotes the word weight after character- and segment-level interaction encoding. Red color indicates candidate weight in the document-level interaction encoding module. (Color figure online)

## 6  Conclusion

In this work, we have investigated the task of document-level sentiment analysis. We design a hierarchical interaction networks (HIN) model to learn a subject-oriented document representation for sentiment classification. The main idea of HIN is to exploit bidirectional interactions between the user-generated summary and document. Furthermore, a sentiment-based rethinking mechanism (SR) refines the weights of document features to learn a more sentiment-aware document representation and alleviate the negative impact of noisy data. Experimental Results on three widely public datasets have demonstrated that HIN-SR outperforms significantly and tackles long documents with the vague semantic links and abundant sentiments effectively. The proposed model is of great significance to DSA and related applications.

For future work, we will consider more interaction information (e.g., interactions between documents) and more detailed theoretical analysis of rethinking mechanism, to further improve the performance.

## References

1. Wang, W., Wang, H.: The influence of aspect-based opinions on user's purchase intention using sentiment analysis of online reviews. Syst. Eng. **36**(1), 63–76 (2016)
2. Hyun, D., Park, C., Yang, M.C., Song, I., Lee, J.T., Yu, H.: Review sentiment-guided scalable deep recommender system. In: SIGIR, pp. 965–968 (2018)
3. Yuan, C., Ma, Q., Zhou, W., Han, J., Hu, S.: Jointly embedding the local and global relations of heterogeneous graph for rumor detection. In: ICDM. IEEE (2019)
4. Yuan, C., Zhou, W., Ma, Q., Lv, S., Han, J., Hu, S.: Learning review representations from user and product level information for spam detection. In: ICDM. IEEE (2019)

5. Kim, Y.: Convolutional neural networks for sentence classification. In: EMNLP, pp. 1746–1751. Doha, Qatar, Oct 2014
6. Tang, D., Qin, B., Feng, X., Liu, T.: Effective LSTMs for target-dependent sentiment classification. In: Proceedings of the 26th International Conference on Computational Linguistics, pp. 3298–3307. Osaka, Japan, Dec 2016
7. Devlin, J., Chang, M.W., Lee, K., Toutanova, K.: BERT: pre-training of deep bidirectional transformers for language understanding. In: NAACL, pp. 4171–4186. Association for Computational Linguistics, Minneapolis, Minnesota, June 2019
8. Sun, C., Qiu, X., Xu, Y., Huang, X.: How to fine-tune BERT for text classification? In: Sun, M., Huang, X., Ji, H., Liu, Z., Liu, Y. (eds.) CCL 2019. LNCS (LNAI), vol. 11856, pp. 194–206. Springer, Cham (2019). https://doi.org/10.1007/978-3-030-32381-3_16
9. Yang, Z., Yang, D., Dyer, C., He, X., Smola, A., Hovy, E.: Hierarchical attention networks for document classification. In: NAACL, pp. 1480–1489. San Diego, California, June 2016
10. Pergola, G., Gui, L., He, Y.: TDAM: a topic-dependent attention model for sentiment analysis. Inf. Process. Manage. **56**(6), 102084 (2019)
11. Remy, J., Tixier, A.J., Vazirgiannis, M.: Bidirectional context-aware hierarchical attention network for document understanding (2019)
12. Vikrant, H., Mukta, T.: Real time tweet summarization and sentiment analysis of game tournament. Int. J. Sci. Res. **4**(9), 1774–1780 (2013)
13. Mane, V.L., Panicker, S.S., Patil, V.B.: Summarization and sentiment analysis from user health posts. In: ICPC, pp. 1–4. IEEE (2015)
14. Ma, S., Sun, X., Lin, J., Ren, X.: A hierarchical end-to-end model for jointly improving text summarization and sentiment classification. In: IJCAI, pp. 4251–4257. International Joint Conferences on Artificial Intelligence Organization, July 2018
15. Wang, H., Ren, J.: A self-attentive hierarchical model for jointly improving text summarization and sentiment classification. In: ACML, pp. 630–645. PMLR, Nov 2018
16. Li, X., Jie, Z., Feng, J., Liu, C., Yan, S.: Learning with rethinking: recurrently improving convolutional neural networks through feedback. Pattern Recognit. **79**, 183–194 (2018)
17. Gui, T., Ma, R., Zhang, Q., Zhao, L., Jiang, Y., Huang, X.: Cnn-based chinese NER with lexicon rethinking. In: IJCAI, pp. 4982–4988. Macao, China (2019)
18. Tang, D., Qin, B., Liu, T.: Document modeling with gated recurrent neural network for sentiment classification. In: EMNLP, pp. 1422–1432. Association for Computational Linguistics, Lisbon, Portugal, Sep 2015
19. Xu, J., Chen, D., Qiu, X., Huang, X.: Cached long short-term memory neural networks for document-level sentiment classification. In: EMNLP, pp. 1660–1669. The Association for Computational Linguistics (2016)
20. Song, J.: Distilling knowledge from user information for document level sentiment classification. In: 35th IEEE International Conference on Data Engineering Workshops, pp. 169–176. IEEE (2019)
21. Rush, A.M., Chopra, S., Weston, J.: A neural attention model for abstractive sentence summarization. In: EMNLP, pp. 379–389. The Association for Computational Linguistics (2015)

22. Bhargava, R., Sharma, Y.: Msats: multilingual sentiment analysis via text summarization. In: 7th International Conference on Cloud Computing, pp. 71–76. IEEE, Jan 2017
23. He, R., McAuley, J.: Ups and downs: modeling the visual evolution of fashion trends with one-class collaborative filtering. In: WWW, pp. 507–517. International World Wide Web Conferences Steering Committee (2016)

# Early Detection of Fake News with Multi-source Weak Social Supervision

Kai Shu[1(✉)], Guoqing Zheng[2], Yichuan Li[3], Subhabrata Mukherjee[2],
Ahmed Hassan Awadallah[2], Scott Ruston[3], and Huan Liu[3]

[1] Department of Computer Science, Illinois Institute of Technology, Chicago, USA
kshu@iit.edu
[2] Microsoft Research AI, Redmond, USA
{zheng,submukhe,hassanam}@microsoft.com
[3] Arizona State University, Tempe, USA
{yichuan1,Scott.Ruston,huanliu}@asu.edu

**Abstract.** Social media has greatly enabled people to participate in online activities at an unprecedented rate. However, this unrestricted access also exacerbates the spread of misinformation and fake news which cause confusion and chaos if not detected in a timely manner. Given the rapidly evolving nature of news events and the limited amount of annotated data, state-of-the-art systems on fake news detection face challenges for early detection. In this work, we exploit multiple weak signals from different sources from user engagements with contents (referred to as weak social supervision), and their complementary utilities to detect fake news. We jointly leverage limited amount of clean data along with weak signals from social engagements to train a fake news detector in a meta-learning framework which estimates the quality of different weak instances. Experiments on real-world datasets demonstrate that the proposed framework outperforms state-of-the-art baselines for early detection of fake news without using any user engagements at prediction time.

**Keywords:** Fake news · Weak social supervision · Meta learning

## 1 Introduction

**Motivation.** Social media platforms provide convenient means for users to create and share diverse information. Due to its massive availability and convenient access, more people seek and receive news information online. For instance, around 68% of U.S. adults consumed news from social media in 2018, a massive increase from corresponding 49% consumption in 2012[1] according to a survey by the Pew Research Center. However, social media also proliferates a plethora of misinformation and fake news, i.e., news stories with intentionally false information [27]. Research has shown that fake news spreads farther, faster, deeper, and more widely than true news [32]. For example, during the 2016 U.S. election,

---

[1] https://bit.ly/39zPnMd.

© Springer Nature Switzerland AG 2021
F. Hutter et al. (Eds.): ECML PKDD 2020, LNAI 12459, pp. 650–666, 2021.
https://doi.org/10.1007/978-3-030-67664-3_39

**Fig. 1.** An illustration of a piece of fake news and related user engagements, which can be used for extracting weak social supervision. Users have different credibility, perceived bias, and express diverse sentiment to the news.

the top twenty frequently-discussed false election stories generated 8.7 million shares, reactions, and comments on Facebook, more than the total of 7.4 million shares of top twenty most-discussed true stories[2]. Widespread fake news can erode the public trust in government and professional journalism and lead to adverse real-life events. Thus, a timely detection of fake news on social media is critical to cultivate a healthy news ecosystem.

**Challenges.** First, fake news is diverse in terms of topics, content, publishing methods and media platforms, and sophisticated linguistic styles geared to emulate true news. Consequently, training machine learning models on such sophisticated content requires *large-scale annotated fake news data* that is egregiously difficult to obtain. Second, it is important to detect fake news early. Most of the research on fake news detection rely on signals that require a long time to aggregate, making them unsuitable for *early detection*. Third, the evolving nature of fake news makes it essential to analyze it with signals from multiple sources to better understand the context. A system solely relying on social networks and user engagements can be easily influenced by biased user feedback, whereas relying only on the content misses the rich auxiliary information from the available sources. In this work, we adopt an approach designed to address the above challenges for early detection of fake news with limited annotated data by leveraging weak supervision from multiple sources involving users and their social engagements – referred to as *weak social supervision*.

**Existing Work.** Prior works on detecting fake news [17,33] rely on large amounts of labeled instances to train supervised models. Such large labeled training data is difficult to obtain in the early phase of fake news detection. To overcome this challenge, learning with weak supervision presents a viable solution [34]. Weak signals are used as constraints to regularize prediction mod-

---

[2] https://bit.ly/39xmXT7.

els [29], or as loss correction mechanisms [6]. Often only a *single* source of weak labels is used.

Existing research has focused either on the textual content relying solely on the linguistic styles in the form of sentiment, bias and psycho-linguistic features [15] or on tracing user engagements on how fake news propagate through the network [33]. In this work, we utilize *weak social supervision* to address the above shortcomings. Consider the example in Fig. 3. Though it is difficult to identify the veracity considering the news content in isolation, the surrounding context from other users' posts and comments provide clues, in the form of opinions, stances, and sentiment, useful to detect fake news. For example, in Fig. 3, the phrase "`kinda agree...`" indicates a positive sentiment to the news, whereas the phrase "`I just do not believe it...`" expresses a negative sentiment. Prior work has shown conflicting sentiments among propagators to indicate a higher probability of fake news [8,27]. Also, users have different credibility degrees in social media and less-credible ones are more likely to share fake news [28]. Although we do not have this information a priori, we can consider *agreement* between users as a weak proxy for their credibility. All of the aforementioned signals from different sources like content and social engagements can be leveraged as weak supervision signals to train machine learning models.

**Contributions.** We leverage weak social supervision to detect fake news from limited annotated data. In particular, our model leverages a small amount of manually-annotated clean data and a large amount of weakly annotated data by proxy signals from multiple sources for joint training in a meta-learning framework. Since not all weak instances are equally informative, the model learns to estimate their respective contributions for the end task. To this end, we develop a Label Weighting Network (LWN) to model the weight of these weak labels that regulate the learning process of the fake news classifier. The LWN serves as a meta-model to produce weights for the weak labels and can be trained by back-propagating the validation loss of a trained classifier on a separate set of clean data. The framework is uniquely suitable for early fake news detection, because it (1) leverages rich weak social supervision to boost model learning in a meta-learning fashion; and (2) only requires the news content during the prediction stage without relying on the social context as features for early prediction. Our contributions can be summarized as:

- **Problem.** We study a novel problem of exploiting weak social supervision for early fake news detection;
- **Method.** We provide a principled solution, dubbed MWSS to learn from Multiple-sources of Weak Social Supervision (MWSS) from multi-faceted social media data. Our framework is powered by meta learning with a Label Weighting Network (LWN) to capture the relative contribution of different weak social supervision signals for training;
- **Features.** We describe how to generate weak labels from social engagements of users that can be used to train our meta learning framework for early fake news detection along with quantitative quality assessment.

- **Experiments.** We conduct extensive experiments to demonstrate the effectiveness of the proposed framework for early fake news detection over competitive baselines.

## 2   Modeling Multi-source Weak Social Supervision

User engagements over news articles, including posting about, commenting on or recommending the news, bear implicit judgments of the users about the news and could serve as weak sources of labels for fake news detection. For instance, prior research has shown that contrasting sentiment of users on a piece of news article, and similarly different levels of credibility or bias, can be indicators of the underlying news being fake. However, these signals are noisy and need to be appropriately weighted for training supervised models. Due to the noisy nature of such social media engagements, we term these signals as *weak social supervision*.

To give a brief overview for the modeling, we define heuristic labeling functions (refer to Sect. 3) on user engagements to harvest such signals to weakly label a large amount of data. The weakly labeled data is combined with limited amount of manually annotated examples to build a fake news detection system that is better than training on either subset of the data. We emphasize that multiple weak labels can be generated for a single news article based on different labeling functions and we aim to jointly utilize both the clean examples as well as multiple sources of weak social supervision in this paper.

In this section, we first formulate the problem statement, and then focus on developing algorithms for the joint optimization of manually annotated clean and multi-source weakly labeled instances in a unified framework.

### 2.1   Problem Statement

Let $\mathcal{D} = \{x_i, y_i\}_{i=1}^n$ denote a set of $n$ news articles with manually annotated clean labels, with $\mathcal{X} = \{x_i\}_{i=1}^n$ denoting the news pieces and $\mathcal{Y} = \{y_i\}_{i=1}^n \subset \{0,1\}^n$ the corresponding clean labels of whether the news is fake or not. In addition, there is a large set of unlabeled examples. Usually the size of the clean labeled set $n$ is smaller than the unlabeled set due to labeling costs. For the widely available unlabeled samples, we can generate weak labels by using different labeling functions based on *social engagements*. For a specific labeling function $g^{(k)} : \mathcal{X}^{(k)} \rightarrow \tilde{\mathcal{Y}}^{(k)}$, where $\mathcal{X}^{(k)} = \{x_j^{(k)}\}_{j=1}^N$ denotes the set of $N$ unlabeled messages to which the labeling function $g^{(k)}$ is applied and $\tilde{\mathcal{Y}}^{(k)} = \{\tilde{y}_j^{(k)}\}_{j=1}^N$ as the resulting set of weak labels. This weakly labeled data is denoted by $\tilde{\mathcal{D}}^{(k)} = \{x_j^{(k)}, \tilde{y}_j^{(k)}\}_{j=1}^N$ and often $n \ll N$. We formally define our problem as:

> **Problem Statement:** Given a limited amount of manually annotated news data $\mathcal{D}$ and $K$ sets of weakly labeled data $\{\tilde{\mathcal{D}}^{(k)}\}_{k=1}^K$ derived from $K$ different weak labeling functions based on weak social signals, learn a fake news classifier $f : \mathcal{X} \rightarrow \mathcal{Y}$ which generalizes well onto unseen news pieces.

## 2.2 Meta Label Weighting with Multi-source Weak Social Supervision

Learning from multiple sources has shown promising performance in various domains such as truth discovery [4], object detection [13], etc. In this work, we have $K + 1$ distinct sources of supervision: clean labels coming from manual annotation and multiple sources of weak labels obtained from $K$ heuristic labeling functions based on users' social engagements.

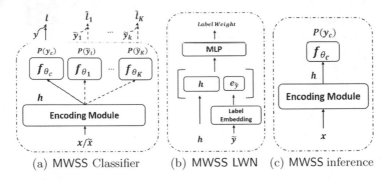

(a) MWSS Classifier    (b) MWSS LWN    (c) MWSS inference

**Fig. 2.** The proposed framework MWSS for learning with multiple weak supervision from social media data. (a) Classifier: Jointly modeling clean labels and weak labels from multiple sources; (b) LWN: Learning the label weight based on the concatenation of instance representation and weak label embedding vector. (c) During inference, MWSS uses the learned encoding module and classification MLP $f_{w_c}$ to predict labels for (unseen) instances in the test data.

Our objective is to build an effective framework that leverages weak social supervision signals from multiple sources in addition to limited amount of clean data. However, signals from different weak sources are intrinsically noisy, biased in different ways, and thus of varying degree of qualities. Simply treating all sources of weak supervision as equally important and merging them to construct a single large set of weakly supervised instances tend to result in sub-optimal performance (as used as a baseline in our experiments). However, it is challenging to determine the contribution of different sources of weak social supervision. To facilitate a principled solution of weighting weak instances, we leverage meta-learning. In this, we propose to treat label weighting as a meta-procedure, i.e., building a *label weighting network* (LWN) which takes an instance (e.g., news piece) and its weak label (obtained from social supervision) as input, and outputs a scalar value as the importance weight for the pair. The weight determines the contribution of the weak instance in training the desired fake news classifier in our context. The LWN can be learned by back-propagating the loss of the trained classifier on a separate clean set of instances. To allow information sharing among different weak sources, for the fake news classifier, we use a shared feature extractor to learn a common representation and use separate functions (specifically, MLPs) to map the features to different weak label sources.

Specifically, let $h_{\theta_E}(x)$ be an encoder that generates the content representation of an instance $x$ with parameters $\theta_E$. Note that this encoder is shared by instances from both the clean and multiple weakly labeled sources. Let $f_{\theta_c}(h(x))$ and $\{f_{\theta_k}(h(x))\}_{k=1,..,K}$ be the $K+1$ labeling functions that map the contextual representation of the instances to their labels on the clean and $K$ sets of weakly supervised data, respectively. In contrast to the encoder with shared parameters $\theta_E$, the parameters $\theta_c$ and $\{\theta_k\}_{k=1,...,K}$ are different for the clean and weak sources (learned by separate source-specific MLPs) to capture different mappings from the contextual representations to the labels from each source.

---

**while** *not converged* **do**
> 1. Update LWN parameters $\alpha$ by descending
> $\nabla_\alpha \mathcal{L}_{val}(\theta - \eta \nabla_\theta \mathcal{L}_{train}(\alpha, \theta))$
> 2. Update classifier parameters $\theta$ by descending $\nabla_\theta \mathcal{L}_{train}(\alpha, \theta)$

**end**

**Algorithm 1:** Training process of MWSS

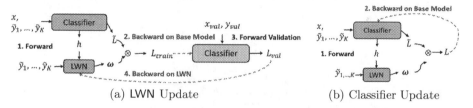

(a) LWN Update                    (b) Classifier Update

**Fig. 3.** The illustration of the MWSS in two phases: (a) we compute the validation loss based on the validation dataset and retain the computation graph for LWN backward propagation; (b) the classifier update its parameters through backward propagation on clean and weakly labeled data. Note that $h$ is the set of hidden representations of the input instances, $\omega$ is the weight for each pair of instances and labels and $\otimes$ is point-wise multiplication. Gray indicates the parameters from the last iteration, blue indicates temporary updates.

For training, we want to jointly optimize the loss functions defined over the (i) clean data and (ii) instances from the weak sources weighted by their respective utilities. The weight of the weak label $\tilde{y}$ for an instance $x$ (encoded as $h(x)$) is determined by a separate Label Weighting Network (LWN) formulated as $\omega_\alpha(h(x), \tilde{y})$ with parameters $\alpha$. Thus, for a given $\omega_\alpha(h(x), \tilde{y})$, the objective for training the predictive model with multiple sources of supervision jointly is:

$$\min_{\theta_E, \theta_c, \theta_1, ..., \theta_k} \mathbb{E}_{(x,y) \in \mathcal{D}} \ell(y, f_{\theta_c}(h_{\theta_E}(x))) + \sum_{k=1}^{K} \mathbb{E}_{(x,\tilde{y}) \in \tilde{\mathcal{D}}^{(k)}} \omega_\alpha(h_{\theta_E}(x), \tilde{y}) \ell(\tilde{y}, f_{\theta_k}(h_{\theta_E}(x)))$$

$$(1)$$

where $\ell$ denotes the loss function to minimize the prediction error of the model. The first component in the above equation optimizes the loss over the clean data, whereas the second component optimizes for the weighted loss (given by $w_\alpha(.)$) of the weak instances from $K$ sources. Figure 2 shows the formulation for both the classifier and LWN.

The final objective is to optimize LWN $w_\alpha(h(x), \tilde{y})$ such that when using such a weighting scheme to train the main classifier as specified by Eq. (1), the trained classifier could perform well on a separate set of clean examples. Formally, the following bi-level optimization problem describe the above intuition as:

$$\min_\alpha \mathcal{L}_{val}(\theta^*(\alpha)) \quad \text{s.t.} \quad \theta^* = \arg\min \mathcal{L}_{train}(\alpha, \theta) \tag{2}$$

where $\mathcal{L}_{train}$ is the objective in Eq. (1), $\theta$ denotes the concatenation of all classifier parameters $(\theta_E, \theta_c, \theta_1, ..., \theta_K)$, and $\mathcal{L}_{val}$ is loss of applying a trained model on a separate set of clean data. Note that $\theta^*(\alpha)$ denotes the dependency of $\theta^*$ on $\alpha$ after we train the classifier on a given LWN.

Analytically solving for the inner problem is typically infeasible. In this paper, we adopt the following one-step SGD update to approximate the optimal solution. As such the gradient for the meta-parameters $\alpha$ can be estimated as:

$$\nabla_\alpha \mathcal{L}_{val}(\theta - \eta \nabla_\theta \mathcal{L}_{train}(\alpha, \theta)) = -\eta \nabla^2_{\alpha,\theta} \mathcal{L}_{train}(\alpha, \theta) \nabla_{\theta'} \mathcal{L}_{val}(\theta') \tag{3}$$

$$\approx -\frac{\eta}{2\epsilon} \left[ \nabla_\alpha \mathcal{L}_{train}(\alpha, \theta^+) - \nabla_\alpha \mathcal{L}_{train}(\alpha, \theta^-) \right] \tag{4}$$

where $\theta^\pm = \theta \pm \epsilon \nabla_{\theta'} \mathcal{L}_{val}(\theta')$, $\theta' = \theta - \eta \nabla_\theta \mathcal{L}_{train}(\alpha, \theta)$, $\epsilon$ is a small constant for finite difference and $\eta$ is learning rate for SGD.

Since we leverage Multiple Weak Social Supervision, we term our method as MWSS. We adopt Adam with mini-batch training to learn the parameters. Algorithm 1 and Fig. 3 outline the training procedure for MWSS.

# 3 Constructing Weak Labels from Social Engagements

In this section, we describe how to generate weak labels from users' social engagements that can be incorporated as weak sources in our model.

## 3.1 Dataset Description

We utilize one of the most comprehensive fake news detection benchmark datasets called FakeNewsNet [25]. The dataset is collected from two fact-checking websites: GossipCop[3] and PolitiFact[4] containing news contents with labels annotated by professional journalists and experts, along with social context information. News content includes meta attributes of the news (e.g., body text),

---

[3] https://www.gossipcop.com/.
[4] https://www.politifact.com/.

whereas social context includes related users' social engagements on the news items (e.g., user comments in Twitter). Note that the number of news pieces in PolitiFact data is relatively small, and we enhance the dataset to obtain more weak labels. Specially, we use a news corpus spanning the time frame 01 January 2010 through 10 June 2019, from 13 news sources including mainstream British news outlets, such as BBC and Sky News, and English language versions of Russian news outlets such as RT and Sputnik, which are mostly related to political topics. To obtain the corresponding social engagements, we use a similar strategy as FakeNewsNet [25] to get tweets/comments, user profiles and user history tweets through the Twitter API and web crawling tools. For GossipCop data, we mask part of the annotated data and treat them as unlabeled data for generating weak labels from the social engagements.

## 3.2 Generating Weak Labels

Now, we introduce the labeling functions for generating weak labels from social media via statistical measures guided by computational social theories.

First, research shows user opinions towards fake news have more diverse sentiment polarity and less likely to be neutral [3]. So we measure the sentiment scores (using a widely used tool VADER [7]) for all the users sharing a piece of news, and then measure the variance of the sentiment scores by computing the standard deviation. We define the following weak labeling function:

> **Sentiment-based:** *If a news piece has a standard deviation of user sentiment scores greater than a threshold $\tau_1$, then the news is weakly labeled as fake news.*

Second, social studies have theorized the correlation between the bias of news publishers and the veracity of news pieces [5]. Accordingly, we assume that news shared by users who are more biased are more likely be fake, and vice versa. Specifically, we adopt the method in [10] to measure user bias (scores) by exploiting users' interests over her historical tweets. The hypothesis is that users who are more left-leaning or right-leaning share similar interests with each other. Following the method in [10], we generate representative sets of people with known public bias, and then calculate bias scores based on how closely a query users' interests match with those representative users. We define the following weak labeling function:

> **Bias-based:** *If the mean value of users' absolute bias scores – sharing a piece of news – is greater than a threshold $\tau_2$, then the news piece is weakly-labeled as fake news.*

Third, studies have shown that less credible users, such as malicious accounts or normal users who are vulnerable to fake news, are more likely to spread fake news [27]. To measure user credibility, we adopt the practical approach in [1]. The hypothesis is that less credible users are more likely to coordinate with each other and form big clusters, whereas more credible users are likely to form small clusters. We use the hierarchical clustering[5] to cluster users based on their meta-information on social media and take the reciprocal of the cluster size as the credibility score. Accordingly, we define the following weak labeling function:

> **Credibility-based:** *If a news piece has an average credibility score less than a threshold $\tau_3$, then the news is weakly-labeled as fake news.*

To determine the proper thresholds for $\tau_1$, $\tau_2$, and $\tau_3$, we vary the threshold values from $[0, 1]$ through binary search, and compare the resultant weak labels with the true labels from the training set of annotated clean data – later used to train our meta-learning model – on GossipCop, and choose the value that achieves the the best accuracy on the training set. We set the thresholds as $\tau_1 = 0.15$, $\tau_2 = 0.5$, and $\tau_3 = 0.125$. Due to the sparsity for Politifact labels, for simplicity, we use the same rules derived from the GossipCop data.

**Table 1.** The statistics of the datasets. Clean refers to manually annotated instances, whereas the weak ones are obtained by using the weak labeling functions

| Dataset | GossipCop | Politifact |
|---|---|---|
| # Clean positive | 1,546 | 303 |
| # Clean negative | 1,546 | 303 |
| # Sentiment-weak positive | 1,894 | 3,067 |
| # Sentiment-weak negative | 4,568 | 1,037 |
| # Bias-weak positive | 2,587 | 2,484 |
| # Bias-weak negative | 3,875 | 1,620 |
| # Credibility-weak positive | 2,765 | 2,963 |
| # Credibility-weak negative | 3,697 | 1,141 |

**Quality of Weak Labeling Functions.** We apply the aforementioned labeling functions and obtain the weakly labeled positive instances. We treat the news pieces discarded by the weak labeling functions as *negative* instances. The statistics are shown in Table 1. To assess the quality of these weakly-labeled instances, we compare the weak labels with the true labels on the annotated clean data in GossipCop – later used to train our meta-learning model. The accuracy of the weak labeling functions corresponding to Sentiment, Bias, and Credibility are 0.59, 0.74, 0.74, respectively. The F1-scores of these three weak labeling functions are 0.65, 0.64, 0.75. We observe that the accuracy of the labeling functions are significantly better than random (0.5) for binary classification indicating that the weak labeling functions are of acceptable quality.

---

[5] https://bit.ly/2WGK6zE.

# 4   Experiments

Now, we present the experiments to evaluate the effectiveness of MWSS. We aim to answer following evaluation questions: (1) **EQ1**: Can MWSS improve fake news classification performance by leveraging weak social supervision; (2) **EQ2**: How effective are the different sources of supervision for improving prediction performance; and (3) **EQ3**: How robust is MWSS on leveraging multiple sources?

## 4.1   Experimental Settings

**Evaluation Measures.** We use F1 score and accuracy as the evaluation metrics. We randomly choose 15% of the clean instances for validation and 10% for testing. We fix the number of weak training samples and select the amount of clean training data based on the *clean data ratio* defined as: clean ratio = $\frac{\#\text{clean labeled samples}}{\#\text{clean labeled samples} + \#\text{weak labeled samples}}$. This allows us to investigate the contribution of clean vs. weakly labeled data in later experiments. All the clean datasets are balanced with positive and negative instances. We report results on the test set with the model parameters picked with the best validation accuracy. All runs are repeated for 3 times and the average is reported.

**Base Encoders.** We use the convolutional neural networks (CNN) [9] and RoBERTa-base, a robustly optimized BERT pre-training model [12] as the encoders for learning content representations. We truncate or pad the news text to 256 tokens, and for the CNN encoder we use pre-trained WordPiece embeddings from BERT to initialize the embedding layer. For each of the $K+1$ classification heads, we employ a two-layer MLP with 300 and 768 hidden units for both the CNN and RoBERTa encoders. The LWN contains a weak label embedding layer with dimension of 256, and a three-layer MLP with $(768, 768, 1)$ hidden units for each with a sigmoid as the final output function to produce a scalar weight between 0 and 1. We use binary cross-entropy as the loss function $\ell$ for MWSS[6].

**Baselines and Learning Configurations.** We consider the following settings: (1) training only with limited amount of manually annotated **clean** data. Models include the following state-of-the-art early fake news detection methods:

- TCNN-URG [17]: This method exploits users' historical comments on news articles to learn to generate synthetic user comments. It uses a two-level CNN for prediction when user comments are not available for early detection.
- EANN [33]: This method utilizes an adversarial learning framework with an event-invariant discriminator and fake news detector. For a fair comparison, we only use the text CNN encoder.

(2) training only with **weakly** labeled data; and (3) training with both the **clean** and **weakly** labeled data as follows:

---

[6] All the data and code are available at: **this clickable link**.

- Clean+Weak: In this setting, we simply merge both the clean and weak sets (essentially treating the weak labels to be as reliable as the clean ones) and use them together for training different encoders.
- L2R [23]: L2R is the state-of-the-art algorithm for learning to re-weight (L2R) examples for training models through a meta learning process.
- Snorkel [21]: It combines multiple labeling functions given their dependency structure by solving a matrix completion-style problem. We use the label generated by Snokel as the weak label and feed it to the classification models.
- MWSS: The proposed model for jointly learning with clean data and multi-sources of weak supervision for early fake news detection.

Most of the above baseline models are geared for single sources. In order to extend them to multiple sources, we evaluated several aggregation approaches, and found that taking the majority label as the final label achieved the best performance result. We also evaluate an advanced multiple weak label aggregation method – Snorkel [19] as the multi-source baseline. Note that our MWSS model, by design, aggregates information from multiple sources and does not require a separate aggregation function like the majority voting.

### 4.2   Effectiveness of Weak Supervision and Joint Learning

To answer **EQ1**, we compare the proposed framework MWSS with the representative methods introduced in Sect. 4.1 for fake news classification. We determine

**Table 2.** Performance comparison for early fake news classification. *Clean* and *Weak* depict model performance leveraging only those subsets of the data; *Clean+Weak* is the union of both the sets.

| Methods | GossipCop | | PolitiFact | |
|---|---|---|---|---|
| | F1 | Accuracy | F1 | Accuracy |
| TCNN-URG (Clean) | 0.76 | 0.74 | 0.77 | 0.78 |
| EANN (Clean) | 0.77 | 0.74 | 0.78 | 0.81 |
| CNN (Clean) | 0.74 | 0.73 | 0.72 | 0.72 |
| CNN (Weak) | 0.73 | 0.65 | 0.33 | 0.60 |
| CNN (Clean+Weak) | 0.76 | 0.74 | 0.73 | 0.72 |
| CNN-*Snorkel* (Clean+Weak) | 0.76 | 0.75 | 0.78 | 0.73 |
| CNN-*L2R* (Clean+Weak) | 0.77 | 0.74 | 0.79 | 0.78 |
| CNN-MWSS (Clean+Weak) | **0.79** | **0.77** | **0.82** | **0.82** |
| RoBERTa (Clean) | 0.77 | 0.76 | 0.78 | 0.77 |
| RoBERTa (Weak) | 0.74 | 0.74 | 0.33 | 0.60 |
| RoBERTa (Clean+Weak) | 0.80 | 0.79 | 0.73 | 0.73 |
| RoBERTa-*Snorkel* (Clean+Weak) | 0.76 | 0.74 | 0.78 | 0.77 |
| RoBERTa-*L2R* (Clean+Weak) | 0.78 | 0.75 | 0.81 | 0.82 |
| RoBERTa-MWSS (Clean+Weak) | **0.80** | **0.80** | **0.82** | **0.82** |

the model hyper-parameters with cross-validation. For example, we set parameters $learning\_rate \in \{10^{-3}, 5 \times 10^{-4}, 10^{-4}, 5 \times 10^{-5}\}$ and choose the one that achieves the best performance on the held-out validation set. From Table 2, we make the following observations:

- Training only on clean data achieves better performance than training only on the weakly labeled data consistently across all the datasets (clean > weak).
- Among methods that only use clean data with CNN encoders, we observe TCNN-URG and EANN to achieve relatively better performance than Clean consistently. This is because TCNN-URG utilizes user comments during training to capture additional information, while EANN considers the event information in news contents (TCNN-URG>CNN-clean, and EANN>CNN-clean).
- On incorporating weakly labeled data in addition to the annotated clean data, the classification performance improves compared to that using only the clean labels (or only the weak labels) on both datasets (demonstrated by clean+weak, L2R, Snorkel > clean > weak).
- On comparing two different encoder modules, we find that RoBERTa achieves much better performance in GossipCop compared to CNN, and has a similar performance in PolitiFact. The smaller size of the PolitiFact data results in variable performance for RoBERTa.
- For methods that leverage both the weak and clean data, L2R and Snorkel perform quite well. This is because L2R assigns weight to instances based on their contribution with a held-out validation set, whereas Snorkel leverages correlations across multi-source weak labeling functions to recover the label.
- In general, our model **MWSS** achieves the best performance. We observe that MWSS > L2R and Snorkel on both the datasets. This demonstrates the importance of treating weak labels differently from the clean labels with a joint encoder for learning shared representation, separate MLPs for learning source-specific mapping functions, and learning to re-weight instances via LWN. To understand the contribution of the above model components, we perform an ablation study in the following section.

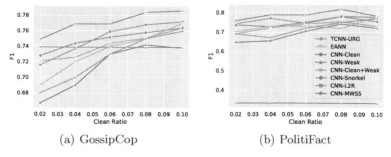

(a) GossipCop    (b) PolitiFact

**Fig. 4.** F1 score with varying clean data ratio from 0.02 to 0.1 with CNN-MWSS. The trend is the similar with RoBERTa encoder (best visualized in color). (Color figure online)

## 4.3    Impact of the Ratio of Clean to Weakly Labeled Data on Classification Performance

To answer **EQ2**, we explore how the performance of MWSS changes with the clean ratio. We set the clean ratio to vary in $\{0.02, 0.04, 0.06, 0.08, 0.1\}$. To have a consistent setting, we fix the number of weakly labeled instances and change the number of clean labeled instances accordingly. In practise, we have abundant weak labels from the heuristic weak labeling functions. The objective here is to figure out how much clean labels to add in order to boost the overall model performance. Figure 4 shows the results. We make the following observations:

- With increasing values of clean ratio, the performance increases for all methods (except *Weak* which uses a fixed amount of weakly labeled data). This shows that increasing amount of reliable clean labels obviously helps the models.
- For different clean ratio configurations, MWSS achieves the best performance compared to other baselines, i.e., MWSS > L2R and Snorkel. This shows that MWSS can more effectively utilize the clean and weak labels via its multi-source learning and re-weighting framework.
- We observe that the methods using Clean+Weak labels where we treat the weak labels to be as reliable as clean ones may not necessarily perform better than using only clean labels. This shows that simply merging the clean and weak sources of supervision without accounting for their reliability may not improve the prediction performance.

**Table 3.** F1/Accuracy on training MWSS on different weak sources with clean data.

| Dataset | Sentiment | Bias | Credibility | All Sources |
|---|---|---|---|---|
| GossipCop | 0.75/0.69 | 0.78/0.75 | 0.77/0.73 | 0.79/0.77 |
| PolitiFact | 0.75/0.75 | 0.77/0.77 | 0.75/0.73 | 0.78/0.75 |

**Table 4.** F1/Accuracy result of ablation study on modeling source-specific MLPs with different clean ratio (C-Ratio). "SH" denotes a single shared MLP and "MH" denotes multiple source-specific ones.

| Model | C-Ratio | L2R | LWN |
|---|---|---|---|
| SH | 0.02 | 0.72/0.68 | 0.73/0.72 |
| | 0.10 | 0.77/0.74 | 0.77/0.73 |
| MH | 0.02 | 0.73/0.71 | 0.75/0.71 |
| | 0.10 | 0.78/0.76 | 0.79/0.77 |

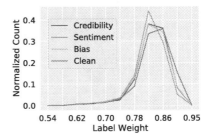

**Fig. 5.** Label weight density distribution among weak and clean instances in GossipCop. The mean of the label weight for *Credibility, Sentiment, Bias* and *Clean* are 0.86, 0.85, 0.86 and 0.87 respectively.

## 4.4   Parameter Analysis

**Impact of Source-Specific Mapping Functions:** In this experiment, we want to study the impact of modeling separate MLPs for source-specific mapping functions (modeled by $f_{\theta_k}$ in Eq. 1) in LWN and L2R as opposed to replacing them with a single shared MLP (i.e. $f_{\theta_k} = f_\theta \; \forall k$) across multiple sources. From Table 4, we observe that MWSS and L2R both work better with multiple source-specific MLPs as opposed to a single shared MLP by better capturing source-specific mapping functions from instances to corresponding weak labels. We also observe MWSS to perform better than L2R for the respective MLP configurations – demonstrating the effectiveness of our re-weighting module.

**Impact of Different Weak Sources**: To study the impact of multi-source supervision, we train MWSS separately with individual weak sources of data along with clean annotated instances with a clean ratio of 0.1. From Table 3, we observe that training MWSS with multiple weak sources achieves better performance compared to that of a single weak source – indicating complementary information from different weak sources help the model. To test whether MWSS can capture the quality of each source, we visualize the label weight distribution for each weak source and clean dataset in Fig. 5. From the weight distribution, we also observe the weight of the sentiment-source (referred as *Sentiment*) to be less than that of other sources. In addition, although the LWN is not directly trained on clean samples, it still assigns the largest weight to the clean source. These demonstrate that our model not only learns the importance of different instances but also learns the importance of the corresponding source.

## 5   Related Work

**Fake News Detection:** Fake news detection methods mainly focus on using *news contents* and with information from *social engagements* [27]. Content-based approaches exploit feature engineering or latent features to capture deception cues [16,24]. For social context based approaches, features are mainly extracted from users, posts and networks. User-based features are extracted from user profiles to measure their characteristics [2]. Post-based features represent users' responses in term of stance, topics, or credibility [2,8]. Network-based features are extracted by constructing specific networks. Recently, deep learning models are applied to learn the temporal and linguistic representation of news [17,33]. Wang *et al.* proposed an adversarial learning framework with an event-invariant discriminator and fake news detector [33]. Qian *et al.* exploited users' historical comments on news articles to learn to generate synthetic user comments for early fake news detection [17].

**Learning with Weak Supervision:** Most machine learning models rely on quality labeled data to achieve good performance where the presence of label noise or adversarial noise cause a dramatic performance drop [22]. Therefore, learning with noisy labels has been of great interest for various tasks [26,34].

Existing works attempt to rectify the weak labels by incorporating a loss correction mechanism [11,14,30]. Patrini *et al.* [14] utilize the loss correction mechanism to estimate a label corruption matrix without making use of clean labels. Other works consider the scenario where a small set of clean labels are available [6,23,35]. For example, Zheng *et al.* propose a meta label correction approach using a meta model which provides reliable labels for the main models to learn. Recent works also consider that weak signals are available from multiple sources [18,20,31] and consider the redundancy and consistency across labels. In contrast, the weak signals in our work are derived from user engagements and we do not make any assumptions about the structure of the label noise.

## 6    Conclusions and Future Work

In this paper, we develop techniques for early fake news detection leveraging weak social supervision signals from multiple sources. Our end-to-end framework MWSS is powered by meta learning with a Label Weighting Network (LWN) to capture the varying importance weights of such weak supervision signals from multiple sources during training. Extensive experiments in real-world datasets show MWSS to outperform state-of-the-art baselines without using any user engagements at prediction time. As future work, we want to explore other techniques like label correction methods to obtain high quality weak labels to further improve our models. In addition, we can extend our framework to consider other types of weak social supervision signals from social networks leveraging temporal footprints of the claims and engagements.

**Acknowledgments.** This work is, in part, supported by Global Security Initiative (GSI) at ASU and by NSF grant # 1614576.

## References

1. Abbasi, M.-A., Liu, H.: Measuring user credibility in social media. In: Greenberg, A.M., Kennedy, W.G., Bos, N.D. (eds.) SBP 2013. LNCS, vol. 7812, pp. 441–448. Springer, Heidelberg (2013). https://doi.org/10.1007/978-3-642-37210-0_48
2. Castillo, C., Mendoza, M., Poblete, B.: Information credibility on twitter. In: WWW (2011)
3. Cui, L., Wang, S., Lee, D.: Same: sentiment-aware multi-modal embedding for detecting fake news (2019)
4. Ge, L., Gao, J., Li, X., Zhang, A.: Multi-source deep learning for information trustworthiness estimation. In: KDD, ACM (2013)
5. Gentzkow, M., Shapiro, J.M., Stone, D.F.: In Handbook of media economics, vol. 1, pp. 623–645. Elsevier (2015)
6. Hendrycks, D., Mazeika, M., Wilson, D., Gimpel, K.: Using trusted data to train deep networks on labels corrupted by severe noise. In: NeurIPS (2018)
7. Hutto, C., Gilbert, E.: Vader: a parsimonious rule-based model for sentiment analysis of social media text. In: ICWSM (2014)
8. Jin, Z., Cao, J., Zhang, Y., Luo, J.: News verification by exploiting conflicting social viewpoints in microblogs. In: AAAI (2016)

9. Kim, Y.: Convolutional neural networks for sentence classification (2014)
10. Kulshrestha, J.: Quantifying search bias: Investigating sources of bias for political searches in social media. In: CSCW, ACM (2017)
11. Li, Y., Yang, J., Song, Y., Cao, L., Luo, J., Li, L.J.: Learning from noisy labels with distillation. In: ICCV (2017)
12. Yinhan, L.: A robustly optimized bert pretraining approach, Roberta (2019)
13. Ouyang, W., Chu, X., Wang, X.: Multi-source deep learning for human pose estimation. In: CVPR, IEEE (2014)
14. Patrini, G., Rozza, A., Krishna Menon, A., Nock, R., Qu, L.: A loss correction approach. In: CVPR, Making Deep Neural Networks Robust to Label Noise (2017)
15. Pennebaker, J.W., Boyd, R.L., Jordan, K., Blackburn, K.: The development and psychometric properties of liwc2015. Technical report (2015)
16. Potthast, M., Kiesel, J., Reinartz, K., Bevendorff, J., Stein, B.: A stylometric inquiry into hyperpartisan and fake news. arXiv preprint arXiv:1702.05638 (2017)
17. Qian, F., Gong, C., Sharma, K., Liu, Y.: Fake news detection with collective user intelligence. In: IJCAI, Neural User Response Generator (2018)
18. Ratner, A., Bach, S.H., Ehrenberg, H., Fries, J., Wu, S., Ré, C.: Snorkel: rapid training data creation with weak supervision. Proc. VLDB Endowment **11**(3), 269–282 (2017)
19. Ratner, A., Bach, S.H., Ehrenberg, H., Fries, J., Wu, S., Ré, C.: Snorkel: rapid training data creation with weak supervision. In: Proceedings of the VLDB Endowment. International Conference on Very Large Data Bases, vol. 11, no. 3, p. 269. NIH Public Access (2017)
20. Ratner, A., Hancock, B., Dunnmon, J., Sala, F., Pandey, S., Ré, C.: Training complex models with multi-task weak supervision. arXiv preprint arXiv:1810.02840 (2018)
21. Ratner, A., Hancock, B., Dunnmon, J., Sala, F., Pandey, S., Ré C.: Training complex models with multi-task weak supervision (2018)
22. Reed, S., Lee, H., Anguelov, D., Szegedy, C., Erhan, D., Rabinovich, A.: Training deep neural networks on noisy labels with bootstrapping. arXiv preprint arXiv:1412.6596 (2014)
23. Ren, M., Zeng, W., Yang, B., Urtasun, R.: Learning to reweight examples for robust deep learning. arXiv preprint arXiv:1803.09050 (2018)
24. Rubin, V.L., Lukoianova, T.: Truth and deception at the rhetorical structure level. J. Assoc. Inf. Sci. Technol. **66**(5), 905–917 (2015)
25. Shu, K., Mahudeswaran, D., Wang, S., Lee, D., Liu, H.: Fakenewsnet: A data repository with news content, social context and dynamic information for studying fake news on social media. arXiv preprint arXiv:1809.01286 (2018)
26. Shu, K., Mukherjee, S., Zheng, G., Awadallah, A.H., Shokouhi, M., Dumais, S.: Learning with weak supervision for email intent detection. In: SIGIR (2020)
27. Shu, K., Sliva, A., Wang, S., Tang, J., Liu, H.: A data mining perspective. KDD exploration newsletter, Fake news detection on social media (2017)
28. Shu, K., Wang, S., Liu, H.: The role of social context for fake news detection. In: WSDM, Beyond news contents (2019)
29. Stewart, R., Ermon, S.: Label-free supervision of neural networks with physics and domain knowledge. In: AAAI (2017)
30. Sukhbaatar, S., Bruna, J., Paluri, M., Bourdev, L., Fergus, R.: Training convolutional networks with noisy labels. arXiv preprint arXiv:1406.2080 (2014)
31. Varma, P., Sala, F., He, A., Ratner, A., Ré C.: Learning dependency structures for weak supervision models. In: ICML (2019)

32. Vosoughi, S., Roy, D., Aral, S.: The spread of true and false news online. Science **359**(6380), 1146–1151 (2018)
33. Wang, Y., et al.: Event adversarial neural networks for multi-modal fake news detection. In: CIKM, Eann (2018)
34. Zhang, Z.Y., Zhao, P., Jiang, Y., Zhou, Z.H.: Learning from incomplete and inaccurate supervision. In: KDD (2019)
35. Zheng, G., Awadallah, A.H., Dumais, S.: Meta label correction for learning with weak supervision. arXiv preprint arXiv:1911.03809 (2019)

# Generating Financial Reports from Macro News via Multiple Edits Neural Networks

Wenxin Hu[1], Xiaofeng Zhang[1(✉)], and Yunpeng Ren[1,2(✉)]

[1] School of Computer Science, Harbin Institute of Technology, Shenzhen, China
zhangxiaofeng@hit.edu.cn, applerenyunpeng@163.com
[2] Qianhai Financial Holdings Co., Ltd., Shenzhen, China

**Abstract.** Automatically generating financial reports given a piece of breaking macro news is quite challenging task. Essentially, this task is a text-to-text generation problem but is to learn long text, i.e., greater than 40 words, from a piece of short macro news. Moreover, the core component for human beings to generate financial reports is the logic inference given a piece of succinct macro news. To address this issue, we propose the novel multiple edits neural networks which first learns the outline for given news and then generates financial reports from the learnt outline. Particularly, the input news is first embedded via skip-gram model and is then fed into Bi-LSTM component to train the contextual representation vector. This vector is used to learn the latent word probability distribution for the generation of financial reports. To train this end to end neural network model, we have collected one hundred thousand pairs of news-report data. Extensive experiments are performed on this collected dataset. The proposed model achieves the SOTA performance against baseline models w.r.t. the evaluation criteria BLEU, ROUGE and human scores. Although the readability of the generated reports by our approach is better than that of the rest models, it remains an open problem which needs further efforts in the future.

**Keywords:** Financial data mining · Text generation model · Natural language generation

## 1 Introduction

Text-to-text generation, one of the most significant tasks in natural language generation (NLG), has attracted extensive research efforts in various sub research domains, e.g., report generation [17,19], machine translation [26,42], dialogue system [14,29] and text summarization [9]. Among these tasks, generating long text report from short text is utmost challenging especially for the generation of financial reports.

Generally, the delicate human efforts are needed to generate financial reports especially given a piece of breaking news. However, this might be quite demanding and the quality of the generated reports seriously relies on the human writers,

© Springer Nature Switzerland AG 2021
F. Hutter et al. (Eds.): ECML PKDD 2020, LNAI 12459, pp. 667–682, 2021.
https://doi.org/10.1007/978-3-030-67664-3_40

and thus are quit e diverse. To alleviate these aforementioned issues, this work is thus motivated to automatically generate financial reports via the proposed deep neural network models. Essentially, this task is also a text generation problem. There exist a good number of successful approaches for text generation such as recurrent neural network (RNN) type models and variational autoencoder (VAE) based approach [2]. To employ existing approaches, each pair of breaking news and the corresponding financial reports are fed into the deep neural networks to train model parameters. However, there exist two challenging difficulties which invalidate existing approaches. First, the length of the input breaking news is rather short, e.g., "The US Federal Reserve on Tuesday announced the establishment of a temporary repurchase agreement facility for foreign central banks and international monetary authorities amid coronavirus uncertainty". And the length of the generated reports is usually much greater than that of the input macro news, which imposes the first difficulty, i.e., long text generation issue. Second, the generation of financial reports usually involves human beings' intellectual efforts, e.g., inferring and reasoning abilities. Apparently, these challenges remain outstanding problems and thus need more research efforts.

To address these issues, we intuitively assume that the report writers may first draft an outline and then write reports based on the outline. The outline is believed to well reflect human beings' reasoning content which cannot be directly generated from the short news. Thus, the appropriate working flow should be that we generate the outline for each piece of macro news, and then the corresponding financial report is generated from the outline. By simulating the working logic of human beings, we propose this multiple edits neural network model in this paper. The proposed approach first learns the outline for given macro news and then generates financial reports from the learnt outline. Without loss of generality, these textual data, e.g., news and reports, are embedded via the skip-gram model. Then in the proposed approach, we revise the Seq2Seq module to train an end to end framework from macro news to the generated reports. The embedded news data, inspired by [27], are fed into the Bi-LSTM component to train the contextual representation vector. This vector is used to learn the latent word probability distribution for generating the outline. For the generation of long text report, a decoder component is employed and a global loss function is proposed to minimize the loss from the input news to the generated financial reports. To the best of our knowledge, this work is among the first attempts to generate domain-oriented long text report. The major contributions of the proposed approache could be summarized as follows.

- We propose the multiple edits neural networks to generate financial reports from a piece of short breaking news. The length of the generated reports is greater than 200 which is about 5 to 6 times larger than the length of the input text.
- We have collected a large-scale Chinese financial report dataset having over one hundred thousand pairs of news-report data crawled from several popular financial Websites. This dataset could be used as the benchmark dataset to

evaluate approaches for text summarization, topic model, and text-to-text generation tasks.

- We have performed extensive experiments on this dataset. Both the basedline model as well as the state-of-the-art approaches, i.e., the Seq2Seq model and the VAE model are evaluated for performance comparison and the proposed model achieves the SOTA performance against these approaches with respect to the evaluation criteria BLEU [25], ROUGE [20] and human scores.

The rest of this paper is organized as follows. Section 2 summarizes related work of text-to-text generation approaches. Section 3 details the proposed Multiple edits Neural Networks and Sect. 4 illustrates how the experiments are evaluated. We conclude this paper in Sect. 5.

## 2  Related Work

In this section, we first review the related text-to-text generation approaches including RNN [32,38] based approaches, variational autoencoder based approaches [2], and genrative adversarial networks (GAN) based approaches [10]. Then, we briefly review several most related long text generation approaches.

In the literature, the RNN type approaches have already achieved the state-of-the-art performance over the conventional techniques, e.g., n-gram based model. The RNN model can well capture the contextual information contained in a sentence [22]. However, with the increase of the text length, the vanishing gradient problem seriously deteriorates the RNN model performance, which also makes long-distance state information difficult to learn. Therefore, various long short-term memory (LSTM) models [7,11,31] have been proposed to capture long term dependencies among contextual words. To further improve model performance, Tran [34] proposed a contextual text generation model which considers contextual information by utilizing the structural information extracted from a set of documents.

The proposed sequence-to-sequence (Seq2Seq) model [3,5,36] has achieved superior model performance in various NLP related tasks such as neural translation. Cho et al. [3] proposed a neural network architecture, which integrates two RNN components acting as an encode-decoder pair. The encoder maps an input sequence with variable-length into a fixed-length semantic vector, then the decoder maps this vector back to a target sequence with variable-length. To generate more informative responses, Chen et al. [37] proposed the topic-aware sequence-to-sequence (TA-Seq2Seq) model through a joint attention mechanism integrated with a biased text generation probability distribution. Feng et al. [7] proposed a multi-topic-aware long short-term memory (MTA-LSTM) network to generate a paragraph-level Chinese essay. Gu et al. [12] incorporated the copying mechanism into the Seq2Seq learning process and proposed the CopyNet model, which already demonstrates a superior performance over the standard Seq2Seq model in text generation task. Then, an attention-enhanced attribute-to-sequence model [4] was proposed to generate review text according to the provided attribute information.

Alternatively, the VAE [2] approach was customized for the text generation task. Samuel et al. [2] introduced an RNN-based VAE generative model by incorporating distributed latent representations of entire sentences. And this VAE can explicitly model holistic properties of sentences such as style, topic, and high-level syntactic features. Miao et al. [24] proposed an inference network on discrete input to estimate the variational distribution, which models a generic variational inference framework for generative and conditional models. Also, Miao et al. [23] modelled language as a discrete latent variables in a variational auto-encoding framework and proved that the model can effectively exploit both supervised and unsupervised data in sequence-to-sequence task. Then a neural network-based generative architecture with latent stochastic variables [30] was proposed to generate diverse text. On top of the standard VAE model, Semeniuta et al. [28] proposed a hybrid architecture that blends fully feed-forward convolutional and deconvolutional components to generate a long sequence of text. Liao et al. [19] proposed the QuaSE model to edit sequence under quantifiable guidance by the revised VAE model. Yang et al. [39] proposed a topic-to-eassy generation approach by using the prior knowledge.

Compared with the one stage text generation approach via the VAE framework, the generative adversarial network (GAN) [10] was proposed to optimize the generated results. It is impossible for the standard GAN model to generate a sequence of discrete variables as the samples from a data distribution on discrete variables are intractable. Matt et al. [18] resolved this problem by using the Gumbel-softmax distribution. Zhang et al. [41] proposed a generic framework to generate text by employing LSTM and convolutional neural network (CNN) through adversarial training. Then by modeling the data generator as a stochastic policy under reinforcement learning settings, Yu et al. [40] proposed the sequence generation framework, called SeqGAN, to directly perform gradient policy updating rules. Furthermore, RankGAN [21] was proposed for generating high-quality textual descriptions by adapting the discriminator to a classifier. Then, MaskGAN [6] approach was proposed to optimize the process of generating text and LeakGAN [13] was further proposed to generate long text within 40 words.

Among these aforementioned approaches, the Seq2Seq based model and the Pointer-Generator Networks [27] for the long text generation task [1] are most related to our approach. Bahdanau et al. [1] further proposed Seq2Seq+Attn model to improve the performance of encoder-decoder process. The employed fixed-length vector allows the proposed model to search for more relevant source sentences. In [27], a hybrid pointer-generator network was proposed. The proposed network first generates words from the source text via pointing sampling technique, which aids accurate reproduction of information. Then, coverage approach was used to keep track of what has been summarized, which discourages repetition.

Although the Seq2Seq based model can generate long text given the short input text data, the inferring and reasoning content cannot be acquired by these approaches. Moreover, despite some reinforcement learning based approaches [39], the length of generated long text is usually less than 40 words [13], which is not suitable for the generation of financial reports. Therefore, this multiple edits neural network is proposed to simulate the inferring and reasoning ability of human beings.

## 3    The Proposed Approach

In this section, the proposed multiple edits neural network is briefly illustrated as follows. The proposed approach consists of two sub components, i.e., outline generation component and text generation component as plotted in Fig. 1. We first generate the outline text $O$ from the input macro news $X$ and then generate the target text $Y'$ from both the input macro news $X$ and the outline text $O$. Especially, the outline text $O$ and the target text $Y'$ have the same length. For the outline generation component, we will generate the outline text $O$ from the input macro news $X$. The macro news data $X$ contains $m$ words and both of the report $Y$ and the outline text $O$ contain $l$ words, where $l$ is far greater than $m$. Both $X$ and $Y$ are embedded into word vectors. The word vectors of $X$ and $Y$ are respectively fed into the bidirectional LSTM (Bi-LSTM) module and the employed LSTM module to co-train the outline $O$. The attention mechanism is introduced to further emphasize the important words of the input $X$. As for the text generation component, the learnt $O$ is used to generate the text $Y'$ via the general VAE model. The proposed approach is trained in a two-stage manner. Each component is separately illustrated in the following subsections.

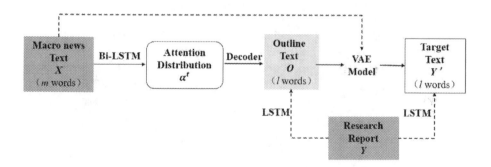

**Fig. 1.** The framework of the proposed multiple edits neural network.

### 3.1    Outline Generation Component

Inspired by the TA-Seq2Seq model [37] and Pointer-Generator Networks [27], we design this outline generation component as plotted in Fig. 2. Let $X = \{x_1, x_2, ..., x_m\}$ represent the macro news data and $O = \{o_1, o_2, ..., o_l\}$ represent the outline to be generated.

In this component, a sequence of $X$ is first embedded through the widely adopted skip-gram model to capture their contextual information. Then, the embedded features of $X$ are fed into the Bi-LSTM module in a one-by-one manner. The output of the Bi-LSTM is a sequence of hidden states $h_t^e$, calculated as

$$h_t^e = f_{encoder}(x_t, h_{t-1}^e) \| f_{encoder}(x_t, h_{t+1}^e), \tag{1}$$

where $f_{encoder}(x_t, h_{t-1}^e)$ and $f_{encoder}(x_t, h_{t+1}^e)$ represent the hidden states of forward network and backward network, respectively. To generate financial report $Y$, the core semantic content of $Y$ are introduced to co-train the generation of $O$. To make use of the position embedding information, similar to transformer model [35], $Y$ is marked with $<start>$ at the beginning and $<end>$ at the end. Then $Y$ is fed into a LSTM model to acquire the hidden states of the decoder. At each step $t$, the decoder receives the word embedding of the previous word as well as the decoder states $s_t^d$ which is computed as

$$s_t^d = f_{decoder}(\hat{o}_{t-1}, s_{t-1}^d). \tag{2}$$

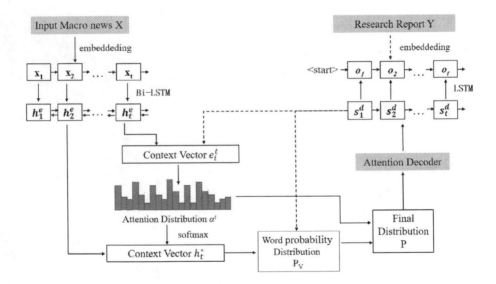

**Fig. 2.** Details of the proposed outline generator.

The attention score $e_i^t$ is calculated as

$$e_i^t = \nu tanh(W_h h_i^e + W_s s_t^d + w_c c_i^t + b_{attn}), \tag{3}$$

where $\nu$, $W_h$, $W_s$, $w_c$ and $b_{attn}$ are learnable parameters. Similarly, $c^t$ is a cover vector which contains the attention information for all previous moments, calculated as follows

$$c^t = \sum_{t=0}^{t-1} \alpha^t, \tag{4}$$

where $\alpha^t$ is the attention probability distribution generally calculated using the softmax function, written as

$$\alpha^t = \frac{exp(e_i^t)}{\sum exp(e_i^t)}. \tag{5}$$

The calculated attention probability distribution $\alpha^t$ is then used to weight the encoder hidden states $h_i^e$ to generate the context vector $h_t^*$, calculated as

$$h_t^* = \sum_i \alpha_i^t h_i^e, \tag{6}$$

The context vector $h_t^*$, which can be seen as a fixed size representation of the input source information at current step, is then concatenated with the decoder hidden state $s_t^d$. After that, $h_t^*$ is fed into a two fully connected layers to generate the vocabulary distribution $P_V$, computed as follows

$$P_V = softmax(V'(V[s_t^d, h_t^*] + b) + b'), \tag{7}$$

where $V$, $V'$, $b$ and $b'$ are learnable parameters of the two linear layers.

The merit of the original pointer generation is to generate unseen words and this is fulfilled by sampling terms from the vocabulary corpus. Therefore, a soft switch $p_{gen}$ is designed to well balance the generation of a word either from the vocabulary corpus by sampling from $P_V$, or from the input sequence by sampling from the attention distribution $\alpha^t$. At time $t$, $p_{gen}$ is calculated using the context vector $h_t^*$, the decoder state $s_t^d$ and the decoder input $o_t$, given as

$$p_{gen} = \sigma(W_{h*}^T h_t^* + W_{s^t}^T s_t^d + W_x^T o_t + b_{gen}), p_{gen} \in [0,1], \tag{8}$$

where vectors $W_{h*}^T$, $W_{s^t}^T$, $W_x^T$ and scalar $b_{gen}$ are learnable parameters and $\sigma$ is the sigmoid function. Further, we obtain the final probability distribution over the extended vocabulary set, given as

$$P(w) = p_{gen}P_V(w) + (1 - p_{gen}) \sum_{t:w_i=w} \alpha^t. \tag{9}$$

If $w$ is an out-of-vocabulary word, the first term $P_V(w)$ becomes 0. Similarly, if $w$ does not appear in the source document, the second term of this Equation is 0. On one hand, Eq. (9) solves the problem of generating out-of-vocabulary words. On the other hand, it makes full use of the attention distribution of the original input text.

To avoid generating repetitive text, we also define a *covloss* function to penalize repeatedly attending to the same location, calculated as

$$covloss^t = \sum_i min(c_i^t, \alpha_i^t). \tag{10}$$

If the generated word at the current moment has been generated at the past, the attention probability of the current word's position is restricted to decrease.

Therefore, the corresponding loss function of the outline generation model $L_{outline}$ could be modeled as the superposition of the primary loss function and *covloss* function, calculated as

$$L_{outline} = -logP(w_t^*) + \lambda \sum_i min(c_i^t, \alpha_i^t), \tag{11}$$

where $w_t^*$ is a generated word and the hyper-parameter $\lambda$ is set to 1.

### 3.2 Text Generation Component

The text generation component is considered as a decoding process and thus the VAE based module is a natural choice to be the decoding component. To generate text $Y'$, we take both the macro news $X$ and the pre-trained outline text $O$ as the model input. Then, we employ the VAE model to learn the posterior distribution of latent variables $P(z|X)$ and samples data from this distribution as depicted in Fig. 1. The $P(z|X)$ can now be rewritten as

$$P(z|X, O) = \frac{P(X, O|z) \cdot P(z)}{P(X, O)}. \tag{12}$$

The log likelihood of sampled data $x$ can be acquired by maximizing the ELBO problem, defined as follows

$$logP(X, O) \geq E_{q(X, O|z)}logP(X, O|z) - KL(q(X, O|z)||P(z)) \tag{13}$$

In this ELBO problem, the first term is the sampled data from $P(X, O|z)$ to calculate the cross entropy loss and the second term, i.e., the KL divergence, is used to ensure that the posterior distribution is close to the prior distribution. In this component, the decoder's hidden state $h_t^d$ is updated as

$$h_t^d = f_{encoder}(\hat{y}_{t-1}, h_{t-1}^d). \tag{14}$$

Therefore, the loss function $L_{report}$ for the text generation component can be written as

$$L_{report} = \sum_{i=0}^{l} -logP(y_i' = y_i|h^d) \tag{15}$$

At last, the overall loss function of the proposed approach can be calculated as the summation of $L_{outline}$ and $L_{report}$, written as

$$Loss = L_{outline} + L_{report}. \tag{16}$$

## 4    Experiments

### 4.1    Dataset and Data Preprocessing

To evaluate the proposed approach, we first carefully collected a large-scale experimental dataset[1] from several popular Chinese financial Websites. We

---

[1] We make available our dataset https://github.com/papersharing/news-and-reports-dataset.

respectively crawled 69,960, 52,360 and 8,017 pairs of news-report data from each data source. Each piece of news is associated with a financial report in the original Website and thus we can form the news-report pair. For data preprocessing, we eliminated the numeric symbols and other special characters from both news data and report data. One sample pair of data is given in Table 1. The left column is the macro news and the right column is the corresponding financial report written by a finance specialist.

To segment the Chinese macro news data, an open source tool ("jieba") is utilized. After word segmentation, the average and median length of the macro news are 28 and 24 words, respectively. Whilst the average and median length of the financial reports are 341 and 331 words, respectively. As the text generation component requires the length of the input data should be the same, we therefore truncated or lengthened the macro news and financial reports to the same length. Now, the length of input data is set to 30 words, and length of financial report is set to 200 words.

**Table 1.** A sample of processed macro news and the corresponding financial report.

| Macro news | Corresponding financial report |
|---|---|
| Shenzhen Expway Co., Ltd. (600548) released the semiannual report for 2018, which achieved the operating income of 2.68 billion yuan. The income increase 16.9% year-on-year and 19.2% month-on-month, which realized the earnings per share of 0.262 yuan. The non-recurring profit and loss in the second quarter was 175 million yuan in compensation for the demolished property of Meiguan Company | Comments: The toll revenue of highways has grown steadily. Overall, the highway traffic and toll revenue are growing steadily and new acquired highway projects will continue to increase the income from main business. The main reason of the increased highway costs was the consolidation of accounting statement of Changsha Ring Road and Yichang highway projects in the middle of last year, which led to the increased labor and depreciation costs this year. The company will further develop the main business of highway projects and promote the construction of Outer Ring Road and GuangShen Highway. The toll project of highway remains stable and the environmental protection projects is developing actively. Also, diversified businesses such as Meilinguan project have a large profit opportunity. The investment rating of this company is identified as recommended |

After word segmentation, the term frequency is quite imbalanced. To further alleviate this issue, we only kept the word in the vocabulary corpus if its term frequency (TF) is higher than 5. By doing so, the vocabulary corpus, used for generating unseen words, contains 63,782 words and we filter out about 0.4%

terms. Moreover, the vocabulary is marked with four tokens, i.e., padding token (PAD), unknown token (UNK), start position token (START) and end position token (END). The PAD is used to fill into the input of the encoder and decoder. The UNK is used to represent words that are not in the vocabulary corpus which are usually rare but meaningful words, such as the entity name (a company or a person). The START and END are added at the beginning and the end of each report, respectively.

## 4.2 Experimental Settings

Experiments are to be evaluated on this collected dataset. The experimental settings as well as the parameter settings are prepared as follows.

To evaluate the proposed approach as well as the baseline models, we randomly chose 90% of the dataset as the training set and the rest 10% as the testing set. The length of the generated report $Y'$ is required to be 200. For the outline generation component, the one-hot vector is generally adopted for the embedding of text data. Without loss of generality, both $X$ and $Y$ are embedded via the high-dimensional one-hot vector and then mapped into a low-dimensional 128 vector. The number of hidden units in the LSTM component is set to 256. The maximum length of encode and decoder are set to 30 and 200, respectively. The learning rate is set to 0.005, the batch size is set to 16 and the weight for repetitive generation loss is set to 1. To improve the efficiency of encoding and decoding process, we use beam search strategies [8] to generate the outline text and the beam parameter is set to 3. For the text generation component, the VAE based component is to be trained. In the proposed VAE component, the dimension of the hidden variables is set to 120, the dropout rate is set to 0.5 to avoid over-fitting problem, the batch size is set to 16 and the parameter of the beam search is also set to 3. We trained the model on two NVIDIA GeForce GTX 1080 Ti GPU and an Intel Core i7-6700 CPU with 64G RAM and 2TB hard disk. The operating system is 64-bit Ubuntu 16.04.

## 4.3 Baseline Models

To evaluate the model performance, several baseline and the state-of-the-art approaches, i.e., the Seq2Seq model [33], the Seq2Seq+Attn model [1], the Pointer-Generator model [27] and the VAE model [16] are performed. Details of these approaches are illustrated as follows.

- The Seq2Seq model is a baseline model for the text generation task. This approach already achieves a superior model performance in various text-to-text generation problem.
- The Seq2Seq+Attn model [1] extends the original Seq2Seq model, and is originally proposed for neural machine translation task. In this model, a fixed-length vector is adopted for both encoder and decoder component, and it allows to automatically soft-search relevant words from the input source sentence to predict the next word to be generated.

- The Pointer-Generator model is first proposed for text summarization task. In this model, a hybrid pointer-generator network is designed to sample words from the input source text data via the so-called pointing process, and then integrates the coverage mechanism to keep track of generated words to penalize repetitive generation.
- The VAE model is one of the most widely adopted generative model. It learns the neural network model by minimizing the error between the reconstructed data and the input data. The learnt decoder component, low-dimensional latent variable space, could then be applied to generate high-dimensional data.

## 4.4   Evaluation Metric

For the evaluation criteria about the text generation task, both the objective and subjective criteria are chosen in the experiments including the Bilingual Evaluation Understudy (BLEU), ROUGE and human score. As most Chinese words consist of less than 5 characters, we chose the BLEU-2, BLEU-3 and BLEU-4 scores as the evaluation metric to evaluate the results generated by all approaches.

## 4.5   Evaluation Results

The evaluation results of the proposed multiple edits neural network model as well as all compared methods are respectively reported in Table 2, Table 3 and Table 4.

**Table 2.** Results of our model and compared methods in terms of BLEU criteria.

| Methods | BLEU-2 | BLEU-3 | BLEU-4 |
|---|---|---|---|
| Seq2Seq | 0.253 | 0.291 | 0.102 |
| Seq2Seq+Attn | 0.289 | 0.332 | 0.112 |
| Pointer-generator | 0.322 | 0.366 | 0.142 |
| VAE | 0.311 | 0.233 | 0.08 |
| **Multiple edits neural network** | **0.432** | **0.389** | **0.164** |

For objective evaluation results, Table 2 reports the evaluation results on the BLEU criteria. The BLEU-2, BLEU-3 and BLEU-4 scores of our proposed approach are 0.432, 0.389 and 0.164 which are 34%, 6%, 15% higher than the scores of the compared methods, respectively. Table 3 reports the evaluation results on the ROUGE criteria. For ROUGE criterion, this experiment adopts ROUGE-1, ROUGE-2 and ROUGE-L scores as the detailed evaluation metric. Obviously, the Pointer-Generator model is more effective with respect to the

**Table 3.** Results of our model and compared methods in terms of ROUGE criteria.

| Methods | ROUGE-1 | ROUGE-2 | ROUGE-L |
|---|---|---|---|
| Seq2Seq | 0.032 | 0.021 | 0.043 |
| Seq2Seq+Attn | 0.066 | 0.053 | 0.095 |
| Pointer-generator | 0.133 | 0.087 | 0.114 |
| VAE | 0.081 | 0.011 | 0.023 |
| **Multiple edits neural network** | **0.141** | **0.084** | **0.127** |

**Table 4.** Results of our model and compared methods in terms of human score.

| Methods | Fluency | Consistency |
|---|---|---|
| Seq2Seq | 1.54 | 0.87 |
| Seq2Seq+Attn | 2.86 | 2.31 |
| Pointer-generator | 3.35 | 3.58 |
| VAE | 2.03 | 2.19 |
| **Multiple edits neural network** | **3.62** | **3.69** |

ROUGH criterion among all compared methods. The proposed model outperforms the Pointer-Generator model w.r.t. ROUGE-1, ROUGE-2 and ROUGE-L. It is also noticed that for ROUGE-2, the score of the Pointer-Generator is close to our proposed model. Moreover, the ROUGE-1 and ROUGE-L scores of our model are respectively 6%, 11% higher than that of the Pointer-Generator model. From the results about the objective evaluation criteria, it is obvious that the proposed approach already achieves a better model performance.

For subjective evaluation results, we chose the human scores to further evaluate the fluency and consistency of the generated reports. In this experiment, we randomly chose 1000 financial reports and manually annotated these reports. This human score annotation is also a common practice in NLP related task. Table 4 reports the evaluation results on the human scores of all approaches. The higher this value, the better the model performance. From this table, it is well noticed that the multiple edits neural network achieves the best human score values. Note that the second best model is the Pointer-Generator model which is also consistent with our expectation.

To summarize, it is well noticed that, from the evaluation results of the BLEU, ROUGE and human score, the model performance of the proposed multiple edits neural network is superior to the baseline model as well as the state-of-the-art approaches. Both these objective and subjective evaluation results already verify the effectiveness of the proposed approach in generating long text from short text. To further evaluate the quality of the generated reports, we also report one generated report extracted from the testing results, given in Table 5, for users to make their own subjective judgement. The right column is the generated reports, and we highlight the correct content in blue color. The coverage ratio of the correctly generated content is acceptable.

**Table 5.** An input macro news and the corresponding generated report.

| Input macro news | Generated reports |
|---|---|
| The European sovereign debt crisis is a "sequelae" phenomenon of the relief policy of financial crisis. The global economy may slow down and fall into a stage of "high debt and low growth" due to the sovereign debt crisis because of the five European countries. Market demand will decrease and the process of going out of the global economic crisis will be prolonged | We believe that the expectations of global economic growth may show up in month, which is the highest level since the beginning of the year. The global economic recovery may get a new height since the month of the year. Furthermore, the economic growth of the United States may get a new height recently. Our view is that the US economic growth is the major factor in global economic growth. On the day of the month, the US Federal Reserve announced its annual expectations of economic growth, but we believe that the expectations of the US Federal Reserve may be the expectations to raise interest rates. Monthly exports increased by a percentage point from the previous month. Judging from the economic growth data, the economic growth rate of the United States will be respectively in the year. And the pressure on the global economic recovery will decrease, economic indexes of European will rise. We believe that the expected economic growth of US Federal Reserve is to be in the year. Economic growth is expected to be in month. Annual economic growth will effect by the period of interest rate cycle since the beginning of the year. The global economy will continue to slow down |

## 4.6 Discussion

Although the Seq2Seq model is one of the most important SOTA approaches for text generation task, its performance is not satisfying for this specific task, i.e., generating long text from short text. From Table 2 to Table 4, it could be seen that the Seq2Seq model only achieves the lowest score especially in the human score measuring the consistency of the generated text and target text. The possible reason is that the lack of sufficient information of the input data. Therefore, it is really challenging to generate a longer text given such insufficient information. Accordingly, the generated text using these models tend to repeat themselves. However, the Seq2Seq model well suits the situation where the length of input data is equal or greater than that of the output data, such as machine translation and dialogue system. The Seq2Seq+Attn model could significantly improve the quality of the generated text. This observation verifies the effectiveness of the attention mechanism with the soft-search strategy. The Pointer-Generator model also employs an attention mechanism and performs well in short-to-long text generation task. Apparently, the attention mechanism truly plays an important role in short-to-long text generation. Compared with the RNN-based neural network models, the VAE-type model directly models the end-to-end generation process. The learnt low-dimensional latent variable layer is thus able to decode (i.e., generate) a high-dimensional output data. This is the possible reason why the VAE type model could also achieve higher evaluation scores. Furthermore, the VAE model with the attention mechanism could emphasize more important words to be generated. The appropriate sequence of the generated words could be guaranteed by the Seq2Seq type model. This is the behind reason why we design our proposed model under the unified framework of combining the VAE structure with the RNN type structure.

## 5   Conclusion and Future Work

In this paper, we propose this novel multiple edits neural network particularly for generating financial reports from short breaking news. This task is a challenging long text generation task. In addition to the existing SOTA long text approaches, the generated text length of our approach could even reach to 200 words whereas the length of the input text is rather limited. To the best of our knowledge, this is among the first attempts to generate the longer text from a comparably short input data. The proposed approach is trained in a two-stage manner. In the first stage, we generate the outline text using the proposed outline generation component analogous to the framework of the Pointer-Generator network with attention mechanism. In the second stage, we generate the financial reports from the outline using the revised VAE component. To evaluate the model performance, a large-scale experimental dataset is collected which consists of over one hundred thousand pairs of news-report data. Extensive experiments are then evaluated on this dataset. From the promising experimental results, it is well noticed that the proposed approach significantly outperforms both the baseline and the state-of-the-art approaches w.r.t. the evaluation criteria, i.e., BLEU, ROUGE and human score. This verifies the effectiveness of the proposed approach, and thus could be applied to largely save the manual cost of finance specialists in the finance industry. In the near future, we will further investigate a better way to improve the readability of the generated reports.

## References

1. Bahdanau, D., Cho, K., Bengio, Y.: Neural machine translation by jointly learning to align and translate. arXiv preprint arXiv:1409.0473 (2014)

2. Bowman, S.R., Vilnis, L., Vinyals, O., Dai, A.M., Jozefowicz, R., Bengio, S.: Generating sentences from a continuous space. arXiv preprint arXiv:1511.06349 (2015)
3. Cho, K., et al.: Learning phrase representations using rnn encoder-decoder for statistical machine translation. arXiv preprint arXiv:1406.1078 (2014)
4. Dong, L., Huang, S., Wei, F., Lapata, M., Zhou, M., Xu, K.: Learning to generate product reviews from attributes. In: Proceedings of EACL, pp. 623–632 (2017)
5. Dušek, O., Jurčíček, F.: Sequence-to-sequence generation for spoken dialogue via deep syntax trees and strings. arXiv preprint arXiv:1606.05491 (2016)
6. Fedus, W., Goodfellow, I., Dai, A.M.: Maskgan: better text generation via filling in the_. arXiv preprint arXiv:1801.07736 (2018)
7. Feng, X., Liu, M., Liu, J., Qin, B., Sun, Y., Liu, T.: Topic-to-essay generation with neural networks. In: Proceedings of the Twenty-Seventh International Joint Conference on Artificial Intelligence, pp. 4078–4084 (2018)
8. Freitag, M., Al-Onaizan, Y.: Beam search strategies for neural machine translation. arXiv preprint arXiv:1702.01806 (2017)
9. Genest, P.E., Lapalme, G.: Framework for abstractive summarization using text-to-text generation. In: Proceedings of the Workshop on Monolingual Text-to-Text Generation, pp. 64–73 (2011)
10. Goodfellow, I., et al.: Generative adversarial nets. In: Advances in Neural Information Processing Systems, pp. 2672–2680 (2014)
11. Graves, A.: Generating sequences with recurrent neural networks. arXiv preprint arXiv:1308.0850 (2013)
12. Gu, J., Lu, Z., Li, H., Li, V.O.K.: Incorporating copying mechanism in sequence-to-sequence learning. arXiv preprint arXiv:1603.06393 (2016)
13. Guo, J., Lu, S., Cai, H., Zhang, W., Yu, Y., Wang, J.: Long text generation via adversarial training with leaked information. In: Proceedings of AAAI (2018)
14. Hao, Y., Liu, H., He, S., Liu, K., Zhao, J.: Pattern-revising enhanced simple question answering over knowledge bases. In: Proceedings of International Conference on Computational Linguistics, pp. 3272–3282 (2018)
15. Hochreiter, S., Schmidhuber, J.: Long short-term memory. Neural Comput. 9(8), 1735–1780 (1997)
16. Kingma, D.P., Welling, M.: Auto-encoding variational bayes. arXiv preprint arXiv:1312.6114 (2013)
17. Kukich, K.: Design of a knowledge-based report generator. In: Proceedings of ACL, pp. 145–150 (1983)
18. Kusner, M.J., Hernández-Lobato, J.M.: Gans for sequences of discrete elements with the gumbel-softmax distribution. arXiv preprint arXiv:1611.04051(2016)
19. Liao, Y., Bing, L., Li, P., Shi, S., Lam, W., Zhang, T.: Quase: sequence editing under quantifiable guidance. In: Proceedings of EMNLP, pp. 3855–3864 (2018)
20. Lin, C.Y.: Rouge: a package for automatic evaluation of summaries. In: Proceedings of the Workshop on Text Summarization Branches Out (WAS 2004) (2004)
21. Lin, K., Li, D., He, X., Zhang, Z., Sun, M.-T.: Adversarial ranking for language generation. In: Advances in Neural Information Processing Systems, pp. 3155–3165 (2017)
22. Lin, R., Liu, S., Yang, M., Li, M., Zhou, M., Li, S.: Hierarchical recurrent neural network for document modeling. In: Proceedings of EMNLP, pp. 899–907 (2015)
23. Miao, Y., Blunsom, P.: Language as a latent variable: Discrete generative models for sentence compression. arXiv preprint arXiv:1609.07317 (2016)
24. Miao, Y., Yu, L., Blunsom, P.: Neural variational inference for text processing. In: Proceedings of ICML, pp. 1727–1736 (2016)

25. Papineni, K., Roukos, S., Ward, T., Zhu, W.-J.: Bleu: a method for automatic evaluation of machine translation. In: Proceedings of ACL, pp. 311–318 (2002)
26. Quirk, C., Brockett, C., Dolan, W.B.: Monolingual machine translation for paraphrase generation. In: Proceedings of EMNLP, pp. 142–149 (2004)
27. See, A., Liu, P.J., Manning, C.D.: Get to the point: Summarization with pointer-generator networks. arXiv preprint arXiv:1704.04368 (2017)
28. Semeniuta, S., Severyn, A., Barth, E.: A hybrid convolutional variational autoencoder for text generation. arXiv preprint arXiv:1702.02390 (2017)
29. Serban, I.V.: Multiresolution recurrent neural networks: an application to dialogue response generation. In: Proceedings of AAAI (2017)
30. Serban, I.V.: A hierarchical latent variable encoder-decoder model for generating dialogues. In: Proceedings of AAAI (2017)
31. Sundermeyer, M., Schlüter, R., Ney, H.: Lstm neural networks for language modeling. In: Thirteenth Annual Conference of the International Speech Communication Association (2012)
32. Sutskever, I., Martens, J., Hinton, G.E.: Generating text with recurrent neural networks. In: Proceedings of ICML, pp. 1017–1024 (2011)
33. Sutskever, I., Vinyals, O., Le, Q.V.E.: Sequence to sequence learning with neural networks. In: Advances in Neural Information Processing Systems, pp. 3104–3112 (2014)
34. Tran, Q.H., Zukerman, I., Haffari, G.: Inter-document contextual language model. In: Proceedings of the 2016 Conference of the North American Chapter of the Association for Computational Linguistics: Human Language Technologies, pp. 762–766 (2016)
35. Vaswani, A.: Attention is all you need. In: Proceedings of the 31st International Conference on Neural Information Processing Systems, pp. 5998–6008 (2017)
36. Vinyals, O., Le, Q.: A neural conversational model. arXiv preprint arXiv:1506.05869 (2015)
37. Xing, C., et al.: Topic aware neural response generation. In: Proceedings of AAAI (2017)
38. Xue, Y., et al.: Multimodal recurrent model with attention for automated radiology report generation. In: Frangi, A.F., Schnabel, J.A., Davatzikos, C., Alberola-López, C., Fichtinger, G. (eds.) MICCAI 2018. LNCS, vol. 11070, pp. 457–466. Springer, Cham (2018). https://doi.org/10.1007/978-3-030-00928-1_52
39. Yang, P., Li, L., Luo, F., Liu, T., Sun, X.: Enhancing topic-to-essay generation with external commonsense knowledge. In: Proceedings of ACL, pp. 2002–2012 (2019)
40. Yu, L., Zhang, W., Wang, J., Yu, Y.: Seqgan: sequence generative adversarial nets with policy gradient. In: Proceedings of AAAI (2017)
41. Zhang, Y., Gan, Z., Carin, L.: Generating text via adversarial training. In: NIPS workshop on Adversarial Training, vol. 21 (2016)
42. Zhao, S., Niu, C., Zhou, M., Liu, T., Li, S.: Combining multiple resources to improve smt-based paraphrasing model. In: Proceedings of ACL, pp. 1021–1029 (2008)

# Continual Learning with Knowledge Transfer for Sentiment Classification

Zixuan Ke[1], Bing Liu[1(✉)], Hao Wang[2], and Lei Shu[1]

[1] University of Illinois at Chicago, Chicago, USA
{zke4,liub}@uic.edu, shulindt@gmail.com
[2] Southwest Jiaotong University, Chengdu, China
cshaowang@gmail.com

**Abstract.** This paper studies continual learning (CL) for sentiment classification (SC). In this setting, the CL system learns a sequence of SC tasks incrementally in a neural network, where each task builds a classifier to classify the sentiment of reviews of a particular product category or domain. Two natural questions are: Can the system transfer the knowledge learned in the past from the previous tasks to the new task to help it learn a better model for the new task? And, can old models for previous tasks be improved in the process as well? This paper proposes a novel technique called KAN to achieve these objectives. KAN can markedly improve the SC accuracy of both the new task and the old tasks via forward and backward knowledge transfer. The effectiveness of KAN is demonstrated through extensive experiments (Code and data are available at: https://github.com/ZixuanKe/LifelongSentClass).

## 1 Introduction

Continual learning (CL) aims to learn a sequence of tasks incrementally [4,18]. Once a task is learned, its training data is typically forgotten. The focus of the existing CL research has been on solving the catastrophic forgetting (CF) problem [4,18]. CF means that when a neural network learns a sequence of tasks, the learning of each new task is likely to change the network weights or parameters learned for previous tasks, which degrades the model accuracy for the previous tasks [17]. There are two main CL settings in the existing research:

**Class Continual Learning (CCL):** In CCL, each task consists of one or more classes to be learned. Only one model is built for all classes seen so far. In testing, a test instance from any class may be presented to the model for it to classify without giving it any task information used in training.

**Task Continual Learning (TCL).** In TCL, each task is a separate classification problem (e.g., one classifying different breeds of dogs and another classifying different types of birds). TCL builds a set of classification models (one per task) in one neural network. In testing, the system knows which task each test instance belongs to and uses only the model for the task to classify the test instance.

© Springer Nature Switzerland AG 2021
F. Hutter et al. (Eds.): ECML PKDD 2020, LNAI 12459, pp. 683–698, 2021.
https://doi.org/10.1007/978-3-030-67664-3_41

In this paper, we work in the TCL setting to continually learn a sequence of sentiment analysis (SA) tasks. Typically, a SA company has to work for many clients and each client wants to study public opinions about one or more categories of its products/services and those of its competitors. The sentiment analysis of each category of products/services is a task. For confidentiality, a client often does not allow the SA company to share its data with or use its data for any other client. Continual learning is a natural fit. In this case, we also want to improve the SA accuracy over time without breaching confidentiality. This presents two key challenges: (1) how to transfer the knowledge learned from the previous tasks to the new task to help it learn better without using the previous tasks' data, and (2) how to improve old models for the previous tasks in the process without CF? In [15], the authors showed that CL can help improve the accuracy of document-level sentiment classification (SC), which is a sub-problem of SA [13]. In this paper, we propose a significantly better model, called KAN (*Knowledge Accessibility Network*). Note that each task here is a two-class SC problem, i.e., classifying whether a review for a product is positive or negative.

A fair amount of work has been done on CL. However, existing techniques have mainly focused on dealing with catastrophic forgetting (CF) [4,18]. In learning a new task, they typically try to make the weights update toward less harmful directions to previous tasks, or to prevent the important weights for previous tasks from being significantly changed. We will detail these and other related work in the next section. Dealing with only CF is far from sufficient for SC. In most existing studies of CL, the tasks are quite different and have little shared knowledge. It thus makes sense to focus on dealing with CF. However, for SC, the tasks are similar because words and phrases used to express sentiments for different products/tasks are similar. As we will see in Sect. 4.4, CF is not a major problem in CL for SC due to the shared knowledge across tasks. Our main goal is thus to leverage the shared knowledge among tasks to perform significantly better than learning individual tasks separately in isolation.

To achieve the goal of leveraging the shared knowledge among tasks to improve the SC accuracy, KAN uses two sub-networks, the *main continual learning* (MCL) network and the *accessibility* (AC) network. The core of MCL is a *knowledge base* (KB), which stores the knowledge learned from all trained tasks. In learning each new task, the AC network decides which part of the past knowledge is useful to the new task and can be shared. This enables *forward knowledge transfer*. Also importantly, the shared knowledge is enhanced during the new task training using its data, which results in *backward knowledge transfer*. Thus, KAN not only improves the model accuracy of the new task but also improves the accuracy of the previous tasks without any additional operations. Extensive experiments show that KAN markedly outperforms state-of-the-art baselines.

## 2  Related Work

Continual learning (CL) has been researched fairly extensively in machine learning (see the surveys in [4,18]). Existing approaches have primarily focused on dealing with catastrophic forgetting (CF). Lifelong learning is also closely related [3,4,22,27], which mainly aims to improve the new task learning through forward knowledge transfer. We discuss them in turn and also their applications in sentiment classification (SC).

**Continual Learning.** Several approaches have been proposed to deal with CF:

*Regularization-based methods,* such as those in [9,12,23], add a regularization in the loss function to consolidate previous knowledge when learning a new task.

*Parameter isolation-based methods,* such as those in [6,16,24], make different subsets of the model parameters dedicated to different tasks. They identify the parts of the network that are important to the previous tasks and mask them out during the training of the new task.

*Gradient projection-based methods,* such as that in [32], ensure the gradient updates occur only in the orthogonal direction to the input of the previous tasks. Then, the weight updating for the new task have little effect on the weights for the previous tasks.

*Exemplar and replay-based methods,* such as those in [2,14,20], retain an exemplar set that best approximates the previous tasks to help train the new task. The methods in [7,8,21,25] instead took the approach of building data generators for the previous tasks so that in learning the new task, they can use some generated data for previous tasks to help avoid forgetting.

As these methods are mainly for avoiding CF, after learning a sequence of tasks, their final models are typically worse than learning each task separately. The proposed KAN not only deals with CF, but also perform forward and backward transfer to improve the accuracy of both the past and the new tasks.

To our knowledge, SRK [15] is the only CL method for sentiment classification (SC). It consists of two networks, a feature learning network and a knowledge retention network, which are combined to perform CL. However, SRK only does forward transfer as it protects the past knowledge and thus cannot do backward transfer as KAN does. More importantly, due to this protection, its forward transfer also suffers because it cannot adapt the previous knowledge but only use it without change. Its results are thus poorer than KAN. The SRK paper also showed that CF is not a major issue for continual SC learning as the SC tasks are highly similar, which also explains why adaption of the previous knowledge in forward transfer in KAN does not cause CF.

KAN is closely related to the TCL system HAT [24] as HAT also trains a binary mask using hard attention. However, HAT's hard attention is for identifying what past knowledge in the network should be protected for each previous task so that the new task learning will not modify this previous knowledge. This is effective for avoiding CF, not appropriate for our SC tasks due to the shared knowledge across tasks in SC. KAN trains an accessibility mask for the current/new task to decide what previous knowledge can be accessed by or shared with the current task to enable both forward and backward knowledge transfer. There is no concept of knowledge transfer in HAT. In terms of architecture, KAN has two sub-networks: the accessibility (AC) network and the main continual learning (MCL) network, while HAT has only one - it does not have the

AC network. KAN's AC network trains the AC mask to determine what knowledge can be shared. The MCL network stores the knowledge and applies the trained AC mask to the knowledge base. This setting enables KAN not only to adapt the shared knowledge across tasks to produce more accurate models, but also to avoid CF. HAT has only one network, and its mask is to block only the knowledge that is important to previous task models.

**Lifelong Learning (LL) for SC.** The authors of [5,28] proposed a Naive Bayes (NB) approach to help improve the new task learning. A heuristic NB method was also proposed in [28]. [30] presented a LL approach based on voting of individual task classifiers. All these works do not use neural networks, and are not concerned with the CF problem. The work in [26,29] uses LL for aspect-based sentiment analysis, which is an entirely different problem than document-level sentiment classification (SC) studied in this paper.

# 3    Proposed Model KAN

To improve the classification accuracy and also to avoid forgetting, we need to identify some past knowledge that is shareable and update-able in learning the new task so that no forgetting of the past knowledge will occur and both the new task and the past tasks can improve.

We solve this problem by taking inspiration from what we humans seem to do. For example, we may "forget" our phone number 10 years ago, but if the same number or a similar number shows up again, our brain may quickly retrieve the old phone number and make both the new and old numbers memorized more strongly. Biological research [10] has shown that our brain keeps track of the knowledge accessibility. If some parts of our previous knowledge are useful to the new task (i.e., shared knowledge between the new task and some previous tasks), our brain sets those parts accessible to enable *forward knowledge transfer*. This also enables *backward knowledge transfer* as they are now accessible and we have the opportunity to strengthen them based on the new task data. For those not useful parts of the previous knowledge, they are set to inaccessible, which protects them from being changed. Inspired by this idea, we design a memory and accessibility mechanism.

We need to solve two key problems: (1) how to detect the accessibility of the knowledge in the memory (which we call *knowledge base* (KB)), i.e., identifying the part of the previous knowledge that is useful to the new task; (2) how to leverage the identified useful/shared knowledge to help the new task learning while also protecting the other part. To address these challenges, we propose the *Knowledge and Accessibility Network* (KAN) shown in Fig. 1.

## 3.1    Overview of KAN

KAN has two components (see Fig. 1), the *main continual learning* (MCL) component in the purple box and the *accessibility* (AC) component in the yellow box. MCL performs the main continual learning and testing (AC is not used in testing

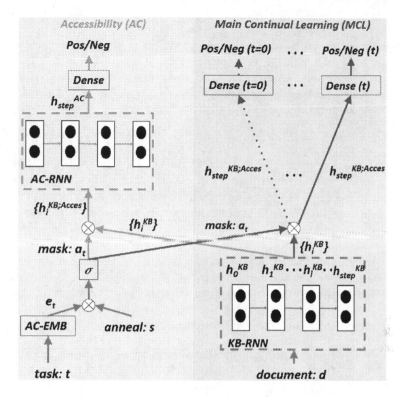

**Fig. 1.** The KAN architecture (best viewed in color). The purple box contains the main continual learning (MCL) component, and the yellow box contains the accessibility (AC) component. The green arrows represent forward paths shared by both components. The yellow and purple arrows represent the forward paths used only in AC and MCL respectively. (Color figure online)

except the mask $a_t$ generated from the task id $t$). We see the sentiment classification heads (pos/neg) for each task at the top. Below them are the dense layers and further down is the *knowledge base* (KB) (the memory) in the green dash-lined box. KB, which is modeled using an RNN (we use GRU in our system) and is called KB-RNN, contains both the task-specific and shared knowledge across tasks. AC decides which part of the knowledge (or units) in the KB is accessible by the current task $t$ by setting a *binary task-based mask* $a_t$. Each task is indicated by its task embedding produced by AC-EMB from the task id ($t$). AC-EMB is a randomly initialized embedding layer. The inputs to KAN are the task id $t$ and document $d$. They are used in training both components via the mask $a_t$ and $\{h_i^{KB}\}$ (hidden states in KB-RNN) links.

**AC Training.** In the AC training phase, only Task Embedding (AC-EMB), AC-RNN and others in the yellow box are trainable. Consider the KB has already retained the knowledge learned from tasks $0...t\text{-}1$. When the new task $t$ arrives, we first train the AC component to produce a binary task-based mask $a_t$

688    Z. Ke et al.

(a matrix with the same size as the KB $\{h_i^{KB}\}$) to indicate which units in KB-RNN are accessible for the new/current task $t$. Since the mask is trained based on the new task data with the previous knowledge in KB-RNN fixed, those KB-RNN units that are not masked (meaning that they are useful to the new task) are the accessible units with their entries in $a_t$ as 1 (unmasked). The other units are inaccessible with their entries in $a_t$ as 0 (masked).

**MCL Training.** After AC training, we start MCL training. In this phase, only KB-RNN and others in the purple box are trainable. The trained binary task-based mask $a_t$ is element-wise multiplied by the output vectors of KB-RNN. This operation protects those inaccessible units since no gradient flows across them while allowing those accessible units to be updated since the mask does not stop gradients for them. This clearly enables *forward knowledge transfer* because those accessible units selected by the mask represent the knowledge from the previous tasks that can be leveraged by the current/new task. It also enables *backward knowledge transfer* and avoidance of CF because (1) if the accessible units are not important to the previous tasks, any modification to them does not degrade the previous tasks' performance, and (2) if the accessible units are useful for some previous tasks, updating them enable them to improve as we now have more data to enhance the shared knowledge.

---

**Algorithm 1:** Continual Learning in KAN

**Input:** Dataset $D = (D_0, ..., D_T)$
**Output:** Parameters of KB-RNN $W^{KB\text{-}RNN}$, of AC-RNN $W^{AC\text{-}RNN}$, and of AC-EMB $W^{AC\text{-}EMB}$

1  **for** $t = 0, .., T$ **do**
2    **if** $t = 0$ **then**
3      $W_0^{KB\text{-}RNN} = MCLtraining(D_0)$
4      $W_0^{AC\text{-}RNN}, W_0^{AC\text{-}EMB} = ACtraining(D_0, W_0^{KB\text{-}RNN})$
5    **else**
6      $W_t^{AC\text{-}RNN}, W_t^{AC\text{-}EMB} = ACtraining(D_t, W_{t-1}^{KB\text{-}RNN})$
7      $W_t^{KB\text{-}RNN} = MCLtraining(D_t, W_t^{AC\text{-}EMB})$

---

**Continual Learning in KAN:** The algorithm for continual learning in KAN is given in Algorithm 1. For each new task, AC training is done first and then MCL training. An exception is at the first task (lines 3–4 in Algorithm 1). At the very beginning, KB has no knowledge, and thus nothing can be used by AC. Therefore, we train MCL (and KB) before AC to obtain some knowledge first. However, after the first task, AC is always trained before MCL (and KB) for each new task (lines 6–7).

## 3.2   Details of Accessibility Training

AC training aims to detect the accessibility of the knowledge retained in the KB given the new task data. As shown in Fig. 1 and Algorithm $2^1$, it takes as inputs a training example $d$ and the trained knowledge base KB-RNN $\boldsymbol{W}_{t-1}^{KB\text{-}RNN}$, which has been trained because AC is always trained after KB (KB-RNN) was trained on the previous task. To generate a binary matrix $\boldsymbol{a}_t$ so that it can work as a mask in both the MCL (and KB) and AC training phases, we borrow the idea of *hard attention* in [1,24,31] where the values in the attention matrix are *binary* instead of a probability distribution as in *soft attention*.

**Hard Attention Training.** Since we can access the task id in both training and testing, a natural choice is to leverage the task embedding to compute the hard attention. Specifically, for a task id $t$, we first compute a task embedding $e_t$ by feeding the task id into the task embedding layer (AC-EMB) where a 1-hot *id* vector is multiplied by a *randomly initialized* parameter matrix. Using the resulting task embedding $e_t$, we apply a gate function $\sigma(x) \in [0,1]$ and a positive scaling parameter $s$ to compute the binary attention $\boldsymbol{a}_t$ as shown in lines 1 and 2 in Algorithm 2. Intuitively, it uses a unit step function as the activation function $\sigma(x)$. However, this is not ideal because it is non-differentiable. We want to train the embedding $e_t$ with back propagation. Motivated by [24], we use a *sigmoid* function with a positive scaling parameter $s$ to construct a pseudo-step function allowing the gradient to flow. This scaling parameter is introduced to control the polarization of our pseudo-step function and the output $\boldsymbol{a}_t$. Our strategy is to anneal $s$ during training, inducing a gradient flow and set $s = s_{max}$ during testing. This is because using a hyperparameter $s_{max} \gg 1$, we can make our sigmoid function approximate to a unit step function. Meanwhile, when $s \to \infty$ we get $\boldsymbol{a}_t \to \{0,1\}$ and when $s \to 0$, we get $\boldsymbol{a}_t \to 0.5$. We start the training epoch with all units being equally active by using the latter and progressively polarize them within the epoch. Specifically, we anneal $s$ as follows:

$$s = \frac{1}{s_{max}} + (s_{max} - \frac{1}{s_{max}})\frac{b-1}{B-1} \tag{1}$$

where $b = 1, ...B$ is the batch index and $B$ is the total number of batches in an epoch.

   To better understand the hard attention training, recall that the resulting mask $\boldsymbol{a}_t$ needs to be binary so that it can be used to block/unblock some units' training in both phases. The task id is used to control the mask to condition the KB. To achieve this, we need to make sure the embedding of the task id is trainable for which we adopt sigmoid as the pseudo gate function. The training procedure is annealing: at first, $s \to 0$ ($s = \frac{1}{s_{max}}$) and the mask is still a soft attention. After certain batches, $s$ becomes large and $\sigma(s \otimes e_t)$ becomes very similar to a gate function. After training, those units with their entries in $\boldsymbol{a}_t$ as 1 are accessible to task $t$ while the others are inaccessible. Another advantage of

---

[1] For simplicity, we only show the process for one training example, but our actual system trains in batches.

training a hard attention is that we can easily retrieve the binary task-based $a_t$ after training: we simply adopt $s_{max}$ to be $s$ and apply $\sigma(s_{max} \otimes e_t)$.

**Apply Hard Attention to the Network.** The fixed KB-RNN takes a training example $d$ as input and produces a sequence of output vectors $\{h_i^{KB}\}$ which have incorporated the previous knowledge (line 3 in Algorithm 2). $i$ denotes the $i$th step of the RNN or the representation of the $i$th word of the input, ranging from 0 to the step size $step$. $\{h_i^{KB}\}$ then performs element-wise multiplication with the task-based mask $a_t$ to get the accessibility representation of the previous knowledge $\{h_i^{KB;Access}\}$ (line 4). Recall that $e_t$ is a task id embedding vector and $e_t \in \mathbb{R}^{dim}$, where $dim$ refers to the dimension size, and therefore $a_t \in \mathbb{R}^{dim}$ ($a_t = \sigma(s \otimes e_t)$). In other words, we expand the vector $a_t$ (repeat the vector $step$ times) to match the size of $\{h_i^{KB}\}$, and then perform element-wise multiplication. This accessibility representation encodes the useful knowledge from the previous tasks. We first feed the representation into AC-RNN to learn some additional new task knowledge (line 5). The last step of the resulting sequence of vectors $\{h_i^{AC}\}$, which is $h_{step}^{AC}$, then goes through a dense layer to reduce the vector's dimension and finally compute the loss based on cross entropy (line 6).

---

**Algorithm 2:** AC Training

**Input:** A training example $d$, scaling parameter $s$, task id $t$, trained KB-RNN $W_{t-1}^{KB\text{-}RNN}$

**Output:** Parameters of AC-RNN $W_t^{AC\text{-}RNN}$ and of AC-EMB $W_t^{AC\text{-}EMB}$

1 $e_t =$AC-EMB$(t)$ // AC-EMB is trainable.
2 $a_t = \sigma(s \otimes e_t)$ // We anneal $s$ as shown in Eq.1
3 $\{h_i^{KB}\} =$KB-RNN$(d)$ // KB-RNN is already trained and is fixed.
4 $\{h_i^{KB;Access}\} = \{h_i^{KB}\} \otimes a_t$
5 $\{h_i^{AC}\} =$AC-RNN$(\{h_i^{KB;Access}\})$ // AC-RNN is trainable.
6 $\mathcal{L}^{AC} = CrossEntropy(d_{label}, Dense(h_{step}^{AC}))$ // Compute the AC loss.

---

### 3.3   Details of Main Continual Learning Training

MCL training learns the current task knowledge and protects the knowledge learned in previous tasks in the KB (KB-RNN). As shown in Algorithm 3 and Fig. 1, it takes an input training example $d$ in the corresponding dataset $D_t$ and encodes $d$ via KB-RNN (line 1), which results in a sequence of vectors $\{h_i^{KB}\}$. Following our training scheme, i.e., AC-RNN and AC-EMB are always trained before KB for a new task $t$ (except for the first one), we already have the trained AC-EMB $W_t^{AC\text{-}EMB}$ when discussing KB. We therefore can compute the mask $a_t$ from $W_t^{AC\text{-}EMB}$ (lines 5–6). Note that we expand $a_t$ to be a binary matrix and $a_t \in \mathbb{R}^{step \times dim}$ where $step$ refers to step size of KB-RNN and $dim$ refers to the dimension size of the task embedding vector. For the first task (i.e., $t = 0$), we simply assume $a_0$ as a matrix of ones (line 3).

**Block the Inaccessible and Unblock the Accessible Units.** Naturally, we want to "remove" those inaccessible units so that only accessible ones can contribute to the training of KB-RNN. An efficient method is to simply element-wise multiply the outputs of KB $\{h_i^{KB}\}$ by $a_t$ (line 7). Since $a_t$ is a binary mask, only those KB with mask 1 can be updated by backward propagation. This is equivalent to modifying the gradient $g$ with the mask $a_t$:

$$g' = a_t \otimes g \qquad (2)$$

The resulting vectors can be seen as the representation of *accessible* knowledge $\{h_i^{KB;Access}\}$. Finally, we take the last step of the accessible knowledge vectors $\{h_i^{KB;Access}\}$, which is $h_{step}^{KB;Access}$, to a fully connected layer to perform classification (line 8). Note that we employ *multi-head* configuration (upper components in the KB training phase in Fig. 1) which means each task is allocated an exclusive dense layer. These dense layers are mapped to the dimension of the number of classes $c$ ($c = 2$) so that we can use different dense layer to perform classification according to different task id.

---

**Algorithm 3:** MCL Training

---

**Input:** A training example $d$, task id $t$, and trained task embedding $W_t^{AC\text{-}EMB}$

**Output:** Parameters of KB-RNN $W_t^{KB\text{-}RNN}$

1   $\{h_i^{KB}\}$ =KB-RNN($d$) // KB-RNN is trainable.

2   **if** $t = 0$ **then**

3      |   Set all values of $a_0$ to 1

4   **else**

5      |   $e_t$ =AC-EMB($t$) // AC-EMB is trainable.

6      |   $a_t = \sigma(s_{max} \otimes e_t)$ // Use $s_{max}$ to retrieve the trained mask.

7   $\{h_i^{KB;Acces}\} = \{h_i^{KB}\} \otimes a_t$

8   $\mathcal{L}^{KB} = CrossEntropy(d_{label}, Dense(h_{step}^{KB;Acces}))$ // Compute the MCL loss.

---

### 3.4 Comparing Accessibility and Importance

Many existing models discussed in Sect. 2 detect the *importance* of units. That is, they identify units or parameters that are important to previous tasks so that in learning the new task, the learner can protect them to avoid forgetting the previous tasks' models. However, it is also stifling the chance for adapting and updating the previous knowledge to help learn the new task in *forward transfer* and for improving previous tasks to achieve *backward transfer*. In contrast, our concept of accessibility is very different, which is for the *current task*. KAN detects the accessibility of units and *safely* update the weights of the accessible units because of the shared knowledge. This enables adaptation in forward transfer. If some units are accessible for the current task, KAN can update them based on the *current task* training data. If those units are also accessible by some previous tasks, it suggests that there is some shared knowledge between the current and the previous tasks. This results in *backward transfer*. One can also see this as *strengthening the shared knowledge* since we now have more data for the shared knowledge training. In short, training the accessible units is helpful to both the current and the previous tasks.

# 4   Experiments

We now evaluate KAN for continual document sentiment classification (SC). We follow the standard continual learning evaluation procedure [11] as follows. We provide a sequence of SC tasks with their training datasets for KAN to learn one by one. Each task learns to perform SC on reviews of a type of product. Once a task is learned, its training data is discarded. After all tasks are learned, we test all task models using their respective test data. In training each task, we use its validation set to decide when to stop training.

## 4.1   Experimental Data

Our dataset consists of Amazon reviews from 24 different types of products, which make up the 24 tasks. Each task has 2500 positive (with 4 or 5 stars) and 2500 negative (with 1 or 2 stars) reviews. We further split the reviews in each task into training, testing and validation set in the ratio of 8:1:1. We didn't use the datasets in [15] as they are all highly skewed with mostly positive examples. Without doing anything, the system can achieve more than 80% of accuracy. We also did not use the commonly employed sentiment analysis datasets from SemEval [19] because its reviews involve only two types of products/services, i.e., laptop and restaurant, which are too few for continual learning.

## 4.2   Baselines

We consider a wide range of baselines: (1) isolated learning of each task; (2) state-of-the-art continual learning methods; (3) existing continual or lifelong sentiment classification models; and (4) a naive continual learning model.

**One task learning (ONE)** builds an isolated model for each task individually, independent of other tasks. There is no knowledge transfer. The network is the same as KAN but without AC and accessibility mask, consisting of word embeddings, a conventional GRU layer and a fully connected layer for classification. The same network is used in the other variant of KAN, i.e., N-CL.

**Elastic Weight Consolidation (EWC)** [9] is a popular regularization-based continual learning method, which slows down learning for weights that are important to the previous tasks.

**Hard Attention to Task (HAT)** [24] learns pathways in a given base network using the given task id (and thus task incremental). The pathways are then used to obtain the task-specific networks. It is the state-of-the-art task continual learning (TCL) model for avoiding catastrophic forgetting (CF).

**Orthogonal Weights Modification (OWM)** [32] is a state-of-the-art class continual learning (CCL) method. Since OWM is a CCL method but we need a TCL method, we adapt it for TCL. Specifically, we only train on the corresponding head of the specific task id during training and only consider the corresponding head's prediction during testing.

**Sentiment Classification by Retained Knowledge (SRK)** [15] is the only prior work on continual sentiment classification. It achieves only limited forward transfer as we discussed in Sect. 2.

**Lifelong Learning for Sentiment Classification (LSC)**, proposed by [5], is a Naive Bayes based lifelong learning approach to SC. It does forward transfer but not continual learning and thus has no CF issue. The main goal of this traditional method is to improve only the performance of the new task.

**Naive continual learning (N-CL)** greedily trains a sequence of SC tasks incrementally without dealing with CF. Note that this is equivalent to KAN after removing AC and mask, i.e., it uses the same network as ONE.

### 4.3  Network and Training

Unless stated otherwise, we employ the embedding with 300 dimensions to represent the input text. GRU's with 300 dimensions are used for both AC-RNN and KB-RNN. We adopt Glove 300d$^2$ as pre-trained word embeddings and fix them during training of KAN. The fully connected layer with softmax output is used as the final layer(s), together with categorical cross-entropy loss. During KAN training, we initialize the hidden state for each element in the batch to 0. We set $s_{max}$ to 140 in the $s$ annealing algorithm, dropout to 0.5 between embedding and GRU layers for both MCL and AC training phases. We train all models with Adam using the learning rate of 0.001. We stop training when there is no improvement in the validation accuracy for 5 consecutive epochs (i.e., early stopping with patience = 5). The batch size is set to 64. During testing, we evaluate the final performance using MCL only. AC is not involved in testing (except the mask $a_t$ generated with the task id $t$ during training). For the baselines SRK, HAT, OWM, and EWC, we use the code provided by their authors (customized for text if needed) and adopt their original parameters.

### 4.4  Results

**Average Results.** We first report the average results of all compared models to show the effectiveness of the proposed KAN. Since the order of the tasks may have an impact on the final results of CL, we randomly choose and run 10 sequences and average their results. We also ensured that the last tasks in the 10 sequences are different. Table 1 gives the average accuracy of all systems. Column 2 gives the average result over 24 tasks for each model after all tasks are learned. Column 3 gives the accuracy result of the last task for each model. Note here that the Last Task results and the All Tasks results are not comparable because

---

$^2$ https://github.com/stanfordnlp/GloVe.

**Table 1.** Average accuracy of different models. #Parameters refers to the number of parameters. LSC is based on Naive Bayes whose number of parameters is the sum of the number of unique words (vocabulary) in each dataset multiplied by the number of classes, which is 2 in our case. Paired *t*-test is used to test the significance of the result of KAN against that of each baseline for ALL Tasks. P-values are given in column 4.

| Models | All tasks | Last tasks | P-value | #Paramters |
|---|---|---|---|---|
| ONE | 0.7846 | 0.7809 | 1.025e−7 | 25.2M |
| LSC | 0.8219 | 0.8246 | 1.581e−2 | — |
| N-CL | 0.8339 | 0.8477 | 1.792e−3 | 25.2M |
| EWC | 0.6899 | 0.7187 | 1.542e−9 | 42.5M |
| OWM | 0.6983 | 0.7337 | 3.219e−15 | 30.0M |
| HAT | 0.6456 | 0.6938 | 1.861e−14 | 42.7M |
| SRK | 0.8282 | 0.8500 | 7.793e−6 | 3.4M |
| **KAN** | **0.8524** | **0.8799** | — | 42.4M |

each Last Task result is the average of the 10 last tasks in the 10 random task sequences, while each ALL Tasks result is the average of all 24 tasks. Column 4 gives the p-value of the significance test to show that KAN outperforms all baselines (more discussion later). Column 5 gives the number of parameters of each model. Note that ONE and LSC are not continual learning (CL) systems and have no CF issue. The rest are continual learning systems.

We first observe that KAN's result is markedly better than that of every baseline. Since the traditional lifelong learning method LSC mainly aims to improve the last task's performance, for the All Tasks column, we give its best result, i.e., putting each of the 24 tasks as the last task. Even under this favorable setting, its result is considerably poorer than that of KAN.

It is interesting to know that comparing to ONE, naive CL (N-CL) does not show accuracy degradation on average even without a mechanism to deal with CF (forgetting). N-CL is actually significantly better than ONE (they use exactly the same network). As mentioned earlier, this is because sentiment analysis tasks are similar to each other and can mostly help one another. That is, CF is not a major issue in CL for sentiment analysis. In fact, N-CL also outperforms SRK on ALL Tasks. This can be explained by the fact that SRK does not adapt the past knowledge or allow backward transfer as discussed in Sect. 2. SRK's main goal was to improve the last task's performance rather than those of all tasks.

Regarding the continual learning (CL) baselines, EWC, OWM and HAT, they perform poorly and are even worse than ONE, which is not surprising because their networks are primarily designed to preserve knowledge learned for each of the previous tasks. This makes it hard for them to exploit knowledge sharing to improve all tasks.

**Significance Test.** To show that our results from KAN are significantly better than those of baselines, we conduct a paired t-test. We test KAN against each of the baselines based on the results of All Tasks from the 10 random sequences. All p-values are far below 0.05 (see Table 1), which indicates that KAN is significantly better than every baseline.

Table 1 also includes the number of parameters of each neural model (Column 4). A large fraction of the parameters for KAN is due to the mask for each task. Note that the similar numbers of parameters of models by no means indicate the models are similar (see Sect. 2).

**Ablation Study.** KAN has two components, AC and MCL. To evaluate the effectiveness of AC, we can remove AC to test KAN. However, the binary mask in KAN needs AC to train. Without AC, it will have no mask and KAN is the same as N-CL. KAN is significantly better than N-CL as shown in Table 1. MCL is the main module that performs continual learning and cannot be removed.

**Individual Task Results.** To gain more insights, we report the individual task results in Table 2 for all continual learning baselines and KAN. We also include ONE for comparison. Note that each task result for a model is the average of the results from the 10 random sequences (except ONE).

Table 2 shows that KAN gives the best accuracy for 21 out of 24 tasks. In those tasks where KAN does not give the best, KAN's performances are competitive. Hence, we can conclude that KAN is both highly accurate and robust.

Regarding SRK, it performs the best in only 2 tasks. However, these two tasks' results are only slightly better than those of KAN. These clearly indicate SRK is weaker than KAN. For the other continual learning baselines: HAT, OWM and EWC, their performances are consistently worse even than ONE. This is expected because their goal is to protect each of the ONE's results and such protections are not perfect and thus can still result in some CF (forgetting).

**Effectiveness of Forward and Backward Knowledge Transfer.** From Tables 1 and 2, we can already see that KAN is able to exploit shared knowledge to improve learning of similar tasks. Here, we want to show whether the forward knowledge transfer and the backward knowledge transfer are indeed effective.

Table 3 shows the accuracy results progressively after every 6 tasks have been learned. In the second row, we give the results after 6 tasks have been learned sequentially. Each accuracy result in the Forward column is the overall average of the 6 tasks when each of them was first learned in each of the 10 random runs, which indicate the forward transfer effectiveness because from the second task, the system can already start to leverage the previous knowledge to help learn the new task. By comparing with the corresponding results of ONE and N-CL, we can see forward transfer of KAN is indeed effective. N-CL also has the positive forward transfer effect, but KAN does better. Each result in the Backward column shows the average test accuracy after all 6 tasks have been learned (over 10 random runs). By comparing with the corresponding result in the Forward column, we can see that backward transfer of KAN is also

**Table 2.** Individual task accuracy of each model, after having trained on all tasks. The number in bold in each row is the best accuracy of the row.

| Task (product category) | SRK | HAT | OWM | EWC | N-CL | ONE | KAN |
|---|---|---|---|---|---|---|---|
| Amazon_Instant_Video | 0.7776 | 0.6297 | 0.6792 | 0.6589 | 0.7989 | 0.7859 | **0.8293** |
| Apps_for_Android | 0.8236 | 0.6351 | 0.7044 | 0.6634 | 0.8431 | 0.8101 | **0.8531** |
| Automotive | 0.8049 | 0.6728 | 0.6849 | 0.6874 | 0.8086 | 0.6917 | **0.8335** |
| Baby | 0.8617 | 0.6947 | 0.7175 | 0.6997 | 0.8703 | 0.8020 | **0.8870** |
| Beauty | 0.8758 | 0.6684 | 0.7333 | 0.6816 | 0.8752 | 0.8081 | **0.8912** |
| Books | **0.8517** | 0.6260 | 0.7078 | 0.7112 | 0.8165 | 0.8101 | 0.8337 |
| CDs_and_Vinyl | 0.7740 | 0.5666 | 0.6487 | 0.6412 | 0.7937 | 0.7152 | **0.7970** |
| Cell_Phones_and_Accessories | 0.8277 | 0.6363 | 0.7163 | 0.6844 | 0.8489 | 0.7899 | **0.8679** |
| Clothing_Shoes_and_Jewelry | 0.8520 | 0.6678 | 0.7221 | 0.7124 | 0.8701 | 0.8727 | **0.8814** |
| Digital_Music | 0.7480 | 0.5656 | 0.6335 | 0.6159 | **0.7717** | 0.7232 | 0.7706 |
| Electronics | **0.8457** | 0.6472 | 0.6895 | 0.6546 | 0.8242 | 0.7636 | 0.8359 |
| Grocery_and_Gourmet_Food | 0.8800 | 0.6664 | 0.7203 | 0.7354 | 0.8686 | 0.8242 | **0.8828** |
| Health_and_Personal_Care | 0.8076 | 0.6095 | 0.6642 | 0.6605 | 0.8235 | 0.7131 | **0.8335** |
| Home_and_Kitchen | 0.8577 | 0.6920 | 0.7289 | 0.7407 | 0.8595 | 0.8081 | **0.8812** |
| Kindle_Store | 0.8500 | 0.6427 | 0.7060 | 0.6966 | 0.8324 | 0.8505 | **0.8584** |
| Movies_and_TV | 0.8377 | 0.6317 | 0.7053 | 0.6986 | 0.8278 | 0.7879 | **0.8527** |
| Musical_Instruments | 0.8142 | 0.7595 | 0.7456 | 0.7570 | 0.8677 | 0.8351 | **0.8851** |
| Office_Products | 0.8180 | 0.6195 | 0.6664 | 0.6592 | 0.8142 | 0.7374 | **0.8346** |
| Patio_Lawn_and_Garden | 0.8130 | 0.6406 | 0.6543 | 0.7033 | 0.8308 | 0.7833 | **0.8363** |
| Pet_Supplies | 0.7936 | 0.6289 | 0.6840 | 0.6720 | 0.8255 | 0.7556 | **0.8481** |
| Sports_and_Outdoors | 0.8420 | 0.6631 | 0.7045 | 0.6953 | 0.8384 | 0.7939 | **0.8696** |
| Tools_and_Home_Improvement | 0.8300 | 0.6530 | 0.6904 | 0.6832 | 0.8408 | 0.7515 | **0.8640** |
| Toys_and_Games | 0.8500 | 0.6700 | 0.7368 | 0.7379 | 0.8644 | 0.8202 | **0.8744** |
| Video_Games | 0.8397 | 0.6502 | 0.7147 | 0.7067 | 0.8381 | 0.7980 | **0.8557** |

**Table 3.** Effects of forward and backward knowledge transfer of KAN. We give progressive results after 6, 12, 18, and 24 tasks have been learned respectively.

| Tasks | ONE | N-CL | | KAN | |
|---|---|---|---|---|---|
| | | Forward | Backward | Forward | Backward |
| First 6 tasks | 0.7846 | 0.7937 | 0.7990 | **0.8068** | **0.8132** |
| First 12 tasks | 0.7865 | 0.8135 | 0.8199 | **0.8314** | **0.8390** |
| First 18 tasks | 0.7870 | 0.8253 | 0.8327 | **0.8424** | **0.8501** |
| First 24 tasks | 0.7846 | 0.8302 | 0.8339 | **0.8471** | **0.8524** |

effective, which means that learning of later tasks can help improve the earlier tasks automatically. The same is also true for N-CL, although KAN does better. Rows 3, 4, and 5 show the corresponding results after 12, 18, and 24 tasks have been learned, respectively.

We also observe that forward transfer is much more effective than the backward transfer. This is expected because forward transfer leverages the previous knowledge first and backward transfer can improve only after the forward transfer has made significant improvements. Furthermore, backward transfer also has the risk of causing some forgetting for the previous tasks because the previous task data are no longer available to prevent it.

## 5   Conclusion and Future Work

This paper proposed KAN, a novel neural network for continual learning (CL) of a sequence of sentiment classification (SC) tasks. Previous CL models primarily focused on dealing with catastrophic forgetting (CF). As we have seen in the experiment section, CF is not a major issue for continual sentiment classification because the SC tasks are similar to each other and have a significant amount of shared knowledge among them. KAN thus focuses on improving the learning accuracy by exploiting the shared knowledge via forward and backward knowledge transfer. KAN achieves these goals using a knowledge base and a knowledge accessibility network. The effectiveness of KAN was demonstrated by empirically comparing it with state-of-the-art CL approaches. KAN's bi-directional knowledge transfer for CL significantly improves its results for SC. Our future work will improve its accuracy and adapt it for other types of data.

**Acknowledgments.** This work was supported in part by two grants from National Science Foundation: IIS-1910424 and IIS-1838770, and a research gift from Northrop Grumman.

## References

1. Aharoni, R., Goldberg, Y.: Morphological inflection generation with hard monotonic attention. In: ACL (2017)
2. Chaudhry, A., Ranzato, M., Rohrbach, M., Elhoseiny, M.: Efficient lifelong learning with A-GEM. In: ICLR (2019)
3. Chen, Z., Liu, B.: Topic modeling using topics from many domains, lifelong learning and big data. In: ICML (2014)
4. Chen, Z., Liu, B.: Lifelong machine learning. Synth. Lect. Artif. Intell. Mach. Learn. **12**(3), 1–207 (2018)
5. Chen, Z., Ma, N., Liu, B.: Lifelong learning for sentiment classification. In: ACL (2015)
6. Fernando, C., et al.: PathNet: evolution channels gradient descent in super neural networks. CoRR abs/1701.08734 (2017)
7. He, X., Jaeger, H.: Overcoming catastrophic interference using conceptor-aided backpropagation. In: ICLR (2018)
8. Kamra, N., Gupta, U., Liu, Y.: Deep generative dual memory network for continual learning. CoRR abs/1710.10368 (2017)
9. Kirkpatrick, J., et al.: Overcoming catastrophic forgetting in neural networks. CoRR (2016)

10. Kornell, N., Hays, M.J., Bjork, R.A.: Unsuccessful retrieval attempts enhance subsequent learning. J. Exp. Psychol. Learn. Memory Cogn. **35**(4), 989 (2009)
11. Lange, M.D., et al.: Continual learning: a comparative study on how to defy forgetting in classification tasks. CoRR abs/1909.08383 (2019)
12. Lee, S., Kim, J., Jun, J., Ha, J., Zhang, B.: Overcoming catastrophic forgetting by incremental moment matching. In: NeurIPS (2017)
13. Liu, B.: Sentiment Analysis: Mining Opinions, Sentiments, and Emotions. Cambridge University Press, Cambridge (2015)
14. Lopez-Paz, D., Ranzato, M.: Gradient episodic memory for continual learning. In: NeurIPS (2017)
15. Lv, G., Wang, S., Liu, B., Chen, E., Zhang, K.: Sentiment classification by leveraging the shared knowledge from a sequence of domains. In: Li, G., Yang, J., Gama, J., Natwichai, J., Tong, Y. (eds.) DASFAA 2019. LNCS, vol. 11446, pp. 795–811. Springer, Cham (2019). https://doi.org/10.1007/978 3 030-18576-3 47
16. Mallya, A., Lazebnik, S.: PackNet: adding multiple tasks to a single network by iterative pruning. In: CVPR (2018)
17. McCloskey, M., Cohen, N.J.: Catastrophic interference in connectionist networks: the sequential learning problem. In: Psychology of Learning and Motivation. Elsevier (1989)
18. Parisi, G.I., Kemker, R., Part, J.L., Kanan, C., Wermter, S.: Continual lifelong learning with neural networks: a review. Neural Netw. **113**, 54–71 (2019)
19. Pontiki, M., et al.: Semeval-2016 task 5: aspect based sentiment analysis. In: SemEval (2016)
20. Rebuffi, S., Kolesnikov, A., Sperl, G., Lampert, C.H.: iCaRL: incremental classifier and representation learning. In: CVPR (2017)
21. Rostami, M., Kolouri, S., Pilly, P.K.: Complementary learning for overcoming catastrophic forgetting using experience replay. In: IJCAI (2019)
22. Ruvolo, P., Eaton, E.: ELLA: an efficient lifelong learning algorithm. In: ICML (2013)
23. Seff, A., Beatson, A., Suo, D., Liu, H.: Continual learning in generative adversarial nets. CoRR abs/1705.08395 (2017)
24. Serrà, J., Suris, D., Miron, M., Karatzoglou, A.: Overcoming catastrophic forgetting with hard attention to the task. In: ICML (2018)
25. Shin, H., Lee, J.K., Kim, J., Kim, J.: Continual learning with deep generative replay. In: NeurIPS (2017)
26. Shu, L., Xu, H., Liu, B.: Lifelong learning CRF for supervised aspect extraction. In: ACL (2017)
27. Silver, D.L., Yang, Q., Li, L.: Lifelong machine learning systems: Beyond learning algorithms. In: AAAI (2013)
28. Wang, H., Liu, B., Wang, S., Ma, N., Yang, Y.: Forward and backward knowledge transfer for sentiment classification. In: ACML (2019)
29. Wang, S., Lv, G., Mazumder, S., Fei, G., Liu, B.: Lifelong learning memory networks for aspect sentiment classification. In: IEEE International Conference on Big Data, Big Data (2018)
30. Xia, R., Jiang, J., He, H.: Distantly supervised lifelong learning for large-scale social media sentiment analysis. IEEE Trans. Affect. Comput. **8**, 480–491 (2017)
31. Xu, K., et al.: Show, attend and tell: neural image caption generation with visual attention. In: ICML (2015)
32. Zeng, G., Chen, Y., Cui, B., Yu, S.: Continuous learning of context-dependent processing in neural networks. Nat. Mach. Intell. **1**, 364–372 (2019)

# Bioinformatics

# Predictive Bi-clustering Trees
# for Hierarchical Multi-label Classification

Bruna Z. Santos[1(✉)], Felipe K. Nakano[2,3], Ricardo Cerri[1], and Celine Vens[2,3]

[1] Department of Computer Science, Federal University of São Carlos,
São Carlos, Brazil
bruna.zamith@hotmail.com, cerri@ufscar.br
[2] Department of Public Health and Primary Care, KU Leuven, Kortrijk, Belgium
{felipekenji.nakano,celine.vens}@kuleuven.be
[3] ITEC, imec Research Group at KU Leuven, Kortrijk, Belgium

**Abstract.** In the recent literature on multi-label classification, a lot of
attention is given to methods that exploit label dependencies. Most of
these methods assume that the dependencies are static over the entire
instance space. In contrast, here we present an approach that dynamically adapts the label partitions in a multi-label decision tree learning
context. In particular, we adapt the recently introduced predictive bi-
clustering tree (PBCT) method towards multi-label classification tasks.
This way, tree nodes can split the instance-label matrix both in a horizontal and a vertical way. We focus on hierarchical multi-label classification (HMC) tasks, and map the label hierarchy to a feature set over the
label space. This feature set is exploited to infer vertical splits, which
are regulated by a lookahead strategy in the tree building procedure. We
evaluate our proposed method using benchmark datasets. Experiments
demonstrate that our proposal (PBCT-HMC) obtained better or competitive results in comparison to its direct competitors, both in terms of
predictive performance and model size. Compared to an HMC method
that does not produce label partitions though, our method results in
larger models on average, while still producing equally large or smaller
models in one third of the datasets by creating suitable label partitions.

**Keywords:** Predictive clustering trees · Hierarchical multi-label
classification · Bi-clustering

## 1 Introduction

Most of the research on machine learning has investigated traditional classification problems whose classes are mutually exclusive, meaning that one instance
may not belong to more than one label (class) simultaneously. Certain applications, however, present more complex learning tasks. In multi-label classification
for instance, instances can be associated to multiple labels at the same time.

Originally, two main approaches were used for multi-label classification [1]:
local and global approaches. The local approach transforms the multi-label setting to a single-label setting, such that traditional classification algorithms can be

© Springer Nature Switzerland AG 2021
F. Hutter et al. (Eds.): ECML PKDD 2020, LNAI 12459, pp. 701–718, 2021.
https://doi.org/10.1007/978-3-030-67664-3_42

applied. Binary relevance and label powerset methods [1] are examples of a local approach. The global approach (also called big bang) adapts classification algorithms, such that they are able to work with the multi-label structure directly. Predictive clustering trees [2] are an example of a global approach. Despite having their respective advantages, the literature does not present a consensus on which strategy is superior. Later, approaches that group labels in different subsets were presented [3–5], resulting in methods in-between the global and local ones. That is, certain subsets of labels are believed to be more correlated among themselves than the label set al.together. In the recent years, a lot of attention in the multi-label literature goes to developing methods to optimally handle such correlations hidden in the label space. Many of these methods require ensemble methods [6–8] or exploit correlations in a pre-processing step [9–11]. In either case, the label dependencies are handled in a static way. In contrast, we hypothesize here that label correlations may differ throughout the instance space. Thus, rather than partitioning the label space as a pre-processing step in the same manner for all instances, models should dynamically detect such correlations, and create suitable partitions during their induction process.

Recently, a method has been introduced in the context of pairwise learning [12], that predicts interactions between two data points by bi-clustering the interaction matrix. Bi-clustering, also called co-clustering or two-way clustering, is the simultaneous clustering of the rows and columns of a matrix. We leverage this idea, which was set in the predictive clustering tree (PCT) framework, to the multi-label classification task, by bi-clustering the label space. This means that the PCT can induce both horizontal and vertical splits in the label matrix, instead of only horizontal ones. The vertical splits allow the model to partition the label set and group labels with a similar interaction pattern. This procedure is performed dynamically during the induction process: at each node to be split, horizontal as well as vertical candidate splits are considered.

Although the bi-clustering method for interaction prediction [12] also employed multi-label classification techniques, the end goal considered here is different, and hence, some adaptations to the method are required. In particular, we generate the vertical splits by incorporating a lookahead strategy in the tree learning procedure. Lookahead [13] is a technique that alleviates the myopia in greedy tree learning algorithms by taking into account the effect that a split has on deeper levels of the tree, when calculating its quality. In order to control computational complexity (which would become a bottleneck when considering all possible $2^{|L|}$ candidate splits of labels, with $L$ the label set for the node under consideration) and maximize interpretability, we employ a feature-based splitting strategy, just as with the horizontal splits.

In this paper, we apply the proposed idea to the task of hierarchical multi-label classification (HMC), i.e., problems where each data instance can be associated to multiple paths of a hierarchy defined over the label set. Such a hierarchy reflects a general-to-specific structure of the labels, and predictions must conform with the hierarchy constraint, that is, if a given label is predicted, all its ancestor labels must be predicted as well. One example is the task of image classification into a topic hierarchy, where an image may be classified as belonging to the paths Nature → Tree → Pine and Nature → Snow.

To evaluate our proposed algorithm, we have performed experiments using 24 benchmark datasets from the domain of functional genomics, email classification and medical X-Ray images. We have compared our approach (which we call PBCT-HMC) to three other approaches: 1) Regular PCTs for HMC (Clus-HMC) [14]; 2) Predictive bi-clustering trees, as originally proposed to interaction prediction tasks [12] and later also applied to multi-label classification tasks [15] (PBCT); and 3) A static approach where label subsets are defined upfront according to the subtrees in the label hierarchy (Clus-Subtree).

The results show that PBCT-HMC outperforms PBCT and Clus-Subtree in terms of predictive performance and model size. Compared to Clus-HMC, while the predictive performance is comparable, our model sizes are larger on average. Nevertheless, for 8 out of 24 datasets, PBCT-HMC yields equally or more interpretable models than Clus-HMC by creating suitable label partitions.

The remainder of this paper is organized as follows: Sect. 2 discusses recent methods from the multi-label classification literature that also focus on exploiting label correlations. Section 3 presents our approach. Section 4 exposes the empirical comparison, with a discussion of the results. Finally, Sect. 5 provides a conclusion of this work and future work directions.

## 2    Related Work

Many multi-label studies have focused on exploiting label relationships, an important aspect for good multi-label predictors. In many of these studies, label relationships are defined by the problem domain, such as in hierarchical multi-label classification (HMC), where a taxonomy organizes the classes involved.

Predictive clustering trees for HMC were investigated in [14]. This work proposes a global method, named Clus-HMC, which induces a single decision tree predicting all the hierarchical labels at once, and shows that it outperforms local variants. The method was later extended to an ensemble approach [16], boosting the performance at the cost of reduced interpretability.

HMC has also been addressed by neural network based methods. In [17], the proposed method associates one multi-layer perceptron (MLP) to each hierarchical level, each MLP being responsible for the predictions in its associated level. The predictions at one level are used to complement the feature vectors of the instances used to train the neural network associated to the next level. Novel architectures for HMC tasks have been proposed in [18], where local and global loss functions are optimized for discovery of local hierarchical class-relationships and global information, being able to penalize violations of the hierarchy.

In [19], the authors propose a component to the output layers of neural networks that is applicable to HMC tasks. It makes use of prior knowledge on the output domain to map the hierarchical structure to the network topology. The so-called adjacency wrapping matrix incorporates the domain-knowledge directly into the learning process of neural networks.

In the absence of explicit label relationships, many studies have focused on extracting relationships from the label vectors of the instances, in order to exploit them during training. In [3], a probabilistic method is presented, that enforces discovery of deterministic relationships among labels in a multi-label model. The study focused on two main relationships: pairwise positive entailment and exclusion. Such relationships are represented as a deterministic Bayesian network.

A label co-occurence graph is created in [5]. Then, community detection methods are applied on the graph, as an alternative to the random partitioning performed by the RAkEL method [20]. A label powerset approach is used on the created label subsets. The results of the community detection methods on label co-occurence graphs outperformed the results obtained by using random partitioning. Another study [10] proposes to modify multi-label datasets by creating new features that represent correlations, in order to exploit instance correlations in the classification task. The features take into account distances between instances in the feature space.

Although the previous methods aim at extracting label relationships from a class hierarchy or label vectors, these relationships are considered to be fixed throughout the dataset. Recently, ensemble methods have been proposed, that allow different relations (modeled as label partitions) in each individual model.

In [7], an ensemble of PCTs is proposed, that consider random output subspaces to multi-label classification: each tree of the model uses a subset of the label space, thereby combining problem transformation and algorithm adaptation approaches into one method. The same approach was also proposed for multi-target regression [8]. Also, this work proposed a new prediction aggregation function that provides best results when used with extremely randomized PCTs. A recent study [21] proposed to use evolutionary algorithms for multi-label ensemble optimization. It selects diverse multi-label classifiers, each focusing on a label subset, in order to build an ensemble which takes into account data attributes, such as label relationship and label imbalance degree.

Differently from these approaches, in the current paper we present a method that learns a single (i.e., interpretable) tree model, that allows different label partitions in different parts of the instance space.

# 3   Predictive Bi-clustering Trees for HMC

## 3.1   Background

A *predictive clustering tree* (PCT) is a generalization of a decision tree, where each node corresponds to a cluster. Depending on the information contained in the leaf nodes, PCTs can be used for different learning tasks, including clustering, classification (single or multi-label) and regression (single or multi-target). They are implemented in the Clus software[1]. PCTs are built in a top-down approach, meaning that the root node corresponds to the complete training set, which is recursively partitioned in each split. The heuristic used to select tests to include

---

[1] https://dtai.cs.kuleuven.be/clus/.

in the tree is the reduction of intra-cluster label variance, i.e., each split aims to maximize the inter-cluster label variance. In the multi-label classification setting, the labels are represented by a binary label vector, and the variance of a set of instances $S$ is defined as:

$$Var(S) = \frac{\sum_i d(\mathbf{h_i}, \overline{\mathbf{h}})^2}{|S|} \tag{1}$$

where $d$ is the distance between each instance's label vector $\mathbf{h_i}$ and the mean label vector from the set, $\overline{\mathbf{h}}$.

In HMC tasks, it is convenient to consider that the similarity of higher levels of the hierarchy are more important than the similarity of lower levels. In order to implement such idea, the HMC implementation of PCTs in Clus (Clus-HMC) uses a weighted Euclidean distance in the function $d$ above [14]. The class weight $w$ decreases exponentially with the depth of the class in the hierarchy. The weight $w$ of a class $c$, is given as:

$$w(c) = w_0^{depth(c)} \tag{2}$$

A *predictive bi-clustering tree* (PBCT) is a PCT that is capable to predict interactions between two node sets in a bipartite network [12]. Interaction prediction is found in many applications, such as predicting drug-protein interactions, user-product interactions, etc. The network can be described by an adjacency matrix where each row and each column corresponds to a node, and each cell receives the value 1 if its corresponding nodes are connected, and 0 otherwise (Fig. 1). Each row and each column is also associated to a feature vector (e.g., chemical characteristics for drugs and structural properties for proteins in a drug-protein interaction network).

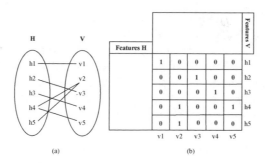

**Fig. 1.** (a) Representation of an interaction network, (b) The same network represented as an interaction matrix.

As PCTs, the PBCT is also built in a top-down approach. That means that in each iteration, a test is applied to one of the features. The test is chosen considering both sets of features for row and column nodes (H and V, as shown in Fig. 1), based on a heuristic and a stop criterion. The heuristic is the same as for multi-label classification (minimize intra-cluster variance of the label vectors), in the sense that for a horizontal split, row-wise label vectors are considered, while for a vertical split, column-wise label vectors.

## 3.2   Proposed Method: PBCT-HMC

In order to leverage PBCTs to the HMC domain, we apply the PBCT construction procedure to the label matrix of the HMC problem. Figure 2 provides an example, where for the label matrix shown in the top left box, there is no proper horizontal split, but if a vertical split is performed first, then it is possible to find two horizontal splits afterwards, resulting in four pure bi-clusters.

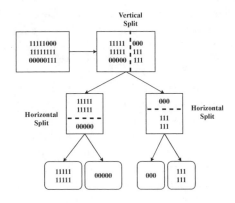

**Fig. 2.** Example in which using a PBCT has advantages compared to using a PCT.

While the PBCT method has been applied to multi-label classification tasks before [15], it was designed for a different task (interaction prediction), and thus, we argue that some additions and adaptations to the algorithm described above are necessary in order to make the method fully accommodated to the HMC context. Our proposed approach is further denoted as PBCT-HMC. We introduce the following notation. H and V, the two node sets from before, now respectively denote the set of instances and the set of labels. Each of these sets is composed by a (usually real-valued) feature matrix ($H^F$ and $V^F$) and a (binary valued) target matrix ($H^T$ and $V^T$), such that $H^T$ is equivalent to transposing $V^T$. Vectors are denoted in boldface and we use the notation $\mathbf{w}[Z]$ to denote the subvector of $\mathbf{w}$, restricted to the components defined by Z.

*Feature Representation of Label Hierarchy.* In order to represent the feature matrix $V^F$, we propose the following simple mapping of the corresponding label hierarchy. We provide a single categorical feature, that denotes for each label $l$ the name of the label just below the root, that is on the path from $l$ to the root. A toy hierarchy is shown in Fig. 3, along with its representation. The proposed mapping takes into account the structural properties of the label hierarchy, it guarantees to produce predictions that fulfil the hierarchy constraint (i.e., the prediction probability for a child label can not exceed that for its parent label), and has the advantages to be small (which is beneficial for the lookahead approach - see further), very simple to model and to be domain independent.

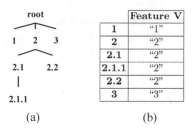

**Fig. 3.** (a) Example of label hierarchy and (b) its resulting feature vector.

*Tree Induction and Split Heuristic.* As in PBCT, tree induction is performed in a top-down manner. Any given node $k$ of the tree can be associated to a bi-cluster defined by a pair $(H_k, V_k)$ with $H_k \subseteq H^T$ and $V_k \subseteq V^T$. The subsets $H_k$ and $V_k$ can be obtained by following the path of split nodes all the way from the root until node $k$. The root node is associated to $(H, V)$.

In order to split node $k$, we first go through each feature in $H^F$, in order to choose the best horizontal test. We apply the variance reduction heuristic, given by Eq. 1, to $H^T$, in order to evaluate the quality of each split. In doing so, we restrict the label vectors $\mathbf{h_i}$ and $\overline{\mathbf{h}}$ to those components that are in $V_k$. As we are dealing with a hierarchical task, we take into account the weights for each class, as exposed in Eq. 2. This results in the following variance definition:

$$Var(H_k, V_k) = \frac{\sum_i d(\mathbf{h_i}[V_k], \overline{\mathbf{h}[V_k]})^2}{|H_k|} \text{ with } \mathbf{h_i} \in H_k \qquad (3)$$

and, consequently, in the following heuristic function for horizontal splits:

$$h_h(s, H_k, V_k) = Var(H_k, V_k) - \left( \frac{|H_{kL}|}{|H_k|} \cdot Var(H_{kL}, V_k) + \frac{|H_{kR}|}{|H_k|} \cdot Var(H_{kR}, V_k) \right) \qquad (4)$$

In Eq. 4, $L$ and $R$ refer to the left and right child nodes that are created for node $k$, after applying horizontal split $s$.

In a regular PBCT [12], the same procedure would be applied to the features in $V^F$, after which the best overall test would be chosen to split the node. However, since the goal here is to perform HMC, we are not interested in the variance reduction over $V^T$. Instead of measuring the quality of a vertical split directly, we need to make sure that the choice of a vertical split will indeed benefit the instance space partitioning, since in the end we want to make predictions for new (unseen) rows. To do so, the algorithm goes through each possible test from $V^F$, defined by taking a subset of its feature values[2], and uses a lookahead approach [13] (illustrated in Fig. 4): For each test, the vertical split is simulated, as well as the next horizontal split (if any) in both resulting child nodes $(H_k, V_{kL})$ and $(H_k, V_{kR})$ of $(H_k, V_k)$. Thus, the heuristic function for vertical splits is based on the best value of the subsequently applied heuristic for horizontal splits. For computational reasons, we only perform a lookahead of depth one. This results in the following function to evaluate the quality of a vertical split $s$:

$$h_v(s, H_k, V_k) = \frac{|V_{kL}|}{|V_k|} \cdot h_h(s_L, H_k, V_{kL}) + \frac{|V_{kR}|}{|V_k|} \cdot h_h(s_R, H_k, V_{kR}) \qquad (5)$$

Splits $s_L$ and $s_R$ are chosen to maximize the values of the horizontal heuristic in the left and right child nodes of node $k$, respectively. The split $s$ that is chosen to split the node $k$, is the one giving the overall maximal value for Eq. 4 or Eq. 5.

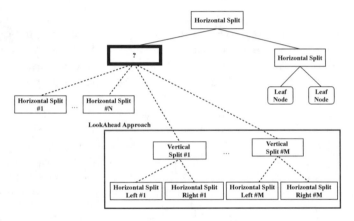

**Fig. 4.** Illustration of lookahead approach: for each test from $V^F$, the vertical split is simulated, as well as the next horizontal split (if any) in the left and right child obtained.

Before applying the split, an F-test is used to check if the variance reduction induced by the split is statistically significant. If the reduction is not significant, a leaf node is created instead. Otherwise, the test is included in the tree, the subsets of $H_k$ or $V_k$ are created to form new bi-clusters and the induction is recursively called until a stopping criterion, e.g. the minimal number of instances in a leaf, is reached. When a vertical split is included, the two subsequent horizontal splits are also included, i.e., a vertical split yields six instead of two new nodes, four of them being nodes to consider for splitting in the next iteration.

Each leaf receives a prototype vector, which corresponds to the vector of classwise (i.e., rowwise) averages. Figure 5(a) shows a small example of a tree resulting from our approach, using the label hierarchy from Fig. 3.

---

[2] In our implementation, we consider a greedy generation of the subsets.

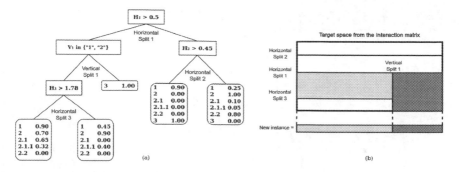

**Fig. 5.** (a) Illustration of an induced PBCT-HMC tree. $H_n$ is the nth feature from $H^F$, and $V_1$ is the (single) categorical feature from $V^F$. (b) Illustration of the prediction procedure for the constructed PBCT-HMC tree. The two bi-clusters where the new instance arrives are colored in gray. The prototypes from these bi-clusters are then copied and concatenated into the prediction vector.

*Stop Criterion.* The smaller the prototype vector gets (after vertical splits), the easier it becomes to find a statistically significant subsequent horizontal split, using a fixed significance level $l_0$. This results in the tree making more splits than necessary and becoming overfitted. For that reason, we apply a correction to the F-test significance level: when checking the significance of a split in a node $k$ defined by $(H_k, V_k)$, we use $l = l_0 \times (|V_k|/|V|)$ as significance level. Thus, the significance criterion becomes stricter, as the target vectors become smaller.

*Making Predictions.* After the tree is built, it is possible to enter a test set and get the predicted vector of probabilities for each test instance. To do this, each test instance is sorted down the tree, starting with the root node. Whenever a horizontal split is encountered, one of the child nodes is followed, according to the outcome of the test. Whenever a vertical split is encountered, however, both child nodes are followed, since a prediction is required for all labels. As such, the test instance may end up in multiple leaf nodes. The final prediction is constructed by concatenating the prototypes from these leaf nodes. Figure 5(b) illustrates the prediction procedure.

*Pseudocode.* Our proposed PBCT-HMC algorithm is presented in Algorithm 1.

## 4    Experiments and Discussion

### 4.1    Datasets

We used 24 datasets in our experiments, from the domain of functional genomics, image and email classification. Twenty datasets contain proteins from the *Saccharomyces cerevisiae* (yeast, 16 datasets) or *Arabidopsis thaliana* (4 datasets,

**Algorithm 1.** PBCT-HMC

---

1: **procedure** INDUCE(DATA H, DATA V, TREE NODE) RETURNS TREE
2:     $bestHorizontal \leftarrow chooseBestTest(H)$
3:     $bestVertical \leftarrow chooseBestTestLookAhead(V)$
4:     **if** $isAcceptable(bestHorizontal, bestVertical)$ **then**    ▷ Stop criterion
5:         **if** $isBetterThan(bestHorizontal, bestVertical)$ **then**    ▷ Horizontal split
6:             $(subsetsH, children) \leftarrow splitHorizontal(node)$
7:             $induce(subsetsH[0], V, children[0])$                ▷ Recursive call
8:             $induce(subsetsH[1], V, children[1])$                ▷ Recursive call
9:         **else**                                              ▷ Vertical split
10:            $adjustFTest()$
11:            $(subsetsV, children) \leftarrow splitVertical(node)$
12:            $induce(H, subsetsV[0], children[0])$             ▷ Recursive call
13:            $induce(H, subsetsV[1], children[1])$             ▷ Recursive call
14:        **end if**
15:    **else**
16:        $includeLeafNode(node)$                    ▷ Create leaf node and stop
17:    **end if**
18:    **return** $node$
19: **end procedure**

---

with suffix '_ara') genome[3]. The datasets differ mainly in how the protein features are represented, e.g., using sequence, cellcycle, or expression information. The label hierarchy employed is the Functional Catalogue (FunCat), a hierarchically structured classification system that enables the functional description of proteins from almost any organism (prokaryotes, fungi, plants and animals). The FunCat annotation scheme covers functions like cellular transport, metabolism and protein activity regulation [22]. The yeast datasets come in 2 versions: the 2007 version datasets[4] were used in the original Clus-HMC publication [14] and the 2018 version datasets[5] come from a recent update of the class labels [23]. The datasets ImgClef07a and ImgClef07d contain annotations of medical X-Ray images, with the attributes being descriptors obtained from EHD diagrams [24]. The Diatoms dataset contains descriptors of image data from diatoms, unicellular algae found in water and humid places [25]. Finally, the Enron dataset specifies a taxonomy for the classification of a corpus of emails [26][6].

Table 1 summarizes the datasets used, presenting the number of instances in each subset (Train, Valid and Test), the number of attributes of each type (Categorical and Numerical), and the number of labels per level of the hierarchy.

---

[3] Available at https://dtai.cs.kuleuven.be/clus/hmc-ens/.
[4] Available at https://dtai.cs.kuleuven.be/clus/hmcdatasets/.
[5] Available at https://itec.kuleuven-kulak.be/supportingmaterial.
[6] Available at http://kt.ijs.si/DragiKocev/PhD/resources/doku.php.

**Table 1.** Summary of datasets

| | Train | Valid | Test | Categorical | Numerical | L1 | L2 | L3 | L4 | L5 | L6 |
|---|---|---|---|---|---|---|---|---|---|---|---|
| Cellcycle2007 | 1628 | 848 | 1281 | 0 | 77 | 18 | 80 | 178 | 142 | 77 | 4 |
| Derisi2007 | 1608 | 842 | 1275 | 0 | 63 | 18 | 80 | 178 | 142 | 77 | 4 |
| Eisen2007 | 1058 | 529 | 837 | 0 | 79 | 18 | 76 | 165 | 131 | 67 | 4 |
| Expr2007 | 1639 | 849 | 1291 | 4 | 547 | 18 | 80 | 178 | 142 | 77 | 4 |
| Gasch1_2007 | 1634 | 846 | 1284 | 0 | 173 | 18 | 80 | 178 | 142 | 77 | 4 |
| Gasch2_2007 | 1639 | 849 | 1291 | 0 | 52 | 18 | 80 | 178 | 142 | 77 | 4 |
| Seq2007 | 1701 | 879 | 1339 | 5 | 473 | 18 | 80 | 178 | 142 | 77 | 4 |
| Spo2007 | 1600 | 837 | 1266 | 3 | 77 | 18 | 80 | 178 | 142 | 77 | 4 |
| Cellcycle2018 | 1628 | 848 | 1281 | 0 | 77 | 20 | 86 | 210 | 171 | 92 | 6 |
| Derisi2018 | 1608 | 842 | 1275 | 0 | 63 | 20 | 86 | 210 | 171 | 92 | 6 |
| Eisen2018 | 1058 | 529 | 837 | 0 | 79 | 19 | 84 | 201 | 159 | 83 | 6 |
| Expr2018 | 1639 | 849 | 1291 | 4 | 547 | 20 | 86 | 210 | 171 | 92 | 6 |
| Gasch1_2018 | 1634 | 846 | 1284 | 0 | 173 | 20 | 86 | 210 | 171 | 92 | 6 |
| Gasch2_2018 | 1639 | 849 | 1291 | 0 | 52 | 20 | 86 | 210 | 171 | 92 | 6 |
| Seq2018 | 1701 | 879 | 1339 | 5 | 473 | 20 | 86 | 210 | 171 | 93 | 6 |
| Spo2018 | 1600 | 837 | 1266 | 3 | 77 | 20 | 86 | 210 | 171 | 92 | 6 |
| Exprindiv_ara | 1579 | 735 | 1182 | 418 | 834 | 15 | 98 | 63 | 35 | – | – |
| Interpro_ara | 1674 | 781 | 1264 | 2816 | 0 | 15 | 101 | 63 | 36 | – | – |
| Scop_ara | 1407 | 648 | 1042 | 1 | 2003 | 15 | 91 | 62 | 32 | – | – |
| Seq_ara | 1674 | 781 | 1264 | 3 | 4448 | 15 | 101 | 63 | 36 | – | – |
| Diatoms | 1032 | 1033 | 1054 | 0 | 371 | 85 | 310 | 3 | – | – | – |
| Enron | 494 | 494 | 660 | 0 | 1001 | 3 | 40 | 13 | – | – | – |
| ImgClef07a | 5000 | 5000 | 1006 | 0 | 176 | 8 | 25 | 63 | – | – | – |
| ImgClef07d | 5000 | 5000 | 1006 | 0 | 158 | 4 | 16 | 26 | – | – | – |

## 4.2 Comparison Methods and Evaluation Measures

We compare PBCT-HMC to its three competitor methods, all set in the PCT framework. Although a lot of recent work on HMC is based on neural networks [17–19], here, next to predictive performance, we also want to assess the interpretability of the models. To do that, we compare the following methods:

- Clus-HMC: Original PCT models proposed for HMC tasks [14][7] (Sect. 3.1);
- PBCT: Original bi-clustering PCT proposed for interaction prediction tasks [12], as described in Sect. 3.1, which has also been applied to multi-label classification tasks before [15];

---

[7] Available at https://dtai.cs.kuleuven.be/clus.

- Clus-Subtree: A static approach where one Clus-HMC model is built for each subtree of the root of the hierarchy;
- PBCT-HMC: Our proposed method[8];

In order to have a fair comparison, we have re-implemented the PBCT method, which was originally implemented in Python, in Clus. For the label feature space, it uses the same mapping of the label hierarchy as PBCT-HMC. The main difference is the split heuristic, which uses rowwise and columnwise variance reduction, without class weights.

The Clus-Subtree method was included to compare against a static variant of our proposed method. In this case, the vertical splits are made upfront: for each subtree of the label hierarchy root, a Clus-HMC model is constructed. Afterwards, all prediction vectors are concatenated. For each model, all available training instances are used, thus including the instances with an all-zeroes label vector (negative instances).

In accordance with the HMC literature [14,18,19], we adopted the Pooled Area Under Precision-Recall Curve (Pooled AUPRC) as the evaluation measure. It is the micro-averaged area under the label-wise precision-recall curves, generated by using threshold values (steps of 0.02) ranging from 0 to 1.

We have optimized the F-test significance level parameter ($l_0$), a parameter used in all compared methods, using the following values: 0.001, 0.005, 0.01, 0.05, 0.1 and 0.125. For each of these values, a model is built using only the training subset, and evaluated using the validation dataset. The value associated to the best performance is selected. Finally, the optimized model is built using the train and validation datasets together, and results are reported using the test dataset. For the parameters minimum of examples perf leaf, we have fixed the value to 5.

### 4.3   Results and Discussion

Table 2 shows the results regarding the Pooled AUPRC for each dataset and each approach. The best results are highlighted. Looking at this table, we can notice that Clus-HMC and PBCT-HMC presented superior results when compared to Clus-Subtree and PBCT. In all datasets, apart from 3 exceptions, either Clus-HMC or PBCT-HMC was superior. When compared between themselves, Clus-HMC and PBCT-HMC yielded very similar results. Although on average, Clus-HMC has a slight advantage, in 7 cases PBCT-HMC outperformed or resulted in the same performance as Clus-HMC. In Fig. 6, we present a critical distance diagram following a Friedman and Nemenyi test, as proposed by [27], comparing the Pooled AUPRC of all methods. The Friedman test resulted in a p-value = 5.441e–07. PBCT-HMC and Clus-HMC present a statistically significant difference when compared to PBCT, but not between themselves. Likewise, PBCT and Clus-Subtree do not present a statistically significant difference. Although PBCT-HMC did not present a statistically significant difference compared to Clus-Subtree, their average ranks are at the limit of the critical difference.

---

[8] Available at https://github.com/biomal/Clus-PBCT-HMC.

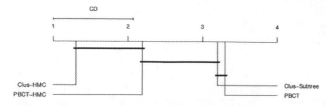

**Fig. 6.** Critical diagram comparing the Pooled AUPRC of all methods (CD = 1.0724)

**Table 2.** Pooled AUPRC using optimal values for the F-test.

|               | Clus-HMC | Clus-Subtree | PBCT  | PBCT-HMC |
|---------------|----------|--------------|-------|----------|
| Cellcycle2007 | **0.172** | 0.170       | 0.154 | 0.165    |
| Derisi2007    | 0.175    | 0.173        | 0.175 | **0.177** |
| Eisen2007     | **0.205** | 0.201       | 0.182 | 0.195    |
| Expr2007      | **0.210** | 0.167       | 0.187 | 0.207    |
| Gasch1_2007   | 0.205    | 0.182        | 0.177 | **0.206** |
| Gasch2_2007   | **0.195** | 0.180       | 0.178 | 0.189    |
| Seq2007       | **0.211** | 0.127       | 0.172 | 0.191    |
| Spo2007       | **0.186** | 0.173       | 0.179 | 0.184    |
| Cellcycle2018 | 0.192    | **0.194**    | 0.181 | 0.192    |
| Derisi2018    | **0.195** | 0.192       | 0.192 | 0.192    |
| Eisen2018     | **0.229** | 0.224       | 0.211 | 0.218    |
| Expr2018      | **0.218** | 0.193       | 0.191 | 0.216    |
| Gasch1_2018   | **0.212** | 0.193       | 0.206 | **0.212** |
| Gasch2_2018   | **0.205** | 0.196       | 0.197 | **0.205** |
| Seq2018       | **0.229** | 0.164       | 0.201 | 0.219    |
| Spo2018       | 0.205    | 0.193        | 0.198 | **0.208** |
| Exprindiv_ara | **0.175** | 0.129       | 0.165 | 0.168    |
| Interpro_ara  | **0.384** | 0.370       | 0.383 | 0.364    |
| Scop_ara      | **0.526** | 0.521       | 0.498 | 0.490    |
| Seq_ara       | **0.280** | 0.217       | 0.271 | 0.272    |
| Diatoms       | 0.220    | 0.217        | 0.217 | **0.234** |
| Enron         | 0.636    | **0.715**    | 0.636 | 0.636    |
| ImgClef07a    | 0.562    | **0.619**    | 0.560 | 0.551    |
| ImgClef07d    | 0.749    | 0.712        | 0.793 | **0.794** |
| Mean          | **0.282** | 0.268       | 0.271 | 0.278    |

The contrasting performance between PBCT-HMC and PBCT is associated to the lookahead procedure and the F-test significance level adjustment. PBCT, which targets interaction prediction and aims to maximize also column-wise

**Table 3.** Model size: Nodes (Leaves)

| | Clus-HMC | Clus-Subtree | PBCT | PBCT_HOR | PBCT-HMC | PBCT-HMC_HOR |
|---|---|---|---|---|---|---|
| Cellcycle2007 | **41 (21)** | 574 (296) | 383 (192) | 290 (192) | 45 (23) | **41 (23)** |
| Derisi2007 | **7 (4)** | 200 (109) | 23 (12) | 20 (12) | 23 (12) | 20 (12) |
| Eisen2007 | **11 (6)** | 414 (216) | 262 (132) | 251 (132) | 209 (105) | 192 (105) |
| Expr2007 | **75 (38)** | 1634 (826) | 289 (145) | 282 (145) | 163 (82) | 156 (82) |
| Gasch1_007 | **67 (34)** | 864 (441) | 521 (261) | 504 (261) | 73 (37) | **67 (37)** |
| Gasch2_007 | 53 (27) | 422 (220) | 71 (36) | 67 (36) | 31 (16) | **29 (16)** |
| Seq2007 | **95 (48)** | 2112 (1065) | 319 (160) | 310 (160) | 269 (135) | 255 (135) |
| Spo2007 | **11 (6)** | 354 (186) | 79 (40) | 74 (40) | 25 (13) | 22 (13) |
| Cellcycle2018 | **35 (18)** | 472 (246) | 47 (24) | 43(24) | 63 (32) | 57 (32) |
| Derisi2018 | 19 (10) | 216 (118) | 35 (18) | 31 (18) | 15 (8) | **13 (8)** |
| Eisen2018 | 55 (28) | 375 (197) | 315 (158) | 287 (158) | 33 (17) | **28 (17)** |
| Expr2018 | **19 (10)** | 1602 (811) | 789 (395) | 761 (395) | 119 (60) | 108 (60) |
| Gasch1_018 | **81 (41)** | 914 (467) | 173 (87) | 165 (87) | 153 (77) | 141 (77) |
| Gasch2_018 | **41 (21)** | 412 (216) | 81 (41) | 76 (41) | 81 (41) | 73 (41) |
| Seq2018 | **23 (12)** | 2162 (1091) | 237 (119) | 225 (119) | 153 (77) | 145 (77) |
| Spo2018 | 39 (20) | 386 (203) | 59 (30) | 54 (30) | 25 (13) | **20 (13)** |
| Exprindiv_ara | **19 (10)** | 1183 (600) | 85 (48) | 83 (43) | 29 (15) | 27 (15) |
| Interpro_ara | 173 (87) | 1119 (568) | 1231 (616) | 1003 (616) | 139 (70) | **131 (70)** |
| Scop_ara | 267 (134) | 814 (415) | 1675 (838) | 1422 (838) | 173 (87) | **165 (87)** |
| Seq_ara | **37 (19)** | 1489 (753) | 207 (104) | 194 (104) | 145 (73) | 140 (73) |
| Diatoms | 427 (214) | 1975 (1038) | 1603 (802) | 1382 (802) | 1269 (635) | 1206 (635) |
| Enron | 55 (28) | **39 (21)** | 63(23) | 59 (23) | 55 (28) | 55 (28) |
| ImgClef07a | **721 (361)** | 1390 (699) | 1971 (986) | 1739 (986) | 1785 (893) | 1755 (893) |
| ImgClef07d | **509 (255)** | 1296 (650) | 595(298) | 593 (298) | 597 (299) | 596 (299) |
| Mean | **120 (60)** | 934 (477) | 462 (206) | 413 (232) | 236 (116) | 217 (116) |

variance reduction, performs vertical splits prematurely, leading to suboptimal splits. As for the F-test, not adjusting its significance level can result in overfitting.

As for model size (Table 3), we see that PBCT-HMC results in models that are 4 and 2 times smaller on average compared to Clus-Subtree and PBCT, respectively. Thus, compared to the methods that also partition the label space, our method can be expected to yield more interpretable models. This finding is expected for Clus-Subtree since it contains multiple models, and thus its size consists of the summed size of all models. As for PBCT, we suspect that the models may be overfitted due to their larger size, but inferior performance. In contrast, when compared to Clus-HMC, on average, our models are twice as large. This can be explained by the fact that Clus-HMC only introduces horizontal splits and that each vertical split in PBCT-HMC results in six new nodes compared to only two nodes for horizontal splits, as explained before. Since at prediction time, the vertical nodes are not used semantically, in the sense that their split condition does not need to be checked for the test instances (they follow the paths to both child nodes), and hence, they do not add to the prediction time, we also report the number of horizontal splits for the bi-clustering-based methods (reported as PBCT_HOR and PBCT-HMC_HOR).

**Table 4.** Induction time of the optimized model, in seconds

| | Clus-HMC | Clus-Subtree | PBCT | PBCT-HMC |
|---|---|---|---|---|
| Cellcycle2007 | **2.62** | 17.51 | 7.80 | 10.64 |
| Derisi2007 | **1.60** | 18.37 | 2.71 | 8.97 |
| Eisen2007 | **4.26** | 11.21 | 2.64 | 15.22 |
| Expr2007 | 29.74 | 309.46 | 25.86 | 94.61 |
| Gasch1_2007 | 20.13 | 46.43 | **10.39** | 29.17 |
| Gasch2_2007 | 3.10 | 23.86 | **2.16** | 4.17 |
| Seq2007 | **20.09** | 216.04 | 23.51 | 75.56 |
| Spo2007 | **2.41** | 24.78 | 3.12 | 14.56 |
| Cellcycle2018 | 3.05 | 16.94 | **2.89** | 18.29 |
| Derisi2018 | **2.98** | 19.90 | 3.36 | 10.47 |
| Eisen2018 | **2.17** | 11.79 | 4.04 | 7.75 |
| Expr2018 | 18.09 | 343.38 | 56.67 | 93.41 |
| Gasch1_2018 | 8.85 | 50.5 | **8.07** | 25.82 |
| Gasch2_2018 | 10.98 | 25.08 | **2.77** | 24.07 |
| Seq2018 | **17.37** | 235.93 | 24.39 | 37.52 |
| Spo2018 | 4.37 | 29.79 | **3.82** | 17.89 |
| Exprindiv_ara | 13.19 | 247.01 | 20.17 | 25.80 |
| Interpro_ara | **2.60** | 15.27 | 7.63 | 10.58 |
| Scop_ara | 55.13 | 136.00 | 102.03 | 85.90 |
| Seq_ara | 90.18 | 1184.37 | 100.44 | 144.04 |
| Diatoms | **3.5** | 311.69 | 45.95 | 62.12 |
| Enron | **0.26** | 3.69 | 1.57 | 1.14 |
| ImgClef07a | **1.08** | 15.76 | 7.32 | 16.50 |
| ImgClef07d | **0.92** | 9.64 | 4.29 | 5.52 |
| Mean | **13.27** | 138.51 | 19.73 | 34.99 |

Although Clus-HMC induced smaller models on average, we see that PBCT-HMC, specially when considering only the horizontal nodes, yielded considerably smaller models for some datasets. Nevertheless, our results raise the question under which conditions label partitions should be considered to exploit correlations.

Finally, Table 4 exposes the induction time for each experiment. Clus-HMC presented lower induction times overall due to its simplicity. On the other hand, Clus-Subtree demands most time since it builds multiple models per dataset. PBCT-HMC comes in third, after PBCT, due to the lookahead strategy. The induction time depends on the number of categorical feature values of the dataset $V$, meaning the hierarchy size has a direct effect on the induction time. However, as the table shows, the run time is still feasible.

# 5    Conclusions

In this work, we have proposed a predictive bi-clustering tree method for hierarchical multi-label classification (HMC). Opposed to traditional approaches for HMC, our method, named PBCT-HMC, automatically partitions the label space during its induction process. To achieve this, a lookahead approach is incorporated in the tree construction, such that a label space partition is introduced if it leads to a better instance space partition in a deeper level of the tree.

Our experiments demonstrated that PBCT-HMC obtained better or competitive predictive performances compared to three related methods, yielding the second smallest models overall. Specially in comparison with the original PBCT and the static method Clus-Subtree, PBCT-HMC performed better in the large majority of the datasets, achieving a great interpretability gain. Still, our results put into question the advantage of considering label partitions to exploit label correlations in multi-label classification, a topic that receives a lot of attention. Further research is needed to identify which type of models can benefit from it.

Future extensions of our work might address more challenging applications, e.g., datasets with a label hierarchy structured as a directed acyclic graph, and the application to non-hierarchical multi-label classification tasks.

**Acknowledgments.** We acknowledge Sao Paulo Research Foundation (FAPESP grants #2017/13218-5 and #2016/25078-0) and Research Fund Flanders (FWO) for financial support.

# References

1. Tsoumakas, G., Katakis, I.: Multi-label classification: an overview. Int. J. Data Warehouse. Min. (IJDWM) **3**(3), 1–13 (2007)
2. Blockeel, H., Raedt, L.D., Ramon, J.: Top-down induction of clustering trees. In: Proceedings of the Fifteenth International Conference on Machine Learning, ICML 1998, pp. 55–63 (1998)
3. Papagiannopoulou, C., Tsoumakas, G., Tsamardinos, I.: Discovering and exploiting deterministic label relationships in multi-label learning. In: Proceedings of the 21th ACM SIGKDD International Conference on Knowledge Discovery and Data Mining, pp. 915–924. Association for Computing Machinery (2015)
4. Madjarov, G., Gjorgjevikj, D., Dimitrovski, I., Džeroski, S.: The use of data-derived label hierarchies in multi-label classification. J. Intell. Inf. Syst. **47**(1), 57–90 (2016)
5. Szymanski, P., Kajdanowicz, T., Kersting, K.: How is a data-driven approach better than random choice in label space division for multi-label classification? CoRR (2016)
6. Joly, A., Geurts, P., Wehenkel, L.: Random forests with random projections of the output space for high dimensional multi-label classification. In: Machine Learning and Knowledge Discovery in Databases, pp. 607–622 (2014)
7. Breskvar, M., Kocev, D., Džeroski, S.: Multi-label classification using random label subset selections. In: Discovery Science (2017)
8. Breskvar, M., Kocev, D., Džeroski, S.: Ensembles for multi-target regression with random output selections. Mach. Learn. **107**(11), 1673–1709 (2018). https://doi.org/10.1007/s10994-018-5744-y

9. Prati, R.C., de França, F.O.: Extending features for multilabel classification with swarm biclustering. In: IEEE Congress on Evolutionary Computation, pp. 2964–2971 (2013)
10. de Abreu, I.B.M., Mantovani, R.G., Cerri, R.: Incorporating instance correlations in multi-label classification via label-space. In: International Joint Conference on Neural Networks (IJCNN), pp. 581–588 (2017)
11. Feng, L., An, B., He, S.: Collaboration based multi-label learning. In: Proceedings of the AAAI Conference on Artificial Intelligence, vol. 33, pp. 3550–3557 (2019)
12. Pliakos, K., Geurts, P., Vens, C.: Global multi-output decision trees forinteraction prediction. Mach. Learn. **107**(8), 1257–1281 (2018). https://doi.org/10.1007/s10994-018-5700-x
13. Elomaa, T., Malinen, T.: On lookahead heuristics in decision tree learning. In: Zhong, N., Raś, Z.W., Tsumoto, S., Suzuki, E. (eds.) ISMIS 2003. LNCS (LNAI), vol. 2871, pp. 445–453. Springer, Heidelberg (2003). https://doi.org/10.1007/978-3-540-39592-8_63
14. Vens, C., Struyf, J., Schietgat, L., Džeroski, S., Blockeel, H.: Decision trees for hierarchical multi-label classification. Mach. Learn. **73**(2), 185 (2008)
15. Pliakos, K., Vens, C., Tsoumakas, G.: Predicting drug-target interactions with multi-label classification and label partitioning. In: IEEE-ACM Transactions On Computational Biology And Bioinformatics, pp. 1–11 (2019)
16. Schietgat, L., Vens, C., Struyf, J., Blockeel, H., Kocev, D., Dzeroski, S.: Predicting gene function using hierarchical multi-label decision tree ensembles. BMC Bioinf. **11**, 2 (2010)
17. Cerri, R., Barros, R.C., de Carvalho, A.C., Jin, Y.: Reduction strategies for hierarchical multi-label classification in protein function prediction. BMC Bioinform. **17**(1), 373 (2016) https://doi.org/10.1186/s12859-016-1232-1
18. Wehrmann, J., Cerri, R., Barros, R.: Hierarchical multi-label classification networks. In: Proceedings of the 35th International Conference on Machine Learning, vol. 80, pp. 5075–5084 (2018)
19. Masera, L., Blanzieri, E.: Awx: an integrated approach to hierarchical-multilabel classification. In: Machine Learning and Knowledge Discovery in Databases, pp. 322–336 (2019)
20. Tsoumakas, G., Katakis, I., Vlahavas, I.: Random k-labelsets for multi-label classification. IEEE Trans. Knowl. Data Eng. **23** 1079–1089 (2011)
21. Moyano, J., Gibaja, E., Cios, K., Ventura, S.: Combining multi-labelclassifiers based on projections of the output space using evolutionary algorithms. Knowl.-Based Syst. **196**, 105770 (2020)
22. Ruepp, A, Zollner, A.M.D.: The funcat, a functional annotation scheme for systematic classification of proteins from whole genomes. Nucleic Acids Res. **32**(18), 5539–5545 (2004)
23. Nakano, F.K., Lietaert, M., Vens, C.: Machine learning for discovering missing or wrong protein function annotations. BMC Bioinf. **20**(1), 485 (2019)
24. Dimitrovski. I., Kocev, D., Loskovska, S., Džeroski, S.: Hierchical annotation of medical images. In: Proceedings of the 11th International Multiconference - Information Society IS 200. IJS, Ljubljana, pp. 174–181 (2008)
25. Dimitrovski, I., Kocev, D., Loskovska, S., Džeroski, S.: Hierarchical classification of diatom images using ensembles of predictive clustering trees. Ecol. Inf. **7**(1), 19–29 (2012)

26. Klimt, B., Yang, Y.: The Enron corpus: a new dataset for email classification research. In: Boulicaut, J.-F., Esposito, F., Giannotti, F., Pedreschi, D. (eds.) ECML 2004. LNCS (LNAI), vol. 3201, pp. 217–226. Springer, Heidelberg (2004). https://doi.org/10.1007/978-3-540-30115-8_22
27. Demšar, J.: Statistical comparisons of classifiers over multiple data sets. J. Mach. Learn. Res. **7**, 1–30 (2006)

# Self-attention Enhanced Patient Journey Understanding in Healthcare System

Xueping Peng[1]($\boxtimes$), Guodong Long[1], Tao Shen[1], Sen Wang[2], and Jing Jiang[1]

[1] FEIT, AAII, University of Technology Sydney, Ultimo, Australia
{xueping.peng,guodong.long,jing.jiang}@uts.edu.au,
Tao.Shen@student.uts.edu.au
[2] School of Information Technology and Electrical Engineering,
The University of Queensland, Brisbane, Australia
sen.wang@uq.edu.au

**Abstract.** Understanding patients' journeys in healthcare system is a fundamental prepositive task for a broad range of AI-based healthcare applications. This task aims to learn an informative representation that can comprehensively encode hidden dependencies among medical events and its inner entities, and then the use of encoding outputs can greatly benefit the downstream application-driven tasks. A patient journey is a sequence of electronic health records (EHRs) over time that is organized at multiple levels: patient, visits and medical codes. The key challenge of patient journey understanding is to design an effective encoding mechanism which can properly tackle the aforementioned multi-level structured patient journey data with temporal sequential visits and a set of medical codes. This paper proposes a novel self-attention mechanism that can simultaneously capture the contextual and temporal relationships hidden in patient journeys. A multi-level self-attention network (MusaNet) is specifically designed to learn the representations of patient journeys that is used to be a long sequence of activities. We evaluated the efficacy of our method on two medical application tasks with real-world benchmark datasets. The results have demonstrated the proposed MusaNet produces higher-quality representations than state-of-the-art baseline methods. The source code is available in https://github.com/xueping/MusaNet.

**Keywords:** Electronic health records · Self-attention network · Medical outcome · Healthcare · Bi-directional

## 1 Introduction

Healthcare system stores huge volumes of electronic health records (EHR) that contain detailed admission information about patients over a period of time [12,29]. The data is organised in multi-level structure: the patient journey level, the patient visit level, and the medical code level. A useful example of this structure shown in Fig. 1. In the example, an anonymous patient visits his/her

© Springer Nature Switzerland AG 2021
F. Hutter et al. (Eds.): ECML PKDD 2020, LNAI 12459, pp. 719–735, 2021.
https://doi.org/10.1007/978-3-030-67664-3_43

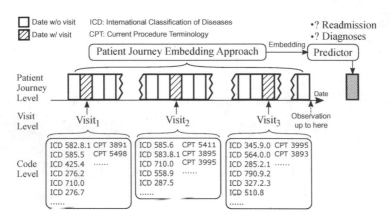

**Fig. 1.** An example of a patient's medical records in multi-level structure – from top to bottom: patient journey level, individual visit level, and medical code level.

doctor and is admitted to hospital on different days. The diagnoses and procedures performed at each of these visits are recorded as industry-standard medical codes. Each medical code at the lowest level records an independent observation while the set of codes at higher levels can depict the medical conditions of a patient at a given time point. At the top level, for each patient, all occurrences of medical events at different time-stamps are chained together as a patient journey, which offers more informative details. Understanding patient activity based on a patient's journey, such as re-admissions and future diagnoses, is a core research task that of significant value for healthcare system. For example, re-admission statistics can be used to measure the quality of care. Diagnoses can be used to more fully understand a patient's problems and medical research [26]. However, researchers have faced many challenges in their attempts to represent patient journeys from EHR data with temporality, high-dimensionality, and irregularity [25].

Recurrent neural network (RNN) has been widely used to model medical events in sequential EHR data for clinical prediction tasks [2,3,5,16,18,25]. For instance, Ref. [2,5] indirectly exploit an RNN to embed the visit sequences into a patient representation by multi-level representation learning to integrate visits and medical codes based on visit sequences and the co-occurrence of medical codes. Other research works have, however, used RNNs directly to model time-ordered patient visits for predicting diagnoses [3,4,16,18,25]. Convolutional Neural Network (CNN) also has been exploited to represent patient journey. For example, Ref. [22] transforms a record into a sequence of discrete elements separated by coded time gaps, and then employ CNN to detect and combine predictive local clinical motifs to stratify the risk. Reference [30] utilises CNN in code level to learn visit embedding. These RNN and CNN based models follow ideas of processing sentences [13] in documents from NLP to treat a patient journey as a document and a visit as a sentence, which only has a sequential relationship, while two arbitrary visits in one patient journey may be separated by different time intervals. The interval between two visits, which has been largely disregarded in the existing studies on patient journey representation, can be modelled as auxiliary information fed into the supervised algorithms.

Recently, research has proposed the integration of attention mechanisms and RNNs to model sequential EHR data [4,16,24,26]. Reference [32] used a sole attention mechanism to construct a sequence-to-sequence model for a neural machine translation task that achieved a state-of-the-art quality score. According to Ref. [27], attention allows for more flexibility in sequence lengths than RNNs, and is more task/data-driven when modelling dependencies. Unlike sequential models, attention is easy to compute. Computation can also be significantly accelerated with distributed/parallel computing schemes. However, to the best of our knowledge, a neural network based entirely on attention has never been designed for analytics with EHR data.

To fill this gap in the literature, and address some of the open issues listed above, we propose a novel attention mechanism called masked self-attention (mSA) for temporal context fusion that uses self-attention to capture contextual information and the temporal dependencies between patient's visits. In addition, we propose an end-to-end neural network, called multi-level self-attention network (MusaNet) to understand medical outcomes using a learned representation of a patient journey based solely on the proposed attention mechanism. MusaNet constructs a multi-level self-attention representation to simultaneously represent visits and patient journeys using attention pooling and mSA layers. It is worth noting that, compared to the existing RNN and CNN-based methods, MusaNet can yield better prediction performance for a long sequence of medical records.

Experiments conducted on two prediction tasks with two public EHR datasets demonstrate that the proposed MusaNet is superior to several state-of-the-art baseline methods.

To summarize, our main contributions are:

- a novel attention mechanism, called mSA, that uses self-attention to capture the contextual information and long-term dependencies between patient's visits;
- an end-to-end neural network called "MusaNet" that predicts medical outcomes using a learned representation of a patient journey based solely on the proposed attention mechanism; and
- an evaluation of the proposed model on two benchmark datasets with two prediction tasks, demonstrating that the MusaNet model is superior to all the compared methods.

## 2 Background

### 2.1 Definitions and Notations

**Definition 1 (Medical Code).** *A medical code is a term or entry to describe a diagnosis, procedure, medication, or laboratory test administered to a patient. A set of medical codes is formally denoted as $X = \{x_1, x_2, \ldots, x_{|X|}\}$, where $|X|$ is the total number of unique medical codes in the EHR data.*

**Definition 2 (Visit).** *A visit is a hospital stay from admission to discharge with an admission time stamp. A visit is denoted as $V_i = <x_i, t_i>$, where $x_i = [x_{i1}, x_{i2}, ..., x_{ik}]$, $i$ is the $i$-th visit, $t_i$ is the admission time of the $i$-th visit, $k$ is the total number of medical codes in a visit.*

**Definition 3 (Patient Journey).** *A patient journey consists of a sequence of visits over time, denoted as $J = [V_1, V_2, ..., V_n]$, where $n$ is the total number of visits for the patient.*

**Definition 4 (Temporal position).** *Temporal position refers to a difference in days between admission time $t_i$ of the $i$-th visit and admission time $t_1$ of the first visit in a patient journey, denoted as $p_i = |t_i - t_1|$, where $i = 1, \ldots, n$.*

**Definition 5 (Task).** *Given a set of patient journeys $\{J_l\}_{l=1,...}$, the task is to predict medical outcomes.*

In the remainder of the paper, a patient's medical data means a stored and chronological sequence of $m$ visits in a patient journey $J_i$. To reduce clutter, we omit the superscript and subscript ($i$) indicating $i$-th patient, when discussing a single patient journey.

Table 1 summarizes notations we will use throughout the paper.

**Table 1.** Notations for BiteNet.

| Notation | Description |
| --- | --- |
| $X$ | Set of unique medical codes |
| $|X|$ | The size of unique medical concept |
| $V_i$ | The $i$-th visit of the patient |
| $v_i$ | The representation of $t$-th visit of the patient |
| $v$ | Sequence of $n$ visit embeddings of the patient |
| $x_i$ | Set of medical codes in $V_i$ |
| $x$ | Sequence of medical codes, $[x_1, x_2, ..., x_n]$ |
| $e_i$ | Set of medical code embeddings in $x_i$ |
| $e$ | A sequence of medical code embeddings, $[e_1, e_2, ..., e_n]$ |
| $J$ | A patient journey consisting of a sequence of visits over time |
| $d$ | The dimension of medical code embedding |

## 2.2 Medical Code Embedding

The concept of word embedding was introduced to medical analytics by Ref. [19,20] as a way to learn low-dimensional real-valued distributed vector representations of medical concepts for downstream tasks instead of using discrete medical codes. Hence, medical code embedding is a fundamental processing unit for learning EHRs in a deep neural network. Formally, given a sequence or set of medical concepts $c = [c_1, c_2, ..., c_n] \in \mathbb{R}^{|C| \times n}$, where $c_i$ is a one-hot vector, and $n$ is the sequence length. In the NLP literature, a word embedding method like word2vec [19,20] converts a sequence of one-hot vectors into a sequence of low-dimensional vectors $e = [e_1, e_2, ..., e_n] \in \mathbb{R}^{d \times n}$, where $d$ is the embedding dimension of $e_i$. This process can be formally written as $e = W^{(e)} c$, where $W^{(e)} \in \mathbb{R}^{d \times |C|}$ is the embedding weight matrix, which can be fine-tuned during the training phase.

A visit can be represented by the set of medical codes embedded with real-valued dense vectors. A straightforward approach to learning this representation $v_i$ is to sum the embeddings of medical codes in the visit, which is written as

$$v_i = \sum_{e_{ik} \in e_i} e_{ik}, \tag{1}$$

where $e_i$ is the set of medical code embeddings in the $i$-th visit, and $e_{ik}$ is the $k$-th code embedding in $e_i$. A visit can also be represented as a real-valued dense vector with a more advanced method, such as self-attention and attention pooling discussed below.

### 2.3   Attention Mechanism

**Vanilla Attention.** Given a patient journey consisting of a sequence of visits $J = [v_1, v_2, ..., v_m]$ and a vector representation of a query $q \in \mathbb{R}^d$, vanilla attention [1] computes the alignment score between $q$ and each visit $v_i$ using a compatibility function $f(v_i, q)$. A softmax function then transforms the alignment scores $\alpha \in \mathbb{R}^n$ to a probability distribution $p(z|J, q)$, where $z$ is an indicator of which visit is important to $q$. A large $p(z = i|J, q)$ means that $v_i$ contributes important information to $q$. This attention process can be formalized as

$$\alpha = [f(v_i, q)]_{i=1}^n, \tag{2}$$

$$p(z|J, q) = softmax(\alpha). \tag{3}$$

The output $s$ is the weighted average of sampling a visit according to its importance, i.e.,

$$s = \sum_{i=1}^n p(z = i|J, q) \cdot v_i. \tag{4}$$

Additive attention is a commonly-used attention mechanism [1], where a compatibility function $f(\cdot)$ is parameterized by a multi-layer perceptron (MLP), i.e.,

$$f(v_i, q) = w^T \sigma(W^{(1)} v_i + W^{(2)} q + b^{(1)}) + b, \tag{5}$$

where $W^{(1)} \in \mathbb{R}^{d \times d}$, $W^{(2)} \in \mathbb{R}^{d \times d}$, $w \in \mathbb{R}^d$ are learnable parameters, and $\sigma(\cdot)$ is an activation function.

Reference [27] recently proposed a multi-dimensional (multi-dim) attention mechanism as a way to further improve an attention module's ability to model contexts through feature-wise alignment scores. So, to model more subtle context and dependency relationships, the score might be large for some features but small for others. Formally, $P_{ki} \triangleq p(z_k = i | \boldsymbol{J}, q)$ denotes the attention probability of the $i$-th visit on the $k$-th feature dimension, where the attention score is calculated from a multi-dim compatibility function by replacing the weight vector $w$ in Eq. (5) with a weight matrix. Note that, for simplicity, we ignore the subscript $k$ if it does not cause any confusion. The attention result can, therefore, be written as $s = \sum_{i=1}^{n} P_{\cdot i} \odot v_i$

**Sophisticated Self-attention Mechanism.** Information about the contextual and temporal relationships between two individual visits $v_i$ and $v_j$ in the same patient journey $\boldsymbol{J}$ is captured with a self-attention mechanism [32] because a context-aware representation probably leads to better empirical performance [9, 27, 32].

Specifically, the query $q$ in an attention compatibility function, such as a multi-dim compatibility function, is replaced with $v_j$, i.e.,

$$f(v_i, v_j) = W^T \sigma(W^{(1)} v_i + W^{(2)} v_j + b^{(1)}) + b. \tag{6}$$

Similar to the probability matrix $P$ in multi-dim attention, each input individual visit $v_j$ is associated with a probability matrix $P_j$ such that $P_{ki}^j \triangleq p(z_k = i | \boldsymbol{J}, v_j)$. The output representation for each $v_j$ is

$$s_j = \sum_{i=1}^{n} P_{\cdot i}^j \odot v_i \tag{7}$$

The final output of self-attention is $\boldsymbol{s} = [s_1, s_2, \ldots, s_n]$, each of which is a visit-context embedded representation for each individual visit.

However, a fatal defect in previous self-attention mechanisms for NLP tasks is that they are order-insensitive and therefore, cannot model temporal relationships between the visits in a patient journey. Even if position embedding [32] is used to alleviate this problem, it is designed for consecutive natural language words rather than patient visits with arbitrary intervals.

**Attention Pooling.** Attention pooling [14,15] explores the importance of each individual visit within an entire patient journey given a specific task. It works by compressing a sequence of visit embeddings from a patient journey into a single context-aware vector representation for downstream classification or regression. Formally, $q$ is removed from the common compatibility function, which is written as

$$f(v_i) = W^T \sigma(W^{(1)} v_i + b^{(1)}) + b. \tag{8}$$

The multi-dim attention probability matrix $P$ is defined as $P_{ki} \triangleq p(z_k = i | \boldsymbol{J})$. The final output of attention pooling, which is used as sequence encoding, is similar to the self-attention mechanism above, i.e.,

$$s = \sum_{i=1}^{n} P_{\cdot i} \odot v_i \tag{9}$$

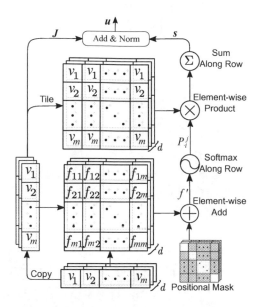

**Fig. 2.** The masked self-attention mechanism (mSA). $f_{ij}$ is defined as $f(v_i, v_j)$ in Eq. (6). $f'$ is defined as $f'(v_i, v_j)$ in Eq. (10). $P^j_{\cdot i}$ is formally defined in Eq. (13), and $s$ is defined in Eq. (7).

## 3  Proposed Model

This section begins by introducing the masked self-attention module, followed by the MusaNet context fusion module for predicting medical outcomes.

### 3.1  Masked Self-attention

Because self-attention was originally designed for NLP tasks, it does not consider temporal relationships (e.g., interval between visits) within inputs. Obviously, it is very important when modeling sequential medical events. Inspired by the work of Ref. [27,28] with masked self-attention, we developed masked self-attention (mSA) to capture the contextual and temporal relationships between visits. The structure of mSA is shown in Fig. 2.

The self-attention mechanism outlined in Eq. (6) is rewritten into a temporal-dependent format:

$$f'(v_i, v_j) = f(v_i, v_j) + M^{pos}_{ij}, \tag{10}$$

where $f(v_i, v_j) = W^T \sigma(W^{(1)}v_i + W^{(2)}v_j + b^{(1)}) + b$ captures the contextual dependency between the $i$-th visit and the target $j$-th visit. In this context, masks can be used to encode temporal order information into an attention output. Our approach also incorporates three masks, such as, forward $M^{fw}$, backward $M^{bw}$ and diagonal mask $M^{diag}_{ij}$, i.e.,

$$M_{ij}^{fw} = \begin{cases} 0, & i < j \\ -\infty, & otherwise \end{cases} \tag{11}$$

$$M_{ij}^{bw} = \begin{cases} 0, & i > j \\ -\infty, & otherwise \end{cases} \tag{12}$$

The forward mask $M^{fw}$, only attends later visits $j$ to earlier visits $i$, and vice versa with the backward one. Here, let $M_{ij}^{pos} \in \{M_{ij}^{fw}, M_{ij}^{bw}\}$.

Given an input patient journey $\boldsymbol{J}$ and a positional mask $M$, $f'(v_i, v_j)$ is computed according to Eq. (10) followed by the standard procedure of self-attention to compute the probability matrix $P^j$ for each $v_j$ as follows:

$$P_{\cdot i}^j \triangleq [p(z_k = i | \boldsymbol{J}, v_j)]_{k=1}^d, \tag{13}$$

The output $\boldsymbol{s}$ is computed as Eq. (7). Then, layer normalization ($Norm$) and rectified linear unit ($ReLU$) is used to generate the output $\boldsymbol{u}$ as follows,

$$\boldsymbol{u} = \text{Norm}(\text{ReLU}(\boldsymbol{J} + \boldsymbol{s})). \tag{14}$$

## 3.2   Multi-level Self-attention Network

We propose a patient journey embedding model called the "Multi-lelvel Self-Attention Network (MusaNet)" with mSA as its major components. The architecture of MusaNet is shown in Fig. 3. In MusaNet, the embedding layer is applied to the input medical codes of visits, and its output is processed by the code-level attention layer to generate visit embeddings $[v_1, v_2, ..., v_m]$ given in Eq. (9). The visit interval encodings are then added to the visit embeddings, which is followed by two parameter-untied mSA blocks with forward mask $M^{fw}$ in Eq. (11) and $M^{bw}$ in Eq. (12), respectively. Their outputs are denoted by $u^{fw}, u^{bw} \in \mathbb{R}^{d \times m}$. The visit-level attention layers in Eq. (9) are applied to the outputs followed by the concatenation layer to generate output $u^{bi} \in \mathbb{R}^{2d}$. Lastly, a feed-forward layer consisting of dense and softmax layers is employed to generate a categorical distribution over targeted categories.

**Interval Encoding.** Although mSA incorporates information on the order of visits in a patient journey, the relative time interval between visits is an important factor in longitudinal studies. We include information on the visit intervals $p = [p_1, p_2, ..., p_m]$ in the sequence. In particular, we add interval encodings to the visit embeddings of the sequence. The encoding is performed by mapping interval $t$ to the same lookup table during both training and prediction. The $d$-dimensional interval encoding is then added to the visit embedding with the same dimension.

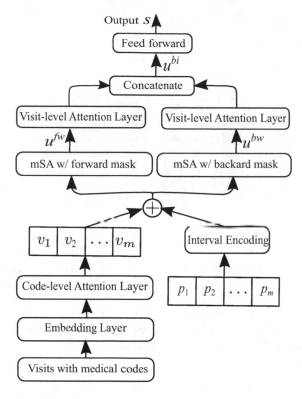

**Fig. 3.** The multi-level self-attention network (MusaNet) framework. The inputs are $m$ sequential visits with medical codes. $m$ is the maximal number of visits in individual patient journeys over the dataset, and padding is applied if the number of visits in a patient journey is less than $m$.

## 4  Experiments

### 4.1  Data Description

**Dataset.** MIMIC-III [11] is an open-source, large-scale, de-identified dataset of ICU patients and their EHRs. The dataset consists of 46k+ ICU patients with 58k+ visits collected over 11 years. In this paper, we consider two sub-datasets: Dx and Dx&Tx, where Dx is a dataset which only includes diagnosis codes for each visit, and Dx&Tx is another dataset which includes diagnosis and procedure codes for each visit.

**Data Pre-processing.** We chose patients who made at least two visits. All infrequent diagnoses codes were removed and the threshold was empirically set to 5. In summary, we extract 7,499 patients with an average of 2.66 visits per patient; the average number of diagnoses and procedures each visit are 13 and 4, respectively.

## 4.2   Experiment Setup

**Baseline Models.** We compare the performance of our proposed model against several baseline models[1]:

- *RNN* and *BRNN*, We directly embed visit information into the vector representation $v_t$ by summation of embedded medical cods in the visit, and then feed this embedding to the GRU. The hidden state $h_t$ produced by the GRU is used to predict the $(t+1)$-th visit information.
- *RETAIN* [4], which learns the medical concept embeddings and performs heart failure prediction via the reversed RNN with the attention mechanism.
- *Deepr* [22]: which is a multilayered architecture based on convolutional neural networks (CNNs) that learn to extract features from medical records and predict future risk.
- *Dipole* [16], which uses bidirectional recurrent neural networks and three attention mechanisms (location-based, general, concatenation-based) to predict patient visit information.
- *SAnD* [30]: which employs a masked, self-attention mechanism, and uses positional encoding and dense interpolation strategies to incorporate temporal order to generate a sequence-level prediction.
- *Transformer Encoder* (TransEnc) [32]: which only considers Transformer with one encoder layer to replace mSA in MusaNet.

**Validation Tasks.** The two tasks we selected to evaluate the performance of our proposed model are to predict re-admission and future diagnosis [26].

- **Re-admission (Readm)** is a standard measure of the quality of care. We predicted unplanned admissions within 30 d following a discharge from an indexed visit. A visit is considered a "re-admission" if admission date is within thirty days after discharge of an eligible indexed visit [26].
- **Diagnoses** reflect the model's understanding of a patient's problems. In the experiments, we aim to predict diagnosis categories instead of the real diagnosis codes, which are the nodes in the second hierarchy of the ICD9 codes[2].

**Evaluation Metrics.** The two evaluation metrics used are:

- **PR-AUC:** (Area under Precision-Recall Curve), to evaluate the likelihood of re-admission. PR-AUC is considered to be a better measure for imbalanced data like ours [7].
- **precision@k:** which is defined as the correct medical codes in top k divided by $\min(k, |y_t|)$, where $|y_t|$ is the number of category labels in the $(t+1)$-th visit [18]. We report the average values of precision@k in the experiments. We vary $k$ from 5 to 30. The greater the value, the better the performance.

---

[1] GRAM [3] and KAME [18] and MMORE [31] are not baselines as they use external knowledge to learn the medical code representations.
[2] http://www.icd9data.com.

**Table 2.** Prediction performance comparison of future re-admission and diagnoses (Dx is for diagnosis, and Tx is for procedure).

| Data | Model | Readm (PR-AUC) | Diagnosis Precision@k | | | |
|------|-------|----------------|--------|--------|--------|--------|
| | | | k = 5 | k = 10 | k = 20 | k = 30 |
| Dx | RNN | 0.3021 | 0.6330 | 0.5874 | 0.6977 | 0.8068 |
| | BRNN | 0.3119 | 0.6362 | 0.5925 | 0.7014 | 0.8128 |
| | RETAIN | 0.3014 | 0.6498 | 0.5948 | 0.6999 | 0.8102 |
| | Deepr | 0.2999 | 0.6434 | 0.5865 | 0.6981 | 0.8113 |
| | Dipole | 0.2841 | 0.6484 | 0.5997 | 0.7034 | 0.8121 |
| | SAnD | 0.2979 | 0.6179 | 0.5709 | 0.6805 | 0.7959 |
| | TransEnc | 0.3116 | 0.6482 | 0.5980 | 0.7037 | 0.8139 |
| | MusaNet | **0.3261** | **0.6507** | **0.6069** | **0.7104** | **0.8227** |
| Dx & Tx | RNN | 0.3216 | 0.6317 | 0.5857 | 0.6973 | 0.8093 |
| | BRNN | 0.3270 | 0.6402 | 0.5961 | 0.7088 | 0.8138 |
| | RETAIN | 0.3161 | 0.6497 | 0.6021 | 0.7061 | 0.8148 |
| | Deepr | 0.3142 | 0.6391 | 0.5947 | 0.7004 | 0.8125 |
| | Dipole | 0.2899 | 0.6515 | 0.6097 | 0.7121 | 0.8149 |
| | SAnD | 0.2996 | 0.6242 | 0.5774 | 0.6878 | 0.8004 |
| | TransEnc | 0.3137 | 0.6559 | 0.6086 | 0.7105 | 0.8144 |
| | MusaNet | **0.3344** | **0.6618** | **0.6127** | **0.7146** | **0.8242** |

**Implementation.** We implement all the approaches with Tensorflow 2.0. For training models, we use RMSprop with a minbatch of 32 patients and 10 epochs. The drop-out rate is 0.1 for all the approaches. The data split ratio is 0.8:0.1:0.1 for training, validation and testing sets. In order to fairly compare the performance, we set the same embedding dimension $d = 128$ for all the baselines and the proposed model.

## 4.3 Results

**Overall Performance.** Table 2 reports the results of the two prediction tasks on the two datasets - future re-admissions and diagnoses. The results show that MusaNet outperforms all the baselines. This demonstrates that the superiority of our framework results from the explicit consideration of inherent hierarchy of EHRs, and the contextual and temporal dependencies which are incorporated into the representations. Furthmore, we note that the performance obtained by the models remains approximately the same (RETAIN, RNN) or even drops by up to 0.43% (Deepr) after adding the procedure to the training data over Precision@5. The underlying reason may be that the future diagnosis prediction is less sensitive to the procedures, thus the relationships among procedures cannot

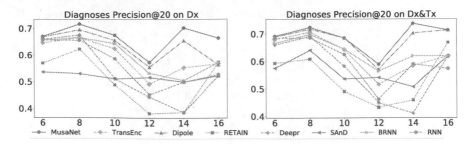

**Fig. 4.** Robustness comparison regarding precision@20 on Dx and Dx&Tx. Length of visit sequence varies from 6 to 16.

be well captured when predicting few diagnoses. Furthermore, by using attention mechanisms, Dipole, SAnD, TransEnc, and our MusaNet model achieve a marginal improvement when comparing the performance of using Dx&Tx for training and that of Dx. This implies that attention could play an important role in the learned representations of the procedures.

**Robustness for Lengthy Sequence of Visits.** We conducted a set of experiments to evaluate the robustness of MusaNet by varying the length of sequential visits of a patient from 6 to 16, which are considered to be long patient journeys in the medical domain. Figure 4 shows the results. Overall, the model MusaNet outperforms the other models as the length of a patient's sequential visits increases in diagnoses prediction both Dx and Dx&Tx. In particular, the performance of Dipole is comparative to our MusaNet and follows a similar trend with an increase in the length of sequential visits. From these results, a conclusion can be drawn that the positional masks in the mSA module and interval encoding in our framework play a vital role in capturing the lengthy temporal dependencies between patient's sequential visits.

**Ablation Study.** We performed a detailed ablation study to examine the contributions of each of the model's components to the prediction tasks. There are three components, which are (Attention) the two attention pooling layers to learn the visits from the embedded medical codes and learn the patient journey from the embedded visits, (PosMask) the position mask in mSA module, and (Interval) the interval encodings to be added to the learned visit embeddings.

- **Attention:** replace each of the two attention pooling layers with a simple summation layer;
- **PosMask:** remove the forward and backward position mask in the mSA module;
- **Interval:** remove the interval encodings module;
- **MusaNet:** our proposed model in the paper.

**Table 3.** Ablation performance comparison.

| Data | Ablation | Readm (PR-AUC) | Diagnosis Precision@k | | | |
|------|----------|----------------|---------|---------|---------|---------|
|      |          |                | k = 5   | k = 10  | k = 20  | k = 30  |
| Dx   | Attention | 0.3170        | 0.6371  | 0.5921  | 0.7044  | 0.8142  |
|      | PosMask   | 0.3033        | 0.6287  | 0.5821  | 0.7044  | 0.8125  |
|      | Interval  | 0.3132        | 0.6318  | 0.5907  | 0.6977  | 0.8080  |
|      | MusaNet   | **0.3261**    | **0.6507** | **0.6069** | **0.7104** | **0.8227** |
| Dx& Tx | Attention | 0.3278      | 0.6459  | 0.5997  | 0.7030  | 0.8104  |
|      | PosMask   | 0.3237        | 0.6457  | 0.5957  | 0.7097  | 0.8174  |
|      | Interval  | 0.3290        | 0.6485  | 0.6050  | 0.7141  | 0.8189  |
|      | MusaNet   | **0.3344**    | **0.6618** | **0.6127** | **0.7146** | **0.8242** |

**Fig. 5.** Importance of visits for patient 1 (Left), and importance of diagnoses in visit 1 (Right).

From Table 3, we see that the full complement of MusaNet achieved superior accuracy to the ablated models. Specifically, we note that the position mask from the mSA module (PosMask) contributes the highest accuracy to re-admission prediction over Dx, which gives us confidence in using the position mask to learn the patient journey representations without sufficient data. Moreover, it is clear that the attention pooling component provides valuable information with the learned embeddings of the patient journey for the performance of diagnoses prediction over Dx&Tx. Specifically, MusaNet predicted re-admissions by 2.83% more accurately on Dx and by 1.07% more accurately on Dx&Tx. Similarly, MusaNet predicted diagnoses precision@5 by 2.2% more accurately on Dx and diagnoses precision@30 by 1.4% more accurately on Dx&Tx.

**Case Study: Visualization and Explainability.** One aspect of this method is that it hierarchically compresses medical codes into visits and visits into patient journeys. At each level of aggregation, the model decides the importance of the lower-level entries on the upper-level entry, which makes the model explainable. To showcase this feature, we visualized two patient journeys. These examples come from the re-admission prediction task with the Dx dataset. From the importance distribution of the patient visits, we analyzed the most important visits to future re-admission. For example, Visit 1 in Fig. 5 was the most influential factor on Patient 1's re-admission. After zooming on Visit 1, we found a vital insight in Diagnosis 3 (ICD 585.6), i.e., end-stage renal disease, which

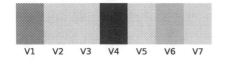 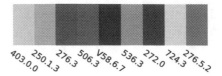

**Fig. 6.** Importance of visits for patient 2 (Left), and importance of diagnoses in visit 4 (Right).

would obviously cause frequent and repeated re-admissions to hospital. As shown in Fig. 6, Patient 2 had 7 visits. Visit 4 contributed most to the re-admissions. Again, the diagnoses reveal the cause: long-term use of insulin (ICD V58.6.7) and pure hypercholesterolemia (ICD 272.0).

# 5    Related Work

## 5.1    Medical Concept Embedding

Borrowing ideas from word representation models [10,19,20,23,33], researchers in the healthcare domain have recently explored the possibility of creating representations of medical concepts. Much of this research has focused on the Skip-gram model. For example, Minarro-Gimnez et al. [21] directly applied Skip-gram to learn representations of medical text, and Vine et al. [8] did the same for UMLS medical concepts. Choi et al. [6] went a step further and used the Skip-gram model to learn medical concept embeddings from different data sources, including medical journals, medical claims, and clinical narratives. In other work [2], Choi et al. developed the Med2Vec model based on Skip-gram to learn concept-level and visit-level representations simultaneously. The shortcoming of all these models is that they view EHRs as documents in the NLP sense, which means that temporal information is ignored.

## 5.2    Patient Journey Embedding

Applying deep learning to healthcare analytical tasks has recently attracted enormous interest in healthcare informatics. RNN has been widely used to model medical events in sequential EHR data for clinical prediction tasks [2,3,5,16,18, 25]. Choi et al. [2,5] indirectly exploit an RNN to embed the visit sequences into a patient representation by multi-level representation learning to integrate visits and medical codes based on visit sequences and the co-occurrence of medical codes. Other research works have, however, used RNNs directly to model time-ordered patient visits for predicting diagnoses [3,4,16–18,25,34]. CNN has been exploited to represent a patient journey in other way. For example, Nguyen et al. [22] transform a record into a sequence of discrete elements separated by coded time gaps, and then employ CNN to detect and combine predictive local clinical motifs to stratify the risk. These RNN- and CNN-based models follow ideas of processing sentences [13] in documents from NLP to treat a patient journey as a document and a visit as a sentence, which only has a sequential relationship, while two arbitrary visits in one patient journey may be separated by different time intervals. Attention-based neural networks have been exploited successfully in healthcare tasks to model sequential EHR data [4,16,26,30] and have been shown to improve predictive performance.

## 6   Conclusion

In this paper, we proposed a novel multi-level self-attention network (**MusaNet**) to encode patient journeys. The model framework comprises a masked self-attention module (mSA) for capturing the contextual information and the temporal relationships, attention poolings for tackling the multi-level structure of EHR data, and interval encoding for encoding temporal information of visits. As demonstrated by experiment results, MusaNet produces better representations that is validated using two downstream healthcare tasks.

**Acknowledgments.** This work was supported in part by the Australian Research Council (ARC) under Grant LP160100630, LP180100654 and DE190100626.

## References

1. Bahdanau, D., Cho, K., Bengio, Y.: Neural machine translation by jointly learning to align and translate. arXiv:1409.0473 (2014)
2. Choi, E., et al.: Multi-layer representation learning for medical concepts. In: SIGKDD, pp. 1495–1504. ACM (2016)
3. Choi, E., Bahadori, M.T., Song, L., Stewart, W.F., Sun, J.: Gram: graph-based attention model for healthcare representation learning. In: SIGKDD, pp. 787–795. ACM (2017)
4. Choi, E., Bahadori, M.T., Sun, J., Kulas, J., Schuetz, A., Stewart, W.: Retain: an interpretable predictive model for healthcare using reverse time attention mechanism. In: NeurIPS, pp. 3504–3512 (2016)
5. Choi, E., Xiao, C., Stewart, W., Sun, J.: Mime: multilevel medical embedding of electronic health records for predictive healthcare. In: NeurIPS, pp. 4552–4562 (2018)
6. Choi, Y., Chiu, C.Y.I., Sontag, D.: Learning low-dimensional representations of medical concepts. AMIA Summits on Translational Science Proceedings 2016, p. 41 (2016)
7. Davis, J., Goadrich, M.: The relationship between precision-recall and roc curves. In: ICML, pp. 233–240. ACM (2006)
8. De Vine, L., Zuccon, G., Koopman, B., Sitbon, L., Bruza, P.: Medical semantic similarity with a neural language model. In: CIKM, pp. 1819–1822. ACM (2014)
9. Hu, M., Peng, Y., Huang, Z., Qiu, X., Wei, F., Zhou, M.: Reinforced mnemonic reader for machine reading comprehension. arXiv:1705.02798 (2017)
10. Jha, K., Wang, Y., Xun, G., Zhang, A.: Interpretable word embeddings for medical domain. In: 2018 IEEE International Conference on Data Mining (ICDM), pp. 1061–1066. IEEE (2018)

11. Johnson, A.E., Pollard, T.J., Shen, L., Li-wei, H.L., Feng, M., Ghassemi, M., Moody, B., Szolovits, P., Celi, L.A., Mark, R.G.: Mimic-iii, a freely accessible critical care database. Sci. Data **3**, 160035 (2016)

12. Kho, A.N., Yu, J., Bryan, M.S., Gladfelter, C., Gordon, H.S., Grannis, S., Madden, M., Mendonca, E., Mitrovic, V., Shah, R., Tachinardi, U., Taylor, B.: Privacy-preserving record linkage to identify fragmented electronic medical records in the all of us research program. In: Cellier, P., Driessens, K. (eds.) ECML PKDD 2019. CCIS, vol. 1168, pp. 79–87. Springer, Cham (2020). https://doi.org/10.1007/978-3-030-43887-6_7

13. Kim, Y.: Convolutional neural networks for sentence classification. In: EMNLP, pp. 1746–1751, October 2014

14. Lin, Z., et al.: A structured self-attentive sentence embedding. arXiv:1703.03130 (2017)

15. Liu, Y., Sun, C., Lin, L., Wang, X.: Learning natural language inference using bidirectional lstm model and inner-attention. arXiv:1605.09090 (2016)

16. Ma, F., Chitta, R., Zhou, J., You, Q., Sun, T., Gao, J.: Dipole: diagnosis prediction in healthcare via attention-based bidirectional recurrent neural networks. In: SIGKDD, pp. 1903–1911. ACM (2017)

17. Ma, F., Gao, J., Suo, Q., You, Q., Zhou, J., Zhang, A.: Risk prediction on electronic health records with prior medical knowledge. In: SIGKDD, pp. 1910–1919. ACM, July 2018

18. Ma, F., You, Q., Xiao, H., Chitta, R., Zhou, J., Gao, J.: KAME: knowledge-based attention model for diagnosis prediction in healthcare. In: CIKM, pp. 743–752. ACM, October 2018

19. Mikolov, T., Chen, K., Corrado, G., Dean, J.: Efficient estimation of word representations in vector space. arXiv:1301.3781 (2013)

20. Mikolov, T., Sutskever, I., Chen, K., Corrado, G.S., Dean, J.: Distributed representations of words and phrases and their compositionality. In: NeurIPS, pp. 3111–3119 (2013)

21. Minarro-Giménez, J.A., Marin-Alonso, O., Samwald, M.: Exploring the application of deep learning techniques on medical text corpora. Stud. Health Technol. Inform. **205**, 584–588 (2014)

22. Nguyen, P., Tran, T., Wickramasinghe, N., Venkatesh, S.: Deepr: a convolutional net for medical records. arXiv:1607.07519 (2016)

23. Peng, X., Long, G., Pan, S., Jiang, J., Niu, Z.: Attentive dual embedding for understanding medical concepts in electronic health records. In: 2019 International Joint Conference on Neural Networks (IJCNN), pp. 1–8. IEEE (2019)

24. Peng, X., Long, G., Shen, T., Wang, S., Jiang, J., Blumenstein, M.: Temporal self-attention network for medical concept embedding. In: 2019 IEEE International Conference on Data Mining (ICDM), pp. 498–507 (2019)

25. Qiao, Z., Zhao, S., Xiao, C., Li, X., Qin, Y., Wang, F.: Pairwise-ranking based collaborative recurrent neural networks for clinical event prediction. In: IJCAI, pp. 3520–3526 (2018)

26. Rajkomar, A., et al.: Scalable and accurate deep learning with electronic health records. NPJ Digital Med. **1**(1), 18 (2018)

27. Shen, T., Zhou, T., Long, G., Jiang, J., Pan, S., Zhang, C.: Disan: directional self-attention network for RNN/CNN-free language understanding. In: AAAI (2018)

28. Shen, T., Zhou, T., Long, G., Jiang, J., Zhang, C.: Bi-directional block self-attention for fast and memory-efficient sequence modeling. arXiv:1804.00857 (2018)

29. Shickel, B., Tighe, P.J., Bihorac, A., Rashidi, P.: Deep EHR: a survey of recent advances in deep learning techniques for electronic health record (EHR) analysis. IEEE J. Biomed. Health Inform. **22**(5), 1589–1604 (2018)

30. Song, H., Rajan, D., Thiagarajan, J.J., Spanias, A.: Attend and diagnose: clinical time series analysis using attention models. In: AAAI (2018)

31. Song, L., Cheong, C.W., Yin, K., Cheung, W.K., Cm, B.: Medical concept embedding with multiple ontological representations. In: IJCAI, pp. 4613–4619 (2019)

32. Vaswani, A., et al.: Attention is all you need. In: NeurIPS, pp. 5998–6008 (2017)

33. Wang, S., Li, X., Chang, X., Yao, L., Sheng, Q.Z., Long, G.: Learning multiple diagnosis codes for ICU patients with local disease correlation mining. ACM Trans. Knowl. Discovery Data (TKDD) **11**(3), 1–21 (2017)

34. Zhang, Q., Wu, J., Zhang, P., Long, G., Zhang, C.: Salient subsequence learning for time series clustering. IEEE Trans. Pattern Anal. Mach. Intell. **41**(9), 2193–2207 (2018)

# MMCNN: A Multi-branch Multi-scale Convolutional Neural Network for Motor Imagery Classification

Ziyu Jia[1,2], Youfang Lin[1,2,3], Jing Wang[1,2,3(✉)], Kaixin Yang[1], Tianhang Liu[1], and Xinwang Zhang[1]

[1] School of Computer and Information Technology, Beijing Jiaotong University, Beijing, China
{ziyujia,yflin,wj,kxyang,lthang,xwzhang1}@bjtu.edu.cn
[2] Beijing Key Laboratory of Traffic Data Analysis and Mining, Beijing, China
[3] CAAC Key Laboratory of Intelligent Passenger Service of Civil Aviation, Beijing, China

**Abstract.** Electroencephalography (EEG) based motor imagery (MI) is one of the promising Brain–computer interface (BCI) paradigms enable humans to communicate with the outside world based solely on brain intentions. Although convolutional neural networks have been gradually applied to MI classification task and gained high performance, the following problems still exist to make effective decoding of raw EEG signals challenging: 1) EEG signals are non-linear, non-stationary, and low signal-noise ratio. 2) Most existing end-to-end MI models utilize single scale convolution which limits the result of classification because the best convolution scale varies with different subject (called *subject difference*). In addition, even for the same subject, the best convolution scale also differs from time to time (called *time difference*). In this paper, we propose a novel end-to-end model, named Multi-branch Multi-scale Convolutional Neural Network (MMCNN), for motor imagery classification. The MMCNN model effectively decodes raw EEG signals without any pre-processing including filtering. Meanwhile, the multi-branch multi-scale convolution structure successfully addresses the problems of subject difference and time difference based on parallel processing. In addition, multi-scale convolution can realize the characterization of information in different frequency bands, thereby effectively solving the problem that the best convolution scale is difficult to determine. Experiments on two public BCI competition datasets demonstrate that the proposed MMCNN model outperforms the state-of-the-art models. The implementation code is made publicly available https://github.com/jingwang2020/ECML-PKDD_MMCNN.

**Keywords:** Motor imagery · Convolutional neural network · EEG signal · Brain–computer interface

© Springer Nature Switzerland AG 2021
F. Hutter et al. (Eds.): ECML PKDD 2020, LNAI 12459, pp. 736–751, 2021.
https://doi.org/10.1007/978-3-030-67664-3_44

# 1    Introduction

Brain–computer Interface (BCI) establishes the connection between the human brain and outside world, thereby can control the external instruments by decoding brain signals. The BCI system is mainly based on two types of brain activities: electrophysiological and hemodynamic [12]. The electrophysiological activity is often measured by the electroencephalography (EEG), electrocorticography (ECoG), and magnetoencephalography (MEG). And the hemodynamic is quantified by neuroimaging methods, such as functional magnetic resonance and near infrared spectroscopy. Because of the low cost of collecting EEG signals and low risk to users, the EEG signals are by far the most widely used data sources in BCI systems.

Motor imagery (MI) is one of the typical EEG based BCI paradigms. The EEG signals of MI are different movement intention signals from the scalp when a person imagines moving different parts of the body or different control commands on the instruments. In this way, human intentions can be recognized by analyzing the EEG signals, which has been attracting increasing attentions. Many researchers use the spatial features and time-frequency features from EEG signals to perform MI classification tasks. For example, the Common Spatial Pattern (CSP) [6] and Filter Bank Common Spatial Pattern (FBCSP) [2], etc. are often utilized to extract the spatial features from the EEG signals. In general, the wavelet transform and fast Fourier transform are applied to extract the time-frequency features. However, the manual features extraction often requires researchers to have a large amount of prior knowledge.

In recent years, the end-to-end deep learning's revolutionary advances in audio and visual signals recognition have gained significant attentions. To avoid manual features extraction, some researchers have successfully applied end-to-end deep learning models for MI classification [9,13,16]. For example, Schirrmeister et al. present three types of convolutional neural networks (CNN) models (Shallow ConvNet, Deep ConvNet, and Hybrid ConvNet) for MI classification [13]. However, due to the non-linearity, non-stationarity and low signal-noise ratio of EEG signals, the input of most existing models still depends on the filtered EEG signals. Specifically, although the existing end-to-end models avoid a large number of features engineering, it is often necessary to filter the raw EEG signals in a specific frequency band (such as 4–40 Hz). This frequency band is generally considered to be EEG signals related to MI tasks. In addition, most existing end to end models apply single convolution scale which limits the result of MI classification. Figure 1 presents the impact of kernel size on different test subjects. For subject A, the accuracy achieves highest when kernel size is 45, while it is 105 for subject B. This phenomenon is called *subject difference*. Meanwhile, for same subject A, the best kernel size for session 1 and session 2 are 15 and 55, respectively. This phenomenon is called *time difference*. Therefore, it is difficult to determine the best convolution scale due to the subject difference and time difference in MI classification, which makes the MI classification challenging.

738    Z. Jia et al.

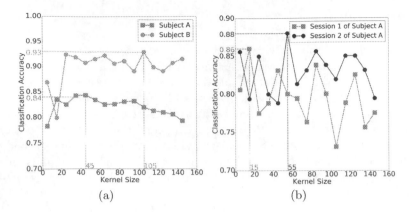

(a)                              (b)

**Fig. 1.** (a) Classification accuracy of two different subjects with different kernel size (*subject difference*). (b) Classification accuracy of two different sessions of Subject A with different kernel size (*time difference*).

To address the above challenge, we propose a novel end-to-end deep learning model: *Multi-branch Multi-scale Convolution Neural Network* (MMCNN). The proposed MMCNN is composed of 5 parallel EEG Inception Networks (EINs), consisting 15 types of convolution kernel size. Overall, the contributions of this paper are three folds. Firstly, we propose the end-to-end MMCNN model effectively decodes raw EEG signals without any pre-processing including filtering and has great practical application prospects. Secondly, we design the multi-branch multi-scale convolution structure, which successfully solves the problems of subject difference and time difference for MI classification based on parallel processing. Thirdly, experiments are conducted on two public BCI Competition datasets and experimental results present that our model outperforms all the baseline models. In addition, our model can achieve higher classification results based on fewer EEG channels.

## 2 Related Work

Time series analysis has always been concerned by researchers [19,20]. The EEG signals based on MI tasks are the typical time series. Researchers usually extract a large number of features for MI tasks from EEG signals, of which Cz, C3, and C4 are the three channels often used. The extracted features mainly describe EEG signals from two aspects: spatial features and time-frequency features. The methods of extracting spatial features mainly include the CSP and its variants [6]. For example, Zhang et al. propose a multi-kernel extreme learning machine model, which utilizes the CSP method to extract features [17]. Ang et al. propose a FBCSP method to solve the problem that the performance of CSP depends on the operational frequency band of EEG [2]. In addition, Ang et al. also apply the FBCSP method to classify the EEG signals of MI [1]. For the extraction of time-frequency features, fast Fourier transform, wavelet transform are often used to analyze EEG signals. For example, Lu et al. present a deep learning scheme based on restricted Boltzmann machines, which utilizes fast Fourier transform to extract the frequency domain features of the EEG signals [10]. In addition, Sun et al. propose a bispectrum features extraction method to classify the EEG signals of MI [14].

In recent years, deep learning has made great achievements in computer vision, natural language processing, and speech recognition. In particular, the end-to-end deep learning framework integrates multiple processing stages (such as data processing and features extraction) into a unified model, and builds direct projection from input to output, which excels in a variety of tasks. Therefore, the end-to-end deep learning framework does not rely on features engineering and has great practical application value. In the field of motor imagery, there are already some successful end-to-end deep learning frameworks. For example, Schirrmeister et al. employ the advantages of convolutional neural networks for end-to-end learning and design three different frameworks of convolutional neural networks: the Deep ConvNet, the Shallow ConvNet, and the Hybrid ConvNet [13]. Zhao et al. propose the WaSF ConvNet to solve the problem that tradition ConvNets requires a large amount of training data [18]. In order to improve the application effect of MI classification, Dose et al. design an end-to-end model based on the convolutional neural networks, which can both be used as a subject-independent classifier, as well as a classifier adapted to a single subject [5]. Lawhern et al. propose an EEG-based compact convolutional network called EEGNet, which is constructed by using depthwise and separable convolutions [9]. Wu et al. propose the parallel filter bank convolutional neural network named MSFBCNN, which is an end-to-end network can extract temporal and spatial features [16].

Due to the EEG signals of MI vary greatly from subject to subject, the best convolution scale is different with different subjects. Moreover, for the same subject, the best convolution scale also varies when time is different. Hence, most of existing models employ single scale convolution which limits the MI classification result. To solve this issue, we design a convolutional neural network with multi-branch multi-scale structure to perform MI classification.

## 3  Preliminaries

The raw EEG signals of motor imagery are defined as $X = [x_1, x_2, \ldots x_n] \in R^{n \times T \times C}$, where $n$ represents the number of EEG samples; $T = t \times f$ represents the number of time points for each EEG signal, where $t$ represents the duration time of the motor imagery segment, and $f$ represents the sampling frequency; $C$ represents the number of EEG channels.

The EEG based MI classification problem can be defined as: learning a mapping function $F$ based on the end-to-end deep convolutional neural network, which maps the raw EEG signals $X$ to the corresponding MI class:

$$Y_{classification} = F(X) \tag{1}$$

where $F$ denotes the mapping function, $Y_{classification}$ denotes the classification result.

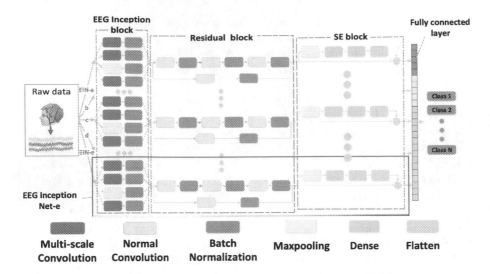

**Fig. 2.** Overall framework of the proposed MMCNN model. Specifically, the MMCNN contains 5 parallel EEG Inception Networks (EINs), one of which is EIN-e marked by a black box. In addition, each EIN is composed of an EEG Inception block (EIB), a Residual block, and a Squeeze and Excitation block (SE block). Each EIB has 3 different convolution kernel sizes (green color). For example, EIB-a (kernel size: $1 \times 5/1 \times 10/1 \times 15$); EIB-e (kernel size: $1 \times 160/1 \times 180/1 \times 200$). (Color figure online)

## 4    Multi-branch Multi-scale Convolutional Neural Network

In this section, we present the structure of the Multi-branch Multi-scale Convolutional Neural Network (MMCNN), which is an end-to-end deep learning model as shown in Fig. 2. The proposed MMCNN model consists of 5 parallel EEG Inception Networks (EIN). Each EIN consists of an EEG Inception block (EIB), a Residual block, and a Squeeze and Excitation block (SE block). As the best convolution scale is different with subject and time, the MMCNN model implements multi-scale convolution to solve this problem: all EIN branches adopt different convolution kernel size (different scales). In addition, we apply the Residual block to avoid the network degradation phenomenon. At the same time, we employ the SE block to adaptively capture attentive features. Our model uses the ELU function instead of the traditional RELU function to improve the anti-noise capability. Overall, the MMCNN model is carefully designed for the MI classification.

### 4.1    EEG Inception Block

Our MMCNN model utilizes 5 EIBs (EIB a-e) with growing convolution kernel size in parallel EINs to implement the multi-scale convolution. We design the

EIBs, which is inspired by the classic Inception network [15]. Each EIB has four parts including the multi-scale 1D convolution and pooling operations. Take the structure of EIB-a as an example, which is shown in Fig. 3. The specific convolution kernel size are as following: $1 \times 5, 1 \times 10, 1 \times 15$ for EIB-a, $1 \times 40, 1 \times 45, 1 \times 50$ for EIB-b, $1 \times 60, 1 \times 65, 1 \times 70$ for EIB-c, $1 \times 80, 1 \times 85, 1 \times 90$ for EIB-d, $1 \times 160, 1 \times 180, 1 \times 200$ for EIB-e. The kernel size gradually increases among all EIBs. The EIB process can be defined as:

$$I_q = \left[ p_q^{j=1,k} * x; p_q^{j=2,k} * x; p_q^{j=3,k} * x; F_{maxpooling}(x) \right] \tag{2}$$

where $I_q$ denotes the output from EIB in different EINs and $I_q \in R^{T' \times C'}$, $q \in [a, b, c, d, e]$ with five branches in EIN, $p_q^{j,k}$ denotes parameters from the $k$-th filter in $j$-th branch, $*$ denotes the convolution operation, and the $x$ denotes the input sample, and the $T'$ denotes the number of time points, the $C'$ denotes the number of channels.

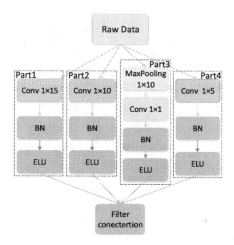

**Fig. 3.** The Structure of the EIB-a. The EIB-a has four parts, of which three parts are multi-scale convolutional layers, and one part is the pooling layer.

**Activation Function.** Most deep learning models usually employ the rectified linear unit (ReLU) function as the activation function, because the ReLU function is simple to implement and converges quickly. The expression of the ReLU function is defined as:

$$f(x) = \begin{cases} x, x > 0 \\ 0, x \leq 0 \end{cases} \tag{3}$$

where $x$ is the input value and the ReLU is used as the activation function to reduce the dependency between the parameters and alleviate over-simulation. However, it also leads to the sparsity of the network, so that the gradient of the negative part is 0, and some neurons may never be activated, and the corresponding parameters are never updated, which limits the learning ability of the network [11]. Considering the characteristics of EEG signals with low signal-noise ratio, our model utilizes the exponential linear unit (ELU) function as the activation function, which is defined as:

$$f(x) = \begin{cases} x, & x > 0 \\ \alpha\left(e^x - 1\right), & x \leq 0 \end{cases} \tag{4}$$

where $\alpha$ is a constant greater than 0, the ELU function is shown in Fig. 4.

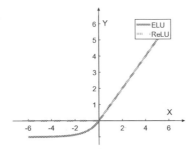

**Fig. 4.** The ELU activation function and the ReLU activation function.

The biggest advantage of the ELU function is that it blends the Sigmoid function and the ReLU function [3]. As shown in Fig. 4, for ELU function, the left part has soft saturation and the right part has no saturation. The linear part on the right makes the ELU function relieve the gradient vanishing, and the soft saturation on the left makes the ELU more robust to input changes or noise. This is why the ELU function is suitable for end-to-end models with low signal-noise raw EEG signals. In addition, the ELU function has better learning characteristics than the ReLU function: there is no negative value in ReLU function, and the ELU function has negative values, which can converge the output mean of the activation unit to a position close to zero, making the normal gradient closer to the unit natural gradient. Meanwhile, the ELU function reduces the computational complexity and speeds up the model's learning speed.

### 4.2   Residual Block

To avoid network degradation as the number of network layers increases [7], the Residual block is utilized, which is presented in Fig. 5.

The Residual block consists of two branches: one branch performs a series of layers interleaved with 1D-convolutional layers and Batch Normalization layers (BN) layers and obtains output $F_{res}\left(I_q\right)$; and the other branch inserts a shortcut connection. The two branches are merged and form the final output, which is defined as:

$$U_q = F_{res}\left(I_q\right) + I_q \tag{5}$$

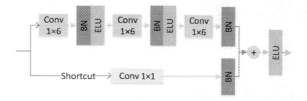

**Fig. 5.** The Structure of the Residual block.

where $U_q$ denotes the output of the Residual block, $I_q$ denotes the input. In this way, the features extracted from the shallow layer are transferred to the deeper layer. Therefore, Residual block largely addresses the degradation problem.

### 4.3 Squeeze and Excitation Block

We design a Squeeze and Excitation block (SE block) for 1D CNN, which is inspired by the SE blocks for 2D CNN in the image field [8], to pay more attention to adaptive extraction of important features. The structure of the SE block is shown in Fig. 6.

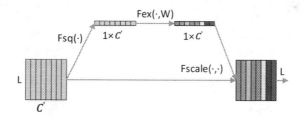

**Fig. 6.** The Structure of the SE block. $C'$ denotes the number of feature maps and $L$ denotes the size of feature maps. The feature maps are defined as the features extracted by the network in deep learning.

The SE block consists of two components: squeeze and excitation. The squeeze tackles the issue of exploiting channel dependencies. The entire channel is compressed into a channel descriptor $m$ by using global average pooling:

$$m_q = F_{sq}(U_q) = \frac{1}{L}\sum_{n=1}^{L} U_q(n) \tag{6}$$

where $m_q$ denotes the channel descriptor, $L$ denotes the size of feature map.

Excitation learns sample-specific activations for each channel through a channel-dependent self-gating mechanism. It is conductive to govern the excitation of each channel. We calculate the weight $Sq$ of the feature map:

$$S_q = F_{ex}(m_q, W) = \sigma(h(m_q, W)) = \sigma(W_2 \delta(W_1 m_q)) \tag{7}$$

where $W_1$ denotes dimensionality-reduction layer, and the dimension of $W_1$ is controlled by ratio to reduce the amount of calculation. $W_2$ denotes $L$ dimensional layer and $\delta$ denotes the ELU function. Then the $m_q$ is assigned weights:

$$f_q = F_{scale}\left(U_q, S_q\right) = U_q \cdot S_q \tag{8}$$

where $f_q$ denotes the output of the SE block, $F_{scale}$ denotes the operation of assigning weight.

# 5    Experiment

## 5.1    Datasets

**Dataset 1.** BCI Competition IV 2b dataset[1] is a binary classification problem dataset with MI of the left hand movement and the right hand movement which has 3 EEG bipolar recordings (C3, Cz, and C4) with a sampling frequency 250 Hz. The dataset includes 9 subjects and each subject has 5 sessions. The first two sessions contain training data without feedback, and each session includes 120 trials. The last three sessions are recorded with smiley feedback, and each session includes 160 trails.

**Dataset 2.** The BCI Contest IV 2a dataset[2] contains data of 9 subjects performing motor imagery classification (left-hand, right-hand, tongue, and foot movements). The EEG signals include 22 EEG bipolar recordings with a frequency 250 Hz. And each subject has 2 sessions, each session includes 288 trials. The duration of the collected MI is the same as the 2b dataset.

## 5.2    State-of-the-Art Models

We compare our model with the following open source end-to-end models.

**Deep ConvNet** [13]: Deep ConvNet has four convolution–max–pooling blocks, with a special first block designed to handle EEG input, followed by three standard convolution–max–pooling blocks and a dense softmax classification layer.

**Shallow ConvNet** [13]: The layers in the Shallow ConvNet is less than the Deep ConvNet, and the kernel size of temporal convolution in the Shallow ConvNet is $1 \times 25$, larger than the kernel size in the Deep ConvNet, which is $1 \times 10$.

**Hybrid ConvNet** [13]: Hybrid ConvNet combines the structure of the Deep ConvNet and the Shallow ConvNet.

**EEGNet** [9]: A deep learning model utilizes a single-scale neural network with deep convolution and separable convolution, which constructs a uniform approach for different BCI tasks.

**MSFBCNN** [16]: A parallel filter bank convolutional neural network has three parts: front feature extraction layer, feature reduction layer, and classification layer.

---

[1] http://www.bbci.de/competition/iv/#dataset2b.
[2] http://www.bbci.de/competition/iv/#dataset2a.

## 5.3   Experiment Settings

In the experiments, the 5-fold cross-validation is used 5 times to evaluate all models. The ELU is used as the activation function. The optimization method is Adam, and the learning rate is set to 0.0001. In addition, we also employ L2 regularization in the EIB and Residual block, where the parameters are set to 0.002 and 0.01, respectively. Besides, the dropout probability is set to 0.8 and the early-stopping is applied in the training stage. The experiment uses 3 EEG bipolar recordings (C3, Cz, and C4), which are related to MI tasks. The detailed parameters of the proposed MMCNN structure are presented in our code.

## 5.4   Comparison with State-of-the-Art Models

We evaluate the performance of all end-to-end models on the two public datasets. Table 1 presents the performance of our method with the baseline models on the BCI Competition IV 2b. It can be seen that our method achieves the highest average accuracy and kappa among all subjects. Meanwhile, Table 2 presents the results on the BCI Competition IV 2a. The average accuracy of 8 subjects among all subjects is much higher than other baseline models except for subject 2. For subject 2, our MMCNN has the second highest average accuracy and kappa, which are slightly lower than those of EEGNet.

The convolution kernel sizes used by traditional Deep ConvNet and Shallow ConvNet are $1 \times 10$ and $1 \times 25$, respectively. The result shows the average accuracy of Deep ConvNet is higher than Shallow ConvNet in BCI Competition IV 2b but is lower than Shallow ConvNet in BCI Competition IV 2a. The same situation happens between EEGNet and MSFBCNN as well. This phenomenon indicates that the results of the end-to-end models with a single convolution scale are different between datasets, which is mainly due to the difference of subjects. Our model has multi-scale convolution which can reduce the limitation of classification results due to the differences between subjects. Hence, our model achieves the best classification accuracy and kappa in almost every subject in two different datasets.

## 5.5   Ablation Study

**Ablation Study on Multi-scale Convolution.** In order to study the effect of increasing convolutional scales on the results in our model, we use different number of EINs in our model. The variant models starts with EIN-a and then gradually adds parallel network. Figure 8 shows the accuracy and kappa of the variant models increase with the increase of parallel networks. The results show that with the increase of convolution scale, our model gradually solves the problem of subject difference and time difference, thus improving the accuracy of MI classification. All parallel networks are used to achieve the highest accuracy. According to signal processing theory [4], the EIN-a with smaller convolution kernel tends to capture high-frequency information. On the contrary,

**Table 1.** Performance comparison of different models on BCI Competition IV 2b.

|   |   | Shallow ConvNet | Deep ConvNet | Hybrid ConvNet | MSFBCNN | EEGNet | MMCNN (ours) |
|---|---|---|---|---|---|---|---|
| S1 | ACC | 0.802 ± 0.009 | 0.843 ± 0.009 | 0.800 ± 0.004 | 0.818 ± 0.002 | 0.717 ± 0.052 | **0.849 ± 0.014** |
|    | Kappa | 0.645 ± 0.079 | 0.684 ± 0.018 | 0.660 ± 0.117 | 0.635 ± 0.006 | 0.434 ± 0.103 | **0.697 ± 0.028** |
| S2 | ACC | 0.671 ± 0.016 | 0.661 ± 0.003 | 0.660 ± 0.008 | 0.627 ± 0.012 | 0.671 ± 0.013 | **0.704 ± 0.005** |
|    | Kappa | 0.341 ± 0.029 | 0.323 ± 0.008 | 0.318 ± 0.014 | 0.253 ± 0.023 | 0.340 ± 0.027 | **0.409 ± 0.007** |
| S3 | ACC | 0.733 ± 0.005 | 0.736 ± 0.011 | 0.716 ± 0.013 | 0.706 ± 0.005 | 0.741 ± 0.009 | **0.755 ± 0.009** |
|    | Kappa | 0.462 ± 0.010 | 0.472 ± 0.023 | 0.431 ± 0.027 | 0.412 ± 0.010 | 0.481 ± 0.017 | **0.510 ± 0.018** |
| S4 | ACC | 0.942 ± 0.007 | 0.953 ± 0.004 | 0.946 ± 0.006 | 0.952 ± 0.003 | 0.950 ± 0.002 | **0.963 ± 0.003** |
|    | Kappa | 0.884 ± 0.014 | 0.906 ± 0.009 | 0.885 ± 0.011 | 0.903 ± 0.006 | 0.899 ± 0.005 | **0.925 ± 0.007** |
| S5 | ACC | 0.901 ± 0.005 | 0.907 ± 0.010 | 0.894 ± 0.008 | 0.891 ± 0.012 | 0.740 ± 0.008 | **0.924 ± 0.005** |
|    | Kappa | 0.801 ± 0.008 | 0.812 ± 0.020 | 0.775 ± 0.023 | 0.781 ± 0.024 | 0.481 ± 0.015 | **0.847 ± 0.010** |
| S6 | ACC | 0.818 ± 0.009 | 0.854 ± 0.006 | 0.831 ± 0.012 | 0.831 ± 0.011 | 0.719 ± 0.007 | **0.863 ± 0.011** |
|    | Kappa | 0.634 ± 0.018 | 0.708 ± 0.011 | 0.662 ± 0.024 | 0.659 ± 0.022 | 0.437 ± 0.014 | **0.725 ± 0.023** |
| S7 | ACC | 0.847 ± 0.014 | 0.852 ± 0.009 | 0.848 ± 0.005 | 0.834 ± 0.002 | 0.798 ± 0.009 | **0.876 ± 0.004** |
|    | Kappa | 0.693 ± 0.029 | 0.703 ± 0.017 | 0.695 ± 0.009 | 0.668 ± 0.004 | 0.592 ± 0.018 | **0.751 ± 0.007** |
| S8 | ACC | 0.811 ± 0.020 | 0.831 ± 0.007 | 0.812 ± 0.014 | 0.827 ± 0.005 | 0.808 ± 0.005 | **0.842 ± 0.012** |
|    | Kappa | 0.622 ± 0.039 | 0.661 ± 0.014 | 0.624 ± 0.029 | 0.652 ± 0.010 | 0.614 ± 0.010 | **0.683 ± 0.024** |
| S9 | ACC | 0.800 ± 0.007 | 0.813 ± 0.007 | 0.782 ± 0.011 | 0.801 ± 0.007 | 0.799 ± 0.010 | **0.818 ± 0.006** |
|    | Kappa | 0.596 ± 0.019 | 0.625 ± 0.015 | 0.562 ± 0.021 | 0.603 ± 0.013 | 0.598 ± 0.018 | **0.636 ± 0.012** |
| Mean | ACC | 0.814 ± 0.076 | 0.828 ± 0.081 | 0.810 ± 0.081 | 0.810 ± 0.090 | 0.771 ± 0.076 | **0.844 ± 0.075** |
|      | Kappa | 0.631 ± 0.153 | 0.655 ± 0.163 | 0.624 ± 0.161 | 0.618 ± 0.179 | 0.542 ± 0.153 | **0.687 ± 0.149** |

**Table 2.** Performance comparison of different models on BCI Competition IV 2a.

|   |   | Shallow ConvNet | Deep ConvNet | Hybrid ConvNet | MSFBCNN | EEGNet | MMCNN (ours) |
|---|---|---|---|---|---|---|---|
| S1 | ACC | 0.767 ± 0.026 | 0.792 ± 0.018 | 0.745 ± 0.026 | 0.734 ± 0.018 | 0.633 ± 0.015 | **0.821 ± 0.013** |
|    | Kappa | 0.531 ± 0.050 | 0.586 ± 0.034 | 0.485 ± 0.054 | 0.464 ± 0.039 | 0.268 ± 0.029 | **0.641 ± 0.022** |
| S2 | ACC | 0.583 ± 0.016 | 0.595 ± 0.014 | 0.578 ± 0.028 | 0.530 ± 0.022 | **0.612 ± 0.012** | 0.598 ± 0.015 |
|    | Kappa | 0.167 ± 0.032 | 0.205 ± 0.019 | 0.152 ± 0.058 | 0.066 ± 0.047 | **0.224 ± 0.023** | 0.196 ± 0.029 |
| S3 | ACC | 0.883±0.024 | 0.881±0.015 | 0.882±0.014 | 0.890 ± 0.017 | 0.873 ± 0.058 | **0.928 ± 0.005** |
|    | Kappa | 0.765 ± 0.046 | 0.763 ± 0.029 | 0.762 ± 0.029 | 0.782 ± 0.026 | 0.746 ± 0.116 | **0.856 ± 0.007** |
| S4 | ACC | 0.654 ± 0.035 | 0.628 ± 0.012 | 0.665 ± 0.014 | 0.649 ± 0.012 | 0.635 ± 0.019 | **0.690 ± 0.017** |
|    | Kappa | 0.306 ± 0.073 | 0.254 ± 0.032 | 0.322 ± 0.034 | 0.297 ± 0.026 | 0.268 ± 0.042 | **0.375 ± 0.030** |
| S5 | ACC | 0.846 ± 0.008 | 0.849 ± 0.012 | 0.836 ± 0.008 | 0.790 ± 0.011 | 0.870 ± 0.009 | **0.873 ± 0.004** |
|    | Kappa | 0.689 ± 0.015 | 0.696 ± 0.023 | 0.666 ± 0.016 | 0.578 ± 0.022 | 0.737 ± 0.019 | **0.742 ± 0.009** |
| S6 | ACC | 0.652 ± 0.022 | 0.648 ± 0.010 | 0.615 ± 0.023 | 0.598 ± 0.020 | 0.560 ± 0.009 | **0.685 ± 0.020** |
|    | Kappa | 0.308 ± 0.039 | 0.290 ± 0.021 | 0.229 ± 0.041 | 0.203 ± 0.034 | 0.119 ± 0.021 | **0.362 ± 0.043** |
| S7 | ACC | 0.872 ± 0.029 | 0.846 ± 0.015 | 0.858 ± 0.003 | 0.827 ± 0.017 | 0.862 ± 0.009 | **0.892 ± 0.007** |
|    | Kappa | 0.742 ± 0.059 | 0.687 ± 0.031 | 0.715 ± 0.008 | 0.655 ± 0.031 | 0.720 ± 0.020 | **0.784 ± 0.014** |
| S8 | ACC | 0.915 ± 0.009 | 0.892 ± 0.021 | 0.888 ± 0.014 | 0.867 ± 0.004 | 0.825 ± 0.008 | **0.916 ± 0.006** |
|    | Kappa | **0.848 ± 0.035** | 0.781 ± 0.041 | 0.772 ± 0.029 | 0.730 ± 0.007 | 0.647 ± 0.015 | 0.831 ± 0.012 |
| S9 | ACC | 0.881 ± 0.014 | 0.892 ± 0.010 | 0.891 ± 0.014 | 0.869 ± 0.006 | 0.905 ± 0.010 | **0.926 ± 0.009** |
|    | Kappa | 0.760 ± 0.032 | 0.782 ± 0.020 | 0.779 ± 0.029 | 0.734 ± 0.012 | 0.807 ± 0.021 | **0.850 ± 0.017** |
| Mean | ACC | 0.784 ± 0.117 | 0.780 ± 0.115 | 0.773 ± 0.118 | 0.751 ± 0.123 | 0.753 ± 0.131 | **0.814 ± 0.117** |
|      | Kappa | 0.568 ± 0.235 | 0.560 ± 0.228 | 0.543 ± 0.237 | 0.501 ± 0.244 | 0.504 ± 0.260 | **0.626 ± 0.272** |

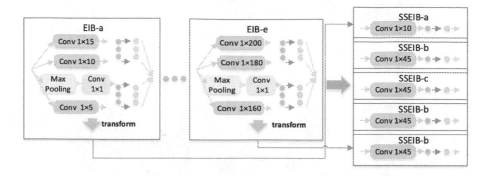

**Fig. 7.** The process of transforming the EIB (a-e) to single-scale EIB (a-e). The model replaces all EIBs with the middle size convolution in each EIB.

EIN-e with larger convolution kernel tends to capture low-frequency information. Multi-scale convolution can help the model to capture a large number of features and improve the classification effect significantly.

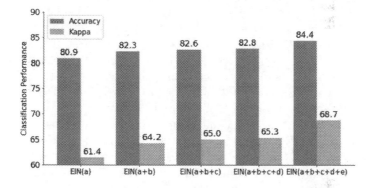

**Fig. 8.** Performance of different number of EIN.

To further evaluate the effectiveness of three different convolutional scales within the EIB, we transform each EIB into the Single Scale EIB (SSEIB) and form the new EIN called Single Scale EIN (SSEIN) as shown in Fig. 7. The variant models start with SSEIN-a and then gradually adds parallel network. Figure 9 presents the accuracy is improved during this process. The results show that increasing the number of parallel networks is helpful for MI classification and significantly enhance the performance of the variant model. However, the single scale block can not extract accurate features, the overall accuracy is lower than the model using EIB on the same scale. For example, the accuracy of the SSEIN-a is 68.1%, but the accuracy of the EIN-a is 80.9%. This shows that the multi-scale structure in EIN (or EIB) is effective.

In summary, a large number of experiments show that not only the multi-scale convolution among different EINs can improve the classification accuracy, but also the multi-scale convolution in a single EIB is effective.

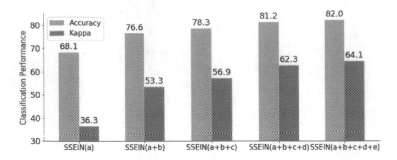

**Fig. 9.** Performance of different number of Single Scale EIN (SSEIN). The model replace the EINs with the Single Scale EIN, which is include the Single Scale EIB, the Residual block and the SE block. The specific transform process is shown in Fig. 7.

**Ablation Study on Different Blocks.** To study the effect of different blocks of the MMCNN, we design two variants of the MMCNN model. One variant is to remove SE block (a-e), the other variant is to remove both the SE block (a-e) and Residual block (a-e) at the same time. We compare these two variants with the MMCNN model on the BCI Competition IV 2b dataset. As shown in Fig. 10, the classification accuracy and kappa are reduced after the SE block is removed. This shows that the SE block can automatically learn valuable features and improve accuracy and kappa of MI classification. In addition, only EIB is used to get the lowest result, which shows that the Residual block effectively avoids the problem of network degradation by connecting the shallow layer and the deep layer of the network. At the same time, it improves the ability of model representation learning, and then improves the classification effect of the model.

**Ablation Study on Activation Functions.** In order to explore the impact of different activation functions on the classification results, two different activation functions (ELU and ReLU) are used to perform experiments on the BCI Competition IV 2b dataset. As shown in Fig. 11, the result shows that the model classification accuracy and kappa can be improved by using the ELU function. Due to the ELU function has better robustness to noise, the accuracy reaches 84.4% and kappa is 0.687. The classification accuracy using the ReLU function can only reach 81.4% and kappa is 0.626. Hence, when the data has a low signal-noise ratio, the ELU function can be preferred as the activation function of the model.

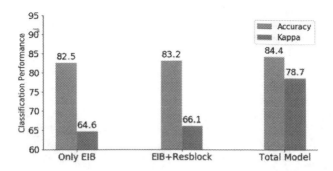

**Fig. 10.** Performance of ablation study on different blocks.

## 5.6  The Impact of Different EEG Channels

Because different EEG channels have different importance for MI classification results, identifying channels that are more important for classification can help reduce signal acquisition costs and develop portable devices. Therefore, we further explore the impact of different EEG channels on the MI classification.

Figure 12 shows that the more channels used, the better the results. Specifically, the classification effect using 3 EEG channels simultaneously is the best. In contrast, the classification accuracy using only one EEG channel is the lowest. It is worth noting that the accuracy of our model based on the combination of C4 and C3 channels reaches 81.0%, which is similar to the accuracy of the traditional models (such as the Hybrid ConvNet and EEGNet) based on 3 EEG channels on BCI Competition IV 2b dataset (see Table 1). The experiments show that when performing MI classification, the information contained in each channel improves the accuracy. However, the importance of different channels to improve the classification is different. When the number of channels is limited, the combination of C4 and C3 can achieve a better result. Our experiment provides a certain reference standard for the selection of fewer EEG channels and helps the development of portable MI equipment.

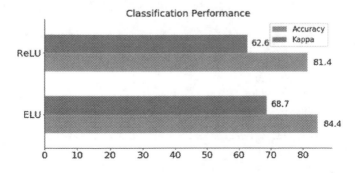

**Fig. 11.** Performance of different activation function.

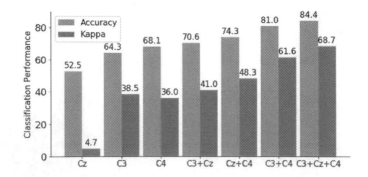

**Fig. 12.** Performance of different EEG channel for classification.

## 6   Conclusion

In this paper, we propose an end-to-end multi-branch multi-scale CNN for MI classification. Our model successfully solves the problem that the best convolution scale varies with the different subject and time. Experiments on two public benchmark BCI competition datasets show that our model is superior to the state-of-the-art models. In addition, our model can achieve similar accuracy to existing models with only a few EEG channels, which has promoted the development of portable BCI systems. The proposed model is a general framework for EEG signal classification, so we can apply it to other practical applications in the future, such as the EEG-based emotion recognition.

**Acknowledgments.** Financial supports by National Natural Science Foundation of China (61603029), the Fundamental Research Funds for the Central Universities (2020YJS025) are gratefully acknowledge.

## References

1. Ang, K.K., Chin, Z.Y., Wang, C., Guan, C., Zhang, H.: Filter bank common spatial pattern algorithm on BCI competition iv datasets 2a and 2b. Front. Neurosci. **6**, 39 (2012)
2. Ang, K.K., Chin, Z.Y., Zhang, H., Guan, C.: Filter bank common spatial pattern (FBCSP) in brain-computer interface. In: 2008 IEEE International Joint Conference on Neural Networks (IEEE World Congress on Computational Intelligence), pp. 2390–2397. IEEE (2008)
3. Clevert, D.A., Unterthiner, T., Hochreiter, S.: Fast and accurate deep network learning by exponential linear units (ELUs). arXiv preprint arXiv:1511.07289 (2015)
4. Cohen, M.X.: Analyzing Neural Time Series Data: Theory and Practice. MIT Press, Cambridge (2014)
5. Dose, H., Møller, J.S., Iversen, H.K., Puthusserypady, S.: An end-to-end deep learning approach to MI-EEG signal classification for BCIs. Expert Syst. Appl. **114**, 532–542 (2018)

6. Fukunaga, K.: Introduction to Statistical Pattern Recognition. Academic Press Professional, Inc., Boston (1990)
7. He, K., Zhang, X., Ren, S., Sun, J.: Deep residual learning for image recognition. In: Proceedings of the IEEE Conference on Computer Vision and Pattern Recognition, pp. 770–778 (2016)
8. Hu, J., Shen, L., Sun, G.: Squeeze-and-excitation networks. In: Proceedings of the IEEE Conference on Computer Vision and Pattern Recognition, pp. 7132–7141 (2018)
9. Lawhern, V.J., Solon, A.J., Waytowich, N.R., Gordon, S.M., Hung, C.P., Lance, B.J.: EEGNet: a compact convolutional neural network for EEG-based brain-computer interfaces. J. Neural Eng. **15**(5), 056013 (2018)
10. Lu, N., Li, T., Ren, X., Miao, H.: A deep learning scheme for motor imagery classification based on restricted Boltzmann machines. IEEE Trans. Neural Syst. Rehabil. Eng. **25**(6), 566–576 (2016)
11. Nair, V., Hinton, G.E.: Rectified linear units improve restricted Boltzmann machines. In: Proceedings of the 27th International Conference on Machine Learning (ICML 2010), pp. 807–814 (2010)
12. Nicolas-Alonso, L.F., Gomez-Gil, J.: Brain computer interfaces, a review. Sensors **12**(2), 1211–1279 (2012)
13. Schirrmeister, R.T., et al.: Deep learning with convolutional neural networks for EEG decoding and visualization. Hum. Brain Mapp. **38**(11), 5391–5420 (2017)
14. Sun, L., Feng, Z., Lu, N., Wang, B., Zhang, W.: An advanced bispectrum features for EEG-based motor imagery classification. Expert Syst. Appl. **131**, 9–19 (2019)
15. Szegedy, C., et al.: Going deeper with convolutions. In: Proceedings of the IEEE Conference on Computer Vision and Pattern Recognition, pp. 1–9 (2015)
16. Wu, H., et al.: A parallel multiscale filter bank convolutional neural networks for motor imagery EEG classification. Front. Neurosci. **13**, 1275 (2019)
17. Zhang, Y., et al.: Multi-kernel extreme learning machine for EEG classification in brain-computer interfaces. Expert Syst. Appl. **96**, 302–310 (2018)
18. Zhao, D., Tang, F., Si, B., Feng, X.: Learning joint space-time-frequency features for EEG decoding on small labeled data. Neural Netw. **144**, 67–77 (2019)
19. Jia, Z., Lin, Y., Jiao, Z., Ma, Y., Wang, J.: Detecting causality in multivariate time series via non-uniform embedding. Entropy **21**(12), 1233 (2019)
20. Jia, Z., Lin, Y., Liu, Y., Jiao, Z., Wang, J.: Refined nonuniform embedding for coupling detection in multivariate time series. Phys. Rev. E **101**(6), 062113 (2020)

# Author Index

Printed in the United States
By Bookmasters